SIGMA MOLECULAR ORBITAL THEORY

A CHEMISTRY–PHYSICS INTERFACE

YALE SERIES IN THE SCIENCES

General Editors, Oktay Sinanoğlu and Edward U. Condon

A CHEMISTRY—PHYSICS INTERFACE

SIGMA MOLECULAR ORBITAL THEORY

by

Oktay Sinanoğlu
Professor of Theoretical Chemistry, Yale University

and

Kenneth B. Wiberg
Professor of Organic Chemistry, Yale University

(including material by a number of other authors on their specialties)

NEW HAVEN AND LONDON, YALE UNIVERSITY PRESS, 1970

Library of Congress catalog card number: 77-89906
Standard book number: 300-01145-8
set in IBM Selectric Press Roman type by Science Press, Inc.,
and printed in the United States of America by
The Murray Printing Company, Forge Village, Mass.
Distributed in Great Britain, Europe, Asia, and
Africa by Yale University Press Ltd., London; in
Canada by McGill University Press, Montreal; and
in Latin America by Centro Interamericano de Libros
Académicos, Mexico City.

Preface

Many of our colleagues have shared our feeling that the time was ripe for a look at the present state of the recently developed "sigma molecular orbital" methods. These methods have already made quantum chemistry broadly useful in a practical way in organic chemistry. It is to be expected that such methods will become more and more everyday tools for the practicing chemist. At the moment, the methods are still in the developmental stage. Much effort is being expended on comparisons and tests of the reliability of the various approximations used. We hope that researchers and graduate students with diverse interests from organic chemistry, theoretical chemistry, molecular biophysics, and inorganic chemistry will find this "state-of-the science" book useful in seeing what methods are available, what the limitations are, and where one may expect them to apply. We thank our many colleagues, themselves the originators and developers of many of the methods, for their encouraging remarks as to the timeliness and desirability of this project and for their taking part in it. We also acknowledge with gratitude the support of the project by the Yale administration through an NSF institutional grant. Our special thanks in this regard go to President Kingman Brewster, Provost Charles Taylor, Dean John Perry Miller, and Dr. Arthur Ross. At the production stage, it was a pleasure to work with Miss Jane Olson and Mrs. Anne Wilde of the Yale University Press, and with Miss Marie Sarkes and Mrs. Miriam Bess.

Oktay Sinanoğlu and Kenneth B. Wiberg

New Haven, Conn.
November 1969

In addition to the material by O. Sinanoğlu and K. B. Wiberg, this book contains original contributions prepared for this book by the following authors:

L. C. Allen	Dept. of Chem., Princeton University, Princeton, N.J.
H. Basch	Bell Telephone Laboratories, Murray Hill, N.J.
B. J. Bertus	Dept. of Chem., Louisiana State University, Baton Rouge, La.
T. D. Bouman	Chem. Division, Argonne National Lab., Argonne, Ill.
A. L. H. Chung	Dept. of Chem., Rutgers University, New Brunswick, N.J.
J. H. Corrington	Dept. of Chem., Xavier University, New Orleans, La.
L. C. Cusachs	Dept. of Chem., Tulane University, New Orleans, La.
E. R. Davidson	Dept. of Chem., University of Washington, Seattle, Wash.
J. W. Faller	Dept. of Chem., Yale University, New Haven, Conn.
A. A. Frost	Dept. of Chem., Northwestern University, Chicago, Ill.
K. Fukui	Dept. of Hydrocarbon Chem., Kyoto University, Kyoto, Japan
T. L. Gilbert	Chem. Division, Argonne National Lab., Argonne, Ill.
G. L. Goodman	Chem. Division, Argonne National Lab., Argonne, Ill.
C. Hollister	Dept. of Chem., New York University, New York, N.Y.
R. G. Jesaitis	Dept. of Chem., University of California, Berkeley, Cal.
G. Klopman	Dept. of Chem., Case Western Reserve Univ., Cleveland, Ohio
S. P. McGlynn	Dept. of Chem., Louisiana State University, Baton Rouge, La.
D. J. Mickish	Dept. of Physics, Oklahoma State Univ., Stillwater, Okla.
J. W. Moskowitz	Dept. of Chem., New York University, New York, N.Y.
H. A. Pohl	Dept. of Physics, Oklahoma State Univ., Stillwater, Okla.
A. Pullman	Institut de Biologie-Physico-chimique, Paris, France
C. Sandorfy	Dept. of Chem., University of Montreal, Montreal, Canada
A. Streitwieser, Jr.	Dept. of Chem., University of California, Berkeley, Cal.
C. Trindle	Chem. Division, Argonne National Lab., Argonne, Ill.
A. C. Wahl	Chem. Division, Argonne National Lab., Argonne, Ill.
M. A. Whitehead	Dept. of Chem., McGill University, Montreal, Canada

A number of reprints by other authors and by some of the authors above are also included. The sources of the reprints are indicated in the table of contents.

Contents

Chapter I

Introduction

Quantum chemistry may be considered to be entering a new phase. There was a time when nonempirical calculations were carried out on the smallest of molecules, and semiempirical ones on the π electrons of conjugated systems alone. Recently, semiempirical molecular orbital (MO) methods have been extended to include all electrons, σ and π, and to apply to saturated molecules as well as conjugated. Purely nonempirical MO calculations, although still in general far from the desired Hartree-Fock MO result, are being made on sizable organic molecules, too. Clearly, for predictions on chemical reactions, σ electrons are essential. Inclusion of σ electrons in conjugated systems shows that although some qualitative conclusions based on π alone remain valid, in general the picture changes. Sigma and π are not as clearly separable as once used to be thought.

In the present book an attempt is made to present an overall view of this rapidly developing and still very fluid field with the help of contributions from our colleagues and with occasional inclusion of articles reprinted from the original literature. Chapters II and III contain the main semiempirical methods with applications to different properties. Throughout, attention is drawn to the limitations of the methods as well. In spite of considerable advance, both the semiempirical and nonempirical methods are still at a rudimentary stage.

Chapter IV includes an annotated bibliography of recent organic chemical applications. In Chapter VI semiempirical methods are compared with each other and with nonempirical calculations. The chapter also includes some theoretical results which attempt to justify the use and approximations of the simpler semiempirical methods. In Chapter IV the reader will find discussions of some experimental problems in organic chemistry, and in Chapter VIII in inorganic chemistry, which are called to the attention of theorists. Streitwieser gives examples of qualitative notions in organic chemistry that could now be tested by nonempirical calculations.

Chapter VII presents a view of what computers can do at their best used as powerful instruments of a priori prediction and how they may even do group theory from scratch. Applicability of MO methods, especially to energetics, depends on the error that would remain even after the MO's are calculated at the Hartree-Fock level, that is, on electron correlation and its behavior. This question is examined in Chapter VII-6 and 7. Sigma-π separation, or rather nonseparation, has been studied in the papers included in Chapter VII-7 and 8. In the last part of the book, Faller gives an inorganic chemist's view of the state of usefulness of σ-MO theory with selected recent applications.

Chapter II

Semiempirical Sigma Molecular Orbital Theories

II-1

Semiempirical methods allow calculations on a large number of molecules at little cost. They have been useful as guides in chemical applications, and have gained more and more importance in both organic and inorganic chemistry. However, these methods often involve drastic and as-yet untested approximations. They need, therefore, to be used judiciously. Although a user may often be interested only in applications to his own chemical problems, he would nevertheless be well advised to look into the foundations of the method that is being used in some detail and to familiarize himself with the limitations.

For a long time, MO theory was applied to only the π electrons of conjugated molecules. For a survey of the π-electron methods and how useful they have been, the reader is referred to the following books: R. G. Parr, *Quantum Theory of Molecular Electronic Structure*, Benjamin, New York, 1961, and O. Sinanoğlu, ed., *Modern Quantum Chemistry,* Vol. 1, Orbitals, Academic Press, New York, 1965. For applications of Hückel-type methods in organic chemistry, the reader is referred to the book by A. Streitwieser, Jr., *Molecular Orbital Theory for Organic Chemists,* Wiley, New York, 1961.

Recently, the analogues of these various semiempirical MO methods have been developed for systems containing σ electrons. This is a very important development, as almost all molecules involve σ-type electrons. As in the π-electron methods, the σ-MO theory has also evolved over various stages of approximations. The Hückel method has been extended in a number of ways, as has the self-consistent-field (SCF) method, the latter to include electron-electron repulsions in an approximate way.

In a historically very important paper (Chapter II-2), Sandorfy applied Hückel-type approximations to saturated hydrocarbons, for the first time including the hydrogens, and concluded that MO methods of the linear combination of atomic orbitals type (LCAO) would be capable of giving as good charge distributions in such molecules as on conjugated systems.

In Hückel-type theories, the total energy is assumed to be a sum of one-electron orbital energies, and only the one-electron Hamiltonian matrix is diagonalized to obtain these. The H'_{rs}, one-, and two-center Hamiltonian matrix elements are obtained either from atomic properties and/or related to overlap integrals S_{rs} in a number of ways. Further, the LCAO may be that of the regular atomic orbitals (AO's) such as the 2s's and 2p's and the hydrogen 1s's, or of hybridized valence atomic orbitals (LCVO). Fukui and co-workers, in their pioneering applications, used especially an LCVO method without overlaps. They obtained charge distributions and σ dipole moments for a large number of saturated and unsaturated molecules, and they studied with this method chemical reactivities and phenomena such as the breakup of hydrocarbons under electron impact. These applications, as well as references to the literature up to 1965, have been surveyed by Fukui in *Modern Quantum Chemistry*, Vol. I, O. Sinanoğlu, ed.,

Academic Press, New York, 1965. More recent applications and a brief discussion of the hybrid-based MO method will be found in Chapter III-3.

A strong impetus to the treatment of σ systems has developed from the "extended Hückel" work of Hoffmann (Chapter II-3). Hoffmann used a Hückel method with all the one-electron matrix elements H'_{rs} and all the overlaps S_{rs} included. The method was applied with the same parametrization to a large class of saturated and unsaturated hydrocarbons. Hoffmann concluded that with such a method the geometries of molecules could be well predicted. In later papers, Hoffmann has used his extended Hückel method as a guide for organic chemistry (see Chapter IV-4). The success obtained from simple MO calculations in predicting geometries, as was the case with Walsh's rules for smaller molecules, has stimulated work on the reasons of this success by a study of more complete and nonempirical methods (see Chapter VI).

It is interesting that another, much simplified method, which, however, differs from the Hückel approach in important aspects, the subminimal *ab initio* method of Frost (Chapter VII-5), also gives remarkable success in the predictions of geometries. In the method of Frost, the full exact Hamiltonian and all the matrix elements that it gives rise to are used, but, in turn, the orbitals themselves are chosen in the simplest possible manner.

The Hückel-type methods do not include electron-electron interactions and the adjustments of the parameters that would result from these. Such changes in the parameters are particularly important in ionic species and with heteroatoms, where straight extended Hückel calculations run into more difficulty.

As in the π-electron theory, one way to go beyond the Hückel methods while retaining their basic simplicity and lack of difficult integrals has been to introduce iterations to the parameters, dependent on charge distribution. Such iterated extended Hückel methods (IEHC) are discussed and compared with other methods by Pullman in Chapter VI-6. Cusachs and McGlynn have been among the main proponents of this method (see Chapter VI-4 and 7).

A very important development in σ-MO theory involved the inclusion of electron-electron repulsions to achieve a self-consistency of the orbitals. A number of versions of approximate SCF have been developed. With σ systems, additional problems concerning parameters and the rotational invariance of atomic orbitals on each center come up which were not present in π-electron theory. Some of these problems and the often drastic approximations they necessitate have been discussed by Pople, Santry, and Segal (Chapter VI-5). A hybrid-based MO theory, but including electron-electron repulsions, has been given by Katagiri and Sandorfy [*Theoret. Chim. Acta* 4, 203 (1966)].

Although most of the other methods have been applied mainly to ground states, Katagiri and Sandorfy have been concerned with applications to electronic spectra as well.

A version of approximate SCF is the CNDO/2 (complete neglect of differential overlap) method of Pople and Segal (Chapter II-4). Computer programs for the CNDO/2 method as well as for the extended Hückel method of Hoffmann are available in the *Quantum Chemistry Program Exchange* (Department of Chemistry, Indiana University, Bloomington, Ind.).

These semiempirical methods give considerable success for molecular charge distributions and geometries. They fail, however, in other properties, when used with the same parameters, such as heats of formation. One of the severest approximations in the CNDO/2 method is that 2s and 2p electrons are treated as equivalent to achieve rotational invariance in Coulomb repulsion parameters.

In later chapters of this book the reader will find extensive comparisons of the various semiempirical methods with each other as well as with nonempirical calculations (see Chapter VI-6). A number of articles and discussions on the nature of the approximations are also included. That often different parametrizations must be used for different properties is a crucial and suggestive aspect. Some discussion on this point will be found in Chapter III, where examples of applications to diverse properties are also given.

II-2 Reprinted from *Canadian Journal of Chemistry* **33**, 1337 (1955)

LCAO MO CALCULATIONS ON SATURATED HYDROCARBONS AND THEIR SUBSTITUTED DERIVATIVES[1]

BY C. SANDORFY[2]

ABSTRACT

The simple semiempirical molecular orbital method was applied to saturated hydrocarbons in three different approximations. Electronic charge distribution diagrams were obtained. Characteristic differences between saturated and conjugated compounds concerning bond localization and the effect of an electronegative substituent can be interpreted on the basis of these diagrams. The 1*s* electrons of the hydrogens are introduced for the first time in semiempirical molecular orbital calculations.

I. INTRODUCTION

The molecular orbital method in its simple form, where the electrons are assigned to one-electron wave functions (ϕ_j) expressed as linear combinations of atomic orbitals

[I] $$\phi_j = c_1\psi_1 + c_2\psi_2 + c_3\psi_3 + \ldots,$$

was applied with success to the evaluation of energy levels and electronic charge distribution in conjugated molecules (19, 26).

The majority of these calculations were carried out by use of the very flexible Hückel approximation (18) where the π-electrons are supposed to live a somewhat independent life in the averaged field of the nuclei and the core and σ-electrons. Objections against this approximate method are easy to find. The separation of π- and σ-electrons needs justification and recently Altmann and Coulson (1, 8) questioned it. However work by Ross (37) and Moser (25) seems to prove that this approximation is a fair one. Other objections arise owing to the unsufficient account taken of spin and the mutual repulsion of the electrons. Partial improvement in these respects might be attained by more elaborate methods, such as the methods of antisymmetrized molecular orbitals (15), configurational interaction (6, 9), and self-consistent LCAO (36). Nevertheless, in spite of a certain success, these improved methods fail to give a decisively better understanding of the molecules than that obtained by the simpler semiempirical LCAO method. The reason for this lies probably in the fact that these methods require an actual knowledge of the wave functions whereas the semiempirical character of the simpler method makes this unnecessary.

Since 1949 the evolution of theoretical chemistry has taken new directions following the critical work of Lennard-Jones (Ref. 20 and subsequent papers), the new ideas of "the most probable configuration" (2, 22, 39, 43) introduced before by Artmann, and Daudel's new approach to the problem of localizability of electrons (la théorie des loges) (11).

The fact remains, however, that the simple semiempirical LCAO MO method succeeded in giving a fair interpretation of the electronic spectra of

[1]*Manuscript received May 2, 1955.*
Contribution from Département de Chimie, Université de Montréal, Montréal, Québec.
[2]*Département de Chimie, Université de Montréal, Montréal, Québec.*

conjugated molecules (5, 27, 33) and of the electronic charge distribution in them (7, 12, 34, 42). The reasons for this surprising success were examined by Mulliken (28) (Roger (35) even tried to deduce this method independently of wave mechanics).

In view of this fact we are compelled to believe that, although apparently very approximate, the semiempirical LCAO MO method constitutes a very useful tool in investigation of molecular structure (see Refs. 13 and 34), and it may be considered as the next stage of evolution after Pauling's resonance theory (14, 30).

II. CALCULATIONS ON SATURATED COMPOUNDS

The LCAO molecular orbital method was first applied to the H_2 molecule, and the H_2^+ ion (see Ref. 31), and there is no reason why its use should be restricted to "mobile" or π-electrons.

There exists however a very serious limitation due to the greater complexity of the problems involved. There is a certain justification in treating π-electrons separately from the others, but, if we turn our attention to the saturated compounds, there is now no reason to neglect any of the orbitals participating in bonds, or the lone pair electrons. All orbitals have to be taken into consideration, except possibly the K orbitals.

Thus, for example, in the case of *n*-butane we have 26 orbitals instead of 4 in the π-electron treatment of 1,3-butadiene.

This fact is not however prohibitive in every case, because, firstly, symmetry considerations enable us to simplify the secular equations, and, secondly, calculating machines can be used to solve systems of equations of a relatively high degree.

The main points to notice are that, in using a wave function of the form of [I], we do not "delocalize" the electrons, and also that when the molecular orbital is a linear combination of atomic orbitals it bears no relation to the "mobility" of the electrons in certain orbitals.

Lennard-Jones (20) has insisted on the fact that the total wave function, which is an antisymmetrized product of the molecular orbitals (the ϕ_j), remains unchanged for any orthonormal transformation of the ϕ_j and especially on formation of linear sums such as, for example,

$$\begin{vmatrix} \phi_1 (1) & \phi_2 (1) \\ \phi_1 (2) & \phi_2 (2) \end{vmatrix} \equiv \begin{vmatrix} d_1 (1) & d_2 (1) \\ d_1 (2) & d_2 (2) \end{vmatrix},$$

$$d_{1,2} = (\phi_1 \pm \phi_2)/N^2,$$

where N is a normalizing factor. Thus, for example, the molecules can be described equally well in terms of either molecular orbitals or "equivalent orbitals" localized in a limited part of the molecule. The electronic charge density at a given point depends on the squared modulus of the total wave function, which is the same in both cases.

Hall (17) proposed a semiempirical (and self-consistent) method of treating saturated compounds, based on group orbitals and using experimental ionization potentials to derive the necessary parameters.

Brown (3), starting from somewhat different principles, has put forward a method very similar to that of Hall. Because of their very nature these methods do not make possible the evaluation of electronic charge distribution (except for ions) or of the spectral properties.

We attempted to obtain at least electronic charge distributions by the semi-empirical LCAO MO method. We tried three different approximations.

III. THE "C" APPROXIMATION

Daudel and the author (38) described this method which is only a tentative one because of the rather unjustified approximations involved in it. It has proved to be useful, however, in increasing our practical knowledge concerning computations on saturated compounds.

Let us consider a paraffinic hydrocarbon. If we suppose tetrahedral hybridation we have four sp^3 hybrids on each carbon atom and the $1s$ orbitals of the hydrogen atoms. In this first approximation we not only disregard the hydrogens but also those hybrids which are linked to the hydrogens. There is no reason to do this but it would be interesting to investigate whether or not consideration of the carbon–carbon linkages alone could explain certain properties of these compounds. We put all the coulombic terms equal to $\alpha_C = \alpha = 0$, we neglect interactions between nonadjacent carbon atoms or orbitals not linked together and overlap integrals. (These latter will be taken into account in approximation "H".) Furthermore, for two carbon atoms bound together, we put $\beta_{C-C} = \beta$. Lacking any information concerning interactions between hybrids on the same carbon atom we varied this parameter $\beta' = m\beta$ in certain cases (*vide infra*).

We define as "orbital charge" the quantity

$$q_r = \sum c_r^2, \qquad \text{[II]}$$

r being an index for the atomic orbitals, the summation being extended over all molecular spin-orbitals occupied by electrons in the ground state.* For n-paraffins we obtain $q_r = 1$ in this approximation using any value of β'. If we write the secular determinant with the above conventions, and if $\beta' = \frac{1}{4}\beta$, we obtain for n-propane the secular determinant given in Table I, where we

TABLE I

C APPROXIMATION. SECULAR DETERMINANT FOR n-PROPANE

	1	2	3	4
1	y	1	0	0
2	1	y	$\frac{1}{4}$	0
3	0	$\frac{1}{4}$	y	1
4	0	0	1	y

have divided everywhere by β and put $|(\alpha-\epsilon)/\beta| = y$, ϵ being the energy corresponding to the molecular orbitals. The system of numbering shown in Fig. 1 was used. This secular equation would be the same as the one for

*All atomic and molecular orbitals are normalized to unity. Thus q_r is expressed in electron-units. "Effective" charges are obtained by subtracting from these values the corresponding positive charges in the nuclei.

butadiene in the π-electron approximation if we had put $\beta' = \beta$. Such analogies may be useful in many cases. In the case of certain branched chain hydrocarbons, the secular equation may be analogous to the one of an odd-membered ring conjugated hydrocarbon and thus we might expect to obtain orbital charges different from unity (10).

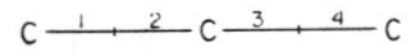

FIG. 1. Numbering system for Table I.

Now we replace one CH_3 group in propane by substituent X, and we suppose that $\beta_{X-C} = \beta_{X-X}$, $\alpha_X = \alpha_C+\beta$, and that no other electron of X but the one bound to the next carbon atom has to be considered. Then X will be an imaginary substituent with an electronegativity higher than carbon.

We carry through the calculation successively with $m = +2, +1, +0.25, 0, -0.25, -1$, and -2 ($\beta' = m\beta$). The result found is that the sign of m does not change the values of the charges in this particular case. The diagrams obtained are shown in Fig. 2. We also computed "n-X-propane" with

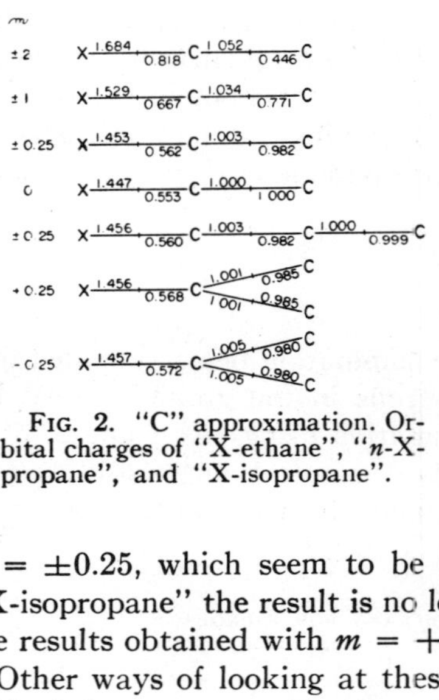

FIG. 2. "C" approximation. Orbital charges of "X-ethane", "n-X-propane", and "X-isopropane".

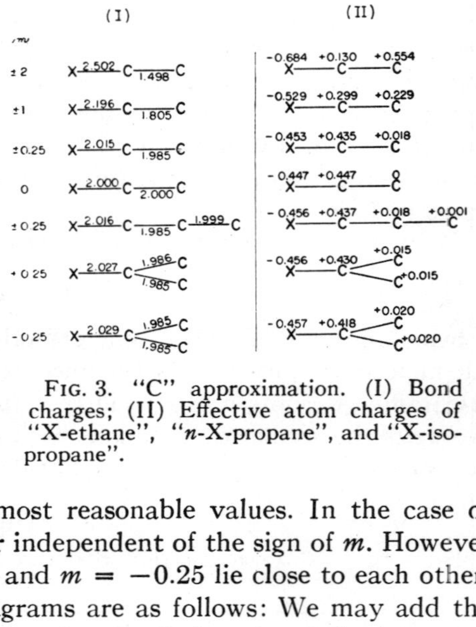

FIG. 3. "C" approximation. (I) Bond charges; (II) Effective atom charges of "X-ethane", "n-X-propane", and "X-isopropane".

$m = \pm 0.25$, which seem to be the most reasonable values. In the case of "X-isopropane" the result is no longer independent of the sign of m. However the results obtained with $m = +0.25$ and $m = -0.25$ lie close to each other.

Other ways of looking at these diagrams are as follows: We may add the orbital charges on one X—C or C—C bond and call them bond charges or we may add the orbital charges on orbitals belonging to one given atom and call them atom charges. (See Fig. 3.) The main conclusions from these diagrams are that, at least in this rough approximation, the electron attraction of the substituent X mainly affects the carbon orbital which contributes to the bond X—C, and that this effect decreases rapidly with increasing distance from X. The orbital charges are alternant as in a conjugated chain but this alternance is masked as soon as we compute the bond changes. There are two orbitals now on each atom instead of one as in pure π-electron calculations.

IV. THE "CH" APPROXIMATION

We here include the hydrogens, although not yet explicitly. The tetrahedral hybrids which bind together two carbons (or one carbon and one X) are taken individually but the hybrids linked to hydrogens are considered as forming a CH group orbital with the 1*s* electron of hydrogen.

This way of setting up the secular equation may be considered, we think, as a compromise between Hall's method of two-atom-orbitals and the straight LCAO treatment, with all the orbitals individualized, as we have attempted in Section V.

In Fig. 4 there are diagrams of *n*-propane, *n*-butane, *n*-pentane, and 2-methyl butane. For computing them we took all the coulombic terms to be equal. We furthermore took $\beta_{C-C} = \beta$ for carbon—carbon bonds, $\beta' = \frac{1}{4}\beta$ for

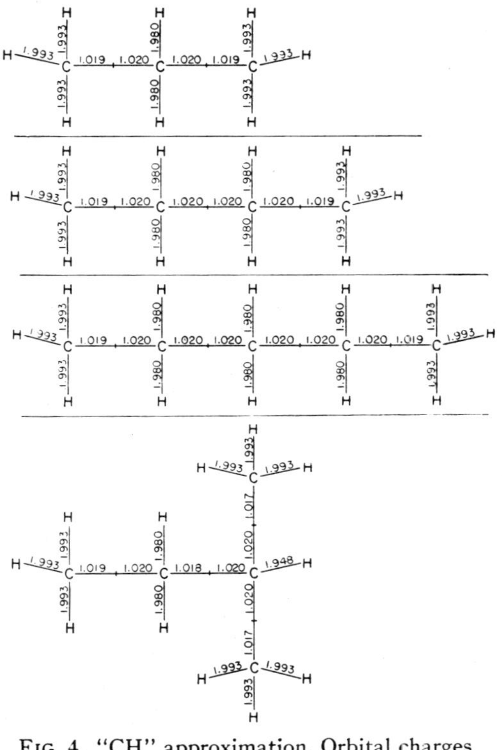

FIG. 4. "CH" approximation. Orbital charges of *n*-propane, *n*-butane, *n*-pentane, and 2-methyl butane.

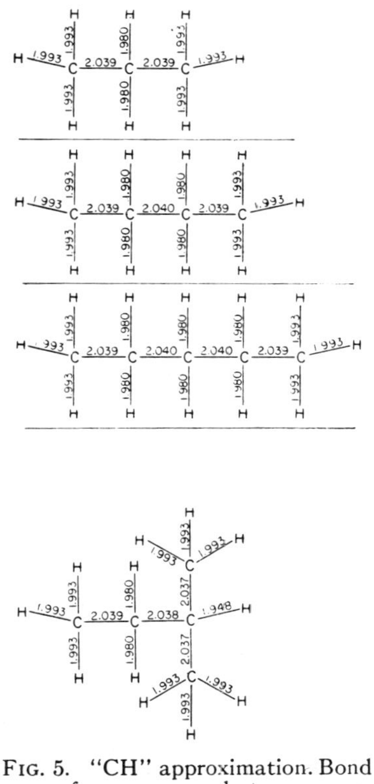

FIG. 5. "CH" approximation. Bond charges of *n*-propane, *n*-butane, *n*-pentane, and 2-methyl butane.

interactions between two tetrahedral hybrids on the same carbon, and also $\beta'' = \frac{1}{4}\beta = \beta'$ for interactions between two C—H bonds on the same carbon. For an interaction between one hybrid and one C—H bond on the same carbon we took $\beta'' = \frac{1}{8}\beta$.

The value 0.25 for β' was chosen according to the preceding approximation and the ratio β'/β'' is about the same as e/d in Lennard-Jones and Hall's (21) treatment of paraffins in their semiempirical method where the choice of parameters is based on ionization potentials. e is the nondiagonal term between two C—C bonds with a common carbon atom, and d the one between a C—C and a C—H bond also having a carbon atom in common. The numerical values given by these authors were $e = 1.46$ ev. and $d = 0.70$ ev. We suppose here that the interaction between two C—C bonds, with a common carbon atom, may be represented by that of the two hybrids on the central carbon. Concerning the parameter representing the interaction between two C—H bonds on the same carbon the value given by Hall (17) was $b = 1.75$ ev. in the case of methane. Seeking only the right order of magnitude we adopted β' for the corresponding parameter.

The (unfactorized) secular determinant for *n*-propane is given in Table II. The numbering system on Fig. 6 was used. The diagrams in Fig. 4 give the orbital charges for four hydrocarbons (these are actually bond charges for the CH bonds).

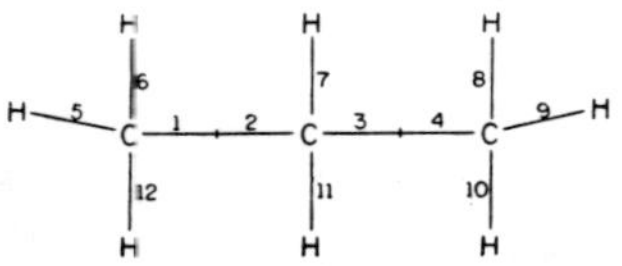

FIG. 6. Numbering system for Table II.

TABLE II

"CH" APPROXIMATION. SECULAR DETERMINANT FOR *n*-PROPANE

	1	2	3	4	5	6	7	8	9	10	11	12
1	y	1	0	0	$\frac{1}{8}$	$\frac{1}{8}$	0	0	0	0	0	$\frac{1}{8}$
2	1	y	$\frac{1}{4}$	0	0	0	$\frac{1}{8}$	0	0	0	$\frac{1}{8}$	0
3	0	$\frac{1}{4}$	y	1	0	0	$\frac{1}{8}$	0	0	0	$\frac{1}{8}$	0
4	0	0	1	y	0	0	0	$\frac{1}{8}$	$\frac{1}{8}$	$\frac{1}{8}$	0	0
5	$\frac{1}{8}$	0	0	0	y	$\frac{1}{4}$	0	0	0	0	0	$\frac{1}{4}$
6	$\frac{1}{8}$	0	0	0	$\frac{1}{4}$	y	0	0	0	0	0	$\frac{1}{4}$
7	0	$\frac{1}{8}$	$\frac{1}{8}$	0	0	0	y	0	0	0	$\frac{1}{4}$	0
8	0	0	0	$\frac{1}{8}$	0	0	0	y	$\frac{1}{4}$	$\frac{1}{4}$	0	0
9	0	0	0	$\frac{1}{8}$	0	0	0	$\frac{1}{4}$	y	$\frac{1}{4}$	0	0
10	0	0	0	$\frac{1}{8}$	0	0	0	$\frac{1}{4}$	$\frac{1}{4}$	y	0	0
11	0	$\frac{1}{8}$	$\frac{1}{8}$	0	0	0	$\frac{1}{4}$	0	0	0	y	0
12	$\frac{1}{8}$	0	0	0	$\frac{1}{4}$	$\frac{1}{4}$	0	0	0	0	0	y

The first observation is that although we used molecular orbitals in these calculations the bonds appear to be very well "localized". A certain displacement of the electrons takes place from the C—H bonds towards the C—C bonds and the carbon chain thus becomes negatively charged and the C—H bonds positively charged.

Other ways of looking at these results can be seen in the diagrams of "bond charges" (Fig. 5) and "group charges". The latter mean the sum of orbital charges (or CH bond charges) of orbitals (or CH bonds) around a given atom. Effective group charges are 0 for *n*-paraffins in this approximation. The same numbers occur in all these diagrams (at least with an accuracy of 0.001) and there does not seem to be any alternance in the chain. CH bonds on primary carbon atoms have an effective charge of +0.007, CH bonds on secondary carbon atoms have +0.020, and CH bonds on tertiary carbon atoms have +0.052. This may be related to the observation that nucleophilic substitution takes place, in order of preference, on $\geqslant$CH, $>CH_2$, and $—CH_3$ groups.

If we substitute X in *n*-butane and *n*-pentane we obtain the results given in Figs. 7 and 8 (X has the same properties as in Section III).

According to these diagrams almost all the charge pulled on the heteroatom because of its electronegativity being greater than carbon comes from the carbon orbital bonded to it and also from the C—H bonds on that carbon. There still is some perturbation in the next C—C bond but after the second carbon the effect of substituent X becomes negligible and the values of charges are the same as in the related hydrocarbons.

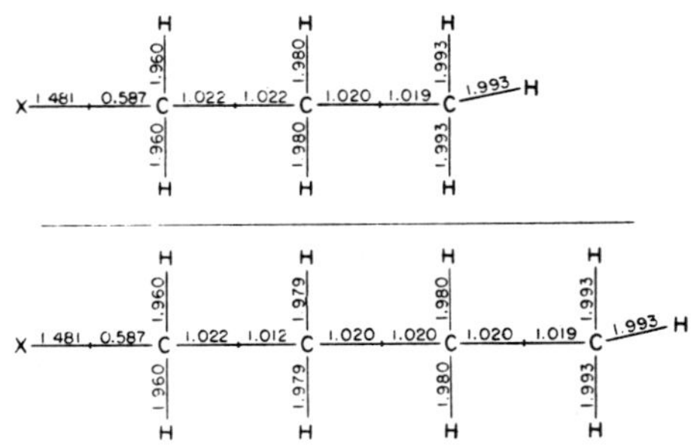

FIG. 7. "CH" approximation. Orbital charges of "*n*-X-propane" and "*n*-X-butane".

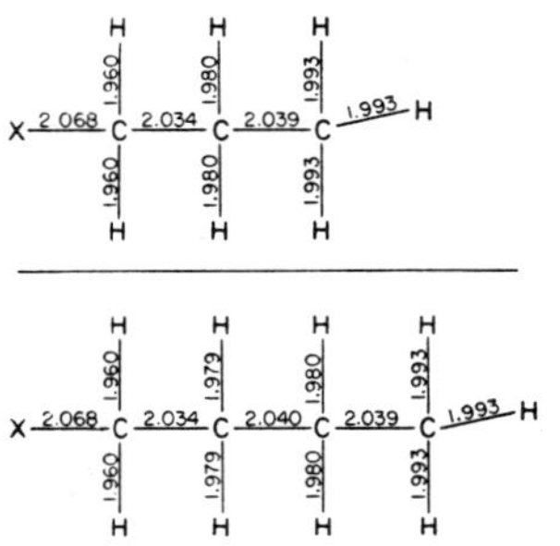

FIG. 8. "CH" approximation. Bond charges of "*n*-X-propane" and "*n*-X-butane".

If we compare these diagrams to the one computed by Gouarné and Yvan (16) (Fig. 9) with comparable parameters for a conjugated chain substituted at carbon one, then we see the significant difference between conjugated and saturated chains. The effect of X reaches much further in conjugated compounds. This result is obtained in spite of the use of similar wave functions and shows that, because of the different arrangement of orbitals, certain electrons appear to be "mobile" and others "localized". This also illustrates the fact that the use of molecular orbitals is just a way of writing wave functions and does not necessarily involve the "delocalization" of the electrons.

FIG. 9. Comparison between saturated and conjugated compounds. (The diagram of the conjugated compound from Gouarné and Yvan.) Effective atom charges. "CH" approximation.

As the coulombic terms of the C—H bonds are likely to be different from those of the orbitals forming C—C bonds we computed *n*-X-butane with $\alpha_{C-H} = \alpha_C + \beta$ and $\alpha_{C-H} = \alpha_C - \beta$ (Fig. 10). There seems to be a slight alternance of more and less negative charges in the carbon chain with these parameters (in the third decimal), but the main picture remains the same.

When we replace X by a "real" substituent we have to face new difficulties. Outside of the need for new parameters one has to find a way of including lone pairs of electrons in semiempirical calculations. Actually, all the usual substituents of organic compounds contain lone pairs of electrons.

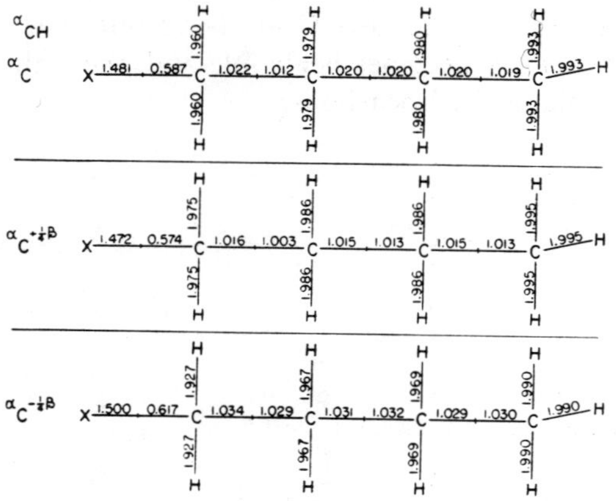

FIG. 10. "CH" approximation. Variation of α_{C-H}. Orbital charges of "*n*-X-butane".

We only attempted to compute a diagram for 1-amino-propane (Fig. 11).

As Pitzer (32) gave good reasons to believe that C—N bonds are weaker than C—C bonds (66 kcal. against 80 kcal.) owing to electronic repulsion, we did not include the electrons of the lone pair directly into the calculation but gave to β_{C-N} a value equal to $0.8\beta_{C-C}$, to represent the relative weakness of the C—N bond. Another parameter used was $\alpha_N = \alpha_C + 0.65\beta$ for the coulombic term of nitrogen (supplement equal to the difference of electronegativities of nitrogen and carbon). Interaction between the two N—H bonds was considered to be like that between two C—H bonds and interaction between H—N and N—C to be like that between H—C and C—C.

Fig. 11 needs little comment.

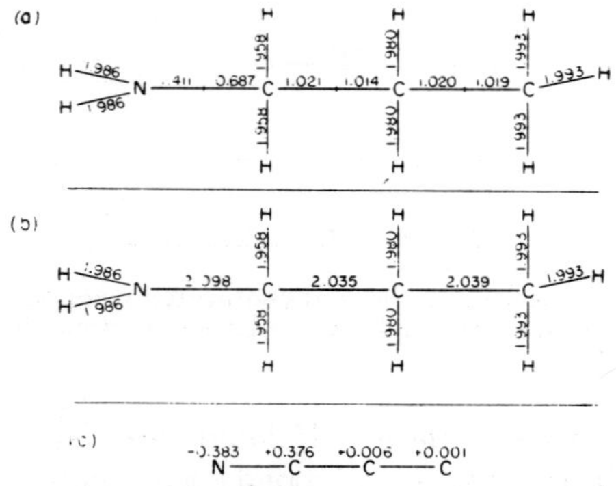

FIG. 11. "CH" approximation. 1-Amino-propane: (*a*) orbital charges, (*b*) bond charges, (*c*) effective atom charges.

V. THE "H" APPROXIMATION

In our third approximation all sp^3 hybrids and also all hydrogen 1*s* orbitals are included separately.

This has never been done before, to our knowledge, in a semiempirical treatment. It is not easy to select the parameters. We included overlap integrals right from the beginning. Overlap integrals are very high in σ-bonds. In spite of this we did not introduce them in approximations "C" and "C—H". We did not because, as Chirgwin and Coulson (4) have shown, this does not alter the values of charges provided all the coulombic terms are equal, and thus it does not affect our calculations related to pure hydrocarbons. It does change the charges in the case of the substituted compounds. It is unlikely however to change the qualitative features. Overlap integrals are fully taken into account in approximation "H".

Mulliken, Rieke, Orloff, and Orloff (29) computed overlap integrals using Slater functions. From their tables we obtain:

$$S_{\text{C-H}} = 0.684,$$
$$S_{\text{C-C}} = 0.647,$$
$$S_{\text{intra}} = -0.500.$$

S_{intra} is the overlap integral between two sp^3 hybrids on the same carbon atom. We put the β's proportional to the overlap integrals and the coulombic integrals equal to the electronegativity of the atoms minus the electronegativity of carbon (C: 2.50; H: 2.00).

Using Wheland's (40, 41) approximation the coulombic term corresponding to substituent X becomes

$$y+\delta_{\text{X}}(1-Sy) = 0.35y+1,$$

with $\delta_{\text{X}} = 1$ and $S = 0.65$.

For a hydrogen atom

$$y + \delta_{\text{H}}(1-Sy) = y-0.50(1-0.65y) = 1.33y-0.50$$

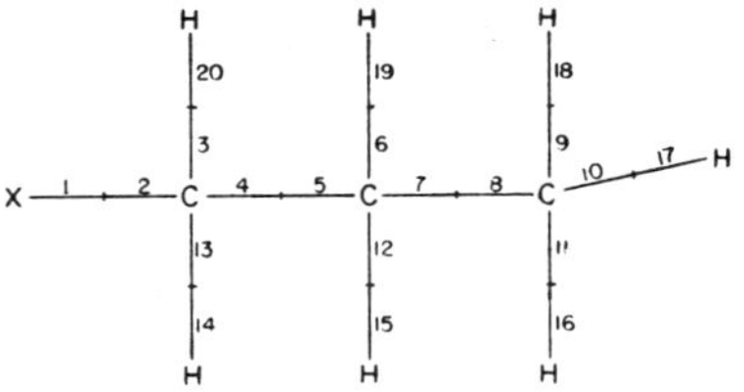

FIG. 12. Numbering system for Table III.

with $\delta_{\text{H}} = -0.50$. The nondiagonal terms become, for an interaction between two adjacent carbon atoms (resonance integrals including overlap), $\gamma_{\text{C-C}} = 1$; between X and C, $\gamma_{\text{X-C}} = 1$; between C and H, $\gamma_{\text{C-H}} = 0.684/0.647 = 1.06$; between two hybrids on the same carbon, $\gamma_{\text{intra}} = -0.500/0.684 = -0.77$. With these values the secular determinant for "*n*-X-propane" (Fig. 12) becomes as shown in Table III.

TABLE III

"H" APPROXIMATION. SECULAR DETERMINANT FOR "*n*-X-PROPANE"

	1	2	3	4	5	6	7	8	9	10	11	12	13	14	15	16	17	18	19	20
1	$(0.35y+1)$	1	0	0	0	0	0	0	0	0	0	0	0	0	0	0	0	0	0	0
2	1	y	−0.77	−0.77	0	0	0	0	0	0	0	0	−0.77	0	0	0	0	0	0	0
3	0	−0.77	y	−0.77	0	0	0	0	0	0	0	0	−0.77	0	0	0	0	0	0	1.06
4	0	−0.77	−0.77	y	1	0	0	0	0	0	0	0	−0.77	0	0	0	0	0	0	0
5	0	0	0	1	y	−0.77	−0.77	0	0	0	0	−0.77	0	0	0	0	0	0	0	0
6	0	0	0	0	−0.77	y	−0.77	0	0	0	0	−0.77	0	0	0	0	0	0	1.06	0
7	0	0	0	0	−0.77	−0.77	y	1	0	0	0	−0.77	0	0	0	0	0	0	0	0
8	0	0	0	0	0	0	1	y	−0.77	−0.77	−0.77	0	0	0	0	0	0	0	0	0
9	0	0	0	0	0	0	0	−0.77	y	−0.77	−0.77	0	0	0	0	0	0	1.06	0	0
10	0	0	0	0	0	0	0	−0.77	−0.77	y	−0.77	0	0	0	0	0	1.06	0	0	0
11	0	0	0	0	0	0	0	−0.77	−0.77	−0.77	y	0	0	0	0	1.06	0	0	0	0
12	0	0	0	0	−0.77	−0.77	−0.77	0	0	0	0	y	0	0	1.06	0	0	0	0	0
13	0	−0.77	−0.77	−0.77	0	0	0	0	0	0	0	0	y	1.06	0	0	0	0	0	0
14	0	0	0	0	0	0	0	0	0	0	0	0	1.06	$(1.33y-0.5)$	0	0	0	0	0	0
15	0	0	0	0	0	0	0	0	0	0	0	1.06	0	0	$(1.33y-0.5)$	0	0	0	0	0
16	0	0	0	0	0	0	0	0	0	0	1.06	0	0	0	0	$(1.33y-0.5)$	0	0	0	0
17	0	0	0	0	0	0	0	0	0	1.06	0	0	0	0	0	0	$(1.33y-0.5)$	0	0	0
18	0	0	0	0	0	0	0	0	1.06	0	0	0	0	0	0	0	0	$(1.33y-0.5)$	0	0
19	0	0	0	0	0	1.06	0	0	0	0	0	0	0	0	0	0	0	0	$(1.33y-0.5)$	0
20	0	0	1.06	0	0	0	0	0	0	0	0	0	0	0	0	0	0	0	0	$(1.33y-0.5)$

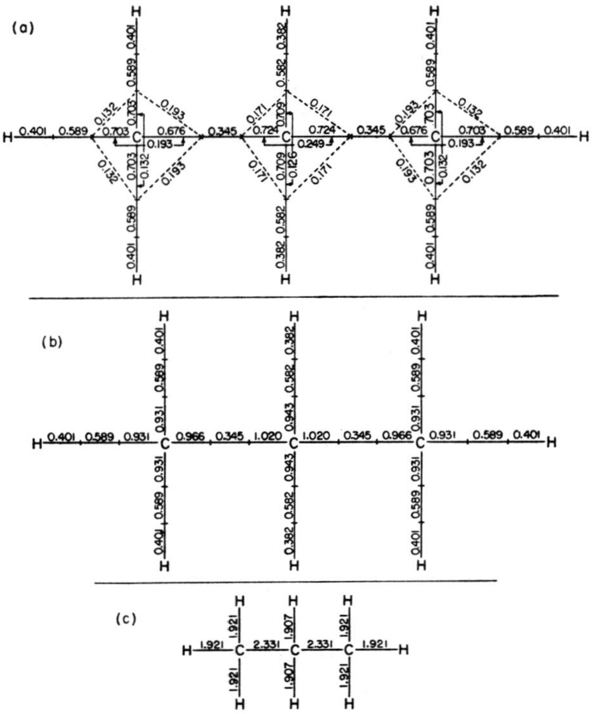

FIG. 13. "H" approximation. *n*-propane (see text).

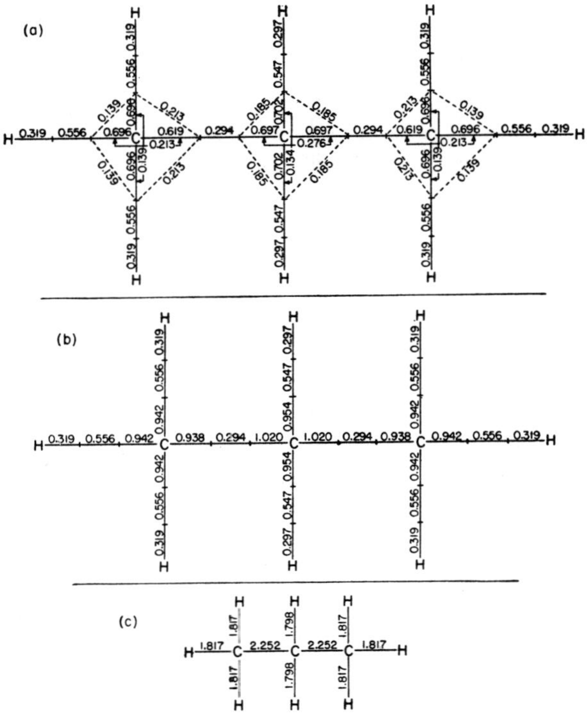

FIG. 14. "H" approximation. (+)-Ion of *n*-propane (see text).

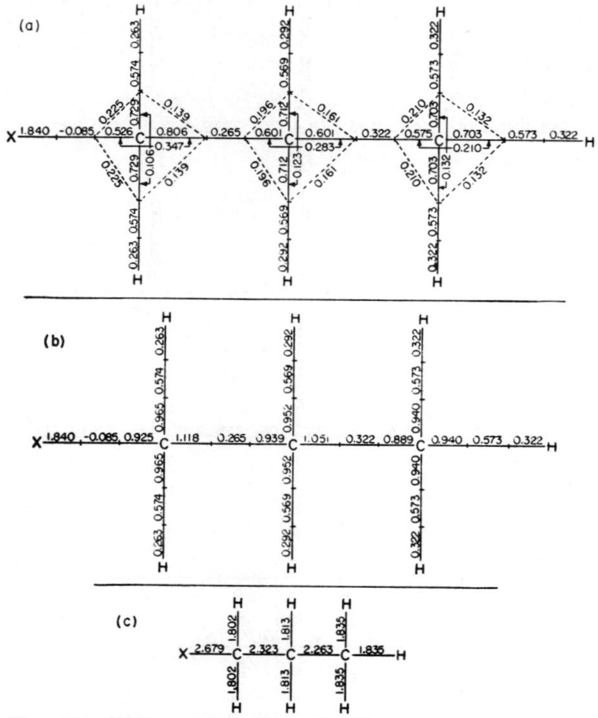

FIG. 15. "H" approximation. "*n*-X-propane" (see text).

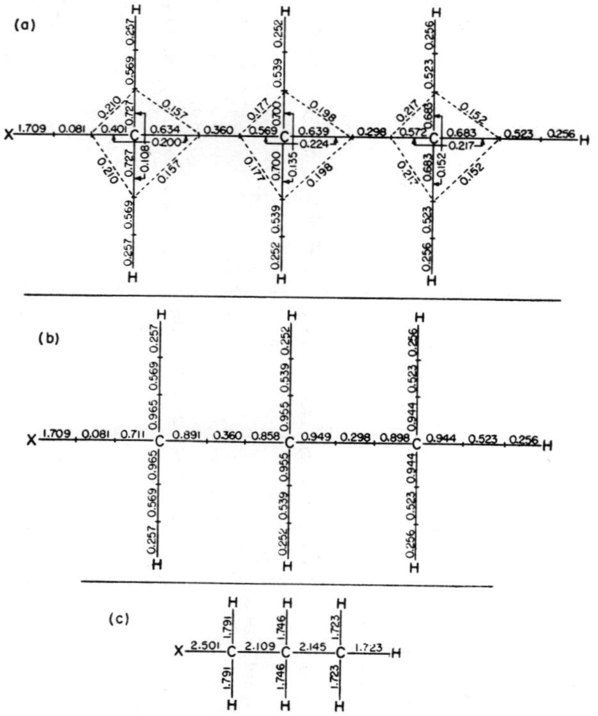

FIG. 16. "H" approximation. (+)-Ion of "*n*-X-propane" (see text).

In Figs. 13–14 we find diagrams for *n*-propane and its (+) ion, which we obtain by taking away an electron from the molecular orbital of highest energy. In Figs. 15 and 16 are diagrams for "*n*-X-propane" and its (+) ion. (Length of lines has no quantitative meaning in these figures.)

It has been long recognized that inclusion of overlap integrals makes definition of charges (and bond orders) more delicate (4, 23).

The most natural way of representing charge distribution in our opinion is the one suggested by MacWeeny (24) for conjugated compounds. He defines *local* charges as

[III] $$q_r = \sum c_r^2$$

and

[IV] $$q_{rs} = S_{rs} \sum c_r c_s.$$

In this manner we obtain diagrams (*b*) in Figs. 13–16, which are comparable with MacWeeny's diagram of naphthalene.

By adding all the charges in each chemical bond we obtain diagrams (*c*).

The coulombic terms we used are very probably too high and effective charges are lower in reality. This does not however affect our conclusions.

In C—H bonds the high value of overlap charge shows the trend of the electrons to approach nearer to the carbon. Again the C—C bonds have effective negative charges and the C—H bonds positive ones. If we compare the diagrams of *n*-propane with those of its (+) ion we find that the electronic charge is mainly taken away from the hydrogens. This is in accord with their lower electronegativity.

Now if we compare *n*-propane to "X-propane" we see that by far the greatest perturbation takes place in the X—C bond. (Effective charges are relatively higher in this approximation.) In the (+) ion of X-propane the relatively greatest share of the missing electron is taken away from the carbon next to the electronegative substituent X.

VI. CONCLUSIONS

It is believed, on the basis of the above results, that the LCAO MO method, with some refinements, will be capable of giving as good an account of electronic distribution in saturated hydrocarbons and their derivatives as in the case of conjugated compounds.

RÉSUMÉ

La simple méthode des orbitales moléculaires a été appliquée en trois approximations différentes à l'évaluation des diagrammes moléculaires des hydrocarbures saturés. Ces diagrammes qui donnent une idée de la répartition des électrons dans ces molécules permettent d'interpréter quelques différences fondamentales entre corps saturés et corps non-saturés à insaturation conjuguée. (Localisabilité des liaisons, l'effet d'un substituant électronégatif.)

Pour la première fois les électrons 1*s* des hydrogènes ont été introduits dans un calcul à l'aide de la méthode semiempirique des orbitales moléculaires.

ACKNOWLEDGMENTS

Work on saturated compounds by the LCAO method was started in Paris at the Centre de Chimie Théorique. The author wishes to express his gratitude to Prof. Raymond Daudel for many valuable suggestions.

We wish to express our thanks to Prof. Lucien Piché, head of the Department of Chemistry of the Université de Montréal, for his constant encouragement.

Secular equations necessary to obtain the diagrams in Sections IV and V of this article are of degrees 14 to 20. Their eigenvalues and eigenvectors were computed on the calculating machines installed by the National Research Council of Canada at the University of Toronto. We wish to thank the National Research Council for having allowed us a certain number of hours of machine time and Dr. C. C. Gotlieb, chief computer, and his collaborators, for having carried through the calculations.

A part of this work was carried through under the tenure of a university postdoctorate fellowship of the National Research Council and for this too the author wishes to express his appreciation.

We thank Mr. David Davies for help in preparation of the manuscript.

REFERENCES

1. ALTMANN, S. L. Proc. Roy. Soc. (London), A, 210: 327. 1951.
2. ARTMANN, K. Z. Naturforsch. 1: 426. 1946.
3. BROWN, R. D. J. Chem. Soc. 2615. 1953.
4. CHIRGWIN, B. H. and COULSON, C. A. Proc. Roy. Soc. (London), A, 201: 196. 1950.
5. COULSON, C. A. Proc. Phys. Soc. (London), 60: 257. 1948.
6. COULSON, C. A. and FISCHER, I. Phil. Mag. 40: 386. 1949.
7. COULSON, C. A. and LONGUET-HIGGINS, H. C. Proc. Roy. Soc. (London), A, 191: 39. 1947.
8. COULSON, C. A., MARCH, N. H., and ALTMANN, S. L. Proc. Natl. Acad. Sci. U.S. 38: 372. 1952.
9. CRAIG, D. P. Proc. Roy. Soc. (London), A, 200: 474. 1950.
10. DAUDEL, R. Private communication.
11. DAUDEL, R., ODIOT, S., and BRION, H. J. chim. phys. 51: 74, 358, 361, 554. 1954.
12. DAUDEL, R., SANDORFY, C., VROELANT, C., YVAN, P., and CHALVET, O. Bull. soc. chim. France, 17: 66. 1949.
13. DEWAR, M. J. S. The electronic theory of organic chemistry. The Clarendon Press, Oxford. 1949.
14. DEWAR, M. J. S. and LONGUET-HIGGINS, H. C. Proc. Roy. Soc. (London), A, 214: 482. 1952.
15. GOEPPERT-MAYER, M. and SKLAR, A. L. J. Chem. Phys. 6: 645. 1938.
16. GOUARNÉ, R. and YVAN, P. Compt. rend. 228: 1345. 1949.
17. HALL, G. G. Proc. Roy. Soc. (London), A, 205: 541. 1951.
18. HÜCKEL, E. Z. Physik, 70: 204. 1931.
19. LENNARD-JONES, J. E. Trans. Faraday Soc. 25: 668. 1929.
20. LENNARD-JONES, SIR J. Proc. Roy. Soc. (London), A, 198: 1, 14. 1949.
21. LENNARD-JONES, SIR J. and HALL, G. G. Trans. Faraday Soc. 48: 355. 1952.
22. LINNETT, J. W. and POË, A. J. Trans. Faraday Soc. 47: 1033. 1951.
23. LÖWDIN, P.-O. J. Chem. Phys. 18: 365. 1950.
24. MACWEENY, R. J. Chem. Phys. 19: 1614. 1951.
25. MOSER, C. M. J. Chem. Phys. 21: 2098. 1953.
26. MULLIKEN, R. S. Phys. Rev. 41: 49. 1932.
27. MULLIKEN, R. S. J. Chem. Phys. 7: 14, 20, 121, 339, 353. 1939.
28. MULLIKEN, R. S. J. chim. phys. 46: 497. 1949.
29. MULLIKEN, R. S., RIEKE, C. A., ORLOFF, D., and ORLOFF, H. J. Chem. Phys. 17: 1248. 1949.
30. PAULING, L. The nature of the chemical bond. Cornell Univ. Press, Ithaca, N.Y. 1945.
31. PAULING, L. and WILSON, E. B. Introduction to quantum mechanics. McGraw-Hill Book Company, Inc., New York and London. 1935.

32. PITZER, K. S. J. Am. Chem. Soc. 70: 2140. 1948.
33. PLATT, J. R. J. Chem. Phys. 18: 1168. 1950.
34. PULLMAN, A. and PULLMAN, B. Les théories électroniques de la chimie organique. Masson et Cie, Paris. 1952.
35. ROGER, F. Compt. rend. 236: 2207. 1953.
36. ROOTHAAN, C. C. J. Revs. Mod. Phys. 23: 69. 1951.
37. ROSS, I. G. Trans. Faraday Soc. 48: 973. 1952.
38. SANDORFY, C. and DAUDEL, R. Compt. rend. 238: 93. 1954.
39. VROELANT, C. Compt. rend. 236: 1666, 1887. 1953.
40. WHELAND, G. W. J. Am. Chem. Soc. 63: 2025. 1941.
41. WHELAND, G. W. J. Am. Chem. Soc. 64: 900. 1942.
42. WHELAND, G. W. and PAULING, L. J. Am. Chem. Soc. 57: 2086. 1935.
43. ZIMMERMANN, H. K. and VAN RYSSELBERGHE, P. J. Chem. Phys. 17: 598. 1949.

II-3 Reprinted from *The Journal of Chemical Physics* 39, 1397 (1963)

An Extended Hückel Theory. I. Hydrocarbons

ROALD HOFFMANN*

Chemistry Department, Harvard University, Cambridge 38, Massachusetts

(Received 10 April 1963)

The Hückel theory, with an extended basis set consisting of 2*s* and 2*p* carbon and 1*s* hydrogen orbitals, with inclusion of overlap and all interactions, yields a good qualitative solution of most hydrocarbon conformational problems. Calculations have been performed within the same parametrization for nearly all simple saturated and unsaturated compounds, testing a variety of geometries for each. Barriers to internal rotation, ring conformations, and geometrical isomerism are among the topics treated. Consistent σ and π charge distributions and overlap populations are obtained for aromatics and their relative roles discussed. For alkanes and alkenes charge distributions are also presented. Failures include overemphasis on steric factors, which leads to some incorrect isomerization energies; also the failure to predict strain energies. It is stressed that the geometry of a molecule appears to be its most predictable quality.

INTRODUCTION

THE Hückel theory has been widely exploited in chemistry, but a glance at any recent textbook will show that the emphasis has been rather one-sided.[1] The vast majority of calculations have been for planar conjugated and aromatic systems. Where hydrogens must be considered, they have been brought in by the artifice of a perturbation or a pseudoheteroatom. Where nonplanarity plays a role, the effects of the violation of the sigma–pi separation have been generally ignored. The few calculations on aliphatics have been limited to the method of linear combinations of bond orbitals,[2] and the resultant parametrization is difficult to relate to that used for aromatics. Indeed, the steady pursuit of correlations between theoretically computed π-electron properties and measurables has unfortunately cast a shadow of unreality on the σ framework.[3]

It is claimed here that the Hückel method, without the assumption of zero differential overlap, allows simply the calculation of the basic properties of all organic systems, aliphatic and aromatic, as well as inorganic structures, with one simple parametrization. The structure and relative importance of sigma and pi orbitals, where a separation exists, may be easily assessed. Conformational predictions, i.e., predictions of what three-dimensional shapes molecules take on in their ground states, are shown to be generally adequate. Indeed, the extended Hückel scheme succeeds, independent of the difficulties in choosing parameters, in just those areas, such as charge distributions, where the π-electron theory works; and performs as miserably in other areas, such as spectral predictions.

* Junior Fellow, Society of Fellows, Harvard University.

[1] A. Streitwieser, *Molecular Orbital Theory* (John Wiley & Sons, Inc., New York, 1961). R. Daudel, R. Lefebvre, C. Moser, *Quantum Chemistry* (Interscience Publishers, Inc., New York, 1959). B. Pullman and A. Pullman, *Les théories électroniques de la chimie organique* (Masson et Cie., Paris, 1952).

[2] (a) G. G. Hall and J. Lennard-Jones, Proc. Roy. Soc. (London) **A202**, 336 (1950); **205**, 357, 541 (1951); (b) R. D. Brown, J. Chem. Soc. **1953**, 2615; (c) M. J. S. Dewar and R. Petitt, *ibid.* **1954**, 1625; (d) C. Sandorfy, Can. J. Chem. **33**, 1337 (1955); (e) K. Fukui, H. Kato, K. Morokuma, A. Imamura, and C. Nagata, Bull. Chem. Soc. Japan **35**, 38 (1962) and references therein.

[3] A recent example of the consequences of this attitude may be seen in the work of O. Sovers and W. Kauzmann, J. Chem. Phys. **38**, 813 (1963).

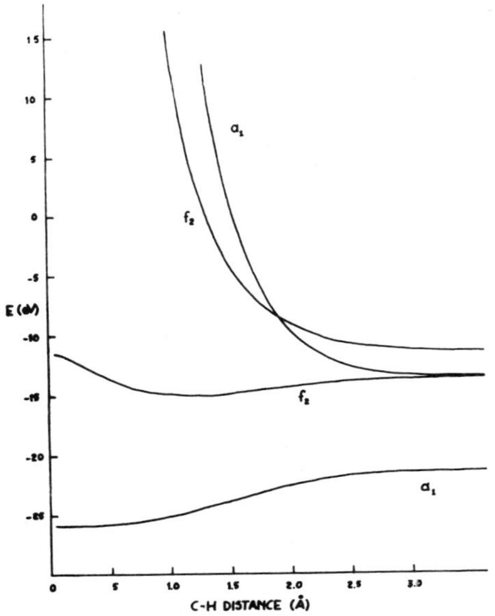

Fig. 1. Energy levels of methane as a function of C–H distance

METHOD OF CALCULATION

Expansion of a molecular orbital as a linear combination of atomic orbitals

$$\Psi_i = \sum_j c_{ij}\phi_j \qquad (1)$$

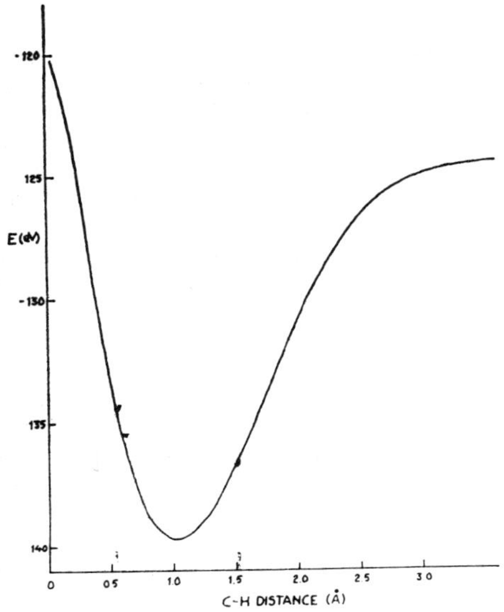

Fig. 2. Total energy of methane as a function of C–H distance.

yields, on minimizing the total energy, the set of Hückel equations

$$\sum_{i=1}^{n}[H_{ij}-ES_{ij}]c_{ij}=0 \qquad j=1, 2, \cdots n. \qquad (2)$$

For a calculation of a molecule C_nH_m we use a basis set consisting of m hydrogen Slater orbitals, exponent 1.0; n $2s$ and $3n$ $2p$ carbon Slater orbitals, exponent 1.625. The order of the resulting secular determinant is $4n+m$. The complete secular determinant is treated, all interactions accounted for, and off-diagonal E's retained. The critical choice is our manner of guessing the matrix elements H_{ij}. The H_{ii} are chosen as valence state ioniza-

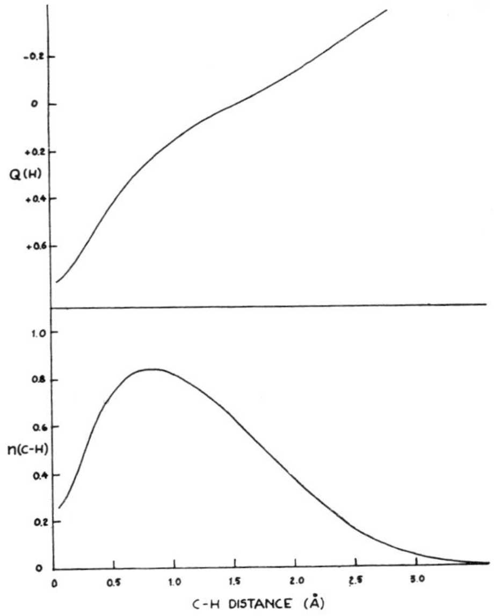

Fig. 3. Net charge on hydrogen (top) and C–H overlap population (bottom) in methane as a function of C–H distance.

tion potentials, and the particular values used are essentially those of Skinner and Pritchard[4] for the carbon sp^3 valence state

$$H_{ii}(C2p) = -11.4 \text{ eV},$$

$$H_{ii}(C2s) = -21.4,$$

$$H_{ii}(H1s) = -13.6.$$

The H_{ij} are approximated as

$$H_{ij}=0.5K(H_{ii}+H_{jj})S_{ij}. \qquad (3)$$

[4] H. A. Skinner and H. O. Pritchard, Trans. Faraday Soc. **49,** 1254 (1953); H. O. Pritchard and H. A. Skinner, Chem. Rev. **55,** 745 (1955).

TABLE I. Calculated total energies of some hydrocarbons.

	Molecule	ΣE_i (eV)		Molecule	ΣE_i (eV)
C_2H_6	ethane	−243.673	C_2H_4	ethylene	−210.484
CH_4	methane	−139.608	C_4H_6	*trans*-butadiene	−386.094
C_3H_8	propane	−347.889	C_3H_6	propylene	−314.895
C_4H_{10}	*n*-butane	−452.095	C_3H_4	allene	−281.438
C_4H_{10}	isobutane	−452.195	C_4H_8	*cis*-butene-2	−419.074
C_5H_{12}	*n*-pentane	−556.300	C_4H_8	*trans*-butene-2	−419.219
C_5H_{12}	isopentane	−556.133	C_4H_8	isobutylene	−419.292
C_5H_{12}	neopentane	−556.534	C_4H_8	butene-1	−419.052
C_6H_{14}	*n*-hexane	−660.504	C_4H_6	methyl allene	−385.686
C_6H_{14}	2,2-dimethylbutane	−660.200			
C_6H_{14}	2,3-dimethylbutane	−660.160	C_2H_2	acetylene	−178.028
C_6H_{14}	2-methylpentane	−660.326	C_3H_4	methyl acetylene	−282.300
C_6H_{14}	3-methylpentane	−660.070	C_4H_6	dimethyl acetylene	−386.447
C_7H_{16}	*n*-heptane	−764.709			
C_8H_{18}	*n*-octane	−868.914	C_6H_6	benzene	−527.068
C_9H_{20}	*n*-nonane	−973.118	C_7H_8	toluene	−631.437
C_6H_{12}	(chair) cyclohexane	−625.463	C_8H_{10}	*ortho*-xylene	−735.699
C_7H_{14}	methyl cyclohexane	−729.756	C_8H_{10}	*meta*-xylene	−735.808
C_8H_{16}	1,1-dimethylcyclohexane	−833.572	C_8H_{10}	*para*-xylene	−735.784
C_8H_{16}	1,2(*a-e*) dimethylcyclohexane	−833.247	$C_{10}H_8$	naphthalene	−843.085
C_8H_{16}	1,2(*e-e*) dimethylcyclohexane	−833.779	$C_{10}H_8$	azulene	−841.681
C_8H_{16}	1,3(*a-e*) dimethylcyclohexane	−833.511	$C_{14}H_{10}$	anthracene	−1158.974
C_8H_{16}	1,3(*e-e*) dimethylcyclohexane	−834.049	$C_{14}H_{10}$	phenanthrene	−1158.790
C_8H_{16}	1,4(*a-e*) dimethylcyclohexane	−833.520	$C_5H_5^-$	cyclopentadienyl anion	−449.624
C_8H_{16}	1,4(*e-e*) dimethylcyclohexane	−834.049	$C_7H_7^+$	cycloheptatrienyl cation	−602.329
$C_{10}H_{18}$	*trans*-decalin	−1007.396	$C_3H_3^+$	cyclopropenyl cation	−252.076
$C_{10}H_{18}$	*cis*-decalin	−1006.592	C_6H_6	fulvene	−525.960
C_3H_6	cyclopropane	−314.019	C_8H_6	pentalene	−665.576
C_4H_8	cyclobutane (planar)	−417.013	$C_{12}H_{10}$	biphenyl	−1019.095
C_5H_{10}	cyclopentane ($\frac{1}{2}$ chair)	−520.892	$C_{12}H_8$	biphenylene	−981.202
C_4H_8	methylcyclopropane	−418.286			

This parametrization, first discussed by Mulliken[5] and used in a molecular calculation by Wolfsberg and Helmholtz,[6] is examined in the Appendix, along with the only remaining choice, that of K in Eq. (3), which in our calculations is taken as 1.75. The computer program which performs our calculations has been described earlier.[7] Input to the program consists of precise atomic coordinates of the various atoms. The overlap matrix is internally computed, and the Hamiltonian matrix constructed from it by the recipe of Eq. (3). The complete set of Eq. (2) is solved with two matrix diagonalizations; for large molecules it is this step which determines the time consumed. The resultant wavefunction is subjected to a Mulliken population analysis,[8] yielding overlap populations and gross atomic populations or effective charges. At present the program is limited to a maximum of 68 orbitals; anthracene (66) and decalin (58) thus about define the limits of present calculations. A complete run on the latter molecule takes about 9 min on an IBM 7090, and a run on one configuration of ethane about 10 sec. This speed allows computations at many different geometries; indeed the approach to stereochemical problems used here will be to perform calculations at a variety of distances and orientations.

In Fig. 1 we show the energy levels of methane calculated by this method at a number of C–H distances, preserving tetrahedral symmetry. In Fig. 2 we plot the total Hückel energy, simply a sum of orbital energies of the eight valence electrons in the four filled orbitals of CH_4. Also shown in Fig. 3 are C–H overlap populations and the variation of the resultant charge on the hydrogens. A clear minimum in the potential curve is apparent at about 1.0 Å. The shape of the curve near the minimum is also more or less correct, as indicated by a rough calculated C–H stretching force constant of 5.5 mdyn/Å compared with the experimental value of 5.0.[9]

For almost all molecules, organic or inorganic, which we have considered, a Hückel calculation of this type, carried out as a function of internuclear distance, gives rise to a potential curve having a minimum not far from the correct experimentally determined geometry of the molecule.[10]

Of course this behavior has in theory been expected of all LCAO–MO functions and has been tested in nonempirical calculations on small molecules. It has been found that the correct internuclear distance has been obtained with wavefunctions which give a rather poor

[5] R. S. Mulliken, J. Chim. Phys. **46**, 497, 675 (1949).

[6] M. Wolfsberg and L. Helmholtz, J. Chem. Phys. **20**, 837 (1952).

[7] R. Hoffmann and W. N. Lipscomb, (a) J. Chem. Phys. **36**, 2179, 3489 (1962); (b) *ibid.* **37**, 2872 (1962).

[8] R. S. Mulliken, J. Chem. Phys. **23**, 1833, 1841, 2338, 2343 (1955).

[9] G. Herzberg, *Molecular Spectra and Molecular Structure* (D. Van Nostrand, Inc., New York, 1959), Vol. II.

[10] The exceptions are some diatomics and triatomics for which this theory fails; for instance the obvious case of the ground state of the hydrogen molecule.

TABLE II. Heats of formation at 0°K.[a]

	Molecule	$\Delta H_f°$		Molecule	$\Delta H_f°$
CH_4	methane	−15.987	C_8H_{16}	1,*trans* 4 dimethylcyclohexane	−31.99
C_2H_6	ethane	−16.517	C_5H_{10}	cyclopentane	−10.68
C_3H_8	propane	−19.482	C_2H_2	acetylene	54.329
C_4H_{10}	*n*-butane	−23.67	C_3H_4	methyl acetylene	46.017
C_5H_{12}	*n*-pentane	−27.23	C_4H_6	dimethyl acetylene	38.09
C_6H_{14}	*n*-hexane	−30.91	C_2H_4	ethylene	14.522
C_7H_{16}	*n*-heptane	−34.55	C_3H_6	propylene	8.468
C_8H_{18}	*n*-octane	−38.20	C_4H_8	butene-1	4.96
C_9H_{20}	*n*-nonane	−41.84	C_4H_8	*cis*-butene-2	3.48
C_4H_{10}	isobutane	−25.30	C_4H_8	*trans*-butene-2	2.24
C_5H_{12}	isopentane	−28.81	C_4H_8	isobutylene	0.98
C_5H_{12}	neopentane	−31.30	C_3H_4	allene	47.70
C_6H_{14}	2-methylpentane	−32.08	C_4H_6	methyl allene	42.00
C_6H_{14}	3-methylpentane	−31.97	C_4H_6	butadiene	29.78
C_6H_{14}	2,2-dimethylbutane	−34.65	C_6H_6	benzene	24.000
C_6H_{14}	2,3-dimethylbutane	−32.73	C_7H_8	toluene	17.500
C_6H_{12}	cyclohexane	−20.01	C_8H_{10}	*o*-xylene	11.096
C_7H_{14}	methylcyclohexane	−26.30	C_8H_{10}	*m*-xylene	10.926
C_8H_{16}	1,1 dimethylcyclohexane	−30.93	C_8H_{10}	*p*-xylene	11.064
C_8H_{16}	1,*cis* 2 dimethylcyclohexane	−28.95			
C_8H_{16}	1,*trans* 2 dimethylcyclohexane	−30.91			
C_8H_{16}	1,*cis* 3 dimethylcyclohexane	−32.02			
C_8H_{16}	1,*trans* 3 dimethylcyclohexane	−30.06			
C_8H_{16}	1,*cis* 4 dimethylcyclohexane	−30.08			

[a] As given by F. D. Rossini, K. S. Pitzer, R. L. Arnett, R. M. Braun, and G. C. Pimentel in *Selected Values of Physical and Thermodynamic Properties of Hydrocarbons and Related Compounds* (Carnegie Press, Pittsburgh, 1953).

binding energy.[11] Unfortunately, the clear implication of these results, that geometries are more easily predictable than energies, was lost in the eagerness with which the deceptively useful resonance energies of early semiempirical theories were accepted. Some other factors which have prevented semiempirical calculations of geometries include a certain inertia against calculating overlap integrals between noncoplanar atoms, and the unavailability until recent times of computers efficient enough to solve the high-order secular equations.

Partial justification of using a simple sum of orbital energies is given in the Appendix. It should be emphasized here that a sound theoretical basis for our model is, however, not as yet available.

The determination of the most favorable arrangement of atoms in even such a small molecule as ethane involves very many calculations at a multitude of geometries. To a limited extent we have carried out this absolute minimization procedure for methane, acetylene, ethylene, and ethane. The results are

methane: tetrahedral, C–H, 1.02 Å;

acetylene: linear; C–C, 0.85 Å; C–H, 1.0 Å;

ethylene: D_{2h} planar; C–C, 1.47 Å; C–H, 0.95 Å; angle HCH 125°;

ethane: (tetrahedral angles assumed) C–C 1.92 Å; C–H 1.0 Å, staggered.

[11] For H_2 the simplest LCAO and VB functions give a poor binding energy but an equilibrium separation within 10% of the correct value. Scaling, i.e., varying the Slater exponent, improves the energy somewhat and predicts the distance to 1% [see the review of H_2 calculations in A. D. McLean, A. Weiss, and M. Yoshimine, Rev. Mod. Phys. **32**, 211 (1960)]. For F_2 the best simple LCAO function gives the internuclear separation to 10%, but fails to predict binding [B. J. Ransil, Rev. Mod. Phys. **32**, 239, 245 (1960)].

All distances quoted are ±0.05 Å. At the indicated distance for ethane, the difference in energy between the staggered and eclipsed form was 0.8 kcal. It will be noted that these minima come at C–C distances which are much too short for acetylene, much too long for ethane. We are faced with the following dilemma: Since the method is evidently useful for crude conformational analysis, should we run a separate absolute minimization for each molecule, or should we process all molecules of a similar chemical nature at some standard distance. Considering the labor involved in the first alternative, the latter choice is clearly indicated. The next problem was to choose the distance at which to perform calculations for a large number of molecules. Should, for instance, aliphatics be processed at a C–C distance giving the most stable ethane (1.92 Å), or should their properties be computed at a more realistic separation, taking the risk of anomalous effects arising from a calculation which we know is not at the equilibrium molecular distance for the given approximate method of computation. The latter choice was made, and, we believe, vindicated to some extent by the results obtained below.

A large variety of hydrocarbons was studied, in various idealized geometries, which in unstrained cases corresponded to tetrahedral angles at aliphatic carbons, 120° HCH angle in olefins; C—C, 1.54 Å; C═C, 1.34 Å; C≡C, 1.21 Å, C—C in aromatics 1.40 Å; C—H, 1.10 Å throughout. Aromatic hydrogens were placed radially out of the ring they were on. The total energies for the most stable conformations of the compounds calculated are given in Table I. In Table II we list some experimental heats of formation at 0°K which will be relevant in the discussion. Where conformational problems arise, several calculations at different orientations were performed and these will be discussed separately. In Figs. 4–8 are shown the net charges and overlap populations for most molecules treated. For the more studied cases of ethane, ethylene, and benzene, the entire wavefunction will be given; for the other molecules these are available from the author.

Barrier to Internal Rotation in Ethane

At C–C, 1.54 Å; C–H, 1.10 Å; tetrahedral angles, the difference in energy between an eclipsed and a stag-

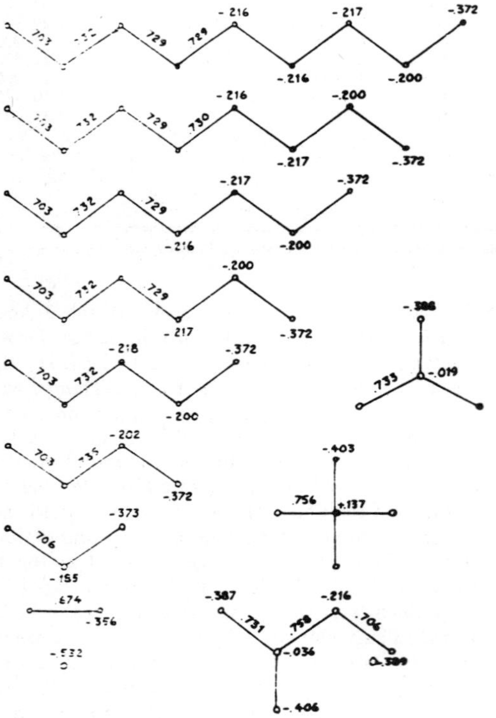

FIG. 4. Population analysis for alkane wavefunctions. Signed quantities are net charges, other numbers are C–C overlap populations.

gered conformation of CH_3CH_3 is computed to be 4.0 kcal/mole. The experimental value is 2.7–3.0 kcal/mole.

The aetiology of barriers is well developed.[12] It is our opinion that too much effort has gone into a search for a simple explanation to this phenomenon; in what follows we discuss only the symptoms accompanying ethane's torsional behavior.

The coordinates, energies, and AO coefficients for the seven filled orbitals in the staggered and eclipsed

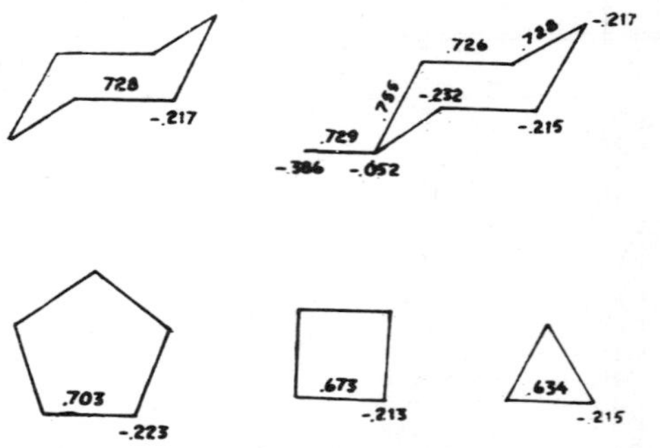

FIG. 5. Population analysis for some cycloalkane wavefunctions: chair cyclohexane, equatorial methyl cyclohexane, planar cyclopentane, planar cyclobutane, cyclopropane. Signed quantities are net charges, other numbers are C–C overlap populations.

conformations are listed in Tables III, IV. In Table V we give the gross atomic populations orbital by orbital, as well as certain subtotal overlap populations. Below we list $E_{ecl}-E_{stag}$ orbital by orbital (for degenerate e levels the energy difference is for one orbital of the pair)

e	0.101 eV,
a_1	−0.005,
e	−0.056,
a_2	0.002,
a_1	0.001.

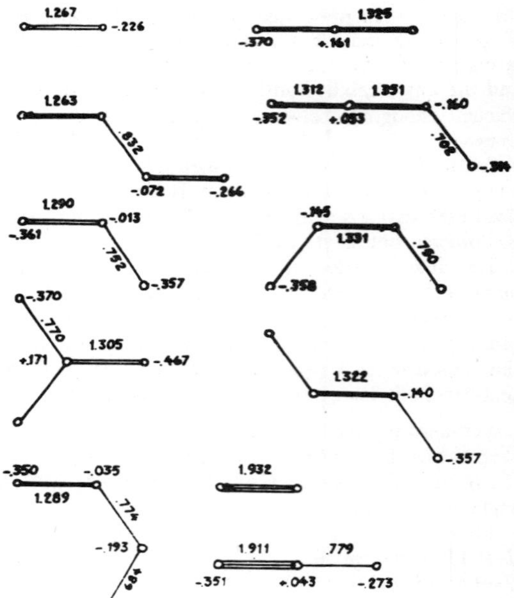

FIG. 6. Population analysis for some ethylenic and acetylenic hydrocarbons. The most stable conformation (see text) is chosen. Signed quantities are net charges, other numbers are C–C overlap populations.

[12] (a) W. G. Dauben and K. S. Pitzer in *Steric Effects in Organic Chemistry*, edited by M. S. Newman (John Wiley & Sons, Inc., New York, 1956), p. 1; (b) E. B. Wilson, Jr., Advan. Chem. Phys. **2**, 367 (1959); (c) D. J. Millen, Progr. Stereochem. **3**, 138 (1962).

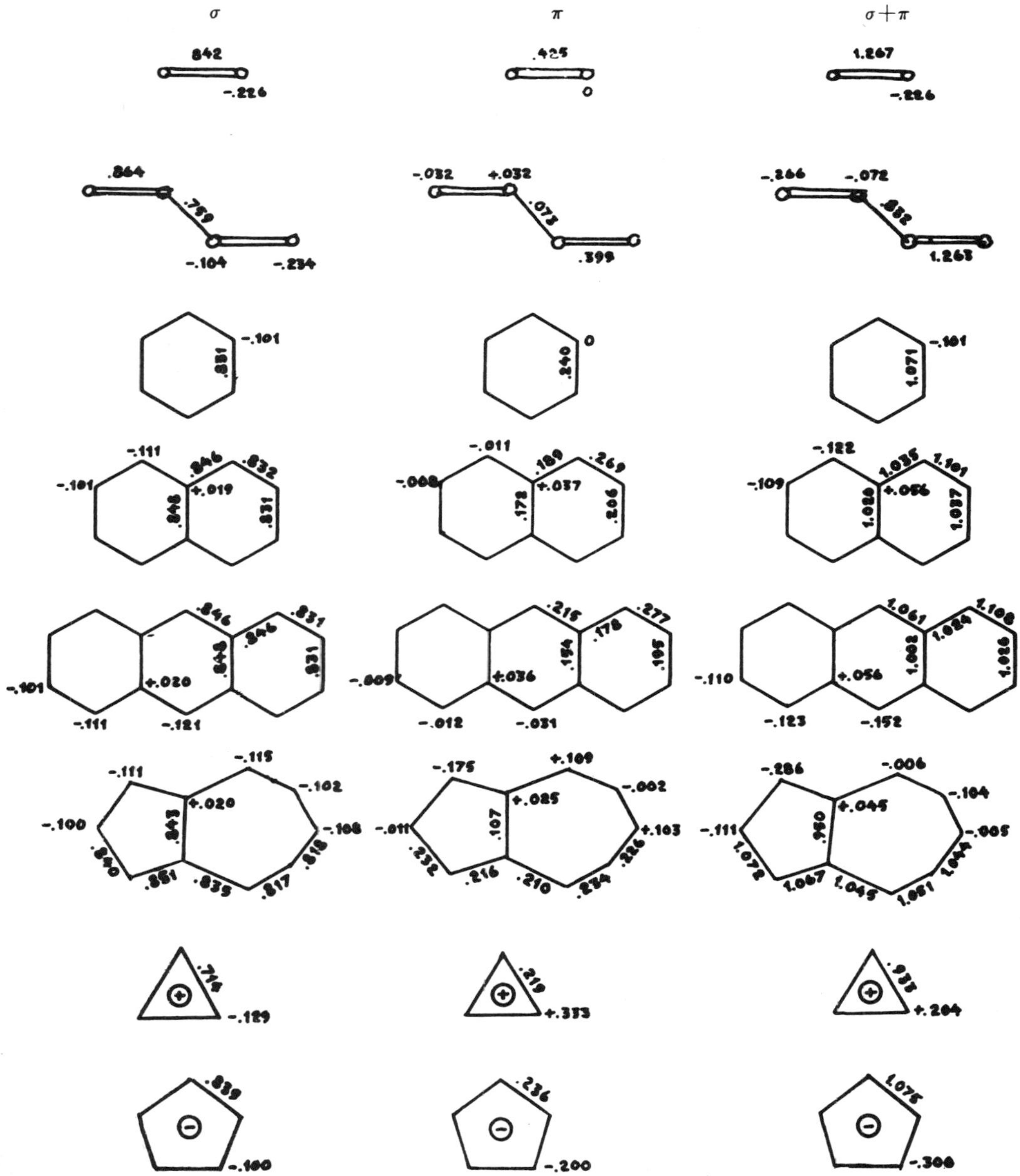

FIG. 7(a). Population analysis for aromatics and some other conjugated molecules. Net charges (signed numbers) and overlap populations (unsigned) are given separately for σ, π, $\sigma+\pi$ orbitals.

It can be noticed that the prevailing contribution to the barrier comes from the doubly degenerate orbitals; the top filled level determines the direction of the barrier. These orbitals are C–H orbitals composed of 1*s* H and $2p_x$, $2p_z$ carbon AO's ($2p_y$ lie along the C–C axis) and an examination of the electron distribution shows that in both cases transfer of electrons from carbons to protons is associated with greater stability of the staggered form. The total C–C overlap population is less for the eclipsed form,[13] while the C–H overlap population is very slightly greater. The charge on the

[13] In the minimization procedure the staggered form stabilized at a slightly shorter C–C distance than the eclipsed.

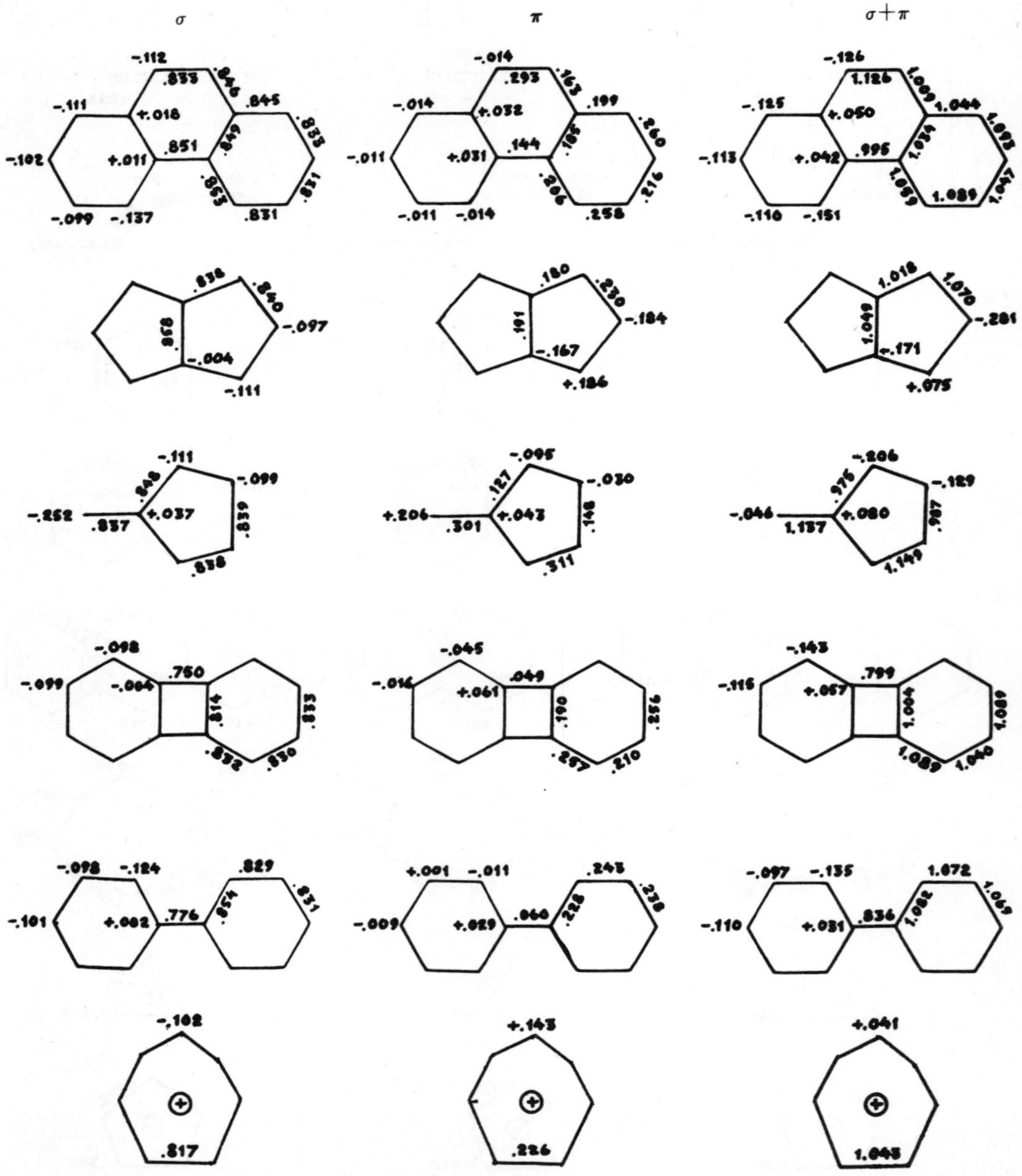

FIG. 7(b). Population analysis for aromatics and some other conjugated molecules. Net charges (signed numbers) and overlap populations (unsigned) are given separately for σ, π, $\sigma+\pi$ orbitals.

proton is slightly less in the staggered form. The nonbonded H–H overlap populations are -0.070 within a methyl group, for closest interactions across the molecule -0.017 staggered, -0.032 eclipsed. While the barrier itself is dependent on K (see Appendix), the qualitative features of the above symptoms, particularly the relative role of the various orbital contributions, does not change. Moreover all the manifestations of the barrier are in good agreement with those obtained in a recent nonempirical calculation by Pitzer.[14]

The shape of the barrier has been investigated. The equation

$$\Delta E = (4.02)0.5(1-\cos 3\theta)\ \text{kcal/mole}$$

[14] R. M. Pitzer (to be published).

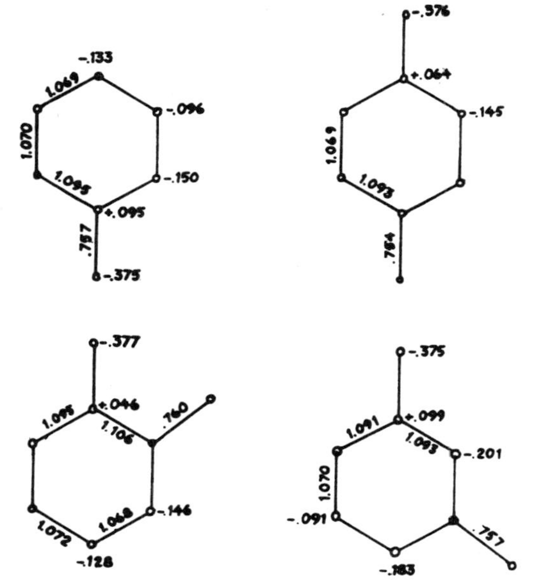

FIG. 8. Population analysis for toluene and the xylenes.

is followed with an absolute deviation of less than 1% over the entire range; the deviation is such that the actual curve is always wider than that given by a cosine form.

Normal and Branched Alkanes

The normal alkanes through *n*-nonane were considered in their completely staggered *trans* configurations. As expected the increment in total energy per CH_2 is constant, being 104.065 eV for methane–ethane, 104.216 for ethane–propane, 104.205±0.001 for all the higher members of the progression. The energy of the highest occupied orbital,

CH_4 −14.977 eV, C_6H_{14} −12.675,
C_2H_6 −13.759, C_7H_{16} −12.561,
C_3H_8 −13.419, C_8H_{18} −12.475,
C_4H_{10} −13.055, C_9H_{20} −12.409,
C_5H_{12} −12.832,

which is presumably related to the molecular ionization potential, decreased uniformly, but the absolute values are about 2 eV greater than the observed ionization potentials. In the population analysis (Fig. 4) note the remarkable uniformity as one progresses in the series. The terminal carbon in a chain carries a charge of −0.372, the penultimate is the most positive of the carbons, with a charge of −0.200, subsequent interior carbons stabilize at −0.216. The charges on the hydrogens and the carbon–hydrogen overlap populations are not shown in this figure, but representative values are given in Tables VI, VII and are discussed separately below.

Three conformations were examined for propane, corresponding to staggered–staggered, eclipsed–staggered, eclipsed–eclipsed arrangements of the terminal hydrogens with respect to the interior atoms. The following energies were computed

stag–stag, −347.889 eV,
stag–ecl, −347.645,
ecl–ecl, −347.351.

The equilibrium conformation predicted is thus the all staggered form, as was found by Lide in a microwave

TABLE III. Geometry of ethane: Cartesian coordinates of atoms.

staggered atom	x	y	z
C_1	0.0	0.77	0.0
C_2	0.0	−0.77	0.0
H_1	1.037089	1.136666	0.0
H_2	−0.518544	1.136666	−0.898146
H_3	−0.518544	1.136666	0.898146
H_4	0.518545	−1.136666	0.898146
H_5	−1.037089	−1.136666	0.0
H_6	0.518544	−1.136666	−0.898146
eclipsed			
C_1, C_2, H_1, H_2, H_3 same as staggered			
H_4	1.037089	−1.136666	0.0
H_5	−0.518544	−1.136666	0.898146
H_6	−0.518544	−1.136666	−0.898146

TABLE IV. Molecular orbitals in ethane.[a]

staggered

	-13.759 e_g		-14.111 a_{1g}	-15.857 e_u		-21.823 a_{2u}	-26.671 a_{1g}
H(1)	-0.3711	0.0630	-0.1177	0.3132	-0.0215	-0.1885	-0.0847
H(2)	0.1310	-0.3529	-0.1177	-0.1379	0.2820	-0.1885	-0.0847
H(3)	0.2401	0.2899	-0.1177	-0.1753	-0.2605	-0.1885	-0.0847
H(4)	0.1310	-0.3529	-0.1177	0.1379	-0.2820	0.1885	-0.0847
H(5)	-0.3711	0.0630	-0.1177	-0.3132	0.0215	0.1885	-0.0847
H(6)	0.2401	0.2899	-0.1177	0.1753	0.2605	0.1885	-0.0847
Cs(1)	-0.0000	0.0000	0.0788	-0.0000	0.0000	-0.4011	-0.4608
Cs(2)	-0.0000	0.0000	0.0788	0.0000	-0.0000	0.4011	-0.4608
Cx(1)	-0.4558	0.0774	-0.0000	0.3979	-0.0274	-0.0000	0.0000
Cx(2)	0.4558	-0.0773	0.0000	0.3979	-0.0274	0.0000	0.0000
Cy(1)	0.0000	-0.0000	-0.5618	-0.0000	0.0000	-0.1036	0.0201
Cy(2)	-0.0000	0.0000	0.5618	-0.0000	0.0000	-0.1036	-0.0201
Cz(1)	0.0773	0.4558	-0.0000	-0.0274	-0.3979	0.0000	0.0000
Cz(2)	-0.0774	-0.4558	0.0000	-0.0274	-0.3979	0.0000	-0.0000

eclipsed

	-13.658 e''		-14.116 a_1'	-15.913 e'		-21.821 a_2''	-26.670 a_1'
H(1)	-0.3711	0.0009	0.1179	-0.3176	0.0100	-0.1885	0.0846
H(2)	0.1848	-0.3218	0.1179	0.1501	-0.2801	-0.1885	0.0846
H(3)	0.1863	0.3209	0.1179	0.1675	0.2700	-0.1885	0.0846
H(4)	0.3711	-0.0009	0.1179	-0.3176	0.0100	0.1885	0.0846
H(5)	-0.1863	-0.3209	0.1179	0.1675	0.2700	0.1885	0.0846
H(6)	-0.1848	0.3216	0.1179	0.1501	-0.2801	0.1885	0.0846
Cs(1)	0.0000	0.0000	-0.0795	0.0000	-0.0000	-0.4015	0.4609
Cs(2)	-0.0000	0.0000	-0.0795	0.0000	0.0000	0.4015	0.4609
Cx(1)	-0.4723	0.0011	-0.0000	-0.3925	0.0124	-0.0000	-0.0000
Cx(2)	0.4723	-0.0011	-0.0000	-0.3925	0.0124	0.0000	-0.0000
Cy(1)	0.0000	-0.0000	0.5615	-0.0000	-0.0000	-0.1038	-0.0201
Cy(2)	0.0000	0.0000	-0.5615	0.0000	-0.0000	-0.1038	0.0201
Cz(1)	0.0011	0.4723	0.0000	0.0124	0.3925	-0.0000	0.0000
Cz(2)	-0.0011	-0.4723	-0.0000	0.0124	0.3925	0.0000	0.0000

[a] p_x, p_y, p_z orbitals are directed along x, y, z axes. Atoms are located at positions given in Table III.

study.[15] The barrier computed here, 0.244 eV, is much too large.

For *n*-butane 12 conformations were examined, three configurations of terminal hydrogens for each of four arrangements of carbons. In each geometry, the staggered–staggered arrangement of the terminal hydrogens was favored; the relative energies of the four carbon chain conformers are shown schematically in Fig. 9. Qualitatively again the picture conforms to what is known about *n*-butane—there being two potential minima, the more stable of which corresponds to the *trans* or *anti* form, the other to the *gauche* configuration.[12c,16] Quantitatively all the energy differences are too big; the difference between the *anti* and *gauche* forms, the barrier to rotation about the central C–C bond, the barrier to rotation about a terminal carbon–carbon bond (computed in the *anti* form to be 0.475 eV). By now it will be realized that this feature will appear throughout our calculations—an apparent overemphasis of what colloquially would be termed steric factors.

Four conformations were considered for isobutane, the most stable being that with all atoms staggered, as found experimentally.[17] The computed barrier to rotation was 0.300 eV. The isomerization energy of *n*-butane and isobutane is seen, from Table I, to be computed as 2.3 kcal/mole, compared to the experimental value of 1.6 kcal/mole.

For the various pentanes and hexanes, the hydrogen arrangements were assumed to be all staggered, and various carbon conformers were examined. In each case the most stable one was found to be that in which the longest possible chain was *anti* throughout. As may be gleaned from Tables I, II, the theory here fails to predict correctly the energetic relationships among the various isomers; thus while neopentane is computed to be more stable than *n*-pentane, isopentane is not. We attribute the failure here to the overemphasis of steric factors which we already noted in the extreme energy difference between *gauche* and *anti* *n*-butane. Since *gauche* conformations are unavoidable in the branched hexanes and isopentane, in our calculation they destabilize these so much as to make them less stable than the corresponding normal alkanes, in contrast to reality.

TABLE V. Ethane population analysis by orbitals.

Gross atomic populations (staggered)	e_g	a_{1g}	e_u	a_{2u}	a_{1g}
H	0.3117	0.0388	0.2844	0.1637	0.0825
C(*s*)	0	0.0090	0	0.4289	0.7446
C(p_x)	0.5324	0	0.5733	0	0
C(p_y)	0	0.8747	0	0.0799	0.0078
C(p_z)	0.5324	0	0.5733	0	0
n(C–H) = 0.8135		n(C–C) = 0.6742		N(H) = 0.8812	

Gross atomic populations (eclipsed)	e''	a_1'	e'	a_2''	a_1'
H	0.3008	0.0390	0.2926	0.1634	0.0826
C(*s*)	0	0.0091	0	0.4296	0.7446
C(p_x)	0.5488	0	0.5611	0	0
C(p_y)	0	0.8739	0	0.0802	0.0078
C(p_z)	0.5488	0	0.5611	0	0
n(C–H) = 0.8149		n(C–C) = 0.6529		N(H) = 0.8784	

TABLE VI. Aliphatic carbons and hydrogens.

Molecule	Q(C)	Q(H)	n(C–H)
methane	−0.532	+0.133	0.794
primary			
ethane	−0.356	+0.119	0.814
neopentane	−0.403	+0.123	0.809
n-nonane	−0.372	+0.120	0.812
isobutane	−0.388	+0.121	0.810
propylene	−0.357	+0.124[a]	0.808
		+0.136	0.791
methyl allene	−0.314	+0.124[a]	0.807
		+0.132	0.796
trans-butene-2	−0.357	+0.133	0.794
		+0.125[a]	0.809
cis-butene-2	−0.358	+0.138[b]	0.793
		+0.123	0.803
methyl acetylene	−0.273	+0.141	0.782
dimethyl acetylene	−0.275	+0.138	0.786
methyl cyclohexane	−0.386	+0.121	0.811
secondary			
cyclohexane	−0.217	+0.109[c]	0.828
		+0.108	0.829
propane	−0.185	+0.105	0.832
n-nonane next to terminal	−0.200	+0.108	0.831
n-nonane inner	−0.216	+0.108	0.830
cyclopropane	−0.215	+0.107	0.821
cyclobutane	−0.213	+0.106	0.828
tertiary			
isobutane	−0.019	+0.092	0.852
isopentane	−0.036	+0.094	0.850
quaternary			
neopentane	+0.137		

[a] Refers to single hydrogen eclipsing double bond, other entry to other two hydrogens.
[b] Refers to the two hydrogens staggered with respect to double bond, other entry to remaining hydrogen.
[c] Axial hydrogen, other entry refers to equatorial hydrogen.

Cycloalkanes

The cyclohexane system has been subjected to considerable stereochemical scrutinization.[12a,e,16] The ring

[15] D. R. Lide, J. Chem. Phys. **33**, 1514 (1960).
[16] E. L. Eliel, *Stereochemistry of Carbon Compounds* (McGraw-Hill Book Company, Inc., New York, 1962).
[17] D. R. Lide, J. Chem. Phys. **33**, 1519 (1960).

TABLE VII. Olefinic and acetylenic carbons and hydrogens.

Molecule	$Q(C)$	$Q(H)$	$n(C–H)$
primary			
ethylene	−0.226	+0.113	0.813
propylene	−0.361	+0.117[a]	0.810
allene	−0.370	+0.145	0.777
methyl allene	−0.352	+0.139[a]	0.781
butadiene	−0.266	+0.116[a]	0.810
isobutylene	−0.467	+0.119	0.807
butene-1	−0.350	+0.116[a]	0.810
secondary			
butadiene	−0.072	+0.107	0.822
trans-butene-2	−0.140	+0.106	0.822
cis-butene-2	−0.145	+0.105	0.826
butene-1	−0.035	+0.103	0.824
propylene	−0.013	+0.102	0.826
methyl allene	−0.160	+0.126	0.798
tertiary			
allene	+0.161		
methyl allene	+0.033		
isobutylene	+0.171		
acetylenic			
acetylene	−0.157	+0.157	0.789
methyl acetylene	−0.351	+0.158	0.788

[a] Average of two nonequivalent hydrogens.

system proper we have calculated in the conformations

chair	−625.463,
boat(C_{2v})	−624.695,
planar	−623.254.

The chair form is preferred, but the boat–chair energy difference is greater than the observed 5.5 kcal/mole[18] (this value should be really compared with the difference between a chair form and a twisted boat). The difference between axial and equatorial methyl cyclohexane is calculated as 0.529 eV in favor of the equatorially substituted conformer—this being much larger than the experimentally inferred value of about 1.8 kcal/mole.[16] The barrier to rotation is here computed as 0.284 eV. The relative stabilities of the dimethyl cyclohexanes are predicted remarkably well. In each case we have in order of decreasing stability: equatorial–equatorial, equatorial–axial, axial–axial.[16] The sterically hindered axial–axial 1, 3 dimethyl cyclohexane conformer has a particularly high energy. Finally *trans*-decalin is more stable than *cis*-decalin, with the computed difference again being much larger than the observed 2.7 kcal/mole.[12a]

For cyclopentane three conformations were examined:

planar	−520.722,
1/2 chair	−520.892,
envelope	−520.819.

[18] F. R. Jensen, D. S. Noyce, C. H. Sederholm, and A. J. Berlin, J. Am. Chem. Soc. **84**, 386 (1962).

The half-chair and envelope geometries are those given by Brutcher and Bauer[19] and are more puckered than the optimum forms described by Pitzer and Donath.[20] Both puckered forms are more stable than the planar arrangement, and for this degree of puckering, the half-chair is slightly preferred, as was also found by Brutcher and Bauer.

For cyclobutane the most stable conformation found was planar, but the potential curve was quite flat for bent forms with a dihedral angle up to 15°. Cyclopropane, as other cyclic and noncyclic paraffins, was studied with a C–C distance of 1.54 Å and tetrahedral exterior HCH angle. The resultant gap between filled and empty orbitals was smaller than in other saturated hydrocarbons. The barrier to internal rotation in methyl cyclopropane was computed as 0.137 eV. In subsequent work we hope to present contour plots of the electron density in these molecules. Some of the cycloalkane charge distributions are shown in Fig. 5.

Olefinic and Acetylenic Hydrocarbons

The coordinates and occupied molecular orbitals of ethylene are given in Tables VIII, IX. The highest filled and lowest empty orbitals are π type, as expected. The highest filled σ orbital, however, is not far below. Some recent work of Berry indicates that such an energy-spectrum is not unreasonable.[21] A calculation on the twisted D_{2d} geometry gives an energy 3.489 eV higher for its orbitally degenerate ground state.[22] The geometry

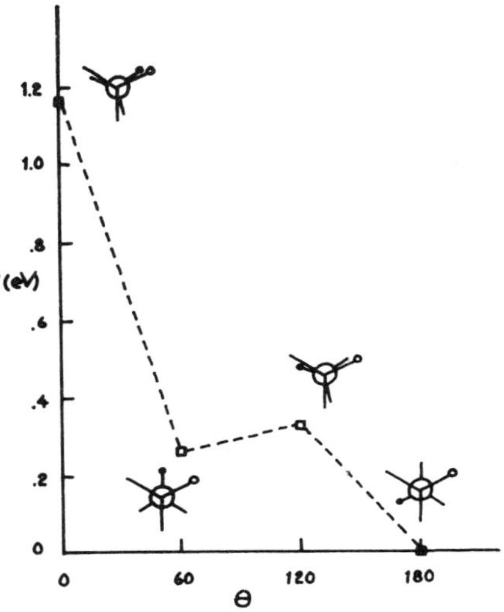

FIG. 9. Calculated energies of four conformations of *n*-butane, referred to the energy of the *trans* or *anti* form.

[19] F. V. Brutcher, Jr., and W. Bauer, Jr., J. Am. Chem. Soc. **84**, 2233 (1962).
[20] K. S. Pitzer and W. E. Donath, J. Am. Chem. Soc. **81**, 3213 (1959).
[21] R. S. Berry, J. Chem. Phys. **38**, 1934 (1963).
[22] R. S. Mulliken and C. C. J. Roothaan, Chem. Rev. **41**, 219 (1947).

is reversed in the first excited state. Allene (constructed with the same C–C distance as ethylene) is favored in the D_{2d} geometry by 1.264 eV over the planar structure. Again in the first excited state the favored geometry is the reverse. For propylene two conformations of the methyl group were examined and the form with a hydrogen eclipsing the double bond was favored over the staggered conformer by 1.1 kcal/mole. The absolute conformation is in agreement with a microwave determination.[23] Butadiene was examined in planar *cis* and *trans* geometries, and the *trans* form was favored by 7.6 kcal/mole.[24] The lowest filled π orbital was here found below a σ level. In methyl allene, as in propylene, the conformation with a hydrogen eclipsing the double bond was more stable by 1.3 kcal/mole. The calculated order of the propylene and methyl allene barriers is thus incorrect, the latter being 1.59 kcal/mole,[25] the former 1.98 kcal/mole.[26] However, from the previous section we have already learned that even fair absolute barriers are not to be expected from these calculations, while predictions of eqailibrium conformations are satisfactory.

The next group of compounds examined includes butene-1, *cis* and *trans* butene-2, and isobutylene. For butene-1, eight conformations were studied and the one of lowest energy had a nonplanar carbon arrangement with one of the methylene hydrogens eclipsing the double bond and the terminal methyl group staggered with respect to the methylene hydrogens. For isobutylene three conformations were examined—the most stable one having both terminal methyl groups arranged as in propylene (calculated barrier here 0.8 kcal/mole). A similar arrangement, again with hydrogens eclipsing the double bond, was found in *trans*-butene-2, where the computed barrier is 1.1 kcal/mole. For *cis*-butene-2 the nonbonded steric factor dominates and the stable conformation comes out with both methyl groups staggered with respect to the double bond. The barrier here is calculated very large, 10.6 kcal/mole. Thermochemically the *cis*-butene-2 barrier is smaller than that of the *trans* isomer,[12a] very clearly our method of calculation fails badly here.

The order of stabilities of the butenes is given correctly, though with some evidence of the steric problem in our model of *cis*-butene-2, as may be seen from the isomerization energies to the most stable isomer, isobutylene

	observed	calculated
trans-butene-2	1.26	1.68 kcal/mole,
cis-butene-2	2.50	5.03,
butene-1	3.98	5.54.

[23] D. R. Herschbach and L. C. Krisher, J. Chem. Phys. **28**, 728 (1958).

[24] R. G. Parr and R. S. Mulliken, J. Chem. Phys. **18**, 1338 (1950), obtained somewhat smaller estimates. See also O. Polansky, Monatsh. Chem. **94**, 23 (1963).

[25] D. R. Lide and D. E. Mann, J. Chem. Phys. **27**, 874 (1957).

[26] D. R. Lide and D. E. Mann, J. Chem. Phys. **27**, 868 (1957).

TABLE VIII. Ethylene atom positions.

atom	x	y	z
C_1	0.0	0.67	0.0
C_2	0.0	−0.67	0.0
H_1	0.952629	1.22	0.0
H_2	−0.952629	1.22	0.0
H_3	0.952629	−1.22	0.0
H_4	−0.952629	−1.22	0.0

TABLE IX. Molecular orbitals in ethylene.[a]

	-13.218 b_{1u}	-13.776 b_{1g}	-14.448 a_g	-16.215 b_{3u}	-20.604 b_{2u}	-26.981 a_g
H(1)	0.	0.3436	0.1904	-0.2668	0.2396	0.0888
H(2)	0.	-0.3436	0.1904	0.2668	0.2396	0.0888
H(3)	-0.	-0.3436	0.1904	-0.2668	-0.2396	0.0888
H(4)	-0.	0.3436	0.1904	0.2668	-0.2396	0.0888
Cs(1)	-0.	0.0000	-0.0741	-0.0000	0.3860	0.4870
Cs(2)	0.	-0.0000	-0.0741	-0.0000	-0.3860	0.4870
Cx(1)	0.	0.4428	-0.0000	-0.3821	0.0000	0.0000
Cx(2)	-0.	-0.4428	0.0000	-0.3821	-0.0000	0.0000
Cy(1)	0.	0.0000	0.5265	0.0000	0.1768	-0.0240
Cy(2)	0.	-0.0000	-0.5265	-0.0000	0.1768	0.0240
Cz(1)	0.6275	0.	0.	-0.	-0.	0.
Cz(2)	0.6275	-0.	-0.	0.	0.	-0.

[a] p_x, p_y, p_z orbitals are directed along x, y, z axes. Atoms are located at positions given in Table VIII.

Calculations were also performed for acetylene,[27] methyl acetylene, and dimethyl acetylene. For the latter the computed barrier is 0.01 kcal/mole in favor of the eclipsed form. Since this magnitude is near the estimated accuracy of our calculation, it is not clear if the number is significant. Methyl acetylene is correctly computed to be more stable than allene; the relative stability of butadiene and methyl allene is also given correctly, but not their relation to dimethyl acetylene.

Charges on Carbons and Hydrogens

The figures illustrating charge distributions in the hydrocarbons do not indicate hydrogen charges of C–H overlap populations. The inclusion of these quantities in the drawings would complicate the latter extensively. Moreover, these numbers show a regularity which may be appreciated from some typical and atypical cases given in Tables VI, VII. The over-all variation in hydrogen charge is small, while that of the carbon charges is considerable. The more carbons are bonded to a given carbon, the more positive it becomes; the associated hydrogens become slightly less positive and the C–H overlap populations rise. Those hydrogens which are sterically unhappy, i.e., which are forced into excessive proximity to other hydrogens, acquire a positive character roughly proportional to their discomfort—this already begins to be seen in the eclipsed ethane and the

[27] The acetylene wavefunction we obtain compares favorably with the SCF functions calculated by A. D. McLean, J. Chem. Phys. **32**, 1595 (1960) [see also A. D. McLean, B. J. Ransil, and R. S. Mulliken, *ibid.* **32**, 1873 (1960)], and L. Burnelle, *ibid.* **35**, 311 (1961). Our orbital energies, in eV, are $2\sigma_g$−27.120, $2\sigma_u$−19.642. $3\sigma_g$−15.186, π_u−13.533. The total C–C overlap population is 1.93, of which 1.00 comes from the π orbitals. The charges and overlap population may be found in Table VII.

TABLE X. Atomic positions in benzene.

	x	y	z
C_1	1.40	0.0	0.0
C_2	−1.40	0.0	0.0
C_3	0.70	1.212436	0.0
C_4	0.70	−1.212436	0.0
C_5	−0.70	−1.212436	0.0
C_6	−0.70	1.212436	0.0
H_1	2.50	0.0	0.0
H_2	−2.50	0.0	0.0
H_3	1.25	2.165064	0.0
H_4	1.25	−2.165064	0.0
H_5	−1.25	−2.165064	0.0
H_6	−1.25	2.165064	0.0

axial hydrogens of cyclohexane, and was very apparent in some of the more sterically unfavored conformations which we examined. It is interesting to note that in the equilibrium conformations of molecules of the propylene type, there was a noticeable difference in the charges on the methyl hydrogens, the one eclipsing the double bond being more or less "normal," the other two more positive.

The carbon charge distributions are in themselves extremely interesting, particularly since we believe that these are the first estimates of these quantities. Though they are undoubtedly too drastic, we think they bear careful study in the interpretation of reactions—but such an undertaking is beyond the present scope of this work. We will only point to some of the more interesting results: the uniformity and charge alternation in the normal paraffin series; the charge distribution in methyl cyclohexanes (the charge alternation reminiscent of that to be discussed below for toluene ; the charge distributions in the allenes and butenes.

Aromatic Hydrocarbons

In Tables X, XI we give the coordinates and occupied molecular orbitals for benzene, calculated at C–C, 1.40 Å; C–H, 1.10 Å. The most interesting feature of this level scheme is that the lowest bonding π orbital is located below some of the σ levels. Indeed, such behavior was noted for all aromatics: the highest filled orbital was π type, as were the first few unoccupied levels, but lower bonding σ and π levels were interspersed. It is difficult to conceive of an experiment which could distinguish our arrangement of energy levels from the conventionally assumed one where all the occupied σ levels lie below the π's[28]; confirmation must await a complete SCF calculation on benzene.

In Fig. 7 we show charge distributions and overlap populations for a number of simple alternant and nonalternant conjugated systems. These are shown separately for the σ and π frameworks and for the composite, $\sigma+\pi$. Fukui *et al.*[29] have recently carried out calculations on σ frameworks in aromatics by taking linear combinations of H 1*s* and C *sp*² hybrid orbitals, with overlap and nonnearest-neighbor interactions neglected. The general features of the charge distributions obtained here agree with the above-quoted work, though there are discrepancies (e.g., order of charges of 1 and 2 positions in naphthalene is reversed). We believe that the great advantage of our calculation is that σ and π orbitals are obtained within one parametrization, while in Fukui's work the relative magnitudes of the Coulomb and resonance integrals used in σ and π calculations

TABLE XI. Molecular orbitals in benzene.[a]

	-12.797		-12.839		-14.297	-14.510	-14.637		-16.576	-16.601	-19.933		-25.785		-29.567
	e_{1g}		e_{2g}		b_{2u}	a_{2u}	e_{1u}		a_{1g}	b_{1u}	e_{2g}		e_{1u}		a_{1g}
H(1)	-0.	-0.	0.1833	-0.1998	-0.0000	0.	-0.2059	0.1588	-0.1803	0.2526	-0.1391	0.1787	0.0019	0.1081	-0.0151
H(2)	0.	0.	0.1833	-0.1998	0.0000	-0.	0.2059	-0.1588	-0.1803	-0.2526	-0.1391	0.1787	-0.0019	-0.1081	-0.0151
H(3)	0.	0.	-0.2647	-0.0589	-0.0000	-0.	0.0345	0.2577	-0.1803	-0.2526	0.2243	0.0311	0.0946	0.0524	-0.0151
H(4)	0.	0.	0.0814	0.2587	0.0000	-0.	-0.2404	-0.0989	-0.1803	-0.2526	-0.0852	-0.2098	-0.0927	0.0557	-0.0151
H(5)	-0.	-0.	-0.2647	-0.0589	0.0000	0.	-0.0345	-0.2577	-0.1803	0.2526	0.2243	0.0311	-0.0946	-0.0524	-0.0151
H(6)	-0.	-0.	0.0814	0.2587	-0.0000	0.	0.2404	0.0989	-0.1803	0.2526	-0.0852	-0.2098	0.0927	-0.0557	-0.0151
Cs(1)	0.	0.	-0.0209	0.0228	0.0000	-0.	0.0722	-0.0557	0.0199	0.1744	-0.1949	0.2504	0.0074	0.4150	-0.2810
Cs(2)	-0.	-0.	-0.0209	0.0228	-0.0000	0.	-0.0722	0.0557	0.0199	-0.1744	-0.1949	0.2504	-0.0074	-0.4150	-0.2810
Cs(3)	-0.	-0.	0.0301	0.0067	0.0000	0.	-0.0121	-0.0904	0.0199	-0.1744	0.3143	0.0436	0.3631	0.2011	-0.2810
Cs(4)	-0.	-0.	-0.0093	-0.0295	-0.0000	0.	0.0843	0.0347	0.0199	-0.1744	-0.1194	-0.2940	-0.3557	0.2139	-0.2810
Cs(5)	0.	0.	0.0301	0.0067	-0.0000	-0.	0.0121	0.0904	0.0199	0.1744	0.3143	0.0436	-0.3631	-0.2011	-0.2810
Cs(6)	0.	0.	-0.0093	-0.0295	0.0000	-0.	-0.0843	-0.0347	0.0199	0.1744	-0.1194	-0.2940	0.3557	-0.2139	-0.2810
Cx(1)	-0.	-0.	0.1935	-0.2109	-0.0000	0.	-0.3017	0.2326	-0.2263	0.1846	-0.0643	0.0826	-0.0003	-0.0148	-0.0045
Cx(2)	-0.	-0.	-0.1935	0.2109	-0.0000	0.	-0.3017	0.2326	0.2263	0.1846	0.0643	-0.0826	-0.0003	-0.0148	0.0045
Cx(3)	0.	0.	-0.2102	0.2859	-0.2875	-0.	0.1438	0.1729	-0.1131	-0.0923	0.0299	0.1649	-0.0297	0.0384	-0.0022
Cx(4)	0.	0.	-0.2666	0.2340	0.2375	-0.	-0.1307	-0.1830	-0.1131	-0.0923	-0.1672	0.0114	0.0311	0.0373	-0.0022
Cx(5)	0.	0.	0.2102	-0.2859	-0.2875	-0.	0.1438	0.1729	0.1131	-0.0923	-0.0299	-0.1649	-0.0297	0.0384	0.0022
Cx(6)	0.	0.	0.2663	-0.2340	0.2875	-0.	-0.1307	-0.1830	0.1131	-0.0923	0.1672	-0.0114	0.0311	0.0373	0.0022
Cy(1)	0.	0.	0.2763	0.2535	-0.3319	-0.	-0.0843	-0.1093	-0.0000	-0.0000	0.1451	0.1129	0.0554	-0.0010	0.0000
Cy(2)	0.	0.	-0.2763	-0.2535	-0.3319	-0.	-0.0843	-0.1093	0.0000	0.0000	-0.1451	-0.1129	0.0554	-0.0010	0.0000
Cy(3)	0.	0.	-0.2013	-0.2368	0.1660	-0.	-0.0246	0.3362	-0.1960	-0.1599	0.1024	-0.0786	0.0022	-0.0304	-0.0039
Cy(4)	-0.	-0.	-0.2533	-0.1302	0.1660	0.	0.3314	0.0617	0.1960	0.1599	-0.0511	0.1185	0.0033	0.0303	0.0039
Cy(5)	0.	0.	0.2013	0.2368	0.1660	-0.	-0.0246	0.3362	0.1959	-0.1599	-0.1024	0.0786	0.0022	-0.0304	0.0039
Cy(6)	-0.	-0.	0.2533	0.1302	0.1660	0.	0.3314	0.0617	-0.1959	0.1599	0.0511	-0.1185	0.0033	0.0303	-0.0039
Cz(1)	0.5101	0.1367	-0.	0.	-0.	0.3257	-0.	0.	-0.	0.	0.	-0.	0.	0.	-0.
Cz(2)	-0.5101	-0.1367	0.	-0.	0.	0.3257	0.	-0.	0.	-0.	-0.	0.	-0.	-0.	0.
Cz(3)	0.3734	-0.3734	0.	-0.	0.	0.3257	0.	-0.	0.	-0.	-0.	0.	-0.	-0.	0.
Cz(4)	0.1367	0.5101	0.	-0.	0.	0.3257	0.	-0.	0.	-0.	-0.	0.	-0.	-0.	0.
Cz(5)	-0.3734	0.3734	-0.	0.	-0.	0.3257	-0.	0.	-0.	0.	0.	-0.	0.	0.	-0.
Cz(6)	-0.1367	-0.5101	-0.	0.	-0.	0.3257	-0.	0.	-0.	0.	0.	-0.	0.	0.	-0.

[a] p_x, p_y, p_z orbitals are directed along *x*, *y*, *z* axes. Atoms are located at positions given in Table X.

[28] M. P. Gouterman has suggested that a careful search for transitions arising from $\sigma\rightarrow\pi^*$ excitations, and thus polarized perpendicular to the aromatic ring plane, would be useful in this respect. Our energy spectrum supports that given schematically by J. C. Slater, *Quantum Theory of Molecules and Solids* (McGraw-Hill Book Company, Inc., New York, 1963), Vol. 1, p. 234.

must be estimated *a posteriori*. The general features of our population analysis may be summarized as follows.

Charges

(1) π charges for alternant hydrocarbons generally small (in the simplest Hückel theory they all vanish) and fall into two groups depending on the number of nearest-neighbor carbons; slightly positive for carbons bonded to three others, slightly negative for carbons bonded to two others. The π charge distributions in naphthalene agree with those calculated by Ruedenberg[29] and McWeeny.[30]

(2) π charges appreciable for nonalternant hydrocarbons.

(3) σ charges, independent of alternant character, fall into three classes depending on the number of nearest-neighbor carbons, and clustered around the following values: -0.23 single neighbor, -0.11 two neighbors, $+0.01$ three neighbors.

(4) Proton charges (not indicated in figures) are all close to $+0.10$ (±0.01), except for the sterically unfavored hydrogens in phenanthrene and planar biphenyl, which are more positive.

(5) Within each σ and π class defined in (1) and (3) above, for alternant hydrocarbons the magnitude of charge variation over the molecule is small, and in the same direction in the σ and π frameworks, therefore in the entire molecule.

(6) For nonalternant hydrocarbons, σ and π variations are not necessarily in the same direction. The π charge variation dominates and is in agreement with calculations from the simple Hückel theory.[1]

Overlap Populations

(1) Quite constant in the σ framework, independent of bond location (except for biphenylene, where the long bonds are clearly indicated even in the σ overlap population; biphenyl was computed with a C–C distance of 1.54 Å for the central bond and thus the small overlap population was built into the calculation).

(2) Quite varied in the π orbitals, and in excellent agreement with ordinary HMO bond orders and thus observed bond lengths. $\sigma+\pi$ dominated by π variation.

Conformations and isomerization energies calculated for the aromatics are generally fair. The sixfold barrier to internal rotation in toluene is computed to be negligible (less than 1 cal). A large barrier is encountered in *ortho*-xylene. This molecule is of particular interest because of the similarity of the barrier geometry to that in *cis*-butene-2; here the underlying rotational barrier should be nearly vanishing and only the nonbonded steric interaction remains, thus permitting an analysis of the relative roles of the two contributions. Unfortunately we have found no estimate of the *ortho*-xylene barrier. The value computed by our method, no doubt too large again, is 0.419 eV with the most stable conformation, similar to that calculated for *cis*-butene-2, being that in which a hydrogen in both methyl groups eclipses the benzene ring plane away from the other methyl group.

TABLE XII. π-electron energies in the aromatics.

ethylene	−26.436 eV
cyclopropenyl cation	−28.414
butadiene	−52.964
cyclopentadienyl anion	−77.064
benzene	−80.208
cycloheptatrienyl cation	−82.002
fulvene	−78.928
pentalene	−105.328
naphthalene	−133.676
azulene	−132.934
biphenylene	−159.964
biphenyl	−160.436
phenanthrene	−187.286
anthracene	−187.020

Internal rotation about the central carbon–carbon bond in biphenyl has attracted much attention, theoretical as well as experimental. The molecule is planar in the crystal[31] and twisted in the vapor.[32] Theoretical computations indicate a small energy difference favoring a nonplanar form.[33] We have performed a calculation at only two conformations: rings coplanar and perpendicular. In both cases normal bond lengths were retained for the phenyl groups, but the bond joining the rings was taken as 1.54 Å. The perpendicular form comes out more stable by 0.449 eV—again we think overbiasing of nonbonded repulsions is involved.

The order of stabilities of the xylenes is in agreement with thermochemical data, though the quantitative differences are once again too large. Naphthalene emerges 32.3 kcal/mole more stable than azulene, in fortuitous agreement with the experimental value of 32.6 kcal/mole.[34] Fulvene is computed to be 25.6 kcal/mole less stable than benzene, the observed difference being about 27 kcal/mole.[35] Anthracene comes out 4.2 kcal/mole more stable than phenanthrene, while in actuality the latter isomer is more stable by 6.9 kcal/mole.[36] Again we think phenanthrene is discriminated against as a result of its two sterically unhappy hydrogens. This is borne out by the fact that the energy of the π orbitals only (Table XII) favors phenanthrene by 5.4 kcal/mole.

[29] K. Ruedenberg, J. Chem. Phys. **34**, 1878 (1961).
[30] R. McWeeny, J. Chem. Phys. **19**, 1614 (1951).

[31] J. Trotter, Acta Cryst. **14**, 1135 (1961).
[32] I. L. Karle and L. O. Brockway, J. Am. Chem. Soc. **66**, 1974 (1944); O. Bastiansen, Acta Chem. Scand. **3**, 408 (1949).
[33] C. A. Coulson, Conference on Quantum Mechanical Methods in Valence Theory, Shelter Island, New York, 1961, p. 42. F. J. Adrian, J. Chem. Phys. **28**, 608 (1958).
[34] E. Heilbronner in *Nonbenzenoid Aromatic Compounds*, edited by D. Ginsburg (Interscience Publishers, Inc., New York, 1959), p. 171.
[35] J. H. Day and C. Oestreich, J. Org. Chem. **22**, 214 (1956).
[36] A. Magnus, H. Hartmann, and F. Becker, Z. Physik. Chem. **197**, 75 (1951).

TABLE XIII. Highest occupied and lowest unoccupied energy levels in the aromatics and other compounds.

Molecule	Highest occupied level (eV)	Lowest unoccupied level (eV)	Gap (eV)
benzene	−12.797	−8.345	4.452
naphthalene	−12.073	−9.338	2.635
anthracene	−11.642	−9.839	1.803
phenanthrene	−12.023	−9.310	2.713
azulene	−11.730	−9.872	1.858
pentalene	−11.492	−10.809	0.683
fulvene	−11.991	−10.338	1.653
biphenylene	−11.555	−9.553	2.002
biphenyl (planar)	−12.284	−8.967	3.317
cyclopropenyl cation	−13.374	−8.636	4.738
cyclopentadienyl anion	−11.991	−6.464	5.527
cycloheptatrienyl cation	−12.922	−9.894	3.028
toluene	−12.502	−8.348	4.154
p-xylene	−12.246	−8.347	3.899
m-xylene	−12.397	−8.247	4.150
o-xylene	−12.382	−8.221	4.161
ethane	−13.759	3.131	16.890
ethylene	−13.218	−8.238	4.980
butadiene	−12.592	−9.031	3.561
acetylene	−13.533	−7.142	6.391

However, it may well be that phenanthrene is not precisely planar.[37]

In Fig. 8 we see the population analysis results for toluene and the xylenes. These are of particular interest since the calculation is accomplished taking into account the hydrogen atoms directly, thus without any *ad hoc* assumptions about hyperconjugation. The toluene charge distribution is in agreement with its *o–p* directing character. It is interesting that the charge differentiation in the ring is accomplished with only a small charge transfer: the C_6H_5 group in benzene contains 29.101 electrons, in toluene 29.021. Incidentally, this is contrary to the accepted picture of a methyl group as an electron donor.[38]

Energies of the top filled and lowest empty orbitals for aromatics and a few other compounds are shown in Table XIII. The absolute values of the gaps are too small if one relates them to the energy of the aromatic *p*-band transitions, but the variation is very similar to that computed by the simple Hückel theory. The calculated ionization potentials are similarly off, in this case being too large. The particularly small gap for pentalene can be relieved by formation of the −2 ion, which indeed was recently synthesized.[39]

A calculation was performed on a number of possible geometries for cyclo-octatetraene. In agreement with recent experimental results,[40] the tub form was favored in the neutral molecule, but the anion and dianion preferred the planar geometry.

Energetic Relationships

In our previous work on the boron hydrides we found that it was possible to make some absolute thermochemical sense from the computed total energies. If we calculate a binding energy for CH_4, i.e., subtract $\Sigma E = -139.608$ eV from ΣE of the atoms at infinity[41] (C at infinity taken in sp^3 valence state), we obtain a binding energy of 29.608 eV, roughly 1.8 times the correct atomization energy. This factor, not very different from K, approximately relates all of the calculated binding energies to the true dissociation energies.[42]

In themselves the binding energies are not very interesting and are not tabulated here—they may easily be computed from the data in Table I. A question of interest is whether an over-all additive scheme of bond energies exists in our calculations. The prognosis must be poor since different atomic separations are used for different classes of compounds. For this purpose we can work with total energies, if we remember not to attribute any direct significance to the magnitudes manipulated. From *n*-octane and *n*-nonane we obtain the contribution to the total energy per bond: C–C, 34.4610 eV; C–H, 34.8715 eV. These reproduce the other normal paraffin energies very well but lead to disagreements already for methane and cyclohexane. Extended to other types of bonds, matters are sometimes good, sometimes bad. Another serious deficiency manifests itself in the energy per CH_2 in the cycloalkane series: C_3H_6, 104.673 eV; C_4H_8, 104.253; C_5H_{10}, 104.178; C_6H_{12}, 104.244. Thus no strain energies are apparent.

CONCLUSIONS

The most important accomplishment of this work is the demonstration that semiempirical molecular orbital theory need not limit itself to planar conjugated molecules. We have been able, with one and the same set of parameters, to gain insight into such diverse properties as the barrier to internal rotation and the relative role of σ and π frameworks in aromatics. In subsequent papers in this series we will show that indeed one need not confine oneself to organic compounds; that this parametrization makes it possible to make a good guess at the wavefunctions of inorganic compounds as well.

Let us review briefly the clear failures of the theory as it was applied with our choice of parameters. There exists a tendency to overemphasize steric repulsions, which finally leads to incorrect isomerization energies for the pentanes and hexanes. This behavior cannot be corrected by two obvious maneuvers: changing K, or

[37] G. Ferguson and J. M. Robertson in *Advances in Physical Organic Chemistry*, edited by V. Gold (Academic Press Inc., New York, 1963), p. 203.

[38] See, however, W. M. Schubert, R. B. Murphy, and J. Robins, Tetrahedron **17**, 199 (1962).

[39] T. J. Katz and M. Rosenberger, J. Am. Chem. Soc. **84**, 865 (1962).

[40] H. L. Strauss and G. K. Fraenkel, J. Chem. Phys. **35**, 1738 (1961). T. J. Katz and H. L. Strauss, *ibid.* **32**, 1873 (1960).

[41] Note that as usual with LCAO–MO calculations, the wavefunction does not have the correct behavior at infinity.

[42] It is interesting in this connection to note that in a calculation in which the Mulliken approximation was used for three- and four-center integrals, the binding energy was also overestimated. (L. Burnelle, Ref. 27).

increasing the hydrogen Slater exponent to a value more appropriate for molecules, about 1.2. These manipulations can reproduce the experimental barrier in ethane, but do not alleviate our difficulties with the hexanes. Using the alternate relation $H_{ij}=K'S_{ij}$ does help somewhat but if we are to deal later with a wide variety of heteroatoms we must reject this alternative (see Appendix). It is our feeling that the absolute minimization procedure carried out in detail for each molecule might lead to improved results, but this process seems much too complicated considering the degree of sophistication of the model. It should be stated that we are fortunate to obtain as good qualitative results as we have, having chosen to process our molecules at distances not corresponding to minima. The other major failure, the absence of strain energies in small rings, remains without an explanation.

As for the successes, they lie in the correct assignment of equilibrium conformations where these are known. Barriers are way off, and even qualitative behavior as one goes from molecule to molecule is sometimes incorrect. We have much greater faith in the charge distributions, since our experience with the simple Hückel theory tells us that these are quite insensitive to the choice of parameters. Indeed, we consider the calculation of charge distributions in aliphatics and the σ and π systems in aromatics as interesting a feature of this method as the fact that the Hückel theory can produce a barrier in ethane.

Finally, we want to issue a plea for a search for that semiempirical parametrization which will improve on these results. For it seems clear to us that a simple wavefunction with an atrocious energy can still nicely predict the geometry of a molecule, and thus answer the chemists prime question of molecular structure. Though the necessary integrals for a reasonable *a priori* calculation are now becoming available, a semiempirical procedure is still necessary, since the number and variety of these integrals increases astronomically with molecular size.

ACKNOWLEDGMENTS

I would like to express my indebtedness to L. L. Lohr, Jr., R. M. Pitzer, W. N. Lipscomb, E. B. Wilson, Jr., E. J. Corey, M. P. Gouterman, and H. Kobayashi for many discussions relating to the problems covered in this paper. Abundant computer time was made available by the Harvard and MIT Computation Centers. This work was supported in part by the National Institutes of Health.

APPENDIX

Choice of H_{ij}

Once one decides not to neglect the off-diagonal overlaps and Hamiltonian matrix elements in the secular determinant $|H_{ij}-ES_{ij}|=0$, one is faced with the

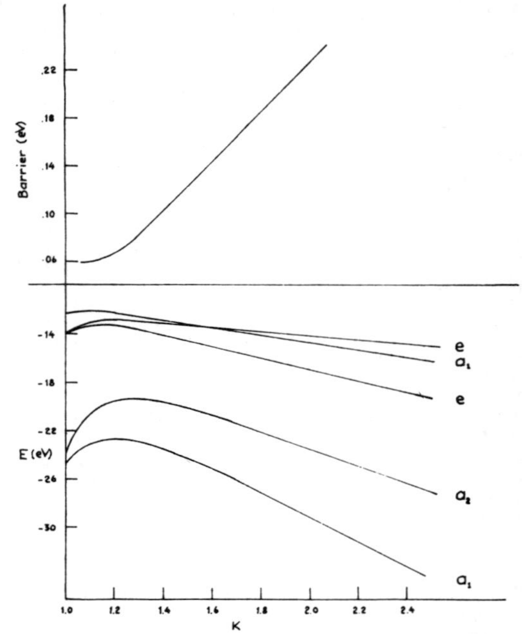

FIG. 10. Barrier to internal rotation (top) and individual bonding energy levels (bottom) for ethane as a function of K.

choice of the distance dependence of the H_{ij}. In our earlier work on the boron hydrides we used the relationship $H_{ij}=K'S_{ij}$, with a value of $K'=-21$ eV. However, if one is to consider a large variety of heteroatoms, one is forced to use an inordinately high magnitude of K' due to the requirement that K' be smaller than any diagonal matrix element, i.e., valence-state ionization potential. The difficulty may be appreciated from the following example. Suppose all $H_{ii}=\alpha$ and we chose K' also equal to α. Then our secular equation becomes $|(\alpha-E)S_{ij}|=0$ which is satisfied by all $E=\alpha$! For $K'>\alpha$ level inversion results, with generally absurd consequences. Thus K' must be strictly less than α.

A better approximation[43] is to take our Eq. (3), $H_{ij}=0.5K(H_{ii}+H_{jj})S_{ij}$. If one makes use of the Mulliken approximation for the product of two charge distributions χ_i and χ_j

$$\chi_i\chi_j=0.5S_{ij}(\chi_i\chi_i+\chi_j\chi_j),$$

one obtains $K=1$. However, this again leads to absurd results for the homonuclear case. One is thus forced to

[43] L. L. Lohr, Jr., has used a similar expression $H_{ij}=K''S_{ij}(H_{ii}H_{jj})^{\frac{1}{2}}$ which differs from ours only in second order and has certain computational advantages. L. L. Lohr, Jr., and W. N. Lipscomb, J. Chem. Phys. **38**, 1607 (1963). T. Jordan, H. W. Smith, L. L. Lohr, Jr., and W. N. Lipscomb, J. Am. Chem. Soc. **85**, 846 (1963). L. L. Lohr, Jr. and W. N. Lipscomb, *ibid.* p. 240. See also C. J. Ballhausen and H. B. Gray, Inorg. Chem. **1**, 111 (1962).

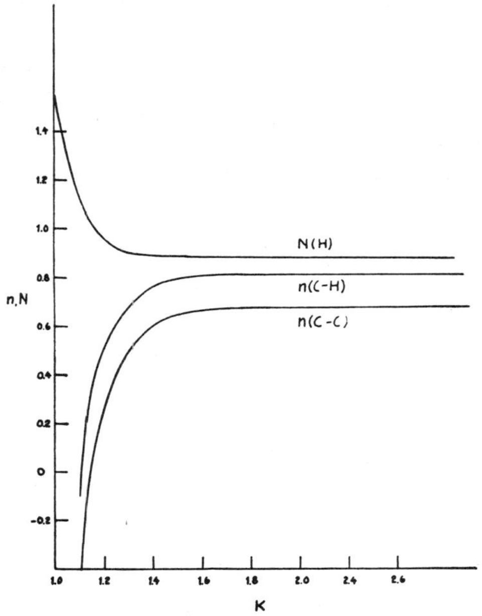

FIG. 11. Gross atomic population on hydrogen, C–H and C–C overlap populations in ethane as a function of K.

use K greater than 1.0. Various other authors have used $K=1.87$ and 2.00.[6,43]

In Fig. 10 we show the variation of the energies of the occupied orbitals of ethane (staggered, C–C, 1.54 Å, C–H, 1.10 Å) with K. Above $K=1.6$ a good linear proportionality holds. A different proportionality, however, holds for the energy levels of the eclipsed conformation, so that the barrier also varies with K (Fig. 10). In Fig. 11 we show the corresponding variation in the overlap populations and charges for the staggered geometry. Note the insensitivity of these quantities once K is again greater than about 1.6; the off-diagonal matrix elements begin to dominate and the wavefunction becomes independent of K.

We conclude that if we are interested in charge distributions in molecules, that these are insensitive to K over a large region. The binding energy, however, becomes proportional to $K-1$ for large K and already exceeds the observed heat of atomization for any K greater than 1.0. The value 1.75 was chosen as a reasonable compromise between the desire to match the experimental barrier in ethane, and the necessity to work in a region where populations are stable.

Total Energy and Electronic Energy

It will be noted that the minimum in Fig. 2 arises when we plot the simple sum of one-electron energies vs internuclear separation. The complete Hamiltonian may be written as the sum of electron–electron, electron–nuclear, and nuclear–nuclear energies,

$$H=\sum_{ee'}H_{ee'}+\sum_{en}H_{en}+\sum_{nn'}H_{nn'}. \qquad (4)$$

In the Hückel theory H is approximated by a sum of one-electron effective Hamiltonians, whose matrix elements we endeavor to guess in some systematic manner

$$H=\sum_{e}H_{\text{eff}}. \qquad (5)$$

The term $H_{nn'}$ in Eq. (4) is a purely classical nuclear–nuclear repulsion, and one has the choice of taking it over to the left side of the expression before approximation by (5); or one can leave it where it is, and in effect include part of the nuclear repulsion in each effective one-electron Hamiltonian. If to the potential curve of Fig. 1 we add the nuclear repulsions of the protons and the carbon shielded by its 1s electrons, the minimum vanishes. This, and the behavior of the simple sum of one-electron energies at small internuclear distances leads us to conclude that our method of guessing H_{ij} simulates within the electronic energies the presence of nuclear repulsions at small distances, and this is what gives us our minimum.

To this operational argument we may add a theoretical one, due to an observation made by Slater.[44] The sum of the one-electron energies of a Hartree–Fock Hamiltonian is equal to the total energy minus the nuclear–nuclear repulsions, plus the electron–electron repulsions. The last two terms cancel, roughly (it is sufficient if their difference varies slowly with distance), and thus the simple sums of one-electron energies behave approximately as the true molecular energies. This parallelism and the accompanying overestimation of binding energies should be investigated further. The procedure of viewing simple sums of one-electron energies was advanced in our previous calculations on boron hydrides and carboranes,[7] and by L. L. Lohr, Jr., in a number of calculations on transition metal ions, noble-gas halides, and sulfones.[43]

---

[44] J. C. Slater, *Quantum Theory of Molecules and Solids* (McGraw-Hill Book Company, Inc., New York, 1963), Vol. I, p. 108.

II-4 Reprinted from *The Journal of Chemical Physics* **44**, 3289 (1966)

Approximate Self-Consistent Molecular Orbital Theory. III. CNDO Results for AB_2 and AB_3 Systems*

J. A. Pople
Carnegie Institute of Technology and Mellon Institute, Pittsburgh, Pennsylvania
AND
G. A. Segal
Carnegie Institute of Technology, Pittsburgh, Pennsylvania
(Received 27 December 1965)

The approximate self-consistent molecular orbital theory with complete neglect of differential overlap (CNDO) presented in earlier papers has been modified in two ways. (a) Atomic matrix elements are chosen empirically using data on both atomic ionization potentials and electron affinities. (b) Certain penetration-type terms, which led to excess bonding between formally nonbonded atoms in the previous treatment, have been omitted. The new method (denoted by CNDO/2) has been applied to symmetrical triatomic (AB_2) and tetratomic (AB_3) molecules, for a range of bond angles. The theory leads to calculated equilibrium angles, dipole moments, and bending force constants which are in reasonable agreement with experimental values in most cases.

1. INTRODUCTION

In two earlier papers[1,2] (subsequently referred to as Parts I and II), a new approximate method was introduced for calculating self-consistent molecular orbitals for all valence electrons in molecules containing no atom heavier than fluorine. This is based on the "complete neglect of differential overlap" (CNDO) approximation in which the overlap distribution $\phi_\mu(1)\phi_\nu(1)$ of any two atomic orbitals ϕ_μ and ϕ_ν is neglected in all electron repulsion integrals. By choosing some parameters in a semiempirical manner to fit atomic data and others to fit more precise calculations on diatomics, it was found possible to obtain wavefunctions and energies which predicted geometries, bending force constants, and rotational barriers in fair agreement with experiment, at least for small polyatomic molecules.

In this paper, we make certain modifications suggested by results of the previous calculations. To distinguish the methods, we refer to the procedure defined previously as CNDO/1 and the new version as CNDO/2. In addition, an extension of the theory to open-shell molecules (with an unequal number of α-spin and β-spin electrons) is described. In Sec. 4, results of calculations on a number of symmetrical triatomic and tetratomic molecules are given and compared with experimental data where possible.

2. MODIFICATION OF THE METHOD

In the CNDO approximation, as described in Parts I and II, the coefficients $c_{i\mu}$ of the LCAO SCF molecular orbitals

$$\Psi_i=\sum_\mu c_{i\mu}\phi_\mu \tag{2.1}$$

are normalized eigenvectors of the Hartree–Fock matrix $F_{\mu\nu}$ with elements

$$F_{\mu\mu}=U_{\mu\mu}+(P_{AA}-\tfrac{1}{2}P_{\mu\mu})\gamma_{AA}+\sum_{B(\neq A)}(P_{BB}\gamma_{AB}-V_{AB}), \tag{2.2}$$

$$F_{\mu\nu}=\beta_{AB}{}^0S_{\mu\nu}-\tfrac{1}{2}P_{\mu\nu}\gamma_{AB}\quad(\mu\neq\nu). \tag{2.3}$$

In these expressions ϕ_μ is an atomic orbital of Atom A and ϕ_ν of Atom B. Other symbols are defined as follows:

$U_{\mu\mu}$ is the diagonal matrix element of ϕ_μ using only the core Hamiltonian of its own Atom A (that is, the kinetic energy plus the potential energy in the field of the nucleus and inner shells). $P_{\mu\nu}$ is the charge density and bond-order matrix

$$P_{\mu\nu}=2\sum_i^{\text{occ}} c_{i\mu}c_{i\nu}, \tag{2.4}$$

the diagonal terms representing atomic-orbital charge densities. P_{AA} is the total charge on Atom A

$$P_{AA}=\sum_\mu^A P_{\mu\mu}. \tag{2.5}$$

γ_{AB} is an average interaction energy between an electron in any valence atomic orbital of A with another in an orbital of B. V_{AB} is the interaction energy of an electron in any valence orbital of A with the core (nucleus + inner shells) of B. $\beta_{AB}{}^0$ is a bonding parameter given by

$$\beta_{AB}{}^0=\tfrac{1}{2}(\beta_A{}^0+\beta_B{}^0), \tag{2.6}$$

where $\beta_A{}^0$ is chosen empirically but depends only on the nature of Atom A. $S_{\mu\nu}$ is the overlap integral between atomic orbitals ϕ_μ and ϕ_ν.

At the same level of approximation, the total energy of the molecule can be written as a sum of one- and two-atom terms,

$$E_{\text{total}}=\sum_A E_A+\sum_{A<B}E_{AB}, \tag{2.7}$$

---

* This paper was presented as part of the International Summer School on Quantum Theory of Polyatomic Molecules organized at Menton, France by the North Atlantic Treaty Organization.

[1] J. A. Pople, D. P. Santry, and G. A. Segal, J. Chem. Phys. 43, S129 (1965).

[2] J. A. Pople and G. A. Segal, J. Chem. Phys. 43, S136 (1965).

where

$$E_A = \sum_{\mu}^{A} P_{\mu\mu}U_{\mu\mu} + \tfrac{1}{2}\sum_{\mu}^{A}\sum_{\nu}^{A}(P_{\mu\mu}P_{\nu\nu} - \tfrac{1}{2}P_{\mu\nu}^2)\gamma_{AA} \tag{2.8}$$

and

$$E_{AB} = \sum_{\mu}^{A}\sum_{\nu}^{B}[2P_{\mu\nu}\beta_{AB}^0 S_{\mu\nu} - \tfrac{1}{2}P_{\mu\nu}^2\gamma_{AB}] + [Z_A Z_B R_{AB}^{-1} - P_{AA}V_{AB} - P_{BB}V_{BA} + P_{AA}P_{BB}\gamma_{AB}]. \tag{2.9}$$

The only additional assumption here is that the interaction energy of the cores of Atoms A and B is approximated by $Z_A Z_B R_{AB}^{-1}$, where Z_A, Z_B are the charges of the cores (in units of $+e$).

One of the principal failures of CNDO/1 was that for diatomic molecules it led to calculated bond lengths which were too short and binding energies which were too large. This feature arises because of a "penetration" effect in which electrons in an orbital on one atom penetrate the shell of another leading to net attraction. If we rewrite (2.2) in the form

$$F_{\mu\mu} = U_{\mu\mu} + (P_{AA} - \tfrac{1}{2}P_{\mu\mu})\gamma_{AA} + \sum_{B(\neq A)}(P_{BB} - Z_B)\gamma_{AB} + \sum_{B(\neq A)}(Z_B\gamma_{AB} - V_{AB}), \tag{2.10}$$

then the last terms may reasonably be described as penetration integral contributions to $F_{\mu\mu}$.

These penetration terms give rise to calculated bonding energies even when the bond orders connecting two atoms are zero. Thus, if the energy of the first triplet state ($^3\Sigma_u^+$) of H_2 is calculated at this level of approximation, the bond order between the hydrogen 1s orbitals is zero and the theoretical interaction energy is

$$E = \gamma_{AB} - 2V_{AB} + R_{AB}^{-1}. \tag{2.11}$$

With the parameters used in CNDO/1, this has a minimum of −0.637 eV at a distance of 0.85 A, whereas accurate calculations show this state to be repulsive[3] (except for weak van der Waals attraction at large distances).

To eliminate these difficulties, the penetration terms in (2.10) are neglected in the new version of the method. This corresponds to putting

$$V_{AB} = Z_B\gamma_{AB} \tag{2.12}$$

throughout the calculation. No complete justification of this can be given, but a study of the H_2^+ problem in Appendix A does suggest that the neglect of overlap distributions introduces errors of a similar kind but of opposite sign to the neglect of penetration, so (2.12) does correct the CNDO/1 procedure in the right direction and has the further merit of simplifying the over-all method.

The second change made concerns the estimation of the local core matrix element $U_{\mu\mu}$. In CNDO/1, this was obtained from the atomic ionization potential I_μ (referred to appropriate average atomic states) by the relation

$$-I_\mu = U_{\mu\mu} + (Z_A - 1)\gamma_{AA}, \tag{2.13}$$

the orbital ϕ_μ belonging to Atom A. An alternative procedure would have been to use atomic electron affinities A_μ and

$$-A_\mu = U_{\mu\mu} + Z_A\gamma_{AA}. \tag{2.14}$$

In molecular orbital theory, we wish to be able to account satisfactorily for the tendency of an atomic orbital both to acquire and lose electrons, so that the new procedure adopted is to use the average of (2.13) and (2.14)

$$-\tfrac{1}{2}(I_\mu + A_\mu) = U_{\mu\mu} + (Z_A - \tfrac{1}{2})\gamma_{AA}. \tag{2.15}$$

Using (2.12) and (2.15), the basic equations for the Fock matrix in the CNDO/2 method can now be written

$$F_{\mu\mu} = -\tfrac{1}{2}(I_\mu + A_\mu) + [(P_{AA} - Z_A) - \tfrac{1}{2}(P_{\mu\mu} - 1)]\gamma_{AA} + \sum_{B(\neq A)}(P_{BB} - Z_B)\gamma_{AB}, \tag{2.16}$$

$$F_{\mu\nu} = \beta_{AB}^0 S_{\mu\nu} - \tfrac{1}{2}P_{\mu\nu}\gamma_{AB}. \tag{2.17}$$

This form for $F_{\mu\mu}$ shows up the self-consistent character of the theory in a very simple manner. The first term is a fundamental electronegativity for the atomic orbital, closely related to the scale introduced by Mulliken.[4] The remaining terms show how this is modified by the actual distribution of charge in the molecule. $F_{\mu\mu}$ reduces to $-\tfrac{1}{2}(I_\mu + A_\mu)$ if the orbital contains one electron ($P_{\mu\mu} = 1$) and if all atoms have zero net charge ($P_{AA} = Z_A$, $P_{BB} = Z_B$).

The values used for $-\tfrac{1}{2}(I_\mu + A_\mu)$ are listed in Table I and their derivation from atomic data is described in Appendix B. Other features of CNDO/2 are the same as CNDO/1. Slater atomic orbitals are used to calculate $S_{\mu\nu}$ (with an effective charge of 1.2 for hydrogen) and the γ_{AB} are obtained theoretically from valence s orbitals. The parameters β_A^0 are identical with those used in CNDO/1.

The calculations using this method have been carried out by means of an ALGOL program, the input data consisting of nuclear charges, the number of electrons and bond lengths and angles. Initial estimates of the LCAO coefficients were obtained from a Hückel-type theory using matrix elements

$$F_{\mu\mu}^{(0)} = -\tfrac{1}{2}(I_\mu + A_\mu), \tag{2.18}$$

$$F_{\mu\nu}^{(0)} = \beta_{AB}^0 S_{\mu\nu}, \tag{2.19}$$

[3] H. M. James and A. S. Coolidge, J. Chem. Phys. **6**, 730 (1938).

[4] R. S. Mulliken, J. Chem. Phys. **2**, 782 (1934).

TABLE I. Matrix elements from atomic data (electron volts).

	H	Li	Be	B	C	N	O	F
$\frac{1}{2}(I_s+A_s)$	7.176	3.106	5.946	9.594	14.051	19.316	25.390	32.272
$\frac{1}{2}(I_p+A_p)$		1.258	2.563	4.001	5.572	7.275	9.111	11.080

and the final solution was approached by an iterative scheme as described in Part II. Self-consistency was achieved in all calculations reported in Sec. 4 to an accuracy of 0.0001 in all coefficients.

3. EXTENSION TO OPEN-SHELL CONFIGURATIONS

The CNDO method is easily extended to open shells of electrons if a single-determinant wavefunction is used with different molecular orbitals for α and β spins. If the number of α electrons exceeds the β electrons by one, this gives a component of the doublet state of a free radical. If there are two extra α electrons, the wavefunction corresponds to a component of the lowest triplet state.

The two sets of LCAO molecular orbitals may be written

$$\psi_i^\alpha = \sum_\mu c_{i\mu}^\alpha \phi_\mu$$

$$\psi_i^\beta = \sum_\mu c_{i\mu}^\beta \phi_\mu, \tag{3.1}$$

and there is corresponding partial charge density and bond-order matrices

$$P_{\mu\nu}^\alpha = \sum_i^{\text{occ}} c_{i\mu}^\alpha c_{i\nu}^\alpha,$$

$$P_{\mu\nu}^\beta = \sum_i^{\text{occ}} c_{i\mu}^\beta c_{i\nu}^\beta. \tag{3.2}$$

The total charge density and bond-order matrix is given by

$$P_{\mu\nu} = P_{\mu\nu}^\alpha + P_{\mu\nu}^\beta \tag{3.3}$$

and we may also define a spin-density matrix

$$Q_{\mu\nu} = P_{\mu\nu}^\alpha - P_{\mu\nu}^\beta. \tag{3.4}$$

The LCAO coefficients $c_{i\mu}^\alpha$ and $c_{i\mu}^\beta$ are eigenvectors of separate F matrices, for which general expressions were given by Pople and Nesbet.[5] Simplification by the CNDO approximations is straightforward and details are not given. The final results are

$$F_{\mu\mu}^\alpha = -\tfrac{1}{2}(I_\mu + A_\mu) + [(P_{AA} - Z_A) - (P_{\mu\mu}^\alpha - \tfrac{1}{2})]\gamma_{AA} + \sum_{B(\neq A)} (P_{BB} - Z_B)\gamma_{AB}, \tag{3.5}$$

$$F_{\mu\nu}^\alpha = \beta_{AB}^0 S_{\mu\nu} - P_{\mu\nu}^\alpha \gamma_{AB}, \tag{3.6}$$

and corresponding expressions for the β matrix. Equations (3.5) and (3.6) reduce to (2.16) and (2.17) if α and β orbitals are identical.

Computational details require little elaboration. Each self-consistent cycle consists of diagonalization of both F matrices using the $P_{\mu\nu}^\alpha$ and $P_{\mu\nu}^\beta$ from the previous cycle, followed by recalculation of both. Calculations on open-shell systems reported in the following sections were also carried out by an ALGOL computer program.

4. SYMMETRICAL TRIATOMIC MOLECULES AB_2

The CNDO/2 method described in the preceding sections has been applied to a series of triatomic AB_2 molecules in linear $D_{\infty h}$ and bent C_{2v} forms. Using fixed bond lengths (chosen from experimental data where possible), the total energy and wavefunctions have been calculated as functions of the BAB angle. Table II gives the resultant theoretical equilibrium angles, dipole moments, atomic charges, and bending force constants together with corresponding experimental data. The wavefunctions and $P_{\mu\nu}$ matrices are not reproduced in full, although certain points about them are mentioned in the following discussion.

It is clear from the table that the theory successfully predicts the bond angle in most cases and provides substantiation for the qualitative rules of Walsh[6] using arguments based on independent-electron molecular orbital theory. In any form of SCF theory, of course, it is not possible to obtain the total energy as a sum of one-electron energies, so that there is no direct analog of the Walsh diagrams, but the results of these calculations do lend support to his rules.

For the dihydride molecules AH_2, the CNDO/2 calculations always order the valence-shell molecular orbitals (with increasing energy) $1a_1$, $1b_2$, $2a_1$, $1b_1$, $3a_1$. In the linear form $2a_1$ and $1b_1$ become a degenerate 1π pair, being concentrated entirely on the central atom, the order of MO's then being $1\sigma_g$, $1\sigma_u$, $1\pi_u$, $2\sigma_g$. This corresponds to the ordering of Walsh; electronic configurations follow by filling these orbitals using an aufbau principle.

BeH_2 has the configuration

$$BeH_2:\ (1\sigma_g)^2(1\sigma_u)^2 \quad {}^1\Sigma_g^+$$

and is predicted to be linear. The wavefunctions and charge distributions are similar to those quoted in Part II.

[5] J. A. Pople and R. K. Nesbet, J. Chem. Phys. **22**, 571 (1954).

[6] A. D. Walsh, J. Chem. Soc. **1953**, 2260, 2266, 2296, 2301.

TABLE II. Summary of results for AB_2 molecules.

No. of valence electrons	Molecule	B–A–B angle calc	B–A–B angle obs	Dipole moment calc[aa] (debyes)	Dipole moment obs[aa] (debyes)	Bending force constant calc (mdyn/Å)	Bending force constant obs (mdyn/Å)	P_{BB}	r_e[a] (Å)
4	BeH_2	180.0	···	0	···	0.07	···	1.144	1.343[s]
5	BH_2 ($^2A_1$)	136.6	···	0.51	···	0.27	···	1.040	1.180[s]
	BH_2 ($^2\Pi-{}^2B_1$)	180.0	···	0	···	0.23	···	1.015	1.180[s]
6	CH_2 ($^1A_1$)	108.6	103.2[b]	2.26	···	0.69	···	0.991	1.094[s]
	CH_2 ($^3\Sigma_g^- - {}^3B_1$)	141.4	180[b]	0.75	···	0.38	···	0.937	1.094[s]
7	NH_2 ($^2B_1$)	107.3	103.3[c]	2.16	···	0.81	···	0.917	1.024
	NH_2 ($^2\Pi-{}^2A_1$)	145.1	144[d]	0.87	···	0.34	···	0.849	1.024[s]
	OH_2^+ ($^2B_1$)	118.7	···	···	···	0.52	···	0.591	0.960[s]
8	OH_2	107.1	104.45[e]	2.08	1.8[y]	0.95[e]	0.69[f]	0.856	0.960[s]
9	FH_2	180.0	···	0	···	0.19	···	0.968	0.920[s]
15	BO_2	180.0	180[g]	0	···	0.36	0.26[g]	6.134	1.250[s]
	CO_2^+	180.0	180[h]	···	···	0.44	···	5.823	1.176[s]
16	CO_2	180.0	180[f]	0	···	0.58	0.57[f]	6.266	1.162
	BeF_2	180.0	180[i]	0	···	0.12	0.73[j]	7.271	1.360[s]
	NO_2^+	180.0	180[k]	···	···	0.44	0.42[l]	5.883	1.154
17	CO_2^-	142.3	134[m]	···	···	0.73	···	6.528	1.200[s]
	NO_2	137.7	132[n]	−0.75	±0.4[o]	0.66	0.40[p]	6.204	1.200
	BF_2	124.6	···	0.05	···	0.62	(0.58)[q]	7.162	1.300[s]
18	NO_2^-	118.3	115.4[r]	···	···	1.13	1.75[s]	6.543	1.236
	O_3	114.0	116.8	−1.26	±0.58[t]	0.76	1.28[f]	6.152	1.278
	CF_2	104.6	(100 or 108)[u]	0.53	···	1.09	(1.28)[q]	7.145	1.320[s]
19	NF_2	102.5	104.2[v]	−0.12	···	0.97	···	7.113	1.350
20	OF_2	99.2	103.8[w]	−0.21	±0.297[x]	0.71	0.55[f]	7.060	1.410

[a] *Tables of Interatomic Distances and Configurations in Molecules and Ions* (The Chemical Society, London, 1965).
[b] G. Herzberg, Proc. Roy. Soc. (London) **A262**, 291 (1961).
[c] K. Dressler and D. A. Ramsay, Phil. Trans. **A251**, 553 (1959).
[d] R. N. Dixon, Mol. Phys. **9**, 357 (1965).
[e] D. W. Posener and M. W. P. Strandberg, Phys. Rev. **95**, 374 (1954).
[f] G. Herzberg, *Infrared and Raman Spectra* (D. Van Nostrand Co. Inc., New York, 1945).
[g] A. Sommer, D. White, M. J. Linevsky, and D. E. Mann, J. Chem. Phys. **38**, 87 (1963).
[h] D. A. Ramsay, Advan. Spectry. **1**, 1 (1959).
[i] L. Brewer, G. R. Somayajulu, and E. Bracket, Chem. Rev. **63**, 111 (1963).
[j] A. Buchler and W. Klemperer, J. Chem. Phys. **29**, 121 (1958).
[k] J. W. M. Steeman and C. H. Macgillavry, Acta. Cryst. **7**, 402 (1954).
[l] R. Teranishi and J. C. Decius, J. Chem. Phys. **22**, 896 (1954).
[m] D. W. Ovenall and D. H. Whiffen, Mol. Phys. **4**, 135 (1961).
[n] S. Claesson, J. Donohue, and V. Schomaker, J. Chem. Phys. **16**, 207 (1948).
[o] C. T. Zahn, Physik Z. **33**, 686 (1932).
[p] I. C. Hisatsune, J. P. Devlin, and S. Califano, Spectrochim. Acta **16**, 450 (1960).
[q] G. Nagarajan, Australian J. Chem. **16**, 717 (1963).
[r] G. B. Carpenter, Acta Cryst. **8**, 852 (1955).
[s] R. E. Weston, Jr., and T. F. Brodasky, J. Chem. Phys. **27**, 683 (1957).
[t] R. H. Hughes, J. Chem. Phys. **21**, 959 (1953).
[u] D. E. Milligan, D. G. Mann, and M. E. Jacox, J. Chem. Phys. **41**, 1199 (1964).
[v] M. D. Harmony and R. J. Myers, J. Chem. Phys. **35**, 1129 (1961).
[w] J. A. Ibers and V. Schomaker, J. Phys. Chem. **57**, 699 (1953).
[x] L. Pierce, R. Jackson, and N. DiCianni, J. Chem. Phys. **35**, 2240 (1961).
[y] A. L. McClellan, *Tables of Experimental Dipole Moments* (W. H. Freeman and Co., San Francisco, Calif. 1963).
[s] Assumed average bond length.
[aa] Dipole moment + in the direction A^-B^+.

BH_2 is a free radical which has two low-lying electronic configurations in the bent form

$$BH_2\text{:}\quad (1a_1)^2(1b_2)^2 2a_1 \quad {}^2A_1,$$

$$(1a_1)^2(1b_2)^2 1b_1 \quad {}^2B_1.$$

These coalesce into a degenerate $^2\Pi$ state in the linear form. By specifying the population of the $2p$ atomic orbital corresponding to the $1b_1$ molecular orbital, it is possible to obtain open-shell CNDO/2 wavefunctions for both states, and it was found that the $^2A_1$ state was below $^2B_1$. In agreement with the Walsh rules, the theory leads to a bent ground state and a linear excited state (Fig. 1).

The lowest singlet state of CH_2 is predicted to be strongly bent with the electronic configuration

$$CH_2\text{:}\quad (1a_1)^2(1b_2)^2(2a_1)^2 \quad {}^1A_1.$$

This is in agreement with the experimental findings of Herzberg.[7] It is interesting to note that a substantial dipole moment in the sense C^-H^+ is predicted for this state, even though the total electronic charge on

[7] G. Herzberg, Proc. Roy. Soc. (London) **A262**, 291 (1961).

hydrogen is near unity. This is because the molecular orbital $2a_1$ is largely a directed lone pair concentrated at the back of the molecule. The triplet state of CH_2

$$CH_2: \quad (1a_1)^2(1b_2)^2(2a_1)(1b_1) \quad {}^3B_1$$

is also predicted to be bent, although less than the singlet. This is in disagreement with the experimental result.[7] However, this is a situation where the CNDO approximation is least satisfactory. For linear CH_2, the electronic configuration is $(1\sigma_g)^2(1\sigma_u)^2(1\pi_u)^2$ and leads to ${}^3\Sigma_g^-$, ${}^1\Delta_g$, and ${}^1\Sigma_g^+$ states. Of these ${}^3\Sigma_g^-$ should be the lowest, but the CNDO method fails to predict this as it neglects the one-center exchange integral causing the splitting. This triplet stabilization (due to the separation of electrons of the same spin into spatially distinct orbitals) is probably a maximum in the linear form and its inclusion would very likely modify the calculated bond angle. In any case, the calculated triplet-state energy at 141.4° is only 0.17 eV below the linear value.

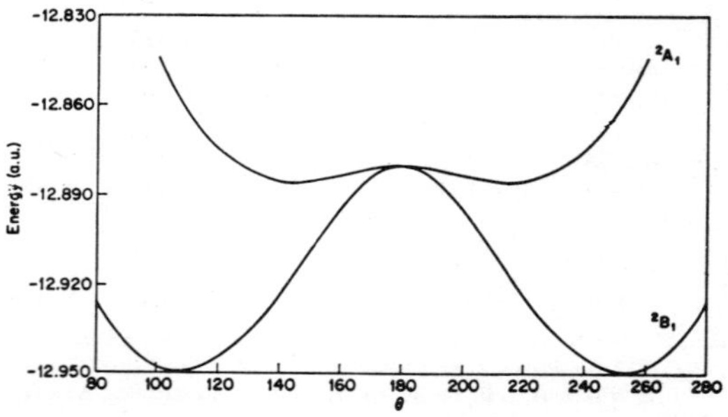

FIG. 1. Energy on bending of NH_2.

NH_2 in the bent form again has the two electronic states

$$NH_2: \quad (1a_1)^2(1b_2)^2(2a_1)^2(1b_1) \quad {}^2B_1,$$
$$(1a_1)^2(1b_2)^2(2a_1)(1b_1)^2 \quad {}^2A_1$$

and is another Renner-type molecule. The calculations predict both states to be bent, with ${}^2B_1$ lowest in energy and furthest from the linear form (Fig. 2). The equilibrium angle for the upper state was calculated to be 145.1° with a barrier height of 1103 cm^{-1}. Early experimental work[8] suggested that the upper state is linear, but a recent reconsideration[9] of this data gave an equilibrium angle of 144°±5° and a barrier height of 777±100 cm^{-1}.

The water molecule has the closed-shell configuration

$$H_2O: \quad (1a_1)^2(1b_2)^2(2a_1)^2(1b_1)^2 \quad {}^1A_1$$

and the calculations predict a bent molecule with a bond angle of the correct order of magnitude. In this

[8] Reference c, Table II.
[9] Reference d, Table II.

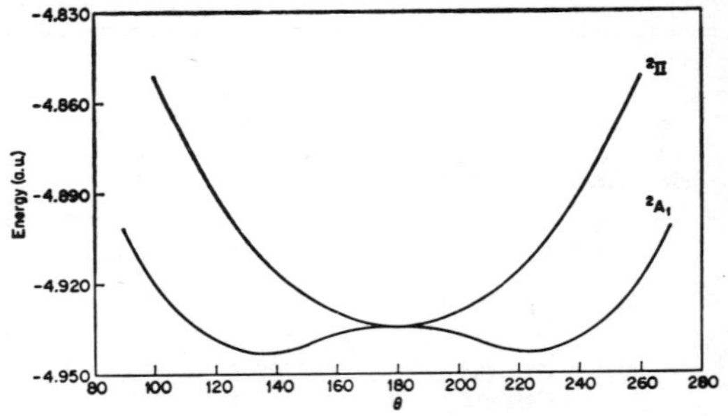

FIG. 2. Energy on bending of BH_2.

case, we have studied the breakdown of the total energy into monatomic and diatomic parts using Eqs. (2.7)–(2.9). The results (relative to corresponding quantities for the linear form) are shown in Fig. 3. From this diagram, it is clear that the monatomic parts of the energy decrease as the molecule bends away from the linear form, but this is partly offset by a rise in the oxygen–hydrogen diatomic part. This corresponds to the accepted qualitative picture. As the molecule deviates from linearity, the population of the oxygen 2*s* orbital increases (one of the lone pairs acquiring *s* character) and this is the primary "driving force" causing the molecule to be bent. On the other hand the oxygen–hydrogen bonding energy is weaker in the bent molecule, primarily because the overlap between digonal hybrids and hydrogen (in the linear form) is more effective than overlap involving hybrids with more *p* character (in the bent form). According to Fig. 3, there is some direct hydrogen–hydrogen

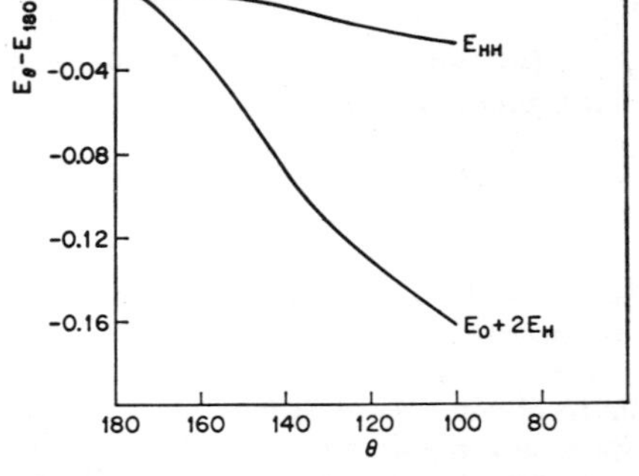

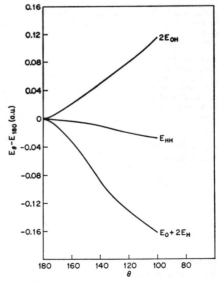

FIG. 3. Angular dependence of energy components for H_2O.

TABLE III. Summary of results for AB_2 molecules.

No. of valence electrons	Molecule	B–A–B angle calc	B–A–B angle obs	Dipole moment calc[n] (debyes)	Dipole moment obs[n] (debyes)	Bending force constant[o] calc (mdyn/Å)	Bending force constant[o] obs (mdyn/Å)	P_{BB}	r_e[a] (Å)
6	BH_3	120.0	···	0	···	0.73	···	1.066	1.180[m]
7	CH_3	120.0	(120)[b]	0	···	0.27	···	0.968	1.094
8	OH_3^+	113.9	117[c]	···	···	0.70	···	0.630	0.960
	NH_3	106.7	106.6[d]	2.08	1.47[e]	0.95	0.6[f]	0.924	1.020[m]
9	OH_3	120.0	···	0	···	1.09	···	0.966	0.970[m]
24	CO_3^-	120.0	120[g]	···	···	1.53	1.46[f]	6.796	1.313
	BF_3	120.0	120[h]	0	···	0.91	0.87[f]	7.233	1.300
	NO_3^-	120.0	120[i]	···	···	1.48	1.47[f]	6.533	1.242
24	CF_3	113.5	111.1[j]	−0.17	···	1.61	···	7.162	1.320[m]
26	NF_3	104.0	102.5[k]	0.05	±0.23[l]	1.33	0.76[k]	7.112	1.370

[a] *Tables of Interatomic Distances and Configurations in Molecules and Ions* (The Chemical Society, London, 1965).
[b] G. Herzberg, Proc. Roy. Soc. (London) **A262**, 291 (1961).
[c] Y. K. Yoon and G. B. Carpenter, Acta. Cryst. **12**, 17 (1959).
[d] A. Almenninger and O. Bastiansen, Acta Chem. Scan. **9**, 815 (1955).
[e] A. L. McClellan, *Tables of Experimental Dipole Moments* (W. H. Freeman and Co., San Francisco, Calif. 1963).
[f] G. Herzberg, *Infrared and Raman Spectra* (D. Van Nostrand Co., Inc., New York, 1945).
[g] N. Elliott, J. Am. Chem. Soc. **59**, 1380 (1937).
[h] A. H. Nielson, J. Chem. Phys. **22**, 659 (1954).
[i] E. Grison, K. Eriks, and J. L. DeVries, Acta. Cryst. **3**, 290 (1950).
[j] R. W. Fessenden and R. H. Schuler, J. Chem. Phys. **43**, 2704 (1965).
[k] V. Schomaker and C. S. Lu, J. Am. Chem. Soc. **72**, 1182 (1950).
[l] S. N. Ghosh, J. Chem. Phys. **21**, 308 (1953).
[m] Assumed average bond length.
[n] Dipole moment+in the direction A^-B^+.
[o] Relative to the angle of deviation from plane for planar molecules, otherwise relative to B–A–B angle.

bonding favoring the bent form, but this is not the main factor involved.

The calculations on H_2O^+ predict an opening of the H–O–H bond angle by about 12° on the ionization of water. This is probably because the increased net charge on the hydrogen atoms leads to hydrogen–hydrogen repulsion. No experimental information seems to be available on this point.

FH_2 is not known experimentally, but calculations were carried out to find the effect of an antibonding ($3a_1$) electron on AH_2 angles. Since this orbital has *more* 2*s* character in the linear form, its occupation tends to linearize the molecule and the calculations suggest that FH_2 would be linear with the bond lengths assumed.

Calculations on AB_2 molecules, where B is oxygen or fluorine, are also listed in Table II, from which it is clear that the theory is fairly successful in predicting equilibrium bond angles along these series. This angle is primarily determined by the number of valence electrons as suggested by Walsh.[6] Dipole moments and bending force constants are in moderate agreement with experimental values with the exception of BeF_2 which is calculated to have a much lower force constant than observed.

Calculations on some AH_3 and AB_3 molecules are summarized in Table III, which again show good agreement with experimental bond angles where available. It is interesting to note that H_3O (assuming the same bond lengths as H_2O) is calculated to be planar so that, as in the AH_2 systems, the addition of an a_1 antibonding electron restores the more symmetrical configuration. Another significant success of the theory is the prediction that CH_3 is planar (experimental evidence suggests that this is probably so[7]) while CF_3 is pyramidal (as found recently by Fessenden and Shuler).[10] At first this is surprising since the carbon in CF_3 is expected to be more positive than in CH_3 (and as found by CNDO/2). However, in the planar configuration of CF_3, the $2p\pi$ atomic orbital of carbon has a population greater than unity (compared to unity in planar CH_3). Thus, although the over-all electron charge transfer is from C to F, in the π system there is a "back donation" from F to C. Since it is the population of the $2p\pi$ orbital of carbon that determines how much stabilization is gained by a nonplanar distortion (with consequent increase of *s* character), it is reasonable to find CF_3 pyramidal.

5. CONCLUSIONS

As a result of the calculations reported in the previous section, the following remarks may be made about the present status of approximate theories of this type.

(1) The CNDO/2 treatment leads to generally good agreement with experimental values for the bond angles of low-lying states of these small molecular species.

[10] Reference j, Table III.

Very few *a priori* treatments of nonlinear polyatomic molecules have been published so we do not yet know whether full LCAO SCF calculations with a comparable basis set would give similar results. However, even if the reasons for the success of the CNDO method are not fully understood, it is clearly a promising model system for studying larger molecules.

(2) It should be emphasized that the theory in this version does not give a satisfactory treatment of the total energy or the dissociation process. The approximations made appear to be effective only in handling angular geometry and long-range nonbonded interactions.

(3) A second unsatisfactory feature of the CNDO/2 method is that it often fails to give a separation of an open-shell molecular orbital configuration into different states (as in the linear form of CH_2). This can be attributed mainly to the neglect of one-center exchange integrals. To improve this aspect of the theory, other less approximate methods, such as the NDDO scheme (neglect of diatomic differential overlap) outlined in Paper I, is necessary.

ACKNOWLEDGMENT

This research was supported by a grant from the National Science Foundation.

APPENDIX A

Application to H_2^+

For the one-electron hydrogen molecule–ion, we may compare the CNDO method with an accurate LCAO treatment. Suppose ϕ_A and ϕ_B are the two 1*s* hydrogen atomic orbitals. Then we may either use these as a basis or, alternately, two equivalent but orthogonal orbitals (Löwdin orbitals) χ_A, χ_B defined by

$$\phi_A=(\cos\epsilon)\chi_A+(\sin\epsilon)\chi_B,$$

$$\phi_B=(\sin\epsilon)\chi_A+(\cos\epsilon)\chi_B, \qquad \text{(A1)}$$

where

$$2\epsilon=\arcsin S, \qquad \text{(A2)}$$

S being the overlap integral.

In the ground state of H_2^+, the occupied LCAO molecular orbital is

$$\Psi=(\chi_A+\chi_B)/\sqrt{2}=(\phi_A+\phi_B)/\{[(2(1+S)]\}^{\frac{1}{2}} \qquad \text{(A3)}$$

and the electronic energy may be written

$$E_{\text{electronic}}=\frac{1}{2(1+S)}\int(\phi_A+\phi_B)[-\tfrac{1}{2}\nabla^2-r_A^{-1}-r_B^{-1}]$$

$$\times(\phi_A+\phi_B)d\tau=\frac{U-V+\beta}{1+S}, \qquad \text{(A4)}$$

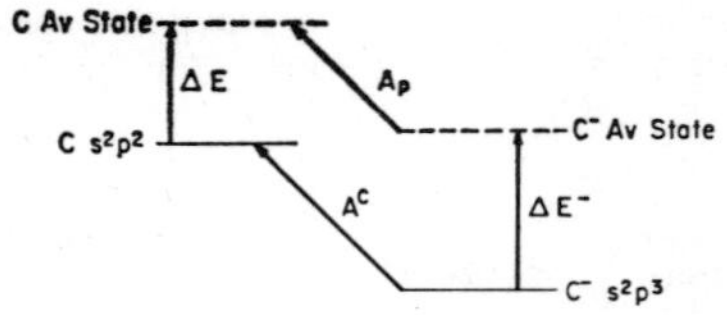

FIG. 4. Correction of electron affinity of carbon to average states.

where

$$U=\int\phi_A[-\tfrac{1}{2}\nabla^2-r_A^{-1}]\phi_A d\tau, \qquad \text{(A5)}$$

$$V=\int\phi_A^2 r_B^{-1}d\tau, \qquad \text{(A6)}$$

$$\beta=\int\phi_A[-\tfrac{1}{2}\nabla^2-r_A^{-1}-r_B^{-1}]\phi_B d\tau. \qquad \text{(A7)}$$

U is the electronic energy of a single atom and V is the electron–other-nucleus potential.

Alternatively, the electronic energy may be written in terms of the χ orbitals,

$$E_{\text{electronic}}=\frac{1}{2}\int(\chi_A+\chi_B)[-\tfrac{1}{2}\nabla^2-r_A^{-1}-r_B^{-1}](\chi_A+\chi_B)d\tau$$

$$=U'-V'+\beta', \qquad \text{(A8)}$$

where

$$U'=\int\chi_A[-\tfrac{1}{2}\nabla^2-r_A^{-1}]\chi_A d\tau, \qquad \text{(A9)}$$

$$V'=\int\chi_A^2 r_B^{-1}d\tau, \qquad \text{(A10)}$$

$$\beta'=\int\chi_A[-\tfrac{1}{2}\nabla^2-r_A^{-1}-r_B^{-1}]\chi_B d\tau. \qquad \text{(A11)}$$

It is easily shown that (if the energy zero is altered so that $U=0$)

$$U'-V'=-(V+S\beta)/(1-S^2), \qquad \text{(A12)}$$

$$\beta'=(\beta+SV)/(1-S^2). \qquad \text{(A13)}$$

The procedure adopted in CNDO/1 is essentially to use (A8), taking $U'-V'=-V$ and estimating β' as $\beta^0 S$, where β^0 is independent of the internuclear distance. But from (A12) and A13)

$$U'-V'+S\beta'=V \qquad \text{(A14)}$$

exactly, so that a more satisfactory procedure would be to take

$$\beta'=\beta^0 S, \qquad \text{(A15)}$$

$$U'-V'=-V-\beta^0 S^2, \qquad \text{(A16)}$$

in the energy expression (A8).

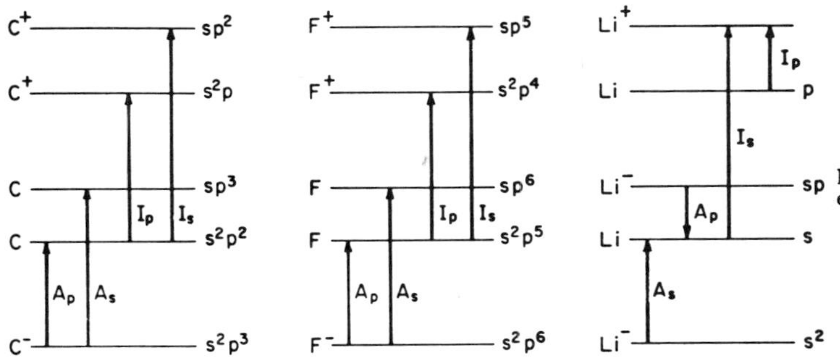

FIG. 5. Relation of A_s and A_p for carbon, lithium, and fluorine (levels represent average energy of configuration).

Since β^0 is negative, this means that V' is too large if taken equal to V, so that in this case, at least, the full allowance for overlap tends to reduce the effective other-nucleus interaction. For many-electron molecules, this corresponds to reducing the magnitude of penetration integrals.

APPENDIX B

Determination of Core Matrix Elements from Atomic Data

The ionization potentials of both s and p electrons for the first-row elements, I_μ, were obtained by averaging over contributing terms in the manner described in Paper II. Their values were taken to be those given in the previous work with the exception of hydrogen for which the observed value, 13.605 eV, was adopted. The evaluation of the matrix elements $(I_\mu+A_\mu)$ then requires a knowledge of the electron affinities of these elements for both s and p electrons based upon the same type of average states.

The majority of the electron affinities of the first-row elements are inexactly known and the determination of the required quantities was therefore based upon the electron affinity of lithium for an s electron and the electron affinity of carbon and fluorine for p electrons. These have been estimated[11] by isoelectronic extrapolation to be

$$A^{\mathrm{Li}}=0.82 \text{ eV} \quad \text{lithium,}$$

$$A^{\mathrm{C}}=1.24 \text{ eV} \quad \text{carbon,}$$

$$A^{\mathrm{F}}=3.50 \text{ eV} \quad \text{fluorine.}$$

These energies represent the transitions $s\rightarrow s^2$ and $s^2p^5\rightarrow s^2p^6$ for lithium and fluorine, respectively, and require no averaging adjustment, each configuration giving rise to but one term. The electron affinity of carbon for a p electron, however, is the energy of the transition $s^2p^2\rightarrow s^2p^3$. The estimated electron affinity, A^{C}, requires correction to a transition between average states, the configurations s^2p^3 giving rise to ${}^2P$, ${}^2D$, and ${}^4S$ terms and s^2p^2 to ${}^1S$, ${}^1D$, and ${}^3P$.

[11] B. Edlen, J. Chem. Phys. **33**, 98 (1960).

The quoted A^{C} is the estimated energy of the transition between the lowest term in the multiplet arising from s^2p^3, that is, ${}^4S$ to the lowest term in the s^2p^2 multiplet ${}^3P_1$. The corrected electron affinity of carbon for a p electron is, as shown in Fig. 4,

$$A_p=A_{\mathrm{exptl}}+(\Delta E-\Delta E^-),$$

where ΔE is the height of the average state of a neutral atom over its most stable contributing term and ΔE^- the height of the analogous average state of a negative ion over its lowest term. ΔE was evaluated from spectroscopic data[12] for carbon and nitrogen and was found to be 0.600 and 2.265 eV for C and N, respectively. ΔE^+ was similarly found to be 3.167 eV for the positive ion O^+. O^+ and N are isoelectronic with C^- and ΔE^- for carbon was estimated to be 1.363 eV by linear extrapolation of ΔE^+ for O^+ and ΔE for N to ΔE^- for C^-. This leads to a corrected affinity of carbon for a p electron, A_p, of 0.477 eV.

The affinity of lithium for a p electron and those of carbon and fluorine for an s electron are related to the corrected experimental values obtained above by the energy intervals shown in Fig. 5, where the energy of each configuration is taken as its average energy. Thus

$$\text{carbon} \quad A_s=A_p+E_{\mathrm{av}}(sp^3)-E_{\mathrm{av}}(s^2p^2),$$

$$\text{fluorine} \quad A_s=A_p+E_{\mathrm{av}}(sp^6)-E_{\mathrm{av}}(s^2p^5),$$

$$\text{lithium} \quad A_p=A_s-E_{\mathrm{av}}(s^2)+E_{\mathrm{av}}(sp),$$

where in the case of lithium it was assumed that

$$E_{\mathrm{av}}(s^2)-E_{\mathrm{av}}(sp)=E_{\mathrm{av}}(p)-E_{\mathrm{av}}(s).$$

The values of (I_s+A_s) and (I_p+A_p) for these three elements were fitted to separate quadratic curves, and the values for the other first-row elements determined by interpolation. The value of (I_s+A_s) for hydrogen was determined from the calculated value[13] $A^{\mathrm{H}}=0.747$ eV.

[12] C. E. Moore, Natl. Bur. Std. (U.S.) Circ. No. 467 (1949).

[13] H. O. Pritchard and H. A. Skinner, Chem. Rev. **55**, 745 (1955).

Chapter III

Energetics with Sigma Molecular Orbital Theory

III-1

As in π-electron theory, semiempirical σ MO methods have been used by and large for charge distributions, dipole moments, and conclusions that can be drawn from charge distributions, such as the inductive effects of substituents and reactivity of sites. A challenge to MO methods is posed by energy properties. Are MO methods as such capable of yielding heats of formation, ionization potentials, activation energies, and the energies of transition in electronic spectra?

Considerable progress on the basic aspects of MO theory as to which properties it should yield to quite good accuracy, and where it is expected to fail, has been made in recent years. A framework for molecular orbitals is provided by the Hartree-Fock method, which yields the best single determinantal wave function made up of spin orbitals, each determined in the SCF of all the electrons in the system. Properties which are expectation values of one-electron operators ("one-electron properties") are expected to come out well when calculated by Hartree-Fock MO theory or reasonable approximations to it. [See O. Sinanoğlu and D. F. Tuan, *Ann. Rev. Phys. Chem.* **15**, 251 (1964).] Equilibrium distances in the geometry, and force constants, have also been related to one-electron properties. Properties of the type of energy differences, such as heats of formation or binding energies, are, however, a different matter. The binding energy of a molecule is the difference between its total energy and its atoms, each in its ground state, separated out of infinity. In obtaining the experimental binding energy, the energy of zero-point vibrations is corrected for.

The exact total energy of the system is given, barring relativistic effects, by

$$E = E_{HF} + E_{corr}$$

where E_{corr}, the correlation energy, is the difference between the exact nonrelativistic N-electron energy and the Hartree-Fock energy. Similarly, the binding energy equals

$$\text{B.E.} = \Delta E_{HF} + \Delta_{corr}$$

Thus the binding energy of a molecule may be described as being due to a *Hartree-Fock binding* and a *correlation binding*.

The correlation binding constitutes a large fraction of the total binding energy as shown in Chapter III-2d. The flourine molecule F_2 is an extreme example, which comes out unstable with respect to two flourine atoms when calculated nonempirically by an accurate Hartree-Fock method. *Correlation binding fractions* of 20 to 30 per cent are more typical figures in other molecules. The paper by Hollister and Sinanoğlu gives, by two semiempirical methods, estimates of correlation binding fractions. It is observed that the correlation binding fractions are quite constant within classes of molecules, which gives the hope that it may be possible to predict the actual binding energy itself using such a ratio, from some molecules of a class and a correlation energy estimate on the unknown molecule, which is obtainable from simple MO coefficients and atomic data. The simple correlation estimates have been compared with experiment and actual Hartree-Fock on small molecules. No accurate Hartree-Fock MO results, however, are available on molecules larger than ethane. It is not clear, therefore, how well the estimates work on larger molecules, as it is as yet not possible to compare with experimental correlation energies, which require a knowledge of accurate Hartree-Fock values. Much more work is needed on the correlation energies of sizable molecules for which additional and more detailed semiempirical methods are under development.

How do the theoretical conclusions above, as to what may be expected from Hartree-Fock molecular orbitals (SCF) and the role of electron correlation in various properties, fit in with the semiempirical σ-MO methods?

The article by Whitehead (Chapter II-2a) studies a wide range of properties using different versions of a CNDO theory. If the parameters are fixed as in the Pople-Santry-Segal CNDO/2 method, the binding energies obtained come out three to eight times too large. The ionization potentials obtained are also not satisfactory when compared with the experimental ones, including the inner ionization potentials obtained by ultraviolet photoelectron spectroscopy (D. Turner et al.). If, on the other hand, the parameters are fitted empirically using valence-state ionization potentials and the binding energies of small diatomic and polyatomic molecules, then on other systems the binding energies obtained come out quite well. With these empirical parameters, however, the orbital populations, charge distributions, and other properties change. Success with parameters obtained from binding properties on heats of formation is obtained also by Dewar and Klopman (Chapter III-2b) and by Wiberg (Chapter III-2c). Wiberg examines the parameters of CNDO/2 theory as to how they need to be modified to yield better bond distances and straight-line correlations with experimental heats of formation. Dewar and Klopman, it may be noted, use a different version of approximate MO-SCF theory, in which 2s and 2p orbitals have different repulsion integrals, as they should, at the expense of rotational invariance not preserved, but this, they show, does not make any practical difference.

It is clear from these results that semiempirical MO theory may be made to yield satisfactory results for a given type of property if its parameters are chosen for that property. But then the success with other properties is spoiled. This apparent incompatibility is understandable, at least qualitatively, on the basis of Hartree-Fock and correlation theory. It has been shown that in ground states electron correlation does not affect charge distributions and the molecular orbitals much [O. Sinanoğlu and D. F. Tuan, *J. Chem. Phys.* **38**, 1740 (1963)]. One should therefore obtain the charge distributions and the molecular orbitals from a semiempiricized theory approximating the Hartree-Fock SCF-MO, in the energy, excluding any portion of correlation from the parameters. (For an analogous situation in π-electron Pariser-Parr-Pople approximate SCF theory, see Chapter VII-7.) When, in a CNDO-type theory, the parameters are fitted from binding-energy data, the orbitals obtained may end up fortuitously altered from what they should be at the Hartree-Fock level. The theory would indicate, therefore, that if one can obtain a simple approximate SCF theory which would yield a satisfactory approximation to just the Hartree-Fock total energy, the molecular orbitals would yield good charge and one-electron properties. To get binding-energy-related properties, one would then add separately correlation-energy corrections obtained by other semiempirical methods, such as the ones given in Chapter III-2d.

One question in calculating binding energies semiempirically, as in the work of Dewar and Klopman, Wiberg, and Whitehead, is: Should the energy be calculated for the known geometry of the molecule, or should one seek the minimum-energy geometry that would result from the method used? If the geometry comes out close to experimental, with parameters appropriate for charge distributions, then a different geometry would be expected to result, with new parameters used appropriate to binding energies. The question, therefore, is probably susceptible to a theoretical analysis, which, however, has not been carried out.

In Chapter III-2e Mickish and Pohl show that properties of the hydrogen bond such as the equilibrium distances and force constants agree satisfactorily with experiment when obtained by semiempirical MO calculations.

Reactivity

To obtain the rate of a chemical reaction, the entropy of activation, $\Delta S^{\ddagger}$, and the enthalpy of activation, $\Delta H^{\ddagger}$, are needed. This requires an educated guess of the geometry of the activated complex. For obtaining the entropy of activation, simple formulas based on classical mechanics with quantum corrections have been developed, at least for classes of reactions such as "abstraction" [O. Sinanoğlu and K. S. Pitzer, *J. Chem. Phys.* **30**, 422 (1959)]. The activation enthalpy, $\Delta H^{\ddagger}$, on the other hand, is even more difficult than heats of formation to predict. In an activated complex, there are stretched-out bonds. As ordinary, closed-shell, MO theory often gives dissociation to the wrong atomic states where there are situations near dissociation, an open-shell MO method must be used. This is a difficulty already at the framework Hartree-Fock MO theory level, compounded by difficulties in semiempiricism. An important attempt to tackle this problem with σ-MO theory has been made by Klopman (Chapter III-3a).

As mentioned in Chapter II, Fukui and co-workers pioneered the use of σ-MO theory for chemical reactivities, using a hybrid-based Hückel approach. (See K. Fukui in *Modern Quantum Chemistry*, Vol. I, O. Sinanoğlu, ed., Academic Press, New York, 1965, for a detailed presentation.) This approach gives a qualitative guide to reactivities, based on the highest occupied and lowest unoccupied molecular orbitals, their shapes, and their charge distributions. The reader interested in reactivity will also find organic chemical applications, along these lines, of σ-MO theory discussed in Chapter IV.

Electronic Spectra

The transition energies in electronic spectra constitute another property difficult to obtain with semiempirical MO theory. The energy differences in general depend strongly on electron correlation. In excited states, both the Hartree-Fock MO theory and the correlation theory are more complex than in ground states. The excited states involve novel and sizable correlation effects not present in the ground, and therefore the correlation energy corrections are in general not expected to cancel out. Such correlation effects and how they affect both the transition energies and the transition probabilities have been studied recently in detail for many atomic states [O. Sinanoğlu in *Atomic Physics*, V. Hughes et al., eds., Plenum Press, New York, 1969, and O. Sinanoğlu and I. Öksüz, *Phys. Rev. Letters* **21**, 507 (1968)].

However, little is known about these effects in molecules, where they may be even more important. It is also well known that in the orbital theory itself one needs to minimize the energy of each state separately to obtain orbitals appropriate to that state. The relatively small amount of data available on transition energies of saturated and nonconjugated, unsaturated compounds, coupled with the difficulty in making assignments of the observed bands, makes it difficult to find experimental data against which to test the calculations. This is an area which clearly needs much more work, both experimentally and theoretically.

In his interesting article (III-4a), Sandorfy calls attention to the need for σ-MO calculations on excited states. Most calculations and methods have dealt with ground-state properties only. He reviews the new experimental results on normal and branched paraffin hydrocarbons and the available but inconclusive interpretations of the UV spectra.

In Chapter III-4b, McGlynn et al. study experimentally and theoretically the electronic spectra and structure of sulfur compounds, where the presence of sulfur poses additional problems in parametrization. A Hückel-type approach is used which works better on single sulfur compounds than on those with several sulfur atoms. McGlynn and co-workers have also studied the spectra of molecules such as pyridine with open-shell calculations of excited states (Chapter VI-7).

III-2

Binding: Heats of Formation, Ionization Potentials, and Hydrogen Bonding

III-2a

Semiempirical All-Valence-Electron SCF-MO-CNDO Theory

M. A. WHITEHEAD

1. Introduction

The CNDO approximation was formulated by Pople and Segal,[1] who established the Hamiltonian matrix elements as[2]

$$F_{kk} = U_{kk} + \left(P_{AA} - \tfrac{1}{2} P_{kk}\right) g_{AA} + \sum_{B \neq A} (P_{BB} g_{AB} + V_{AB}) \tag{1}$$

and

$$F_{k\ell} = -\tfrac{1}{2} (\beta^0_A + \beta^0_B) S_{k\ell} - \tfrac{1}{2} P_{k\ell}\, g_{AB} \qquad k \neq 1 \tag{2}$$

where

$$P_{k\ell} = 2 \sum_i C^*_{ki} C_{\ell i} \tag{3}$$

and

$$P_{AA} = \sum_k^A P_{kk} \tag{4}$$

The orbitals ϕ_k and ϕ_ℓ are on atoms A and B, respectively. U_{kk} is the diagonal matrix element of the kth orbital on atom A with respect to the kinetic energy and to the potential energy of the core of atom A; P_{AA} is the total valence-shell electronic charge on atom A; P_{kk} is the kth valence-shell orbital electronic charge; g_{AB} represents "an average repulsion between an electron in a valence orbital on atom A and another in a valence orbital on B"[3]; V_{AB} is the interaction of an orbital on atom A with the core of B and has the same value for all orbitals on atom A, so as to preserve the invariance of the Roothaan equations to atomic transformations.[3] The bonding parameters β^0_A and β^0_B are characteristic of the atoms A and B, respectively, while the overlap integral $S_{k\ell}$ is defined by

$$S_{k\ell} = \int \phi^*_k (1)\, \phi_\ell (1)\, dV_1 \tag{5}$$

The only problem, therefore, lies in finding the parameters to put into these equations.[1,2,4,5]

2. Atomic Parameters

(For values of some semiempirical atomic parameters, see Table 1.)

Now, if the full SCF-LCAO-MO equations could be simplified by the CNDO approximation without any restrictions, the diagonal matrix elements of the total Hamiltonian for the kth orbital on atom A would be

$$F_{kk} = U_{kk} + \sum_{\ell} P_{\ell\ell}\, g_{\ell\ell}(1 - \delta_{k\ell}) + \sum_{m} P_{mm} g_{km} - \sum_{B \neq A} V_{kB} \tag{6}$$

However, this has to be made invariant to local transformations of the atomic orbital basis functions; hence the electron repulsion matrix $g_{k\ell}$ between the kth and ℓ^{th} orbitals on atoms A and B, respectively, has to be given the same value, g_{AB}, for all pairs of atomic orbitals on the two atoms.[3] A basis set of s and p orbitals gives four distinct electron repulsion parameters, g_{ss}, g_{sp}, g_{pp}, and $g_{pp'}$, in which p and p′ are two different p orbitals on the same atom; these must be averaged to give g_{AA}', the average repulsion between two valence electrons on the same atom. This is equivalent to assuming that the energy of any valence state $(ns)^p(np)^q$ is the same, which is true within 2 to 3 eV.[6,7] Thus equation (6) becomes

$$F_{kk} = U_{kk} + g_{AA} \sum_{\ell} P_{\ell\ell}(1 - \delta_{k\ell}) + \sum_{m} P_{mm} g_{AB} - \sum_{B \neq A} V_{AB} \tag{7}$$

These two equations can only be made equivalent for a specific charge distribution, by equating the diagonal matrix elements; we chose to make them equivalent when the valence-shell electron

Table 1. Semiempirical Atomic Parameters (in eV)

Element	$\bar{U}_{ss}$	$\bar{U}_{pp}$	g_{AA}	g^*_{AA}	C
H	-13.595	—	12.848	12.848	13.595
Li	-4.999	-3.673	3.469	3.458	4.999
Be	-15.543	-12.280	5.935	5.953	25.151
B	-30.371	-24.702	8.000	8.048	61.444
C	-50.686	-41.530	10.207	10.333	123.517
N	-70.093	-57.848	11.052	11.308	204.291
O	-101.306	-84.284	13.625	13.907	335.908
F	-129.544	-108.933	15.054	15.233	487.697
Na	-4.502	-3.247	2.982	3.031	4.502
Mg	-13.083	-9.603	4.623	4.656	21.544
Aℓ	-22.828	-18.592	5.682	5.680	47.203
Si	-36.494	-30.375	6.964	7.015	92.438
P	-58.610	-50.940	9.878	9.886	172.095
S	-66.796	-58.008	9.205	9.260	227.860
Cℓ	-86.774	-75.681	10.292	10.366	335.847
K	-3.170	-3.115	3.702	3.560	3.170
Ca	-9.842	-7.696	3.977	3.979	15.707
Ga	-25.032	-19.807	5.936	5.942	52.063
Ge	-35.844	-29.973	6.608	6.634	92.527
As	-50.151	-44.485	8.399	8.361	150.653
Se	-66.005	-57.927	9.121	9.156	227.686
Br	-76.413	-65.412	8.823	8.838	294.760
Rb	-3.555	-2.804	2.495	2.384	3.555
Sr	-9.430	-7.074	3.749	3.761	15.110
In	-23.056	-17.663	5.530	5.582	47.185
Sn	-26.981	-21.869	4.297	4.304	72.317
Sb	-47.427	-40.923	7.657	7.761	141.347
Te	-64.464	-57.144	8.985	9.039	223.174
I	-76.905	-69.091	9.448	9.382	301.030

population for all the atoms in the molecule is equally distributed among the one s and three p orbitals. Then

$$F_{kk} = U_{kk} + \frac{P_{AA}}{4} \sum_{\ell} g_{k\ell}(1 - \delta_{k\ell}) + \tfrac{1}{4} \sum_{m} P_{BB}\, g_{km} - \sum_{B \neq A} V_{kB} \tag{6'}$$

and

$$F_{kk} = U_{kk} + \tfrac{7}{8}\, g_{AA} P_{AA} + \sum_{B \neq A} (P_{BB} g_{AB} - V_{AB}) \tag{7'}$$

The atomic terms are equal when

$$g_{AA} = g_{AA}^{k} = \tfrac{2}{7} \sum_{\ell} g_{k\ell}\, (1 - \tfrac{1}{2} \delta_{k\kappa}) \tag{8}$$

where the superscript k indicates that the matrix element of the kth orbital is used to evaluate g_{AA}. However, there is no reason to prefer any one orbital, and consequently equation (8) should be averaged over all the orbitals on atom A to give

$$g_{AA} = \tfrac{1}{4} \sum_{k} g_{AA}^{k} = \tfrac{1}{14} \sum_{k} \sum_{\ell} g_{k\ell}\, (1 - \tfrac{1}{2} \delta_{k\ell}) \tag{9}$$

or in the chosen basis,

$$g_{AA} = \tfrac{1}{28}\, (g_{ss} + 12 g_{sp} + 3 g_{pp} + 12 g_{pp'}) \tag{10}$$

This procedure destroys the equivalence of the atomic terms in equations (6′) and (7′), and it is impossible to restore the equality for all values of P_{AA}. The best procedure is to restore the equality for an exactly neutral atom, thus eliminating errors due to the CNDO approximation in calculations on homopolar bonding. When all valence electrons are considered, P_{AA} for an exactly neutral atom equals the core charge Z_A, so that the adjusted parameter $\overline{U}_{kk}$ is

$$\overline{U}_{kk} = U_{kk} + \frac{Z_A}{4} \left[\sum_{\ell} g_{k\ell}(1 - \delta_{k\ell}) - \tfrac{7}{2}\, g_{AA} \right] \tag{11}$$

and for s and p orbitals this becomes

$$\overline{U}_{ss} = U_{ss} + \frac{Z_A}{4} \left[\tfrac{1}{2}\, g_{ss} + 3 g_{sp} - \tfrac{7}{2}\, g_{AA} \right]$$

and

$$\overline{U}_{pp} = U_{pp} + \frac{Z_A}{4} \left[\tfrac{1}{2}\, g_{pp} + g_{sp} + 2 g_{pp'} - \tfrac{7}{2}\, g_{AA} \right]$$

respectively. These parameters were evaluated from atomic spectra using the equation due to Oleari et al.,[4]

$$E = C + \sum_{k} n_k U_{kk} + \tfrac{1}{2} \sum_{k} \sum_{\ell \neq k} n_k n_\ell g_{k\ell} + \tfrac{1}{2} \sum_{k} n_k (n_k - 1) g_{kk} \tag{12}$$

in which the parameters C, U_{ss}, U_{pp}, g_{sp}, g_{pp}, and $g_{pp'}$ are determined so that the E in equation (12) best fits certain atomic valence-state energies as a function of n_k. These parameters vary markedly with atomic charge[4] and hence must be evaluated for valence states close to neutrality. The constant C occurs because the core state is not used in evaluating the parameters. The necessary valence-state energy data were obtained from Hinze and Jaffé,[6,7] and the parameters for any atom were evaluated entirely from valence-state energies; the $g_{k\ell}$ were averaged to give g_{AA}; the core matrix elements U_{ss} and U_{pp} were adjusted after the averaging process for g_{AA}.

Equation (12) contains seven atomic parameters and the additive constant C, so seven valence states were needed to solve for the parameters; these were chosen so that the states (a) differed from electroneutrality by *one* electron, (b) formed a set sufficient to evaluate the seven parameters, (c) were unipositive rather than uninegative, and (d) were preferred if they had low promotion energies. Thus for carbon the parameters were calculated from the following equations:

$$\begin{aligned}
g_{ss} &= E(C^-, s^2ppp) - 2E(C, sppp) + E(C^+, ppp)\\
g_{sp} &= [E(C^-, s^2ppp) - E(C, sppp)] - [E(C, s^2pp) - E(C^+, spp)]\\
g_{pp} &= E(C^-, sp^2pp) - 2E(C, sppp) + E(C^+, spp)\\
g_{pp'} &= [E(C^-, sp^2pp) - E(C, sp^2p)] - [E(C, sppp) - E(C^+, spp)]\\
U_{ss} &= E(C, sppp) - E(C^+, ppp) - 3g_{sp}\\
U_{pp} &= E(C, sppp) - E(C^+, spp) - g_{sp} - 2g_{pp'}
\end{aligned}$$

The only remaining atomic parameters to define are the atomic limits of the interatomic repulsion integrals g_{AB}. If equations (6′) and (7′) are equated and averaged over all the orbitals on atom A, then

$$g_{AB} = \frac{1}{16}\sum_k \sum_m g_{km} \tag{13}$$

The two-center repulsion integral for any two valence orbitals can be expressed by [8-10]

$$\lim_{R \to O} g_{k\ell} = \tfrac{1}{2}\,(g^A_{k\ell'} + g^B_{k'\ell}) \tag{14}$$

where $g^A_{k\ell}{}'$ is the repulsion integral between an electron in the kth orbital of atom A and one in the ℓ'^{th} orbital of atom A, which is of the same type as the ℓth orbital on atom B. Thus from equations (13) and (14) the atomic limits for g_{AB}, $g_{AB'}$ (g^*_{AA} and g^*_{BB}) are given by

$$\lim_{R \to O} g_{AB} = \tfrac{1}{2}\,(g^*_{AA} + g^*_{BB}) \tag{15}$$

where

$$g^*_{AA} = \frac{1}{16}\sum_k \sum_\ell g_{k\ell} \tag{16}$$

and for a basis of s and p orbitals,

$$g^*_{AA} = \frac{1}{16}\,(g_{ss} + 6g_{sp} + 3g_{pp} + 6g_{pp'}) \tag{17}$$

If one of the atoms in the molecule is hydrogen, then g^*_{AA} equal to g_{ss} was used for hydrogen and equation (17) for the other atom.[11] Thus the parameters were chosen so that for any specific charge distribution the diagonal Hamiltonian matrix elements would have the same values as they would if the use of a common value for all the g^i_{AA} were not required; but they cannot have the same change with charge distribution as they would without such a restriction.

From atomic spectra it was found that $g_{ss} > g_{sp} \cong g_{pp} > g_{pp'}$ so that $g^s_{AA} > g_{AA} > g^p_{AA}$, and equating these raises the electronegativity of the s orbital and lowers it for the p orbital when the electron population exceeds $\frac{1}{4}\,Z_A$, and vice versa.

The accuracy of these parameters can be measured by comparing the empirical ionization potentials and empirical electron affinities with those calculated from

$$E = C^0 + \sum_k n_k \overline{U}_{kk} + \frac{1}{2}\left(\sum_k n_k\right)\left(\sum_k n_k - 1\right) g_{AA} \tag{18}$$

the valence-state energy relative to the ground state of the neutral atom, chosen as zero, rather than the core with all the valence electrons removed, because the higher ionization potentials are uncertain.[12] C° is calculated by equating the stable neutral valence state E to its promotion energy.[6,7] It was found that the ionization potentials and electron affinities for the common valence states of carbon (homopolar) and fluorine (heteropolar bonds) were accurate within 1 eV.[11] Thus the parameters were calculated for a charge distribution in which the valence-shell electrons on each atom were equally distributed among the valence orbitals. When only s and p orbitals are included, this approximates the charge distribution in real molecules, and the parameters shoule be reasonable for molecular calculations, but inclusion of d orbitals in a similar way for elements with slight d-orbital bonding would be quite inaccurate.

Of course, equation (10) is not itself invariant to a change in basis set, but it was found[11] to vary by only a few tenths of an electron volt if a hybrid basis set was used, so that a value of g_{AA} determined using s-p valence-state energies is valid for MO calculations using hybrid orbitals. It is better to use the nonhybrid basis to estimate g_{AA} because the valence-state promotion energies are more accurate, and the core Hamiltonian matrix off-diagonal elements between pure orbitals vanish by symmetry.

3. Interatomic Parameters

Having calculated the atomic parameters of the diagonal matrix elements of F_{kk}, it was necessary to estimate the interatomic parameters.

Interatomic Electron-Electron Repulsion Integrals

The repulsion between an electron in the kth orbital of atom A and one in the rth orbital of atom B is

$$g_{kr} = (kk|rr) = \int \phi_k^*(1)\phi_k(1)\frac{1}{r_{12}}\phi_r^*(2)\phi_r(2)\,dV_1 dV_2 \tag{19}$$

and this can be calculated for orbitals of a specified form, and averaged to give g_{AB}: the interatomic electron repulsion integral. Therefore, we calculated both the atomic and interatomic electron repulsion integrals using equation (19) for valence-shell Slater s orbitals using formulas listed by Roothaan[13] following the procedure of Pople and Segal.[1,2]

However, the CNDO approximation is analogous to the ZDO approximation in π-electron theory, which gives better agreement with experiment when both the atomic and interatomic electron repulsion integrals are evaluated empirically, which allows for electron correlation and lowers the integrals below their theoretical values. Two of the empirical formulas proposed in π-electron calculations are still valid in CNDO, and both were used for calculations together with the empirical assessment of the atomic electron repulsion integrals described above.

a. The interatomic repulsion integrals were calculated by the Mataga formula[9] adapted to calculations involving all valence electrons:

$$g_{AB} = \frac{1}{R_{AB} + a} \tag{20}$$

where

$$a = \frac{2}{g^*_{AA} + g^*_{BB}}$$

where g^*_{AA} and g^*_{BB}, the atomic limits, replace the g_{kk} and g_{rr} of the original π-theory formula, and g_{AB} replaces the g_{kr}, in agreement with the arguments above.

b. They were calculated using Ohno's formula,[10] similarly modified to

$$g_{AB} = \frac{1}{\sqrt{R^2_{AB} + a^2}} \tag{21}$$

These two equations are a representative sample of the values of g_{kr} used in semiempirical MO calculations on π systems,[14] adapted to CNDO. The use of such semiempirical methods results in much better agreement between experiment and theory than the theoretical methods, justifying their use. The similarity between the Ohno formula and the theoretical formula due to Paoloni has been demonstrated recently.[75]

Core Attraction Integrals V_{AB}

When the interatomic terms of equations (6′) and (7′) are equated and averaged over all orbitals on atom A,

$$V_{AB} = \frac{1}{4} \sum_i V_{iB} \tag{22}$$

but since semiempirical values of V_{iB} are unknown, this cannot be used to obtain V_{AB}. Consequently, they are evaluated from

$$V_{AB} = - Z_B g_{AB} \tag{23}$$

as in the CNDO/2 theory, [1] in which case the penetration contributions to F_{kk} vanish; this is very important when semiempirical values of g_{AB} are used, because an analytical assessment of V_{AB} would result in large penetration contributions.[2]

Overlap Integrals

The overlap integrals for Slater orbitals of principal quantum number 1, 2, 3, or 5 were evaluated analytically by the methods of Mulliken et al.[15] The orbital exponents Z' and effective quantum numbers n' are given by Slater's rules.[16] Since in accurate molecular calculations using the Roothaan equations with a variable Z_H' it was found that $Z_H' = 1.2$ lead to lower energies than $Z_H' = 1.0$, $Z_H' = 1.2$ was preferred and, in fact, demonstrated to be more effective than $Z_H' = 1.0$ in this semiempirical SCF-MO-CNDO theory.[17] This is because contraction of the orbital leads to more stable bonding.[18]

For $n = 4$, overlap integrals cannot be evaluated analytically because of the nonintegral value of r present as a factor. Consequently, it was decided to use an approximate orbital, made by assuming that the radial function $R_4(r)$ could be interpolated between $R_3(r)$ and $R_5(r)$ in the same way as n' is interpolated. Two forms were tested:

$$\begin{aligned} R_4'(r) &= N_4^*(0.3R_3 + 0.7R_5) \\ &= N_4'(0.3r^2 e^{-Z'r/3a_0} + 0.7r^3 e^{-Z'r/4a_0}) \end{aligned} \tag{24}$$

and

$$R_4''(r) = N_4''(0.3r^2 + 0.7r^3)e^{-Z'r/3.7a_0} \tag{25}$$

where N is the normalizer. The accuracy of these was tested by computing their overlap with the Slater orbital,

$$R_4(r) = N_4 r^{2.7} e^{-Z' r/3.7 a_0} \tag{26}$$

These overlaps are one-center integrals and were computed using gamma functions, giving

$$S_{R_4 R_4'} = 0.99651$$

and

$$S_{R_4 R_4''} = 0.99979$$

Thus $R_4''(r)$ was used in all the calculations; N_4'' was 1.01384531.

Bonding Parameters

These were empirically evaluated for each element (see Tables 2 and 2a). Preliminary calculations showed that bonding energies were most sensitive to changes in the bonding parame-

Table 2. Bonding Parameters β_A^0 (in eV)

Code[a]	M1	M2	01	02	R1	R2	PS[b]
Atomic parameters	Empirical[c]	Empirical	Empirical	Theoretical	Theoretical	Theoretical	Theoretical
Interatomic g_{AB}	Mataga[e]	Mataga	Ohno[f]	Ohno	Theoretical[d]	Theoretical	Theoretical
Hydrogen exponent Z_H'	1.0	1.2	1.0	–1.2	1.0	1.2	1.2
Evaluation of β_A^0	Empirical[g]	Empirical[g]	Empirical[g]	Empirical[g]	Empirical[g]	Empirical[g]	Pople and Segal[h]
H	4.9	5.4	3.9	4.3	5.4	5.2	9
Li	0.4	0.7	-0.9	-0.8	2.5	3.8	9
Be	3.8	4.0	3.2	3.4	4.3	5.2	13
B	5.8	5.6	5.2	5.0	6.2	6.5	17
C	8.7	8.2	7.8	7.3	9.1	9.0	21
N	9.6	8.8	8.0	7.3	11.2	10.6	25
O	14.2	12.8	11.7	10.5	16.1	14.7	31
F	19.2	17.2	15.7	14.1	22.6	20.4	39
Si	5.0	5.2	4.6	4.7			
P	6.0	6.0	5.3	5.3			
S	6.7	6.5	5.8	5.6			
Cℓ	9.3	8.9	8.1	7.8			
Ge	4.3	4.4	3.8	4.0			
As	4.6	4.7	4.0	4.1			
Se	5.7	5.7	5.0	4.9			
Br	7.3	7.2	6.4	6.3			
Sn	3.4	3.6	1.9	2.1			
Sb	4.5	4.7	3.9	4.2			
Te	5.7	6.1	5.1	5.4			
I	6.5	6.7	5.8	6.0			

[a] Arbitrary code for parameter set.
[b] CNDO/2 Method of Pople and Segal.
[c] From atomic spectra.
[d] From theoretical integral formulas of Roothaan—first row only.
[e] From equation (20).
[f] From equation (21).
[g] From hydride bonding energies as described in the text.
[h] By comparison with minimal-basis set calculations by Roothaan method.

Table 2a. Bonding Parameters β_A^0 (in eV)

Element	O2 "hydrides"[a]	O2 "diatomics"[b]
H	4.3	4.83
Li	-0.8	-1.34
Be	3.4	2.89
B	5.0	4.41
C	7.3	6.91
N	7.3	6.84
O	10.5	9.64
F	14.1	13.86
Si	4.7	4.22
P	5.3	4.73
S	5.6	5.09
Cl	7.8	7.24
Ge	4.0	3.44
As	4.1	3.59
Se	4.9	4.46
Br	6.3	5.77
Sn	2.1	1.55
Sb	4.2	3.65
Te	5.4	4.86
I	6.0	5.43

[a]From Table 2; calculated using the β of H_2 and the hydride of the element.
[b]β_A^0 calculated using the β of the diatomic molecules of each element.

ters, compared to ionization potentials, dipole moments, and nuclear quadrupole coupling constants (Table 2b). Therefore, the β_A^0 for hydrogen was initially chosen to give the correct dissociation energy of the hydrogen molecule.[17] The β_A^0 for other elements were then chosen to give correct bonding energies for the binary hydrides AH_n, using

$$\beta_{AB}^0 = \frac{-(\beta_{AH}^0 + \beta_{BH}^0)}{2} \tag{A}$$

Therefore, all bonding parameters depend on the atomic wave function used for hydrogen and the value of β_H^0; $Z_H' = 1.2$ gave the better fit.

These empirical parameters are all smaller than the theoretical values[17] and more consistent with the values of core resonance integrals in π-electron theories.[17,19]

The bonding parameters were thus evaluated for the interaction of the valence-shell orbitals on atom A with the ls orbital on hydrogen. However, in general, an A–B bond does not use ls orbitals and this may be a poor approximation. Later[62] β_{AB}^0 was approximated by

$$\beta_{AB}^0 = \frac{-(\beta_{AA}^0 + \beta_{BB}^0)}{2} \tag{B}$$

where β_{AA}^0 is chosen to give the experimental dissociation energy of the homonuclear diatomic A_2. This would be expected to be a better parameterization for an A–B bond not involving hydrogen.

For many homonuclear diatomic molecules there is a great deal of controversy about tne experimental dissociation energies. Therefore, the parameterization was based on an acceptable set of experimental data for a *few* elements, and new parameters for the other elements were obtained by a systematic correction of those derived from the hydrides.

Table 2b. Variation of Molecular Properties of HF and HCℓ with Halogen Bonding Parameter β_A^0 [a]

Physical property	β_A^0 for HF					β_A^0 for HCℓ			
	20	19	18	17	Exptl.	10	9	8	Exptl.
Bonding energy, eV	7.30	6.87	6.44	6.03	6.11[c]	5.22	4.66	4.11	4.61[d]
First ionization potential (π), eV	16.27	16.24	16.21	16.18	16.06[c]	13.19	13.18	13.17	12.80[c]
Second ionization potential (σ), eV	16.78	16.64	16.49	16.35	16.48[c]	14.27	14.09	13.89	16.28[c]
Third ionization potential (σ), eV	37.90	37.79	37.69	37.58	–	24.98	24.88	24.79	–
Dipole moment, D	1.897	1.897	1.898	1.898	1.8195[e]	2.058	2.025	1.991	1.12[f]
^{35}Cℓ quadrupole coupling constant, MHz						85.72	86.02	86.29	67.3[g]

[a] SCF-MO calculations with CNDO approximation. Parameters: g_{AA} from atomic spectra, g_{AB} from Mataga formula, $Z_H' = 1.2$, $\beta_H^0 = 5.4$ eV.
[b] Ref. 24.
[c] D. C. Frost, C. A. McDowell, and D. A. Vroom, *J. Chem. Phys.* **46**, 4255 (1967).
[d] Ref. 24.
[e] R. Weiss, *Phys. Rev.* **131**, 659 (1963).
[f] C. A. Burrus, *J. Chem. Phys.* **31**, 1270 (1959).
[g] M. Cowan and W. Gordy, *Phys. Rev.* **111**, 209 (1958).

The first-row elements, B, C, N, O, and F, were chosen for the initial parameterization. The bonding parameters were calibrated to give the experimental equilibrium dissociation energies, D_e, at the equilibrium internuclear distance r_e.[62] These bonding parameters were then used to determine the value of β_H^0 required to calculate the experimental bonding energies for the binary hydrides of the above elements. The average value of the bonding parameter for hydrogen was 4.83 ± 0.3 for BH_3, NH_3, H_2O, and HF. In the case of CH_4, the value obtained for β_H^0 was 5.74, owing to uncertainty in the experimental dissociation energy of C_2.[22,75] The calibration of β_C^0 to give the experimental bonding energy of CH_4 using the average value for β_H^0 gave a large range of errors for the energies of atomization of the hydrocarbons studied. Consequently, the value of β_C^0 was chosen to minimize the average deviation from the experimental bonding energy for methane and the small hydrocarbons in Table 3a with $\beta_H^0 = 4.83$. The bonding parameters of the other elements were calibrated from the bonding energies of the corresponding binary hydrides. This approach is necessary for the second-, third-, and fourth-row elements because insufficient data are available for their homonuclear diatomics. Molecules such as $C\ell_2$, Br_2, and I_2, for which accurate data are available, were used to test the above procedure.

The new bonding parameters for the semiempirical SCF-MO-CNDO theory with interatomic repulsion integrals calculated by the Ohno formula, and an orbital exponent of 1.2 for hydrogen, are compared with the previous parameters in Table 2a. The new parameters are 0.5 ± 0.1 eV *lower* than the "hydride" parameters with the exception of β_H^0, which is correspondingly larger.

The exact electronic energy of a molecule is[20]

$$E_{e\ell^0} = E_{HF} + E_c + V_{nn} \tag{27}$$

where E_{HF} is the Hartree-Fock energy for the best single determinant wave function and E_c is the correlation energy. The parameters were adjusted to give the correct bonding energies, including E_c, despite the single determinant wave function.[20,21]

For a closed-shell molecule with a single-determinant wave function

$$E_{e\ell} = \frac{1}{2} \sum_k \sum_\ell P_{k\ell} H_{k\ell} + \sum_i E_i + V_{nn} \tag{28}$$

where E_i is the orbital-energy eigenvalue given by $\Sigma_k \Sigma_\ell C_{k\ell}^* C_{\ell i} F_{k\ell}$ and where V_{nn} is the expectation value of the internuclear potential energy for a fixed nuclear configuration. In order to find bonding (dissociation) energies, the zero of this equation must be considered.[17] In an all-valence-electron SCF-MO theory, the right-hand side of equation (28) vanishes when (a) the molecule has no valence-shell electrons, or (b) the atomic cores are at infinity.

Now, chemically the bonding energy E_B is that energy required to separate the molecule into neutral ground-state atoms at infinity; hence

$$E_B = \sum_A E_A - E_{e\ell} \tag{29}$$

where E_A is the valence-shell electronic energy of atom A, and in a diatomic molecule E_B equals D_e (the heat of dissociation) measured from the minimum of the potential-energy curve.[22] The theoretical bonding energy is therefore

$$E_B = \sum_A E_A - E_e - V_{nn} \tag{30}$$

since $E_{e\ell} = E_e + V_{nn}$.[17] The atomic energy E_A is thus the energy to remove all the valence-shell electrons from the atom A, just as E_e is the energy to remove them from the molecule. Successive ionization from atoms or molecules requires greater energy. The decrease in electrostatic repulsion of the electron to be ionized is incorporated in any SCF theory, but the decrease in the screening of the electron to be ionized from the nucleus is not included when a minimum basis set is used because the orbital parameters have fixed values. The parameters evaluated from atomic spectra [11] are only valid for valence states near electroneutrality, and consequently the molecular energies must be calculated with the same parameters so that there is an error cancellation in calculating E_B.

Thus in the CNDO approximation the atomic state energy is given by equation(18). Since the energy must now be expressed relative to the core state, with all the valence electrons removed, C^0 becomes zero and for n_s s electrons and n_p p electrons,

$$E = n_s\bar{U}_{ss} + n_p\bar{U}_{pp} + (n_s + n_p)(n_s + n_p^{-1})g_{AA} \tag{31}$$

Internuclear Potential Energy

Finally it is necessary to define V_{nn}. If the atomic cores are point charges, then

$$V_{nn} = \sum_{A>B} Z_A Z_B R_{AB}^{-1} \tag{32}$$

in atomic units.[1,3] The net electrostatic repulsion between two atoms is therefore

$$E_{AB} = V_{nn} + P_{AA}V_{AB} + P_{BB}V_{BA} + P_{AA}P_{BB}g_{AB} \tag{33}$$

which for neutral atoms where $P_{AA} = Z_A$ gives

$$E_{AB}^0 = Z_A Z_B (R_{AB}^{-1} - g_{AB}) \tag{34}$$

No choice of parameters will give accurate bonding energies with this equation,[17] and it was found necessary to assume that the electrostatic interaction between two neutral atoms E_{AB}^0 vanishes, and hence

$$V_{nn} = \sum_{A>B} Z_A Z_B g_{AB} \tag{35}$$

parallelling π-system assumptions.[23] Hence between two nonneutral atoms the net electrostatic repulsion is

$$E_{AB} = (P_{AA} - Z_A)(P_{BB} - Z_B)g_{AB} \tag{36}$$

Equation (35) was found to give good bonding energies at experimental bond lengths.[17]

Thus the final equation used to calculate the bond energies by the semiempirical all-valence-electron SCF-MO-CNDO theory is

$$E_B = \sum_A E_A - \tfrac{1}{2}\sum_k \sum_\ell P_{k\ell}H_{k\ell} - \sum_i E_i - \sum_{A>B} Z_A Z_B g_{AB} \tag{37}$$

combining equations (28), (35), and (30), and E_A from (31).

4. Bonding Energies

For comparison purposes the bonding energies (see Tables 3 and 3a) were all extrapolated to 0°K, assuming ideal gas capacities and ignoring anharmonicity corrections, because the effect on E_B is only 0.004 eV for H_2, less for other diatomics, and unknown for most polyatomics.[24] The bonding energies calculated from the SCF-MO-CNDO theory with empirical parameters for simple diatomics (N_2, CO, LiF, IBr, etc.), polyatomics (BrCN, OCS, SO_2, etc.), and hydro-

Table 3. Bonding Energies Calculated by SCF-MO Theory with CNDO Approximation and Empirical Bonding Parameters (in eV)

Parameter set	M1	M2	O1	O2	R1	R2	Exptl.
N_2	12.325	10.480	12.402	10.839	12.997	11.566	9.903
CO	13.838	11.931	13.798	12.166	14.136	12.579	11.225
CS	8.093	7.414	8.660	7.970			7.190
CO_2	22.032	18.981	21.310	18.592	22.578	20.160	16.856
OCS		16.142	17.968	15.984			14.417
CS_2	14.154	13.012	14.262	13.090			11.980
NNO			17.401	14.986	18.899	16.634	11.724
SO_2	11.085	9.494	11.255	9.761			11.177
O_3	11.995	9.804	11.009	9.034			6.345
C_2H_2	19.333	17.724	19.969	18.264	20.003	19.448	17.530
C_2H_4	25.290	24.250	25.609	24.366	25.169	24.831	24.357
C_2H_6	31.603	31.032	31.650	30.799	31.077	30.867	30.818
C_3H_8	46.431	45.230	46.250	44.705	45.699	45.373	43.563
B_2H_6	28.657	27.706	27.523	26.580	28.398	27.652	26.004
LiF	5.970	5.551	6.557	6.290	4.101	3.822	5.940
F_2	2.887	2.060	2.627	1.983	0.991	0.064	1.653
Cl_2	3.491	3.178	3.398	3.170			2.508
Br_2	2.767	2.687	2.774	2.695			1.991
I_2	1.677	1.837	1.750	1.907			1.557
ClF	3.882	3.191	3.604	3.079			2.668
BrF	3.369	2.799	3.216	2.778			2.682
BrCl	2.997	2.804	2.977	2.826			2.334
IF	1.524	1.153	1.570	1.295			2.91
ICl	2.382	2.307	2.405	2.368			2.190
IBr	2.162	2.201	2.212	2.251			1.928
CH_3F	19.558	18.634	19.126	18.234	18.813	17.847	18.384
CH_3Cl	17.680	17.329	17.669	17.211			17.154
CH_3Br	16.808	16.608	16.891	16.532			16.640
CH_3I	15.789	15.733	15.944	15.727			15.931
HCN	16.113	14.705	16.367	14.996	16.410	15.603	13.537
CH_3CN			30.866	28.710			26.586
FCN	19.703	17.167	19.265	17.113	20.114	18.199	13.529
ClCN		15.262	17.159	15.468			12.310
BrCN	16.051		16.183	14.589			
ICN	14.868	13.333	15.094	13.658			11.159

carbons (C_2H_2, C_3H_8, etc.), as well as substituted hydrocarbons (CH_3X), compared exceedingly well with the experimental measurements,[17] showing that the theory includes correlation energy quite accurately, because the use of theoretical bonding parameters gives a poor comparison of predicted and experimental energies. In all cases the use of $Z'_H = 1.2$ gave better agreement than $Z'_H = 1.0$. However, it is not all clear whether the Mataga or Ohno formula is more accurate in assessing the electron repulsion integrals, although they are both more accurate than the theoretical g_{AB}, because both give more accurate E_B for half the molecules considered.[17] Thus it seems that the exact values of g_{AB} do not matter, provided the atomic integrals and the atomic limits of the interatomic integrals are calculated from atomic spectra. The theoretical parameters are so unsuitable that they predict bonding energies three to eight times the correct value[17]; the error cannot be due to the omission of correlation energy because the results are too high; it must be due to the large values of the bonding parameters.[2] From Table 3a it is clear that the new bonding parameters yield bonding energies which, in general, agree better with experiment than those from bonding parameters based on the dissociation energy of hydrogen.

Table 3a. Bonding Energies Calculated by SCF-MO-CNDO Theory (in eV)

Molecule	02 "hydrides"	Deviation	02 "diatomics"	Deviation	Exptl.
LiF	6.290	0.350	6.158	0.218	5.940
CO	12.166	0.941	10.986	-0.239	11.225
CS	7.970	0.780	7.092	-0.098	7.190
$C\ell_2$	3.170	0.662	2.747	0.239	2.508
Br_2	2.695	0.704	2.278	0.287	1.991
I_2	1.907	0.350	1.459	-0.098	1.557
ClF	3.079	0.411	2.861	0.193	2.668
BrF	2.778	0.096	2.581	-0.101	2.682
$BrC\ell$	2.826	0.492	2.270	-0.064	2.334
IF	1.295	-1.615	1.136	-1.774	2.910
$IC\ell$	2.368	0.178	1.955	-0.235	2.190
IBr	2.251	0.323	1.826	-0.102	1.928
CO_2	18.592	1.736	16.659	-0.197	16.856
OCS	15.984	1.567	14.295	-0.122	14.417
CS_2	13.090	1.110	11.636	-0.344	11.980
NNO	14.986	3.262	13.345	1.621	11.724
O_3	9.034	2.689	7.696	1.351	6.345
SO_2	9.761	-1.416	8.385	-2.792	11.177
C_2H_2	18.264	0.734	17.316	-0.214	17.530
C_2H_4	24.366	0.009	23.940	-0.417	24.357
C_2H_6	30.799	-0.019	30.690	-0.128	30.818
C_3H_8	44.705	1.142	44.263	0.700	43.563
CH_3F	18.234	-0.150	18.158	-0.226	18.384
$CH_3C\ell$	17.211	0.057	16.940	-0.214	17.154
CH_3Br	16.532	-0.108	16.282	-0.358	16.640
CH_3I	15.727	-0.194	15.486	-0.445	15.931
HCN	14.996	1.459	14.181	0.644	13.537
CH_3CN	28.710	2.124	27.519	0.933	26.586
FCN	17.113	3.584	16.042	2.513	13.529
$C\ell CN$	15.468	3.158	14.090	1.780	12.310
BrCN	14.589	2.919	13.223	1.553	11.670
ICN	13.658	2.499	12.299	1.140	11.159

Approximately a threefold improvement for the halogen diatomics has been achieved. This provides considerable empirical justification for the scheme used. For the interhalogen diatomics the new bonding parameters are in general better, although both parameterizations are inadequate for IF. A possible explanation is that the approximations used for the resonance integrals are less valid for interactions between first- and fourth-row elements, where the direct proportionality between $H_{k\ell}$ and $S_{k\ell}$ may be less accurate, owing to orbital mismatch.[63]

The bonding energies of the four methyl halides are predicted more accurately by the original parameters, although the deviation observed with the new parameters are quite acceptable. This result is not surprising, because the bonding parameters for carbon and the halides have been chosen from the corresponding hydrides.[17] Consequently, the original parameters are expected to be very successful for the methyl halides because only the resonance integrals between orbitals on carbon and the halide of interest have not been calibrated to experimental bonding energies directly.

5. Ionization Potentials

The calculated orbital energies were compared with the ionization potentials measured by photoelectron spectroscopy, when available, or by threshold methods[25-27] (see Tables 4 and

Table 4. Comparison of the SCF-MO Orbital Energies (in eV) of First-Row Molecules Calculated Using the CNDO Approximation and Empirical Bonding Parameters, with the Experimental Ionization Potentials

		(1)	(2)	(3)	(4)	(5)	(6)	(7)
		Parameter set						
Molecule	Symmetry	M2	O2	R2	MP	OP	RP	Exptl. I.P.
$H_2(D_{\infty h})$	σ_g	-14.69	-15.44	-18.62	-17.11	-18.60	-20.09	15.45
$LiH(C_{\infty v})$	σ	- 8.09	- 7.90	- 9.94	-12.34	-13.21	-13.15	(7.81–7.91)
$BeH_2(D_{\infty h})$	σ_u	-11.41	-11.77	-13.43	-16.27	-17.26	-17.89	
	σ_g	-12.74	-12.83	-14.79	-17.15	-18.07	-18.85	
$BH_3(D_{3h})$	e'	-12.11	-12.85	-14.89	-16.80	-17.94	-19.16	11.4
	a_1'	-17.73	-17.78	-20.54	-24.56	-25.63	-26.91	
$CH_4(T_d)$	t_2	-12.70	-13.50	-15.64	-17.05	-18.34	-19.78	12.99
	a_1	-23.51	-23.40	-26.16	-32.32	-33.56	-34.68	(24)
$NH_3(C_{3v})$	a_1	-12.45	-13.30	-15.47	-13.26	-14.00	-16.30	10.35
	e	-13.56	-14.20	-16.58	-17.75	-19.22	-20.21	14.95
	a_1	-27.07	-26.71	-30.50	-34.59	-35.89	-37.08	
$H_2O(C_{2v})$	b_1	-14.11	-14.33	-17.90	-14.11	-14.72	-17.83	b_1 12.61
	a_1	-14.03	-14.88	-17.25	-16.10	-17.11	-19.38	14.23
	b_2	-14.68	-15.36	-17.85	-18.64	-20.22	-21.44	18.02
	a_1	-32.69	-32.51	-35.84	-37.87	-39.03	-40.46	
$HF(C_{\infty v})$	π	-16.19	-16.67	-20.85	-16.70	-17.33	-21.28	π 16.06
	σ	-16.38	-17.30	-20.14	-19.48	-20.75	-23.14	σ 16.48
	σ	-37.60	-37.88	-42.91	-40.87	-41.84	-45.55	
$N_2(D_{\infty h})$	σ_g	-13.57	-14.25	-16.49	-15.34	-15.66	-18.30	σ_g 15.58
	π_u	-13.62	-14.44	-16.34	-18.16	-19.40	-20.38	π_u 16.70
	σ_u	-23.15	-23.46	-25.94	-22.03	-22.03	-25.16	σ_u 18.80
	σ_g	-30.23	-29.65	-34.00	-40.95	-41.87	-43.27	
$CO(C_{\infty v})$	σ	-13.20	-13.81	-15.94	-14.56	-14.78	-17.26	σ 14.00
	π	-13.76	-14.40	-16.88	-18.25	-19.52	-21.09	π 16.54
	σ	-20.83	-20.53	-22.97	-21.62	-22.00	-24.67	σ 19.65
	σ	-33.40	-33.15	-36.50	-43.03	-44.09	-45.33	
$CO_2(D_{\infty h})$	π_g	-13.78	-14.55	-17.34	-12.87	-13.88	-15.70	π_g 13.79
	π_u	-16.26	-16.83	-20.51	-21.18	-22.29	-24.81	π_u 17.59
	σ_u	-13.58	-14.55	-16.38	-17.24	-18.23	-20.43	σ_u 18.07
	σ_g	-20.69	-21.01	-23.26	-21.03	-21.81	-24.42	σ_g 19.36
	σ_u	-33.73	-33.81	-37.43	-41.31	-42.34	-43.99	
	σ_g	-34.67	-34.46	-38.18	-43.09	-44.06	-45.40	
$NNO(C_{\infty v})$	π	–	-13.58	-15.81	-11.83	-13.14	-14.44	π 12.90
	σ	–	-15.35	-17.88	-17.28	-18.21	-20.72	σ 16.40
	π	–	-16.63	-20.18	-21.82	-22.96	-25.44	π 18.14
	σ	–	-23.83	-26.22	-21.62	-22.67	-24.98	σ 20.08
	σ	–	-29.54	-33.93	-38.01	-39.16	-41.03	
	σ	–	-34.33	-38.82	-45.14	-46.22	-48.00	
$O_3(C_{2v})$	a_2	-14.35	-14.59	–	-12.05	-13.13	-14.68	12.3
	a_1	-14.72	-15.20	–	-13.32	-13.83	-17.35	12.52
	b_2	-14.75	-15.20	–	-13.73	-14.29	-17.71	13.52
	b_1	-16.18	-16.85	–	-20.31	-21.40	-23.54	16.4–17.4
	b_2	-18.11	-18.19	–	-21.17	-22.14	-24.49	
	a_1	-18.67	-18.34	–	-21.73	-22.36	-25.73	19.24
	a_1	-30.10	-30.70	–	-28.14	-28.40	-32.10	
	b_2	-33.24	-33.32	–	-36.72	-37.57	-39.90	
	a_1	-38.78	-37.88	–	-47.55	-48.35	-50.49	
$C_2H_2(D_{\infty h})$	π_u	-10.49	-11.38	-13.09	-15.14	-16.40	-17.46	π_u 11.40
	σ_g	-12.33	-12.94	-14.88	-18.12	-19.35	-20.62	σ_g 16.44
	σ_u	-18.20	-18.63	-20.59	-24.03	-25.27	-26.55	σ_u 18.42
	σ_g	-24.20	-24.19	-26.57	-34.98	-36.17	-36.84	
$C_2H_4(D_{2h})$	b_{3u}	-10.30	-11.11	-12.72	-13.86	-15.01	-16.60	$b_{3u}(\pi)$ 10.48
	b_{3g}	-11.10	-11.95	-13.61	-13.40	-14.54	-15.82	12.50

Table 4 (*continued*)

Molecule	Symmetry	(1) M2	(2) O2	(3) R2	(4) MP	(5) OP	(6) RP	(7) Exptl. I.P.
	Parameter set							
	a_g	−11.63	−12.36	−14.21	−16.66	−17.84	−19.15	14.39
	b_{2u}	−14.73	−15.25	−17.77	−22.33	−23.68	−25.11	15.63
	b_{1u}	−19.14	−19.48	−21.61	−24.97	−26.19	−27.49	(19.13)
	a_g	−25.43	−25.12	−27.90	−37.16	−38.30	−39.06	
$C_2H_6(D_{3d})$	e_g	−11.32	−12.26	−14.06	−13.74	−14.93	−16.32	11.49
	a_{1g}	−11.32	−12.07	−13.76	−15.94	−17.02	−18.21	
	e_u	−14.18	−14.81	−17.26	−20.63	−21.93	−23.41	14.74
	a_{2u}	−20.22	−20.52	−22.80	−26.57	−27.79	−29.07	(20.13)
	a_{1g}	−26.04	−25.60	−28.60	−38.13	−39.27	−40.08	
$C_3H_8(C_{2v})$	b_1'	−10.93	−11.93	−13.63	−12.72	−13.88	−15.28	11.07
	b_2	−10.64	−11.47	−13.00	−13.78	−14.89	−16.12	
	a_1	−10.96	−11.83	−13.44	−13.42	−14.55	−15.86	
	a_2	−12.29	−13.17	−15.25	−16.16	−17.42	−18.91	13.17
	b_2	−12.72	−13.52	−15.67	−18.10	−19.28	−20.67	
	a_1	−14.17	−14.81	−17.19	−20.76	−22.00	−23.40	15.17
	b_1	−15.06	−15.60	−18.24	−22.67	−23.98	−25.51	
	a_1	−19.38	−19.76	−21.98	−25.94	−27.18	−28.55	(19.8)
	b_2	−22.76	−22.76	−25.36	−31.80	−32.98	−34.05	
	a_1	−27.62	−26.99	−30.21	−41.86	−42.97	−43.68	
$B_2H_6(D_{2h})$	b_{3g}	− 8.82	−10.02	−11.22	− 9.14	−10.25	−10.94	12.0
	b_{2u}	−10.65	−11.54	−13.39	−14.56	−15.96	−16.70	
	b_{3u}	−12.72	−13.63	−15.50	−19.39	−20.72	−21.20	
	a_g	−14.07	−14.47	−16.75	−21.66	−23.04	−23.64	
	b_{1u}	−17.20	−17.36	−19.96	−25.45	−26.87	−27.55	
	a_g	−21.28	−21.16	−24.38	−35.01	−36.40	−36.84	
$LiF(C_{\infty h})$	π	−10.10	−10.88	−11.02	−14.11	−15.05	−14.99	
	σ	−10.14	−10.88	−11.20	−14.17	−15.01	−15.37	
	σ	−31.22	−31.75	−33.20	−37.40	−38.38	−38.61	
$F_2(D_{\infty h})$	π_g	−17.80	−17.94	−22.97	−16.77	−16.77	−22.09	π_g 15.63
	σ_g	−17.02	−17.66	−19.68	−20.22	−21.19	−22.72	σ_g 17.35
	π_u	−19.42	−19.27	−24.89	−20.45	−20.45	−25.77	π_u 18.46
	σ_u	−37.56	−37.82	−43.20	−36.12	−36.12	−41.99	
	σ_g	−41.26	−40.91	−47.43	−44.31	−44.52	−49.73	
$CH_3F(C_{3v})$	e	−12.65	−13.68	−16.02	−13.96	−14.77	−17.57	e 12.85
	a_1	−14.11	−15.06	−16.84	−18.61	−19.70	−21.27	a_1 14.10
	e	−15.87	−16.34	−20.38	−20.42	−21.39	−24.12	(a_1) 16.89
	a_1	−23.15	−23.33	−26.16	−28.82	−29.83	−32.05	
	a_1	−36.87	−37.16	−42.04	−43.47	−44.40	−47.10	
$HCN(C_{\infty v})$	π	−12.44	−13.33	−15.31	−16.40	−17.71	−18.58	13.91
	σ	−12.81	−13.72	−15.31	−15.32	−15.94	−18.15	
	σ	−20.33	−20.69	−22.74	−23.81	−24.90	−26.43	
	σ	−27.86	−27.60	−31.31	−37.55	−38.64	−39.62	
$CH_3CN(C_{3v})$	e	–	−12.49	–	−13.81	−15.09	−15.89	12.22
	a_1	–	−12.32	–	−14.61	−15.29	−17.28	
	e	–	−14.90	–	−20.54	−21.97	−23.21	
	a_1	–	−18.68	–	−20.43	−21.53	−22.88	
	a_1	–	−24.75	–	−34.24	−35.45	−36.46	
	a_1	–	−27.62	–	−39.07	−40.19	−40.83	
$FCN(C_{\infty v})$	π	−13.01	−13.97	−16.18	−14.25	−15.20	−17.14	
	σ	−14.85	−15.90	−17.45	−15.92	−16.51	−19.15	
	π	−16.94	−17.54	−21.40	−21.55	−22.49	−25.51	
	σ	−20.63	−21.04	−23.50	−22.87	−23.87	−26.26	
	σ	−28.67	−28.41	−32.53	−37.37	−38.31	−39.92	
	σ	−38.59	−38.86	−43.60	−46.08	−47.14	−49.85	

Table 4a. Ionization Potentials Calculated by SCF-MO-CNDO Theory (in eV)

Molecule	O2 "hydrides"	O2 "diatomics"	Exptl.
LiF	10.88	10.80	
N_2	14.25	14.16	15.58
CO	13.81	13.67	14.00
CS	11.82	11.72	
F_2	17.94	17.51	15.63
Cl_2	13.16	13.21	11.50
Br_2	11.83	11.88	10.71
I_2	11.86	11.68	9.65
ClF	14.34	14.37	⩽12.70
BrF	12.94	12.96	⩽11.90
BrCl	12.35	12.41	⩽11.10
IF	13.45	13.47	⩽10.50
ICl	12.46	12.51	10.55
IBr	11.85	11.89	10.23
CO_2	14.55	14.32	13.79
OCS	12.26	12.31	11.27
CS_2	11.33	11.38	10.11
NNO	13.58	13.62	12.90
O_3	14.59	14.81	12.30
SO_2	12.69	12.71	12.32
B_2H_6	10.02	9.92	12.00
C_2H_2	11.38	11.24	11.40
C_2H_4	11.11	10.99	10.48
C_2H_6	12.26	11.96	11.49
C_3H_8	11.93	11.36	11.07
CH_3F	13.68	13.68	12.85
CH_3Cl	12.41	12.42	11.42
CH_3Br	11.61	11.61	10.69
CH_3I	11.64	11.50	9.80
HCN	13.33	13.27	13.91
CH_3CN	12.49	12.22	12.22
FCN	13.97	13.93	
ClCN	13.08	13.10	
BrCN	12.40	12.40	
ICN	12.36	12.22	

4a). This comparison depends on the validity of Koopmans' theorem,[28] by which

$$I_i = -E_i$$

where I_i is the energy to remove an electron from the ith orbital of a molecule, leaving the nuclei fixed and the orbitals unchanged, and E_i is the eigenvalue of the ith orbital. Peters[29] has shown that these E_i do correspond more closely to the experimental ionization potentials than in the case of localized orbitals. Thus the first ionization potential corresponds to removal of an electron from the highest occupied molecular orbital or set of degenerate molecular orbitals and formation of a ground-state ion. Higher I_i belong to excited ions formed by removing electrons from more stable molecular orbitals.[25]

The orbital-energy levels should be compared with *vertical ionization potentials*, but these have only recently become available from photoelectron spectroscopy. Luckily for molecules such as CO_2, SCO, CS_2, and NNO, the vertical and adiabatic ionization potentials differ by only a few tenths of an electron volt.[30] Thus, where possible, vertical ionizations are used; otherwise adiabatic. For molecules with many distinct orbital energies in the photoelectron spectroscopy range, MO theory often predicts groups of orbitals close together in energy, corresponding to a single I_i; it has been assumed[31,32] that the observed I_i corresponds to several unresolved ionic states.

The different ionization potentials of a molecule are identified by the irreducible representation of the molecular symmetry group of the ionic state which is formed and of the orbital from which the electron is removed. Thus in a closed-shell molecule the ground state and ionic state belong to the totally symmetric representation and have the same symmetry.[33] Cotton's notation has been used[34] in this article.

For molecules containing only hydrogen and the first-row elements, the orbital energies calculated with electron repulsion integrals evaluated from atomic spectra are higher and in better agreement with experiment than those calculated with theoretical g_{AA}. The theoretical g_{AA} lead to ionization potentials which are too large, just as they did in atoms.[17] Thus, if the g_{AA} are adjusted to give good atomic ionization potentials and electron affinities, they give more accurate molecular ionizations. However, the actual form of g_{AB}—Mataga or Ohno—is less important, because the choice affects the orbital energies by less than 1 eV. The Ohno g_{AB} gives better agreement in slightly more cases. The choice of Slater exponent has even less effect, so that no clear choice of Z'_H could be made.

The energies of lone pair or nonbonding orbitals such as the a_1 and b_1 orbitals of NH_3 and H_2O, respectively, are relatively insensitive to changes in the bonding parameters. This is because

$$E_i = \sum_k \sum_\ell C^*_{ki} C_{\ell i} F_{k\ell} \tag{38}$$

and the only terms in the summation dependent upon the bonding parameters are those $F_{k\ell}$ between orbitals on different atoms, and these terms are small unless the orbital is bonding or antibonding.

In the Pople-Segal approach,[1,2] only the local terms of the diagonal-core Hamiltonian matrix elements U_{ss} and U_{pp}, are evaluated from atomic spectra,[2] while g_{AA} and g_{AB} were derived theoretically. The results of this theory are in very poor agreement with experiment, showing that the degree of empiricism is not sufficient to permit accurate calculation of orbital energies.[17]

The semiempirical orbital energies of some small molecules (CH_4, H_2O, F_2, HF, etc.) were compared to the Hartree-Fock orbital energies calculated from exact SCF-MO calculations with extended basis sets.[35-40] If Koopmans' theorem is valid, the Hartree-Fock results should be closer to the ionization energy; they are not, showing the inexactness of Koopmans' theorem. However, the semiempirical and Hartree-Fock results are in excellent agreement, and when the semiempirical result is closer to the experimental ionization potential, it is due to the errors in the approximate solution of Roothaan's equations, canceling those due to Koopmans' theorem.

In the case of molecules containing elements heavier than neon, the SCF-MO-CNDO calculations have been made with empirical bonding parameters and electron repulsion integrals only because these were most accurate for molecules containing light atoms. The agreement with experiment was very good.[17]

Some specific molecules can be considered which show discrepancies.

Hydrogen

The Hamiltonian matrix elements for H_2 are

$$F_{11} = U_{11} + \tfrac{1}{2} g_{11} \tag{39}$$

and

$$F_{12} = -\beta_H^0 S_{12} - \frac{1}{2} g_{12} \quad (40)$$

For all parameter choices F_{11} equals the negative of the ls orbital electronegativity of the neutral hydrogen atom (7.171 eV).[6] Hence the ionization energy of H_2 is

$$I_{H_2} = 7.171 + \beta_H^0 S_{12} + \frac{1}{2} g_{12} \quad (41)$$

and the dissociation energy is[17]

$$E_B = 2\beta_H^0 S_{12} + \frac{1}{2}(g_{12} - g_{11}) \quad (42)$$

so that if g_{11} is evaluated by Pariser's formula,[41] $g_{12} = 10.766$ eV and $\beta_H^0 S_{12} = 2.896$ eV, these are correctly predicted. However, empirical parameters are not predicted this way, because in other molecules symmetry is absent, and I_{AB} and E_B are not simply expressible in terms of the parameters. Instead, the g_{AB} were calculated by Ohno's or Mataga's formula, and the bonding parameters were chosen to fit the atomization energies alone. Thus $g_{AB}^{H_2}$ is 7.731 (Mataga) and 10.713 (Ohno), and as the Ohno value is close to the g_{12} required to give an exact I_{H_2}, it gives the more accurate prediction of I_{H_2}.

Furthermore, equations (41) and (42) illustrate that for a constant g_{11} and g_{12}, the overlap S_{12} will vary with the Slater exponent Z'_H, but the β_H^0 varies so as to maintain the E_B, and consequently the I_{H_2} is independent of Z'_H.

When the g_{11} is evaluated from the theoretical integral,[11] the g_{12} required to fit both E_B and I_{H_2} is 3.206 eV, compared to the theoretical g_{12} of 15.896 eV, so that theoretical electron repulsion integrals lead to an inaccurate I_{H_2}.

LiH

In the absence of experimental ionization potentials, the results were compared with the rigorous upper bound of LiH^+ calculated by Browne[42] using a generalized valence-bond function. Together with a probably lower bound and D_e for LiH, this upper bound permits calculation of I_{LiH}, which is in good agreement with the SCF-MO-CNDO prediction.

CH_4

The predicted value of 23.4 eV supports the idea that the second ionization potential of 24 eV determined by Collin and Deliviche[27] is correct, because no second ionization potential is found below 21.21 eV by photoelectron spectroscopy.[43] The first ionization potential is predicted at 12.70 eV, in good accord with the 12.99-eV experimental result.

NH_3, etc.

The highest occupied molecular orbital is an a_1 N lone pair insensitive to changes in bonding parameters and very close to the e level; the separation of these levels is seriously underestimated. The separation of the a and e levels in PH_3, AsH_3, and SbH_3 is predicted to be almost zero, but this could be an underestimation.

H_2O, etc.

The order of the computed orbital energies depends on the choice of parameters, Ohno predicting the b_1 lone pair on oxygen to be the highest, in agreement with experiment.[43] In H_2S, H_2Se, and H_2Te, all the parameters predict b_1 as highest.

CO_2, etc.

The first ionization potential is π and correctly predicted, but all the calculations predict that the highest σ_u orbital is higher than one or both the π orbitals, in disagreement with experiment and with the results from complete SCF-MO calculations using an extended basis set but in agreement with the results from a minimum basis set. Thus the energy decrease due to greater wave-function flexibility is greatest for the σ_u orbital, and this may be why the semiempirical method fails.

In OCS and CS_2 the order of energy levels is again in disagreement with experiment, but in NNO, which is isoelectronic with CO_2, the second ionization potential has been found to be σ, in agreement with the semiempirical, and complete, SCF-MO theory. The inversion of σ and π from CO_2 to NNO may be due to the loss of the center of symmetry, thus permitting u and g mixing, causing a greater separation of the closer σ orbitals than of the π orbitals.

SO_2 and O_3

The experimental results have not been assigned to specific orbitals, but the order correlates quite well with a_2, a_1, b_2, b_1, b_2, and a_1 for the first six levels of O_3, and in SO_2 a good fit is obtained if the first two observed lines are three incompletely resolved lines and the third line is three unresolved lines. The fourth observed I_{SO_2} at 20.07 eV[44] is confirmed by the calculated 21.6-eV line.

C_2H_4

The b_{3u} π orbital is the highest occupied orbital[44] and is so predicted by the semiempirical parameters but not by the Pople-Segal parameters. The fifth ionization potential is predicted at 19.14 eV and measured at 19.13 eV[45] and is b_{1u} by empirical parameters and Mataga.

B_2H_6

Assuming a B—B z axis and a bridge H—H x axis, the bridge hydrogen atomic orbitals participate in the a_g and b_{3u} orbitals corresponding to "three-center bonds" with energies of -12.72 and -14.07 eV.

Halogens

The σ_g orbital is calculated to be the highest occupied molecular orbital in F_2 and I_2, and the second highest in $C\ell_2$ and Br_2 according to SCF-MO-CNDO empirical theory; photoelectron spectroscopy shows the σ_g orbital second highest in F_2 and third highest in the other halogens.[46] Either an extended or minimum basis set SCF-MO treatment predicts the correct order, so the error in the empirical method is either in the method or parameters; using Pople-Segal parameters gives the correct order.

The interhalogens, methyl fluoride, hydrogen cyanide, and methyl cyanide, show similar correlations and anomalies. However, in a total of over 200 energy levels, the above are the only discrepancies.[45]

Thus the SCF-MO-CNDO theory does predict fairly accurate ionization potentials when empirical electron repulsion integrals and bonding parameters are used; it is a drastic approximation to the complete SCF-MO theory with extended basis set and explicit treatment of all electron interaction integrals; but even in this theory, errors in Koopmans' theorem prevent exact ionization-potential prediction.

The new parameterization of β_A^0 from diatoms does not lead to a significant quantitative improvement between calculated and experimental *ionization potentials*, but the order of first ionization potentials, for the methyl halides and halogen diatomics, is correctly predicted, whereas it was not correctly predicted by the original parameterization.[45]

The difference between values calculated by the two parameterizations is only a few tenths of an electron volt; for many molecules it is less than 0.1 eV.

6. Dipole Moments

(For some values of dipole moments, see Tables 5, 5a, and 5b.)

Since the dipole moment of a neutral molecule is invariant with respect to translation of axes, invariance must be preserved in the CNDO approximation. In a closed-shell molecule, with doubly occupied molecular orbitals, the dipole moment, μ, is

$$\mu = -2e \sum_i \int \psi_i^* \dot{x} \psi_i dv + e \sum_A Z_A X_A \tag{43}$$

where the dipole lies along the X axis, $\dot{x}$ is the one-electron position operator, and X_A is the position of nucleus A. The matrix elements $X_{k\ell} = \int \phi_k \dot{x} \phi_\ell \, dv$ are *dipole integrals* and ϕ_k and ϕ_ℓ are the basis atomic orbitals; hence

$$\mu = -e \operatorname{tr}(PX) + e \sum_A Z_A X_A \tag{44}$$

The invariance can be preserved in three ways.

a. μ can be calculated by assuming the electron population of each atom to be a point charge at the nucleus, when

$$\mu = e \sum_A (Z_A - P_{AA}) X_A \tag{45}$$

This is invariant to translation for a neutral molecule and is consistent with the CNDO approximation in which the dipole moments between different atomic orbitals vanish, and

$$\mu = -e \sum_k P_{kk} X_{kk} + e \sum_A Z_A X_A \tag{46}$$

Table 5. Comparison of Methods of Calculation of Dipole Moments[a] for First-Row Molecules from SCF-MO-CNDO Theory with Empirical Atomic and Bonding Parameters (in D)

Calculation of μ: Interatomic repulsion integral[b]:	Point charge M2	Point charge O2	Pople-Segal M2	Pople-Segal O2	Dixon M2	Dixon O2	Exptl.
LiH	4.083	6.030	6.699	7.218	7.024	7.479	5.882
H_3N	0.465	0.962	1.973	2.221	0.643	0.991	1.468
H_2O	0.876	1.417	1.803	2.175	0.929	1.391	1.87
HF	1.397	1.866	1.899	2.265	1.578	2.003	1.8195
CO	1.549	2.097	0.789	1.372	1.308	2.017	−0.112
NNO		0.838		0.601		0.862	0.166
O_3[c]	0.734	0.893	0.876	0.909	0.429	0.583	0.58
C_3H_8[d]	0.183	0.106	0.084	0.041	0.151	0.082	0.083
LiF	6.052	6.818	6.602	7.098	6.833	7.222	6.328
CH_3F	2.721	2.812	2.647	2.687	2.391	2.408	1.8555
HCN	1.582	1.689	3.015	2.946	1.547	1.501	2.985
CH_3CN		2.440		3.709		2.173	3.92
FCN	−0.667	−0.427	0.482	0.648	−0.364	−0.165	2.17

[a] Positive sign indicates polarity A^+B^-, where A is the first atom written.

[b] M2 and O2 are retained to make table consistent with preceding tables and signify that the formulas of Mataga and Ohno with $Z'_H = 1.2$ have been used in the calculations.

[c] Positive sign indicates central oxygen at positive end.

[d] Positive sign indicates central carbon at positive end.

Table 5a. Calculated Dipole Moments[a] for Molecules Including Non-First-Row Elements from SCF-MO Theory and SCF-MO-CNDO Theory (in D)

Calculation of μ: Interatomic repulsion integral:	Point charge		SCF-MO-CNDO including sp polarization		Dixon		
	M2	O2	M2	O2	M2	O2	Exptl.
H_3P	-0.127	-0.344	2.379	1.906	0.827	0.260	0.578
H_3As	-0.064	-0.259	2.908	2.427	1.286	0.712	0.22
H_3Sb	-0.351	-0.796	2.587	1.863	0.707	-0.120	0.116
H_2S	0.408	0.576	2.152	2.100	0.985	0.932	0.974
H_2Se	0.648	0.961	2.565	2.627	1.340	1.435	0.24, 0.62
H_2Te	0.662	0.930	2.789	2.817	1.503	1.550	
$HC\ell$	1.006	1.296	2.022	2.180	1.453	1.650	1.12
HBr	− 0.727	0.919	1.813	1.865	1.050	1.122	0.83
HI	0.548	0.659	1.979	1.934	1.154	1.109	0.445
CS	1.234	1.410	1.614	1.460	1.406	1.869	1.97
OCS	-0.891	-1.108	0.241	0.028	-0.448	-0.766	0.7124
SO_2	1.633	2.919	1.273	2.190	0.179	1.548	1.59
$C\ell F$	1.412	1.780	0.962	1.386	0.977	1.401	0.881
BrF	2.160	2.710	1.636	2.260	1.668	2.292	1.29
$BrC\ell$	0.744	0.925	0.646	0.835	-3.644	-3.143	0.57
IF	2.859	3.481	2.113	2.841	2.248	2.969	
$IC\ell$	1.188	1.441	0.736	1.021	0.833	1.123	0.65
IBr	0.452	0.552	0.109	0.233	0.126	0.253	
$CH_3C\ell$	2.101	2.104	2.667	2.588	2.327	2.235	1.869
CH_3Br	1.697	1.649	2.436	2.304	1.994	1.847	1.797
CH_3I	1.387	1.287	2.496	2.297	1.998	1.781	1.647
$C\ell CN$	0.383	0.498	0.922	1.014	-0.109	-0.036	2.802
BrCN		1.094		1.412		0.506	2.94
ICN	1.287	1.462	1.210	1.365	0.415	0.598	3.71

[a] Sign convention as in Table 5.

The center of charge for a p or s orbital is at the nucleus and equations (46) and (45) are identical, but this is not true for hybrid orbitals, and especially with lone pairs the displacement of the center of the electron population from the nucleus contributes substantially to the molecular dipole moment.[2,47-49]

b. If all the diatomic dipole integrals in equation (44) are ignored, but the dipole integrals for different orbitals on the same atom included, then[2]

$$\mu' = -e \sum_{k} \sum_{\ell} P_{k\ell} X_{k\ell} \theta_{k\ell} + e \sum_{A} Z_A X_A \tag{47}$$

where[3] $\theta_{k\ell} = 1$ if ϕ_k and ϕ_ℓ are on the same atom, and zero otherwise. μ' is invariant to translation and to rotation and hybridization.[3] The only nonzero $X_{k\ell}$ is $X_{s\text{-}p_x}$, and the term containing this integral represents the atomic polarization effect, which was absent from the point-charge formula (45). For Slater orbitals[16] this is

$$X_{s\text{-}p_x} = \int \phi_s X \phi_{p_x} \, dv = \frac{n'(n' + \frac{1}{2})\, a_0}{\sqrt{3}\, Z'} \tag{48}$$

c. Formal justification of CNDO depends on considering the basis orbitals as approximations to Löwdin orbitals,[50]

$$\bar{\phi} = \phi S^{-1/2} \tag{49}$$

Table 5b. Dipole Moments Calculated by SCF-MO-CNDO Theory (in D)

Molecule	O2 "hydrides"	O2 "diatomics"	Exptl.
LiH	7.218	7.220	5.882
H_3N	2.221	2.219	1.468
H_2O	2.175	2.165	1.87
HF	2.265	2.259	1.8195
H_3P	1.906	1.916	0.578
H_3As	2.427	2.458	0.22
H_3Sb	1.863	1.872	0.116
H_2S	2.100	2.111	0.974
H_2Se	2.627	2.646	0.62
H_2Te	2.817	2.825	
HCℓ	2.180	2.180	1.12
HBr	1.865	1.866	0.83
HI	1.934	1.931	0.445
CO	1.372	1.555	-0.112
CS	1.460	1.610	1.97
NNO	0.601	0.621	0.166
O_3	0.909	0.897	0.58
OCS	0.028	0.185	0.7124
SO_2	2.190	2.148	1.59
LiF	7.098	7.139	6.328
CℓF	1.386	1.426	0.881
BrF	2.260	2.318	1.29
BrCℓ	0.835	0.898	0.57
IF	2.841	2.912	
ICℓ	1.021	1.102	0.65
IBr	0.233	0.277	
C_3H_8	0.041	0.061	0.083
CH_3F	2.687	2.759	1.8555
$CH_3Cℓ$	2.104	2.680	1.869
CH_3Br	1.649	2.372	1.797
CH_3I	1.287	2.318	1.647
HCN	2.946	2.919	2.985
CH_3CN	3.709	3.610	3.92
FCN	0.648	0.558	2.17
CℓCN	1.014	0.801	2.802
BrCN	1.412	1.198	2.94
ICN	1.365	1.195	3.71

in which case the X matrix becomes[50]

$$\overline{X} = S^{-1/2} X S^{-1/2} \tag{50}$$

and μ becomes

$$\overline{\mu} = -e \operatorname{tr}(P\overline{X}) + e \sum_A Z_A X_A \tag{51}$$

which is invariant to origin change.

All three methods were used to calculate dipole moments from the SCF-MO-CNDO theory; they were compared with the experimental results from microwave spectroscopy.

First-Row Atom Molecules

The point-charge formula (method a above) gives poor agreement with experiment, owing to the neglect of atomic polarization effects. The inclusion of these effects in method b leads to a great improvement in correlation with experiment for most molecules, but the use of Löwdin orbitals makes the correlation worse. Consequently, equation (47) is most accurate in calculating μ, and errors may be due to errors in the wave equation rather than inaccuracies in equation (47).

The computed μ are sensitive to the choice of g_{AB}, and the Mataga form is more accurate in all molecules other than FCN. Further, the choice of g_{AA} is important because the g_{AA} determine the potential energy for a given charge distribution, which determines the self-consistent charge distribution. However, the choice of Z'_H does not effect μ.

For some molecules the dipole moment computed using Pople-Segal bonding parameters differs appreciably from that calculated with empirical parameters. This is due to the dependence of the molecular energy on the bonding parameters, where

$$E_e = \frac{1}{2} \sum_k \sum_\ell P_{k\ell} (F_{k\ell} + H_{k\ell}) \tag{52}$$

When the bonding parameters are increased, so are $H_{k\ell}$ and $F_{k\ell}$. Hence E_e can be lowered by increasing $P_{k\ell}$ by transferring electron density from lone-pair into bonding orbitals. Such transfer occurs in LiF and CH_3F from the fluorine lone pairs, decreasing the dipole moment, but in HF this effect is canceled by increased atomic polarization of the fluorine atom.[51] In F^+CN^-, the dipole moment is increased by transfer of electron density away from fluorine; in CO the increased bonding parameters cause enough charge transfer to reverse the polarity of the computed dipole moment:

Recent calculations[52] support the accuracy of the computed dipole moments, because the dipole moments of several organic molecules were calculated by the SCF-MO-CNDO theory with theoretical g_{AA} and Pople-Segal bonding parameters and had an overall accuracy comparable with those for calculations on first-row molecules.

Molecules Containing Non-First-Row Atoms

Using equation (47), dipole moments which are much higher than the experimental values are obtained; the calculations include only s and p valence-shell orbitals. Recently[49] d orbitals have been included in the SCF-MO-CNDO theory for second-row elements, using theoretical g_{AA} and g_{AB} similar to those developed for first-row elements.[2] This showed that pd polarization terms involving $X_{k\ell}$ between p and d orbitals on the same atom were, apparently, important in calculating dipole moments. This effect was opposite to sp polarization and about equal, so that use of $X_{s\text{-}p_x}$ alone exaggerates the net atomic polarization. An exact SCF-MO calculation on PH_3[53] showed that inclusion of d orbitals reduced μ from 2.34 to 0.86 D, still higher than the experimental value of 0.578 D. This effect is due to the transfer of electron density from the lone pair into the bond, reducing the atomic polarization of phosphorus.

Thus the omission of d orbitals in the SCF-MO-CNDO calculations on molecules containing second- and higher-row atoms results in the dipole moments being very much less accurate than those calculated by the same theory for molecules containing first-row atoms, because the sp and pd polarizations cancel.

The point charge μ^1 is a much better fit with experiment because it ignores the polarization terms completely, while the Löwdin basis (method c) also predicts better dipole moments, but this is difficult to explain.

The difference between the calculated dipole moments by the two methods of parametrizing β_A^0 is small compared to the difference between calculated and experimental values.

7. Quadrupole Coupling Constants

(For some values of quadrupole coupling constants, see Table 6, 6a, 6b, 6c, and 6d.)
The effective field gradient at the nucleus is

$$eq = -e\,\mathrm{tr}(Pq_A) + e \sum_{A \neq B} Z_B q_{AB} \tag{53}$$

in which q_A is the electric field gradient at nucleus A, having elements

$$(q_A)_{k\ell} = \int \phi_k^* \frac{3\cos^2\theta - 1}{r^3} \phi_\ell \, dv \tag{54}$$

Table 6. Nuclear Quadrupole Coupling Constants for $^{35}C\ell$ (in MHz)

	From semiempirical SCF-MO-CNDO theory; $Z_H' = 1.2$		
	Mataga g_{AB} M2	Ohno g_{AB} O2	Exptl.
$HC\ell$	- 88.25	- 82.02	- 67.3
$CH_3C\ell$	- 78.63	- 79.01	- 74.77
$C\ell F$	-126.81	-132.78	-146.00
$C\ell_2$	-106.16	-106.88	-108.95
$BrC\ell$	- 98.27	- 97.04	-103.6
$IC\ell$	- 94.47	- 92.64	- 82.5
$C\ell CN$	- 73.96	- 69.48	- 83.2

Table 6a. Nuclear Quadrupole Coupling Constants for ^{79}Br (in MHz)

	From semiempirical SCF-MO-CNDO theory; $Z_H' = 1.2$		
	Mataga g_{AB} M2	Ohno g_{AB} O2	Exptl.
HBr	657.9	642.7	530.5
CH_3Br	607.6	616.7	577.15
BrF	952.4	1007.3	1089.0
$BrC\ell$	806.9	824.5	876.8
Br_2	750.8	754.7	765
IBr	721.2	718.4	722
BrCN		588.9	686.5

Table 6b. Nuclear Quadrupole Coupling Constants for ^{127}I (in MHz)

	From semiempirical SCF-MO-CNDO theory; $Z_H' = 1.2$		
	Mataga g_{AB} M2	Ohno g_{AB} O2	Exptl.
HI	-2020	-2009	-1823.3
CH_3I	-1907	-1949	-1934
IF	-2920	-3078	
$IC\ell$	-2463	-2529	-2930.0
IBr	-2305	-2338	-2731
I_2	-2214	-2227	-2153
ICN	-1995	-1962	-2420

Table 6c. Nuclear Quadrupole Coupling Constants for ^{14}N (in MHz)

	From SCF-MO-CNDO theory with empirical bonding parameters; $Z_H' = 1.2$			With Pople–Segal bonding parameters; $Z_H' = 1.2$			
	Mataga g_{AB} M2	Ohno g_{AB} O2	R2	MP	OP	RP	Exptl.
NH_3	1.090	0.674	-0.502	-4.111	-3.930	-4.256	-4.0842
N_2	-0.663	-0.482	-0.911	-2.290	-2.290	-2.317	-4.65
$\underline{N}NO$		6.104	5.817	0.612	0.840	0.437	-0.792
$N\underline{N}O$		-3.007	-4.792	0.152	-0.015	-0.528	-0.238
HCN	6.664	6.280	6.108	-1.923	-1.903		-4.58
CH_3CN		6.676		-1.527	-1.482	-1.908	-4.214
FCN	8.823	9.208	8.150	-0.579	-0.468	-0.995	-2.67
CℓCN	7.956	7.897					-3.63
BrCN		7.220					-3.83
ICN	7.048	6.680					-3.80

and q_{AB} is the electric field gradient per unit charge, of nucleus B at A,

$$q_{AB} = \frac{3\cos^2\theta_B - 1}{R_{AB}^3} \tag{55}$$

Thus the direct computation of eq involves evaluating[54] the three-center integrals $(q_A)_{k\ell}$.

Equation (53) can be simplified to apply to the SCF-MO-CNDO theory.

a. The off-diagonal matrix elements of $(q_A)_{k\ell}$ which contain the differential overlap of two orbitals as a factor in the integrand are ignored. The matrix elements between two different orbitals on A vanish by symmetry, while the others refer to overlap regions not on atom A and are small compared to the diagonal matrix elements for orbitals on A. Thus

$$eq = -e\sum_k P_{kk}(q_A)_{kk} + e\sum_{A \neq B} Z_B q_{AB} \tag{56}$$

b. It is assumed that the electric field gradient due to the electron population of each atom in the molecule cancels that due to the core,

$$\sum_k^B P_{kk}(q_A)_{kk} = Z_B q_{AB} \tag{57}$$

This has been shown to be true in accurate calculation of field gradients. The net effects of other atoms is small, owing to the r^{-3} dependence of the field gradient operator; hence

$$eq = -e\left[\sum_k^A P_{kk}(q_A)_{kk}\right] \tag{58}$$

in the SCF-MO-CNDO theory. This is equivalent to the expression for eq in simple MO treatments of coupling constants,[55-57] and consequently the quadrupole coupling constant is given by

$$q = q_0\left(P_{zz} - \frac{P_{xx} + P_{yy}}{2}\right) \tag{59}$$

in which P_{kk} represents the electron population of the kth orbital and is equivalent to the n_k of previous theory.[56] The q_0 is the quadrupole coupling constant per p electron evaluated from atomic spectra.

Table 6d. Nuclear Quadrupole Coupling Constants by SCF-MO-CNDO Theory (in MHz)

	Molecule	O2 "hydrides"	O2 "diatomics"	Exptl.[a]
$^{35}C\ell$	$HC\ell$	- 82.02	- 82.02	- 67.3
	$CH_3C\ell$	- 79.01	- 79.18	- 74.77
	$C\ell F$	-132.78	-133.30	-146.00
	$C\ell_2$	-106.88	-107.22	-108.95
	$BrC\ell$	- 97.04	- 96.91	-103.6
	$IC\ell$	- 92.64	- 92.51	- 82.5
	$C\ell CN$	- 69.48	- 67.26	- 83.2
^{79}Br	HBr	642.7	642.7	530.5
	CH_3Br	616.7	618.2	577.15
	BrF	1007.3	1011.8	1089.0
	$BrC\ell$	824.5	830.8	876.8
	Br_2	754.7	756.9	765.0
	IBr	718.4	719.3	722.0
	BrCN	588.9	574.0	686.5
^{127}I	HI	-2009	-2009	-1823.3
	CH_3I	-1949	-1960	-1934
	IF	-3078	-3091	
	$IC\ell$	-2529	-2549	-2930
	IBr	-2338	-2352	-2731
	I_2	-2227	-2238	-2153
	ICN	-1962	-1930	-2420
^{14}N	NH_3	0.674	1.240	-4.042
	N_2	-0.482	-0.433	-4.65
	$\underline{N}NO$	6.104	6.249	-0.792
	$N\underline{N}O$	-3.007	-3.323	0.238
	HCN	6.280	6.969	-4.58
	CH_3CN	6.676	7.430	-4.214
	FCN	9.208	9.702	-2.67
	$C\ell CN$	7.897	8.481	-3.63
	BrCN	7.220	7.822	-3.83
	ICN	6.680	7.279	-3.80

[a]Obtained from Ref. 57.

Halogens

A comparison of the predicted q for HX, CH_3X, XY, X_2, XZ, and XCN for $^{35}C\ell$, ^{79}Br, and ^{127}I as X or Y with the experimental values measured in the gas phase showed that the q, unlike the μ, were in good agreement with experiment despite the absence of d orbitals. The q are not sensitive to the values of q_{AB} or Z_H' chosen, and the results for bromine are as accurate as those for chlorine and iodine, showing that the approximate form for R_4'' (r), equation (25), does not affect the accuracy of the theory.[57]

Nitrogen

The calculated q are in poor agreement with experiment, and the q calculated from SCF-MO-CNDO with empirical parameters bear little relation to the experimental values; many q have the wrong sign, corresponding to smaller electron populations along the bond than perpendicular to it, in contradiction to the experimental results. The SCF-MO-CNDO theory with theoretical parameters does predict the correct sign, but the magnitude of the q is incorrect except in NH_3.

Ab initio calculations[58] on N_2 using equation (53) show that q is very sensitive to the basis set, because eq_0, the field gradient at atom A per p electron, is proportional to the cube of Z',

$$q_0 = \frac{4}{5} \frac{Z'^3}{n'(n' - \frac{1}{2})(n' - 1)a_0^3} \qquad (60)$$

for a Slater orbital,[16] while the other terms in equation (53) are insensitive to changes in the basis set, owing to a rough cancellation of the electron and nuclear terms. The Sternheimer effect is not responsible for the failure because it has only a 10 per cent effect in N_2.[58]

In semiempirical assessments of q, it is assumed that all errors inherent in *ab initio* calculations, and any error in the quadrupole moment of ^{14}N, will be eliminated by evaluating q_0 from atomic spectra.[59] This obviously is true in the halogens, but not for nitrogen; equation (59) is expected to be less accurate for ^{14}N, because it is a smaller atom with more complex bonding.[60] While the main failure may lie in equation (59), comparison of the SCF-MO-CNDO orbital populations with those obtained from complete minimum-basis-set SCF-MO calculations shows that part of the error is in the CNDO wave functions. Thus in HCN the Mataga μ is accurate to within 1 per cent but has populations $s = 1.93$, $p_\sigma = 0.50$, and $p_\pi = 1.40$, in contrast to the gross orbital populations of $s = 1.77$, $p_\sigma = 1.37$, and $p_\pi = 0.97$ from the complete SCF-MO theory.[61]

The calculated orbital populations are sensitive to the chosen bonding parameters and the Pople-Segal parameters give a better q, an example of the situation in which parameters chosen for the accurate prediction of one molecular property does not lead to the best prediction of another property.

The new β_A^0 parameters from diatomics are more accurate for most halogen nuclear quadrupole coupling constants, but they do not improve the nitrogen results.

8. Conclusion

The success of the theory can be judged by the accuracy of the physical properties it can predict accurately. The semiempirical electron repulsion parameters and bonding parameters predict quite accurate ionization potentials, in some cases more accurate than those from exact SCF-MO calculations, if Koopmans' theorem is valid. For a few molecules the orbital-energy order is incorrectly predicted, but generally the agreement is good. The bonding energies of molecules not used in the calibration are also accurate. However, quantities dependent on the molecular charge distribution are less accurately predicted. The dipole moments are accurate for molecules containing first-row atoms, and the quadrupole coupling constants are accurate for the halogens. The principal failures of the theory are the dipole moments for molecules containing atoms not in the first row of the periodic table and the quadrupole coupling constants of nitrogen.

These comments are true of both the parametrization from H_2 and the hydrides, and of the parametrization from the diatomics, of β_A^0. In the second case, although it has been necessary to compromise between the parameterizations suggested by equations (A) and (B), the general improvement in the accuracy obtained for the bonding energies of $C\ell_2$, Br_2, and I_2, and the fact that the accuracy is comparable for *all rows* of the periodic table, justify the parameterization from diatomics. Further improvements may be achieved by calibration of the bonding parameters for specific molecules. For example, the bonding parameters based on the dissociation energy of hydrogen yield more accurate bonding energies for the methyl halides. No single value for the carbon bonding parameter correctly predicted the bonding energies of the hydrocarbons studied. Wiberg,[72] using the parameters due to Pople and Segal,[2] found it impossible to find a value of the constant which would exactly fit both the C—C and C=C bond lengths in C_2H_6 and C_2H_4, respectively. The calculated bonding energies may be improved by introduction of separate bonding parameters for the resonance integrals between σ and π orbitals. A modification of this form has been used successfully to discuss the electronic spectra of aromatic compounds.[73] This modification may also lead to improved results for the cyanide molecules considered above. As a result of the introduction of separate parameters, the Pople-Santry-Segal approximation would *not be invariant* to hybridization of the AO basis set.

The results for the semiempirical theory have also been compared with those calculated with Pople-Segal parameters, which were shown to overestimate molecular energy quantities; the predicted ionization potentials were too large when calculated with theoretical electron repulsion integrals and/or with Pople-Segal parameters. The Pople-Segal bonding parameters predicted too-large bonding energies, and even if the parameters were calibrated to give the correct energies for some molecules, the predictions for other molecules were not good.

The relative merits of the various parameters is less evident in the predicted values of properties dependent on molecular charge distribution. The overall accuracy of the dipole moments is approximately the same for each parameter set, and the prediction of ^{14}N quadrupole coupling constants favors theoretical bonding parameters.

The advantage of semiempirical theory is therefore that it predicts more accurate orbital energies and total molecular energies.

At present the CNDO theory is not capable of predicting accurate bond lengths, and these are equilibrium input parameters. Recent work[74] suggests that the CNDO theory can be made bond-length-dependent when suitable values of V_{nn} are developed.

Appendix: Off-Diagonal Core Hamiltonian Matrix Elements, H_{kl}

The matrix element, $H_{k\ell}$, the resonance integral, is defined as the matrix element of the one-electron Hamiltonian, including both kinetic and potential energy, in the electrostatic field of the core. Thus

$$H_{k\ell} = \langle \phi_k | \overset{\circ}{H}_{core} | \phi_\ell \rangle = \left\langle \phi_k \middle| -\tfrac{1}{2}\nabla^2 - \sum_A V_A \middle| \phi_\ell \right\rangle \qquad \text{(A1)}$$

where V_A is the potential due to the core of atom A. The resonance integrals, which have been termed the most controversial parameters in semiempirical SCF theories,[64] have been evaluated by numerous approximations.

The Wolfsberg-Helmholtz approximation[65] has been used extensively to evaluate resonance integrals in independent electron theories. In this approximation

$$H_{k\ell} = \frac{k(H_{kk} + H_{\ell\ell})}{2} S_{k\ell} \qquad \text{(A2)}$$

where $k \leqslant 2$ is a constant and H_{kk} and $H_{\ell\ell}$ are evaluated as the valence-state ionization potentials for the atomic orbitals ϕ_k and ϕ_ℓ.

Another approximation used in independent electron theories is due to Ballhausen and Gray[63]; here

$$H_{k\ell} = k'(H_{kk} \times H_{\ell\ell})^{1/2} S_{k\ell} \qquad \text{(A3)}$$

Equation (A3) does not have a theoretical basis. However, it has the property that the interaction between two orbitals, $H_{k\ell}$, decreases as the difference $|H_{kk} - H_{\ell\ell}|$ increases, if the sum $(H_{kk} + H_{\ell\ell})$ is constant, in accordance with the concept of orbital matching,[66] in which orbitals are best suited for bonding if their energies are not too different. (The SCF-MO-CNDO theory automatically includes this feature. If the two orbitals are not well matched, the bond order $P_{k\ell}$ is small and does not contribute significantly to the matrix element $F_{k\ell}$.)

Cusachs[67] assumed a quadratic dependence of $H_{k\ell}$ on overlap,

$$H_{k\ell} = \frac{H_{kk} + H_{\ell\ell}}{2} S_{k\ell}(2 - |S_{k\ell}|) \tag{A4}$$

However,[32] the kinetic integrals for many pairs of orbitals are poorly approximated by equation (A4). Other approximations have been used in independent electron theories, including an attempt to improve systematically the matrix elements to reproduce the results of an SCF calculation by a noniterative procedure.[32]

If the resonance integral, $H_{k\ell}$, is chosen to be proportional to the corresponding overlap integral, $S_{k\ell}$,[3] then the constant of proportionality must be independent of the type of orbital so that the calculation is invariant to a transformation of the atomic basis set. The approximations discussed above are not invariant. Therefore, to preserve invariance, Pople et al.[3] have introduced the approximation

$$H_{k\ell} = \beta^{\circ}_{AB} S_{k\ell} \tag{A5}$$

It is assumed that the parameter, β°_{AB}, depends only on the atoms A and B and not on the atomic orbitals. It is given by

$$\beta^{\circ}_{AB} = \frac{-(\beta^{\circ}_{A} + \beta^{\circ}_{B})}{2} \tag{A6}$$

The bonding parameters, β°_{A}, were chosen by Pople and Segal[2] to give results comparable to those obtained by accurate solution of the Roothaan equations for small molecules.

Wratten[64] has pointed out that this parameterization suffers from the defect that $S_{k\ell}$ has been neglected in the simplification of the Roothaan equations.[33] But in equation (A1), the off-diagonal core Hamiltonian matrix elements between orbitals on different atoms contain the differential overlap of two orbitals as a factor in their integrand, and should therefore be neglected to be consistent with the CNDO approximation. However, they cannot be neglected, because the theory would not then predict bonding.

A further relationship has been proposed by Linderberg,[68] which within the zero-differential-overlap approximation avoids this contradiction:

$$H_{k\ell} = \frac{27.21}{R} \frac{dS_{k\ell}}{dR} \tag{A7}$$

Equation (A7) has been used to reproduce the empirical bonding parameters of Pople et al. for σ systems, when the Fischer-Hjalmars exponential form[69] for the overlap integral was used.

Dewar and Klopman[31] in a semiempirical SCF-MO theory, which includes the two-center exchange integrals and is therefore intermediate between the CNDO and NDDO approximations,[3] adopted the following expression:

$$H_{k\ell} = (\beta_{AA} \times \beta_{BB})^{1/2} (I_k + I_\ell) [r^2_{k\ell} + (\rho_k + \rho_\ell)^2]^{-1/2} S_{k\ell} \tag{A8}$$

which was chosen on the basis of a physical interpretation of equation (A4) together with the assumption that the one-electron resonance integral was proportional to each of the factors in equation (A8). I_k and I_ℓ are the valence-state ionization potentials, $r_{k\ell}$ is the interatomic distance between the atoms of which ϕ_k and ϕ_ℓ are atomic orbitals, and ρ_k and ρ_ℓ are constants characteristic of the two atoms.[31] The first factor is an empirically determined parameter which is assumed to have a common value for all orbitals on atoms A and B. The assumption of a geometric mean can only be justified empirically. However, the arithmetic mean may be justified theoretically by arguments similar to those suggested by Mulliken[47] for the evaluation of difficult integrals in MO theory.[70]

As Rudenberg[71] pointed out, an orbital ϕ_ℓ can be expanded over the complete set of orbitals, ϕ_k, of atom A,

$$\phi_\ell = \sum_{k'}^{A} \phi_{k'} S_{k'\ell} \tag{A9}$$

The overlap distribution is then

$$\phi_k^* \phi_\ell = \sum_{k'}^{A} \phi_k^* \phi_{k'} S_{k'\ell} \tag{A10}$$

Similarly, on expanding ϕ_k in terms of the complete set of orbitals on atom B,

$$\phi_k^* \phi_\ell = \sum_{\ell'}^{B} \phi_\ell \phi_{\ell'} S_{k\ell'} \tag{A11}$$

Averaging equations (A10) and (A11) leads to the identity

$$\phi_k^* \phi_\ell = \frac{1}{2} \left(\sum_{k'}^{A} \phi_k^* \phi_{k'} S_{k'\ell} + \sum_{\ell'}^{B} \phi_\ell \phi_{\ell'} S_{k\ell'} \right) \tag{A12}$$

In the Mulliken approximation,[47] the dominant terms are assumed to be the ones which are not themselves overlap distributions; hence

$$\phi_k^* \phi_\ell = \frac{1}{2} (\phi_k^* \phi_k + \phi_\ell^* \phi_\ell) S_{k\ell} \tag{A13}$$

Equation (A13) appears to be a drastic approximation; however, on integration over all space, the errors cancel, and the definition of the overlap integral is obtained.

Substitution of equation (A13) into equation (A1) gives the result

$$H_{k\ell} = \frac{H_{kk} + H_{\ell\ell}}{2} S_{k\ell} \tag{A14}$$

To obtain the form used by Pople, Santry, and Segal it is necessary to introduce the approximation that the diagonal matrix elements for all orbitals may be replaced by a single bonding parameter, which is independent of the orbital involved and only characteristic of the atom.

The theoretical basis given above and the derivation of the Pople-Segal bonding parameters[2] from equation (A7) provide considerable justification for the calculation of resonance integrals from equations (A5) and (A6). To preserve the invariance properties of the Roothaan equations and to have a theoretically based form for the resonance integral, the Pople-Santry-Segal formulation has been used in this article. An empirical bonding parameter permits at least a partial correction of the errors introduced by the assumption of the CNDO approximation. The choice of bonding parameter is, therefore, very important. The calculation of bonding energies[17] has shown that bonding parameters which differ greatly from the Pople and Segal bonding parameters[2] must be used with the atomic parameters described previously.[11]

Acknowledgment

This research was supported by the National Research Council of Canada.

References

1. J. A. Pople and G. A. Segal, *J. Chem. Phys.* **44**, 3289 (1966).
2. J. A. Pople and G. A. Segal, *J. Chem. Phys.* **43**, S136 (1965).
3. J. A. Pople, D. P. Santry, and G. A. Segal, *J. Chem. Phys.* **43**, S129 (1965).
4. L. Oleari, L. Di Sipio, and G. De Michelis, *Mol. Phys.* **10**, 97 (1966).
5. G. Klopman, *J. Am. Chem. Soc.* **86**, 1463 (1964).
6. J. Hinze, Ph.D. Thesis, University of Cincinnati, Cincinnati, Ohio, 1962.
7. J. Hinze and H. H. Jaffé, *J. Am. Chem. Soc.* **84**, 540 (1962).
8. R. Pariser and R. G. Parr, *J. Chem. Phys.* **21**, 767 (1963).
9. N. Mataga, *Bull. Chem. Soc. Japan* **31**, 453 (1958).
10. K. Ohno, *Theoret. Chim. Acta* **2**, 219 (1964).
11. J. M. Sichel and M. A. Whitehead, *Theoret. Chim. Acta* **7**, 32 (1967).
12. C. E. Moore, Atomic Energy Levels, *N.B.S. Circ. 467,* Vols. 1–3, Washington, 1949, 1952, 1958.
13. C. C. J. Roothaan, *J. Chem. Phys.* **19**, 1445 (1951).
14. K. Ohno, in *Advances in Quantum Chemistry*, Vol. 3, Academic Press, New York, 1967, pp. 240–323.
15. R. S. Mulliken, C. A. Rieke, D. Orloff, and H. Orloff, *J. Chem. Phys.* **17**, 1248 (1949).
16. J. C. Slater, *Phys. Rev.* **36**, 57 (1930).
17. J. M. Sichel and M. A. Whitehead, *Theoret. Chim. Acta* **11**, 220 (1968).
18. K. Ruedenberg, *J. Chem. Phys.* **34**, 326 (1962).
19. J. A. Pople, *Trans. Faraday Soc.* **49**, 1375 (1953).
20. P.-O. Löwdin, *Advances in Chemical Physics*, Vol. 2, Wiley-Interscience, New York, pp. 207–322.
21. C. Hollister and O. Sinanoğlu, *J. Am. Chem. Soc.* **88**, 13 (1966).
22. G. Herzberg, *Spectra of Diatomic Molecules*, Van Nostrand, Princeton, N.J., 1950.
23. A. L. H. Chung and M. J. S. Dewar, *J. Chem. Phys.* **42**, 756 (1965).
24. *JANAF Interim Thermochemical Tables*, The Dow Chemical Co., Midland, Mich., 1960.
25. D. W. Turner and M. I. Al-Joboury, *J. Chem. Phys.* **37**, 3007 (1962).
26. B. L. Kurbatov, F. I. Vilessov, and A. N. Terenin, *Soviet Phys. Doklady* **6**, 883 (1962).
27. F. H. Field and J. L. Franklin, *Electron Impact Phenomena*, Academic Press, New York, 1957.
28. T. Koopmans, *Physica* **1**, 104 (1933).
29. D. Peters, *J. Chem. Phys.* **45**, 3474 (1966).
30. D. W. Turner and D. P. May, *J. Chem. Phys.* **46**, 1156 (1967).
31. M. J. S. Dewar and G. Klopmen, *J. Am. Chem. Soc.* **89**, 3089 (1967).
32. M. D. Newton, F. P. Boer, and W. N. Lipscomb, *J. Am. Chem. Soc.* **88**, 2353, 2361, 2367 (1967).
33. C. C. J. Roothaan, *Rev. Mod. Phys.* **23**, 69 (1951).
34. F. A. Cotton, *Chemical Applications of Group Theory*, Wiley-Interscience, New York, 1963.
35. A. D. McLean and M. Yoshimme, *Tables of Linear Molecular Wave Functions*, IBM, San Jose, Calif., 1967.
36. M. Krauss, *J. Chem. Phys.* **38**, 564 (1963).
37. R. Moccia, *J. Chem. Phys.* **40**, 2176, 2186 (1964).
38. P. E. Cade, K. D. Sales, and A. C. Wahl, *J. Chem. Phys.* **44**, 1973 (1966).
39. J. W. Moskowitz and M. C. Harrison, *J. Chem. Phys.* **42**, 1726 (1965).
40. A. C. Wahl, *J. Chem. Phys.* **41**, 2600 (1964).
41. R. Pariser, *J. Chem. Phys.* **21**, 568 (1953).
42. J. C. Browne, *J. Chem. Phys.* **41**, 3495 (1964).
43. M. I. Al-Joboury and D. W. Turner, *J. Chem. Soc.* **1967**, 373.
44. M. I. Al-Joboury and D. W. Turner, *J. Chem. Soc.* **1964**, 4434.
45. J. M. Sichel and M. A. Whitehead, *Theoret. Chim. Acta* **11**, 239 (1968).
46. D. C. Frost, C. A. McDowell, and D. A. Vroom, *J. Chem. Phys.* **46**, 4255 (1967).
47. R. S. Mulliken, *J. Chim. Phys.* **46**, 497, 675 (1949).
48. C. A. Coulson, *Valence,* Oxford University Press, New York, 2nd ed., 1961.
49. D. P. Santry and G. A. Segal, *J. Chem. Phys.* **47**, 158 (1967).
50. P.-O. Löwdin, *J. Chem. Phys.* **18**, 365 (1950).
51. J. M. Sichel and M. A. Whitehead, *Theoret. Chim. Acta* **11**, 254 (1968).
52. J. A. Pople and M. Gordon, *J. Am. Chem. Soc.* **89**, 4253 (1967).
53. D. B. Boyd and W. N. Lipscomb, *J. Chem. Phys.* **46**, 910 (1967).
54. C. W. Kern and M. Karplus, *J. Chem. Phys.* **42**, 1062 (1965).
55. C. H. Townes and B. P. Dailey, *J. Chem. Phys.* **17**, 782 (1949).

56. W. Gordy, W. V. Smith, and R. F. Trambarulo, *Microwave Spectroscopy*, Wiley, New York, 1953.
57. J. M. Sichel and M. A. Whitehead, *Theoret. Chim. Acta* **11**, 263 (1968).
58. J. W. Richardson, *Rev. Mod. Phys.* **32**, 461 (1960).
59. T. P. Das and E. L. Hahn, *Nuclear Quadrupole Resonance Spectroscopy*, Academic Press, New York, 1958.
60. C. H. Townes, *Discussions Faraday Soc.* **19**, 281 (1955).
61. D. A. McLean, *J. Chem. Phys.* **37**, 627 (1962).
62. R. J. Boyd and M. A. Whitehead, submitted to *J. Chem. Soc.*
63. C. J. Ballhausen and H. B. Grey, *Inorg. Chem.* **1**, 111 (1962).
64. R. J. Wratten, *Chem. Phys. Letters* **1**, 667 (1968).
65. M. Wolfsberg and L. Helmholtz, *J. Chem. Phys.* **20**, 837 (1952).
66. M. A. Whitehead and H. H. Jaffe, *Trans. Faraday Soc.* **57**, 1854 (1961).
67. L. C. Cusachs, *J. Chem. Phys.* **43**, S157 (1965).
68. J. Linderberg, *Chem. Phys. Letters* **1**, 39 (1967).
69. O. Fischer-Hjalmars, *Arkiv Fysik* **21**, 123 (1962).
70. J. M. Sichel, Ph.D. Thesis, McGill University, Montreal, Canada, 1967.
71. K. Rudenberg. *J. Chem. Phys.* **19**, 1433 (1951).
72. W. B. Wiberg, *J. Am. Chem. Soc.* **90**, 59 (1968).
73. J. D. Bene and H. H. Jaffe, *J. Chem. Phys.*, **48**, 1807 (1968).
74. R. J. Boyd and M. A. Whitehead, unpublished results.
75. G. Herzberg, private communication; recent experiments on C_2 suggest 6.25 for D_0^0.

III-2b Reprinted from *The Journal of the American Chemical Society* **89**, 3089 (1967)

Ground States of σ-Bonded Molecules. I. A Semiempirical SCF MO Treatment of Hydrocarbons[1]

Michael J. S. Dewar and Gilles Klopman[2]

Contribution from the Department of Chemistry, The University of Texas, Austin, Texas 78712. *Received January 28, 1967*

Abstract: In a recent series of papers[3] it was shown that the heats of formation of conjugated molecules can be calculated with surprising accuracy by the Pople SCF MO method, the σ bonds being treated as localized. Here we describe an extension of the method to include all the valence electrons in a molecule, using procedures similar to those suggested by Pople, Santry, and Segal[4] and by Klopman.[5] Preliminary calculations for a number of hydrocarbons are reported; the agreement between the calculated and observed heats of formation is already very satisfactory, implying that this approach should ultimately give results of sufficient accuracy to be of value in predicting the structures and reactivities of molecules.

In a recent series of papers[3] it was shown that the heats of formation from atoms of conjugated molecules can be calculated with quite unexpected accuracy (±0.1%), using the localized bond model for the σ bonds and calculating the π binding energy by the Pople method. However, although this approach represents a very considerable advance over anything previously reported, it is still of limited chemical value; it cannot be applied to reactions even of conjugated molecules since transition states do not normally have the symmetry necessary for the π approximation to be applicable, nor can it be applied to many problems concerning the behavior of unconjugated molecules, *e.g.*, conformational equilibria and steric hindrance.

However, in view of the unexpected success of the π calculations, it seems reasonable to hope that an analogous treatment of σ bonds might prove equally successful; if so, we would have a complete solution of the basic problems of chemistry. Preliminary calculations of this kind for diatomic molecules have indeed proved very promising,[5] and Pople, Santry, and Segal[4] have reported preliminary calculations for larger molecules. Here we describe our own initial efforts in this direction, which already seem to have achieved a degree of accuracy almost in the "chemical" zone.

Theoretical Approach

The Pople SCF MO method is now familiar, and the problems involved in its extension to σ-bonded systems have been discussed in a formal manner by Pople, Santry, and Segal. The following pictorial representation has the advantage of clarifying these problems and will also help to illustrate our own approach.

In the original Pople treatment of conjugated systems, the π MOs ψ_μ are written as linear combinations of p AO's ϕ_i of the participating atoms (eq 1). The choice of

$$\psi_\mu = \sum_i a_{\mu i}\phi_i \qquad (1)$$

basis set functions ϕ_i is unambiguous, since the orientation of each ϕ_i is determined by the geometry of the π system. In calculations for a three-dimensional, σ-bonded system, the situation is more complicated.

(1) This work was supported by the Air Force Office of Scientific Research through Grant No. AF-AFOSR-1050-67.
(2) Robert A. Welch Postdoctoral Fellow.
(3) A. L. H. Chung and M. J. S. Dewar, *J. Chem. Phys.*, **42**, 756 (1965); M. J. S. Dewar and G. J. Gleicher, *J. Am. Chem. Soc.*, **87**, 685, 692, 3255 (1965); *J. Chem. Phys.*, **44**, 759 (1966); *Tetrahedron*, **21**, 1817, 3423 (1965); *Tetrahedron Letters*, **50**, 4503 (1965); M. J. S. Dewar and C. C. Thompson, Jr., *J. Am. Chem. Soc.*, **87**, 4414, (1965); M. J. S. Dewar, G. J. Gleicher, and C. C. Thompson, Jr., *ibid.*, **88**, 1349 (1966).
(4) J. A. Pople, D. P. Santry, and G. A. Segal, *J. Chem. Phys.*, **43**, S129 (1965); J. A. Pople and G. A. Segal, *ibid.*, **43**, S136 (1965); **44**, 3289 (1966).
(5) G. Klopman, *J. Am. Chem. Soc.*, **86**, 4550 (1964); **87**, 3300 (1965).

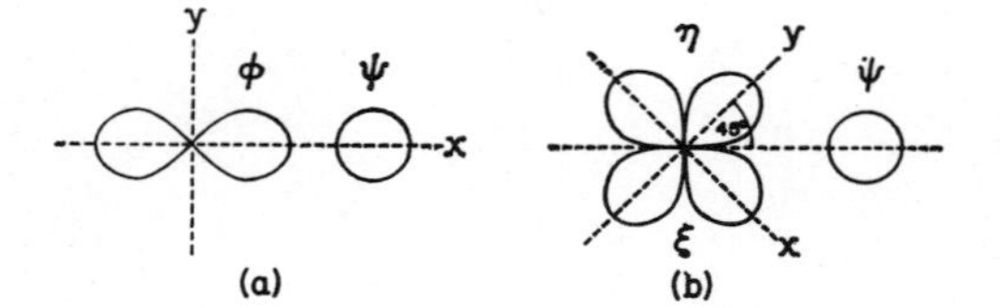

Figure 1. Illustrating the interactions between an electron in an s AO (ψ) of atom N, and an electron in a p AO (ϕ) of atom M.

Each atom other than hydrogen will contribute at least three p AO's; the orientation of these orbitals is arbitrary, since there will normally be no reference frame to fix the choice of coordinate axes. Since the choice of axes is arbitrary, the results of the calculation must be independent of it. While this condition is automatically met in a complete Roothaan SCF MO treatment, the same is not necessarily the case in the simplified version of Pople. The following example illustrates the difficulty.

Consider the interactions between two electrons, one occupying an s AO (ϕ) of atom M, and the other, a p AO (ψ) of atom N, ψ having its axis along the line joining the two nuclei (Figure 1a). In a coordinate system where this line is one of the coordinate axes, ψ will be represented as a single p AO. The terms in the total energy that represent the interactions between the two electrons are of two kinds; first there will be a one-electron resonance integral β, given by

$$\beta = \int \phi \mathrm{H}^{\mathrm{C}} \psi d\tau \quad (2)$$

where H^{C} is the core operator; secondly there will be an electron repulsion term γ, given by

$$\gamma = \int\int \phi^2(1)\frac{e^2}{r_{12}}\psi^2(2)\, d\tau_1 d\tau_2 \equiv (\phi\phi,\psi\psi) \quad (3)$$

in the usual notation for such integrals.

Suppose now that we calculate these interactions in a coordinate system rotated through 45° about the z axis (Figure 1b). The AO ϕ must now be written as a linear combination of the p_x AO (ξ) and the p_y AO (η); *i.e.*

$$\phi = \frac{1}{\sqrt{2}}(\xi + \eta) \quad (4)$$

The corresponding one- and two-electron interaction terms, β' and γ', are then given by eq 5 and 6. If our

$$\beta' = \int \phi \mathrm{H}^{\mathrm{C}} \psi\, d\tau = \frac{1}{\sqrt{2}}\left(\int \xi \mathrm{H}^{\mathrm{C}} \psi\, d\tau + \int \eta \mathrm{H}^{\mathrm{C}} \psi\, d\tau\right) \quad (5)$$

$$\gamma' = \int\int \phi^2(1)\frac{e^2}{r_{12}}\psi^2(2)\, d\tau_1 d\tau_2 = \frac{1}{2}\int\int \xi^2(1)\frac{e^2}{r_{12}}\psi^2(2)\, d\tau_1 d\tau_2 + \frac{1}{2}\int\int \eta^2(1)\frac{e^2}{r_{12}}\psi^2(2)\, d\tau_1 d\tau_2 + \int \xi(1)\eta(1)\frac{e^2}{r_{12}}\psi^2(2)\, d\tau_1 d\tau_2 \equiv \frac{1}{2}[(\xi\xi,\psi\psi) + (\eta\eta,\psi\psi) + 2(\xi\eta,\psi\psi)] \quad (6)$$

calculation is to be independent of the choice of coordinate axes, it is necessary that

$$\beta \equiv \beta' \quad (7)$$

$$\gamma \equiv \gamma' \quad (8)$$

Equation 7 will be automatically satisfied if we make the usual assumption that one-electron resonance integrals are proportional to overlap integrals. The second condition will be met in a full Roothaan treatment where all electron repulsion integrals are included; however, in the standard Pople treatment, where integrals involving overlap between different AO's are neglected, complications will arise, since the final integral in eq 6 will be set equal to zero. Since the remaining integrals in eq 6 are equal, from symmetry, eq 8 will be satisfied only if

$$(\phi\phi,\psi\psi) = (\xi\xi,\psi\psi) = (\eta\eta,\psi\psi) \quad (9)$$

Since these integrals represent the mutual repulsions of two clouds of charge, one representing the distribution of an electron occupying a p AO, the other the distribution of an electron occupying an s AO, the condition implied in eq 8 is equivalent to the assumption that such clouds of charge are spherically symmetrical. This is the CNDO approximation of Pople, *et al.*[4] (complete neglect of differential overlap). In it one assumes that electron repulsion integrals of the type $(\chi\chi,\omega\omega)$ depend on the nature of the orbitals χ and ω and on their distance apart, but not on their orientation.

Our experience indicates, however, that this approximation is too severe. In spite of very extensive trials, we have been unable to devise any satisfactory scheme for calculating heats of formation of molecules based on the CNDO approximation. This is not surprising, for the directivity of valence probably depends at least partly on the variation with orbital orientation of repulsion integrals involving p AO's. If so, one would not expect to be able to calculate heats of formation accurately, using an approximation in which these variations are neglected.

If such variations in the repulsion integrals are to be included in our treatment, we must then include three- and four-orbital repulsion integrals, involving overlap of orbitals on a common center. In other words, *all* two-center repulsion integrals must be included. We can still, of course, neglect three- and four-center integrals, involving overlap between AO's of different atoms, for neglect of these does not affect the invariance of our calculations to choice of coordinate axes. Thus in the notation of Figure 1, and with χ representing an AO of a third atom, we can set

$$(\chi\phi,\psi\psi) = (\chi\xi,\psi\psi) = (\chi\eta,\psi\psi) = 0 \quad (10)$$

without affecting the invariance to rotation, for the contribution of such integrals will be zero no matter what axes we choose.

This is the NDDO approximation of Pople, *et al.*[4] (neglect of diatomic differential overlap); it involves obvious technical difficulties, and no calculations have as yet been reported in which the full NDDO scheme has been adopted. Not only are there a large number of additional integrals to be evaluated, but it is also difficult to estimate them by the kind of semiempirical approach we have used for the two-orbital integrals.[1,5] While the NDDO approximation may prove essential, and while we are at present developing an appropriate program for applying it, we decided first to try the following intermediate approximation in the hope that it might combine simplicity with adequate accuracy.

Consider the integrals (km,ln) between AO's of two atoms M and N. First we transform the AO's of the atoms into the coordinate system of Figure 2. The repulsion integrals between the original AO's can at once be expressed in terms of corresponding integrals between the transformed AO's. In this new system, all three-orbital integrals involving overlap between pairs of p AO's vanish through symmetry, as also do most of the corresponding four-orbital integrals. We now assume that the remaining three- and four-orbital integrals can be neglected. The neglect of integrals involving overlap between an s AO and p AO of a given center can be shown to have no effect on the invariance of the calculations to choice of coordinate axes. The neglect of integrals involving four distinct p AO's can in principle affect this invariance; however, various arguments indicate that such effects are negligible (see below). With these assumptions our problem is greatly simplified, for the remaining two-center integrals are now of the standard two-orbital type, *i.e.*, (kk,ll). This approach might be termed the PNDDO approximation (partial neglect of diatomic differential overlap).

There are ten distinct integrals of this kind to be considered for the orbitals indicated in Figure 2, *viz.*

$$s\sigma_M\text{–}s\sigma_N,\ s\sigma_M\text{–}p\sigma_N,\ p\sigma_M\text{–}s\sigma_N,\ p\sigma_M\text{–}p\sigma_N,\ s\sigma_M\text{–}p\pi_N,\ p\pi_M\text{–}s\sigma_N,\ p\pi_M\text{–}p\pi_N,\ p\sigma_M\text{–}p\pi_N,\ p\pi_M\text{–}p\sigma_N,\ p\pi_M\text{–}p\pi_N^* \quad (11)$$

The last integral, $p\pi_M\text{–}p\pi_N^*$, is one between the p_y AO of M and the p_z AO of N, or conversely. For n atoms, there are therefore $5n(n-1)$ different integrals; in our computer program, each set of integrals is stored in one-half of an $n \times n$ matrix, five such matrices being required.

Our treatment also involves repulsion integrals between orbitals of a single center; here the three- and four-orbital integrals vanish through symmetry, only two-orbital integrals of the type (kk,mm) and (km,km) remaining. Integrals of the latter type must be retained since otherwise we could not distinguish between singlet and triplet states of atoms; thus the difference in energy between the singlet and triplet configurations $(1s)^2(2s)^2(2p)^2$ of carbon arises from an integral of this type where ϕ_k and ϕ_m are different 2p AO's.

We neglect inner electrons, *e.g.*, the 1s electrons in carbon; we treat the valence electrons as moving in the field of a set of cores, each composed of a nucleus and a set of occupied inner AO's. Thus the core of carbon is an ion C^{4+}, consisting of the nucleus and a pair of 1s electrons.

The one-center repulsion integrals (kk,mm) and (km,km) are estimated from spectroscopic data for the corresponding atom by a procedure considered in detail below. In this it is assumed that the repulsion between a pair of electrons in the valence shell of a given atom has a value independent of the orbitals occupied by the electrons and depending only on their relative spins, *i.e.*

$$\text{repulsion between electrons of like spin} = A_M^+ \quad (12)$$

$$\text{repulsion between electrons of opposite spin} = A_M^- \quad (13)$$

The repulsion integrals can be expressed in terms of

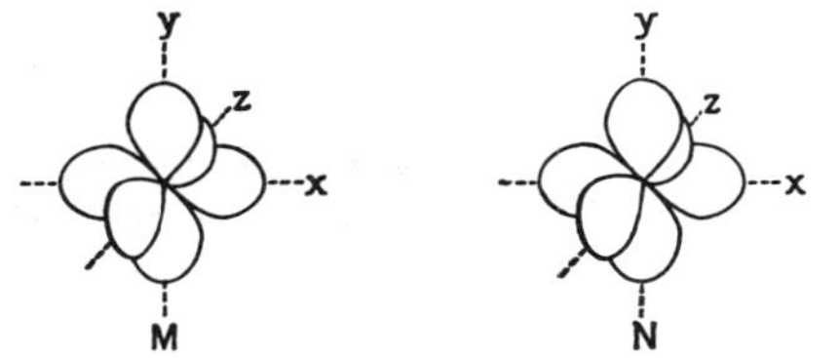

Figure 2. Illustrating calculation of electron repulsion integrals.

these quantities as

$$(kk,mm) = A_M^- \qquad (k = m \text{ or } k \neq m) \quad (14)$$

$$(km,km) = A_M^- - A_M^+ \qquad (k \neq m) \quad (15)$$

In order to make the treatment as general as possible, we derived an expression for the total electronic energy (E_{el}) for an open-shell SCF MO treatment. Here q_i^α, q_j^β are respectively the densities of α-spin and β-spin electrons in the AO's ϕ_i and ϕ_j given by

$$q_i^\alpha = \sum_\mu n_\mu^\alpha a_{\mu i}^2 \qquad q_j^\beta = \sum_\nu n_\nu^\beta a_{\nu j}^2 \quad (16)$$

where n_μ^α and n_μ^β are the numbers of electrons (zero or unity) occupying the corresponding MO's ψ_μ^α and ψ_ν^β. Likewise p_{ij}^α and p_{ij}^β are the corresponding bond orders, defined by eq 17. In the case of a closed-shell

$$p_{ij}^\alpha = \sum_\mu n_\mu^\alpha a_{\mu i} a_{\mu j}$$
$$p_{ij}^\beta = \sum_\nu n_\nu^\beta a_{\nu i} a_{\nu j} \quad (17)$$

molecule, the net charge densities q_i and bond orders p_{ij} are given by

$$q_i = 2q_i^\alpha = 2q_i^\beta \quad p_{ij} = 2p_{ij}^\alpha = 2p_{ij}^\beta \quad (18)$$

The expression for E_{el} is shown in eq 19. Here

$$E_{el} = \sum_k^{(M)}\Big((q_k^\alpha + q_k^\beta)W_k + \sum_m^{(M)}\Big\{\sum_{N\neq M}\Big[k_x m_x V_{kN}^\sigma + (k_y m_y + k_z m_z)V_{kN}^\pi\Big](p_{km}^\alpha + p_{km}^\beta) + \frac{1}{2}\Big[q_k^\alpha q_m^\alpha - (p_{km}^\alpha)^2 + q_k^\beta q_m^\beta - (p_{km}^\beta)^2\Big]A_M^+ + \frac{1}{2}(q_k^\alpha q_m^\beta + q_k^\beta q_m^\alpha)A_M^-\Big\}\Big) + \sum_k^{(M)}\sum_{l>k}^{(N)}\Big\{2(p_{kl}^\alpha + p_{kl}^\beta)\Big[k_x l_x \beta_{kl}^\sigma + (k_y l_y + k_z l_z)\beta_{kl}^\pi\Big]\Big\} + \sum_m^{(M)}\sum_n^{(N)}\Big\{(p_{km}^\alpha p_{ln}^\alpha + p_{km}^\beta p_{ln}^\beta + p_{km}^\alpha p_{ln}^\beta + p_{km}^\beta p_{ln}^\alpha - p_{kn}^\alpha p_{lm}^\alpha - p_{kn}^\beta p_{lm}^\beta)\Big[k_x l_x m_x n_x \gamma_{kl}^{\sigma\sigma} + (k_y l_y m_y n_y + k_z l_z m_z n_z)\gamma_{kl}^{\pi\pi} + k_x m_x(l_y n_y + l_z n_z)\gamma_{kl}^{\sigma\pi} + (k_y m_y + k_z m_z)l_x n_x \gamma_{kl}^{\pi\sigma} + (k_y m_y l_z n_z + k_z m_z l_y n_y)\gamma_{kl}^{\pi\pi *}\Big]\Big\} \quad (19)$$

ϕ_k and ϕ_m are AO's of one atom M, while ϕ_l and ϕ_n are AO's of some other atom N; the summations are labeled accordingly to avoid confusion.

The integral W_k represents a sum of the kinetic energy of an electron occupying the AO ϕ_k and its potential energy due to attractions by the core of the corresponding atom (M). The integrals V_{kN}^σ and V_{kN}^π represent respectively the attractions between an electron in a σ-type AO (s or pσ) or a π-type AO, ϕ_k, and the core of atom N. The quantities A^+ and A^- have already been defined (eq 14 and 15). The remaining one-electron integrals β_{kl}^{σ} and β_{kl}^{π} are resonance integrals between AO's of atoms M and N in the local coordinate system of Figure 2. There are five nonvanishing combinations of this type, *i.e.*

$$s\sigma\text{–}s\sigma,\ s\sigma\text{–}p\sigma,\ p\sigma\text{–}s\sigma,\ p\pi\text{–}p\pi,\ p\sigma\text{–}p\sigma \qquad (20)$$

where the first symbol designates the orbital ϕ_k, the second ϕ_l.

The quantities γ_{kl} are two-orbital repulsion integrals for orbitals of atoms M and N, again in the local coordinate system of Figure 2; *i.e.*

$$\gamma_{kl} \equiv (kk,ll) \qquad (21)$$

The integrals are labeled with superscripts to indicate the types of orbital involved; the distinction between sσ and pσ types follows automatically from the nature of the AO's. Thus if ϕ_k and ϕ_m are p AO's, the integral $\gamma_{kl}^{\sigma\pi}$ is of the pσ–pπ type, while if $k = m$ and ϕ_k is a σ AO, the integral is of the sσ–pπ type. (Note that in our scheme $\gamma_{kl}^{\sigma\pi}$ vanishes if ϕ_k is an s AO and $k \neq m$, since three- and four-orbital integrals involving s–p overlap are neglected.)

The quantities $k_x, k_y \ldots n_z$ are involved in the transformation of the original basis set of AO's into the locally oriented sets of Figure 2. They are defined in (22), where $i = k, l, m$, or n, and other terms are defined

	Type of orbital ϕ_i: s	p_x	p_y	p_z
$i_x =$	1	X	Y	Z
$i_y =$	0	$-rY/R$	rX/R	0
$i_z =$	0	$-rXZ/R$	$-rYZ/R$	R/r

(22)

$$X = |X_M - X_N|/r;\ Y = |Y_M - Y_N|/r;\ Z = |Z_M - Z_N|/r;\ r = (X^2 + Y^2 + Z^2)^{1/2};\ R = (X^2 + Y^2)^{1/2} \qquad (23)$$

in eq 23. X_p, Y_p, and Z_p are the coordinates of atom P in the original coordinate system used to specify the positions of the atoms in the molecule.

The elements of the F matrix for the α-spin electrons are given in eq 24–26. Equation 25 refers to off-

$$F_{kk} = W_k^{(M)} + \sum_{N\neq M} [k_x^2 V_{kN}^{\sigma} + (k_y^2 + k_z^2)V_{kN}^{\pi}] + q_k^{\beta}A_M^- + \sum_{m\neq k}^{(M)} (q_m^{\alpha}A_M^+ + q_m^{\beta}A_M^-) + \sum_l^{(N)}\sum_n^{(N)} (p_{ln}^{\alpha} + p_{ln}^{\beta})(kk,ln) \qquad (24)$$

$$F_{km}^{(M)} = \sum_{N\neq M} [k_x m_x V_{kN}^{\sigma} + (k_y m_y + k_z m_z)V_{kN}^{\pi}] - p_{km}^{\alpha}A_M^+ + \sum_l^{(N)}\sum_n^{(N)} (p_{ln}^{\alpha} + p_{ln}^{\beta})(lm,ln) \qquad (25)$$

$$F_{kl}^{(M,N)} = k_x l_x \beta_{kl}^{\sigma} + (k_y l_y + k_z l_z)\beta_{kl}^{\pi} - \sum_m^{(M)}\sum_n^{(N)} p_{mn}(km,ln) \qquad (26)$$

diagonal matrix elements between AO's ϕ_k and ϕ_m of the same atom M, while eq 26 refers to corresponding elements between AO's of two different atoms M and N. Here the electron repulsion integrals have been left in their original form, over AO's set up in the original coordinate system; in order to evaluate them, the AO's ϕ_k, ϕ_l, ϕ_m, and ϕ_n for each pair of atoms M and N are transformed into the local coordinate system of Figure 2.

The calculations were carried out at the Computation Center of The University of Texas, using first a CDC 1604 digital computer, and later a CDC 6600. The program, while somewhat complex, followed a fairly conventional path. The integrals β_{km} and γ_{km}, and the quantities k_x etc. of eq 22, are first computed and stored. An initial F matrix is then set up, using assumed values for the q's and p's; in the case of hydrocarbons, we set the charge density (q) equal to unity for each valence orbital, and each bond order (p) equal to zero. The F matrix is then diagonalized, new q's and p's are computed, and the process is continued until the sum of the energies of the occupied orbitals converges to within a predetermined limit. The total electronic energy is then computed (eq 19) and the total bonding energy found from it by adding the core repulsion and subtracting the total energy of the isolated atoms. The following section indicates the procedure we have followed in estimating the various integrals and other quantities appearing in the treatment.

Calculations of Integrals Etc.

The quantities appearing in this treatment are of five types: (a) valence-shell ionization potentials, W_k; (b) one-center repulsion integrals, (kk,mm) and (km,km); (c) one-electron resonance integrals, β_{kl}; (d) two-center repulsion integrals, γ_{kl}; (e) the core repulsion and core-electron attraction. For reasons indicated above, we are prepared if necessary to treat any or all of these quantities as parameters, our sole purpose being to develop a reliable and general method for calculating heats of formation of molecules of all kinds (including transition states) with "chemical" accuracy. However, one must obviously try to minimize the number of arbitrary parameters in a treatment such as this; we have accordingly adopted the course of introducing parametric functions for the calculation of the various quantities, these functions containing the minimum number of parameters and being chosen on the basis of physical intuition. In this connection, empirical data for atoms can be regarded as free information, for we are concerned only with the heats of formation of molecules, not with their total binding energies. The true parameters in our treatment are those whose values must be fixed by reference to properties of specific molecules; the number of such "molecular" parameters must be kept as small as possible if the method is to be useful and convincing. The procedure we followed in the calculations reported here was as follows.

(a and b) Valence-Shell Ionization Potentials and One-Center Repulsion Integrals. These were estimated for carbon from spectroscopic data by the following[5] procedure. We represent the core of the carbon atom as the ion C^{4+}, consisting of the nucleus and two 1s electrons. The quantities W_s or W_p should then represent the energy of a 2s or 2p electron, respectively, moving in the field of this core, and one might therefore try to equate them to the appropriate fourth ionization potentials of carbon, *i.e.*

$$C^{3+}(1s)^2(2s) \rightarrow C^{4+}(1s)^2 \quad \Delta H = W_s$$
$$C^{3+}(1s)^2(2p) \rightarrow C^{4+}(1s)^2 \quad \Delta H = W_p \qquad (27)$$

However, the orbitals in C^{3+} are smaller than those in neutral carbon and their binding energies correspondingly greater; the quantities W_s and W_p were therefore chosen to give a best fit to the energies of various states of neutral and singly ionized carbon.

We assume that the repulsion between two valence-shell electrons has the same value, regardless of the orbitals occupied by the electrons; for electrons of parallel spin, the repulsion is A^+, and for electrons of opposite spin, A^-. Thus the energy (E^1 or E^3) of a carbon atom in the singlet or triplet states $(1s)^2(2s)^2(2p)^2$, represented by single determinants with $S_z = 0$ or $\pm\hbar$, respectively, are given in eq 28 and 29. The

Singlet $(1s)(\overline{1s})(2s)(\overline{2s} \quad (\overline{2p})$

$$E^1 = 2W_s + 2W_p + 2A^+ + 4A^- \qquad (28)$$

Triplet $(1s)(\overline{1s})(2s)(\overline{2s})(2p_x)(2p_y)$

$$E^3 = 2W_s + 2W_p + 3A^+ + 3A^- \qquad (29)$$

quantities A^+ and A^- are related to the one-center repulsion integrals by eq 14 and 15.

In order to estimate A^+ and A^- it is necessary to determine E^1 and E^3 from spectroscopic data. The single determinants in eq 28 and 29 do not of course correspond to true states of carbon. There are 15 possible configurations, corresponding to the possible partitions of two electrons between the three 2p AO's; these are indicated in (30) and (31). In our approximation, all

States with $S_z = 0$

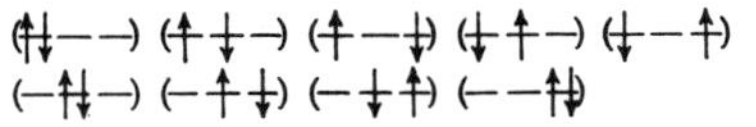

(30)

States with $S_z = \pm\hbar$

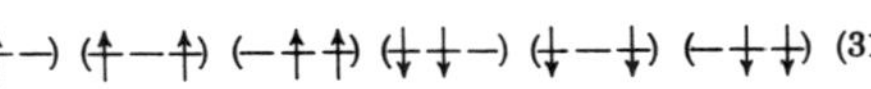

(31)

the "singlet" configurations of (30) have the same energy E^1, while all those of (31) have energy E^3. In practice these 15 configurations correspond to the following 15 real states of carbon.

^1S, one state

^1D, five substates ($M = 2, 1, 0, -1, -2$)

^3P, nine substates ($M = 1, 0, -1;\ S = 1, 0, -1$) (32)

In order to obtain correct representations of these states, we should construct appropriate linear combinations of the configurations indicated in (30) and (31). Since configurations of different multiplicity do not mix, the six triplet states of (31) lead only to six substates of ^3P (those with $S_z = \hbar$ or $-\hbar$). Since these substates have the same energy (that of the state ^3P), we can equate this energy to A^+. The remaining states are the ^1S and ^1D states, and the three substates of ^3P with $S_z = 0$. These are represented by a set of nine orthogonal linear combinations of the configurations of (30) and (31). Now it is easily shown that the *total* energy of such a set of linear combinations is the same as the sum of the individual energies of the original configurations; the total energy of the nine configurations is of course $9E^1$, while that of the nine real states and substates is $[(^1S) + 5(^1D) + 3(^3P)]$, where (^1S), (^1D), and (^3P) are the energies of the corresponding states. Hence

$$E^1 = \frac{1}{9}[(^1S) + 5(^1D) + 3(^3P)] \qquad (33)$$

The configuration thus appears as a weighted mean or *barycenter*[5] of the appropriate states. The energies of barycenters can thus be calculated from spectroscopic data, and the results can then be used to determine the various one-center integrals A^+, A^-, W_s, and W_p.

This treatment of atoms may seem rather primitive, but it is fully justified by its practical success.[5] The number of appropriate barycenters for a given atom is usually greater than the number of parameters; the energies of all the barycenters are given well by this approach for a wide variety of different atoms.

In calculating the binding energy of a molecule, we naturally compare its calculated total energy with a sum of the energies of ground-state barycenters of the component atoms, for, since we use a single Slater determinant to describe the molecule, it would be inconsistent not to use a similar description for its component atoms.

(c) The one-electron resonance integral β_{kl} can be interpreted physically as the energy of an electron occupying the overlap cloud between the AO's ϕ_k and ϕ_l, and moving in the field of the core and remaining electrons. We would therefore expect β_{kl} to be proportional (a) to the magnitude of the overlap cloud, *i.e.*, to the overlap integral S_{kl}; (b) to some mean of the binding energies of the AO's ϕ_k and ϕ_l; (c) and to the distance between the overlap cloud and the nuclei of the atoms of which ϕ_k and ϕ_l are AO's. The last two conditions follow since the potential field in the overlap region arises mainly from the two atoms. We therefore adopted the following expression for β_{kl}.

$$\beta_{kl} = (\beta_{kl})_0 S_{kl}(I_k + I_l)[r_{kl}^2 + (\rho_k + \rho_l)]^{-1/2} \qquad (34)$$

Here I_k and I_l are the valence-state ionization potentials of the AO's ϕ_k and ϕ_l, calculated for the appropriate barycenters by the method of ref 5; r_{kl} is the internuclear distance between the atoms of which ϕ_k, ϕ_l are AO's; ρ_k and ρ_l are quantities appearing in the expressions for two-center repulsion integrals (see below); $(\beta_{kl})_0$ is a parameter to be determined empirically, being the same for all valence orbitals of a given atom. The overlap integrals S_{kl} were calculated in the usual way using Slater–Zener orbitals ($Z = 3.25$ for carbon). In order to reduce the number of parameters in the treatment, we assumed that β_0 has a common value β_{xy} for orbitals of two atoms x and y, regardless of the type of orbitals (s or p) and mode of overlap (σ or π), and that

$$\beta_{xy} = \sqrt{\beta_{xx}\beta_{yy}} \qquad (35)$$

Equation 34 is more complicated than the corresponding expressions used by other authors; we have tried a number of such simpler expressions, but with less success. Thus omission of the terms I_k and I_l leads to results for unsaturated molecules such as ethylene in which the orbitals appear in the wrong order of energy; it is essential to use different values of β for s and p AO's. Again, omission of the term in r gave heats of formation for acetylene that were much too low; it is apparently necessary to use for β an expression that increases more rapidly with decreasing bond length than does the corresponding overlap integral.

(d) The two-center integrals $(\phi\phi,\psi\psi)$ were estimated in two different ways, one for the CNDO calculations and one for the PNDDO approximation. The integral $(\phi\phi,\psi\psi)$ must in each case obey two boundary conditions. As the internuclear distance r tends to zero, the integral should approximate to a one-center repulsion integral, while, when $r \rightarrow \infty$, the integral should approximate to e^2/r.

In the CNDO calculations, we adopted expression 36, one which has been used successfully in previous work.[3,5] Here ρ_k and ρ_l are constants characteristic

$$\gamma_{kl} = e^2[r_{kl}^2 + (\rho_k + \rho_l)^2]^{-1/2} \quad (36)$$

of the two atoms, chosen to make γ_{kl} approach the corresponding one-center integral as $r_{kl} \rightarrow 0$; *i.e.*

$$2\rho_k = e^2/A_k^- \qquad 2\rho_l = e^2/A_l^- \quad (37)$$

In the second approach, different values were assumed for the integral γ_{kl}, depending on the nature of the orbitals involved and their mode of overlap. Two arguments guided us in choosing a suitable expression for the integrals. First, an analysis of the role of electron correlation, using a model[5] similar to that invoked in the SPO approach,[6] suggested that the integrals should fall into three distinct groups, *i.e.*, (38)–(40).

I. correlation large: sσ:sσ; pπ:pπ; sσ:pπ (38)

II. correlation medium: sσ:pσ; pσ:pπ (39)

III. correlation small: pσ:pσ (40)

Secondly, this subdivision of the integrals also appears in the values estimated theoretically, using Slater–Zener orbitals;[7] these are shown in Table I.

Table I. Values for Carbon–Carbon Two-Center Repulsion Integrals

Class	Type	Value of integral, ev Calcd using Slater–Zener AO's	Calcd from (41)–(43)
1	sσ:sσ	9.28	
	sσ:pπ	9.12	7.13
	pπ:pπ	8.98	
	p$\pi(x)$:p$\pi(y)$	8.98	
2	sσ:pσ	9.61	7.81
	pπ:pσ	9.41	
3	pσ:pσ	9.99	8.45

Our object was to duplicate this pattern, subject to the condition that the integral (ii,kk) between orbitals of two identical atoms should converge to the corresponding one-center integral (ii,ii) at zero internuclear separation. The expressions we adopted are given in eq 41–43, where in class II, the orbital ϕ_k is the one of

Class I

$$(ii,kk) = e^2[r_{ik}^2 + (\rho_i + \rho_k)^2]^{-1/2} \quad (41)$$

Class II

$$(ii,kk) = e^2[r_{ik}^2 + (\rho_i + \rho_k T_{ik})^2]^{-1/2} \quad (42)$$

Class III

$$(ii,kk) = e^2[r_{ik}^2 + (\rho_i T_{ik} + \rho_k T_{ik})^2]^{-1/2} \quad (43)$$

the pσ type, and where

$$T_{ik} = e^{-r_{ik}/2(\rho_i+\rho_k)} \quad (44)$$

The values calculated in this way for carbon atoms at an internuclear distance of 1.55 A are listed in the last column of Table I.

(e) Core Repulsion. Having calculated the total electronic energy (E_{el}), we can then find the total energy of a molecule by adding to this the core repulsion. Our last problem is to decide how to calculate this.

In the π calculations, the repulsion between two cores M and N was set equal to the corresponding two-center repulsion integral; however, if we try to do this in the present case, we find that the molecule collapses to a point. The repulsion between point charges (*i.e.,* the nuclear repulsion) is much greater at short distances than is the corresponding repulsion between clouds of charge representing occupied orbitals; this enhanced repulsion is one of the factors that keeps the atoms in a molecule at bond's length. In the π calculations, this difficulty did not arise since we assumed Morse functions for the σ components of bonds; here our calculations include all the valence electrons, so there is no escape.

Nor is it satisfactory to set the core repulsion equal to a point charge potential, *i.e.*, to $Z_M Z_N e^2/R$, where Z_M and Z_N are the nuclear charges, for in this case the calculated binding energy is too small. The reason for this is implied in the literature. Consider for example H_2. The potential field in which the electrons move is greater than that in an isolated hydrogen atom; consequently, the orbitals of H_2 are more compact than one would expect for a combination of ordinary 1s hydrogen AO's. Indeed, if we carry out an orbital treatment, regarding the nuclear charges (Z) as variation parameters, we find the best agreement with experiment given by a value of Z considerably greater than unity. In our treatment, where the "atomic" parameters are fixed from spectroscopic data for isolated atoms, we assume in effect that the effective nuclear charge is the same for each atom in isolation as it is when the atom forms part of a molecule. In order to get realistic binding energies, either we must abandon this assumption or we must make some allowance for it by compensating changes in the other parameters. In this case the changes are best made in the nuclear repulsion, because this does not affect calculations of the electron distribution or orbital energies.

We therefore calculated the core repulsion from an appropriate parametric function. The function chosen must satisfy two boundary conditions. For large r_{ik}, it must approach the corresponding interelectronic repulsion between neutral atoms in order that the net potential due to a neutral atom should vanish at large distances, while for small r_{ik} it must have a value between this and that calculated for point charges. We

(6) See M. J. S. Dewar and N. L. Sabelli, *J. Phys. Chem.*, **66**, 2310 (1962).

(7) We are grateful to Dr. F. A. Matsen for these values.

have tried a large number of possible one-parameter functions of this type; the most successful was that of

$$C_{MN} = E_{MN} + [Z_M Z_N e^2/r_{MN} - E_{MN}]e^{-\alpha_{MN} r_{MN}} \quad (45)$$

eq 45. Here C_{MN} is the core repulsion between atoms M and N; E_{MN} is the corresponding electronic repulsion between neutral atoms M and N (*i.e.*, (*kk,mm*) summed over the valence orbitals); Z_M and Z_N are the formal core charges in units of *e* (*i.e.*, the number of valence electrons) of the two atoms; α_{MN} is a parameter. In order to reduce the number of parameters in the treatment, we assumed (*cf.* eq 35) that the value of α_{MN} for two dissimilar atoms M and N is given in terms of the parameters α_{MM} and α_{NN} for pairs of similar atoms by

$$\alpha_{MN} = \sqrt{\alpha_{MM}\alpha_{NN}} \quad (46)$$

Our treatment contains very few "molecular" parameters, *i.e.*, parameters whose value has to be determined from data for molecules rather than atoms. There are just two molecular parameters for each kind of atom X, *i.e.*, the parameter β_{XX} that appears in the expression for one-electron resonance integrals involving orbitals of X, and the parameter α_{XX} that appears in the expressions for corresponding core repulsion.

The attraction between an electron in an AO *i* of one atom M and the core of atom N, was set equal to *minus* the sum of repulsions between the electron and the valence electron of N (*cf.* the corresponding approximation in the π treatment[3]).

Application to Hydrocarbons

The method outlined above has been applied to a variety of hydrocarbons. The general procedure was as follows: (a) the parameters β_{HH} and α_{HH} were chosen to give the correct internuclear distance and bond energy in H_2; (b) assuming various values for β_{CC}, α_{CC} was chosen to give the correct heat of formation for CH_4; (c) heats of formation were then calculated for acetylene, ethane, and propane, using the values of β_{HH} and α_{HH} from step a, and with the various pairs of values for β_{CC} and α_{CC} from step b; (d) having thus established optimum values for the parameters, calculations were carried out for a number of other saturated and unsaturated hydrocarbons.

In order to apply this treatment, it is necessary to know the Cartesian coordinates of the atoms in a molecule; these must be calculated from the known (or assumed) bond lengths and bond angles. In our case the positions of the atoms are specified by the bond lengths, bond angles, and dihedral angles of the bonds in the molecule; we have written a program whereby the coordinates of the atoms are calculated from these data. In the calculations reported below, we assumed tetrahedral geometry for sp^3 carbon and trigonal geometry for sp^2 carbon (bond angles, 120°). The assumed bond lengths are shown in Table II.

The values for the parameters in the treatment are listed in Table III, while Table IV compares calculated and observed heats of formation for the various compounds. Except when otherwise stated, saturated C_2 units were assumed to have the conformation observed for propene (*i.e.*, one sp^3 CH bond eclipsing the C=C bond). Other calculated quantities will be found below (see Discussion).

Table II. Bond Lengths for CC and CH Bonds

Bond	Hybridization	Length, nm
C—C	sp^3–sp^3	0.1534
	sp^3–sp^2	0.1520
	sp^3–sp	0.1459
	sp^2–sp^2	0.1483
C⋯C	(Aromatic)	0.1397
C=C	. . .	0.1337
C≡C	. . .	0.1205
C—H	sp^3	0.1093
	sp^2	0.1083
	(Benzene)	0.1084
	sp	0.1059

Table III. Parameters for Carbon and Hydrogen

Atom	β_{XX}, pm	α_{XX}, nm^{-1}
C	45.84	47.08
H	27.87	14.8

Table IV. Comparison of Calculated and Observed Heats of Formation (ΔH_f) of Hydrocarbons from Atoms in the Gas Phase at 25°

Compound	ΔH_f, kcal/mole Obsd[a]	Calcd	$\delta\Delta H_f$[b]
Ethane	674.6	677.3	2.7
Ethane (eclipsed)	671.7[c]	676.5	4.8
Propane	954.3	957.3	3.0
n-Butane	1234.7	1237.5	2.8
n-Pentane	1514.7	1517.6	2.9
Isobutane	1236.7	1236.8	0.1
Isopentane	1516.6	1516.3	−0.3
Cyclopropane[d]	812.6	809.1	−3.5
Cyclohexane (chair)	1680.0	1680.2	0.2
Cyclohexane (boat)	1678.0[e]	1674.9	3.1
Ethylene	537.7	540.1	2.4
Propene	820.4	822.6	2.2
cis-2-Butene	1102.0	1101.4	−0.6
trans-2-Butene	1103.0	1103.4	0.4
trans-1,3-Butadiene	969.8	971.6	1.8
cis-1,3-Butadiene	967.5[f]	971.3	3.8
Benzene	1318.1	1314.0	−4.1
Allene	675.2	697.2	22.0
Acetylene	391.8	414.0	22.2
Methylacetylene	676.8	704.4	27.6

[a] The thermochemical data are taken from Rossini except where otherwise stated. [b] Difference between calculated and observed heats of formation in kcal/mole. [c] Calculated from the value for staggered ethane, using the experimentally determined height (2.9 kcal/mole) of the rotational barrier; see D. R. Lide, *J. Chem. Phys.*, **29**, 1426 (1958). [d] Heat of formation calculated for structure with D_{3h} symmetry; see H. A. Skinner and G. Pilcher, *Quart. Rev.* (London), **20**, 264 (1966). [e] Calculated from the value for chair conformation, using the experimental value (5.2 kcal/mole) for the heat of conversion to the boat conformation; see E. L. Eliel, "Stereochemistry of Carbon Compounds," McGraw-Hill Book Co., Inc., New York, N. Y., 1962. [f] Calculated from the observed difference in energy between the *cis* and *trans* isomers; see Table V.

Invariance to Rotation

As has been pointed out, the treatment used here is not strictly invariant to choice of coordinate axes, due to the neglect of repulsion integrals involving four p AO's of two different atoms. Four lines of argument suggested, however, that variations of this kind should be small. Firstly, integrals of this type represent quadrupole–quadrupole repulsions and are consequently much smaller than the charge–charge repulsions corresponding to "normal" repulsion integrals; even in the case of adjacent carbon atoms, the integrals have values[7] of only about 0.1 ev. Secondly, the effect of changing the coordinate axes appears only as secondary changes in the values of these integrals, and the net effect in the sum of the integrals between a given pair of atoms should consequently be small. Thirdly, the integrals between a given pair of atoms do not all have the same sign; the resulting cancellations will further reduce their net contribution to the total energy and so likewise to its variation with choice of axes. And finally, the integrals in question, representing as they do higher multipole repulsions, decrease very rapidly with distance;. integrals between nonadjacent atoms are essentially negligible.

Obviously, however, these arguments needed to be checked experimentally. We therefore repeated the calculations for a number of molecules in various orientations relative to the coordinate axes; in each case the eigenvalues, total energies, charge densities, and bond orders were identical with the accuracy with which they are printed (seven significant figures in the total energy, four in the other quantities). As a further check, we carried out calculations for carbon monoxide and hydrogen cyanide in various orientations; here again the results were quite unaffected by choice of coordinate axes, although these molecules contain heteroatoms and the neglected quadrupole–quadrupole integrals should be greatest for triple bonds since these are so short. It seems clear from these results that our procedure is for all practical purposes invariant to rotation of the coordinate axes, any variations being entirely negligible.

Discussion

The agreement between the calculated and observed heats of formation in Table IV is rather remarkable, the differences in most cases being less than 4 kcal/mole (0.15 ev). The only serious discrepancies (~1 ev) occur in the case of allene and the acetylenes; these are probably due to our use of a nuclear potential which does not increase sufficiently rapidly at short distances. Thus our method correctly predicts the heat of formation (and so by implication the strain energy) of cyclopropane, in which the bonds are single; on the other hand, attempts to calculate bond lengths, by minimizing the total energy of a molecule with respect to them, have given values which are much too small.

Several other qualitative checks also seem satisfactory. Thus ethane is correctly predicted to be most stable in the staggered conformation, cyclohexane in the chair conformation, and 2-butene and 1,3-butadiene in *trans* configurations; previous SCF MO calculations for 1,3-butadiene had incorrectly predicted the *cis* form to be more stable.[8]

Admittedly the differences in energy are not predicted exactly; this is clear from the data listed in Table V. In one case our procedure even leads to a qualitatively incorrect prediction, *i.e.*, that normal paraffins should be more stable than their branched isomers, while the predicted barrier to rotation in ethane is too small. Nevertheless, the over-all picture is very encouraging, given that the work described here represents only a preliminary approach to the problem and given that the errors in the calculated heats of formation are less by two orders of magnitude than those derived from other recent SCF MO calculations.[4,9]

Table V. Comparisons of Energies of Isomeric Hydrocarbons

Reaction	Energy change, kcal/mole Calcd	Obsd
Ethane (staggered → eclipsed)	0.8	2.9
1,3-Butadiene (*trans* → *cis*)	0.3	2.2[a]
2-Butene (*trans* → *cis*)	2.0	1.0
Cyclohexane (chair → boat)	2.0	5.3
n-Butane → isobutane	−2.0	0.7
n-Pentane → isopentane	−1.9	1.3

[a] J. G. Aston, *Discussions Faraday Soc.*, **10**, 73 (1951).

The objective of these calculations was admittedly different from ours. Both Pople and Segal[4] and Lipscomb, *et al.*,[9] were trying to devise some simple semiempirical MO procedure that would reproduce the results to be expected from an *a priori* Roothaan-type approach. The parameters were therefore chosen in such a way as to make the results of the two calculations agree for small molecules where *a priori* calculations have, or could, be made. This procedure of course ensured that the semiempirical treatments would give poor estimates of heats of formation, seeing that heats of formation calculated by the Hartree–Fock method are known to be very inaccurate.

Another check on our work is provided by the photoionization potentials measured for various hydrocarbons by Al-Joboury and Turner.[10] The ionization potentials of a molecule should, according to Koopman's theorem, be approximately equal to the calculated orbital energies; Table VI shows that this parallel exists in a remarkable way for a variety of hydrocarbons and for all measured ionization potentials up to about 19 ev. The photoionization spectra show numerous peaks in this region, due to the possibility of producing ions in vibrationally excited states; our calculations suggest that in several cases Turner, *et al.*,[10] may have mistaken multiple peaks as being due to different vibrational states of a single ion, rather than to two or more different ions of similar energy. Similar difficulties arise in attempts to correlate observed electronic spectra of molecules with calculated excitation energies.

The last column of Table VI shows orbital energies calculated by Palke and Lipscomb[9b] by an *a priori* SCF LCAO MO method, using the POLYATOM program. It will be seen that their orbital energies run parallel to ours but are in general greater; the correlation with measured photoionization potentials is clearly poor. Another comparison of this kind is provided by the population analyses shown in Table VII; here again our values run parallel to those given by the *a priori* procedure and also to those reported by Pople and Segal.

Our method predicts small dipole moments for several of the compounds studied; values are listed in Table VIII. The available experimental evidence (last column

(8) R. G. Parr and R. S. Mulliken, *J. Chem. Phys.*, **18**, 1338 (1950).

(9) M. D. Newton, F. P. Boer, and W. N. Lipscomb, *J. Am. Chem. Soc.*, **88**, 2353, 2361, 2367 (1966); (b) W. E. Palke and W. N. Lipscomb, *ibid.*, **88**, 2384 (1966).

(10) M. I. Al-Joboury and D. W. Turner, *J. Chem. Soc.*, 5141 (1963); 4434 (1964); 616 (1965).

Table VI. Comparisons of Ionization Potentials with Orbital Energies

Compound	Ionization potential, ev	Orbital energy, ev This paper	Ref 9b
Methane	12.99	13.88	14.74
Ethane	11.49	12.51	13.08
		13.04	13.38
	14.74	14.98	16.15
	19.18	20.80	22.51
Ethylene	10.48	10.86	10.09
		12.76	13.77
	12.50	12.94	15.28
	14.39	15.25	17.51
	15.63		
	19.13	19.18	21.28
Acetylene	11.36	11.06	11.03
	16.27	13.63	17.85
	18.33	18.08	20.44
Propane	11.07	12.01	
		12.49	
		12.85	
	13.17	13.73	
		13.90	
	15.17	14.68	
	15.70	15.46	
	18.57	19.67	
n-Butane	10.50	11.63	
		12.39	
	12.36	12.78	
		13.07	
		13.21	
		14.13	
	14.13	14.36	
		14.47	
	15.69	15.79	
Isobutane	10.78	11.88	
		12.54	
	12.54	13.48	
		13.79	
	14.51	14.59	
		15.40	
	18.63	18.68	
Cyclohexane	9.79	11.51	
	11.33	12.23	
		12.59	
	12.22	12.69	
		13.48	
	14.37	15.10	
		15.51	
Benzene	9.25	10.15	
	11.49	11.54	
	12.19	12.72	
		12.86	
	13.67	13.45	
	14.44	15.67	
	16.73	16.07	
	18.75	18.98	
1,3-Butadiene (*trans*)	9.08	10.16	
	11.25	11.70	
		11.83	
	12.14	12.58	
		13.09	
	13.23	14.39	
		14.71	
	18.78	17.99	
		19.24	

of Table VIII) suggests that the calculated moments are of the right order of magnitude.

Summary

While the results reported here are preliminary in nature,[11] they are sufficient to suggest that this kind of approach has exciting possibilities. It seems very likely that it may be improved to a point where heats of formation, etc., of molecules of all kinds may be predicted with an accuracy comparable with that already achieved for conjugated hydrocarbons, using the Hückel approximation. If so, the impact on chemistry would be considerable, for not only would one be able to calculate heats of formation and reaction with "chemical" accuracy, but it would also be possible to predict reaction mechanisms and rates of reaction.

Our results represent a considerable improvement over those of previously reported investigations. The main factors responsible for this seem to be the following: (a) our treatment of the internuclear repulsion as a parameter to allow for the effects of orbital contraction (if the repulsion is treated as one between point charges, the calculated heats of formation must inevitably be too small); (b) our use of different integrals for s and p AO's of a given center, together with a sufficient inclusion of integrals involving one-center differential overlap to make the calculations effectively invariant to choice of coordinate axes; (c) our use of thermochemical data to fix the parameters in our treatment, rather than the results of *a priori* calculations.

There are several obvious ways in which this general approach could be modified and extended, in particular the use of Hartree–Fock AO's in the calculation of over-

Table VII. Population Analyses for Hydrocarbons

Compound	Orbital	Population This paper	Ref 9b	Ref 3
Methane	H	1.077	0.876	0.965
	C2s	1.136	1.274	1.081
	C2p	0.852	1.088	1.020
Ethane	H	1.064	0.876	0.967
	C2s	1.199	1.248	1.042
	C2pσ	0.872	0.981	1.007
	C2pπ	0.865	1.074	1.044
Ethylene	H	1.033	0.860	0.954
	C2s	1.243	1.197	...
	C2pσ	0.844	1.013	
	C2pπ	0.845	1.072	...
	C2pπ*(a)	1.000	1.000	...
Acetylene	H	0.941	0.812	0.893
	C2s	1.263	1.105	
	C2pσ	0.797	1.086	...
	C2pπ	1.000	1.000	...

Table VIII. Calculated and Observed Dipole Moments of Hydrocarbons

Compound	Dipole moment, D. Calcd	Obsd
Propane	0.03	0.08[a]
Isobutane	0.05	0.13[b]
cis-2-Butene	0.08	...
Propyne	0.24	0.75[c]
cis-1,3-Butadiene	0.04	...

[a] D. R. Lide, *J Chem. Phys.*, **33**, 1879 (1960). [b] A. A. Maryott and G. Birnbaum, *ibid.*, **24**, 1022 (1956); D. R. Lide and D. E. Mann, *ibid.*, **29**, 914 (1958). [c] F. J. Krieger and H. H. Wenzek, *J. Am. Chem. Soc.*, **60**, 2115 (1938).

(11) For this reason we have not reported the results (*e.g.*, eigenvalues, eigenvectors, bond orders, etc.) in detail; we will be happy to communicate them to anyone interested.

lap integrals and the use of separate parameters for different types of bonds in place of the approximations of eq 35 and 46. We are studying these and other analogous possibilities, and we are also extending our treatment to include all integrals involving one-center differential overlap (NDDO approximation[5]) in case this should prove necessary in treating molecules containing heteroatoms.

III-2c Reprinted from *The Journal of the American Chemical Society* **90,** 59 (1968)

Application of the Pople–Santry–Segal Complete Neglect of Differential Overlap (CNDO) Method to Some Hydrocarbons and Their Cations[1]

Kenneth B. Wiberg

Contribution from the Department of Chemistry, Yale University, New Haven, Connecticut. *Received February 2, 1967*

Abstract: The CNDO method has been applied to hydrocarbons, hydrocarbon cation radicals, free radicals, and carbonium ions. If the original Pople parameters are modified somewhat, it is possible to calculate structures (both bond lengths and bond angles) which are in good agreement with experimental data. In addition, if the energies calculated are considered to be in arbitrary units, a semiempirical scaling factor may be determined which permits the heats of atomization and heats of formation to be satisfactorily calculated. The CNDO method appears to have considerable promise in the investigation of organic chemical phenomena.

Much of the recent progress in the study of unsaturated organic molecules has its origin in the molecular orbital calculations based on the π-electron approximation. A similar treatment of saturated compounds would allow an examination of a wider range of phenomena. A number of approaches to such a treatment have been suggested,[2-5] and one, the extended Hückel method,[2] has been applied to a number of cases. More recently, Pople, Santry, and Segal,[4,5] have developed an approximate SCF treatment (CNDO, complete neglect of differential overlap) which has the virtue of being independent of the coordinate system used. This paper will explore the application of the method to some hydrocarbons and to the cations derived from them.

The CNDO treatment begins with extended Hückel molecular orbitals derived from the 2s and 2p atomic orbitals on carbon and the 1s atomic orbital on hydrogen. The proper linear combinations of these orbitals are obtained by constructing and diagonalizing the Hückel secular determinant in which the on-diagonal elements are the average ionization potentials and the off-diagonal elements are made proportional to the overlap integral between the atomic orbitals. The bond orders and electron densities are calculated using the occupied molecular orbitals, and from this the approximate Hartree–Fock matrix is constructed in a fashion similar to that previously used by Pople for π-electronic systems.[6] Diagonalization of the latter leads to a new set of molecular orbitals. The procedure is repeated until there is no longer a significant change in the coefficients.

The hybridization at a given carbon is not specified at the beginning of the calculation; only the coordinates of the atoms, the average ionization potential of each type of atomic orbital used, the core charge, the β-proportionality constant, and the exponent for the Slater-type atomic orbital are required. The values suggested by Pople, *et al.*, are summarized in Table I. The final hybridization is determined by the geometry initially supplied and by the difference in the above parameters between 2s and 2p orbitals.

Table I. Parameters Used in CNDO Calculations

Orbital	Av ionization potential, V	Core charge	Slater exponent	β-Proportionality constant
C, 2s	14.051	4	1.625	21
C, 2p	5.572	4	1.625	21
H, 1s	7.1761	1	1.2	9

Pople and Segal[5] have shown that the CNDO procedure, although quite good for bond angles and charge distributions, does not give the correct energy or equilibrium bond length. The parameters used are not uniquely determined from any available data, and it would appear that modification of the β-proportionality constants and the carbon average ionization potentials could be made without doing violence to the general scheme.

It seemed appropriate first to determine the sensitivity of the calculated structures to variations in the parameters used. Ethane and ethylene were chosen as test cases and the equilibrium C—C and C═C bond lengths were obtained for a number of sets of parameters. For each calculation, each of the parameters was chosen within a given range by the use of a random number generator. Thus, there was no correlation between one set of parameters and another. The results of a number

(1) This investigation was supported by the U. S. Army Research Office, Durham.
(2) R. S. Mulliken, *J. Chim. Phys.*, **46**, 497, 675 (1949); M. Wolfsberg and L. Helmholtz, *J. Chem. Phys.*, **20**, 837 (1952); R. Hoffmann, *ibid.*, **39**, 1397 (1963); **40**, 2480 (1964); *J. Am. Chem. Soc.*, **86**, 1259 (1964); *Tetrahedron Letters*, 3819 (1965); J. A. Pople and D. P. Santry, *Mol. Phys.*, **7**, 269 (1964).
(3) G. Klopman, *J. Am. Chem. Soc.*, **86**, 4450 (1964); H. A. Pohl, R. Appel, and K. Appel, *J. Chem. Phys.*, **41**, 3385 (1964); J. J. Kaufman, *ibid.*, **43**, 5152 (1965).
(4) J. A. Pople, D. P. Santry, and G. A. Segal, *ibid.*, **43**, S129 (1965).
(5) J. A. Pople and G. A. Segal, *ibid.*, **43**, 5136 (1965); **44**, 3289 (1966).
(6) J. A. Pople, *J. Phys. Chem.*, **61**, 6 (1957).

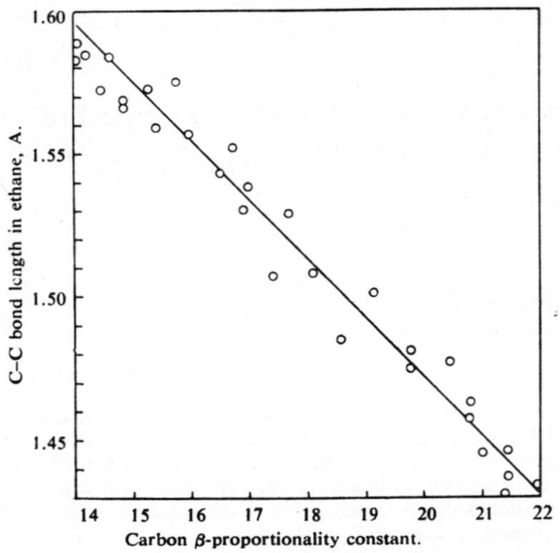

Figure 1. Effect of variation in parameters on the equilibrium C–C bond length for ethane. Although the lengths are plotted against the carbon β-proportionality constants, each of the five parameters were varied in each case. The scattering is due to the small effect of the other parameters on the bond length.

effect of a variation in the Slater exponent, Z, is also indicated. The two dotted lines indicate the observed bond lengths.

It is not possible to find a value of the constant which will exactly fit both the C—C and C═C bond lengths. However, a value may be chosen (17.5) which gives 1.51 A for the C—C length and 1.36 A for the C═C length; both are close to the experimental values (1.53 and 1.33 A).[7] It was not possible to significantly improve the fit by varying Z, and thus the original values were used in all of the remaining calculations.

Having values which give reasonable bond lengths, it was now of interest to try to obtain a linear correlation between calculated and observed energies. The calculated values are the total energies for all bonding electrons. In order to convert the energies to those which may be compared with experimental data, they were converted to heats of atomization by subtracting the energies of the appropriate number of hydrogen atoms and carbon atoms. These are obtained in the same way as are the energies of the molecules being considered. Similarly, the heats of atomization may be calculated from the heats of formation knowing the heats of formation of carbon and hydrogen atoms.[8]

The calculated energies do not include zero-point

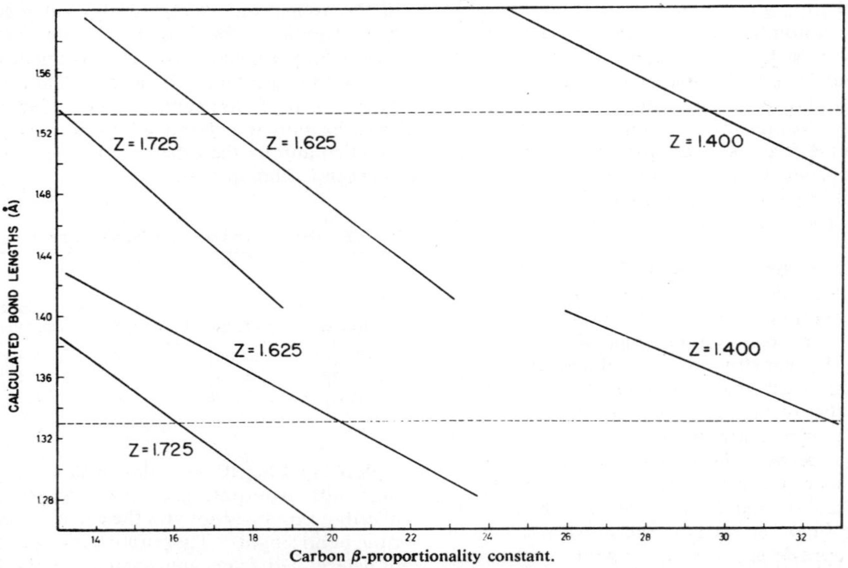

Figure 2. Effect of the carbon β-proportionality constant on the equilibrium bond lengths for ethane (upper lines) and ethylene (lower lines).

of such calculations are indicated in Figure 1. Here, the equilibrium C—C bond length in ethane is plotted as a function of the carbon β-proportionality constant. It can be seen that this constant largely determines the bond length, and that the other parameters (each of which was varied in each calculation) have relatively little effect.

The effect of the β-proportionality constant itself is shown in Figure 2. Here, the other parameters are kept constant, and the equilibrium bond lengths for ethane and ethylene are plotted against the constant. In each case, a very good linear relation is found. The energies. In order to make a valid comparison, it is necessary to use the heats of formation calculated for 0°K and to correct for zero-point energies.[9] Although the calculated energies are in atomic units, we shall for the present purpose consider them to be in arbitrary

(7) *Cf.*, D. R. Linde, Jr., *Tetrahedron*, **17**, 125 (1962).

(8) "Selected Values of Chemical Thermodynamic Properties," National Bureau of Standards Circular 500, U. S. Government Printing Office, Washington, D. C., 1952.

(9) The vibrational frequencies were taken from R. G. Snyder and H. J. Schachtschneider, *Spectrochim. Acta*, **19**, 85 (1963).

Table II. Effect of a Variation in the MO Parameters on the Geometry and Energies of Methane, Ethane, and Ethylene

					Methane	Ethane			Ethylene			$\Delta H/\Delta E$				Ethane
β_H	β_C	I(1s)	I(2s)	I(2p)	R(CH)	R(CH)	R(CC)	HCC	R(CH)	R(CC)	HCC	Methane	Ethane	Ethylene	R	H chg
9.0	21.0	7.176	14.051	5.572	1.114	1.121	1.452	112.1	1.113	1.306	124.3	228.14	213.84	213.96	0.880	1.002
9.0	17.5	7.176	14.051	5.572	1.141	1.146	1.516	111.5	1.136	1.366	123.3	279.87	270.76	281.46	1.254	1.000
10.0	17.5	7.176	14.051	5.572	1.131	1.137	1.520	111.4	1.128	1.367	123.1	263.23	257.08	268.76	1.345	0.998
10.0	17.5	6.5	14.051	5.572	1.131	1.136	1.523	111.2	1.127	1.369	123.0	247.40	243.78	256.27	1.443	0.975
10.0	17.5	7.176	13.0	5.572	1.130	1.135	1.516	111.4	1.126	1.364	123.3	257.04	250.02	259.65	1.266	1.002
10.0	17.5	7.176	14.051	6.0	1.131	1.135	1.520	111.3	1.127	1.367	123.1	260.37	254.02	264.76	1.312	0.985
9.0	17.5	6.151	11.445	7.257	1.131	1.136	1.509	111.4	1.127	1.359	123.2	230.42	222.97	227.99	1.114	0.928
9.0	17.5	6.5	11.445	7.257	1.131	1.137	1.508	111.5	1.127	1.358	123.2	237.70	228.86	233.18	1.071	0.940
9.0	17.0	6.151	11.445	7.257	1.135	1.139	1.518	111.4	1.130	1.367	123.0	236.68	230.10	236.54	1.174	0.927
9.0	17.5	6.151	11.75	7.257	1.131	1.136	1.510	111.4	1.128	1.359	123.1	231.84	224.62	230.07	1.132	0.927
9.0	17.5	6.151	11.455	7.60	1.130	1.135	1.509	111.4	1.126	1.358	123.1	228.21	220.72	225.30	1.099	0.918
10.0	17.5	6.151	11.445	7.257	1.122	1.127	1.513	111.3	1.119	1.360	123.1	218.48	213.20	219.25	1.190	0.928
10.0	17.5	7.176	11.0	7.4	1.122	1.127	1.508	111.5	1.119	1.358	123.3	235.37	226.08	229.43	1.034	0.960
10.0	17.5	7.176	11.445	7.257	1.123	1.128	1.510	111.5	1.120	1.359	123.2	238.58	229.62	233.78	1.064	0.962
10.0	17.5	7.176	11.3	7.4	1.123	1.128	1.509	111.5	1.119	1.358	123.2	236.94	227.84	231.61	1.050	0.958
10.0	17.5	7.176	10.0	6.5	1.123	1.128	1.506	111.5	1.119	1.356	123.3	235.51	225.83	229.01	1.024	0.990
10.0	17.5	7.176	10.3	6.4	1.124	1.129	1.506	111.5	1.120	1.357	123.3	237.67	228.22	231.95	1.045	0.991
10.0	17.5	7.176	10.1	6.4	1.124	1.128	1.506	111.5	1.119	1.356	123.3	236.60	227.03	230.48	1.034	0.992

units, and we shall attempt to determine a conversion factor from these units to kilocalories/mole.

The results of a number of calculations are summarized in Table II. Here, the parameters on which the calculated energies depend are systematically varied and the equilibrium geometry is calculated for methane, ethane, and ethylene. In addition, the ratio of the experimental and calculated heats of atomization are given. The parameter R refers to the calculated heats of atomization and is given by

$$R = \frac{\Delta E(\text{ethane}) - \Delta E(\text{ethylene})}{\Delta E(\text{ethylene}) - \Delta E(\text{methane})}$$

The experimental value of this ratio is 1.034. Finally, the table also gives the calculated charge on a hydrogen in ethane.

The first set of parameters in the table are the original Pople values. A decrease of the carbon β-proportionality constant to 17.5 results in a reasonable carbon–carbon bond length. However, the value of R is not satisfactory. Small changes in the other parameters do not lead to a markedly improved value of R. All of the average ionization potentials were then considerably varied giving the second group of parameters. Here, R is considerably more satisfactory. The data suggest that there is little value in modifying the average ionization potential for hydrogen, and, since this is an experimentally determined quantity, it was returned to its original value.

The values of the β-proportionality constants were fixed by the necessity of fitting the experimental structural data. The values of the carbon ionization potentials were then set by requiring that the correct value of R be obtained, and that the charge on hydrogen be close to unity. This led to the final set of parameters in the table.

Although the correct value of R is obtained, the ratios of the observed and calculated heats of atomization are not constant, indicating that if one is plotted against the other, a small intercept would be found. It did not appear possible to eliminate the intercept.

In order to see if the new set of parameters was reasonable, the equilibrium C—C bond length and the equilibrium C—C—C and H—C—H bond angles were calculated for propane, benzene, acetylene, and allene (Table III). In each case, there is good agreement between the calculated and observed values. The bond angles agree within about 1°, and the bond lengths agree to within about 0.03 A. A further increase in the value of the hydrogen β-proportionality constant might be desirable. It should be noted that the changes in the parameters which resulted in improved agreement with the observed bond lengths also gave an improved agreement with the observed bond angles.

Table III. Observed and Calculated Geometrical Parameters

		Length, A			Value, deg		
Compd	Bond	Calcd	Obsd	Angle	Calcd	Obsd	Ref
Methane	C—H	1.124	1.106				a
Ethane	C—H	1.128	1.107	H-C-C	111.5	110.5	b
	C—C	1.506	1.536				
Propane	C—C	1.513	1.526	C-C-C	113.3	112.4	c
				H-C-H	105.6	106.1	
Ethylene	C—H	1.120	1.084	C-C-H	123.3	122.3	d
	C=C	1.356	1.332				
Benzene	C—C	1.423	1.397				e
Acetylene	C—H	1.103	1.059				f
	C≡C	1.239	1.205				
Allene	C—H	1.120	1.082	C-C-H	122.8	120.2 121.5	g
	C=C	1.342	1.312				

[a] L. S. Bartell, K. Kuchitsu, and R. J. de Neui, *J. Chem. Phys.*, **35**, 1211 (1961). [b] A. Almenningen and O. Bastiansen, *Acta Chem. Scand.*, **9**, 815 (1955). [c] D. R. Linde, Jr., *J. Chem. Phys.*, **33**, 1514 (1960). [d] L. S. Bartell and R. A. Bonham, *ibid.*, **31**, 400 (1959). [e] A. Almenningen, O. Bastiansen, and L. Fernholt, *Kgl. Norske Videnskab. Selskabs Skrifter*, No. 3 (1958); A. Langseth and B. P. Stoicheff, *Can. J. Phys.*, **34**, 350 (1956). [f] J. H. Callomon and B. P. Stoicheff, *ibid.*, **35**, 373 (1957). [g] A. Almenningen, O. Bastiansen, and M. Traetteberg, *Acta Chem., Scand.*, **13**, 1699 (1959).

In comparing the experimental and observed atomization energies, the bond lengths were rounded off to the nearest 0.01 A, and the parameters observed for propane were assumed also to apply to n-butane. The energies were calculated, and it was found advantageous to make an additional small correction in the carbon average ionization potentials (2s = 10.3, 2p = 6.3). The values thus obtained are summarized in Table IV. A

Table IV. Calculated and Observed Energies[a]

Compd	ΔH_f, kcal/mole	ΔH (atom.)	Zero-pt energy	Classical ΔH (atom.)	E, au	ΔE (atom.)[b]	Calcd ΔH (atom.)[c]
Methane	−15.99	392.86	27.11	419.97	−9.8155	1.7629	419.61
Ethane	−16.52	667.02	45.18	712.20	−18.1617	3.1114	713.60
Propane	−19.48	943.61	62.46	1006.07	−26.5048	4.4568	1006.91
Butane	−23.67	1221.43	79.77	1301.20	−34.8406	5.7949	1298.62
Ethylene	14.52	532.74	30.70	563.44	−16.4167	2.4213	563.14
Benzene	24.00	1308.06	62.84	1370.90	−44.9582	6.1367	1373.14
Allene	44.70	672.95	33.23	706.18	−23.0351	3.0969	710.44
Acetylene	54.33	389.69	16.18	406.50	−14.6883	1.7479	416.33

[a] I(2s) = 10.3 ev, I(2p) = 6.3 ev. [b] E(C) = −5.9428, E(H) = −0.5275. [c] ΔH (atom.) = 218.01 × ΔE (atom.) + 35.28.

Table V. Structures and Energies of Hydrocarbon Derivatives

Species	r (CH)	r (CC)	A (CCH)	A (CCC)	E, au	ΔE (atom.)	ΔH_f (0°K), ev	ΔH (atom.), ev	Zero-pt energy, ev	Class. ΔH (atom.)	Calcd ΔH
					A. Radical Cations						
CH_4^+	1.151				−9.1487	1.6236	12.29[a]	17.64	1.18	18.82	18.84
$C_2H_6^+$	1.144	1.435	114.8		−17.6101	3.0873	10.93	30.86	1.92	32.78	32.72
$C_3H_7^+$	(1.14)	1.475		121.0	−26.0242	4.5037	10.23	43.42	2.71	46.13	46.14
					ΔH (atom.) = 9.48ΔE + 3.45 ev (9.48 ev = 218.7 kcal/mole)						
					B. Radicals						
$CH_3\cdot$	1.127				−8.8581	1.3329	33.4[b]	291.9	20.0	311.9	311.7
$C_2H_5\cdot$	1.118	1.468	115.5		−17.2311	2.7083	27.9	571.0	38.1	609.1	609.4
2-$C_3H_7\cdot$	1.122	1.480		128.3	−25.5941	4.0736	22.9	849.6	55.4	905.0	904.8
					ΔH (atom.) = 216.4ΔE + 23.2 kcal/mole						
					C. Cations						
CH_3^+	1.128				−8.3849	1.3873	263[c]	376	20	396	396
$C_2H_5^+$	1.126	1.428	114.1		−16.8364	2.8411	230	682	38	720	721
2-$C_3H_7^+$	1.128	1.456		128.3	−25.2431	4.2501	205	981	55	1036	1036
					ΔH (atom.) = 223.6ΔE + 85.5 kcal/mole						

[a] The values are based on the ionization potentials of the hydrocarbons (R. I. Reed, "Ion Production by Electron Impact," Academic Press Inc., London, 1962) and the heats of formation of the hydrocarbons. [b] Kcal/mole; values based on the C–H bond dissociation energies (T. L. Cottrell, "The Strengths of Chemical Bonds," Academic Press Inc., New York, N. Y., 1954) and the heats of formation of the hydrocarbons. [c] The values are based on the ionization potentials of the free radicals (F. P. Lossing, P. Kebarle, and J. B. DeSousa, *Advan. Mass Spectry.*, **1**, 431 (1959)).

least-squares fit of calculated and observed atomization energies for methane through benzene gave

$$\Delta H\ (\text{atom.}) = 218.01 \times \Delta E\ (\text{atom.}) + 35.28$$

The last column gives the heats of atomization calculated form this equation, and it can be seen that the fit is quite good. The heats of atomization of allene and acetylene were also calculated using the same equation. The result was satisfactory for allene, but was in error by 2.5% for acetylene.

The calculations summarized above suggest that it should be possible to calculate the geometries and heats of atomization of most alkanes and alkenes with only a relative small error. The fact that the energy of benzene was correctly calculated is particularly significant since this indicates the method can accommodate the energy changes due to electron delocalization.

It is quite possible that since the parameters were adjusted to fit data on hydrocarbons, they may not be appropriate for free radicals, cations, and similar species. This was tested by calculating the equilibrium geometry and energy for a group of radical cations, free radicals, and cations, giving the data shown in Table V. Both the radicals and cations were found to prefer the planer conformation. Little is known about the detailed structures of these molecules; however, the calculated bond lengths and angles appear reasonable. When the experimental energies were compared with the cal- culated energies, in each case the slope of the line was close to 220 kcal/mole/energy unit. Considering the errors associated with measuring the heats of formatio of these species, and the fact that the ionization potentia measurements lead to nonequilibrium geometries fo the cations, the slopes are the same as that for the hydro carbons within the uncertainty associated with the experi mental measurements.

The intercept did vary considerably. The hydrocar bons and the free radicals gave about the same intercept and the radical cations and cations gave about the sam intercept. Thus, the magnitude of the intercept appear to be associated with the charge on the molecule. I can be seen that the CNDO method should be useful i estimating equlibrium geometry, and the difference i energy between molecules of the same charge type. I should be noted that although the Pople paramete were adjusted to give a reasonable fit to the experiment data, no attempt was made to determine the set of pa rameters which would give a "best fit" to a body of dat It is probable that the accuracy of the predictions coul be further improved by small variations in the param ters.

Although the CNDO procedure appears very usefu one should not gain the impression that it is free fro difficulties. The energies of eclipsed and staggered ethanes were calculated giving a difference of 0.0034. Using the conversion factor obtained herein, this is 0.7 kcal/mole whereas the observed value is 3 kcal/mole. Thus, for cycloalkanes and similar molecules, the energy due to the torsional barrier must be added to the calculated energy. Further, in connection with the data in Table V, the energy of the ethylene ion-radical was calculated. Unlike all the other cases, two minima were found. The first had r (CH) = 1.139, r (CC) = 1.300, A (CCH) = 138.0 and ΔE (atom.) = 2.4114. The second had r (CH) = 1.127, r (CC) = 1.450, A (CCH) = 121.7 and ΔE (atom.) = 2.4293. Using these energies, one calculates a ΔH (atom.) which is 1 ev too low. It is possible that the two conformations and the error in calculated energy are related, and that, for example, the ethylene radical cation is nonplanar. The vinyl radical and cation were not considered since the energies of these species appear not to be well established.

Experimental Section

The energies were calculated using the program written by G. A. Segal which was made available through the Quantum Chemistry Program Exchange at Indiana University. Initially, the program was modified so that the carbon and hydrogen β-proportionality constants and average ionization potentials were chosen within a predetermined range (about $\pm 20\%$ of the original value) using a random number generator. For each set of parameters, the energies of three ethane and three ethylene structures differing in C–C bond length, were calculated and the equilibrium geometry was obtained assuming a parabolic function. The data thus obtained are shown in Figure 1.

The data shown in Table II were obtained by calculating the energies of a number of structures for each molecule. For methane, three C–H bond lengths were used, and a parabolic function was assumed. For ethane and ethylene, three C–H bond lengths, three C–C bond lengths, and three C–C–H bond angles were taken. This set leads to 27 structures. Initially, the energies of each of the structures was obtained and the energies were fitted to a general quadratic function of the three geometrical parameters. A very good fit was obtained. From this, the equilibrium geometry and energy were easily obtained. The fitting procedure was repeated using only the odd-numbered structures, giving a set of 14. No significant difference in result was noted, and subsequent calculations were based on a set of 14 structures.

The data in Table III were obtained in a similar fashion, using the parameters on the last line of Table II. In the case of propane, the methyl group geometry found in ethane was assumed, and only the geometry about the central carbon was varied. The C–H bond length was taken as 1.12 A. The energies shown in Table IV were obtained using β_H = 10.0, β_C = 17.5, I (1s) = 7.176, I (2s) = 10.3, and I (2p) = 6.3. The CH bond lengths were taken as 1.12 A except for acetylene for which 1.10 A was used. The C–C lengths were rounded to the nearest 0.01 A, and the calculated bond angles were used. Butane was assumed to have the geometrical parameters calculated for propane. The heat of formation of carbon was taken as 170.39 kcal/mole and that for hydrogen was 51.62 kcal/mole.

The results summarized in Table V were obtained in the same fashion as that of Table III. The heats of atomization were calculated for the process leading to a neutral carbon atom, hydrogen atom, and, in the case of the cations, a proton. The heat of formation of the latter was taken as 365.14 kcal/mole. The zero-point energies of the radical cations were assumed to be the same as that for the hydrocarbons. The difference in zero point energy between a hydrocarbon and its radical cation might be expected to be roughly constant, leading to relatively little error in the calculated slope. The radicals were assumed to have a zero-point energy equal to that of the parent hydrocarbon less one C–H stretching vibration (3000 cm^{-1}) and two bending vibrations (1000 cm^{-1}). The cations were assumed to have the same zero-point energy as the radicals.

III-2d Reprinted from *The Journal of the American Chemical Society* **88**, 13 (1966)

Molecular Binding Energies[1]

C. Hollister[2] and O. Sinanoğlu[3]

Contribution from the Sterling Chemistry Laboratory, Yale University, New Haven, Connecticut. *Received, August 23, 1965*

Abstract: The effect of the correlation energy on dissociation energies and binding energies of molecules is studied. Molecular correlation energies are obtained semiempirically using only simple molecular orbital calculations and atomic correlation energies. Calculations are done for homonuclear and heteronuclear diatomic molecules and for polyatomic molecules, including some hydrocarbons. The agreement with available experimental data is good. "Experimental" Hartree-Fock molecular orbital energies are predicted. The fractional contribution of correlation to the binding energy is examined and systematic trends are observed, which allow prediction of the binding energy where it is not known.

In the past few years, noteworthy progress in calculation of energies of molecules and atoms has been made. The availability of accurate total energies raises the question of determining binding energies of various types: the dissociation energy, D_e, of diatomic molecules, the energy of atomization of polyatomic molecules, bond-dissociation energies, and energies of fragmentation. Independent calculations of such quantities would be useful for a quantum theoretical basis of bond energies, intramolecular, nonbonded forces, etc.[4a] This is especially important for transient species, radicals, and other species for which the experimental determination of binding energies is difficult. This would also enable one to predict whether a species will exist, as in high-energy fuel and oxidizer research.

Molecular orbital (MO) theory is most widely used for energy calculations. The prototype for this sort of calculation is the Hartree-Fock (H.F.) MO approach. For practical use on larger molecules, we must approximate the H.F. wave function by constructing the MO's as linear combinations of atomic orbitals (LCAO-MO theory) or use a still more approximate method such as the extended Hückel MO (HMO) methods.[5] Sufficient progress has now been made so that these methods are applicable to many systems.[6] Approximate MO treatments are adequate for estimating charge distributions and total energies. However, binding energies are small differences between two large quantities, and in order to estimate them we need to include correlation between electrons, which even the H.F. treatment omits. Fortunately, the correlation energy is insensitive to the exact details of the orbitals.[4,7] Thus, using simple MO's, we can get the correlation energy quite accurately. Theory[4,7] and calculations[8,9] for atomic correlations have shown which effects are physically important. With these results for guidance, we can get molecular correlation energies rather simply.

Two semiempirical methods for estimating molecular correlation energies from atomic data are developed here and applied to molecules of various types, *e.g.*, F_2, C_2N_2, and C_6H_6. For some diatomic molecules, the results have already been compared with H.F. calculations.[10,11] For larger molecules, the results give a sizable portion of the energy of atomization and predict "experimental" H.F. energies to which H.F. calculations may be compared. Systematic trends are observed from which dissociation energies and binding energies may be estimated.

The Effect of Electronic Environment on Correlation

H.F. MO theory treats each electron in a molecule as if it were moving in the average field of all the other electrons. Such a procedure neglects instantaneous collisions between electrons, and this is the source of the energy defect known as the correlation energy. The H.F. wave function is antisymmetric in accordance with the Pauli principle, and this keeps electrons with the same spin apart. However, two electrons with opposite spin may occupy the same space orbital. The repulsion between electrons in the same region of space raises the calculated energy. The correlation part of the wave function has been studied extensively[4,7] and the contribution of pair correlations and three- and many-body correlations to the energy have been examined.[4,8,9] It was found that mainly pair correlations are significant for closed shell systems. The correlation energy, E_{corr}, may be written (for a single determinantal state)

(1) This paper is based on the dissertation submitted by C. Hollister in partial fulfillment of the requirements for the Ph.D. degree to Yale University, Sept. 1965.

(2) U. S. Public Health Service Trainee, Grants 748-04 and 748-05, 1963-1965.

(3) Alfred P. Sloan Fellow.

(4) (a) See, for example, O. Sinanoğlu, *Advan. Chem. Phys.*, **6**, 315 (1964). (b) For a review of recent quantum chemical methods and work in the field see O. Sinanoğlu and D. F. Tuan, *Ann. Rev. Phys. Chem.*, **15**, 251 (1964).

(5) For a review, see, for example, K. Fukui in "Modern Quantum Chemistry—Istanbul Lectures, Part I, Orbitals," O. Sinanoğlu, Ed., Academic Press Inc., New York, N. Y., 1965. For a discussion of the relationship of various MO methods, see O. Sinanoğlu, *J. Phys. Chem.*, **66**, 2283 (1962).

(6) See, for example, the ARPA symposium on fuel and oxidizer research, 149th National Meeting of the American Chemical Society, Detroit, Mich., April 1965.

(7) O. Sinanoğlu, *J. Chem. Phys.*, **36**, 706, 3198 (1962).

(8) O. Sinanoğlu and D. F. Tuan, *ibid.*, **38**, 1740 (1963).

(9) V. McKoy and O. Sinanoğlu, *ibid.*, **41**, 2689 (1964); D. F. Tuan and O. Sinanoğlu, *ibid.*, **41**, 2677 (1964); some pair correlations are also available for second-row atoms: see V. McKoy and O. Sinanoğlu in "Modern Quantum Chemistry—Istanbul Lectures, Part II, Interactions," O. Sinanoğlu, Ed., Academic Press Inc., New York, N. Y., 1965.

(10) C. Hollister and O. Sinanoğlu in "Modern Quantum Chemistry–Istanbul Lectures," ref. 9.

(11) Extensive calculations of H.F. wave functions for diatomic molecules are presently underway at the Laboratory of Molecular Structure and Spectra of the University of Chicago. References to calculations on specific molecules are given in Tables II and III. The "calcd." H.F. results kindly supplied by the Chicago group are tentative and should be treated as such. Values in Tables II and III should be compared with the final H.F. values published.

$$E_{corr} \cong \sum_{i>j}^{N} \epsilon_{ij} \quad (1)$$

where ϵ_{ij} is the energy contribution due to the correlation of two electrons in MO spin orbitals i and j, respectively. Qualitatively one can easily see why eq. 1 is a good approximation. The effect of antisymmetry is to surround each electron with a Fermi hole which no electron of the same spin can penetrate. It has been shown[7] that the "fluctuation potential" between a pair of electrons is expected to be appreciable only over a range which is smaller than the extent of the Fermi hole. But with three or more electrons, at least two of them are parallel so that they cannot all get within the range of one another's correlation potentials simultaneously. For a review of the formalism and the manner in which these two-, three- and many-electron effects arise, see ref. 4a.

Each pair correlation may be treated independently. This means molecular or atomic correlation can be built up from individual ϵ_{ij}'s. We must now examine the effect of environment on the pair energies, *viz.*, exclusion effects and the fields of the other electrons. The pair energies may be determined by the variational method.[4a] The variational expression for ϵ_{ij} is

$$\tilde{\epsilon}'_{ij} = [2\langle B(ij), g_{ij}\hat{u}_{ij}\rangle + \langle \hat{u}_{ij}(e_i + e_j + m_{ij})\hat{u}_{ij}\rangle]/[1 + \langle \hat{u}_{ij}, \hat{u}_{ij}\rangle] \quad (2)$$

with

$$\tilde{\epsilon}'_{ij} \rightarrow \epsilon_{ij} = \langle B(ij), g_{ij}\hat{u}_{ij}\rangle$$

at the minimum and where i and j are the H.F. spin orbitals of the two electrons which are correlating, B is the two-electron antisymmetrizer, $\hat{u}_{ij}$ is the pair function which represents the correlation and is to be determined, e_i is the H.F. operator minus the ith orbital energy, $g_{ij} = 1/r_{ij}$, and m_{ij} is the fluctuation potential. The trial pair function is varied, subject to the conditions

$$\langle \hat{u}_{ij}(\mathbf{x}_i, \mathbf{x}_j), k(\mathbf{x}_i)\rangle_{x_i} = \int \hat{u}_{ij}(\mathbf{x}_i, \mathbf{x}_j)k(\mathbf{x}_i)d\mathbf{x}_i = 0 \quad (3)$$

for all occupied orbitals, k, until a minimum for eq. 2 is found, *i.e.*, until $\delta\epsilon_{ij}[\hat{u}_{ij}] = 0$. The other electrons affect the correlation of a given pair in two ways: the H.F. operator appearing in e_i and e_j contains the average potential of the electrons in all the occupied orbitals, and eq. 3 requires that the pair function be orthogonal to all the occupied orbitals. This orthogonality is indicated by the caret above $\hat{u}_{ij}$ and reflects the "exclusion effect."[12]

Calculations on first-row atoms show[9] that pair energies may be roughly divided into two types: dynamical correlations which are transferable from system to system, and nondynamical correlations which are not. An example of a transferable pair energy is $\epsilon(1s^2)$. One may take the value of $\epsilon(1s^2)$ calculated for the He-like ion of an atom over into the neutral atom, because both types of environmental effects are small. However, $\epsilon(2s)^2$ is not transferable, so that $\epsilon(2s^2)$ determined for the Be-like ion of an atom may not be taken over for use in the neutral atom.[9] In the Be-like ion, $\epsilon(2s^2)$ is almost entirely due to excitations of the 2s electrons to the nearby 2p orbitals (2s and 2p are exactly degenerate in the limit of infinite Z). (For the many references to the $1s^22s^2$ case in the literature see ref. 4b.) However, $\hat{u}_{ij}$ must be orthogonal to all occupied orbitals, as given by eq. 3. For Be-like ions, $\hat{u}_{ij}$ will contain all the 2p orbitals, but in the neutral atom which has some of the 2p orbitals occupied, $\hat{u}_{ij}$ cannot contain the occupied 2p orbitals. Thus, $\epsilon(2s^2)$ will decrease as one adds 2p electrons to a Be-like ion. One would expect that the correlation energies of transferable pairs, such as $1s^2$, might be taken over into the molecule without much change. However, nontransferable pairs must be treated differently.

Consider the nitrogen atom (^4S) with the ground-state configuration, $1s^22s^22p_+^{\alpha}2p_0^{\alpha}2p_-^{\alpha}$. $\epsilon(2s^2)$ determined from a configuration interaction (CI) function will be zero as all the 2p orbitals are occupied. Using the LCAO method for nitrogen molecule, one can construct ten space orbitals from the atomic orbitals, 1s, 2s, and 2p. The ground-state configuration of N_2 ($^1\Sigma_g^+$) is $1\sigma_g^21\sigma_u^22\sigma_g^22\sigma_u^23\sigma_g^21\pi_u^4$, leaving $1\pi_g$ and $3\sigma_u$ unoccupied. One might expect to find some "nondynamical" correlation in N_2 associated with the low-lying unoccupied $1\pi_g$ and $3\sigma_u$ orbitals, which will be analogous to unoccupied 2p orbitals in the atom. To test this idea, a 2 × 2 CI was performed on N_2 using the orbitals determined by Ransil.[13] The pair function chosen was

$$\hat{u}_{ij} = c(2)^{-1/2}[B(rs) + B(\bar{s}\bar{r})] + c'B(pq) \quad (4)$$

where

$$r = 1\pi_{g+}^{\alpha},\ s = 1\pi_{g-}^{\beta},\ \bar{r} = 1\pi_{g+}^{\beta},\ \bar{s} = 1\pi_{g-}^{\alpha},\ p = 3\sigma_u^{\alpha},\ \text{and}\ q = 3\sigma_u^{\beta} \quad (5)$$

The variational coefficients, c and c', were determined using eq. 2. Three calculations were performed, taking (ij) of eq. 2 as combinations of $2\sigma_g$, $3\sigma_g$, and $2\sigma_u$ with α or β spins.

It is always possible to find a unitary transformation, t, which will express the localized "core" orbitals in terms of the molecular orbitals (MO's) without changing the energy. The transformation used here is that given by Peters.[14] If there were no mixing in of $2p\sigma$, the transformation would give atomic 2s orbitals exactly since $(\sigma_g2s)^2(\sigma_u2s)^2$ will transform to $(2s_a)^2(2s_b)^2$, where a and b number the nuclei. However, t including $3\sigma_g$ is approximately equal to t for equivalent orbitals, so that the amount of $2p_\sigma$ in the localized orbital (LO) is slight. Thus, $(LO_a)^2$ is very close to $2s_a^2$ of the free atom. The pair energy will then transform as[4,7]

$$\epsilon_{\rho\nu} = \langle \sum_{\nu\pm\rho} t_{\rho i}t_{\nu j}B(ij), g_{ij} \sum_{\nu\pm\rho} t_{\rho i}t_{\nu j}\hat{u}_{ij}\rangle \quad (6)$$

where $\rho\nu$ represent LO^2 and i,j run over $2\sigma_g$, $3\sigma_g$, and $2\sigma_u$. $\epsilon(LO^2)$ per atom in N_2 is computed to be -0.3 e.v. The core energy is lowered by about 0.3 e.v. in the molecule compared to the corresponding core of the free atom. Thus the effect contributes about 0.6 e.v. to the binding energy.

The above calculation indicates the magnitude of the exclusion effect. For more accurate values of the

(12) O. Sinanoğlu, *J. Chem. Phys.*, 33, 1212 (1960).

(13) B. J. Ransil, *Rev. Mod. Phys.*, 32, 245 (1960).

(14) D. Peters, *J. Chem. Soc.*, 2003 (1963).

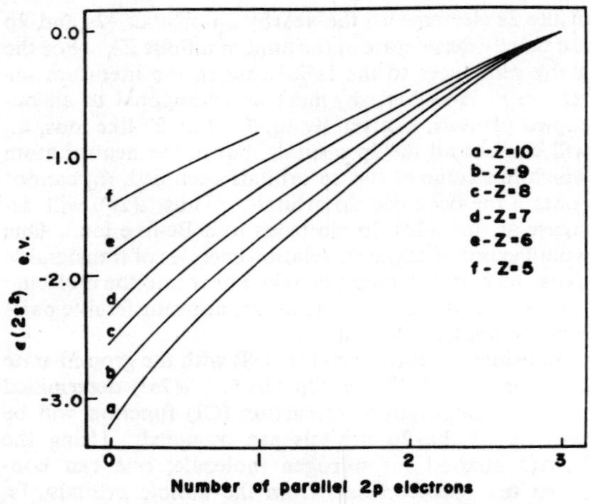

Figure 1. Dependence of $\epsilon(2s^2)$ on the occupation of the 2p shell.

correlation energy, CI with more complete H.F. orbitals may be needed. However, a much simpler method, which does not require any molecular H.F. results, suggests itself for estimates.

For the different ionized species of a given atom, one may plot $\epsilon(2s^2)$ *vs.* the number of parallel spin p electrons, using values of $\epsilon(2s^2)$ given in ref. 9. Figure 1 shows the data for first-row atoms and ions. This allows graphical interpolation to determine $\epsilon(2s^2)$ for nonintegral p occupation numbers. Nonintegral p occupation often arises in molecules owing to the displacement of electrons into bonding regions, and the p occupation in the molecule will generally be less than in the corresponding free atom. Such an interpolation for N_2 gives an $\epsilon(2s^2)$ per atom of -0.5 e.v., about the same magnitude as that determined by the CI calculation above. Interpolated values are used in the rest of this work where $\epsilon(2s^2)$ is required.

Two semiempirical methods for estimating molecular correlation energies from atomic data are described below. Method I uses only the total correlation energy (E_{corr}) of atoms and ions. Total atomic or ionic values of E_{corr} are from Clementi.[15] Method II uses atomic pair correlation energies,[9] modified where necessary by environmental effects. Table I gives values of ϵ_{ij}

Table I. Pair Correlation Energies for First-Row Atoms

Z	$\epsilon(1s^2)$,[a] e.v.	$\epsilon(1s \to 2s)$,[b] e.v.	$\epsilon(2s \to 2p)$,[b] e.v.	$\epsilon(2p_z{}^2)$,[b] e.v.
3	1.184			
4	1.205	0.058		
5	1.219	0.105	0.379	
6	1.227	0.099	0.579	1.0
7	1.233	0.075	0.680	1.0
8	1.238	0.107	0.780	1.0
9	1.241	0.110	0.762	1.0

[a] Data from ref. 15. [b] Data from ref. 9.

for first-row atoms and ions. Table I together with Figure 1 [for interpolated values of $\epsilon(2s^2)$] provides all the data necessary for estimating correlation using method II in molecules containing only first-row atoms.[9]

(15) E. Clementi, *J. Chem. Phys.*, **38**, 2248 (1963); **39**, 175 (1963).

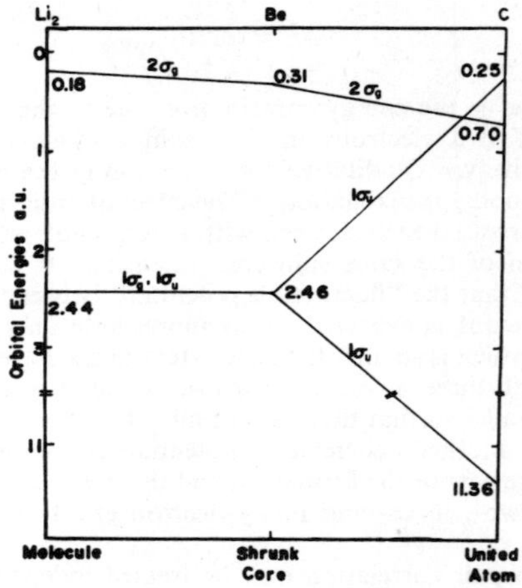

Figure 2a. Orbital energies in united atom, shrunk-core atom' and molecule. Li_2 molecule.

Method I. The "Shrunk-Core" Model

In early work, diatomic molecules were studied by considering two limiting atomic cases: the united atom and the separated atoms. MO diagrams[16] were constructed connecting related atomic orbitals of the two atomic limits with the orbitals of the diatomic molecule lying between the two extremes. Historically, this approach developed from H_2^+. The united atom model is reasonable for diatomic hydrides and has been used by Stanton[17] for estimating dissociation and correlation energies of such molecules. The correspondence between MO energies and orbital energies of the united atom is good in the case of hydrides, but large differences occur if both atoms in a diatomic molecule have core electrons. As the molecule contracts until the nuclei unite, we must consider the repulsion of the inner electronic shells of the constituent atoms. Consequently, while the $\sigma_g 1s$ orbital of the molecule goes over, in the united-atom limit, to 1s, and the $\sigma_g 2s$ goes to 2s, the $\sigma_u 1s$ goes to $2p_\sigma$.

An examination of the coefficients in an LCAO treatment of a molecule indicates that the lowest MO's are made up chiefly of atomic core electrons centered about each nucleus; their orbital energies bear little resemblance to the corresponding ones in the united atom. For a general molecule, there are two factors which will affect the orbital energy. The first is the effective nuclear charge seen by the electrons, which will be important mainly for the 1s electrons. The other is the exclusion effect of the inner electrons (resulting in the orthogonality requirements on the orbitals), keeping the outer electrons at a certain distance from the nucleus. In the united-atom model, the innermost electrons see too high an effective Z in comparison to what they would see in the molecule if they are centered about their respective nuclei; and since the inner electrons are so tightly bound, they do not exclude the outer electrons from a large enough region of space. A much closer analogy to the mole-

(16) See, for example, C. A. Coulson, "Valence," Oxford University Press, Oxford, 1952.
(17) R. E. Stanton, *J. Chem. Phys.*, **36**, 1298 (1962).

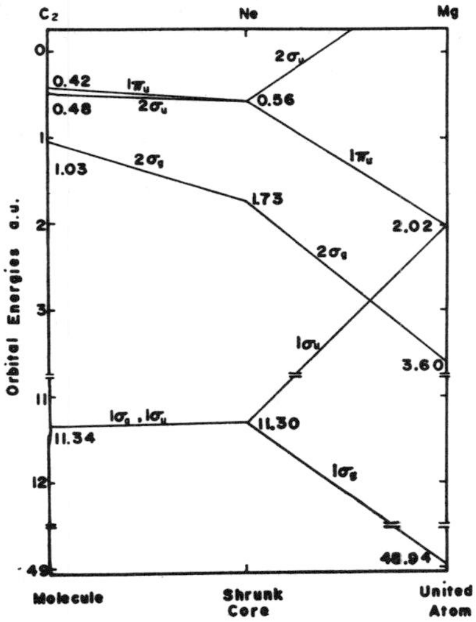

Figure 2b. Orbital energies in united atom, shrunk-core atom, and molecule. C_2 molecule.

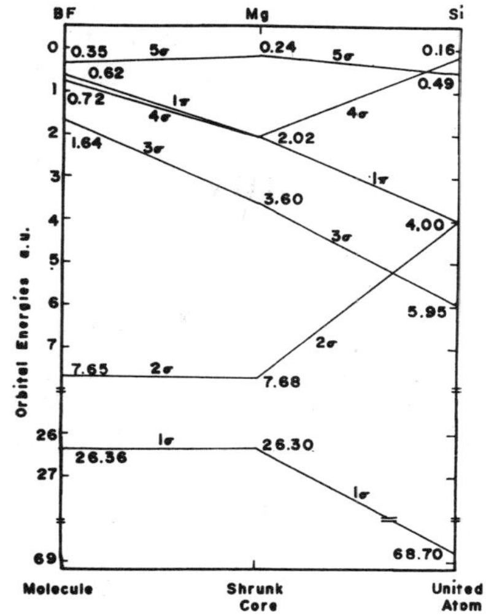

Figure 2c. Orbital energies in united atom, shrunk-core atom, and molecule. BF molecule.

cule is provided by what is called here the "shrunk-core"[18] model. In this model, the core of the resulting atom is very similar to the cores that outer MO's see in the molecule, so that the orbitals and orbital energies of the shrunk-core atom are in much better correspondence to the MO's and their energies. Figure 2 gives MO diagrams for three typical diatomic molecules. The united atom and the shrunk-core atom are shown, but not the separated atoms. The lines connect the MO's with their corresponding AO's in both model atoms. It is clear that the shrunk-core atom does indeed bear a much closer resemblance to the molecule than does the united atom. The same correspondence should apply to the spin-orbital pair correlation energies and to exclusion effects on them. This is important for the cases in which some pair correlation energies are strongly Z-dependent.

An example will make this model clear. Consider the C_2 molecule which has twelve electrons and the symmetry ${}^1\Sigma_g{}^+$. An examination of the LCAO coefficients[13] shows that the molecular core has four 1s electrons still centered on their respective nuclei. Thus, the core is essentially $K_a(C^{+4};\ 1s_a{}^2)K_b(C^{+4};\ 1s_b{}^2)$. The remaining eight electrons (in MO's) see a "1s-like" core, at least in the radial direction from the internuclear axis, with a net positive charge about like the sum of two C^{+4} ions. The shrunk-core model now replaces these eight electrons moving in expanded MO's with the outer part of an atom with a similar $1s^2$ core and net charge. The atom is neon. Thus, C_2 is represented by $K_aK_b\mathrm{Ne}(2s^22p^6;\ {}^1S)$, and the correlation energy is given by

$$E_{\mathrm{corr}}(C_2) = 2E_{\mathrm{corr}}(C^{+4}) + E_{\mathrm{corr}}(\mathrm{Ne}) - E_{\mathrm{corr}}(\mathrm{Ne}^{+8}) \quad (7)$$

This method is a simple one to use, for only total correlation energies of atoms and ions are required. However, there are two points of some importance to be noted in applying this method, both connected with the exclusion effects discussed in the previous section. First, one must choose the shrunk-core atom to have the appropriate symmetry. For example, the shrunk-core atom for $Be_2(1\Sigma_g{}^+)$ is carbon. However, the ${}^3P$ carbon atom does not have the correct number of paired electrons, so the exclusion effects of the occupied 2p orbitals on the inner shells will not be taken into account properly. One must take an appropriate linear combination of E_{corr} for the ${}^1D$ and ${}^1S$ states of carbon to produce a situation which resembles that in Be_2. Second, one could, not necessarily with justification, choose for the core of C_2: $K_a'(C^{+2};\ 1s_a{}^22s_a{}^2)K_b'(C^{+2};\ 1s_b{}^22s_b{}^2)$. However, one may not use the value of E_{corr} for the C^{+2} ion directly. As discussed previously (see also ref. 9), the value of $\epsilon(2s^2)$ for the Be-like ion does not take proper account for the exclusion effects of the occupied 2p orbitals, or the corresponding MO's in the molecule, on this nondynamical pair. The same problem does not arise if the core contains only the $1s^2$ electrons since this pair is transferable. If an examination of the LCAO coefficients shows that the best choice of a core contains 2s electrons, one must use the proper values of E_{corr} as calculated in ref. 9 according to the number of outer shell electrons in the molecule. In C_2, for our second choice of core, $\epsilon(K_a';\ C_2)$ is similar to $\epsilon(K_a';\ C\ {}^3P)$ or $\epsilon(K_a';\ O\ p_x{}^2p_y{}^2)$. In this work, the cores chosen for all diatomic molecules contain only $1s^2$ electrons.

Note that this method does not include the bond distance explicitly. However, the shrunk-core atom should be approximately the same size as the molecule. This is all that may be necessary. Fischer-Hjalmars[19] recently showed that in the H_2 molecule E_{corr} (H_2) remains almost constant over a range of bond distances from $R = 0$ to $R = 2R_e$. The same should hold for pair correlations in molecules with many more electrons since the correlation energy is not very sensitive to the details of the orbitals.

(18) Not to be confused with the shrunk-core model appearing in the early literature of MO theory.

(19) I. Fischer-Hjalmars in "Modern Quantum Chemistry–Istanbul Lectures," ref. 5.

Table II. Homonuclear Diatomic Molecules

Species[a]		$-\Delta E_{corr}$, e.v. Method I	Method II	D_e, e.v.[b]	E_{HF}, a.u. Predicted[c]	Calcd.[p]	$-\Delta E_{corr}(II)/D_e$
Li_2	$^1\Sigma g^+$ [d]	1.26	0.60	1.05[g]	−14.8820	−14.87152[k]	0.57
B_2	$^3\Sigma g^-$ [e]	1.42	1.18	2.90 ± 0.24[h]	−49.1213 ± 0.0088	−49.09088[k]	0.4
C_2	$^1\Sigma g^+$ [d]	3.31	2.99	6.36 ± 0.22[i]	−75.5011 ± 0.0081	−75.40619[l]	0.4
N_2	$^1\Sigma g^+$ [d]	3.20	3.93	9.902[g]	−109.0213	−108.9922[m]	0.4
O_2	$^3\Sigma g^-$ [f]	1.39	1.69	5.178[g]	−149.7470	−149.6659[n]	0.33
F_2	$^1\Sigma g^+$ [d]	2.01	1.82	1.68[j]	−198.8135	−198.7683[o]	1.1

[a] Letters (*d, e, f*) refer to the limited basis wave function used in method II. [b] All zero-point corrections from ref. 29 unless otherwise specified. [c] Using ΔE_{corr} of method II. Note that error refers only to the error in D_e. [d] See ref. 13. [e] A. A. Padgett and V. Griffing, *J. Chem. Phys.*, **30,** 1286 (1959). [f] G. L. Malli and P. E. Cade, unpublished results. [g] See ref. 29. [h] G. Verhaegen and J. Drowart, *J. Chem. Phys.*, **37,** 1367 (1962). [i] L. Brewer, W. T. Hicks, and O. H. Krikorian, *ibid.*, **36,** 182 (1962); E. A. Ballik and D. A. Ramsay, *Astrophys. J.*, **137,** 84 (1963) [G_0]. [j] R. Iczkowski and J. Margrave, *J. Chem. Phys.*, **30,** 403 (1959). [k] J. B. Greenshields, to be published. [l] P. E. Cade, K. D. Sales, A. C. Wahl, and C. C. J. Roothaan, to be published. [m] P. E. Cade, W. Huo, and C. C. J. Roothaan, to be published. [n] G. L. Malli and P. E. Cade, to be published. [o] A. C. Wahl, *J. Chem. Phys.*, **41,** 2600 (1964). [p] See ref. 11.

Table III. Diatomic Hydride Molecules

Species[a]		$-\Delta E_{corr}$, e.v. Method I	Method II	D_e, e.v.[b]	E_{HF}, a.u. Predicted[c]	Calcd.[l]	$-\Delta E_{corr}(II)/D_e$
LiH	$^1\Sigma^+$ [d]	1.88	0.93	2.515[f]	−7.9910	−7.98687[j]	0.37
BH	$^1\Sigma^+$ [d]	1.46	0.80	3.14 ± 0.4[g]	−25.1150 ± 0.0147	−25.02906[j]	0.25
CH	$^2\Pi$ [e]	1.68	1.21	3.64[h]	−38.2779	−38.27935[j]	0.33
NH	$^3\Sigma^-$ [e]	1.90	1.35	3.9 ± 0.5[g]	−54.9579 ± 0.0184	−54.97806[j]	0.33
OH	$^2\Pi$ [e]	1.79	1.45	4.68 ± 0.2[g]	−75.4280 ± 0.0074	−75.42083[j]	0.27
HF	$^1\Sigma^+$ [d]	1.87	1.55	6.06 ± 0.2[g,i]	−100.0751 ± 0.0074	−100.07030[k]	0.25

[a] Letters (*d, e*) refer to the limited basis set wave function used in method II. [b] All zero-point corrrections from ref. 29 unless otherwise specified. [c] Using ΔE_{corr} of method II. Note that error refers only to the error in D_e. [d] See ref. 13. [e] M. Krauss, *J. Chem. Phys.*, **28,** 1021 (1958). [f] R. Velasco, *Can. J. Phys.*, **35,** 1204 (1957). [g] See ref. 29. [h] See ref. 28. [i] See ref. 29; G. A. Kuipers, D. F. Smith, and A. H. Nielsen, *J. Chem. Phys.*, **25,** 275 (1956) [G_0]. [j] P. E. Cade, W. Huo, and C. C. J. Roothaan, to be published. [k] K. D. Sales, P. E. Cade, and A. C. Wahl, to be published. [l] See ref. 11.

Table IV. Heteronuclear Diatomic Molecules

Species[a]	$-\Delta E_{corr}$, e.v. Method I	Method II	D_e, e.v.[b]	E_{HF}, a.u., predicted[c]	$-\Delta E_{corr}(II)/D_e$
CO $^1\Sigma^+$ [d]	2.36	2.45	11.242[g]	−112.8211	0.23
BF $^1\Sigma^+$ [d]	1.21	1.65	8.58 ± 0.5[g]	−124.1931 ± 0.0184	0.2
BeO $^1\Sigma^+$ [e]	2.00	1.66	6.66 ± 0.1[h]	−89.5662 ± 0.0037	0.25
LiF $^1\Sigma^+$ [d]	1.83	1.33	5.99 ± 0.5[i]	−107.0133 ± 0.0184	0.25
NO $^2\Pi$ [f]	2.20	2.96	6.605[j]	−129.3443	0.4

[a] Letters in this column refer to the limited basis wave function used in method II. [b] All zero-point corrections from ref. 29 unless otherwise specified. [c] Using ΔE_{corr} of method II. Note that error refers only to the error in D_e. [d] See ref. 13. [e] M. Yoshimine, *J. Chem. Phys.*, **40,** 2970 (1964). [f] H. Brion, C. Moser, and M. Yamazaki, *ibid.*, **30,** 673 (1959). [g] See ref. 29. [h] W. A. Chupka, J. Berkowitz, and C. F. Giese, *J. Chem. Phys.*, **30,** 827 (1959). [i] See ref. 29; R. Braunstein and J. W. Trischka, *Phys., Rev.*, **98,** 1021 (1955) [G_0]. [j] See ref. 28.

Tables II, III, and IV give values of ΔE_{corr} calculated by this method. ΔE_{corr} is defined by

$$\Delta E_{corr} = E_{corr}(\text{molecule}) - E_{corr}(\text{atoms}) \quad (8)$$

and is the quantity of interest in examining the dissociation energy. Values of atomic and ionic correlation energies have been taken from Clementi.[15]

Method II. "Pair Populations" Method

Method I may only be used for diatomics and certain small polyatomic hydrides. A more generally applicable method is desirable. An "atoms-in-molecules"[20] type of model might be appropriate where dealing with the correlation energy alone, since the fluctuation potential is short range in character making it a "local" property. For a ground-state molecule (closed shell or single determinantal state),[9,22] eq. 1 gives the correlation energy approximately as the sum of the MO pair correlations. In the LCAO–MO approximation, this sum can be expressed as a sum of inter- and intraatomic pair correlations. Though there seem to be several ways of relating molecular correlations to atomic data and of dealing with the overlap charge regions, only one of these ways is developed here into a method capable of useful results.

The starting point will be a minimal basis set LCAO–MO wave function. This means that the basis set used is limited to those atomic orbitals occupied in the free atom, and the atomic orbitals are usually Slater-type orbitals (STO's). Let us examine the approximate wave function by means of LCAO–MO

(20) The "atoms-in-molecules" approach was developed by Moffitt[21] and others. They dealt with the *total* energy of the molecule. The H.F. potential changes considerably going from atoms to the molecule, and the long-range character of the H.F. potential is the source of drastic distortion of atoms in the molecule.

(21) W. Moffitt, *Proc. Roy. Soc.* (London), **A210,** 245 (1951); *Rept. Progr. Phys.*, **17,** 173 (1954). For a review of later work see, *e.g.*, R. G. Parr, "Quantum Theory of Molecular Electronic Structure," W. A. Benjamin, Inc., New York, N. Y., 1963.

(22) H. J. Silverstone and O. Sinanoğlu in "Modern Quantum Chemistry–Istanbul Lectures," ref. 9.

population analysis.[23] Consider a normalized MO in a diatomic molecule

$$\varphi_i = c_a\chi_a + c_b\chi_b \quad (9)$$

where χ_a and χ_b are individually normalized AO's (and may be linear combinations of STO's) on centers a and b, respectively. The total number of electrons, N, in this MO is given by

$$N(\varphi_i)^2 = N(c_a{}^2 + 2c_ac_b S_{ab} + c_b{}^2) = N \quad (10)$$

where

$$S_{ab} = \int\chi_a(1)\ \chi_b\ (1)\mathrm{d}v_1$$

The first term in eq. 10 represents the net population on atom a and gives a measure of the time the electrons spend there. The second term represents the overlap population and gives a measure of the time the electrons spend in the bonding region. The overlap population is usually[23] divided equally between centers. The sum of the net and overlap populations, in the ith MO, on center a, gives the gross atomic population

$$n(a_i) = c_a{}^2 + c_ac_bS_{ab} \quad (11)$$

The sum over the MO's of the gross populations, $n(a_i)$, yields the total gross atomic population for a given AO.

Those populations representing sums over all MO's are invariant[24] with respect to any orthogonal transformation among the occupied LCAO–MO's in a given configuration. The gross atomic population divides the electrons present among the AO's and takes into account, through the overlap populations, the distribution of electrons between different centers. One should then be able to use atomic pair correlation energies, taken in the fractions prescribed by the population analysis, to compute the molecular correlation energy.

Consider the C_2 molecule. Population analysis of the minimal basis set LCAO–MO wave function[13] gives, for each atom[25]: $1s^{2.0002}2s^{1.6702}2p_z{}^{0.3292}2p_\pi{}^{2.000}$. A plausible assumption is that, in singlet molecules and closed shells, these populations represent paired electrons; *i.e.*, that there are half as many pairs of electrons in each AO as the MO pair is apportioned *between* the two centers. Then contributions from distributions when the α spin electron is all on center a, while the β spin electron is all on center b do not become explicit.

Note that in C_2 some of the 2s electrons have been promoted to 2p. Thus, under the pair populations assumption, there are 1.164 *pairs* of 2p electrons, while in the free atom there are none. $|\epsilon(2s^2)|$ is expected to increase owing to nearby unoccupied orbitals, and to the fact that the fractional 2p occupation (of parallel spin p electrons) will often decrease. Atomic pair correlation energies are taken from Figure 1 and Table I.

By comparing the gross atomic population to the populations in the free atom, ΔE_{corr} may be written down directly

$$\Delta E_{\mathrm{corr}} = 2[0.835\epsilon(2s^2)_{\mathrm{molecule}} - \epsilon(2s^2)_{\mathrm{atom}} + \frac{(1.67-2)}{2}\epsilon(1s^2 \rightarrow 2s^2) + 0.1645\epsilon(2p_z{}^2) + \epsilon(2p\pi^2) + \frac{(1.67 \times 2.329 - 4)}{4}\epsilon(1s^22s^2 \rightarrow 2p^2)] \quad (12)$$

Note that in eq. 12 the free-atom value of the correlation energy of a doubly occupied pair such as $\epsilon(1s^2)$ or $\epsilon(2p_z{}^2)$ simply gets multiplied by the difference of molecular and free-atom pair populations, if it can be assumed to be transferable. For $\epsilon(2s^2)$ different molecular and atomic values must be used as discussed above. Interorbital correlations are obtained by first finding the value for one orbital of each type, each with random spin, and then multiplying this by the number of each kind of electron as given by the population analysis. For example, for the $2s \rightarrow 2p$ case,[26] atomic data[9] give directly the value of $\epsilon(1s^22s^2 \rightarrow 2p^2)$ for carbon. Assuming $\epsilon(1s^22s^2 \rightarrow 2p^2) \cong \epsilon(2s^2 \rightarrow 2p^2)$. the random spins value of $\epsilon(2s \rightarrow 2p)$ is one-fourth of this. In the "molecular atom" there are 2.33 random-spin 2p electrons and 1.67 2s electrons. Thus the $2s \rightarrow 2p$ correlation in the molecule, per atom, is: $(1.67 \times 2.33/4)\epsilon(2s^2 \rightarrow 2p^2)$. Using the values of the pair correlations[9] we find

$$\Delta E_{\mathrm{corr}} = 2[(0.835 \times 1.1) - 0.457 - (0.165 \times 0.39) + 0.1645 + 1.0 - (0.028 \times 2.317)] = 2.99\ \mathrm{e.v.} \quad (13)$$

The arithmetic is written out in order to show the contribution of each effect to the binding energy. This method is very easily carried out for any size molecule so long as a reasonable set of LCAO–MO coefficients is available.

Results

Values of ΔE_{corr} computed by both methods are given in Tables II through VI. The "shrunk-core" model was used only for diatomic molecules and small polyatomic hydrides. Also tabulated are experimental values of the binding energy (B.E.) and calculated nonempirical H.F. energies for the molecules where available. ΔE_{corr}(exptl.) is defined as

$$\Delta E_{\mathrm{corr}}(\mathrm{exptl.}) = D_e - E_{\mathrm{HF}}(\mathrm{molecules}) + E_{\mathrm{HF}}(\mathrm{atoms}) \quad (14)$$

The predicted "experimental" H.F. energies were computed as

$$E_{\mathrm{HF}}(\mathrm{molecules}) = E_{\mathrm{HF}}(\mathrm{atoms}) - \Delta E_{\mathrm{corr}} + \mathrm{B.E.} \quad (15)$$

where ΔE_{corr} from method II was used and ΔE_{rel}, the difference in molecular and atomic relativistic energy corrections, was assumed to be small. Finally, the fractional contribution of ΔE_{corr} to the binding energy, ΔE_{corr}/B.E., is given where, again, ΔE_{corr} from method II was used.

Diatomic Molecules. An examination of Tables II–VI shows that the two methods developed here, although they approach the idea of a molecule from two entirely different points of view, give consistent results in most

(23) R. S. Mulliken, *J. Chem. Phys.*, 23, 1833 (1955).
(24) C. Scherr, *ibid.*, 23, 569 (1955).
(25) S. Fraga and B. J. Ransil, *ibid.*, 34, 727 (1961).

(26) This correlation is not completely dynamical. In nonclosed shell states it contains some "semiinternal" correlation (see ref. 22) from mixing such as $2s2p \rightarrow 2p'3d$. Thus, it changes somewhat from system to system.

Table V. Small Polyatomic Molecules

Species[a]	$-\Delta E_{corr}$, e.v. Method I	Method II	B.E., e.v.[b]	E_{HF}, a.u., predicted[c]	$-\Delta E_{corr}$(II) / B.E.
HCN [d]	4.01	3.33	13.53[o]	−92.9644	0.25
CO_2 [e]		3.99	16.869[o]	−187.7807	0.24
H_2O [f]	3.67	2.51	10.08[o]	−76.0876	0.25
NH_2 [g]	3.69	2.62	8.2 ± 0.5[p]	−55.7060 ± 0.0191	0.32
NH_3 [h]	5.58	3.55	13.57[q]	−56.2692	0.26
CH_4 [i]		3.69	18.18[r]	−40.2212	0.2
C_3 [j]		3.74	14.12 ± 0.22[s]	−113.4473 ± 0.0081	0.26
C_4 [k]		3.80	19.25 ± 0.5[t]	−151.3223 ± 0.0191	0.2
O_3 [l]		4.74	6.818[o]	−224.5046	0.7
C_2N_2 [m]		5.28	21.7 ± 0.6[r]	−187.7826 ± 0.0220	0.24
$H_2C{=}O$ [n]		3.74	16.24[o]	−114.0309	0.23

[a] Letters in this column refer to the limited basis wave function used in method II. [b] All zero-point corrections from G. Herzberg, "Infrared and Raman Spectra," D. Van Nostrand Co., New York, N. Y., 1945, unless otherwise specified. [c] Using ΔE_{corr} of method II. Note that error refers only to the error in B.E. [d] A. D. McLean, *J. Chem. Phys.*, **37**, 627 (1962). [e] See ref. 33. [f] F. Ellison and H. Shull, *J. Chem. Phys.*, **23**, 2348 (1955). [g] J. Higuchi, *ibid.*, **24**, 535 (1956). [h] See ref. 27. [i] B. J. Woznick, *J. Chem. Phys.*, **40**, 2860 (1964). [j] E. Clementi, *ibid.*, **34**, 1468 (1961). [k] E. Clementi, *J. Am. Chem. Soc.*, **83**, 4501 (1961). [l] See ref. 31. [m] E. Clementi and A. D. McLean, *J. Chem. Phys.*, **36**, 563 (1962). [n] P. L. Goodfriend, F. W. Birss, and A. B. F. Duncan, *Rev. Mod. Phys.*, **32**, 307 (1960). [o] G. N. Lewis and M. Randall, "Thermodynamics," revised by K. S. Pitzer and L. Brewer, McGraw-Hill Book Co., Inc., New York, N. Y., 1961. [p] T. L. Cottrell, "The Strengths of Chemical Bonds," Academic Press Inc., New York, N. Y., 1953. [q] S. R. Gunn and K. G. Green, *J. Phys. Chem.*, **65**, 779 (1961). [r] See ref. 29. [s] L. Brewer and J. L. Engelke, *J. Chem. Phys.*, **36**, 992 (1962). W. Weltner and D. McLeod, *ibid.*, **40**, 1305 (1964) [G_0]. [t] J. Drowart, R. P. Burns, G. de Maria, and M. G. Inghram, *ibid.*, **31**, 1131 (1959). This is D_0 as there is no spectroscopic data avail- to determine G_0.

Table VI. Larger Molecules

Species[a]	$-\Delta E_{corr}$, e.v. Method II	B.E.,[b] e.v.	E_{HF}, a.u., predicted	$-\Delta E_{corr}$/ B.E.
HCCH[c]	2.64	17.53	−76.9245	0.15
HBNH[d]	3.48			
H_2CCH_2	4.00	24.37	−78.1259	0.16
H_2BNH_2	4.35			
H_3BNH_3	5.05			
H_2CCCH_2	4.87	30.38	−116.0034	0.14
H_3BNCH_2	4.98			
H_2BCNH_2	5.08			
H_2CBNH_2	5.62			
Benzene	8.42	61.084 ± 0.028	−231.0672 ± 0.0010	0.13
Borazine	10.04			

[a] Letters in this column refer to the wave function used in method II. [b] All dissociation energies from G. N. Lewis and M. Randall, "Thermodynamics," revised by K. S. Pitzer and L. Brewer, McGraw-Hill Book Co., Inc., New York, N. Y., 1961. All zero-point corrections from G. Herzberg, "Infrared and Raman Spectra," D. Van Nostrand Co., New York, N. Y., 1945. [c] See ref. 33. [d] This and all following molecules were treated using intermediate unpublished results kindly supplied by Dr. R. Hoffmann. See ref. 32.

cases. The agreement between the molecular H.F. energies predicted here and those calculated at the University of Chicago[11] is good. For almost all cases the values are within 1 e.v. The trends exhibited by the fractional contribution of ΔE_{corr} to D_e are consistent with what is known about the correlation energy. For multiply bonded homonuclear diatomics, this fraction is between 0.33 and 0.4. For singly bonded molecules, the fraction is greater than 0.5 except for B_2 which is a triplet. This is to be expected as the H.F. treatment takes into account "Fermi correlations" between parallel spin electrons and where Fermi correlation is large, coulombic correlation is small. In heteronuclear diatomics the fractions are somewhat less, falling between 0.25 and 0.35 for hydrides and between 0.2 and 0.4 for the others. The polarity of a heteronuclear bond increases the contribution of other factors to the binding energy and, thus, decreases the per cent contribution from the correlation energy.

Small Polyatomic Molecules. Comparison of the two methods developed here given in Table V for polyatomic hydrides shows that the agreement is not as good as for diatomics. This is not surprising as the approximation of a single center being seen by the outer electrons will naturally become poorer in polyatomic molecules, even if the other centers are all hydrogens. Moccia[27] has recently estimated the H.F. energy of H_2O, NH_3, and CH_4 independently, including relativistic corrections. In all cases, his estimates are within 1 e.v. of the values predicted here. The fractional contribution of ΔE_{corr} to B.E. falls between 0.2 and 0.3 for most of the small polyatomics considered (Table V). Although we have no independent estimates of the molecular H.F. energies for polyatomics other than hydrides, let alone calculated values, these fractions would appear to be consistent with the trends noted in diatomics.

Larger Molecules. Table VI contains results for some hydrocarbons and their boron–nitrogen analogs. The fractional contribution of ΔE_{corr} to B.E. is seen to be about constant in those cases where the binding energy is available.

Isoelectronic Series. In Table VII we have grouped results for some isoelectronic series and other related molecules for ease of reference. The two series of

(27) R. Moccia, *J. Chem. Phys.*, **40**, 2176 (1964).

diatomics isoelectronic with N_2 and C_2 show a marked decrease in the contribution of ΔE_{corr} to the binding energy upon the introduction of any polarity, and a slower decrease after that. For the hydrides isoelectronic with HF and OH, all values of a given series fall within 5% of one another. This is also true for the series of molecules related to OH and NH by successive addition of hydrogen atoms.

Table VII

Species	$-\Delta E_{corr}$, e.v. Method II	B.E., e.v.	$-\Delta E_{corr}$/ B.E.
C_2	2.99	6.36 ± 0.2	0.4
BeO	1.66	6.66 ± 0.1	0.25
LiF	1.33	5.99 ± 0.5	0.25
N_2	3.93	9.902	0.4
CO	2.45	11.242	0.23
BF	1.65	8.58 ± 0.5	0.2
HF	1.55	6.06 ± 0.2	0.25
H_2O	2.51	10.08	0.25
NH_3	3.55	13.57	0.26
CH_4	3.69	18.18	0.2
OH	1.45	4.68 ± 0.2	0.27
NH_2	2.62	8.2 ± 0.5	0.32
OH	1.45	4.68 ± 0.2	0.27
H_2O	2.51	10.08	0.25
NH	1.35	3.9 ± 0.5	0.3
NH_2	2.62	8.2 ± 0.5	0.32
NH_3	3.55	13.57	0.26

Discussion

Lithium Molecule. Table II shows that Li_2 is an exception to the generally good agreement between methods I and II for diatomics. Population analysis[25] of an LCAO–MO wave function[13] shows that the binding in Li_2 is due almost entirely to the 2s electrons. In method I, $\epsilon(2s^2)$ of Be ($Z = 4$) is used in computing ΔE_{corr}, while in method II, $\epsilon(2s^2)$ of Li^- ($Z = 3$) is used. However, $\epsilon(2s^2)$ is strongly Z-dependent,[4b] so that a sizable discrepancy appears. In other molecules, however, binding is due largely to the 2p electrons, with E_{corr} almost independent of Z, and the 2s electrons are chiefly nonbonding.

Carbon and Oxygen Molecules. There is a rather large discrepancy (2.6 and 2.2 e.v., respectively) between the predicted and calculated H.F. energies for C_2 and O_2. It is not clear just what gives rise to this. While it seems unlikely that the calculated values[11] are as 1.5 e.v. from the true H.F. energy, the fractional contribution of the reported H.F. energy to D_e is anomalously low, especially for C_2 (about 12%). The trends found here in the fractional contribution of ΔE_{corr} to D_e are consistent with what is known about the correlation energy. We will tentatively assume that the H.F. limit has not yet been reached in these two cases. Only for N_2 and F_2 have absolutely final production runs for the H.F. wave function been made as of this writing. In method II the correlation energy is taken as the sum of approximate pair energies. The predicted H.F. energy should be lower than, or equal to, that calculated, since nonempirical H.F. calculations are carried out by minimization, and the experimental H.F. minimum is not known beforehand; in fact, it is for all the cases treated here.

Nitrogen Molecule. The desirability of having ΔE_{corr} even approximately is strikingly shown in the case of N_2. For many years there was considerable controversy regarding $D_e(N_2)$. From spectroscopic data, two choices were possible. Herzberg[28] and others held the value of 7.5 e.v., while Gaydon[29] supported a value of 9.9 e.v. This controversy has since been resolved in favor of the latter. Note, however, that the H.F. contribution to D_e is only about 5.5 e.v. and thus both values of D_e would still be possible. With the addition of ΔE_{corr} (=4 e.v.), it is clear that the larger value of D_e must be the correct one.

Ozone Molecule. The fractional contribution of ΔE_{corr} to the binding energy of O_3 is 0.7, in contrast to other small polyatomics with values between 0.2 and 0.3. It is impossible to decide if this large value is real for the reasons discussed below. Mulliken[30] has discussed the choice of basis sets for LCAO–MO calculations. The population analysis can give very different results for different choices of basis set. We have consistently chosen the same type of approximate wave function so that comparisons could be made. For most of the molecules treated here, the wave functions used were constructed from STO's occupied in the free atoms. However, the only wave function available for ozone[31] is made up of atomic H.F. orbitals, and the 1s electrons are not included explicitly. Population analysis shows that there is very little 2s–2p promotion in the ozone molecule with this wave function. This tends to increase ΔE_{corr} (see eq. 12 and 13, where the third and last terms are negative owing to decreased 2s population). Fischer-Hjalmars[31] remarks that 2s–2p promotion increases when STO's are used, but she does not indicate by how much, nor whether 1s electrons are included. No similar treatments are available for other molecules, so one cannot determine the effect of using H.F. AO's and of excluding 1s electrons on the gross atomic populations.

Larger Molecules. The wave functions for the molecules given in Table VI are not of the LCAO–MO variety, but were determined using Hoffmann's[32] extended Hückel method. There is also available an LCAO–MO wave function[33] for acetylene, so that one may compare populations from the two wave functions. There is less 2s–2p promotion in the Hückel wave function and a somewhat larger hydrogen population. The 2p population is, thus, considerably smaller than from the LCAO–MO wave function. In the case of acetylene, the decreased contribution of the 2p electrons is almost cancelled by the increased contribution from the 2s electrons (see eq. 12 and 13). The difference in ΔE_{corr} is only 0.3 e.v. However, there is no guarantee that this cancellation will also occur in the larger molecules. One may probably predict heats of formation with a fair amount of confidence, however, since all the molecules are treated in the same approximation. The fractional contribution of ΔE_{corr} to the binding energy is about constant in those cases where experimental values are available. No reliance should be

(28) G. Herzberg, "Spectra of Diatomic Molecules," D. Van Nostrand Co., New York, N. Y., 1950.
(29) A. G. Gaydon, "Dissociation Energies," Chapman-Hall, Ltd., London, 1953.
(30) R. S. Mulliken, *J. Chem. Phys.*, **36**, 3428 (1962).
(31) I. Fischer-Hjalmars, *Arkiv Fysik*, **11**, 529 (1957).
(32) R. Hoffmann, *J. Chem. Phys.*, **39**, 1397 (1963); **40**, 2474 (1964).
(33) A. D. McLean, *ibid.*, **32**, 1595 (1960).

placed on the magnitude of ΔE_{corr} with the limited amount presently known about the wave functions of these large systems.

Limitations and Extensions. The methods given here are applicable to closed shell or single determinantal states. This includes the ground states of most molecules. The extension of these methods to non-closed shell states is being studied. This will allow the treatment of potential energy surfaces as well as of electronic spectra.

For larger molecules, the approximations involved in the "pair populations" method need further basic study. The application of this method is limited by the need for simple MO wave functions and atomic pair correlation energies. The accuracy of pair correlation energies may improve. It is encouraging that ΔE_{corr} may be obtained quite easily from simple MO wave functions. This, of course, increases the applications of H.F. calculations. The agreement with available H.F. results is good, and calculations on larger systems would help test both these correlation methods and various ways of calculating approximate H.F. MO's.

Acknowledgments. This work was supported in part by a grant from the National Science Foundation. Thanks are due to Dr. J. B. Greenshields and, particularly, to Dr. Paul E. Cade for providing many H.F. results prior to publication.

III-2e

Simplified SCF Calculations for Sigma-Bonded Systems: Extension to Hydrogen Bonding

DANIEL J. MICKISH AND HERBERT A. POHL

1. Introduction

During the past forty years, the nature of the hydrogen bond and the cause of its stability has been the subject of much study, both experimental and theoretical. The physicochemical properties of materials are in many cases strongly affected by hydrogen bonding. It is often a key factor in the ferroelectricity of solids,[1,2] in the hyperfine structure of NMR spectra,[3-5] and in the electronic and vibrational spectra of solutions and solids. It is of much biological importance in the problems of protein, DNA, and RNA structure and behavior.[6] Bratož has recently made a critical survey[7] on interpretations of hydrogen bonding. In addition to the early classical electrostatic calculations, recent theoretical treatments of hydrogen bonding have used a variety of quantum mechanical models.

Perhaps the most extensive calculations on hydrogen-bonded systems are those for FHF^-, the bifluoride ion, by Clementi and McLean[8] and for $NH_4C\ell$ by Clementi[9] and by Clementi and Gayles.[10] The quantum mechanical treatments considered all electrons in an LCAO-MO calculation without empirical parameters. Such "all-electron" calculations are so demanding of effort and expense that suitable shortcuts are much sought for. In particular, a limitation on the number of electrons expressly handled is often made. This is in accord with chemical intuition. Attention was, in such cases, focused on the "bonding electrons."

An early series of such treatments of hydrogen bonding used the valence-bond approach with empirical parameters (e.g., Refs. 11–13). An interesting variant of this, using the charge-transfer trial wave function, was examined by Puranik and Kumar[14] and by Bratož.[15] The MO approach has only been used rather recently.

The LCAO-MO technique in which the number of electrons expressly treated was restricted to those of the "bond" was apparently first used by Paoloni[16] with the aid, however, of adjustable parameters. Weissmann and Cohan[17] made a four-electron LCAO-MO calculation for the hydrogen bond of water. Rein[18] and Harris treated the hydrogen bonding of the guanine-cytosine base pair using SCF-LCAO-MO with semiempirical evaluation of many integrals and a single adjustable parameter (the Wolfsberg-Helmholtz parameter, K).[19] The same procedure was used by Lunell and Sperber[20] to consider adenine-thymine, adenine-cytosine, and guanine-thymine base pairs. The Wolfsberg-Helmholtz parameter, K, used in Refs. 18 and 20 (see Ref. 19) has proved useful to discuss homologous series of compounds while using simplified quantum mechanical methods.[21] Furthermore, Cusachs[22] has suggested that this parameter, K, has a useful interpretation. Its use in the base-pair hydrogen-bond calculations[18,20] is therefore not without justification.

Recently several calculations [23,24] have been reported on hydrogen-bonded groups using an all-valence-electron scheme and the extended Hückel procedure.[25] The calculation for $(FHF)^-$ gave reasonable correlation with the observed linear enthalpy-wave number shift relation for hydrogen bonds. Again, such calculations entail much use of the semiempirical evaluation of integrals. Several all-valence-electron calculations were reported on hydrogen bonding in formic acid dimer[26,27] and on the hydrogen bis(trifluoroacetate) ion,[26] in which the CNDO/2 procedure was used.[28,29] Quite reasonable correlation with experimental energies was obtained.

A stimulating approach to the bonding behavior of large and complex molecules was examined by Pollak and Rein.[30] They discussed hydrogen bonding between purines and pyramidines, using second-order quantum mechanical perturbation formulas for intermolecular interactions as had been developed by Haugh and Hirschfelder[31] and by Coulson and Davies.[32] The scheme calculates the energy in terms of direct electrostatic, polarization, and dispersion energies. It was shown to be useful in explaining preferences by base pairs for certain configurations. The input requires prior knowledge of the molecular coordinates and of the σ- and the π-electron distributions of the molecules.

By the use of semiempirical potential functions such as Morse-type potentials, Itoh was able, using a reduced one-dimensional Schrödinger equation, to obtain an interesting model of the hydrogen bond. It yielded binding energies, transition probabilities, tunneling coefficients, and chemical-kinetic constants of interest.

We have seen that hydrogen bonding can be described in a variety of models. For comparative purposes one would prefer a model which is not too laborious to carry out on a number of systems, yet is reasonably precise, descriptive, and needs no adjustable parameters.

Not long ago Pohl et al.[33] made a semiempirical molecular orbital SCF study of the halogen hydrides. The study was extended by Pohl and Raff to calculate physical properties of the 10 possible diatomic interhalogen molecules.[21] They explicitly considered the pair of electrons presumed to be responsible for bonding, making their calculations with approximations parallel to those involved in the Pariser-Parr-Pople method for conjugated π-electron systems.[34] In addition, the methods used by these authors included an allowance for the penetration of the halogen core by the other atom of the diatomic molecules. Their calculations yielded gratifyingly good correspondence to experiment in bond energy, equilibrium bond distance, dipole moment, and vibrational force constant. However, their calculated potential curves behaved poorly at large internuclear separations, mainly because the molecular orbitals used are incapable of providing good electron correlation needed in the limit of describing dissociated hydrogen and halogen atoms.[35]

The failure of the simple MO scheme to yield good potential curves at long internuclear separations was found by Harris and Pohl[36] to be largely removed by making a "split-shell" calculation, i.e., by using different spatial orbitals for the two explicitly considered electrons. Their calculations, otherwise like those of Ref. 33, gave reasonably faithful descriptions of the energy at all internuclear distances for the hydrogen halides.

Extension of the split-shell molecular orbital formulation of wave functions to triatomic molecules was done by Harris et al.[37] for the case of H_3. They calculated the interaction potential surface for $H + H_2 \longrightarrow H_2 + H$ using but few calculational approximations.

Because of the success of the work on "two-electron" systems using the semiempirical SCF molecular orbital scheme[21,33] at small moderate internuclear separations, we undertook the extension of the method to hydrogen bonding. It can be viewed as one having *four* electrons to be treated explicitly.

The specific purpose of this study was to investigate the $X^{(1)}—H^{(2)} \cdots Y^{(3)}$ hydrogen-bonded system using the closed-shell LCAO-MO-SCF technique and various appropriate integral approximations. We treated explicitly only four electrons associated with the hydrogen bond and the valence bond. The remaining nuclei and electrons were treated as appropriate atomic "cores" of formal charge 1, 1, and 2, respectively. The technique is essentially an extension of earlier studies of σ-bonded systems where only two or three electrons were considered explicitly.[21, 33, 36, 37] It is now called the restricted domain of atomic orbitals method (RDAO).

It should be remarked that the molecular orbital approximation is expected to be reasonably faithful for large R_{23} [the internuclear distance from $H^{(2)}$ to $Y^{(3)}$] as long as R_{12} is approximately that for chemical binding, i.e., the range in which the MO approximation is usually valid.[33-36] When R_{23} is large, two electrons will occupy a molecular orbital about $X^{(1)}$ and $H^{(2)}$, and two electrons will occupy a molecular orbital about $Y^{(3)}$ [i.e., the latter becomes an atomic orbital centered on $Y^{(3)}$].[38]

When the semiempirical SCF-MO scheme (but not the split-shell scheme) was applied to molecules with more than two centers, it was found that the description of the energy was reasonably faithful only in the region around the equilibrium bonding distances. The wave functions in this region also showed concomitant reasonable behavior. Both the energy and the molecular orbitals exhibited unusual behavior outside this region. We found it convenient to use the wave function obtained as a guide to trustworthiness of the energy values. We regard this good behavior only within certain geometric limits as a characteristic of the simple molecular orbitals used, and call attention to the phenomenon to indicate the importance of considering trial wave functions having the maximum feasible electron correlation.

2. Methods of Calculation (RDAO-MO Method)

The molecular orbitals were assumed to be linear combinations of atomic orbitals; i.e., in Pople's notation,[39]

$$\psi_i = \sum_{\mu}^{3} c_{i\mu} \phi_\mu \tag{1}$$

Unless otherwise stated, a hybrid atomic orbital was assumed to be located at center 1, directed toward centers 2 and 3 of the linear core configuration; a ls orbital at center 2; and a hybrid orbital at center 3, directed toward centers 1 and 2. The total wave function for the system was formed by taking an antisymmetrized product of the molecular spin orbitals:

$$\Psi = \mathcal{A}\left\{\psi_1(1)\alpha(1)\psi_1(2)\beta(2)\psi_2(3)\alpha(3)\psi_2(4)\beta(4)\right\} \tag{2}$$

The most convenient manner of handling affairs is to use the variational method to optimize the coefficients so as to minimize the energy. To do this we use the nonlinear Roothaan[40] equations:

$$\sum_{\nu}^{3} [F_{\mu\nu} - S_{\mu\nu}E_i]\, C_{i\nu} = 0 \tag{3}$$

where

$$F_{\mu\nu} = H_{\mu\nu} + \sum_{\lambda\sigma}^{3} P_{\lambda\sigma} \left\{\langle\mu\lambda|G|\nu\sigma\rangle - \tfrac{1}{2}\langle\mu\lambda|G|\sigma\nu\rangle\right\} \tag{4}$$

$$H_{\mu\nu} = \int \phi_\mu^*(1)\left\{-\tfrac{1}{2}\nabla^2 - \sum_{\alpha} Z_\alpha/r_{\alpha 1}\right\}\phi_\nu(1)\, d\tau_1 \tag{5}$$

$$\langle\mu\lambda|G|\nu\sigma\rangle = \iint \phi_\mu^*(1)\,\phi_\lambda^*(2)\,(1/r_{12})\,\phi_\nu(1)\,\phi_\sigma(2)\, d\tau_1\, d\tau_2 \tag{6}$$

$$S_{\mu\nu} = \int \phi_\mu^*(1)\,\phi_\nu(1)\, d\tau_1 \tag{7}$$

$$P_{\lambda\sigma} = 2 \sum_{i}^{2} c_{i\lambda}^{*} c_{i\sigma} \tag{8}$$

and Z_α = formal charge on core α. Greek suffixes have been used for atomic orbitals and English suffixes for molecular orbitals. The two lower roots of the secular equation

$$\det | F_{\mu\nu} - S_{\mu\nu} E_i | = 0 \tag{9}$$

were solved to yield the MO-theory approximations for the ionization potentials of the occupied orbitals. The total electronic energy of the four-electron system is then

$$\mathcal{E} = \frac{1}{2} \sum_{\mu\nu}^{3} P_{\mu\nu}(H_{\mu\nu} + F_{\mu\nu}) \tag{10}$$

The total system energy is the sum of the electronic energy and the core repulsion energies.

3. Approximations (RDAO-MO Method)

Reduction of the many-centered integrals was done with the Mulliken[41] approximation:

$$\phi_\mu(1)\phi_\nu(2) \cong (S_{\mu\nu}/2)\ [\phi_\mu(1)\phi_\mu(2) + \phi_\nu(1)\phi_\nu(2)] \tag{11}$$

The general electron repulsion integral (6) can then be written

$$\langle \mu\lambda | G | \nu\sigma \rangle \cong \tfrac{1}{4} S_{\mu\nu} S_{\lambda\sigma}\ [\langle \mu\lambda | G | \mu\lambda \rangle + \langle \mu\sigma | G | \mu\sigma \rangle + \langle \nu\lambda | G | \nu\lambda \rangle + \langle \nu\sigma | G | \nu\sigma \rangle] \tag{12}$$

and is therefore in terms of one- and two-center integrals. The one-center integrals were evaluated using Pariser's suggestion[42]:

$$\langle \mu\mu | G | \mu\mu \rangle = I_\mu + A_\mu \tag{13}$$

The two-center integrals were evaluated at times using the suggestion of Mataga and Nishimoto[43]; i.e.,

$$\langle \mu\lambda | G | \mu\lambda \rangle \cong (R_{\mu\lambda} + a_{\mu\lambda})^{-1} \tag{14}$$

where I_μ is the valence-state ionization potential of center μ, A_μ is the valence-state electron affinity of center μ, and

$$a_{\mu\lambda}^{-1} = \tfrac{1}{2}\ [\langle \mu\mu | G | \mu\mu \rangle + \langle \lambda\lambda | G | \lambda\lambda \rangle] \tag{15}$$

In some parts of the study, $a_{\mu\lambda}$ was taken as zero, producing, in effect, the point-charge approximation for the electron integral as suggested by Pople.[39] The numerical values for the valence-state ionization potentials and electron affinities were calculated from the tables by Hinze and Jaffé[44] (see Table 1).

The diagonal core integrals were evaluated using the Goeppert-Mayer and Sklar[45] suggestion,

$$\langle \mu | -\tfrac{1}{2}\nabla^2 - (1/r_\mu) | \mu \rangle \cong -I_\mu \qquad Z_\mu = 1 \tag{16}$$

and the further assumptions that for $Z_\mu = 2$,

$$\langle \mu | -\tfrac{1}{2}\nabla^2 - (2/r_\mu) | \mu \rangle \cong -I_\mu - \langle \mu\mu | G | \mu\mu \rangle \tag{17}$$

and

$$\langle \mu | (-Z_\nu/r_\nu) | \mu \rangle \cong -Z_\mu \langle \mu\nu | G | \mu\nu \rangle \tag{18}$$

Table 1. Physical Constants and Parameters

Valence state[a]	Orbital exponent	Ground-state ionization potential I_g, a.u.	Valence-state ionization potential I_v, a.u.	Valence-state electron affinity A, a.u.	Penetration-energy parameters	
					R_{HS}, a.u.	Z_{HS}
H (1s)	1.000	0.5000	0.5000	−0.0276	–	–
$\mathrm{N(te^2\,tete\underline{te})}$	1.950	0.5345	0.6962	−0.1490	0.54613	3.20
$\mathrm{N(\underline{te}^2\,tetete)}$	1.950	0.5345	0.5163	−0.1490	0.54613	3.20
$\mathrm{N(s^2\,pp\underline{p})}$	1.950	0.5345	0.5126	−0.0309	0.54613	3.20
$\mathrm{N(\underline{s}^2\,ppp)}$	1.950	0.5345	0.9407	−0.0309	0.54613	3.20
$\mathrm{N(trtr\underline{tr}\pi^2)}$	1.950	0.5345	0.7252	−0.1809	0.54613	3.20
$\mathrm{N(trtrtr\underline{\pi}^2)}$	1.950	0.5345	0.4398	−0.1809	0.54613	3.20
$\mathrm{O(te^2\,te^2\,te\underline{te})}$	2.275	0.5007	0.8969	−0.2247	0.30987	4.66
$\mathrm{O(te^2\,\underline{te}^2\,tete)}$	2.275	0.5007	0.6879	−0.2247	0.30987	4.66
$\mathrm{O(s^2\,p^2\,p\underline{p})}$	2.275	0.5007	0.6355	−0.0739	0.30987	4.66
$\mathrm{O(s^2\,\underline{p}^2\,pp)}$	2.275	0.5007	0.5374	−0.0739	0.30987	4.66
$\mathrm{F(s^2\,p^2\,p^2\,\underline{p})}$	2.600	0.6406	0.7671	−0.1287	0.46819	3.93
$\mathrm{F(s^2\,p^2\,\underline{p}^2\,p)}$	2.600	0.6406	0.6660	−0.1287	0.46819	3.93
$\mathrm{F(te^2\,te^2\,te^2\,\underline{te})}$	2.600	0.6406	0.9230	−0.2200	0.46819	3.93
$\mathrm{F(te^2\,te^2\,\underline{te}^2\,te)}$	2.600	0.6406	0.7470	−0.2200	0.46819	3.93

[a]The underlined symbol indicates the atomic orbital used.

The off-diagonal core integrals were evaluated similarly, and by requiring the identity to hold,

$$H_{\mu\nu} = \tfrac{1}{2}\,[H_{\mu\nu} + H_{\nu\mu}] \tag{19}$$

while it was assumed that (for $Z_\nu = 1$)

$$\langle\mu\,|-\tfrac{1}{2}\,\nabla^2 - (1/r_\nu)|\nu\rangle \cong -\,I_\nu S_{\mu\nu} \tag{20}$$

and (for $Z_\nu = 2$)

$$\langle\mu\,|-\tfrac{1}{2}\nabla^2 - (2/r_\nu)|\,\nu\rangle \cong S_{\mu\nu}\;[-\,I_\nu - \langle\nu\nu\,|G|\,\nu\nu\rangle] \tag{21}$$

The remaining one-center core Coulomb integrals were evaluated analytically using Slater orbitals,

$$\rho_\nu = \langle\nu\,|(-\,Z_\nu/r_\nu)|\,\nu\rangle \cong -\,Z_\nu\,\mu_n/n^* \tag{22}$$

where μ_n is the orbital exponent evaluated according to Slater's recipe[46] and n^* is the effective quantum number[46] of the atomic orbital. The final form, for example, of the one-electron matrix element H_{23} is

$$\begin{aligned} H_{23} = -\,(S_{23}/2)\;[&I_2 + I_3 + \langle 33\,|G|33\rangle + (\rho_2/2) + \rho_3 \\ &+ \langle 12\,|G|12\rangle + (3/2)\,\langle 23\,|G|23\rangle + \langle 13\,|G|13\rangle] \end{aligned} \tag{23}$$

The present study was also carried out in an alternative way by assuming the off-diagonal core integrals to be evaluatable in an extended Mulliken approximation,[18]

$$H_{\mu\nu} = (S_{\mu\nu}/2)\;[H_{\mu\mu} + H_{\nu\nu}] \tag{24}$$

The overlap integrals were evaluated analytically using Slater orbitals.[47] Penetration of the cores by the proton was considered, along the lines found useful earlier,[33,36] and the energy estimated with the aid of the Herman-Skillman tables.[48] The effective charge of a core as seen by a proton was assumed to be that of a shielded potential,

$$Z_{eff} = Z_{core} + (Z_{nuc} - Z_{core})e^{-bR} \tag{25}$$

where

$$b = \frac{1}{R_{HS}} \ln \left[\frac{Z_{nuc} - Z_{core}}{Z_{HS} - Z_{core}} \right]$$

and Z_{nuc} = true nuclear charge of the "core," Z_{core} is, as before, the formal charge of the core, and R is the internuclear distance of the core and proton. The function (i.e., R_{HS} and Z_{HS}) was evaluated from the Herman-Skillman results[48] at a distance of about 0.5 a.u. The nuclear repulsion energy due to the core and proton is then simply (in a.u.)

$$E_n = Z_{eff}/R \tag{26}$$

4. Numerical Detail

The nonlinear eigenvalue problem, equation (9), was solved by an iterative process using the Quantum Chemistry Program Exchange subroutines, CEIG, written by Michels et al.[49] and GIVENS, written by Prosser.[50] The iterative process consisted of three steps that were repeated until three significant figures in the eigenvector components were obtained. The first two steps were simple eigenvector substitutions. The third step, however, comprised a simple resubstitution followed by a second-order extrapolation in which the extrapolation assumed the three previous trial values for an eigenvector were equally spaced on a parabola. The extrapolated value of the component was then computed using the relation

$$c_{i\nu}^{(3)*} = 3c_{i\nu}^{(3)} - 3c_{i\nu}^{(2)} + c_{i\nu}^{(1)} \tag{27}$$

where the asterisk indicates the extrapolated value and $c_{i\nu}^{(n)}$ indicates the nth trial value of the νth component of the ith eigenvector. Convergence was obtained in less than 20 iterations for all but a few isolated points, such as where the component of the eigenvector was nearly zero. This procedure proved to be more efficient than one using a variable parameter such as the Wolfsberg-Helmholtz parameter[18,21] to aid in convergence.

5. Results and Discussion

In searching for the attributes of hydrogen bonding which can be expressed in quantum mechanical terms, we seek a model which preferably does not use adjustable parameters, yet is simple and direct enough to be useful. To begin with, we examined several different combinations of approximations as applied to the unhybridized $F^{(1)}$– $H^{(2)}$ + $F^{(3)}$ system [$F^{(3)}$ assumed to be off at infinity]. The combinations of approximations examined were

a. $a_{\mu\nu} = 0$; $H_{\mu\nu}$ done explicitly as, e.g., in equation (23), and so forth as in Ref. 33.
b. $a_{\mu\nu}$ as given by equation (15); $H_{\mu\nu} \cong (S_{\mu\nu}/2)\,[H_{\mu\mu} + H_{\nu\nu}]$ and so forth as in Ref. 18a.
c. $a_{\mu\nu}$ as given by equation (15); $H_{\mu\nu}$ done explicitly as, e.g., in equation (23).
d. $a_{\mu\nu} = 0$; $H_{\mu\nu} \cong (S_{\mu\nu}/2)\,[H_{\mu\mu} + H_{\nu\nu}]$.

The (constant) energy of the two electrons on $F^{(3)}$ is, in this case, $-3\,I_3 - A_3 = -1.8694$ a.u. The results of the calculation for this system [$F^{(1)}$—$H^{(2)}$ + $F^{(3)}$] showed approximation set a to give both binding and a reasonable bonding distance as well as realistic A0 coefficients for the ground state. Approximation set d gave equally good values for the binding energy and bonding distance but gave unrealistic A0 coefficients of mixed sign for this ground state. Approximation sets b and c did not yield binding and therefore gave no reasonable bonding distance. We feel that sets b and c could serve, however, if one wished to use an adjustable parameter such as the Wolfsberg-Helmholtz parameter. We did not carry this point further. A similar comparison of the approximations sets (a, b, c, and d) on the unhybridized $O^{(1)}-H^{(2)} \cdots O^{(3)}$ system with the oxygens at the normal hydrogen-bonding distance (2.76 Å) gave parallel results. Approximation set a was therefore adopted for further calculations. The calculations were made on a number of differently assigned orbital combinations as trial functions. Those studied are shown in Table 2 and were selected to compare various chemically intuitive ideas regarding hydrogen bonding.

Table 2. Orbital Functions

System	AO basis[a]	Symbol
N–H · · · N	te^2 tetete-1s · · · te^2 tetete	te-te
	π^2 trtrtr-1s · · · π^2 trtrtr	tr-π
	s^2 ppp-1s · · · π^2 trtrtr	p-π
	s^2 ppp-1s · · · s^2 ppp	p-s
O–H · · · O	te^2 te^2 tete-1s · · · te^2 te^2 tete	te-te
	p^2 p^2 pp-1s · · · s^2 p^2 pp	p-p
F–H · · · F	te^2 te^2 te^2 te-1s · · · te^2 te^2 te^2 te	te-te
	s^2 p^2 p^2 p-1s · · · s^2 p^2 p^2 p	p-p

[a]The underlined symbol indicates the atomic orbital used; e.g., π^2 trtrtr-1s . . . π^2 trtrtr means that a trigonal hybrid (of 2s and 2p functions) was used as the AO on $N^{(1)}$ and a directed 2p atomic orbital was used to represent the AO for the lone pair.

The calculated SCF binding energies, equilibrium internuclear distances, and force constants are shown in Table 3. The assumed internuclear distance, R_{13}, for the two outermost atoms is given in the table. The calculation proceeded in several stages. First the total four-electron electronic energy was calculated for a conformation. To this was algebraically added the nuclear repulsions of the cores and the penetration energies. The equilibrium distances were determined at the energy minimum with the aid of a graphical proceedure. The force constants, k_e, were calculated from the best parabola fitted to the region close about the equi-

Table 3. Binding Energies, Internuclear Distances, and Force Constants of Hydrogen-Bonded System

$X^{(1)}-H^{(2)} \cdots Y^{(3)}$	N–H · · · N				O–H · · · O		F–H · · · F	
	te-te	tr-π	p-π	p-s	te-te	p-p	te-te	p-p
R_{13}(a.u.); assumed	5.700	5.700	5.700	5.700	5.215	5.215	4.700	4.700
R_{XH} (a.u.)								
No H bonding (calc.)	2.200	2.176	2.252	2.258	1.891	2.008	1.878	1.872
No H bonding (exptl.)	1.93[b]				1.82[a]		1.72[b]	
R_{XH} (a.u.)								
During H bonding (calc.)	2.353	2.389	2.332	2.180	1.969	2.076	1.948	1.890
During H bonding (exptl.)	2.00				1.91[a]		2.08	
k_e° (10^5 dyn/cm)								
No H bonding (calc.)	3.95	3.75	3.59	3.59	4.37	4.65	5.68	5.15
No H bonding (explt.)[b]	6.1				7.7		9.65	
k_e (10^5 dyn/cm);								
during H bonding (calc.)	1.95	1.54	3.09	5.08	3.39	3.42	5.58	5.10
U_b (kcal/mole)								
Binding energy of H bond (calc.)	9.2	9.8	0.4	-11.5	6.2	4.2	8.5	4.7
Binding energy of H bond (exptl.)	3–5				3–6		6-10	
S_{13} (overlap)	0.026	0.021	0.017	0.012	0.015	0.011	0.012	0.008

[a]Ref. 51, pp. 260–262.
[b]Ref. 53, Vol. II, p. 458.

librium distance. For infinitesimal amplitudes the potential, V, followed by the vibrating proton follows the parabolic law

$$V - V_e = \tfrac{1}{2} k_e (R - R_e)^2$$
$$k_e = 4\pi^2 m_p c^2 \omega_e^2 \tag{28}$$

where m_p is the mass of the proton, c is the velocity of light, and ω is the vibrational frequency measured in cm^{-1}. For the present work, the values of k_e were evaluated graphically from the binding energy versus distance curves, using the conversion factor 1.5549×10^6 to multiply the calculated k_e (Hartrees/Bohr) to obtain k_e (dyn/cm). The spectrally observed values of k_e were computed from data in Ref. 53.

The binding energy, U_b, for the reaction producing the hydrogen bond,

$$X^{(1)}\text{—}H^{(2)} + Y^{(3)} \longrightarrow X^{(1)}\text{—}H^{(2)} \cdots Y^{(3)}; \quad U_b \tag{29}$$

was computed from the difference of the total energies at $R_{xy} = \infty$ and R_{xy} equal to R_{13}, the assumed (experimental) X—Y separation, when the hydrogen atom was at its "equilibrium" separation in each conformation.

By and large, the most satisfying results are obtained using properly hybridized orbitals, in agreement with chemical intuition. Both te-te and tr-π orbital bases for N—H $\cdots$ N give reasonable results for the bond energies, distances, and force constants. The p-π and p-s bases for N—H $\cdots$ N are definitely inferior. Both the te-te and p-p bases appear to give reasonable results for the O—H $\cdots$ O and F—H $\cdots$ F hydrogen-bond systems. Overall, the te-te basis seems adequate in all three systems.

The calculated binding energies for the three hydrogen-bond systems are in reasonable agreement with experiment. It is interesting to note that whereas binding is calculable using the te-te and tr-π orbital bases for N—H $\cdots$ N, the system exhibits "nonbonding" for the more artificial p-π set and is even antibonding when a still more artificial attempt is made to use the deep-lying $2s^2$ orbital as the lone-pair orbital in the p-s basis set. The values cited for experimentally observed binding energies of hydrogen bonds are necessarily broadened by the wide spectrum of compounds (and experimental methods) from which they were obtained. As Coulson,[35] Pimentel and McClellan,[51] and others have shown, the binding energy is much affected by local molecular circumstance. It is a strong function of steric strain, for instance (as reflected, for example, in the X—Y distance of X—H $\cdots$ Y). Furthermore, in view of the pessimism regarding the possibility of one's ever being able to calculate the energy of the hydrogen bond by wave-mechanical methods, as expressed by Coulson only a decade ago (Ref. 54), the present results, which use no adjustable parameter, are to be regarded as gratifying. The present rather satisfying results on the binding energies of the hydrogen bonds concur with earlier experience with the simplified MO approach on di- and triatomic bonding,[21,33,36,37] and with the experience of Parr and Pariser,[55] who made a study of the excitation energies of the H_2 molecule. They used the MO scheme and various approximation methods, finding good results at the experimental equilibrium internuclear distance using zero-differential-overlap and Mulliken approximations.

The values calculated for the X—H internuclear distance in both the unbonded and bonded forms are shown in Table 3. The calculated X—H distances in the absence of the hydrogen bond (i.e., X—H + a Y atom at infinity) are in all cases in reasonable agreement with experiment. They are about 10 per cent too large. The calculated X—H distance is found to increase as hydrogen bonding occurs, and this again is in the direction and magnitude observed experimentally. It is interesting to note that in the series N—H $\cdots$ N, O—H $\cdots$ O, F—H $\cdots$ F, the X—H distance decreases (i.e., for the te-te calculations) in that order as observed experimentally and as suggested by chemical intuition based on ideas about the increase of electronegtivity across the first-row elements.

The vibrational force constants calculated from the curvature of the energy–distance plot are given in Table 3 for the protonic vibration along the X–Y axis. Values were obtained both with and without the formation of the hydrogen bond to the Y atom. The calculated values are in tolerable agreement in both magnitude and trend with the results of spectroscopic measurements. Comparison of the calculated force constants shows that the presence of the third (Y) atom softens the X–H bond, causing a downshift in the IR spectrum.

It may be concluded that the model is fairly successful in describing the potential curves of hydrogen-bonded systems, in the region of equilibrium binding. Where the system is highly distorted from equilibrium, the calculated wave functions (eigenvectors) begin to show indications of unreliability and the model is less reliable. This typical failure of the MO theory at large separations is well discussed by Coulson,[35] Coulson and Fischer,[56] and Löwdin,[57,58] for example. It is the result of undercorrelation of the electrons in the model.

6. Summary and Conclusions

The present study using a simplified LCAO-MO model yields calculations of hydrogen-bonding energies, equilibrium bond distances, vibrational (stretching) force constants, and changes of the latter two quantities as hydrogen bonding occurs. These results are in fair agreement with experiment. Moreover, the trend in the calculated properties, such as bond distance and force constant and of their shifts on bonding within the homologous series N, O, F, is, in all cases examined, in agreement with experimental trend and with chemical experience. From the theoretical point of view, this indicates that the different approximations of the model have the same or a similar effect within a series of related molecules. From the fact that the binding energies are predicted to within a few kilocalories per mole without arbitrariness in the theory, it may be concluded that the different approximations in the model are compatible and suit well for the description of σ-bonded systems. From the applications viewpoint, these features of the model are rather encouraging, because the relative physical and chemical properties of an homologous series can be studied by using reasonable trial wave functions (e.g., hybrids, where indicated) and only a modest computational effort.

Acknowledgments

We should like to acknowledge with thanks the support of a portion of this study by the National Institutes of Health, Public Health Service, and a part by the Paint Research Institute. One of us (D.J.M.) expresses his appreciation to the National Science Foundation for a Traineeship during the period of this study. We wish to thank Dr. L. M. Raff and Mr. Jerry M. Cantril for fruitful discussion of the problem.

References

1. R. Blinc and D. Hadži, *Mol. Phys.* **1,** 391 (1958).
2. R. Blinc, *J. Phys. Chem. Solids* **13,** 204 (1960).
3. J. J. Arnold and M. E. Packard, *J. Chem. Phys.* **19,** 1608 (1954).
4. R. Off, I. Weinberg, and J. R. Zimmerman, *J. Chem. Phys.* **23** 748 (1956).
5. J. A. Pople, W. G. Schneider, and H. F. Bernstein, *High-Resolution Nuclear Magnetic Resonance,* McGraw-Hill, New York, 1959.
6. P.-O. Löwdin, *Biopolymers Symp.* **1,** 161 (1963).
7. S. Bratož, in *Advances in Quantum Chemistry,* Vol. 3, Academic Press, New York, 1967, p. 207.
8. E. Clementi and A. D. McLean, *J. Chem. Phys.* **36,** 745 (1962).
9. E. Cleminti, *J. Chem. Phys.* **47**, 2323 (1967).
10. E. Clementi and J. N. Gayles, *J. Chem. Phys.* **47**, 3837 (1967).
11. C. A. Coulson and U. Danielsson, *Arkiv Fysik* **8**, 245 (1954).

12. N. D. Sokolov, *Zh. Eksperim. i Teor. Fiz.* **23**, 315 (1952).
13. H. Tsubomura, *Bull. Chem. Soc. Japan* **27**, 445 (1954).
14. P. G. Puranik and V. Kumar, *Proc. Indian Acad. Sci.* **58**, 29, 327 (1963).
15. S. Bratož, *Symp. Forces Intermoleculaires*, Bordeaux, 1965 (in press).
16. L. Paoloni, *J. Chem. Phys.* **30**, 1045 (1959).
17. M. Weissmann and N. V. Cohan, *J. Chem. Phys.* **43**, 119 (1965).
18. R. Rein and F. E. Harris, (a) *J. Chem. Phys.* **41**, 3393 (1964); (b) **42**, 2177 (1965); (c) **43**, 4415 (1965).
19. M. Wolfsberg and L. Helmholtz, *J. Chem. Phys.* **20**, 837 (1952).
20. S. Lunell and G. Sperber, *J. Chem. Phys.* **46**, 2119 (1967).
21. H. A. Pohl and L. M. Raff, *Intern. J. Quantum Chem.* **1**, 577 (1967).
22. L. C. Cusachs, *J. Chem. Phys.* **43**, 1575 (1965).
23. K. Morokumo, H. Kato, T. Yonezawa, and K. Fukui, *Bull. Chem. Soc. Japan* **38**, 1263 (1965).
24. K. F. Purcell and R. S. Drago, *J. Am. Chem. Soc.* **89**, 2874 (1967).
25. R. Hoffman, *J. Chem. Soc.* **39**, 1397 (1963).
26. A. Ocvirk, H. Ažman, and D. Hadži, *Theoret. Chim. Acta* **10**, 187 (1968).
27. A. Pullman and H. Berthod, *Theoret. Chim. Acta* **10**, 461 (1968).
28. J. A. Pople, D. P. Santry, and G. A. Segal, *J. Chem. Phys.* **43**, S129 (1965).
29. J. A. Pople and G. A. Segal, *J. Chem. Phys.* **43**, S136 (1965).
30. M. Pollak and R. Rein, *J. Theoret. Biol.* **11**, 490 (1966).
31. E. F. Haugh and J. O. Herschfelder, *J. Chem. Phys.* **23**, 1778 (1958).
32. C. A. Coulson and P. L. Davies, *Trans. Faraday Soc.* **48**, 747 (1952).
33. H. A. Pohl, R. Rein, and K. Appel, *J. Chem. Phys.* **41**, 3385 (1964).
34. See, for example, R. G. Parr, *Quantum Theory of Molecular Electronic Structure,* Benjamin, New York, 1963.
35. C. A. Coulson, *Valence,* Oxford University Press, New York, 1958, p. 149.
36. F. E. Harris and H. A. Pohl, *J. Chem. Phys.* **42**, 3648 (1965).
37. F. E. Harris, D. A. Micha, and H. A. Pohl, *Arkiv Fysik* **30**, 259 (1965) (in English).
38. J. C. Slater, *Quantum Theory of Molecules and Solids,* Vol. I, McGraw-Hill, New York, 1963, p. 112.
39. J. A. Pople, *Trans. Faraday Soc.* **49**, 1375 (1953).
40. C. C. J. Roothaan, *Rev. Mod. Phys.* **23**, 61 (1953).
41. R. S. Mulliken, *J. Chim. Phys.* **46**, 497, 675 (1949).
42. R. Pariser, *J. Chem. Phys.* **21**, 568 (1953).
43. N. Mataga and K. Nishimoto, *J. Phys. Chem.* **13**, 140 (1954).
44. J. Hinze and H. H. Jaffé, *J. Am. Chem. Soc.* **84**, 540 (1962).
45. M. Goeppert-Mayer and A. L. Sklar, *J. Chem. Phys.* **6**, 645 (1938).
46. H. Eyring, J. Walter, and G. E. Kimball, *Quantum Chemistry,* Wiley, New York, 1944, p. 163.
47. C. C. J. Roothaan, *J. Chem. Phys.* **19**, 1445 (1951).
48. F. Herman and S. Skillman, *Atomic Structure Calculations*, Prentice-Hall, Englewood Cliffs, N.J., 1963.
49. H. H. Michels, C. P. Van Dine, and P. Elliott, Quantum Chemistry Program Exchange No. 97, CEIG, Department of Chemistry, Indiana University, Bloomington, Ind.
50. F. Prosser, Quantum Chemistry Program Exchange No. 62.1, GIVENS, Department of Chemistry, Indiana University, Bloomington, Ind.
51. G. C. Pimentel and A. L. McClellan, *The Hydrogen Bond,* Freeman, San Francisco, 1960.
52. S. W. Peterson and H. A. Levy, *Acta Cryst.* **10**, 70 (1957).
53. G. Herzberg, *Spectra of Diatomic Molecules,* Van Nostrand, Princeton, N. J., 1961, Table 39.
54. C. A. Coulson, *Hydrogen Bonding*, D. Hadži and H. W. Thompson, eds., Pergamon Press, Oxford, 1959, p. 339.
55. R. G. Parr and R. Pariser, *J. Chem. Phys.* **23**, 711 (1955).
56. C. A. Coulson and I. Fischer, *Phil. Mag.* **40**, 386 (1949).
57. P.-O. Löwdin, *Adv. Chem. Phys.* **2**, 207 (1959).
58. P.-O. Löwdin, *Rev. Mod. Phys.* **34**, 80 (1962).

III-3

Transition States and Chemical Reactivity

III-3a

The Chemical Reactivity of Sigma-Bonded Molecules—A Generalized Perturbation Treatment for Transition States

G. KLOPMAN

1. Introduction

The theoertical interpretation of chemical reactivity has been restricted mostly to that of conjugated organic molecules.[1] In these cases, more or less quantitative correlations have been obtained between rates of electrophilic, nucleophilic, or even radical reactions and various reactivity indices, such as the free valence index, charge index, localization energies, superdelocalizability, etc. These indices, which were obtained by means of the semiempirical quantum mechanical methods for π-conjugated species, characterize essentially the change in π conjugation that occurs in one of the reagents under the influence of the other.

The correlations usually take the form of a straight line obtained by plotting the rate constants for several similar conjugated species engaged in a specific reaction under the same conditions versus one of the reactivity indices obtained by the Hückel method or by more elaborate techniques such as the SCF methods.[2] The general trend shows a continuous and more or less regular increase in reactivity with increase of the index of the reacting position. This is illustrated in Fig. 1 for several reactions for various polycyclic hydrocarbons where such a pattern is observed. The success of this type of interpretation might lead one to believe that if one were able to describe a satisfactory quantum mechanical method for σ-bonded molecules, a set of indices could similarly be characterized to account for their reactivity.

That this is not the case can already be seen from Fig. 1, when one compares the reactivity of the various electrophiles toward the various hydrocarbons. Thus, the reactivity toward benzene, naphthalene, and anthracene follows the order $2 > 3 > 1$, $3 > 2 > 1$, and $1 \geqslant 3 > 2$, respectively.

It is seen therefore that one single set of indices for the electrophiles would not provide us with a reasonable description of their chemical properties as they would have to change with the nature of the nucleophile. The specific nature of the transition state and the experimental conditions, solvent, temperature, etc., equally affect the sequence. This situation is commonly encountered in reactions of saturated systems and probably reflects the fact that when the rate-determining step depends on the formation of a σ bond, rather than a change in conjugation energy (constant), the rate is controlled by at least two different energy terms that can either

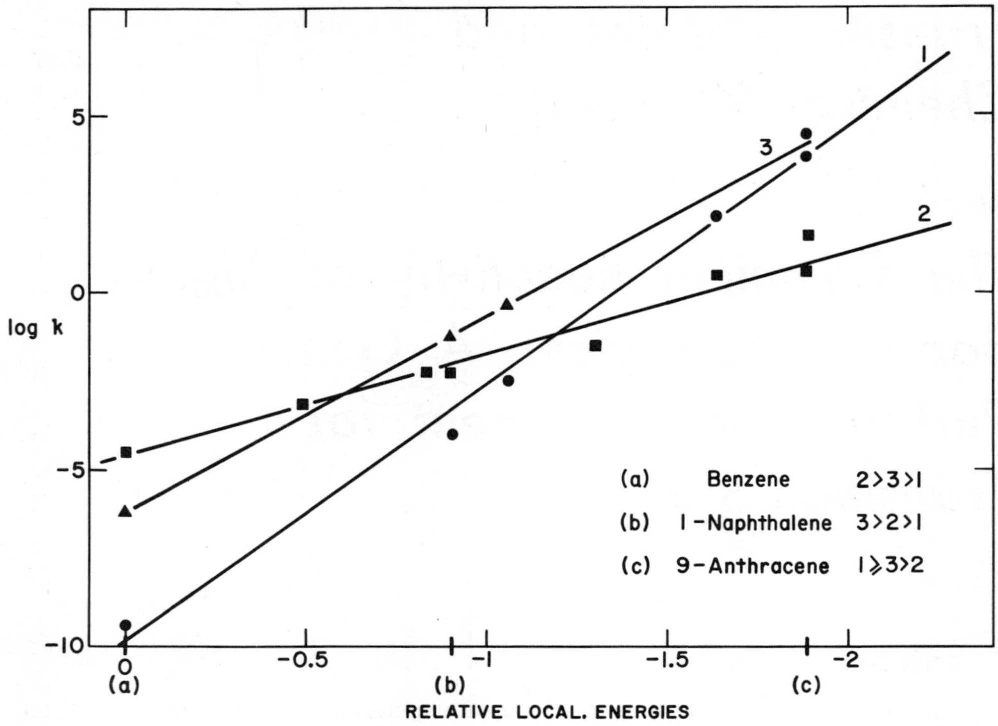

Fig. 1. Rates of reaction of polycyclic aromatic hydrocarbons: 1, basicity in HF at 0°C[3]; 2, chlorination in acetic acid at 20°C[4]; 3, deutero protonation in $CF_3COOD-H_2SO_4-CC\ell_4$.[5]

reinforce or oppose each other. Additional evidence for the above view is found in the field of general nucleophilic reactivity, where it was shown experimentally that the nucleophilicity of bases follows several different patterns, depending on the nature of the electrophile with which they are being reacted.[6]

Thus the sequence of nucleophilic orders toward a saturated carbon is found to be

$$CN^- > I^- > HO^- > Br^- > C\ell^- > F^-$$

whereas that toward a proton (basicity) is

$$OH^- > CN^- > F^- > C\ell^- > Br^- > I^-$$

Similarly, it was found that some metallic cations, such as $A\ell^{3+}$ and Ba^{2+}, form their most stable complexes with various donors and in the sequence OH^- > amines > phosphines > arsines, whereas $T\ell^{3+}$ or Ag^+ follow approximately the reverse sequence.[7]

It therefore appears that the only possible way of describing satisfactorily such chemical behavior would be to calculate the energy change involved in the rate-determining step. To do this, one should be able not only to calculate the electronic properties and essentially the heats of formation of both reactants in their ground states but also that of the transition state.

We have previously described an SCF method capable of providing reasonable values for the heats of formation and charge distribution of σ-bonded species.[8] This method is based on the use of the following approximations:

a. LCAO-MO procedure in its SCF approximation.

b. Separation of spins. Removal of all remaining interactions between configurations having the same azimuthal quantum numbers. Besides the simplification of the calculations of atoms that it provides,[9] this procedure enables us to calculate open-shell structures without interaction of configuration. Moreover, long-distance bonds can be calculated in a straightforward way, in contrast with the restricted procedure, which does not provide the correct dissociation limits. Consequently, this method is particularly suitable for our purpose, because transition states involve such partially dissociated bonds.

c. Neglect of all interaction integrals involving more than two spin orbitals and subsequent neglect of overlap integrals. This procedure is more general and versatile than the commonly used neglect of differential overlap.[10] It maintains the important atomic K integral resulting from the exchange of electrons. In addition, the value of the two-center electron-electron Coulomb interactions remain a function of the respective orientation of their orbitals. To maintain invariance in space, it is, however, necessary to compute a large number of interactions terms, but this has been achieved satisfactorily.

d. Point-charge approximation between the various atoms. The nuclear-nuclear and nuclear-electron interactions are assumed to have the same absolute value as that of the electron-electron interaction. In later versions, however, more sophisticated models have been introduced to overcome some of the difficulties inherent in this approximation. The original point-charge approximation was such that

$$\Gamma = \frac{e^2}{[R^2 + (\rho_x + \rho_y)^2]^{1/2}}$$

where R is the bond distance, ρ_x and ρ_y are the radii of the interacting orbitals such that

$$\rho = \frac{e^2}{2 < ii/ii >}$$

for an s or a p_π orbital, and

$$0 \leqslant \rho < \frac{e^2}{2 < ii/ii >}$$

for a p_σ orbital.

The results are obtained after successive iterations until self-consistency is reached. The procedure involves only one disposable molecular parameter for each type of atom—the resonance integral β. Here, also, several approximations have been tried, from which the following modified form of the Wolfsberg-Helmholz approximation appears to lead to the best results:

$$\beta^{xy} = \frac{S(IP_x + IP_y)\,\beta_0^{xy}}{[R^2 + (\rho_x + \rho_y)^2]^{1/2}}$$

The values of β_0 are determined for each type of atom by adjusting its value so that the treatment leads to a reasonable heat of formation of a standard compound, usually the diatomic molecule. This value being carried through in all subsequent calculations. For an XY bond,

$$\beta_0^{xy} = (\beta_0^{xx}\beta_0^{yy})^{1/2}$$

With these approximations, the calculations were carried through for a large number of compounds and the results are presented for some of them in Table 1.

Now, to be able to deal with the problem of chemical reactivity, we should also be able to calculate the energy of the transition state. As we do not really know the structure of the transition state, we should calculate a large number of possible configurations and determine the reaction path. This, however, requires a large number of computations and is not practical for large systems. Moreover, even if we could achieve these calculations, there would not even be

Table 1. Experimental and Calculated Heats of Formation of Compounds

	Observed, kcal	Calculated, kcal
$HC\ell$[a]	102	100
HI[a]	71	71
$LiC\ell$[a]	115	110
$C\ell Br$[a]	52	51
H_2O[b]	219	219
$HOC\ell$[b]	160	157
H_2O_2[b]	252	255
Ethane[c]	675	677
Propane[c]	954	957
Isobutane[c]	1237	1237
Cyclopropane[c]	813	809
Butadiene[c]	968	972
Benzene[c]	1318	1314

[a] Ref. 8a.
[b] G. Klopman, *J. Am. Chem. Soc.* **87**, 3300 (1965).
[c] Ref. 8b.

any certainty that the results are significant because there is no guarantee that the method provides the exact bond distance-bond energy relationships. Also, most of the reactions of σ-bonded species occur in solution and involve ions. The variations in solvation energy which cannot at this point be included in our SCF procedure certainly affect to a large extent the chemical reactivity and probably invalidate any conclusion drawn from a theory which neglects it.

However, one may describe a perturbation treatment consistent with all the previous approximations but which will be simpler to handle and therefore may include additional features, such as the solvation energy. This treatment, which provides a first approximation of the change in energy produced by a small interaction between the two interacting systems, would lead to the same result as the full SCF if the calculations were carried out until self-consistency.

An additional advantage of such a procedure is that it allows us to gain more insight into the factors which are controlling the reaction and may help us to design the ultimate conditions of an experiment by using qualitative information, which otherwise would be obscured by the mathematical complexity of the full SCF method.

The procedure we followed has been described elsewhere.[11] It is a polyelectronic perturbation treatment involving the hypothesis outlined above and includes the change in solvation energy. The perturbation, however, was restricted to the atoms which are binding in the transition state. This is not a drastic approximation because the two species whose interaction is being calculated are still far apart. The change in solvation which is also restricted to the two reacting atoms is included by means of the solvaton theory,[12] which is a convenient method of handling the Born equation.

The result can be written under the form of the following equation,[13] giving the total change in energy produced by the interaction of the various orbitals m of R and n of S through atom r of R and s of S:

$$\Delta E_{\text{total}} = -q_r q_s \frac{\Gamma}{\epsilon} + \left(\sum_{\substack{m \\ \text{occ}}} \sum_{\substack{n \\ \text{unocc}}} - \sum_{\substack{m \\ \text{unocc}}} \sum_{\substack{n \\ \text{occ}}} \right)$$

$$\left\{ b^2 \left[(E_m^* - E_n^*)_{b^2} - (E_m^* - E_n^*)_{b^2=0} \right] + (E_n^* - E_m^*) + \left[(E_m^* - E_n^*)^2 + 4 (c_r^m)^2 (c_s^n)^2 \beta^2 \right]^{1/2} \right\} \quad (1)$$

where

$$E_m^* = IP_m + q_s (c_r^m)^2 \frac{\Gamma}{\epsilon}$$

$$- a^2 \left[IP_m - EA_m - (c_r^m)^2 (c_s^n)^2 \Gamma \left(\frac{2}{\epsilon} - 1 \right) \right] - \sum_r \frac{x_r (c_r^m)^2}{R_r} \left(1 - \frac{1}{\epsilon} \right) [q_r + 2b^2 x_r (c_r^m)^2]$$

$$E_n^* = IP_n + [q_r + 2(c_r^m)^2] (c_s^n)^2 \frac{\Gamma}{\epsilon}$$

$$- b^2 \left[IP_n - EA_n - (c_r^m)^2 (c_s^n)^2 \Gamma \left(\frac{2}{\epsilon} - 1 \right) \right] - \sum_s \frac{x_s (c_s^n)^2}{R_s} \left(1 - \frac{1}{\epsilon} \right) [q_s - 2b^2 x_s (c_s^n)^2]$$

,

All the quantities involved in equation (1) are known or can be calculated by the full SCF treatment for the isolated reactants in the gas phase. Quantitative results can thus be obtained, but even qualitatively this equation may provide interesting observations, as two main cases may occur.

2. Charge-Controlled Reactions

When the energy difference between the set of occupied orbitals m of the donor and that of the unoccupied n of the acceptors is large, compared to the resonance integral between r and s, then b^2 tends to 0 for each interaction and equation (1) reduces to[14]

$$\Delta E_{total} = -q_r q_s \frac{\Gamma}{\epsilon} + \sum_{occ}^{m} \sum_{unocc}^{n} \frac{2(c_r^m)^2 (c_s^n)^2 \beta^2}{E_m^* - E_n^*}$$

and since $E_m^* - E_n^*$ is large we may approximate it by the average of all pairs such that

$$\gamma = \frac{\beta^2}{(E_m^* - E_n^*)_{av}}$$

and equation (1) becomes

$$\Delta E_{total} = -q_r q_s \frac{\Gamma}{\epsilon} + 2 \sum_{occ}^{m} (c_r^m)^2 \sum_{unocc}^{n} (c_s^n)^2 \gamma$$

The total change in energy is thus associated essentially in such cases with the values of the total charge on the two interacting atoms. Such a situation occurs when the acceptor's empty orbitals are very high; that is, the acceptor has low electron affinity, is highly solvated, and has a low tendency to form covalent bonds (β is small). Similarly, the donor's occupied orbitals should be low, which requires that it has a high electronegativity together with a low tendency to form covalent bonding. It should also be highly solvated and therefore have a small atomic radius.

Under these circumstances, the interaction between the two reactants is very favorable and the reaction involves no or very little charge transfer. We call such a reaction a charge-controlled reaction because the rate is a direct function of the total charges on the reagents. Examples of such reactions are those involving highly electronegative nucleophiles such as fluoride, hydroxide, or even amines and electrophiles such as Ti^{4+}, $A\ell^{3+}$, acid chlorides, etc.

3. Frontier-Controlled Reactions

When the difference in energy $E_m^* - E_n^*$ between the highest occupied orbital of the base and the lowest unoccupied orbital of the acid is very small (degeneracy), then their interaction becomes extremely large and one can practically neglect all other interactions.

Under these circumstances, equation (1) can be approximated by

$$E_{total} = 2c_r^m c_s^n \beta$$

where m and n are the frontier orbitals. Such a case occurs for reactions between atoms whose tendency to form covalent bonding is large. The acceptor has a high tendency to add an electron (high electronegativity), has a large radius, and is poorly solvated. The donor has a low electronegativity, a large radius, high polarizability, and is also poorly solvated.

Therefore, the reaction is now activated by a large density in the frontier orbitals and the reaction is called a frontier-controlled reaction. Examples of nucleophiles following these patterns are iodides, sulfides, and phosphines in reaction with electrophiles such as mercury, silver, and saturated halides. Several applications of this treatment have already been achieved, including a quantitative description of nucleophilicity and of the hard and soft character of acids and bases.[13]

The results seem to be particularly useful for determining the product of ambident reactions and in principle permit prediction of the influence of the structure of the compounds, that of the nucleophile, of the leaving group, and the solvent on the reactivity.

Calculations are presently being carried out for several specific systems where various factors influencing the reactivity are systematically and quantitatively studied.

Acknowledgment

This work has been supported in part by the National Science Foundation (Grant GP-8513).

References and Notes

1. (a) M. J. S. Dewar, in *Advances in Chemical Physics*, Vol. VIII, Wiley-Interscience, New York, 1965, p. 65; (b) A. Streitwieser, Jr., *Molecular Orbital Theory for Organic Chemists*, Wiley, New York, 1961.
2. M. J. S. Dewar and C. C. Thompson, Jr., *J. Am. Chem. Soc.* 87, 4414 (1965).
3. E. L. Mackor, A. Hofstru, and J. H. van der Waals, *Trans. Faraday Soc.* **54**, 66 (1958).
4. S. F. Mason, *J. Chem. Soc.* **1959**, 1233.
5. G. Dalinga, A. A. V. Stuart, P. J. Smith, and E. L. Mackor, *Z. Elecktrochem.* **61**, 1019 (1957).
6. J. O. Edwards, *J. Am. Chem. Soc.* **76**, 1540 (1954).
7. R. G. Pearson, *J. Am. Chem. Soc.* **85**, 3533 (1963).
8. (a) G. Klopman, *J. Am. Chem. Soc.* **86**, 4550 (1964); (b) M. J. S. Dewar and G. Klopman, *J. Am. Chem. Soc.*, **89**, 3089 (1967).
9. G. Klopman, *J. Am. Chem. Soc.* **86**, 1463 (1964).
10. J. A. Pople and G. A. Segal, *J. Chem. Phys.* **43**, 5136 (1965).
11. G. Klopman and R. F. Hudson, *Theoret. Chim. Acta* 8, 165 (1967).
12. G. Klopman, *Chem. Phys. Letters* **1**, 200 (1967).
13. G. Klopman, *J. Am. Chem. Soc.* **90**, 223 (1968).
14. The difference between the unoccupied orbital E_m of the donor and the occupied orbital E_n of the acceptor is even larger, as, for example, between the 3s orbital of F^- and the 1s of Li^+, and their interaction can be neglected.

III-3b

Hybrid-Based Molecular Orbitals and Their Chemical Applications

KENICHI FUKUI

1. Introduction

It is well established that the Hartree-Fock method for many-electron polynuclear systems makes the orbital concept unequivocal. Here we have to concern ourselves with the uniconfigurational wave function in which the sense of electron configuration is definite. The spin-eigenstate uniconfigurational wave function can be a single-determinant function (i.e., an antisymmetrized product of spin orbitals) only in the following cases:

a. Closed-shell wave functions.

b. Open-shell wave functions with maximum multiplicity.

c. Wave functions with a closed-shell structure with additional open-shell structure of maximum multiplicity.

The "best" uniconfigurational wave functions obtained by the Hartree-Fock variational method are composed of "best" orbitals (Hartree-Fock molecular orbitals). In cases a and b, the set of orbitals is in general definitely determined, except arbitrary numerical factors of which the absolute value is unity, taken as being mutually orthogonal and having a definite "orbital energy." The concept of "electron occupation" of orbitals is thus unambiguous.

The Hartree-Fock orbitals are obtained by solving what we call the Hartree-Fock equations.[1] Besides the occupied orbitals, these equations possess solutions corresponding to actually unoccupied, "virtual"'orbitals. Some of them happen to possess negative orbital energies (corresponding to "bound one-electron states"), whereas others have positive energies.

2. Basis Functions in Orbital Formulation

The Hartree-Fock orbital is a one-electron spatial function. This can be expanded in terms of any complete set of three-dimensional functions.

The Hartree-Fock orbital in atoms is of course merely a special case of that in molecules. A whole assembly of atomic Hartree-Fock orbitals belonging to the atoms composing a molecule does not make a complete set, or other sets of approximate atomic orbitals like those of the Slater type. But if we utilize, for instance, a larger and larger number of Slater-type atomic orbitals, belonging to inner shells and valence shells and even those in "vacant" shells, as the basis functions of orbital expansion, the approximation would become better and better (extended-basis approach). Such a result is easily expected from the fact that the molecular potential field is essentially multicentric, localizing at each nucleus.

Several ways may be possible in regard to the selection of basis functions of approximate construction of Hartree-Fock molecular orbitals. One possibility is to use those atomic orbitals which are based on the space-fixed (Cartesian) coordinates. A set of a limited number of occupied (and unoccupied) Hartree-Fock atomic orbitals of component atoms may be taken. Or more simply, as being approximate to them, the Slater type or other atomic orbitals such as the Gaussian type are employed. These are in general nonorthogonal between different atoms. If one wants to avoid the inconvenience that results from this nonorthogonality, one can use Löwdin's orthogonalized orbitals, which essentially localize at a single atom and can be employed instead of the atomic orbitals mentioned above.

Consider a set of orbitals which are originally represented by linear combinations of limited numbers of such atomic orbitals. These can be employed as an approximation to the Hartree-Fock orbitals. The process of optimization in which the atomic orbitals themselves are fixed and only the coefficients of the linear combinations are determined so as to give the minimum energy is due to Roothaan.[2] Frequently, the parameters involved in atomic orbitals, such as orbital exponents in Slater-type orbitals, are simultaneously optimized. Also a technique to use two or more Slater-type atomic orbitals with different exponents in the place of the function of the same quantum number is applied (double or multiple ζ technique).

Another possibility is to use "hybridized" orbitals in place of atomic orbitals associated with the space-fixed coordinates as the basis functions constructing molecular orbitals. Such a hybridized orbital is, for example, taken as a linear combination of those Slater-type orbitals, for example, which belong to the same atom. We can in principle always make a set of n linearly independent hybridized orbitals from n independent Slater-type orbitals with different angular parts and, usually, with the same principal quantum number. The molecular orbitals can thus be approximated as linear combinations containing such hybridized orbitals.

3. Optimization of Hybrid-Based Linear Combinations

Two ways of obtaining a hybrid-basis best orbital are possible. One is to transform the optimized Slater-basis linear combination into the hybrid-basis one. This case will be discussed in Section 4.

The other is to optimize the linear combination originally based on hybridized orbitals. The process of optimization of such molecular orbitals is essentially not different from the Roothaan process. The best orbitals obtained should theoretically be the same as those starting from the linear combination of Slater orbitals, provided the various integrals are properly estimated. In this sense, the two approaches may be said to be equivalent. But in a practical sense, they are not equivalent, as far as we apply conventionalism in the evaluation of integrals.

The best molecular orbitals to be calculated in this way are more closely connected to the bond concept than the Slater-orbital bases. Hybridized orbitals are frequently suitable for describing the directed valence if they are so taken as to possess the maximum extension in the direction of bonds. The electron population of such hybridized orbitals is convenient for discussing electronic behavior *along the bond*.

Such hybrid-basis molecular orbitals were first used essentially in the frame of the Hückel approach,[3] and then in a scheme taking the electron repulsion into account explicitly.[4] The applicability to various chemical problems was also discussed.[5]

In these articles the main interest was in hydrocarbons. The basis atomic orbitals were taken to be the fixed p, sp^3, sp^2, and sp carbon hybrids and, in addition, hydrogen 1s atomic orbitals. This treatment is particularly suitable for the molecules composed of tetrahedral, trigonal, and linear carbon atoms where the hybridized state of carbon atoms is clear cut. These molecules are methane, methyl, and methylene. Next to these come ethane, ethylene, acetylene, and so on. The method was extended to other molecules than hydrocarbons.[3d] But such extension evidently receives a restriction in the molecules in which the concept of directed bond is obscure, as in boranes. For the treatment of these molecules, the use of a basis set associated with space-fixed coordinates is obviously more suitable.

4. Transformation to Hybrid Bases from Slater Bases

As mentioned in Section 3, an approximate best orbital once obtained in the form of a linear combination of Slater orbitals can always be transformed into a linear combination of bases containing hybrids in general. Here is possible another use of the hybrid-based molecular orbital.

This is mathematically a simple problem of linear transformation. Among such transformations, some may be physically significant. One of these is the transformation into the "valence atomic orbitals" (VAO) which was developed by McWeeny and Ruedenberg.[6] The concept of VAO is given as a set of real atomic orbitals (generally hybridized), belonging to one atom and mutually orthogonal, which diagonalizes the bond-order matrix of a molecule within the same atom. It is easily shown [6b] that VAO is the atomic orbital whose set maximizes or minimizes the diagonal elements of the bond-order matrix.

Ruedenberg has shown that VAO is useful for understanding the directed valence, explaining the bond angle in H_2O molecule successfully.[6b] The VAO concept is really convenient for discussing the physical nature of lone-pair and inner-shell orbitals or σ odd-electron orbitals which have no atom in their direction of extension. But it is not appropriate to discuss, for instance, the hybridized state of a carbon atom in a molecule by the VAO criteria. Making mention of a simplest example, VAO's of ethane do not give the state of hybridization correctly. Such a defect may result from the fact that the VAO is so defined as to most or least participate in bonding only within one atom.

5. Orbital Extension and Chemical Reactivity

One of the merits of the use of hybrid-based molecular orbitals is convenience in the discussion of chemical reactivity in the framework of the orbital concept. The particularly important role of the highest occupied (HO) and the lowest unoccupied (LU) molecular orbitals (MO) was stressed in varieties of chemical interactions.[5,7] The extension of HO determines the orientation and stereoselection in the interaction with electron acceptors, and that of LU controls the reactivity with electron donors. In homolytic reactions both HO and LU are controlling. This principle applies widely in the chemistry of both saturated and unsaturated compounds.

The hybrid-based molecular orbitals of propane are shown in Fig. 1. These are taken from the optimized orbitals obtained by Katagiri and Sandorfy.[4a] The larger extension of MO at the hydrogen atoms in the methylene region than the methyl regions can be interpreted as reflecting the higher reactivity of secondary hydrogens than primary ones. A comparison of hydrogen reactivity toward metathetical reactions in various hydrocarbons was also discussed.[8]

Also, an interesting theoretical interpretation of the reactivity of some ring compounds toward ring-opening polymerization was carried out[9] by the use of hybrid-based MO theory of the Hückel type. The extension of LU-MO of β-propiolactones in Fig. 2 explains the position of attack of nucleophiles in this reaction.

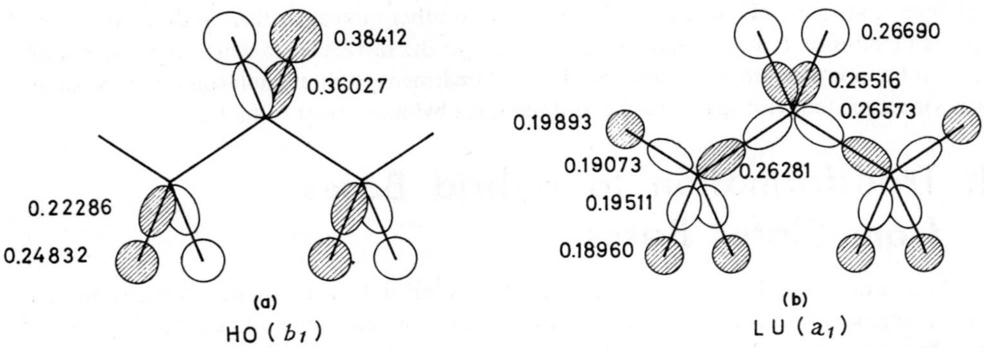

Fig. 1. (a) HO (b_1) and (b) LU (a_1) orbitals in C_3H_8. The figures are orbital coefficients (absolute value), and the shaded and unshaded parts indicate the MO regions with opposite signs.

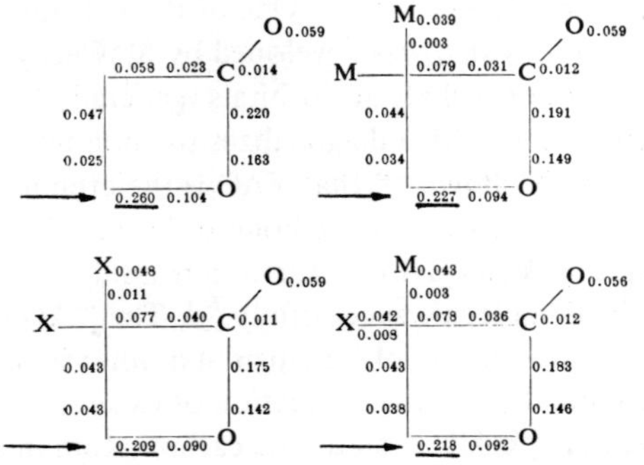

Fig. 2. Extension of LU-MO and the position of attack of nucleophiles in β-propiolactones. M and X denote an electron-repelling and an electron-attracting substituent, the figures indicate the electron population on hybrids, and arrows show the direction of attack of nucleophiles.

In Fig. 3 examples of the σ-LU-MO distribution in several molecules are indicated to interpret their chemical reactivity. Figure 3(a) shows that the LU-MO of methyl hydroperoxide largely locates on the oxygen-oxygen bond, and the partial bond order on that bond is negative and large in absolute value. This corresponds to the fact that the fission of the

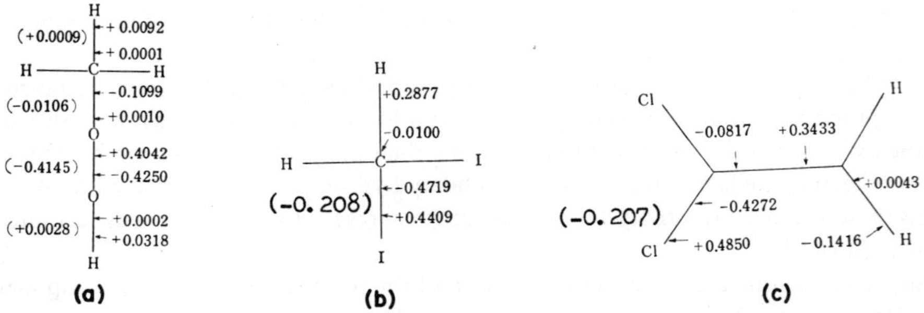

Fig. 3. σ-LU-MO of several compounds. The figures are the AO coefficients, and those in parentheses are the partial bond-order values. (a) Methyl hydroperoxide; (b) diiodomethane; (c) 1,1-dichloroethylene.

O—O bond takes place upon attack by an electron donor. The distribution of the LU level of diiodomethane is highly localized on the carbon-iodine bonds, and the bond order on that bond is largely negative, as indicated in Fig. 3(b). This explains successfully the result in polarographic reduction, in which iodines are reduced to hydrogen through the C–I bond fission of the electron-donated molecule. A similar analysis is possible with respect to the polarographic reduction of 1,1-dichloroethylene. The LU distribution is illustrated in Fig. 3(c).

6. Physical Properties of Halogenated Hydrocarbons

The treatment of halogenated hydrocarbons by the hybrid-based MO theory brings about several interesting results. The electronic structure of halomethanes and other halogenated hydrocarbons was discussed in relation to their chemical reactivity in earlier papers[3d3,3d4] and very briefly in Section 5.

Also, the n-σ^* absorption, PQR coupling constants, and proton chemical shifts were interpreted[10] by this treatment. The polarographic half-wave reduction potential of various halogen compounds was correlated with their LU level.[11] A possibility was suggested

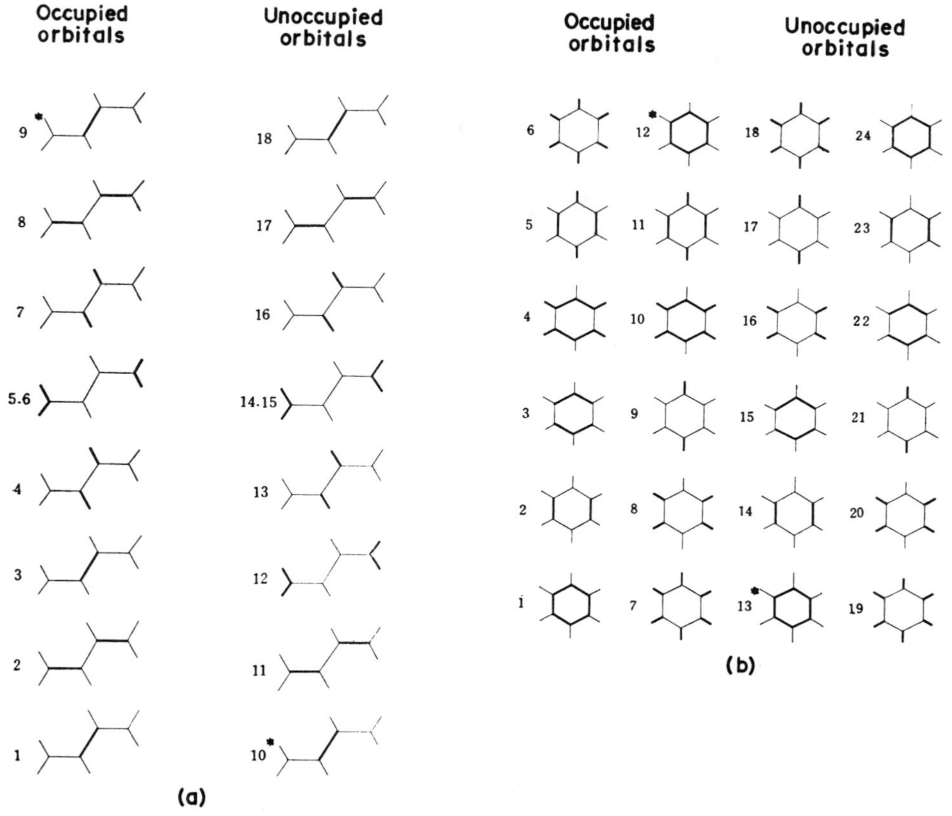

Fig. 4. σ-MO distribution: (a) butadiene; (b) benzene. Thick lines denote the most densely populated bonds. The number specifies the orbitals in ascending order of energy. An asterisk represents HO-MO or LU-MO.

that in some polyhalogenated hydrocarbons the σ-LU level appears below the π-LU level. This conjecture received support from the data of reduction products.

The hybrid-based MO theory is also suitable for discussing the σ skeleton of planar conjugated molecules. Besides the problem which will be treated in Section 7, several interesting results have been obtained with regard to halogenated-conjugated hydrocarbons. The σ dipole moment and the coupling constant of halogen in the PQR spectra were obtained theoretically.[12]

7. Sigma Skeleton of Conjugated Molecules

Physicochemical studies of conjugated molecules, such as high-resolution NMR spectroscopy, PQR frequency shift, dipole moment, and so on, suggest the importance of the investigation of the electronic structure of the σ skeleton.

A result of the hybrid-based MO study is introduced here.[13] Figure 4 indicates the mode of localization of each orbital in certain particular bonds. For instance, the σ-HO and LU-MO's are highly localized on carbon-carbon bonds. The net charge of atoms in several compounds is illustrated in Fig. 5.

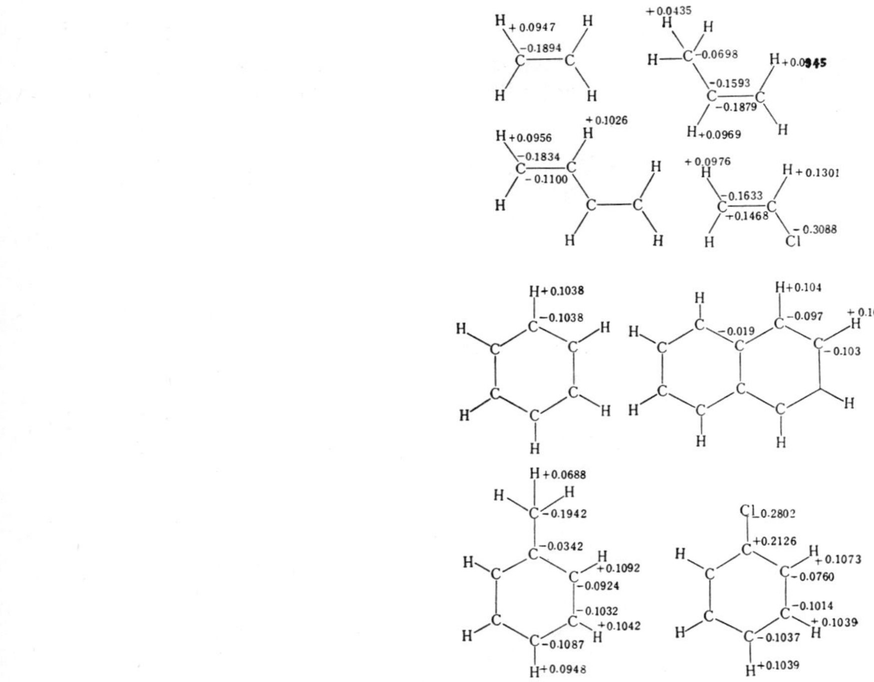

Fig. 5. Net charge on atoms in several conjugated molecules.

The HO and LU σ-MO's of ethylene and naphthalene are indicated in Fig. 6. Such a separate consideration of particular σ orbitals will prove (Section 8) to be important in connection with the chemical reaction of conjugated molecules in which the carbon atoms change their state of hybridization.

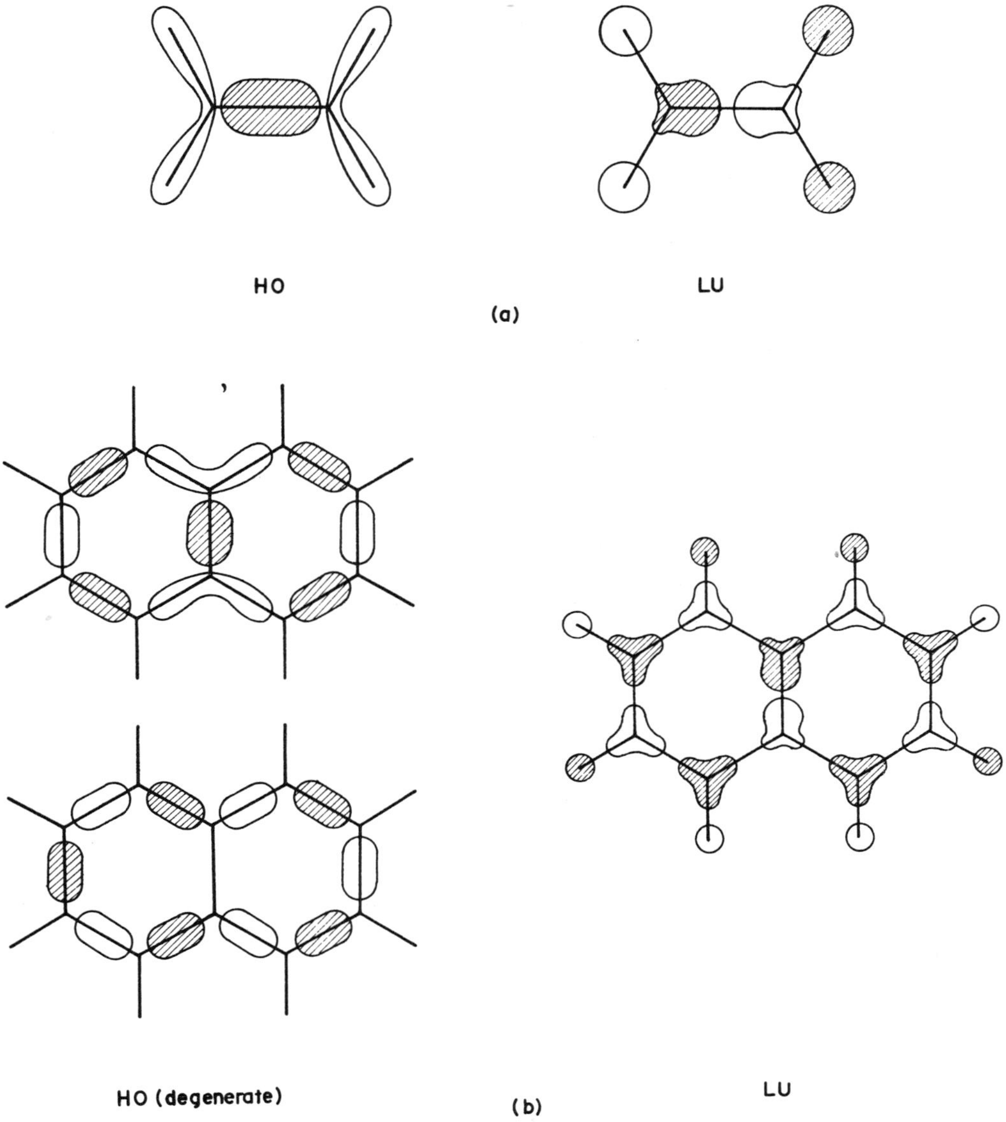

Fig. 6. Schematic diagram of HO and LU orbitals of σ skeletons: (a) ethylene; (b) naphthalene. The shaded and unshaded parts indicate the MO regions with opposite signs.

8. Sigma-Pi Interaction Associated with Nuclear Configuration Change

Many chemical reactions take place with the change of hybridization of carbon atoms. They may be additions, eliminations, substitutions, and some rearranging reactions. These can be regarded as the interaction between the σ part and the π part.

For instance, consider the noncyclic two-center concerted addition of a reagent to a planar unsaturated molecule. Two reaction paths are possible, as shown in Fig. 7. Here we take as an illustration the case of the ethylenic double bond, which is shown in Fig. 8. We

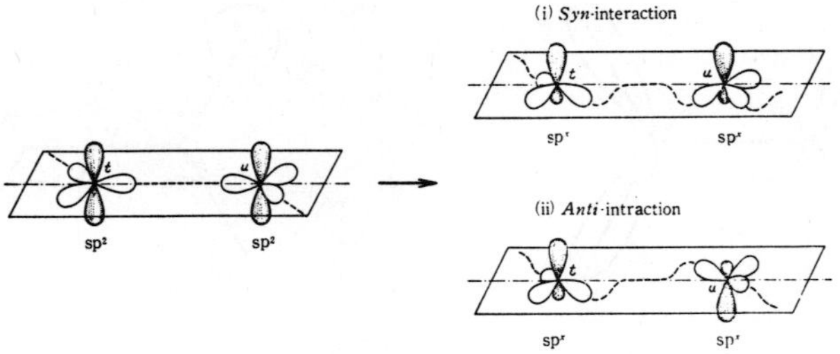

Fig. 7. Schematic representation of the mode of two-center interaction between σ and π parts. The number x in sp^x lies between 2 and 3. The terms *syn* and *anti* are used to avoid confusion with *cis* and *trans* configurations with regard to double bonds.

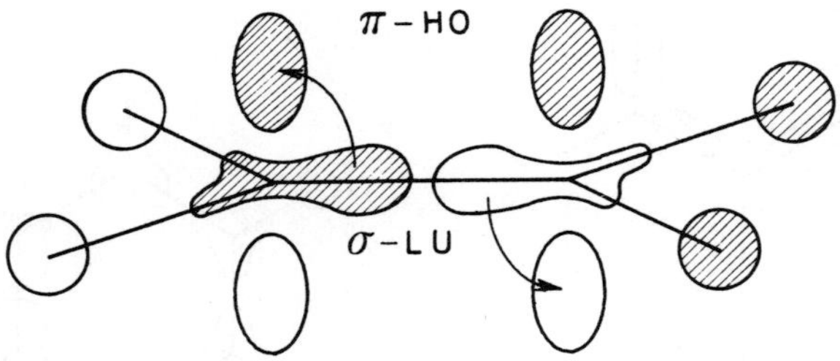

Fig. 8. Steric pathway in α, β-noncyclic interaction in ethylene [see Fig. 6(a)]. The arrows represent the direction of orbital overlapping, which leads to the nuclear configuration change illustrated in Fig. 7(ii).

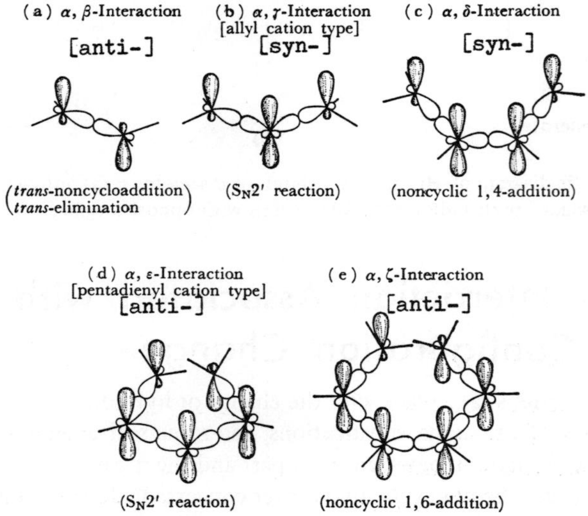

Fig. 9. Theoretical result of the favorable mode of two-center σ-π interaction in various reactions.

may recollect in this connection the discussion of HO and LU orbital extensions in Section 5. It is easily understood that the σ part will behave as an electron-acceptor molecule and the π part as an electron donor. The symmetry consideration of the LU of the σ part and that of the HO of the π part easily leads to the result that the *anti* interaction should be favorable, as is seen in the figure. This conclusion is consistent with experience.

Similar theoretical results concerning other two-center interactions are indicated in Fig. 9 for comparison with experiments.

Many other problems of stereoselection in relation to the σ-π interaction were also discussed by means of hybrid-based MO theory.[14]

9. Conclusion

It is obvious from the instances given above that the hybrid-based MO theory has its reason for existence. This reason is of little theoretical character. It is rather practical in nature, so that the main applications were attempted in the interpretation of chemical problems.

In these circumstances, a major part of the preceding calculations had recourse to the Hückel-type MO approach. However, the theoretical conclusions are expected to be essentially unchanged with or without consideration of electron repulsion, in the light of the results of the comparisons made so far.

References and Notes

1. For instance, see the recent article of Wahl et al. [A. C. Wahl, P. J. Bertoncini, G. Das, and T. L. Gilbert, *Intern. J. Quant. Chem.* **1**, S123 (1967)].
2. C. C. J. Roothaan, *Rev. Mod. Phys.* **23**, 69 (1951).
3. (a) C. Sandorfy, *Can. J. Chem.* **33**, 1337 (1955); (b) G. W. Wheland and P. S. K. Chen, *J. Chem. Phys.* **24**, 67 (1956); (c) Y. Yoshizumi, *Trans. Faraday Soc.* **53**, 125 (1957); (d1) K. Fukui, H. Kato, and T. Yonezawa, *Bull. Chem. Soc. Japan* **33**, 1197 (1960); (d2) *ibid.* **33**, 1201 (1960); (d3) *ibid.* **34**, 442 (1961); (d4) *ibid.* **34**, 1111 (1961); (d5) *ibid.* **35**, 1475 (1962); (e1) J. A. Pople and D. P. Santry, *Mol. Phys.* **7**, 269 (1964); (e2) *ibid.* **9**, 311 (1965).
4. (a) S. Katagiri and C. Sandorfy, *Theoret, Chim. Acta* **4**, 203 (1966); (b) A. Imamura, M. Kodama, Y. Tagashira, and C. Nagata, *J. Theoret. Biol.* **10**, 356 (1966); (c) D. B. Cook, P. C. Hollis, and R. McWeeny, *Mol. Phys.* **13**, 553 (1967); (d) D. B. Cook and R. McWeeny, *Chem. Phys. Letters* **1**, 588 (1968).
5. K. Fukui, in *Modern Quantum Chemistry*, Vol. 1, O. Sinanoğlu, ed., Academic Press, New York, 1965, p. 49, and references cited therein.
6. (a) R. McWeeny, *Rev. Mod. Phys.* **32**, 335 (1960); (b) K. Ruedenberg, *ibid.* **34**, 326 (1962).
7. See many references cited in Fukui's review article (K. Fukui, in *Molecular Orbitals in Chemistry, Physics, and Biology*, P.-O. Löwdin and B. Pullman, eds., Academic Press, New York, 1964, p. 513; compare also the following papers: R. B. Woodward and R. Hoffmann, *J. Am. Chem. Soc.* **87**, 395 (1965); G. Klopman, *J. Am. Chem. Soc.* **90**, 223 (1968); L Salem, *ibid.* **90**, 543, 553 (1968).
8. H. Kato, K. Fukui, and T. Yonezawa, *Bull. Chem. Soc. Japan* **38**, 189 (1965).
9. K. Fukui, H. Kato, T. Yonezawa, and S. Okamura, *Bull. Chem. Soc. Japan* **37**, 904 (1964).
10. T. Yonezawa, H. Kato, H. Saito, and K. Fukui, *Bull. Chem. Soc. Japan* **35**, 1814 (1962).
11. K. Fukui, K. Morokuma, H. Kato, and T. Yonezawa, *Bull. Chem. Soc. Japan* **36**, 217 (1963).
12. K. Morokuma, K. Fukui, T. Yonezawa, and H. Kato, *Bull. Chem. Soc. Japan* **36**, 47 (1963).
13. K. Fukui, H. Kato, T. Yonezawa, K. Morokuma, A. Imamura, and C. Nagata, *Bull. Chem. Soc. Japan* **35**, 38 (1962).
14. K. Fukui and H. Fujimoto, *Bull. Chem. Soc. Japan* **39**, 2116 (1966).

III-4

Electronic Spectra

III-4a

The Electronic Spectra of Sigma-Electron Systems

C. SANDORFY

1. Introduction

The electronic spectra of saturated hydrocarbons are located in the far ultraviolet and this is probably the reason why, until recently, they received so little attention. Methane seems to be the only exception. The most complete works relating to this molecule are those of Ditchburn[1] and Sun and Weissler,[2] who also summarized the previous literature. More general works on aliphatic hydrocarbons started appearing four or five years ago. Among these we mention those of Okabe and Becker,[3] Partridge,[4] Schoen,[5] and Raymonda and Simpson.[6] The absorption spectra of a number of gaseous normal and branched paraffins were measured in the author's laboratory on a McPherson Model 225 vacuum ultraviolet monochromator under approximately 0.2 Å resolution from 2000 to 1150 Å using a double-beam attachment, a hydrogen light source, and photoelectric recording. Our discussion will be based on these spectra.[7,8]

The interest of quantum chemists in σ-electron problems is also relatively recent. The obvious explanation is that the number of valence electrons are much greater than the number of π electrons. Although a number of works are now available on ground-state properties of saturated hydrocarbons, we know about only four works concerning their electronic spectra. These are Mulliken's united atom treatment,[9,10] the Pariser-Parr-Pople type of calculations of Katagiri and Sandorfy[11] and of Brown and Krishna,[12] and the excitonic approach of Raymonda and Simpson.[6]

2. Results and Discussion

a. The bands are usually diffuse and no vibrational structure is observed. Ethane is a significant exception to this.

b. There are one or two weak bands, or at least a pronounced inflection between 1630 and 1575 Å, in all the spectra except that of methane, in which a similar feature is found near 1425 Å. These shift gradually to longer wavelengths.

c. At shorter wavelengths strong bands follow with molecular extinction coefficients in the 10^3 to 10^4 range. The bands exhibit much larger shifts toward lower frequencies, amounting to 4200 cm^{-1} from ethane to propane for the first strong band, then becoming gradually less, and reaching an approximate limit of about 1420 Å for n-pentane. At the same time, the intensities increase gradually (Table 1 and Figs. 1 and 2).

Table 1. Wavelengths and Molecular Extinction Coefficients of the Centers of the First Two Strong Bands of Normal Paraffins

	I		II	
Compound	λ, Å	ε, liters/mole/cm	λ, Å	ε, liters/mole/cm
Methane	1,277	5,839		
Ethane	1,318	8,751		
Propane	1,395	11,240	1,285	13,010
n-Butane	1,410	15,420	1,335	17,950
n-Pentane	1,415	17,330	1,345	20,360
n-Hexane	1,425	19,560	1,335	24,080
n-Heptane	1,430	20,950	1,340	25,980
n-Octane	1,415	24,550	1,340	28,960

According to Mulliken, the united atom configuration of CH_4 is, omitting carbon ls orbitals,

$$[sa_1]^2\,[pf_2]^6\ {}^1A_1$$

where a_1 and f_2 are the usual group theoretical symbols under T_d symmetry. The lowest excited orbitals then would be Rydberg-type, large atomic orbitals 3s, 3p, 3d, . . . with 3s lowest. Mulliken pointed out, however, that antibonding localized C–H orbitals are qualitatively very similar to those of 3s and 3p orbitals of an atom. Thus in this united atom approach the first excited state of methane would have configuration $[sa_1]^2\,[pf_2]^5\,[3sa_1]\ {}^1F_2$ and the correspond-

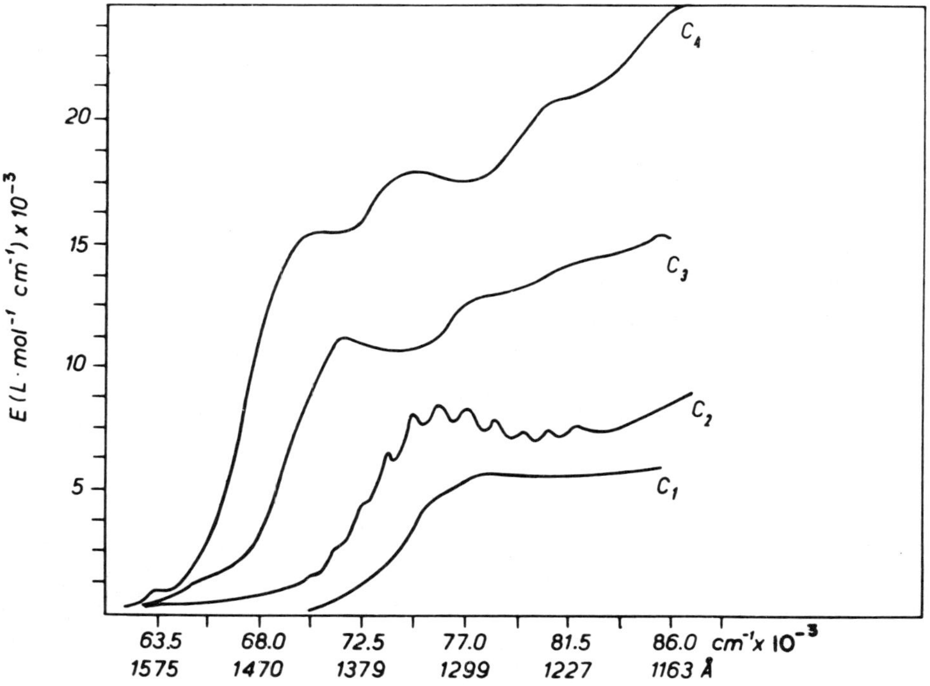

Fig. 1. Far-ultraviolet absorption spectra of methane, ethane, propane, and n-butane. Molecular extinction coefficients against wave numbers.

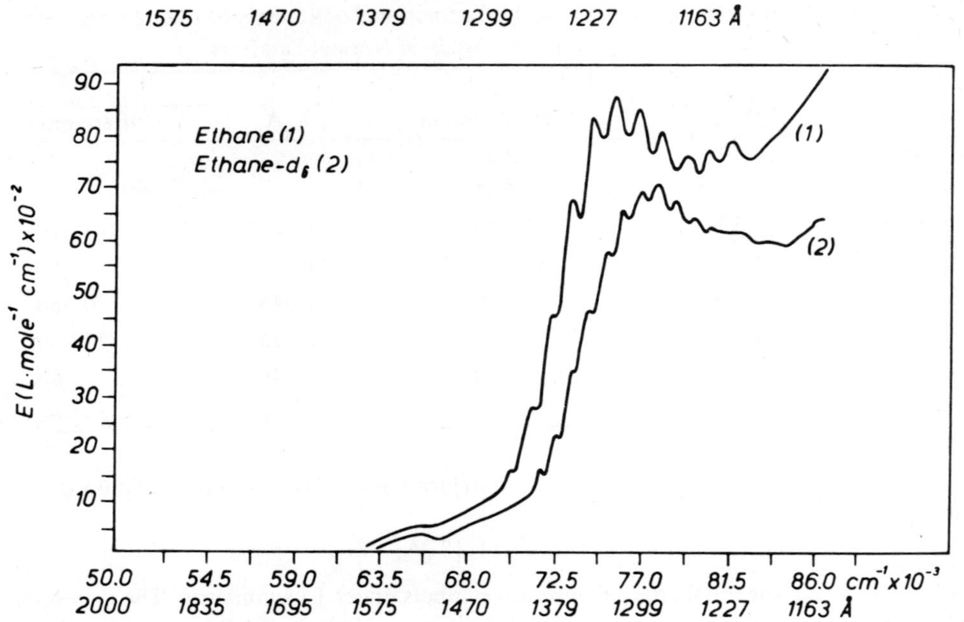

Fig. 2. Far-ultraviolet absorption spectra of ethane and ethane-d$_6$. Molecular extinction coefficients against wave numbers.

ing transition would be of type $^1F_2 \longleftarrow {}^1A_1$ and allowed. The observed intensity ($\epsilon = 5800$, oscillator strength f = 0.26) seems to justify this assignment, although the possibility that the band is a forbidden one borrowing intensity from a stronger transition at higher frequencies is not ruled out. In fact, we know from Schoen's work[5] that the 1277-Å band of methane is followed by even stronger bands.

In Katagiri and Sandorfy's scheme, which is based on the C2s, C2p, and H1s orbitals only, the molecular orbitals are, in order of increasing energy, a_1, f_2, a_1, f_2, so that the transition of lowest energy is $a_1 \longleftarrow f_2$, that is, again $^1F_2 \longleftarrow {}^1A_1$. The sequence of the excited orbitals, however, depends in a delicate way on the choice of certain parameters. If the order was a_1, f_2, f_2, a_1, the first transition would be $f_2 \longleftarrow f_2$, yielding $F_2 \times F_2 = A_1 + E + F_1 + F_2$, the transition to F_2 being allowed and the others forbidden. The latter could be made allowed by vibronic interactions and the band we observed may be due to these. It is also possible that the shoulder at 1425 Å represents a separate electronic band due to one of these forbidden transitions. We have no means of checking upon these tentative assignments, however.

The ground state of ethane in Mulliken's united atom treatment had the following configuration:

$$\underset{CH_3}{[sa_1]^2}\ \underset{CH_3}{[sa_1]^2}\ \underset{CH_3}{[\pi e]^4}\ \underset{CH_3}{[\pi e]^4}\ \underset{C{-}C}{[\sigma + \sigma]^2}\ {}^1A_1$$

Here the two methyl groups are treated as two separate united atoms except that a C–C molecular orbital is formed for the two 2pσ electrons forming that bond. This yields $[\sigma \pm \sigma]$, the plus sign applying to the orbital of lower energy. For the lowest excited state an electron would go to an orbital formed by the two carbon 3s atomic orbitals, [3s + 3s]. Then we obtain the configuration $[sa_1]^2 [sa_1]^2 [\pi e]^4 [\pi e]^4 [\sigma + \sigma]\ [3s + 3s]\ {}^1A_1$ if the electron is taken from the C–C bond. The transition to this state would be $^1A_1 \longleftarrow {}^1A_1$ and forbidden. From considerations based on ionization potentials, Mulliken predicted that the corresponding band should be at about 1600 Å. No such band was known at the time, but we are now making the suggestion that the weak band we find in the spectrum of ethane corresponds to this transition. It is better seen in Fig. 2.*

The first strong band would then be due to the transition of an electron in a C—H bond to the same excited orbital:

$$[sa_1]^2 [sa_1]^2 [\pi e]^4 [\pi e]^3 [\sigma + \sigma]^2 [3s + 3s]\ {}^1E$$

The ${}^1E \longleftarrow {}^1A$ transition is allowed.

In Katagiri and Sandorfy's Pariser-Parr type of calculations the first transition is of the ${}^1E \longleftarrow {}^1A$ type, but we find no equivalent for Mulliken's low-lying ${}^1A_1 \longleftarrow {}^1A_1$ transition.

More extended 3s + 3s + 3s + · · · types of orbitals may account for the bathochromic shift which is observed when the number of carbon atoms increases.

More elaborate calculations using a basis of atomic orbitals including excited ones will probably be needed before we can pass the speculative stage in interpreting these spectra.

It is a matter of some interest to know if the first ionization leaves the "hole" in the C—C bonds or in the C—H bonds or if it is distributed over both. In Mulliken's united atom scheme the highest orbital filled in the ground state is of C—C character in ethane. The Hückel-type calculations of Fukui et al.,[13] Klopman,[14] and Hoffman[15] all seem to favor the C—C.

Raymonda and Simpson[6] assume that, in first approximation, the C—C electrons can be treated separately (like the π electrons in conjugated systems but with less justification), so that the first ionization is again C—C in their method. However, the Pariser-Parr-Pople type of calculations of Katagiri and Sandorfy[11] yield electron densities mainly concentrated in the C—H bonds in the uppermost occupied orbitals, for at least ethane and propane.

From their photoelectron spectra Al-Joboury and Turner[16] also concluded the involvement of C—H rather than C—C bonds. Their bands were broad, however, and they have allowed for the possibility that they result from ionization from more than one close-lying orbital.

The matter does not seem to be definitely settled. Changes from C—H to C—C might occur in going from one paraffin molecule to the other and, in particular, between isomers with different degrees of branching.

The diffuse character of the bands in the spectra of the saturated hydrocarbons is usually attributed to dissociation or predissociation in the excited states, which is indeed very probable in view of the high excitation energies involved. In relatively small molecules where the states are not very crowded, the chances for predissociation are expected to be lower, and this may be the reason ethane and cyclopropane[6] exhibit some vibrational fine structure.

Comparison of the spectra of ethane and ethane-d_6 shows that the spacing in the main apparent progression is related to a bending vibration, which is essentially C—H in character. The suggestion was made[7] that the existence of two maxima in this progression and a certain alternance between a higher and a lower value in the spacing may indicate a Jahn-Teller effect.

Despite the important role played by dissociation, we should be cautious in attributing to these phenomena all the diffuseness that is observed. We have to remember in this respect that because of rotational isomerism our spectra are actually spectra of mixtures and that the number of rotational isomers, as well as the number of totally symmetrical vibrations, increases rapidly with increase in the number of atoms in the molecules.

There are some significant differences between the spectra of the branched-chain paraffins, especially those of the highly branched ones, and those of the normal paraffins.

Typical examples are shown in Figs. 3 and 4. The first bands are much stronger than in the spectra of the normal paraffins and these are followed by a fairly deep minimum toward higher

**Note added in proof*: According to more recent measurements the existence of this band is doubtful.

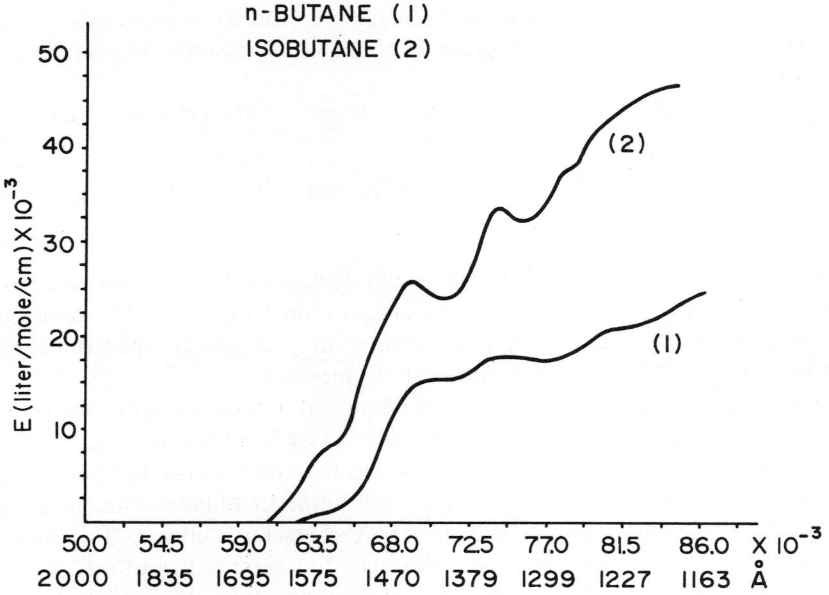

Fig. 3. Far-ultraviolet absorption spectra of n-butane and isobutane. Molecular extinction coefficients against wave numbers.

frequencies. We have no quantum chemical calculations available to help in the interpretation of these spectra. It is possible that in this case the first bands correlate to the first *strong* bands of the normal paraffins. This would again mean that there has been a change in the order of the highest occupied molecular orbitals in the ground state.

Cyclopropane was studied by Brown and Krishna,[12] who used a Pariser-Parr type of approximation, including some configuration interaction. The success of their method based on the C–C bonding orbitals only seems to be in keeping with the more delocalized character of these orbitals. The spectra of cyclopropane and cyclobutane are shown in Fig. 5.

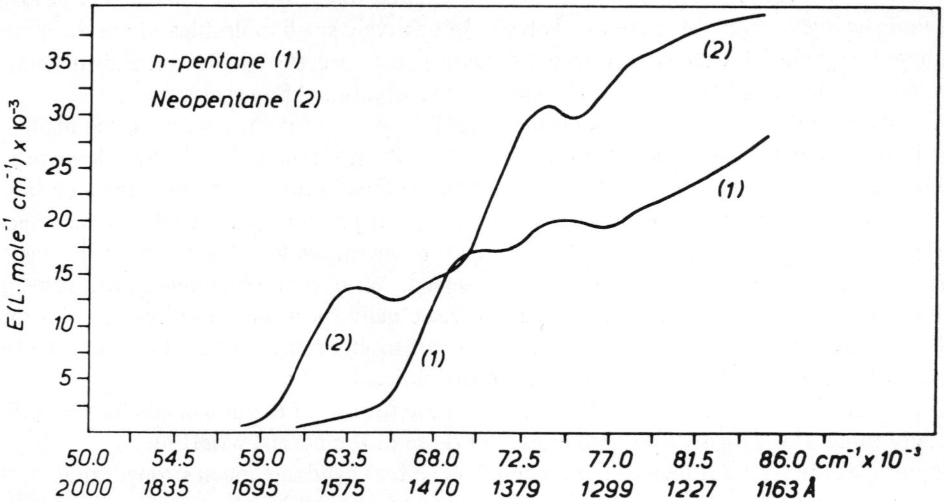

Fig. 4. Far-ultraviolet absorption spectra of n-pentane and neopentane. Molecular extinction coefficients against wave numbers.

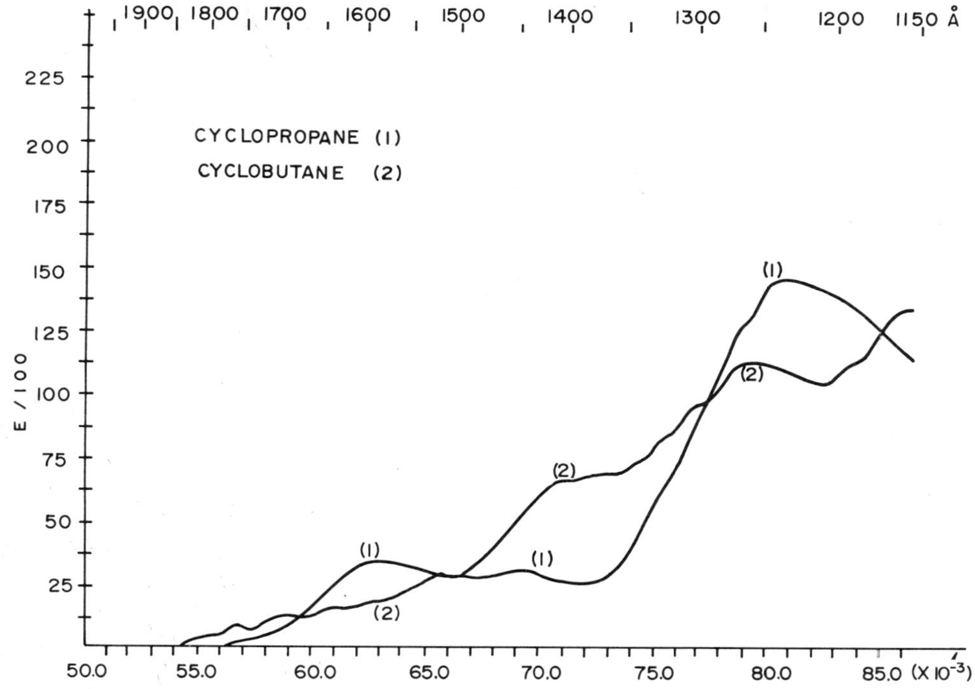

Fig. 5. Far-ultraviolet absorption spectra of cyclopropane and cyclobutane. Molecular extinction coefficients against wave numbers.

All electronic transitions mentioned in this communication were singlet-singlet. The first singlet-triplet transitions were predicted by Katagiri and Sandorfy[11] to lie by about 1 eV to lower frequencies from the first singlet-singlet bands. None of them has been found up to the present time.

We conclude by saying that much remains to be done. The solid theoretical basis for the theoretical discussion of σ-electron spectra is yet to be created.

Acknowledgments

The results presented in this lecture were based on the theses of Dr. B. A. Lombos, P. Sauvageau, and G. Bélanger. We are indebted to Professor W. Lüttke and Dr. J. De Meijere from the University of Göttingen for a generous sample of cyclobutane.

References

1. R. W. Ditchburn, *Proc. Roy. Soc. (London)* **A229**, 44 (1955).
2. H. Sun and G. L. Weissler, *J. Chem. Phys.* **23**, 1372 (1955).
3. H. Okabe and D. A. Becker, *J. Chem. Phys.* **39**, 2549 (1963).
4. R. H. Partridge, *J. Chem. Phys.* **45**, 1685 (1966).
5. R. I. Schoen, *J. Chem. Phys.* **37**, 2032 (1962).
6. J. W. Raymonda and W. T. Simpson, *J. Chem. Phys.* **47**, 430 (1967).
7. B. A. Lombos, P. Sauvageau, and C. Sandorfy, *J. Mol. Spectry.* **24**, 253 (1967).
8. B. A. Lombos, P. Sauvageau, and C. Sandorfy, *Chem. Phys. Letters* **1**, 42,221 (1967).
9. R. S. Mulliken, *J. Chem. Phys.* **3**, 517 (1935).
10. R. S. Mulliken, *J. Am. Chem. Soc.* **86**, 3183 (1964).
11. S. Katagiri and C. Sandorfy, *Theoret. Chim. Acta* **4**, 203 (1966).

12. R. D. Brown and V. G. Krishna, *J. Chem. Phys.* **45**, 1482 (1966).
13. K. Fukui, H. Kato, and T. Yonezawa, *Bull. Chem. Soc. Japan* **34**, 442 (1961).
14. G. Klopman, *Sur la structure électronique des molécules saturées*, Cyanamid European Research Institute, Geneva, 1962.
15. R. Hoffman, *J. Chem. Phys.* **40**, 2047 (1964).
16. M. I. Al-Joboury and D. W. Turner, *J. Chem. Soc.* **B1967**, 373.

III-4b Reprinted from *The Journal of Chemical Physics* **45**, 1367 (1966)

Electronic Spectra and Structure of Sulfur Compounds*

S. D. Thompson, D. G. Carroll, F. Watson, M. O'Donnell, and S. P. McGlynn†
Coates Chemical Laboratories, The Louisiana State University, Baton Rouge, Louisiana
(Received 7 February 1966)

The lower-energy electronic states of a series of saturated sulfur compounds have been investigated theoretically and experimentally. The compounds investigated fell into two categories: the monosulfide series ABS in which the substituents A and B were varied, and the polysulfide series R_2S_n, where $n=1, 2, 3$, and 4 and where R was usually CH_3. The data collected included ultraviolet vapor spectra, solution spectra, solvent shifts, and vibrational data. Computations followed an approach related to that of Wolfsberg and Helmholz. This theoretical method closely approximates the experimentally observed energies in the compounds containing a single sulfur atom but encounters difficulties when two or more sulfurs are present because of the large $3d_S$–$3d_S$ and $4s_S$–$4s_S$ interactions which are predicted. It is indicated that the lower-energy excited electronic states of all the monosulfide compounds investigated are of a molecular nature in which the $4s_S$ and $3d_S$ atomic orbitals of sulfur play particularly dominant but by no means exclusive roles. The lower-energy excited states of the molecules R_2S_n, $n=2, 3$, and 4, also contain significant $4s$ and $3d$ character; additionally, however, they are of S–S antibonding nature.

I. INTRODUCTION

SEVERAL excellent investigations have been conducted[1–3] on Rydberg transitions of systems similar to those investigated here, but the study of the lower-energy electronic excitations of these molecules seems to have been neglected. These latter transitions include excitations to sigma antibonding levels and to the first few low-energy members of the Rydberg series. The low-energy Rydbergs undergo a sufficient interaction with their molecular environment to loes a considerable amount of their "atomic" character.

The compounds investigated were selected to exhibit the effects of varying two parameters. First, hydrogen sulfide, methyl mercaptan, and methylsulfide demonstrate the effects of varying the substituent attached to the sulfur. Second, dimethylmono-, di-, tri-, and tetrasulfides show the changes produced as the length of the sulfur chain is increased and as the effects of the methyl end groups are thereby diminished.

Many of the compounds investigated here absorb at such high energies in the ultraviolet that it is difficult to excite them with available sources. They fail to give any emission at the sensitivities available to us. Spectral shift studies were also of limited use: high-energy transitions were not readily studied because of the transmission limitations of available solvents. A study of the photoinduced scission of S–S linkages has been presented elsewhere.[4]

II. EXPERIMENTAL

A. Instrumentation and Technique

Ultraviolet absorption spectra were measured on Cary recording spectrophotometers 14R and 15. Quartz absorption cells of various path lengths (0.1 to 100 cm) were used. The molar extinction coefficient (ϵ) of $(CH_3)_2S$ was calculated from the equation[5]

$$\epsilon=760\times22.4\times DT/273bp,$$

where D is optical density; T, temperature degrees Kelvin; b, path length (centimeters); and p, pressure

* Supported by contract between the Atomic Energy Commission—Biology Branch and The Louisiana State University.
† Sloan Foundation Fellow.

1 W. C. Price, J. Chem. Phys. **4**, 147 (1936).

2 W. C. Price, J. P. Teegan, and A. D. Walsh, Proc. Roy. Soc. (London) **A201**, 600 (1950).

3 (a) L. B. Clark, Ph.D. dissertation, University of Washington, Seattle, Wash., 1963; (b) L. B. Clark and W. T. Simpson, J. Chem. Phys. **43**, 3666 (1965).

4 V. Ramakrishnan, S. D. Thompson, and S. P. McGlynn, Photochem. Photobiol. **4**, 907 (1965).

5 G. R. A. Brandt, H. J. Emeleus, and R. N. Haszeldine, J. Chem. Soc. **1952**, 2249.

TABLE I. Coulomb integrals H_{ii} (in electron volts).

AO[a,b]	$-H_{ii}$	AO[a,b]	$-H_{ii}$
$3s_S$	20.72	$1s_H$	13.53[f]
$3p_S$	11.61[e]	σ_{1sH}	19.04
$3p_{xS}(H_2S)$	10.47[c,d]	σ_{1sH}^*	8.02
$3p_{xS}(CH_3SH)$	9.44[c,d]	sp^3_C	9.95[g]
$3p_{xS}[(CH_3)_2S]$	8.68[c,e]	σ_{sp^3C}	12.40
$3d_S$	3.67	σ_{sp^3C}	7.50
$4s_S$	3.76	$4p_S$	3.02

[a] Orbital definitions are given in Figs. 1, 2, and 3.
[b] Subscripts S, H, and C denote, respectively, sulfur, carbon, and hydrogen.
[c] Coulomb terms for $3p_{xS}$ orbitals were assumed equal to molecular ionization potentials in H_2S, CH_3SH, and $(CH_3)_2S$; otherwise, their values would also have been 11.61 eV.
[d] Reference 2.
[e] Reference 3(a).
[f] *Handbook of Chemistry and Physics*, C. D. Hodgman *et al.*, Eds. (The Chemical Rubber Publishing Co., Cleveland, Ohio, 1956–1957), 38th ed., p. 2347.
[g] F. P. Lossing, K. U. Ingold, and I. H. S. Henderson, J. Chem. Phys. **22**, 621 (1954).

(millimeters). The vapor pressure was obtained from the expression[6]

$$\log p = 16.51798 - 1876.37/T - 3.0427 \log T.$$

For $T = 293°K$, p is found to be 395 mm. No information was available regarding the vapor pressure of most compounds investigated; consequently extinction coefficients for their vapor spectra are not presented. However, a comparison of absorption intensities may be made using the extinctions calculated from solution studies.

Possible emission spectra of H_2S, $(CH_3)_2S$, $(CH_3)_2S_2$, and $(CH_3)_2S_3$ were investigated using an Aminco-Kiers spectrophosphorimeter. Measurements were made in EPA solvent (ether:isopentane:ethanol = 5:5:2) at ambient and liquid-nitrogen (77°K) temperatures. It is probable that weak emissions are present, but more sensitive apparatus is needed to prove their existence conclusively.

The gas chromatograph used for purification of some compounds was a Beckman Model GC-2. The column was packed with 40% tritolyl phosphate on commercial 42–60-mesh firebrick. Repeated injections of up to 0.5 ml of the substance to be purified were made and the desired component was collected at −76°C.

B. Chemicals

The dimethylsulfide and dimethyldisulfide were the commercially available White Label grade reagents

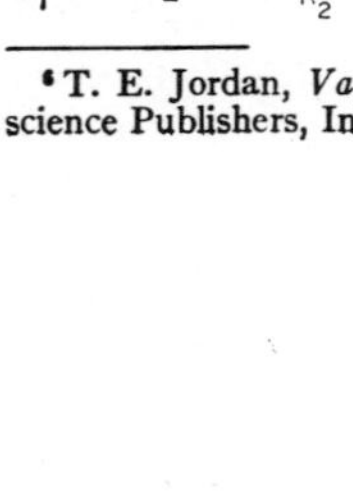

FIG. 1. Coordinates of hydrogen sulfide, methyl mercaptan, and dimethylsulfide. (R=H or CH_3.)

[6] T. E. Jordan, *Vapor Pressure of Organic Compounds* (Interscience Publishers, Inc., New York, 1954), p. 224.

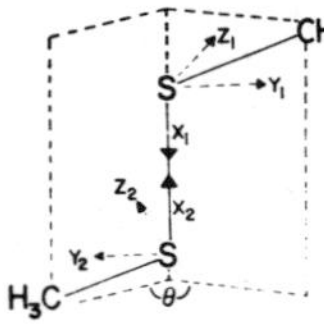

FIG. 2. Coordinates of dimethyldisulfide.

produced by Eastman Organic Chemicals. Both were purified by triple distillation followed by preparative gas chromatography.

Hydrogen sulfide and deuterium sulfide were used as obtained from the Matheson Company and from Volk Radiochemical Company, respectively.

The dimethyltrisulfide was prepared by the method of Strecker.[7] Purification was accomplished by vacuum sublimation and gas chromatography. Physical constants were $b_{15} = 74°–76.8°C$; $n_D^{24.2} = 1.6044$; $d_{24} = 1.197$.

The procedure of Feher, Krause, and Vogelbruch was used to prepare dimethyltetrasulfide.[8] The compound was repeatedly vacuum distilled until the final distillation showed a change of only 0.0002 units in the refractive index. The properties of the dimethyltetrasulfide were $b_{0.3} = 70.5° - 72°C$; $n_D^{24} = 1.6590$.

The analysis is:

	C	H	S
Calc	15.19	3.80	81.01
Found	15.53	3.81	80.85.

The solvents used were Harleco fluorimetric grade ethyl alcohol, methyl alcohol, cyclohexane, and ultrapure water. Hexane was purified by drying over sodium wire, distilling and passing through a 6-ft column of activated silica.

III. MOLECULAR ORBITAL CALCULATIONS

The calculation of energy levels is similar to that of Wolfsberg and Helmholz.[9] The non-valence-shell electrons are assumed to be unaffected by bonding and, with the nuclei, they are taken to constitute an effective core field in which the valence or optical electrons are supposed to move. Consequently, for sulfur it was necessary to consider only the $3s_S$, $3p_S$, $3d_S$, and $4s_S$ atomic orbitals (AO's). The $4p_S$ orbitals were included

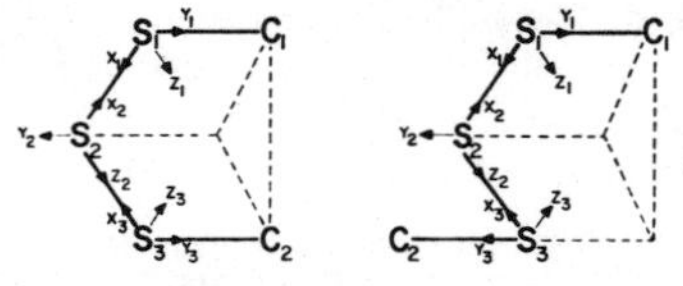

cis *trans*

FIG. 3. Coordinates of dimethyltrisulfide.

[7] W. Strecker, Chem. Ber. **41**, 1105 (1908).
[8] T. Feher, G. Krause, and K. Vogelbruch, Chem. Ber. **90**, 1570 (1957).
[9] M. Wolfsberg and L. Helmholz, J. Chem. Phys. **20**, 837 (1952).

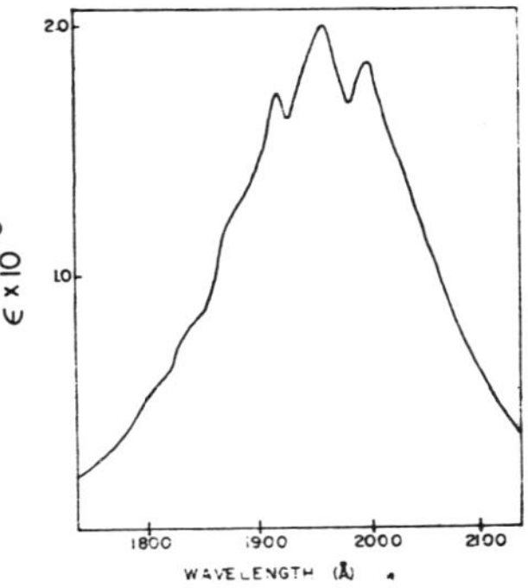

FIG. 4. Structure of the hydrogen sulfide 2000-Å vapor band obtained at −80°C.

in some initial computations; consideration of these orbitals made no significant difference to the lower-energy monosulfide levels, and computations which included them were discontinued. The effects of $4p_S$ inclusion for the molecules R_2S_n, $n=2, 3, 4, \cdots$, were not investigated. Since the VSIE of $4p_S$ is 3.02 eV, these orbitals will certainly be involved in the lower excited states of polysulfides; however, the lowest excited states are still expected to consist mainly of $4s$, $3d$, $3p$, and $3s$ atomic orbitals of sulfur and appropriate atomic orbitals of the attached atoms. For hydrogen and carbon the orbitals of interest were considered to be the $1s_H$ AO, and the $2s_C$ AO and $2p_C$ AO's, respectively.

Coulomb integrals, H_{ii}, were equated to the neutral atom valence-state ionization potentials (VSIP). The Coulomb terms for the highest energy nonbonding electrons were assumed to be equal to their respective molecular ionization potentials. The final values obtained for the various Coulomb integrals are presented in Table I.

Resonance integrals were obtained from the equation[10]

$$H_{ij}=(2-|S_{ij}|)\tfrac{1}{2}(H_{ii}+H_{jj})S_{ij}.$$

The orbitals used were Slater type; this caused certain problems involving the use of $3d_S$ and $4s_S$ orbitals. The effective charges Z for these orbitals are predicted by

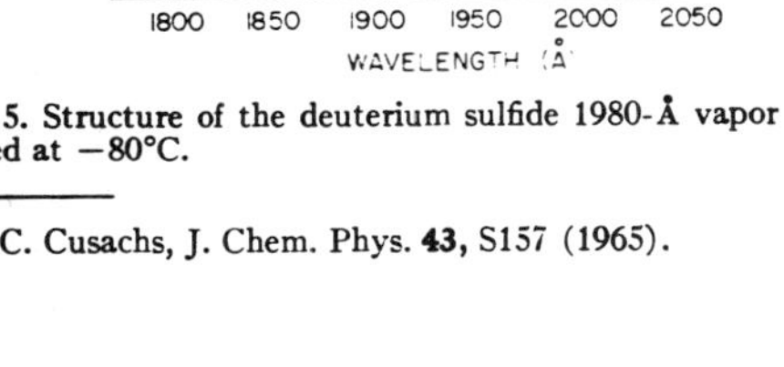

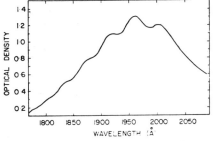

FIG. 5. Structure of the deuterium sulfide 1980-Å vapor band obtained at −80°C.

[10] L. C. Cusachs, J. Chem. Phys. **43**, S157 (1965).

TABLE II. Vibrational intervals in the 2000-Å bands of H_2S and D_2S.

Vibrational interval[a]	H_2S (cm^{-1})	D_2S (cm^{-1})
2–1	1145	804
3–2	1144	817
4–3	1112	804
5–4	1101	871
6–5	1086	828
7–6		794
8–7		834

[a] The first vibrational peak in each spectrum is assigned as the 0–0 band.

Slater's recipe[11] to be zero; this does not appear to be realistic. Consequently, calculations were performed using Z values of 1.0, 0.5, and 0.0 for the $3d_S$'s and 0.9 and 0.0 for the $4s_S$. Although there is little doubt that the charges used do not represent the true situation, they probably do bracket the limits within which it lies.

In the computations, no initial hybridization of AO's was assumed other than considering carbon hybrid orbitals to be sp_C^3's. The coordinates chosen for the various molecules are illustrated in Figs. 1, 2, and 3. In solving the secular determinant, the approximation was made that $ES_{ij}=E\delta_{ij}$, where δ is the Kronecker delta. However, in the evaluation of H_{ij} the overlap S_{ij} was in all instances computed. The molecular geometries used are given in the Appendix.

IV. RESULTS AND DISCUSSION

A. Hydrogen Sulfide (H_2S) and Deuterium Sulfide (D_2S)

The long-wavelength absorption bands of H_2S and D_2S gases were investigated over the temperature range −80° to 25°C in an attempt to sharpen the vibrational structure and to resolve the observed long-wavelength tail. Moderate success was achieved in sharpening the structure but the investigations failed to reveal any distinct transition in the absorption tail. The observed vapor spectra are shown in Figs. 4 and 5. The analysis of Table II indicates that all vibrational peaks are members of a single progression. The frequency differences between successive vibrational peaks of both spectra are given in Table II; the average differences are 1118±20 cm^{-1} and 822±20 cm^{-1} in H_2S and D_2S, respectively. The observed ground-state fundamental vibrations are listed in Table III. The

TABLE III. H_2S and D_2S fundamental vibrational frequencies.

Molecule	$\bar{\nu}_1$ (cm^{-1})	$\bar{\nu}_2$ (cm^{-1})	$\bar{\nu}_3$ (cm^{-1})
H_2S	2611	1290	2684
D_2S	1892	934	1999

[11] J. C. Slater, Phys. Rev. **36**, 57 (1930).

TABLE IV. Results of H_2S computations ($Z_{3d\mathrm{S}}=1.0$; $Z_{4s\mathrm{S}}=0.9$).

Symmetry[a]	Eigenvectors										Eigenvalues[b] (eV)
	σ_{1s_H}	$3s_\mathrm{S}$	$3p_{z\mathrm{S}}$	$4s_\mathrm{S}$	z^{2}[c]	x^2-y^2[c]	$\sigma^*_{1s_\mathrm{H}}$	$3p_{y\mathrm{S}}$	yz[c]	$3p_{x\mathrm{S}}$	
$6a_1$	0.6488	−0.4523	−0.5072	−0.0605	−0.2496	−0.2266	0	0	0	0	2.26
$3b_2$	0	0	0	0	0	0	−0.7189	0.5439	0.4327	0	−0.68
$5a_1$	0.0000	0.0000	0.0000	0.0000	0.6722	−0.7404	0	0	0	0	−3.67
$4a_1$	0.0058	−0.0055	−0.0080	−0.9785	0.1524	0.1383	0	0	0	0	−3.76
$2b_2$	0	0	0	0	0	0	−0.2921	0.3285	−0.8982	0	−4.26
$3a_1$	0.1743	−0.1696	−0.2570	0.1964	0.6770	0.6147	0	0	0	0	−4.26
$1b_1$	0	0	0	0	0	0	0	0	0	1.000	−10.47[d]
$2a_1$	0.2219	−0.5936	0.7710	0.0113	0.0458	0.0415	0	0	0	0	−14.73
$1b_2$	0	0	0	0	0	0	0.6307	0.7722	0.0773	0	−18.37
$1a_1$	−0.7067	−0.6436	−0.2869	−0.1114	−0.0465	−0.0422	0	0	0	0	−38.31

[a] Refer to the C_{2v} character table and Fig. 1 for representation definition.
[b] Interactions of the $3d(xy)$S and $3d(xz)$S are symmetry forbidden. Consequently, these AO's are not included in the table; the corresponding eigenvalues are −3.67 eV. These MO's transform as $2b_1$ and $1a_2$.
[c] $3d_\mathrm{S}$ AO's.
[d] Experimental ionization potential of H_2S.

vibration ν_1 is the symmetrical stretch; ν_2 is the symmetric bend; ν_3 is the antisymmetric stretch. The ratio of the observed D_2S/H_2S excited-state vibrational frequencies is 0.735 and compares favorably with the 0.718 calculated for the unsymmetrical stretch from ground-state data.[12] However, the frequency difference necessitated if the observed excited-state vibrations are members of a ν_3 progression would appear to be much too large. Moreover, theoretical considerations indicate that a symmetrical shape change occurs upon excitation to the 2000-Å state; consequently it is most reasonable to suppose that the active vibrations belong to totally symmetric species.

This then leaves a choice between the symmetrical stretch and the bending mode, both of which are of the a_1 species. Once again, comparison with the ground-state vibrational frequencies would indicate that ν_2 is the more reasonable assignment since it occurs at 1290 and 934 cm^{-1} in the ground states of H_2S and D_2S, respectively. As further evidence for assignment of the progression to a bending mode, it is observed in Table II that the changes in vibrational frequencies in going to successive members of the progressions are relatively minor. Consequently, the anharmonicity χ of the excited state is very small ($\chi<10$ cm^{-1}), indicating that this excited state is a tightly bound state such as might arise from excitation to a $3d_\mathrm{S}$ or $4s_\mathrm{S}$ AO. The fact that the progression which appears seems to be one in ν_2 supports the hypothesis given, because if the excited level were of sigma antibonding character, as has been proposed by a number of authors,[3,13,14] one would expect both a preference for a stretching mode and considerable anharmonicity.

The results of the computations previously outlined are given in Table IV. The lowest energy excited state is predicted to be doubly degenerate, to occur at 6.21 eV, and to possess 75%–80% $3d_\mathrm{S}$ character. The next lowest energy excited state is predicted at 6.71 eV, and is supposed to be mostly $4s_\mathrm{S}$ in nature. Three other excited states occur at 6.80 eV; these are pure $3d_\mathrm{S}$ in character. It is predicted then that the lowest energy excitation is that of an electron from $1b_1$ to either $3a_1$, $4a_1$, or $2b_2$ MO's, to be significantly atomic in nature and to occur in the energy range 6.21–6.80 eV; the observed lowest energy excitation is at 6.51 eV. It is clear that an assignment of the sort given explains the small anharmonicity of the 2000-Å band and the absence of either a ν_1 or ν_3 vibrational progression. The experimental oscillator strength of the 2000-Å band is approximately $f=0.04$. The calculated oscillator strength, while not zero, is indeed very small for all the allowed configurational transitions indicated. However, Carroll, Armstrong, and McGlynn[15] have recently carried out open-shell calculations on H_2S, and have iterated to self-consistency of the density matrix. They have found that increase of μ_{4s_S} from 0.24 to 0.32 increases the oscillator strength of the $3p_{z\mathrm{S}}\rightarrow 4a_1$ electronic transition from 4×10^{-6} to 0.006. Because of this extreme sensitivity of f to the value of the orbital exponent μ, it is quite clear that the observed extinction can readily be accounted for by further increase of μ_{4s_S}. Similar conclusions apply to the allowed excitations to the $3d_\mathrm{S}$ excited configurations, but of these only the 6.21-eV excitations could possibly possess sufficient intensity.

The two excitations to the dominantly S–H antibonding configurations are predicted to occur at 9.79 and 12.73 eV, respectively. It is not possible to effect sufficient lowering of these configurational energies to make them candidates for the upper level of the 2000-Å absorption band. It is possible, however, that these

[12] G. Herzberg, *Molecular Spectra and Molecular Structure. Infrared and Raman Spectra of Polyatomic Molecules* (D. Van Nostrand Co., Inc., Princeton, N.J., 1945), Vol. 2, p. 161.
[13] G. Bergson, Ph.D. dissertation, University of Uppsala, Uppsala, Sweden, 1962.
[14] H. P. Koch, J. Chem. Soc. **1952**, 387.

[15] D. G. Carroll, A. T. Armstrong, and S. P. McGlynn, J. Chem. Phys. **44**, 1865 (1966).

TABLE V. Results of $(CH_3)_2S$ computations ($Z_{3d_S}=1.0$; $Z_{4s_S}=0.9$).

Symmetry[a]	Eigenvectors										Eigenvalues[b] (eV)
	$\sigma^*_{sp^3}{}_C$	$3s_S$	$3p_{zS}$	z^2 [c]	x^2-y^2 [c]	$4s_S$	$\sigma^*_{sp^3}{}_C$	$3p_{yS}$	yz[c]	$3p_{xS}$	
a_1	−0.7458	0.4418	0.4981	0.0151	0.0151	0.0055	0	0	0	0	0.28
b_2	0	0	0	0	0	0	0.7925	−0.6092	−0.0300	0	−1.55
a_1	0.0000	0.0000	0.0000	−0.7071	0.7071	0.0000	0	0	0	0	−3.67
a_1	−0.0136	0.0099	0.0136	−0.7069	−0.7069	0.0046	0	0	0	0	−3.67
b_2	0	0	0	0	0	0	−0.0211	0.0211	−0.9995	0	−3.67
a_1	0.0036	−0.0026	−0.0036	0.0032	0.0032	1.0000	0	0	0	0	−3.76
b_1	0	0	0	0	0	0	0	0	0	1.0000	−8.68[d]
a_1	0.2619	−0.4933	0.8295	0.0020	0.0020	0.0008	0	0	0	0	−14.12
b_2	0	0	0	0	0	0	−0.6095	−0.7927	−0.0035	0	−17.55
a_1	0.6124	0.7492	0.2522	0.0018	0.0018	0.0007	0	0	0	0	−30.89

[a] Taken to be of C_{2v} symmetry species. Consult Fig. 1 and the character tables for representation designations.
[b] Interactions of the $3d(xy)$S and $3d(xz)$S are symmetry forbidden. Consequently, they have not been included in the table; their eigenvalues are −3.67 eV.
[c] $3d_S$ AO's.
[d] Experimental ionization potential of $(CH_3)_2S$.

excitations may account for broad background found[3] in the vacuum-ultraviolet spectra at 88 000 cm^{-1}.

The calculations of Carroll *et al.*[15] are particularly relevant to the present work. Firstly, they substantiate the general conclusions of the calculational scheme which we have used for H_2S; it is thus not unreasonable to hope that this same calculational scheme will yield relatively satisfactory descriptions of all other sulfur molecules so treated. Secondly, they predict the $4s_S$ *nominal* Rydberg orbital to be the lowest excited one, and they conclude that the intensity of the electronic transition to this orbital increases as its Rydberg nature decreases.

B. Dimethylsulfide

The vapor spectrum of dimethylsulfide is presented in Fig. 6. Figure 7 shows the long-wavelength absorption edge in solution. The inflection at approximately 2350 Å is so weak that it is detectable only in solution spectra.

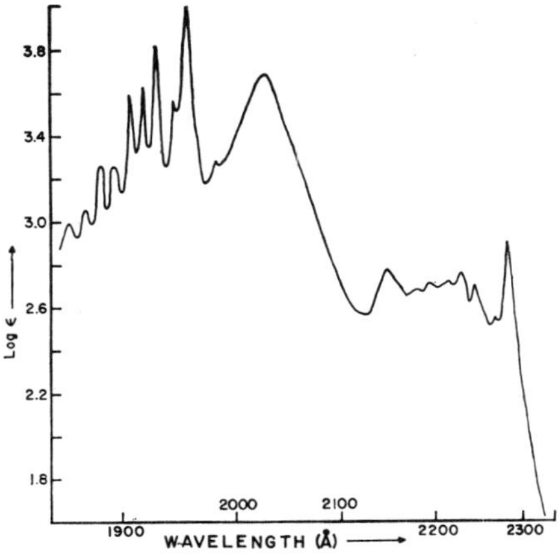

FIG. 6. Vapor spectrum of dimethylsulfide at 23°C.

The MO eigenvectors and eigenvalues are listed in Table V and the observed and calculated transition energies are given in Table VI.

The systems at 5.39 and 6.33 eV exhibit considerable structure and bear a remarkable resemblance to one another. In fact, it is possible to completely analyze both of these bands in terms of progressions and combinations of the symmetrical CSC stretch and the parallel methyl rock, with one exception: the weak second vibrational electronic band which appears to be a torsional vibration. The vibrational assignments are presented in Table VII; they would appear to be quite reasonable but they are by no means unambiguous. The similarity of the two systems is further illustrated by comparing the vibrational frequencies as in Table VIII. The decrease of vibrational frequency of approximately 100 cm^{-1} observed in the 6.33-eV transition relative to the 5.39-eV band would appear reasonable since the former corresponds to a higher-energy excited electronic level.

The calculations predict that the lowest energy transition should be largely $4s_S$ in character while the next three degenerate transitions should be primarily $3d_S$. It does not appear unlikely, in view of the very small energy differences predicted, that the 5.01 A_2 level becomes the lowest energy excited state and the cause of the very weak inflection at 5.20 eV. The $A_2 \leftarrow A_1$ transition is symmetry forbidden, thus the low intensity. The molar decadic extinction coefficient for this transition is in the range 15–20. The assignment of this

TABLE VI. $(CH_3)_2S$ transition energies (electron volts).[a]

Calculated	4.92(B_1)	5.01(B_1) 5.01(B_1) 5.01(A_2)	7.13(A_2)	8.96(B_1)
Observed	5.20	5.39	6.15	6.33

[a] Quantities in parentheses indicate assigned symmetry of excited state.

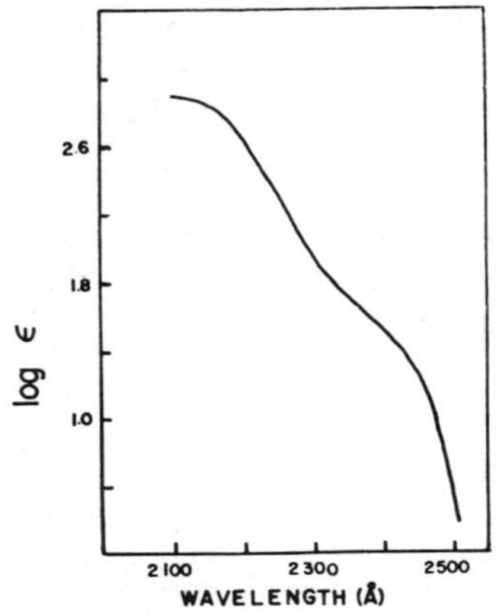

FIG. 7. Solution spectrum of the ~2350-Å inflection of dimethylsulfide.

TABLE VIII. Comparison of the vibrational frequencies in the $(CH_3)_2S$ 5.39- and 6.33-eV bands.

Band (eV)	$\bar{\nu}$(cm^{-1})				
5.39	272	742	1067	1373	1787
6.33	237	637	962	1250	1624
5.39	2018	···	2748		
6.33	1933	2274	2649		

band as symmetry forbidden is substantiated by its strong appearance in *tert*-butylmethylsulfide.[3] The extreme asymmetry of the latter compound makes the transition more allowed.

The system at 5.39 eV has an oscillator strength of 0.016 and is assigned as arising from an excitation to the second of the essentially $3d_S$ levels, while the third essentially $3d_S$ level is assumed to be related to the 6.33 band. This leaves the transition to the predominantly $4s_S$ level to correspond to the diffuse system at 6.15 eV. The similarity of the 5.39- and 6.33-eV bands offers some justification for the assignment of both as involving excited states that are largely $3d_S$ in character. Furthermore, the increase in intensity of the 0–0 band relative to the other vibrations in the short-wavelength packet as compared to the intensity of the 0–0 to the other vibrational members in the long-wavelength packet is what would be expected to occur as the atomic nature of the excited state increases (i.e., as it becomes more Rydberg).

The spectra of dimethylsulfide given in Fig. 8 exhibit large over-all changes in appearance in going from the vapor to solution. First, the weak inflection at 2330 Å shown in Fig. 7 and which is unobservable in the vapor, appears in the solution spectra. Apparently because of its very low intensity this band is masked in the vapor spectrum by the much stronger nearby 2125–2290-Å system. In solution, the vibrational structure of the 2125–2290-Å system is suppressed and the entire band is blue shifted to the extent that it now appears as a shoulder on the side of the strong 2000-Å excitation rather than as a separate well-defined band. This elimination of structure and the increased separation due to the blue shift combine to increase the resolution of the weaker 2300-Å band in solution. However, the ~2300-Å band is still too ill defined to allow any definite statements concerning solvent effects.

TABLE VII. Vibrational assignments in the $(CH_3)_2S$ 2150–2300 Å and 1850–1960 Å bands.

Band max (Å)	$\bar{\nu}$ (cm^{-1})	ir[a] (cm^{-1})	Symmetry	Assignment[b]
1954	0			
1943	237	284 (R)[c]	a_2	Torsional vibr.
1928	637	692.5	a_1	I
1916	962	1028	a_1	II
1905.5	1250	1385	a_1	2×I
1892	1624	1720.5	a_1	I+II
1881	1933	2056 or 2077.5	a_1	2×II
			a_1	3×I
1869	2274	2413	a_1	(2×I)+II
1856	2649	2748.5 or 2770	a_1	I+(2×II)
			a_1	4×I
2276	0			
2262	272	284 (R)[c]	a_2	Torsional vibr.
2238.3	742	692.5	a_1	I
2222	1067	1028	a_1	II
2207	1373	1385	a_1	2×I
2187	1787	1720.5	a_1	I+II
2176	2018	2056 or 2077.5	a_1	2×II
			a_1	3×I
2142	2748	2748.5 or 2770	a_1	I+(2×II)
			a_1	4×I

[a] J. P. Perchard, M. F. Forel, and M. L. Josien, J. Chem. Phys. **61**, 645 (1964).
[b] I, symmetric CSC stretch; II, parallel CH_3 rock.
[c] R means Raman active.

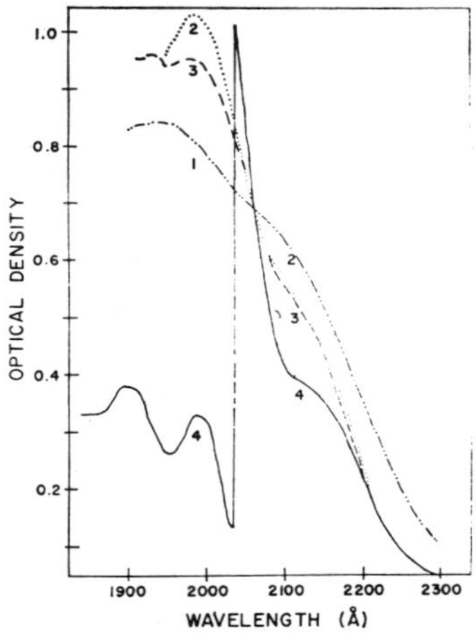

FIG. 8. Solution spectra of dimethylsulfide. The vertical dropoff in Curve 4 indicates a change of optical density scale by one unit. Solvent: 1, H_2O; 2, EtOH; 3, MeOH; 4, hexane.

TABLE IX. Results of CH_3SH computations ($Z_{3d_S}=1.0$; $Z_{4s_S}=0.9$).

	Eigenvectors											Eigenvalues
$1s_H$	sp^3_C	$3s_S$	$3p_{xS}$	$3p_{yS}$	$3p_{zS}$	$4s_S$	z^2 [a]	x^2-y^2 [a]	yz [a]	xy [a]	xz [a]	
0.6554	−0.3638	−0.1922	−0.5034	0.2707	0.0000	−0.0393	0.0268	−0.2152	−0.1608	0.0000	0.0000	1.56
0.2453	0.6833	−0.4575	−0.1602	−0.4599	0.0000	−0.0283	−0.0039	−0.1169	−0.1092	0.0000	0.0000	−0.42
0.0000	0.0000	0.0000	0.0000	0.0000	0.0000	0.0000	−0.6617	−0.4998	0.5589	0.0000	0.0000	−3.67
0.0001	0.0154	−0.0086	0.0015	−0.0154	0.0000	−0.0039	0.7457	0.3635	0.5578	0.0000	0.0000	−3.67
0.0077	−0.0023	−0.0040	−0.0097	0.0034	0.0000	−0.9819	−0.0172	0.1501	0.1138	0.0000	0.0000	−3.76
−0.1814	−0.0168	0.1375	0.2309	−0.0077	0.0000	−0.1829	0.7043	−0.7296	−0.5690	0.0000	0.0000	−4.09
0.0000	0.0000	0.0000	0.0000	0.0000	1.0000	0.0000	0.0000	0.0000	0.0000	0.0000	0.0000	−9.44 [b]
−0.1384	−0.1885	0.4857	−0.5541	−0.6337	0.0000	−0.0050	0.0007	−0.0222	−0.0191	0.0000	0.0000	−13.73
0.3869	−0.4577	−0.1398	0.5622	−0.5497	0.0000	0.0078	−0.0067	0.0443	0.0317	0.0000	0.0000	−18.94
0.5554	0.3939	0.6923	0.2107	0.1061	0.0000	0.0060	−0.0020	0.0289	0.0235	0.0000	0.0000	−35.48
0.0000	0.0000	0.0000	0.0000	0.0000	0.0000	0.0000	0.0000	0.0000	0.0000	1.0000	0.0000	−3.67
0.0000	0.0000	0.0000	0.0000	0.0000	0.0000	0.0000	0.0000	0.0000	0.0000	0.0000	1.0000	−3.67

[a] $3d_S$ AO's. [b] Experimental ionization potential of CH_3SH.

TABLE X. CH_3SH transition energies (electron volts).

Calculated	5.35	5.68	5.77 5.77	9.02
Observed	5.39	6.08	6.75	

The blue shifts of the 2125–2290-Å band continue to increase as solvent polarity increases until in water it has become an inflection. The concomitant increase in intensity is peculiar in that all of the other bands in this as well as other compounds exhibit decreases in intensity in similar circumstances. Indeed, it is more likely that the excitation is actually decreasing in intensity as the solvent polarity is increasing. The *apparent* increase in absorption may be reasonably accounted for if it is assumed that gains from increased overlapping with the much stronger 2000-Å band more than compensate for solvent-induced intensity losses. Solvent effects on the 2000-Å transition cause this transition to decrease in intensity with increasing solvent polarity. Solvent transmission limitations plus a distinct blue shift prevented complete investigation of the band observed in the monosulfide vapor spectrum between 1860–1950 Å. The large slitwidths required for solvents other than hexane below ~1950 Å render data collected below this point somewhat dubious. However, in hexane the structure present in the 1860–1950-Å vapor system has vanished and a definite blue shift has occurred much as in the 2125–2290-Å band. The similar behavior of both the 1860–1950- and 2125–2290-Å bands tends to lend further support to the assignment of both as resulting from dominantly $3d_S \leftarrow 3p_S$ transitions.

C. Methyl Mercaptan

The MO eigenfunctions and eigenvalues are presented in Table IX, and the predicted and observed transitions for CH_3SH are compared in Table X. As in H_2S, the lowest-lying level is predicted to be predominantly $3d_S$, followed in order by essentially pure $4s_S$ and degenerate $3d_S$ levels. The agreement between the calculated and the observed values is surprisingly good.

D. Dimethyldisulfide

The vapor spectrum of dimethyldisulfide is shown in Fig. 9. The MO eigenvectors and eigenvalues are not

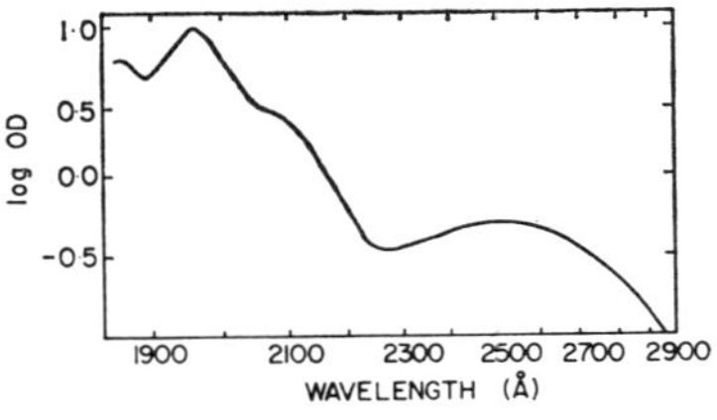

FIG. 9. Vapor spectrum of dimethyldisulfide.

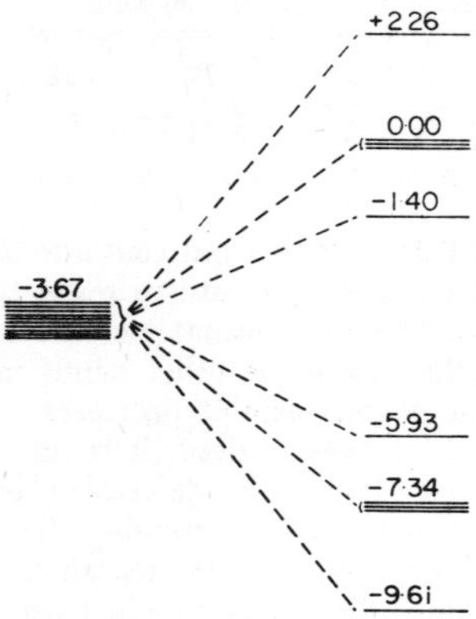

FIG. 10. Splitting of $3d_S(Z=0)$ levels in dimethyldisulfide at $\theta=104°$; units in electron volts. This splitting is maximized because of neglect of ES_{ij} off-diagonal terms.

given here; computed and observed transition energies are presented in Table XI. There appears to be a considerable disparity of experimental and theoretical results in view of the transitions predicted to occur between 2.0 and 3.5 eV. The error apparently stems from the very large overlaps obtained for $S_{4s_S,4s_S}$ (~0.9) and $S_{3d_S,3d_S}$ (0.6–0.9), the neglect of ES_{ij} in the off-diagonal terms of the energy matrix, and the consequent large splittings produced. The magnitude of this effect may be gauged from Fig. 10 in which the $3d_S$ splitting for the case where the dihedral angle is 104 deg and the effective charge is zero has been graphed. Under these conditions the $3d_S$ orbitals do not undergo any interaction with the other available orbitals but the $3d_S$–$3d_S$ interactions are maximized. A number of levels are generated sufficiently close to a predominantly $4s_S$ level at −7.49 eV and two essentially nonbonding levels at −8.54 and −9.80 eV that a large amount of mixing would be anticipated. Increasing the value of Z_{3d_S} to 1.0 has relatively little effect on $S_{3d_S,3d_S}$ values and, therefore, should exert very little influence on the splitting. However, the additional charge does result in considerable mixing of the lower lying $3d_S$'s with the $4s_S$ and nonbonding levels. Consequently the two highest filled MO levels are formed from a combination of $3d_S$, $3p_S$, and $\sigma_{sp^3_C}$ orbitals, but more importantly the first four excited levels are, in order: combinations of $4s_S$, $3d_S$, $3p_S$, and $\sigma_{sp^3_C}$ (predominantly $3d_S$); $4s_S$ and $3d_S$ (very predominantly $4s_S$); $4s_S$ and $3d_S$ (primarily $3d_S$); $3d_S$ only. The contributions of the $3d_S$'s to the next several excited levels are much smaller and the transition energies correspond quite closely to experimentally observed values. Thus the questionable transitions may possibly be due to the large overlaps used. Although the use of smaller overlaps cannot be theoretically justified by us, they almost undoubtedly would produce better correlation with experiment.

TABLE XI. $(CH_3)_2S_2$ transition energies (electron volts).

i	$n_1\to\phi_i$ $\theta=90°$	$n_1\to\phi_i$ $\theta=104°$	$n_2\to\phi_i$ $\theta=90°$	$n_2\to\phi_i$ $\theta=104°$	$\sigma_1\to\phi_i$ $\theta=90°$	$\sigma_1\to\phi_i$ $\theta=104°$
			Predicted energies[a–c]			
1	2.43	2.09	2.72	3.12	7.05	6.99
2	3.30	2.98	3.59	4.01	7.92	7.88
3	3.45	3.15	3.74	4.18	8.07	8.05
4	5.21	4.89	5.50	5.92	9.83	9.79
5	5.29	5.07	5.58	6.10	9.91	9.97
6	5.57	5.19	5.86	6.22	10.19	10.09
7	9.70	9.39	9.99	10.42	14.32	14.29
8	9.71	9.53	10.00	10.56	14.33	14.43
9	10.63	10.36	10.92	11.39	15.25	15.26
10	10.70	10.39	10.99	11.42	15.32	15.29
11	10.71	10.43	11.00	11.46	15.33	15.33
12	11.00	10.69	11.29	11.72	15.62	15.59
13	11.30	11.15	11.59	12.18	15.92	16.05
14	11.84	11.26	12.13	12.29	16.46	16.16
15	13.70	13.44	13.99	14.47	18.32	18.34
			Observed energies			
Energy		4.96		5.89	6.32	6.65
Oscillator strength f		0.0312		0.0283	0.303	...

[a] n_1 is the highest-energy filled MO; it is of $3p_S$ nature; n_2 is the next-highest-energy filled MO; it is also of $3p_S$ nature; σ_1 is the highest-energy filled σ-bonding MO; ϕ_i is the ith excited MO level, where ϕ_1 is the lowest-energy vacant MO.
[b] θ is the dihedral angle defined in Fig. 2.
[c] The numbers quoted here refer to computations for which $Z_{3d_S}=1.00$ and $Z_{4s_S}=0.9$.

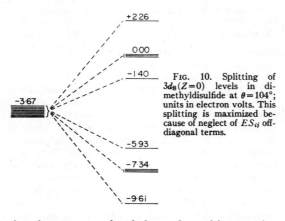

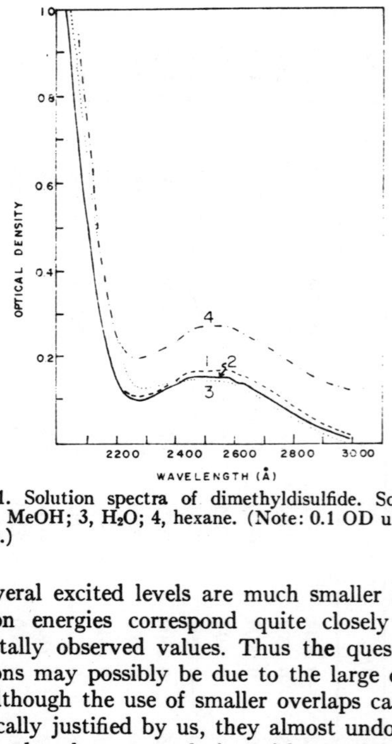

FIG. 11. Solution spectra of dimethyldisulfide. Solvent: 1, EtOH; 2, MeOH; 3, H_2O; 4, hexane. (Note: 0.1 OD unit added to hexane.)

The general features of the disulfide solution spectra are essentially the same as those of the vapor—a broad unresolved band with a maximum in the region of 2500 Å, a shoulder between 2000 and 2075 Å and a comparatively sharp band at ~1950 Å. However, there are certain qualitative differences. A particularly interesting one is in the 2500-Å band, solution spectra of which are shown in Fig. 11. The preceding considerations as well as others not yet published[16] indicate that this band may be assigned as an excitation to a

[16] M. O'Donnell, S. D. Thompson, D. G. Carroll, and S. P. McGlynn (to be published).

TABLE XII. $(CH_3)_2S_3$ transition energies (electron volts).[a]

i	cis $n_1 \to \phi_i$	cis $n_2 \to \phi_i$	cis $n_3 \to \phi_i$	cis $\sigma_1 \to \phi_i$	trans $n_1 \to \phi_i$	trans $n_2 \to \phi_i$	trans $n_3 \to \phi_i$	trans $\sigma_1 \to \phi_i$
				Predicted energies				
1	1.90	2.69	3.91	6.45	1.03	1.78	2.88	5.22
2	7.13	7.92	9.14	11.68	7.21	7.96	9.06	11.40
3	8.20	8.99	10.21	12.75	8.23	8.98	10.08	12.42
4	9.10	9.89	11.11	13.65	9.20	9.95	11.05	13.39
5	9.22	10.01	11.23	13.77	9.31	10.06	11.16	13.50
6	9.49	10.28	11.50	14.04	9.59	10.34	11.44	13.78
7	10.03	10.82	12.04	14.58	10.32	11.07	12.17	14.51
				Observed energies				
	4.42	5.00	5.81	6.17	6.36	6.65		

[a] n_i, ith filled level, n_1 being the highest-energy filled level; ϕ_i, ith excited level, ϕ_1 being the lowest energy vacant level.

level which contains considerable $4s_S$ character. It is further predicted,[16] in apparent agreement with experiment, that this transition will shift to longer wavelengths as the number of sulfur atoms in the chain increases. Yet, in addition to the longer wavelength excitations in the tri- and tetrasulfides (*vide infra*), there is still a band at 2500 Å. It is felt that the solution to this dilemma lies in there being two transitions beneath the 2500-Å absorption envelope of the disulfide. The logical choice for the second excited level would appear to be $\sigma^*_{S-S} \leftarrow n_1$. Certainly the band is sufficiently broad to accommodate both transitions. Consideration of an angular molecular model predicts that the $n \to \sigma^*_{S-S}$ should be more or less independent of the number of sulfur atoms in the chain. Such a conclusion is also justifiable on the basis of available experimental data: the energy of the bonding and antibonding levels should be a function of the sulfur–sulfur distances, and these separations are found to be essentially constant in the di- and trisulfides. Data on the geometry of the tetrasulfide are not available, but S–S distances should be virtually the same in this molecule as in the others. If the 2500-Å transition is indeed independent of the number of sulfur atoms then we should anticipate its occurrence in the same region in the S_8 molecule. Bass has, in fact, reported[17] the presence of a broad, unresolved band at 2550 Å in sulfur vapor spectra at 250°C. As final evidence there is the efficiency of 2500-Å radiation in effecting S–S scission in photolytic experiments. The rather peculiar flattening of this band which is observed in these solution studies could then be attributed to the achievement of a very slight amount of resolution of the two possible transitions.

The shoulder at 2095 Å in the vapor spectrum has become submerged in the stronger 1950-Å band to such an extent as to make an accurate determination of its position impossible. The 1950-Å transition is found to blue shift and decrease in intensity as the solvent polarity increases.

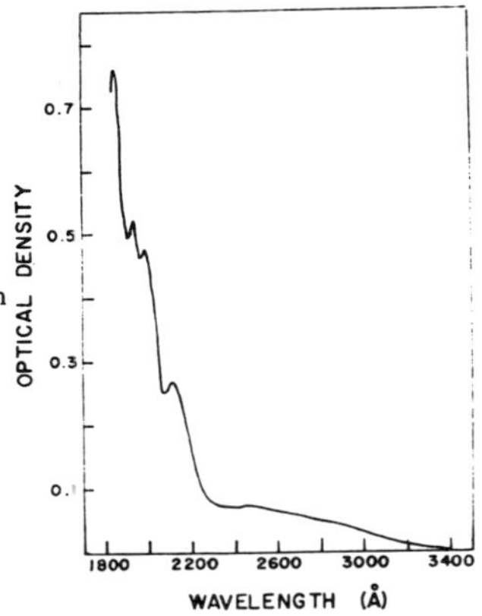

FIG. 12. Vapor spectrum of dimethyltrisulfide.

E. Dimethyltrisulfide

The vapor spectrum of dimethyltrisulfide is presented in Fig. 12; solution spectra are given in Fig. 13. The vapor and solution spectra are virtually identical; the longest wavelength band occurs at 2850 Å and is merely an inflection; the 2500-Å band of the disulfide recurs in the trisulfide.

The computational and experimental results are compiled in Table XII. The lack of agreement is not at all surprising in view of the fact that $3d_S$ AO's were not included in computations of the trisulfide. In calculations on all of the previous molecules it was found that the excited levels corresponding to the observed transitions were generated only when $3d_S$ and $4s_S$ AO's were considered. There is no reason to expect that the inclusion of $3d_S$'s in the computations on this molecule would not have the same effect. Of course, one would also expect, as was found in the disulfide, a number of transitions to be predicted at lower energies (assuming overlaps and resonance integrals were calculated by the same procedure used for the other molecules). Similarly, the computed energy of the lowest energy electronic

[17] A. M. Bass, J. Chem. Phys. **21**, 80 (1953).

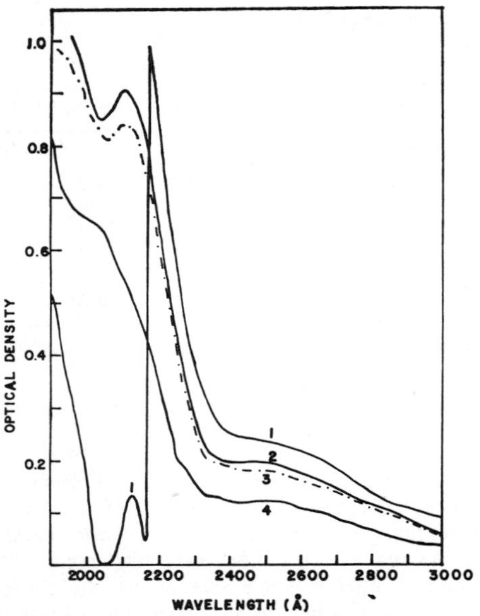

FIG. 13. Solution spectra of dimethyltrisulfide. The vertical dropoff in Curve 1 indicates a change of optical density scale by one unit. Solvent: 1, hexane; 2, EtOH; 3, MeOH; 4, H_2O.

transition of 1.90 eV probably arises from the large value of $S_{4s,4s_S}$ (=0.9) used.

F. Dimethyltetrasulfide

The vapor and solution spectra of dimethyltetrasulfide are given in Figs. 14 and 15, respectively. The 3000-Å absorption is particularly obvious in the solution spectra, whereas the 2500-Å absorption is most obvious in the vapor. A number of other absorption maxima also are observed. No calculations were performed on this molecule.

V. CONCLUSIONS

A. Sulfur–Sulfur Bonding and the Effects of Inclusion of $3d_S$ and $4s_S$ AO's

A general picture of the effect of varying the effective charges of the $3d_S$ and $4s_S$ AO's is shown in Fig. 16 where the energy levels obtained by using various Z

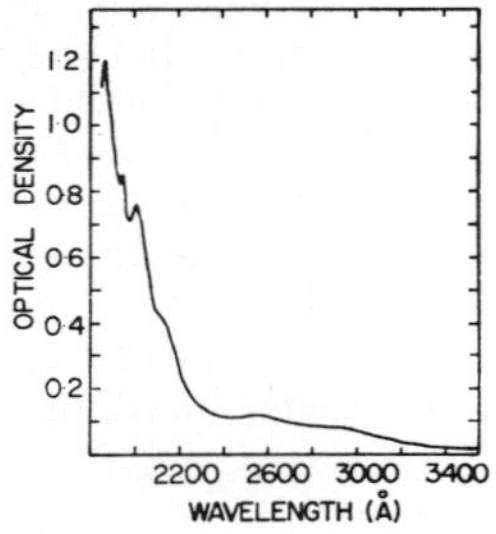

FIG. 14. Vapor spectrum of dimethyltetrasulfide.

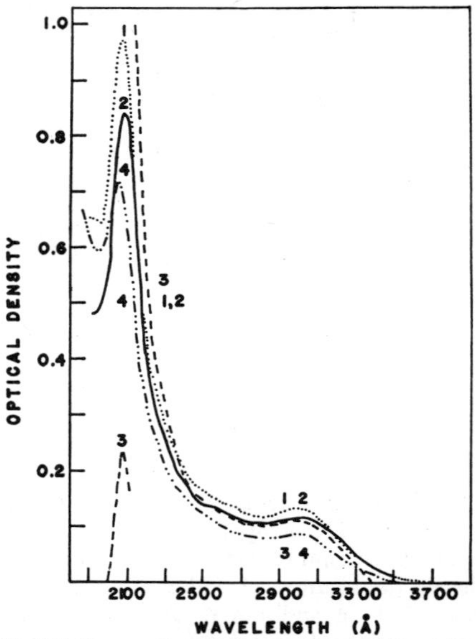

FIG. 15. Solution spectra of dimethyltetrasulfide. Solvent: 1, MeOH; 2, EtOH; 3, cyclohexane; 4, H_2O.

values are plotted for the case of dimethyldisulfide with a dihedral angle of 104°. These results are more or less typical of those found in all of the molecules considered.

Inclusion of $4s_S$ AO's had essentially no effect upon

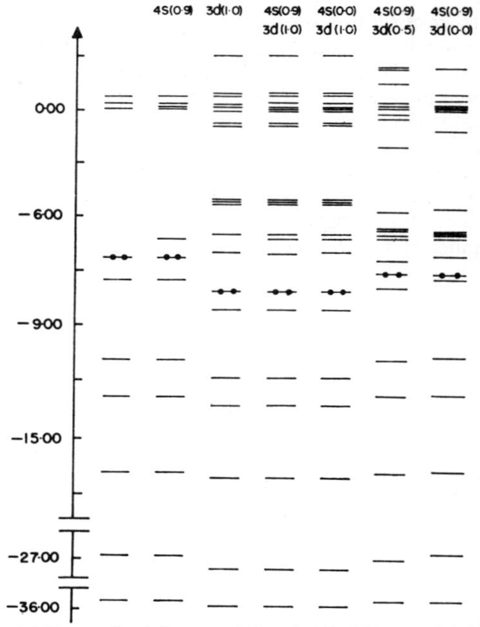

FIG. 16. Energy-level diagram of dimethyldisulfide at $\theta = 104°$ at various $3d_S$ and $4s_S$ effective charges. The highest energy populated MO in the ground-state configuration is indicated by two dots.

TABLE XIII. Hydrogen sulfide molecular orbital energies as functions of orbital exponents $\mu(\equiv Z/n)$ of $3d_S$ and $4s_S$ AO's.

MO	$4s(Z=0.9)$ $3d(Z=0)$	$4s(Z=0.9)$ $3d(Z=0.5)$	$4s(Z=0.9)$ $3d(z=1.0)$	$4s(Z=0)$ $3d(Z=1)$	$3d(Z=1)$ no $4s$	no $3d$ $4S=0.9$	no $4s$ no $3d$
$6a_1$	+1.51	+1.55	+2.26	+2.23	+2.23	+1.51	+1.99
$3b_2$	−1.35	−1.32	−0.68	−0.68	−0.68	−1.35	−1.35
a_2	−3.67	−3.67	−3.67	−3.67	−3.67		
$2b_2$	−3.67	−3.70	−4.26	−4.26(3d)	−4.26		
$2b_1$	−3.67	−3.67	−3.67	−3.67	−3.67		
		−3.67	−3.67	−3.67			
4, $5a_1$	−3.67				−3.67		
		−3.69[0.25(4s)]	−3.76[0.97(4S) +0.29(3d)]	−3.76(4s)			
$3a_1$	−3.78[a]	−3.79[a]	−4.26[d]	−4.24[b]	−4.24	−3.78[a]	
$1b_1$	−10.47[c]	−10.47[c]	−10.47[c]	−10.47[c]	−10.47[c]	−10.47[c]	−10.47[c]
$2a_1$	−14.69	−14.69	−14.73	−14.73	−14.73	−14.69	−14.69
$1b_2$	−18.28	−18.28	−18.37	−18.37	−18.37	−18.28	−18.28
$1a_1$	−38.18	−38.18	−38.31	−38.31	−38.31	−38.18	−38.17

[a] Pure or mostly $4s_S$. [b] Pure or mostly $3d_S$. [c] Highest-energy occupied MO (HOMO).

the energy of the levels generated in their absence. The maximum displacements of existing levels produced when they were added (with Z set equal to 0.9) were of the order of 0.02 eV. Consequently, it is not surprising that reduction of the effective charge, Z_{4s_S}, from 0.9 to 0.0 leaves the levels virtually unchanged. Nonetheless, it is observed that the $4s_S$ AO's undergo quite large splittings in both the di- and trisulfides. For example, the $4s_S$ level of dimethylsulfide at −3.76 eV splits into two levels in dimethyldisulfide which lie at energies −7.46 and −0.05 eV, respectively, and into three levels in dimethyltrisulfide which lie at −11.21, −0.10, and +0.02 eV, respectively. Those levels are widely split and are fairly insensitive to variation of Z_{4s}, exhibiting a maximum variation of approximately 0.1 eV when Z_{4s_S} is decreased from 0.9 to 0.0.

Z values of 0.0, 0.5, and 1.0 were used for the $3d_S$ AO's. The $3d_S$ AO's when inserted with effective charges

TABLE XIV. Solution spectra of the series $(CH_3)_2S_n$ (in angstroms).

Medium	n	λ(max)[a]	logε	λ(max)[a]	logε	λ(max)[a]	logε	λ(max)[a]	logε	λ(max)[a]	logε
Vapor	1			...	...	2125–2290B	...	2025B	...	1850–1955B	
Hexane	1			2330i	~1.9	2140	2.80	2000	3.38	1905	3.40
EtOH	1			2330i	~2.0	2120	2.95	1990	3.29		
MeOH	1			2330i	~1.9	2135	2.94	1990	3.24		
H_2O	1			2340i	~2.0	2100	3.08	1935	3.17		
Vapor	2			2515		2095s[b]				1958	...
Hexane	2			2530	2.62					1945	3.84
EtOH	2			2530	2.59					1945	3.73
MeOH	2			2525	2.52					1930	3.72
H_2O	2			2510	2.52					1890	3.73
Vapor	3	~2900i	...	2480	...	2130	...	2010s	...	1950	...
Hexane	3	~2850i	~2.5	2530s	2.61	2120	4.08	~2000s			
EtOH	3	~2900i	~2.5	2530s	2.59	2105	3.98	...	...		
MeOH	3	~2900i	~2.5	2525s	2.58	2105	3.97	~1980			
H_2O	3	~2900i	~2.5	2510s	2.53	2025s	3.83	...	...		
Vapor	4	2975	...	~2475s	...	2120s	...	2012	...	1948	...
Cyclohexane	4	3015	3.27	~2575i	3.30	2070	4.27	2000	...		
EtOH	4	3000	3.28	~2575i	3.32	2070	4.11	...	...		
MeOH	4	3000	3.30	~2520i	3.34	2065	4.17	...	...		
H_2O	4	2980	...	~2540i	...	2045	...	1980	...		

[a] B, band; s, shoulder; i, inflection.
[b] G. R. A. Brandt, H. J. Emeleus, and R. N. Hazeldine, J. Chem. Soc. **1952**, 2549, have reported absorption at 2030–2070 Å which presumably would correspond to absorption in this region in view of their reported extinction coefficients of logε~3.2.

TABLE XV. Structural parameters for molecules considered in this work.

Molecule	Reference	C–H (Å)	C–S (Å)	S–S (Å)	S–H (Å)	<HCS	<CSC	<CSS	<SSS	Dihedral angle θ	<CSH	<HSH
H_2S[a]	b, c	...	...	...	1.35	...	...	...	...	...	...	92.2
H_2S	d	...	...	...	1.323	...	...	...	...	...	...	92.6
CH_3SH	e	...	1.808	...	1.345	...	...	...	...	...	99.26	...
$(CH_3)_2S$	f	...	1.802	...	...	...	98.87	...	...	...	...	...
$(CH_3)_2S_2$	g	...	1.78±0.03	2.04±0.03	...	...	...	107±3	...	...	...	...
$(CH_3)_2S_3$	h, i	1.09	1.78±0.04	2.04±0.02	...	112	...	104±5	104±5	93	...	...
S_6	j	...	...	2.07	...	...	...	...	105	...	...	...
S_8[a]	k	...	...	2.03±0.005	...	...	...	...	108±0.5	99.3	...	...

[a] Sets of structural parameters used in these computations.
[b] C. C. Price and S. Oae, *Sulfur Bonding* (Ronald Press Co., New York, 1962).
[c] L. N. Ferguson, *Electron Structures of Organic Molecules* (Prentice-Hall, Inc., Englewood Cliffs, N. J., 1952).
[d] C. A. Burrus and W. Gordy, Phys. Rev. **92**, 274 (1953).
[e] I. Solimene and B. P. Dailey, Phys. Rev. **91**, 464 (1953).
[f] H. Dreizler and H. D. Rudolph, Z. Naturforsch. **17a**, 712 (1962).
[g] D. P. Stevenson and J. Y. Beach, J. Am. Chem. Soc. **60**, 287 (1938). The dihedral angle θ was taken at 90° and 104° (see Table XI).
[h] J. Donohue and V. Schomaker, J. Chem. Phys. **16**, 92 (1948).
[i] J. Donohue, J. Am. Chem. Soc. **72**, 2701 (1950).
[j] C. S. Lu and J. Donohue, J. Am. Chem. Soc. **66**, 818 (1944).
[k] S. C. Abrahams, Acta Cryst. **8**, 661 (1955).

of 0.0 or 0.5 remain degenerate and fail to perturb the levels already present in those molecules which contain a single sulfur atom. However, in the disulfide the $3d_S$ AO's exhibit considerable splitting even with a Z value of zero. New levels appear at 2.26, 0.00 (triply degenerate), −1.40, −5.93, −7.34 (triply degenerate), and −9.61 eV. The triply degenerate levels correspond to bonding and antibonding combinations of the xy, xz, and yz AO's while the remaining four levels consist of various combinations of the z^2 AO's and x^2-y^2 AO's. A rather interesting point here is that the lowest of these new levels (−9.61 eV) is approximately 1 eV lower in energy than one of the previously filled levels (−8.54 eV) which presumably is primarily nonbonding in character. This situation continues to exist as the $3d_S$ Z value is increased. This might provide some insight into the role of the $3d_S$ AO's in octet expansion in polysulfide compounds.

Increasing Z_{3d_S} to 0.5 removes the remaining d-orbital degeneracy and results in considerable mixing of the $3d_S$ AO's with nearby levels. The levels are consequently shifted by amounts ranging from 0.1 to 1.5 eV, with the smaller shifts being more common.

The effect of the $4s_S$ levels on sulfur–sulfur bonding is not particularly significant, as may be gauged from Fig. 16. The contribution of $3d_S$ levels to S–S bonding is extensive, as may be seen by comparing Columns 1, 2, and 3 of Fig. 16; this contribution decreases as Z_{3d_S} decreases. The important d AO's are $3d_{z^2}$, $3d_{x^2-y^2}$, and $3d_{yz}$.

B. Sulfur–Hydrogen Bonding

The energies of the H_2S MO's as a function of μ_{4s_S} and μ_{3d_S} are given in Table XIII. Comparison of twice the sum of the filled bonding orbitals (i.e., the four of lowest energy) shows that inclusion of the $4s_S$ AO at $Z=0.9$ and neglect of the $3d_S$ AO's has no effect on ground-state bonding; a total stabilization of only 0.02 eV is predicted. Inclusion of $3d_S$ AO's and neglect of the $4s_S$ AO produces a total stabilization of the ground state of ~0.2 eV which, while not negligible, is certainly not massive. Since the quoted Z values are supposedly maximal and, therefore, produce the situation most contributory to bonding in H_2S, it seems reasonable to assume that $3d_S$ and $4s_S$ contributions to H–S bonding are small.

C. Sulfur–Carbon Bonding

Tabulations similar to Table XIII in the case of CH_3SH and $(CH_3)_2S$ indicate that the contribution of $4s_S$ orbitals to C–S bonding is quite small. The contribution of $3d_S$ AO's to C–S bonding, while larger than that for the $4s_S$ AO, is always less than 0.4 eV in our calculations.

D. Excited States of Sulfur-Containing Compounds

The lower-energy excited electronic states of H_2S, D_2S, CH_3SH, and CH_3SCH_3 are best described as the initial members of Rydberg series. They appear to be heavily constituted of $4s_S$ and $3d_S$ atomic orbitals on sulfur; they do, however, possess considerable molecular character by virtue of their admixture with hydrogen and carbon atomic orbitals. It is this admixture which is primarily responsible for the observed oscillator strengths of the low-energy transitions of these compounds. The lowest energy absorption band of H_2S is assigned to an orbital excitation which is *qualitatively* described as $4s_S \leftarrow 3p_S$. This assignment is in accord with the observed oscillator strength of the 2000-Å H_2S and D_2S absorption bands, with the energy of these bands, with the small observed anharmonicity and with the dominance of the vibrational progression in the bending mode. The other two observed low-energy transitions are assigned to orbital excitations *qualitatively* described by $3d_S \leftarrow 3p_S$; the excitations to the highly antibonding H–S and D–S MO's are expected to occur at energies $\sim$10 eV.

The lower-energy transitions of the disulfide are assigned again to the initial members of Rydberg series. Some evidence of resolution in the 2500-Å band has been obtained and it is suggested that this band envelope contains two transitions which are assigned to the qualitative AO–MO excitations $4s_S \leftarrow 3p_S$ and $\sigma^*_{S-S} \leftarrow 3p_S$ where σ^*_{S-S} is antibonding between the sulfur atoms and primarily constituted of $3d_S$ and $3p_S$ AO's.

The lowest energy absorption bands of the dimethyl-tri- and tetrasulfides at $\sim$2800 and $\sim$3000 Å, respectively, contain considerable $4s_S \leftarrow 3p_S$ character. The persistence of the 2500-Å band throughout the polysulfide series, including S^8, indicates the non-conjugative nature of the MO in which the corresponding MO excitation terminates; this band is supposed to contain considerable $\sigma^*_{S-S} \leftarrow 3p_S$ character.

E. Solvent Effects

Vapor and solution band spectra are collected in Table XIV. Examination of the data in this table reveals that the high-frequency electronic transitions which occur at approximately 2150 and 2000 Å shift to the blue in solvents of large hydrogen-bonding ability. This tendency is further confirmed by examining the shifts which are observed in 10% vol:vol water:ethanol containing different amounts of sodium hydroxide. The presence of alkali induces a red shfit.

If the general expression for solvents shifts derived by Ooshika[18] and McRae[19] is modified according to Ito *et al.*,[20] the following expression is obtained for the solvent-induced frequency shift $\Delta\bar{\nu}$:

$$\Delta\bar{\nu} = B\frac{n_D^2-1}{2n_D^2+1} + C\left[\frac{D-1}{D+2} - \frac{n_D^2-1}{n_D^2+2}\right],$$

where

$$B = (1/hc)[(M_{00})^2 - (M_{ii})^2/A^3]$$

$$C = (2/hc)[M_{00}(M_{00} - M_{ii})/A^3],$$

and where M_{00} and M_{ii} are the dipole-moment vectors of the solute in the ground and ith excited states, respectively, A is the Onsager reaction radius, D is the dielectric constant of the solvent, and where n_D is the refractive index for the sodium line. In nonpolar solvents $D \sim n_D^2$ and the second term vanishes. Consequently, the sign of $\Delta\bar{\nu}$ is determined by the sign of B (i.e., the relative magnitude of M_{ii} and M_{00}). If the ground state has a greater dipole moment than the ith excited state, B is positive and vice versa. By comparing the values of $(n_D^2-1)/(2n_D^2+1)$ for the solvents n-hexane and cyclohexane:

n-Hexane:

$$n_D(25°) = 1.372; \ (n_D^2-1)/(2n_D^2+1) = 0.185;$$

Cyclohexane:

$$n_D(25°) = 1.423; \ (n_D^2-1)/(2n_D^2+1) = 0.203,$$

and the frequencies of the transitions in the two solvents, B is found to be positive; thus $M_{00} > M_{ii}$. We take this to mean that the terminal states of the 2150- and 2000-Å absorption bands are more linear than is the ground state, although this conclusion is by no means unique.

APPENDIX

The structural parameters which were used in the computations carried out in this work are collected in Table XV.

[18] Y. Ooshika, J. Phys. Soc. Japan **9**, 594 (1954).
[19] E. G. McRae, J. Phys. Chem. **61**, 862 (1957).
[20] M. Ito, K. Inuzuka, and S. Imanaishi, J. Am. Chem. Soc. **82**, 1317 (1960).

Chapter IV

Sigma Molecular Orbital Theory and Organic Chemistry

IV-1

Pi-electronic molecular orbital methods have had considerable influence on the development of modern organic chemistry, and σ-electron methods undoubtedly will have an even greater influence. The first two articles in this chapter describe some of the problems which are of interest to organic chemistry and which may be capable of solution using σ-MO methods. Streitwieser points out areas which may be elucidated by nonempirical computer calculations, such as the π-inductive and σ-inductive effects of substituents, structure of transition states, nature of nonclassical carbonium ions, and carbanion stability.

A large number of papers already have appeared dealing with applications of the method to organic chemistry. It would not be possible to include all of them, and for this reason only a few illustrative applications are presented. We add here a brief annotated bibliography of papers which may be of further interest to the reader.

1. Cations, Radicals, and Nonclassical Ions

a. "Toward an Understanding of Non-classical Ions," R. Hoffmann, *J. Am. Chem. Soc.* **86**, 1259 (1964). The 7-norbornyl, 7-norbornenyl, and 7-norbornadienyl carbonium ions were studied with respect to ion stability, movement of the 7-carbon, and charge distribution. The results are in accord with the concept of nonclassical ions.

b. "Extended Hückel Theory: IV. Carbonium Ions," R. Hoffmann, *J. Chem. Phys.* **40**, 2480 (1964). The conformations, relative stabilities, and electronic distributions in many carbonium ions are calculated. Nonclassical ions and protonated cyclopropanes are included.

c. "Methyl Cation and Model Systems for Carbonium Ions," R. E. Davis and A. Ohno, *Tetrahedron* **23**, 1015 (1967). Formaldehyde and borane have been taken as models of a methyl cation, and the properties of these systems were calculated using the extended Hückel method. The differences in bonding force constants are discussed.

d. "Approximate Self-consistent M. O. Theory: V. Intermediate Neglect of Differential Overlap," J. A. Pople, D. L. Beveridge, and P. A. Dobosh, *J. Chem. Phys.* **47**, 2026 (1967). The INDO method, which is an improvement on the CNDO method, is described and applied to the calculation of the unpaired spin density of methyl and ethyl radicals.

e. "Spirarenes," R. Hoffmann, I. Imamura, and G. D. Zeiss, *J. Am. Chem. Soc.* **89**, 5215 (1967). Allylic radicals held in a spiro system may under certain conditions interact so that a stabilized singlet could become the ground state. Conditions for the existence of such interactions are discussed.

f. "Application of the Pople-Santry-Segal CNDO Method to the Cyclopropylcarbinyl and Cyclobutyl Cation and to Bicyclobutane," K. B. Wiberg, *Tetrahedron* **24**, 1083 (1968).

CNDO calculations are presented for cyclopropylcarbinyl and cyclobutyl cations, giving results which are in good agreement with experiment. The bond index is introduced and applied to ^{13}C—H coupling constants and the thermal rearrangement of bicyclobutane to butadiene.

g. "1-Methylcyclobutyl Cation and Cyclobutyl Cation as Classical Ions," R. E. Davis, A. S. N. Murthy, and A. Ohno, *Tetrahedron Letters* **1968**, 1595. Extended Hückel calculations were performed on cyclobutyl cations and it was concluded that a planar conformation was preferred.

h. "The Question of Transannular Bonding in Cyclobutanone, Boretane and the Cyclobutyl Cation," R. E. Davis and A. Ohno, *Tetrahedron* **24**, 2063 (1968). The extended Hückel method was applied to the title species. The cyclobutyl cation was calculated to prefer a planar conformation. (However, see j below.)

i. "Molecular Orbital Theory of the Electronic Structure of Organic Compounds: II. Spin Densities in Paramagnetic Species," J. A. Pople, D. L. Beveridge, and P. A. Dobosh, *J. Am. Chem. Soc.* **90**, 4201 (1968). The INDO method is used to calculate electron-spin-density distributions and nuclear hyperfine constants for a number of paramagnetic organic radicals and ions. The results are generally in reasonable agreement with experimental values.

j. "Extended Hückel Calculations on Bicyclobutonium and Related Cations," J. E. Baldwin and W. D. Foglesong, *J. Am. Chem. Soc.* **90**, 4311 (1968). The energies of several conformations of cyclopropylcarbinyl and cyclobutyl cations are calculated using the extended Hückel method. A puckered cyclobutyl cation was calculated to be more stable than the planar species.

k. "Extended Hückel Calculations of Hydrogen and Nitrogen Electron Paramagnetic Resonance Coupling Constants for σ-Radicals Containing Carbon, Hydrogen, Nitrogen and Oxygen Atoms," R. E. Cramer and R. S. Drago, *J. Am. Chem. Soc.* **90**, 4790 (1968). Satisfactory values of the EPR coupling constants could be calculated.

l. "Molecular-Orbital Theory of Geometry and Hyperfine Coupling Constants of Fluorinated Methyl Radicals," D. L. Beveridge, P. A. Dobosh, and J. A. Pople, *J. Chem. Phys.* **48**, 4802 (1968). The INDO method was applied to the methyl radicals and the results are compared with experimental values.

m. "A Perturbation Molecular Orbital Approach to the Interpretation of Organic Mass Spectra. The Relationship between Mass Spectrometric, Thermolytic and Photolytic Fragmentation Reactions," R. C. Dougherty, *J. Am. Chem. Soc.* **90**, 5780 (1968). Mass-spectrometric reactions are divided into three classes. The application of the PMO method to these reactions is described.

n. "Application of the Perturbation Molecular Orbital Method to the Interpretation of Organic Mass Spectra. The Hexahelicene Rearrangement and Other Electrocylic Mass Spectrometric Reactions," R. C. Dougherty, *J. Am. Chem. Soc.* **90**, 5788 (1968). The use of the PMO method in dealing with electrocyclic reactions is illustrated with examples of rearrangements occurring in a mass spectrometer.

o. "Methyl Group Inductive Effect in Toluene Ions. Comparison of Hückel and Extended Hückel Theory," D. Purins and M. Karplus, *J. Am. Chem. Soc.* **90**, 6275 (1968). A comparison of the results of the two calculations suggests that the overlap between C_1 and the methyl group is the source of the HMO inductive effect for that group.

p. "The Electronic Structure of the Six-Membered Cyclic Transition State in Some γ-Hydrogen Rearrangements," F. P. Boer, T. W. Shannon, and F. W. McLafferty, *J. Am. Chem. Soc.* **90**, 7239 (1968). The course of the γ-hydrogen rearrangement in 2-pentanone is followed by MO calculations assuming a six-membered cyclic activated complex. The question of stepwise and concerted reactions is discussed.

2. Chemical Reactivity

a. "Polyelectron Perturbation Treatment of Chemical Reactivity," G. Klopman and R. F. Hudson, *Theoret. Chim. Acta* **8**, 165 (1967). The formation of an activated complex is treated as a mutual perturbation of the molecular orbitals of both reagents and the relative reactivity of various reacting centers is shown to vary with the magnitude of the perturbation.

b. "General Perturbation Treatment of Chemical Reactivity," R. F. Hudson and G. Klopman, *Tetrahedron Letters* **1967**, 1103. This is closely related to the above paper.

c. "Solvations–Semi-empirical Procedure for Including Solvation in Quantum Mechanical Calculations of Large Molecules," G. Klopman, *Chem. Phys. Letters* **1**, 200 (1967). A mathematical procedure is described which allows the inclusion of the solvent interaction in semiempirical SCF treatments.

d. "Reactivity and the Concept of Charge- and Frontier-Controlled Reactions," G. Klopman, *J. Am. Chem. Soc.* **90**, 223 (1968). The perturbation method reproduces the known qualitative features of the concept of hard and soft Lewis acids and bases, of nucleophilicity order, and other reactivity indices.

3. Excited States

a. "Extended Hückel Theory: VI. Excited States and Photochemistry of Diazirines and Diazomethanes," R. Hoffmann, *Tetrahedron* **22**, 539 (1966). Approximate calculations are presented for the excited state of diazirane and diazomethane. The isomerization of diaziranes to diazomethanes is also considered.

b. "Excited Electronic States of Cyclopropane," R. D. Brown and V. G. Krishna, *J. Chem. Phys.* **45**, 1482 (1966). Treated as system of delocalized σ electrons with respect to carbon skeleton by a method similar to the Pariser-Parr treatment of π-electronic states.

c. "Complete Neglect of Differential Overlap M. O. Theory of Molecular Spectra: I. Virtual Orbital Approximation to Excited States," H. W. Kroto and D. P. Santry, *J. Chem. Phys.* **47**, 792 (1967). An extension of the CNDO method to transition energies and excited-state geometries is described. Semiquantitative results are obtained for formaldehyde, acetylene, and other molecules.

d. "Semi-empirical M. O. Theory of Molecular Spectra: II. Approximate Open-Shell Theory," H. W. Kroto and D. P. Santry, *J. Chem. Phys.* **47**, 2736 (1967). An approximate open-shell theory, based on the CNDO method, is described and is applied to the calculations of the excited-state molecular parameters of several molecules. The method optimizes the MO's for the excited-state configuration, in contrast to methods using ground-state orbitals.

e. "Electronic Excited States of Benzene and Ethylene," P. A. Clark and J. L. Ragle, *J. Chem. Phys.* **46**, 4235 (1967). The application of the CNDO method to the calculation of transition energies was examined and a modification was found which led to good agreement with experiment.

f. "Electronic Structure and Photochemistry of Flavins: IV. σ-Electronic Structure and the Lowest Triplet Configuration of a Flavin," P. S. Song, *J. Phys. Chem.* **72**, 536 (1968). The results of CNDO and extended Hückel calculations were compared with experimental results. The former was the more satisfactory. The results suggest an opposite polarization for the σ core and the π system.

g. "Use of the CNDO Method in Spectroscopy: I. Benzene, Pyridine and the Diazines," J. D. Bene and H. H. Jaffe, *J. Chem. Phys.* **48**, 1807 (1968). The CNDO method was modified by introducing a new empirical parameter k to differentiate resonance integrals between σ orbitals from these between π orbitals. Good results were obtained.
h. "Some Effects of Protonation on the Electronic Structure of Pyridine and Cytosine," A. Denis and M. Gilbert, *Theoret. Chim. Acta* **11**, 31, (1968). The effects of protonation on the charge distribution of pyridine and cytosine were investigated using the iterative extended Hückel technique. The results are used in a PPP π-electron calculation of transition energies.

4. Optical Rotatory Power

a. "Molecular Orbital Theory of Optical Rotatory Strengths of Molecules," Y.-H. Pao and D. P. Santry, *J. Am. Chem. Soc.* **88**, 4157 (1966). The excited states of cyclohexanone and its methyl derivatives are considered and related to the octant rule.

5. Orbital Symmetry—Selection Rules for Concerted Reactions

a. "Stereochemistry of Electrocyclic Reactions," R. B. Woodward and R. Hoffmann, *J. Am. Chem. Soc.* **87**, 395 (1965). This is the paper in which the stereochemistry of electrocyclic reactions is first considered in terms of orbital symmetries. The rules which indicate whether the reaction should be conrotatory or disrotatory are derived.
b. "Selection Rules for Concerted Cycloaddition Reactions," R. Hoffmann and R. B. Woodward, *J. Am. Chem. Soc.* **87**, 2046 (1965). The selection rules for concerted cycloaddition reactions are derived based on orbital symmetry arguments.
c. "Selection Rules for Sigmatropic Reactions," R. B. Woodward and R. Hoffmann, *J. Am. Chem. Soc.* **87**, 2511 (1965). The selection rules for the migration of a σ bond (as in thermal hydrogen shifts) are derived and compared with the available experimental data.
d. "Orbital Symmetries and Endo-Exo Relations in Concerted Cycloaddition Reactions," R. Hoffmann and R. B. Woodward, *J. Am. Chem. Soc.* **87**, 4388 (1965). The preference for endo addition in the Diels-Alder reaction and similar reactions is considered from the viewpoint of orbital symmetry.
e. "Orbital Symmetries and Orientation Effects in a Sigmatropic Reaction," R. Hoffmann and R. B. Woodward, *J. Am. Chem. Soc.* **87**, 4389 (1965). The application of orbital symmetry relationships to the Cope rearrangement is described.
f. "Stereospecificity-Cyclic Reactions," K. Fukui, *Tetrahedron Letters* **1965**, 2009. A discussion of the stereochemistry of electrocyclic reactions which reaches essentially the same conclusions as Woodward and Hoffmann.
g. "Stereoselectivity Associated with Noncycloaddition to Unsaturated Bonds," K. Fukui, *Tetrahedron Letters* **1965**, 2427. The predominance of *trans* addition to olefins is considered by the application of the frontier electron concept. It is predicted that 1,2-addition should take place by a *trans* mechanism but that 1,4-addition should proceed via *cis* addition.
h. "Theoretical Accounting for Stereoselective E2 Reactions," K. Fukui and H. Fujimoto, *Tetrahedron Letters* **1965**, 4303. The frontier electron method was applied to the E2 elimination and accounts for the preferential *trans* stereochemistry. The observed *cis* elimination with the norbornyl systems also is accommodated.

i. "Electronic Structure of Some Intermediates and Transition States in Organic Reactions," R. Hoffmann, *Trans. N.Y. Acad. Sci.* **28**, 475 (1966). Molecular orbital correlation diagrams are used to predict the elimination of hydrogen in diimide reductions and the elimination of carbon monoxide from cyclopentanone. Other systems are discussed also.

j. "Transition States of the Claisen and the Cope Rearrangements," K. Fukui and H. Fujimoto, *Tetrahedron Letters* **1966**, 251. Extended Hückel calculations are presented for the activated complexes in the Claisen and Cope rearrangements.

k. "Molecular Orbital Theoretical Illumination for the Principle of Stereoselection," K. Fukui, *Bull. Chem. Soc. Japan* **39**, 498 (1966). A simplified method for determining the direction of overlap stabilization between interacting atomic orbitals is presented. This approach does not overemphasize the effects of highest occupied or lowest vacant molecular orbitals. A number of common reactions are used to illustrate the method.

l. "Sigma-Pi Interaction Accompanied by Stereoselection," K. Fukui and H. Fujimoto, *Bull. Chem. Soc. Japan* **39**, 2116 (1966). A simple MO treatment is presented for chemical reactions of planar conjugated molecules and for the stereoselectivity which is observed in the reactions. A variety of processes, including the Diels-Alder reaction, osmium tetroxide addition, and the Claisen and Cope rearrangements, are considered.

m. "M. O. Theory of Organic Chemistry: VIII. Aromaticity and Electrocyclic Reactions," M. J. S. Dewar, *Tetrahedron Suppl.*, No. **8**, 75 (1966). The perturbational MO method is found to be in accord with SCF-MO calculations. The method provides a simple interpretation of both thermal and photochemical electrocyclic reactions following the rule that such reactions take place via aromatic activated complexes.

n. "Stereoselection Rules for Electrocyclic Interactions," K. Fukui and J. Fujimoto, *Bull. Chem. Soc. Japan* **40**, 2018 (1967). A set of stereoselection rules governing electrocyclic reactions are proposed. Many examples are given.

o. "Conservation of Orbital Symmetry," R. Hoffmann and R. B. Woodward, *Accounts Chem. Res.* **1**, 17 (1968). A review which illustrates the application of simple concepts in MO theory to virtually every concerted organic reaction.

p. "Intermolecular Orbital Theory of the Interactions between Conjugated Systems: I. Theory," L. Salem, *J. Am. Chem. Soc.* **90**, 543 (1968). The interaction between two conjugated molecules with overlapping p orbitals is described in terms of the π electrons of the separated systems. New orbitals are built out of the interacting molecular orbitals. Applications to concerted thermal and photochemical processes are described.

q. "Intermolecular Orbital Theory of the Interactions between Conjugated Systems: II. Thermal and Photochemical Cycloadditions," L. Salem, *J. Am. Chem. Soc.* **90**, 553 (1968). Application of the above method to cycloaddition reactions is described, giving results in agreement with the Woodward-Hoffmann selection rules. The method is easily applied to large systems.

6. Physical Properties of Molecules

a. "Extended Hückel Theory: I. Hydrocarbons," R. Hoffmann, *J. Chem. Phys.* **39**, 1397 (1963). This paper presents calculations of geometries, barriers to internal rotation, charge distribution, and other properties of saturated and unsaturated hydrocarbons.

b. "Molecular Orbital Treatment of Paraffin Molecules: II. H Approximation," G. Klopman, *Helv. Chim. Acta* **46**, 1967 (1963). The method of Sandorfy was applied to saturated hydrocarbons using a modification which took into account nonbonded interactions. Good agreement was obtained between calculated and observed heats of formation and ionization potentials of C_1–C_5 alkanes.

c. "Distribution of Electronic Levels in Alkanes," R. Hoffmann, *J. Chem. Phys.* **40**, 2047 (1964). Energy levels were calculated for the *trans* staggered C_nH_{2n+2} molecules.
d. "Extended Hückel Theory: III. Compounds of Boron and Nitrogen," R. Hoffmann, *J. Chem. Phys.* **40**, 2474 (1964). Compounds such as borazine and aminoborane are considered, as well as the heteroatomic B-N compounds.
e. "Extended Hückel Theory: II. σ-Orbitals in the Azines," R. Hoffmann, *J. Chem. Phys.* **40**, 2745 (1964). The charge distribution in pyridine is calculated using both σ and π electrons.
f. "Molecular Orbital Theory of Hydrocarbons: II. Ethane, Ethylene and Acetylene," J. A. Pople and D. P. Santry, *Mol. Phys.* **9**, 301 (1965).
g. "Approximate Self-consistent Molecular Orbital Theory: II. Calculations with Complete Neglect of Differential Overlap," J. A. Pople and G. A. Segal, *J. Chem. Phys.* **43**, S136 (1965). Properties of small molecules (including ethane, methylamine, and methanol) are calculated. Bond angles, bending force constants, and barriers to internal rotation are in agreement with experiment, but bond lengths and dissociation energies are not as satisfactory.
h. "Theoretical Observations on Cyclopropane," R. Hoffmann, *Tetrahedron Letters* **1965**, 3819. The potential energy for rotating about the single bond in RCHO was calculated for R = cyclopropyl, vinyl, phenyl, isopropyl, and cyclobutyl. The conformational preference for cyclopropyl is similar to phenyl and vinyl, whereas cyclobutyl is similar to isopropyl. Other calculations on cyclopropyl derivatives also are calculated.
i. "Extended Hückel Theory: V. Cumulenes, Polyenes, Polyacetylenes and C_n," R. Hoffmann, *Tetrahedron* **22**, 521 (1966). Calculations are performed for the ground states and excited states of the title species. Bond-length variation, torsional barriers, and conformations are considered.
j. "Modified Hückel Molecular Orbital Calculations of Nuclear Spin Coupling Constants in Simple Hydrocarbons and Aldehydes," S. Polezzo, P. Cremaschi, and M. Simonetta, *Chem. Phys. Letters* **1**, 357 (1967). An iterative modified Hückel method is used to calculate contact contribution to H—H, ^{13}C—H, and ^{13}C—^{13}C coupling constants in small organic molecules. The results are in fair agreement with experiment. Long-range coupling is not always well predicted.
k. "Ground States of σ-Bonded Molecules: I. Semi-empirical S. C. F. M. O. Treatment of Hydrocarbons," M. J. S. Dewar and G. Klopman, *J. Am. Chem. Soc.* **89**, 3089 (1967). Calculations on heats of formation of hydrocarbons are presented and are found to agree fairly well with experiment.
l. "Ground States of σ-Bonded Molecules: II. Strain Energies of Cyclopropanes and Cyclopropenes," N. C. Baird and M. J. S. Dewar, *J. Am. Chem. Soc.* **89**, 3966 (1967). The SCF-MO method described in Part I is applied to a variety of compounds containing three-membered rings. Generally good agreement with experimental results are found. Tetrahedrane is calculated to be unstable, whereas bicyclobutadiene is predicted to be stable.
m. "Ground States of σ-Bonded Molecules: III. Valence-Shell and π-Electron S. C. F. M. O. Calculations for Conjugated Hydrocarbons," N. C. Baird and M. J. S. Dewar, *Theoret. Chim. Acta* **9**, 1 (1967). A comparison is made between π-electron and valence-electron SCF-MO methods. Good agreement is found. Hyperconjugative interactions were calculated to be small in neutral molecules but substantial in cations.
n. "Valence Shell Calculations on Polyatomic Molecules: I. CNDO SCF Calculations on Nitrogen and Oxygen Heterocycles," J. E. Bloor and D. L. Breen, *J. Am. Chem. Soc.* **89**, 6835 (1967). The five- and six-membered heterocycles were studied and the calculated dipole moments were found to agree well with the experimental values. The total charge densities agreed with ^{13}C chemical shifts.
o. "Valence Shell Calculations on Polyatomic Molecules: II. CNDO SCF Calculations on

Monosubstituted Benzenes," J. E. Bloor and D. L. Breen, *J. Phys. Chem.* **72**, 716 (1968). The CNDO method gave results which agreed well with dipole moments and ^{13}C chemical shifts. There was not a good correlation between orbital energies and ionization potentials.

p. "Application of the Pople-Santry-Segal Complete Neglect of Differential Overlap (CNDO) Method to Some Hydrocarbons and Their Cations," K. B. Wiberg, *J. Am. Chem. Soc.* **90**, 59 (1968). A set of semiempirical parameters were obtained which gave correct equilibrium geometries and dipole moments, and which gave calculated energies which correlated well with the experimental values.

q. "All Valence-Electrons Calculations of the Biological Purines and Pyrimidines: I. CNDO Calculation," C. Giessner-Prettre and A. Pullman, *Theoret. Chim. Acta* **9**, 279 (1968). Calculations were carried out and compared with results obtained using other methods.

r. "Zur Wechselwirkung von π- und σ- Elekronen in der Theorie ungesättigter Moleküle I. Acrolein and Furan," M. Jungen and H. Labhart, *Theoret. Chim. Acta* **9**, 345 (1968). An extended PPP method was applied to furan and acrolein. The results are compared with experimental data and with other calculations.

s. "Trimethylene and the Addition of Methylene to Ethylene," R. Hoffmann, *J. Am Chem. Soc.* **90**, 1475 (1968). Extended Hückel calculations on a distorted cyclopropane indicates the presence of a singlet trimethylene intermediate with a CCC angle of 125°, trigonal terminal methylene groups coplanar with the carbon skeleton. The consequences of these calculations are discussed, as well as those for the potential surface for the addition of methylene to ethylene.

t. "Benzynes, Dehydroconjugated Molecules and the Interaction of Orbitals Separated by a Number of Intervening σ-Bonds," R. Hoffmann, A. Imamura, and W. J. Hehre, *J. Am. Chem. Soc.* **90**, 1499 (1968). Significant and specific interactions among radical lobes in the same molecule separated by a number of σ-bonds have been deduced from calculations. The important factors determining the interaction are discussed.

u. "Electronic Structure of Methylenes," R. Hoffmann, G. D. Zeiss, and G. W. Van Dine, *J. Am. Chem. Soc.* **90**, 1485 (1968). Extended Hückel calculations are carried out on methylene and many of its derivatives, including those having π-electrons. Equilibrium geometries were obtained and preferred modes of bond bending were calculated.

v. "Hydrogen Bonding in Pyridine," W. Adam, A. Grimson, R. Hoffmann, and C. Z. de Ortiz, *J. Am. Chem. Soc.* **90**, 1509 (1968). The extended Hückel theory is applied to the hydrogen bond formed between pyridine and water or methanol.

w. "A Molecular Orbital Study of the Isomerization Mechanism of Diazacumulenes," M. S. Gordon and H. Fischer, *J. Am. Chem. Soc.* **90**, 2471 (1968). Inversion barriers and geometries of AB_3 molecules calculated using the INDO method are in satisfactory agreement with experiment. The methyl cation and anion, diimide, and carbodiimide are considered.

x. "A Self-consistent Field Molecular Orbital Treatment, Including Excited States, of Cyclopropane, Ethylene Oxide and Ethylenimine," D. T. Clark, *Theoret. Chim. Acta* **10**, 111 (1968). The electronic structure of cyclopropane, ethylene oxide, and ethylenimine have been investigated using the CNDO method. The results give a good account of ground-state charge distributions and a reasonable interpretation of the spectra.

y. "A Self-consistent Field Molecular Orbital Treatment of Furan, Including All Valence Electrons," D. T. Clark, *Tetrahedron* **24**, 3285 (1968). A comparison is made between π-electron and valence-electrons SCF-MO calculations for furan. The latter gives satisfactory values for the dipole moment and first ionization potential.

z. "Application of CNDO/II to Some Hydrogen-Bonded Systems," A. Ocvirk, A. Azman, and D. Hadži, *Theoret. Chim. Acta* **10**, 187 (1968). Calculations were carried out on formic acid and other carboxylic acids.

aa. "Extended Hückel Calculations on the Conformation and Structure of Thymine Photodimers," F. Jordon and B. Pullman, *Theoret. Chim. Acta* **10**, 423 (1968). Extended Hückel calculations were carried out on dihydropyrimidines and were related to the structures of the photodimers.

bb. "The Isocyanide-Cyanide and Isoelectronic Rearrangements," G. W. Van Dine and R. Hoffmann, *J. Am. Chem. Soc.* **90**, 3227 (1968). Potential surfaces for the isocyanide-cyanide rearrangement of a hydrogen, methyl, and phenyl are constructed from extended Hückel calculations. The reaction paths for methyl and phenyl are calculated to be different. Application to other rearrangements is discussed.

cc. "Semi-empirical All Valence Electrons SCF-MO-CNDO Theory: II. Interatomic Parameters and Bonding Energies," J. M. Sichel and M. A. Whitehead, *Theoret. Chim. Acta* **11**, 220 (1968). Modified parameters for a CNDO treatment were obtained which give bonding energies in reasonable agreement with experimental values.

dd. "Theoretical Calculations of the Structure and Electronic Properties of Dihydrothymines and Dihydrouracils," F. Jordan, *Theoret. Chim. Acta* **11**, 390 (1968). Extended Hückel calculations were performed on a series of substituted 5, 6-dihydropyrimidinediones. The results were compared with experimental data on conformations and NMR chemical shifts.

ee. "Extended Hückel Method: Calculation of the Ethylene Force Field," J. Paldus and P. Hrabe, *Theoret. Chim. Acta* **11**, 401 (1968). All quadratic force constants for ethylene were calculated using the extended Hückel method. Good agreement with experiment were found for bending modes but not for stretching modes.

ff. "A CNDO Study of Steric Effects in Biphenyl," B. Tinland, *Theoret. Chim. Acta* **11**, 452 (1968). The conformation of biphenyl was studied using the CNDO method. Using a fixed benzene-benzene bond length, no minimum energy was found between 0 and 90°

IV-2

Molecular Orbital Calculations and Organic Chemistry

KENNETH B. WIBERG

Organic chemists usually are interested in three quantities by which molecules, transient species, and activated complexes may be described: (a) the heat of formation (or heat of atomization); (b) the geometry; and (c) the charge distribution. If these three quantities could be calculated, chemical transformation could be discussed in detail and the necessity for a variety of "effects" would in many cases be eliminated. The organic chemist then asks: What is the possibility of calculating these quantities with a precision which would be useful in discussing chemical transformations? This would require a precision of ± 1 kcal/mole (±0.04 eV or ±0.0015 a.u.!) in the heat of formation, ±0.02 Å in bond lengths, ±1 to 2° in bond angles, and ±0.05 in electron density.

If one were asked to calculate the quantities with this precision by a present-day *ab initio* method, the proposition would clearly be absurd. The *ab initio* calculations are simply not that good, and as long as the Hartree-Fock approach is used, they cannot be that good for the correlation energy and correlation effects on the wave functions are not included. As discussed in Chapter III, the correlation energy is an appreciable part (~20 per cent) of the heat of atomization.

In addition to being incapable of realizing chemically useful precision, the *ab initio* methods also suffer from complexity in calculation and high cost in computer time. Neither is desirable for calculations which are to be carried out frequently.

It then appears that the semiempirical methods provide the best possibility for obtaining chemically useful results. This was early seen with the Hückel π-MO calculations which use as primitive a semiempirical method as can be imagined. Yet, the method when properly calibrated could give "resonance energies" for condensed aromatic systems to ±1 kcal/mole and could give the positions of the first and second $\pi \rightarrow \pi^*$ transitions with good accuracy. However, it must be noted that good results with spectra are obtained *only* if the calculated and observed transition energies are plotted against each other. The resultant plot has a nonzero intercept corresponding to a finite transition energy for a zero calculated energy. Abnormal intercepts should be expected in these treatments because part of the error in the treatment (i.e., terms which were neglected) may be absorbed in the intercept.

The π-MO methods are not particularly useful in studying organic chemical phenomena because they include only a small and special part of the molecule. The more recent semiempirical σ-MO methods have provided the possibility of examining all types of chemical transformations and potentially provide a tool which cannot be neglected by the organic chemist.

I believe that cooperation between physical and organic (inorganic) chemists is needed to make real progress in the use of σ-MO methods. On the part of the physical chemists, there is a need for an increased willingness to solve the problems of how to properly calculate repulsion integrals and still maintain rotational invariance,[1] of how to include correlation effects in the semiempirical framework, and of whether or not various degrees of neglect of differential overlap may be maintained. In addition, when the methods are tested against experimental data, it seems only reasonable that both data sets be on the same basis and correspond to minimum-energy geometries.[2]

The organic chemist should take the responsibility of thoroughly testing new methods as they become available, to learn enough about the methods to be able to use them intelligently, and to be willing to think in terms of molecular orbitals rather than localized bonds when considering carbonium ions and other delocalized species. The often-strange discussions of hyperconjugation result from an attempt to stretch valance-bond language too far. The charge alternation in carbonium ions certainly requires an MO description and the same is true of spin densities in free radicals.

If the organic chemist is not willing to do the above, he stands a good chance of being bypassed in the future interesting developments in the field. Further, the organic chemist has a wealth of knowledge concerning the transformations of organic compounds, and this can provide many useful tests of the theoretical methods. Finally, it is unlikely that any group other than the organic chemists will have the motivation to make many varied tests of the theories.

References and Notes

1. Pople et al. [*J. Chem. Phys.* **43**, S129 (1965) (reprinted as Chapter VI-5)] maintain rotational invariance using only one averaged repulsion integral between a pair of atoms and ignoring the distinction among s, p_σ, and p_π orbits. Klopman [*J. Am. Chem. Soc.* **86**, 4550 (1964)] uses a different, but still essentially artificial method for handling the problem. It is probable that part of the difficulty in obtaining good agreement between calculated and observed bond lengths for the series single, double, and triple bonds [see Wiberg, *J. Am. Chem. Soc.* **90**, 59 (1968)] results from the use of inappropriate repulsion integrals.
2. For example, Baird and Dewar [*J. Am. Chem. Soc.* **89**, 3966 (1967)] carried out σ-MO calculations of the energies of a number of molecules using observed geometries as input. However, the parameters used would not give the observed geometries if the energies were minimized with respect to geometry. Further, they compared the calculated energies with the observed energies without correcting the latter for zero-point energies or for the change in energy on going from 25°C to 0°K.

IV-3

Ab initio Calculations and Organic Chemistry

A. STREITWIESER, JR.

Quantum organic chemistry is entering a period of dramatic change. Semiempirical methods have dominated the field, formerly on π systems alone with the Hückel MO method followed by the Pople-Pariser-Parr (PPP) SCF method, and more recently with the all-valence-electron methods of the type of extended Hückel theory (EHT) and the Pople-Santry CNDO method. HMO calculations that were once very laborious and time consuming[1] have been replaced by more meaningful SCF calculations that take only seconds or minutes of computer time. But already, Hartree-Fock calculations with large basis sets have been published for such small organic molecules as methane, methyl cation and anion, etc.[2] Complete nonempirical calculations with limited basis sets have been published for larger molecules such as pyridine.[3] With the advent of still larger and faster computers Clementi and Davis could say in 1966, "We feel confident in stating that within the next 2 to 4 years 'a priori' computations for molecules with about 100–150 electrons and 5–20 atoms will be considered 'routine' effort in theoretical chemistry."[4] It is pertinent to ask at this time what effect this new era of quantum chemistry will have on organic chemistry, especially since larger and faster computers do not come cheaply. Calculations on molecules of reasonable size will take significant amounts of computer time and will be expensive. Indeed, common complaints during private discussions among participants at the recent Yale Seminar on molecular orbitals related to budget limitations—many researchers had more ideas than they had computer funds to implement them.

One comment is indicated—*exact calculations on molecules of fair size by themselves will be of limited significance.* Long lists of orbital energies and pages of wave functions of high quality will not be of much use to the organic chemist. Certainly the total energy will have some use, especially for comparing a series of related molecules—considerable effort has been and still is being made in establishing quantitative correlations between semiempirical calculated energies and approximate experimental results. Many such correlations between reactivities of aromatic systems and HMO calculations have been presented[5]; still better correlations are often obtained with PPP π methods.[6,7] Both Klopman and Wiberg point up similar correlations with σ-MO calculations. But the usefulness of such correlations for prediction is less significant than the fact of a correlation or the deviations of specific systems; that is, such correlations are generally significant not only to provide means for interpolating or extrapolating experimental data to other systems but to establish general concepts—an example is the interpretation of the slopes of linear correlations of rates of related carbonium ion reactions with calculated π-energy differences.[5,8] Of still greater significance and impact on organic chemistry have been the nonquantitative concepts that have come from molecular orbital theory—for example, the Hückel 4n + 2 rule and the Woodward-Hoffman rules. The former generalization has led, starting with Dewar's tropolone structure for stipitatic acid,[9] to a continuing period of brilliant and fertile *synthetic* organic chemistry involving new aromatic systems. The Woodward-Hoffman rules are receiving constant verification and development from new experimental results that are appearing almost daily. The coming period of more exact quantum chemical calculations on organic molecules will present an opportunity to test many of the simplifying concepts so important to the practising organic chemist.

The organic chemist deals with ever larger and more complex systems. We will not soon see thorough calculations of steroids with large basis sets. To use results of calculations of smaller molecules the organic chemist must use the logic of analogy he has used so successfully in building structural organic chemistry. The concept of the functional group emphasizes this limitation. To be useful to the chemist, the quantum mechanical calculations must be dissected into independent or nearly independent units of the localized orbital or group-equivalent type. In principle, delocalized MO's can be transformed to localized MO's, but there are already several definitions for optimum localization and these definitions are generally not easy to implement for other than small molecules or highly symmetrical molecules. It is encouraging to see further research on localized orbitals such as that of Trindle and Sinanoğlu, which is applicable to larger molecules.

The spectacular success of the Woodward-Hoffman rules in utilizing the symmetry of delocalized σ-MO's to derive the stereochemistry of electrocyclic reactions does not negate the need for reduction to more localized units. The rules apply equally well to less symmetrical molecules in which the bonds involved in the reactions can be considered to be perturbations of the fully symmetrical models. An additional example of the chemical deductions from the symmetries of parts of molecules was demonstrated by Hoffmann for benzyne-type compounds during the Yale Seminar.

Geometric structure is a growing product of modern quantum calculations useful to the organic chemist. Better potential functions for conformational energies are a prospect that should enhance the already rather successful classical calculations of conformations.[10,11] But here also we should expect contributions to concepts—an example is provided by Hartree-Fock calculations of the rotation barrier in ethane that show that correlation makes but a minor contribution to the barrier.[12]

Focusing on regions or parts of molecules is also important to the physical organic chemist, who often relates the quantitative behavior of one system as a perturbation of another. The perturbation is treated as the result of an "effect"—for example, an "inductive effect," a "steric effect," a "field effect," etc. Because of the complexity of the systems treated, the application of the various "effects" is often rather qualitative. When the existing "effects" do not suffice to interpret a specific result a new "effect" is postulated, frequently in an *ad hoc* manner. The proliferation of "effects" and disagreements about their magnitudes have led to much polemic in the literature. In the next few years, these "effects" should be analyzed in quantum mechanical terms, and reliable calculations should increasingly provide validation and establishment of the range of application. Some of these concepts reduce to an estimate of an energy change, whereas others relate to electron density changes; with appropriate definitions, both types of concept are subject to computations and more rigorous analyses. The concept of the chemical bond will undoubtedly remain with us, even with good wave functions for small systems. Such wave functions are giving rise to new understanding of bonding, as in Clementi and Gayles' discussion of the electronic reorganization during the NH_3—$HC\ell$ reaction[13] and the treatments of Kern and Karplus[14] and of Bader and Jones[15] of electron density changes from atoms to molecule. But many of our old concepts of bonding should now be subject to scrutiny with precision. An example is Pauling's proposal of some years ago that a difference in electronegativity serves to increase the strength of the bond joining two atoms.[16] This concept has been used frequently in the last three decades and even as recently as 1965 was invoked to interpret various bond strengths to carbon.[17] However, Pearson[18] has recently summarized many examples for which Pauling's proposal gives incorrect predictions. We can add a further case, important because the bonded atoms remain the same, although their electronegativity differences change: the C—Cℓ and C—Br bond energies are essentially the same in $CH_3C\ell$ and CH_3Br as in $CF_3C\ell$ and CF_3Br, respectively,[19] and not higher, as would be expected by application of Pauling's proposal. A direct test of the effect on bond strength of various definitions of electronegativity should be possible, perhaps with the use of fictitious atoms in which all parameters are held constant except those associated with the defined concept whose effect is to be examined.

Another effect used frequently to interpret bond strengths and structure is that of lone-pair repulsion; this "effect" presumably contributes to the low bond strength of F_2[20] and is a factor in the relative stabilities of some fluorocarbanions.[21] It should now be possible to make real calculations of the magnitude of this effect and of the validity of the proposed applications.

The various types of "hyperconjugation" provide an ideal arena for quantum mechanical test. This set of concepts is well defined in quantum mechanical terms,[22] has received much use in the interpretation of organic chemistry,[23,24] and has provoked considerable controversy and dissent.[25] The type of hyperconjugation derived from conjugation or overlap of a methyl group with a π system has had weak experimental support. However, recent SCF calculations, albeit with a minimum basis set of Slater orbitals, gave substantial magnitude to this effect and indicated that even second-order hyperconjugation in ethane,

$$\begin{matrix} \overset{\displaystyle H}{|} \ \overset{\displaystyle H}{|} \\ H_2C-CH_2 \end{matrix} \quad \longleftrightarrow \quad \begin{matrix} H\text{-}H \\ H_2C=CH_2 \end{matrix}$$

is in the kilocalorie range.[26] This recent result will undoubtedly not quell the controversy by itself, but results of more complete calculations may be decisive.

Much more experimental support has been adduced for hyperconjugation as an important factor in the relative stabilities of carbonium ions,

$$\begin{matrix} \overset{\displaystyle H}{|} \\ C-C^+ \end{matrix} \quad \longleftrightarrow \quad \begin{matrix} H^+ \\ C=C \end{matrix}$$

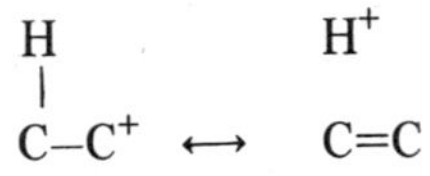

Even for such electron-deficient systems, in which one would expect a driving force for charge delocalization, some controversy has developed; an argument such as the following shows that some hyperconjugation *must* occur. In Fig. 1, the interaction of a doubly occupied σ orbital of the appropriate symmetry with an empty carbon 2p orbital must give an energy lowering. The amount of energy lowering depends in the usual way on the overlap between the two orbitals and on their energy separation and interaction. The real question is: Are these inter-actions sufficient to give an energy lowering of significant magnitude—a magnitude equivalent to the interpretations proposed? Numerous semiempirical calculations, both π and σ, have indicated that hyperconjugation does have the postulated magnitude in carbonium ions, but more complete treatments are certainly called for.

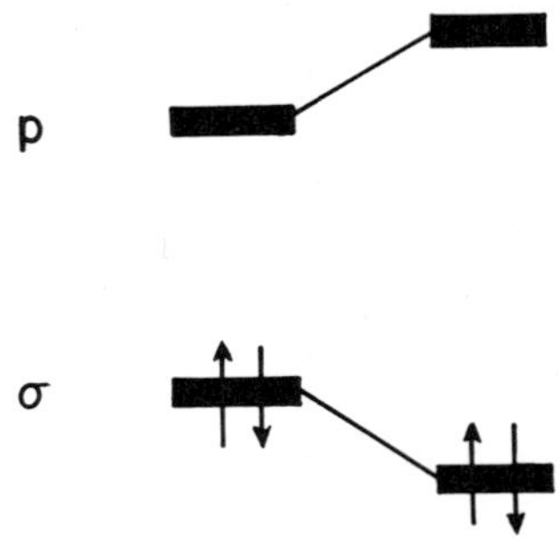

Fig. 1. Hyperconjugation in a carbonium ion.

Anionic hyperconjugation has also been suggested as a significant effect, particularly for the interaction of C—F bonds with a lone pair[27]:

$$
\begin{array}{ccc}
\text{F} & & \text{F}^- \\
| & & \\
\text{C—C}^- & \longleftrightarrow & \text{C=C}
\end{array}
$$

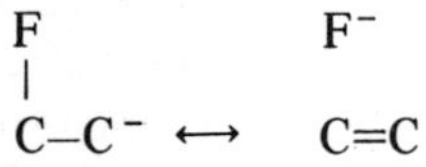

A glance at Fig. 2 shows that this situation involves a fundamental difference from the carbonium ion case. Because both the p orbital and the interacting σ orbital are doubly occupied, the total energy is unchanged to a first approximation and will even increase from electron

$\sigma^\star$

p

σ

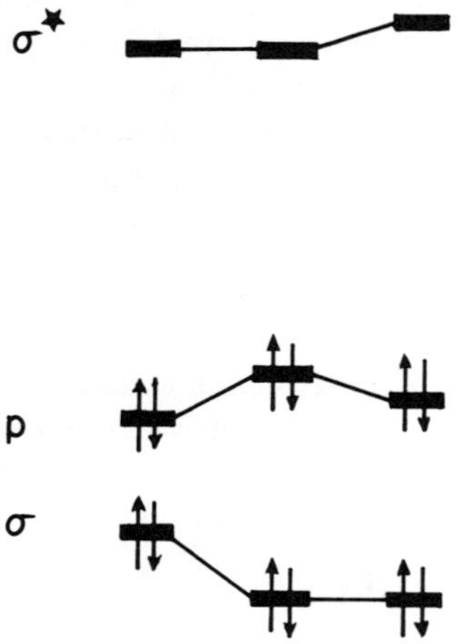

Fig. 2. Anionic hyperconjugation by interaction of a doubly occupied p orbital with a doubly occupied σ orbital of appropriate symmetry and then with the corresponding empty σ^* orbital.

repulsion. Only a significant interaction with σ^* will cause an energy lowering. For this effect to be appreciable, one must argue that the energy of σ^* is accessibly low. The mass spectral observation of the negative ion of perfluoromethylcyclohexane[28] may be taken to suggest that such antibonding σ orbitals have a significant electron affinity, although even here it is possible that bond breaking to an open-chain radical anion has occurred. Recent chemical results indicate that anionic hyperconjugation is not a significant "effect" in organic chemistry.[28,29] The foregoing discussion exemplifies the arguments used in the past—a combination of simple quantum chemical concepts and experimental results. Good calculations on appropriate compounds (e.g., β-fluoroethyl anion in various conformations) are now feasible and should be decisive.

A related type of hyperconjugation has also been used to explain the comparatively short bonds in carbon tetrafluoride[31]:

$$
\text{F—CF}_2\text{—F} \quad \longleftrightarrow \quad {}^+\text{F=CF}_2\ \text{F}^-
$$

Applications of this concept have been thoroughly explored and reviewed by Hine,[32] but again, there is considerable question whether the proposed effect is of sufficient magnitude to account for the observed phenomena. Here also appropriate calculations (e.g., methylene fluoride) should allow isolation of any π component to the overall bonding.

If hyperconjugation is not of sufficient magnitude to explain the C–F bond shortening in the series CH_3F, CH_2F_2, CHF_3, and CF_4, it is important to formulate alternative concepts that are subject to test.[2] One such alternative is that electron withdrawal from carbon by fluorine shrinks the carbon orbitals. It should be possible now to compare calculations of different carbon compounds and compare the "sizes" of the carbon basis orbitals in different compounds by a simple comparison of optimized orbital exponents. Further comparisons should also provide a critical test of another concept commonly used in organic chemistry: localized bonds formed by overlap of appropriately hybridized atomic orbitals. The validity of criteria for the degree of hybridization (for example, exocyclic H–C–H bond angles, as in cyclopropane, correlations involving $J_{^{13}C-H}$, etc.) are now subject to test by *ab initio* calculations.

The charge distributions in organic compounds have long been of interest and many concepts have been suggested that involve charge displacements; examples are the several theories of alternating polarity proposed fifty years ago and the later electronic theory of the English school.[33] The electronic structure of conjugated cations and anions resting on the apparently secure foundation of resonance theory and HMO theory of a few years ago has received some uncomfortable shocks from recent SCF calculations that, although still semiempirical, should be of better quality. For example, resonance structures for benzyl cation (I) agree with HMO

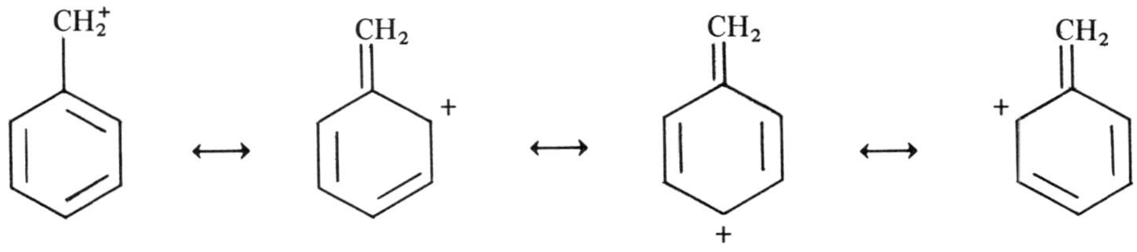

(I)

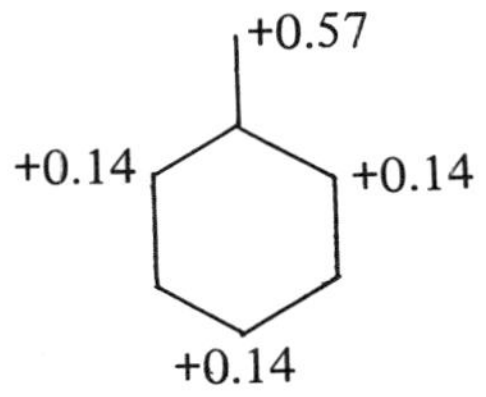

(II)
HMO

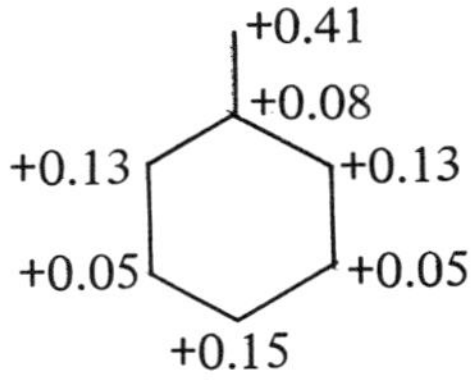

(III)
ω-HMO

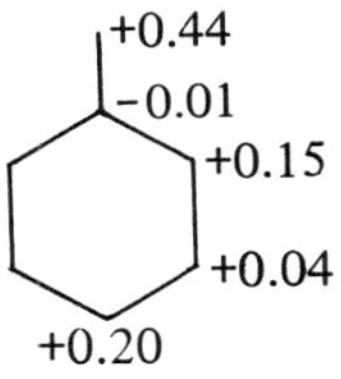

(IV)
SCF-π

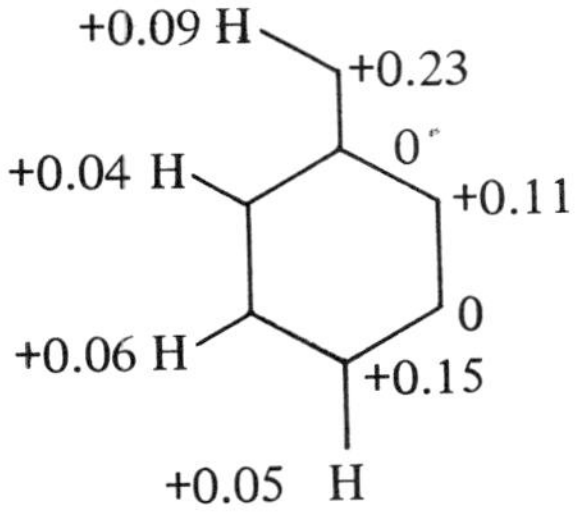

(V)
CNDO

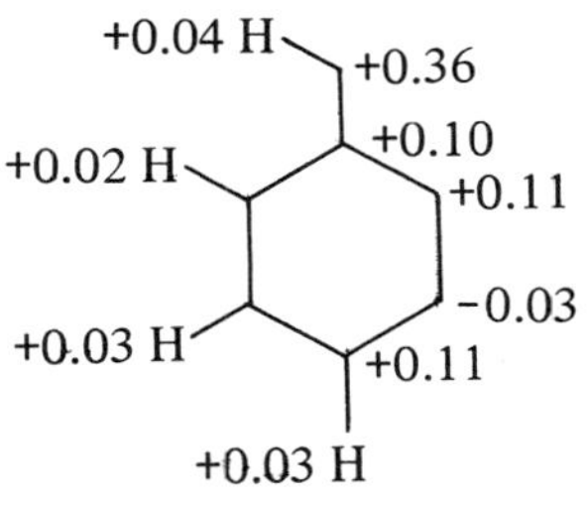

(VI)
EHT

calculations (II) in putting positive charge only on the o, p, and α carbons. The ω modification of the Hückel method spreads the charge somewhat, as in (III). A Pople SCF π calculation with Mataga integrals gives the result[34] in (IV), in which the branching carbon actually has a negative charge—a common result for such calculations. The Pople-Santry CNDO method gives the result in (V) and Hoffmann's EHT method gives (VI),[35] in which the σ systems put substantial positive charge on the ring hydrogens. Which of these results comes closest to the truth? The comparisons of charge distributions in the nucleotide bases as discussed by Madame Pullman in this book show the different results obtained by semiempirical methods; recent results as in the nonempirical calculations by Clementi in pyrrole,[36] pyridine,[13] and pyrazine[37] cause us to reevaluate our assumptions about the separability of σ and π effects.

A knowledge of charge distributions is also required to understand several aspects of the Hammett $\sigma\rho$ relation. Although it is clear that the linear free-energy approximation embodied in the Hammett technique treats the effect of the substituent as a perturbation with a limited number of interaction mechanisms,[38] the interaction mechanisms themselves have been subject to considerable controversy with various substituents. An example is the dissection of the σ parameter itself into σ_I and σ_R components; several definitions have been suggested that result in somewhat different numbers.[39,40] An important distinction is exemplified by the nitro group—to what extent does this group conjugate with a benzene ring? Taft assumes a significant conjugation, whereas Exner deduces that conjugation is negligible. The relative importance of direct field effects, the π-inductive effect, the σ-inductive effect, and conjugation effects have all been proposed to contribute to substituent effects in benzenoid systems. Their relative magnitudes should soon be susceptible to sound quantum mechanical test. Related to these questions are the relative importance of bond-induction effects and field effects in saturated molecules.

These examinations of concepts of bonding and of substituent effects will also have an impact on our ideas of reactions. A better understanding of bonds will certainly affect our thoughts about transition states. Much use has been made, for example, of the Hammond postulate,[41] which states, in brief, that a transition state resembles an unstable intermediate of comparable energy. Some rules advanced by Swain and Thornton[42] on the structure of transition states often give apparently opposite predictions from the application of the Hammond postulate. The apparent discrepancies can usually be resolved by recognizing that the Hammond postulate applies generally to geometric structure—the arrangement of nuclei—whereas the Swain-Thornton rules usually apply to electronic reorganizations. Nevertheless, the validity of these postulates is subject to test when relatively complete calculations of reactions become possible. Clementi's treatment[13,43] of the reaction of $HC\ell$ with NH_3 is a forerunner of more complex studies in the future. The potential-energy reaction profiles so common in discussions of organic reaction mechanisms will doubtless depend on a real theoretical base in the future.

Many of our concepts of transition-state structure have provoked considerable controversy. The current dialogue on nonclassical carbonium ions had its predecessor two decades ago in the concept of the carbonium ion itself as a reaction intermediate. The current extended Hückel theory and CNDO treatments of nonclassical carbonium ion systems will undoubtedly give way before long to less empirical but more expensive calculations. Many of the present controversies may not be settled until then!

In the foregoing I have tried to indicate some of the areas where theoretical chemists can contribute to organic chemistry. The list is far from exhaustive—many other concepts could have been cited, such as s character and carbanion stability, the polar character of free radicals, and specific solvation effects. As the theoretical studies unfold, it will be increasingly incumbent on the physical organic chemist to formulate his concepts and proposals with exact definitions that are subject to a quantum mechanical analysis. We will soon reach the time when we will no longer be able to maintain a safe haven behind hazy concepts and sharp polemic. Continued communication with theoretical chemists and with those doing *ab initio* calculations is essential; the development and efficient use of the large, complex programs used in such calculations has become a research specialty in itself, and such cooperation will undoubtedly be an important component in the future testing of physical organic concepts.

Acknowledgment

This work was supported in part by AFOSR Grant 68-1364.

References and Notes

1. For my first independent MO paper [A. Streitwieser, Jr., *J. Am. Chem. Soc.* **74**, 5288 (1952)] I had occasion to expand a 21-order determinant by hand. The first two expansions gave different results and a third was required. This calculation took a ream of paper and many hours of work.
2. Reviewed by M. Krauss, Compendium of *ad initio* calculations of Molecular Energies and Properties, *N.B.S. Tech. Note 438*, Dec. 1967.
3. E. Clementi, *J. Chem. Phys.* **46**, 4731 (1967).
4. E. Clementi and D. R. Davis, *J. Computational Phys.* **1**, 223 (1966).
5. Summarized in A. Streitwieser, Jr., *Molecular Orbital Theory for Organic Chemists*, Wiley, New York, 1961.
6. M. J. S. Dewar and C. C. Thompson, Jr., *J. Am. Chem. Soc.* **87**, 4414 (1965).
7. A. Streitwieser, Jr., H. A. Hammond, and A. Lewis, unpublished results.
8. M. J. S. Dewar and R. J. Sampson, *J. Chem. Soc.* **1957**, 2946, 2952.
9. M. J. S. Dewar, *Nature* **155**, 50 (1945).
10. K. B. Wiberg, *J. Am. Chem. Soc.* **87**, 1070 (1965).
11. J. B. Hendrickson, *J. Am. Chem. Soc.* **89**, 7036 (1967).
12. E. Clementi and D. R. Davis, *J. Chem. Phys.* **45**, 1374 (1966). See also O. Sinanoğlu, Harvard Lecture Notes, 1962, where this was expected from the theory of electron correlation but no quantitative argument was given.
13. E. Clementi and J. N. Gayles, *J. Chem. Phys.* **47**, 3837 (1967).
14. E. W. Kern and M. Karplus, *J. Chem. Phys.* **40**, 1374 (1964).
15. R. F. W. Bader and G. A. Jones, *J. Chem. Phys.* **38**, 2791 (1963); *Can. J. Chem.* **41**, 586, 2251 (1963).
16. L. Pauling, *The Nature of the Chemical Bond*, Cornell University Press, Ithaca, N.Y., 1948, p. 58; 3rd ed., 1960, p. 79.
17. J. Hine and R. D. Weimar, Jr., *J. Am. Chem. Soc.* **87**, 3387 (1965).
18. R. G. Pearson, *Chem. Commun.* **65** (1968).
19. A. Lord, C. A. Goy, and H. O. Pritchard, *J. Phys. Chem.* **71**, 2701 (1967).
20. Ref. 16, 3rd ed., p. 143.
21. A. Streitwieser, Jr., and F. Mares, *J. Am. Chem. Soc.* **90**, 2444 (1968).
22. R. S. Mulliken, C. A. Rieke, and W. G. Brown, *J. Am. Chem. Soc.* **63**, 41 (1941).
23. G. W. Wheland, *Resonance in Organic Chemistry*, Wiley, New York, 1955.
24. J. W. Baker, *Hyperconjugation*, Oxford University Press, New York, 1952.
25. M. J. S. Dewar, *Hyperconjugation*, Ronald Press, New York, 1962.
26. M. D. Newton and W. N. Lipscomb, *J. Am. Chem. Soc.* **89**, 4261 (1967).
27. J. D. Roberts, R. L. Webb, and E. A. McElhill, *J. Am. Chem. Soc.* **72**, 408 (1950).
28. R. K. Asundi and J. D. Craggs, *Proc. Phys. Soc. (London)* **83**, 611 (1964).
29. A. Streitwieser, Jr., and D. Holtz, *J. Am. Chem. Soc.* **89**, 692 (1967).
30. A. Streitwieser, Jr., A. P. Marchand, and A. H. Pudjaatmaka, *J. Am. Chem. Soc.* **89**, 693 (1967).
31. Ref. 20, p. 314.
32. J. Hine, *J. Am. Chem. Soc.* **85**, 3239 (1963).
33. A. E. Remick, *Electronic Interpretation of Organic Chemistry*, Wiley, New York, 1943, Chap. V.
34. A. Streitwieser, Jr., and J. R. Wright, unpublished results.
35. A. Streitwieser, Jr., and R. G. Jesaitis, unpublished results.
36. E. Clementi, H. Clementi, and D. R. Davis, *J. Chem. Phys.* **46**, 4725 (1967).
37. E. Clementi, *J. Chem. Phys.* **46**, 4737 (1967).
38. J. E. Leffler and E. Grunwald, *Rates and Equilibria of Organic Reactions*, Wiley, New York, 1963.
39. C. D. Ritchie and W. F. Sager, *Progr. Phys. Org. Chem.* **2**, 323 (1964).
40. O. Exner, *Coll. Czech. Chem. Commun.* **31**, 65 (1966).
41. G. S. Hammond, *J. Am. Chem. Soc.* **77**, 334 (1955).
42. C. G. Swain and E. R. Thornton, *Tetrahedron Letters*, **1961**, 211.
43. E. Clementi, *J. Chem. Phys.* **46**, 3151 (1967).

 Reprinted from *The Journal of the American Chemical Society* **90**, 1475 (1968)

Trimethylene and the Addition of Methylene to Ethylene

Roald Hoffmann

Contribution from the Department of Chemistry, Cornell University, Ithaca, New York 14850. *Received May 31, 1967*

Abstract: Extended Hückel calculations on a distorted cyclopropane indicate the presence of a singlet trimethylene intermediate with a CCC angle of 125°, trigonal terminal methylene groups coplanar with the carbon skeleton. This molecule has a high barrier to internal rotation and a low barrier to conrotatory reclosure to cyclopropane. The first excited configuration of trimethylene and cyclopropane is a floppy molecule with no rotational barriers. The electronic structure of trimethylene is unusual with a symmetric π-type level above an antisymmetric combination. Similar level orderings, implying conrotatory closing and concerted 1,2 addition, are found in other "1,3 dipoles." The potential surface for the addition of methylene to ethylene is explored in detail. The most symmetrical approach is symmetry forbidden, and the reaction path is unsymmetrical. It begins as a π approach and terminates as σ. Because of the electronic structure of trimethylene it is possible for this unsymmetrical approach to be stereospecific. The specificity of singlet and triplet methylene additions is attributed not to the difference in spin, but to the difference in the spatial part of the wave function. The ring-opened form of cyclopropanone has an electronic structure different from that of trimethylene and other 1,3 dipoles. It is consistent with the valence-bond formulation of an oxy anion of allyl cation. A consequence of this electronic structure is a disrotatory closure back to the cyclopropanone and propensity to concerted 1,4 addition. The extended Hückel calculations make cyclopropanone and allene oxide unstable with respect to oxyallyl (the ring-opened form). In fact they give no stability for cyclopropanone with respect to conversion to oxyallyl. A π-electron SCF–CI calculation has the ground state of oxyallyl, a triplet, with a singlet only 0.1 eV above.

In this paper two aspects of cyclopropane chemistry are discussed: the question of the existence and electronic structure of a trimethylene intermediate CH_2-CH_2CH_2, and the detailed transition-state geometry and specificities observed in the addition of methylenes to ethylenes. An important and connected problem, the *cis–trans* thermal isomerization of substituted cyclopropanes and the competing rearrangement to propylenes, has not yet been considered in detail.

Trimethylene

The stimulus for a series of calculations on trimethylene arose from the following: (1) the observation of stereospecific 1,2 addition of tetracyanoethylene oxide to olefins (I),[1] (2) the discovery by several groups of similar cycloadditions of aziridines (II),[2] (3) the observation of some novel specificities in the pyrolysis of labeled pyrazolines (III),[3] (4) the 1,4 cycloaddition of cyclopropanones to cyclic dienes (IV),[4] and (5) a general desire to learn something about 1,3-dipolar additions, of, for example, O_3.[5]

(1) W. J. Linn and R. E. Benson, *J. Amer. Chem. Soc.*, **87**, 3657 (1965).
(2) A. Padwa and L. Hamilton, *Tetrahedron Letters*, 4363 (1965); J. E. Dolfini, *J. Org. Chem.*, **30**, 1298 (1965); H. W. Heine and R. Peavy, *Tetrahedron Letters*, 3123 (1965); H. W. Heine, R. Peavy, and A. J. Durbetaki, *J. Org. Chem.*, **31**, 3924 (1966); R. Huisgen, W. Scheer, G. Szeimes, and H. Huber, *Tetrahedron Letters*, 397 (1966).
(3) R. J. Crawford and A. Mishra, *J. Amer. Chem. Soc.*, **87**, 3768 (1965); **88**, 3963 (1966); D. E. McGreer, N. W. K. Chiu, M. G. Vinje, and K. C. K. Wong, *Can. J. Chem.*, **43**, 1407 (1965).
(4) N. J. Turro, P. A. Leermakers, H. R. Wilson, D. C. Neckers, G. W. Byers, and F. F. Wesley, *J. Amer. Chem. Soc.*, **87**, 2613 (1965); A. W. Fort, *ibid.*, **84**, 2620, 4979 (1962); R. C. Cookson and M. J. Nye, *Proc. Chem. Soc.*, 129 (1963).
(5) R. Huisgen, R. Grashey, and J. Sauer in "The Chemistry of the Alkenes," S. Patai, Ed., John Wiley and Sons, Inc., New York, N. Y., 1965, p 808; R. Huisgen, *Angew. Chem. Intern. Ed. Engl.*, **2**, 633 (1963).

In all of the above cases there existed a possibility of a ring-opened intermediate, usually classified as a 1,3-dipolar molecule. While the electronic structure of

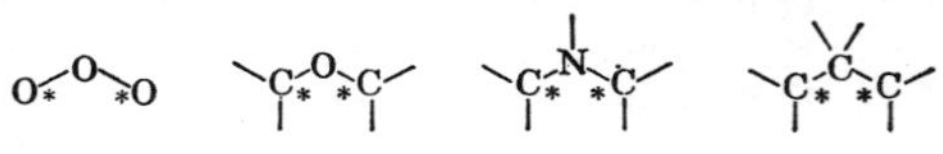

*=+,·,−

the ozone molecule is qualitatively well known, the structures of the other, much less stable, and never isolated, molecules are not obvious. With a good deal of theoretical license the simplest molecule, trimethylene, $CH_2CH_2CH_2$, was chosen for detailed study. The results will be shown to apply to the other molecules in the series, as well as to a number of other unsymmetrical, 1,3-dipolar molecules.

The calculations undertaken were of the extended Hückel type.[6a] The number of degrees of freedom in the general $CH_2CH_2CH_2$ potential surface was reduced to three by fixing C–C distances at 1.54 Å, C–H at 1.10, a tetrahedral central HCH angle, and trigonal terminal CH_2 groups. The remaining degrees of freedom are the CCC angle and the rotations of the terminal methylene groups out of the plane defined by the three carbon atoms. Three geometries defined by the last two angles being 0 or 90° will be important in the subsequent discussion and are drawn in Figure 1. Note in particular that the 90,90 geometry for a small CCC angle goes over to a true cyclopropane structure if the terimnal groups change from a trigonal to an almost tetrahedral local geometry.

(6) (a) R. Hoffmann, *J. Chem. Phys.*, **39**, 1397 (1963), and subsequent papers. The parameters used here are the same except for a H Slater exponent of 1.3. (b) R. B. Woodward and R. Hoffmann, *J. Amer. Chem. Soc.*, **87**, 395 (1965); (c) R. Hoffmann and R. B. Woodward, *ibid.*, **87**, 2046 (1965).

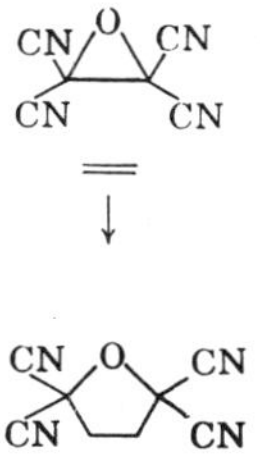

I

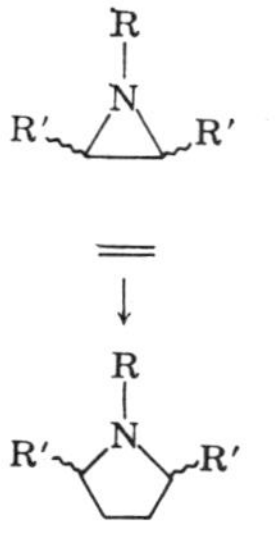

II

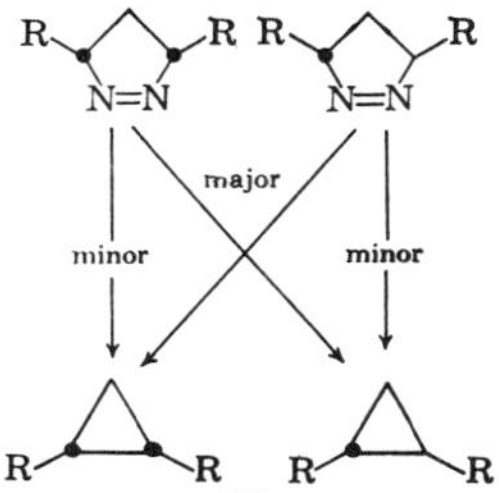

III

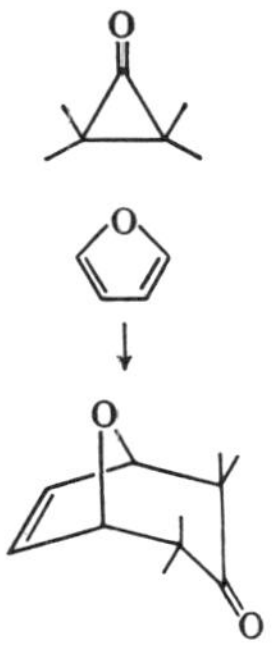

IV

Table I gives several linear sections through the multidimensional potential surface of $CH_2CH_2CH_2$ in ground and excited electronic configurations. The column headings are the geometries illustrated in Figure 1 and the rows are CCC angles. The entries are energies in electron volts relative to a cyclopropane.

Table I. Sections through a $CH_2CH_2CH_2$ Potential Surface[a]

CCC, deg	Ground configuration 0,0	0,90	90,90	Excited configuration 0,0	0,90	90,90
80	4.99	2.76	0.24	5.82	4.31	4.53
100	2.65	2.28	1.08	2.70	2.60	2.62
110	2.10	2.26	1.58	2.40	2.36	2.34
120	1.93	2.36	2.11	2.42	2.37	2.33
130	2.00	2.53	2.31	2.64	2.56	2.51
150	2.65	3.29	2.44	3.56	3.34	3.29

[a] The entries are energies relative to a cyclopropane with C–C 1.54 Å, tetrahedral HCH angles. The geometries refer to Figure 1.

There is here as in the case of methylenes an unavoidable confusion in terminology (in specifying "ground" and "excited" electronic configurations and ground and excited states) which must be kept in mind. Whenever one obtains in some calculation two one-electron

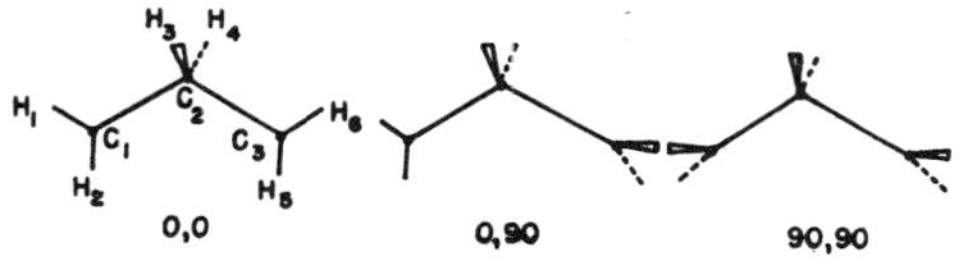

Figure 1. Definition of geometries for trimethylene.

energy levels close in energy (separated by say less than 2 eV), then there are four possible states arising from the placement of two electrons in these two levels. If A is lower in energy than B, then we call the configuration 1 the ground electronic configuration, the singlet and

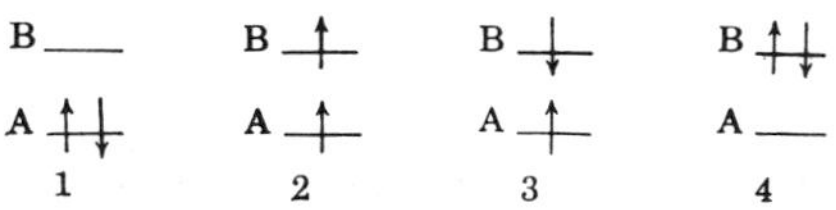

1 2 3 4

triplet states 3 and 2 first excited or simply excited configurations, and the state 4 a doubly excited configuration. Now in fact configuration interaction may strongly mix states 1 and 4 and electron interaction will greatly stabilize the triplet state 2 over the average one-electron energy of its configuration. If the A–B splitting is small, then very likely the triplet state 2 arising from an "excited" electronic configuration will be the true ground state of the molecule while the singlet "ground" electronic configuration 1 will be above it in energy. This is what occurs in methylene and CH_2 and most likely happens in trimethylene as well. The extended Hückel calculations, as other Hückel calculations, unfortunately do not take account of electron interaction, and so when a singlet and triplet arise from a given electronic configuration the calculations presumably yield some average energy of the configuration. Thus in this paper whenever the terms ground and excited appear without the explicit designation "state" they should be taken to mean the ground and excited electronic configurations actually computed, *i.e.*, one-electron energies for the occupation schemes 1 and 2, 3 (the latter two indistinguishable with our deficient method), respectively. The assumption is further made that the potential surfaces of a singlet and triplet of the same orbital symmetry (*e.g.*, 2 and 3) will not differ greatly. This seems to be true for diatomic molecules[7] and the few well-studied polyatomic cases such as formaldehyde.[8]

With the above caution kept in mind, Table I shows sections through the potential surface for ground and excited configurations.

One immediately notes that the most stable point on the ground configuration potential surface is for a 90,90 geometry at a small CCC angle. This, of course, will become a cyclopropane when the geometrical restraints are further relaxed. The most interesting aspect of the ground energy surface is, however, not the anticipated cyclopropane minimum, but the indication of another subsidiary minimum for a 0,0 structure (*i.e.*, terminal methylenes coplanar with the three-carbon chain) with a large CCC angle of approximately 125°. This minimum will be referred to in the subsequent discussion as the trimethylene intermediate. That it is indeed a potential minimum is not proved by the energy sections shown in the table, but is a consequence of a more extended examination of asymmetric distortions of the molecule from this geometry. The easiest passage from the valley of the trimethylene intermediate to the much deeper valley of the cyclopropane is *via* a conrotatory

(7) See the tables in G. Herzberg, "Molecular Spectra and Molecular Structure. I. Spectra of Diatomic Molecules," D. Van Nostrand Co., Inc., Princeton, N. J., 1950.

(8) G. W. Robinson and V. E. DiGiorgio, *Can. J. Phys.*, **36**, 31 (1958), and references therein.

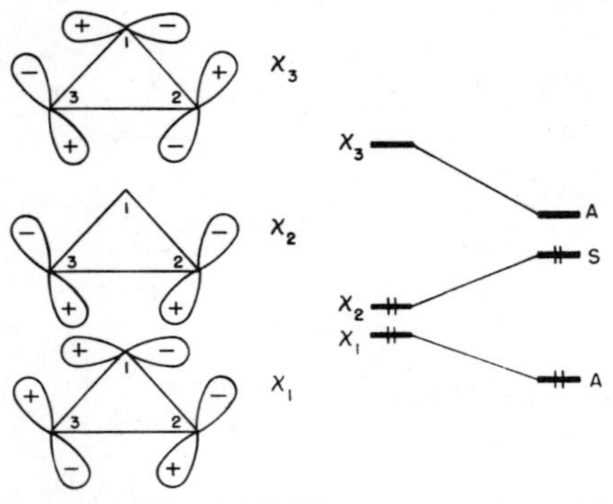

Figure 2. The Walsh orbitals for cyclopropane drawn schematically to indicate their nodal structure. χ_1 and χ_2 are degenerate. The motion of the levels as the 2–3 bond is stretched is shown at right.

motion[6b] of both methylene groups. The activation energy for this motion is calculated to be quite small, about 1 kcal. While the extended Hückel calculations are not to be trusted for such small energy magnitudes, the shape of the potential surface is believed to be qualitatively reliably reproduced. We will return shortly to a consideration of the trimethylene intermediate. Another matter of interest in the ground configuration potential surface is that for a simply opened cyclopropane, *i.e.*, a 90,90 geometry with a large CCC angle, there is retained a sizable barrier to rotation of a methylene group. This barrier is the difference in energy between a 90,90 and a 0,90 geometry and is 2.52 eV at CCC 80°, 1.20 eV at 100°, 0.25 eV at 120°, and increases again at larger CCC angles. The shape of the potential surface is such that a trimethylene which should by chance find itself in a 90,90 geometry at any CCC angle would prefer to collapse directly to a cyclopropane without rotating a terminal methylene group.

The excited configuration potential surface is quite different. Instead of two fairly sharp minima there now appears a very broad valley ranging over all possible orientations of the terminal CH_2 groups and over a CCC angle range from 100 to 130°. This excited trimethylene is thus a molecule "floppy" in most of its degrees of freedom, with no barriers to internal rotation at all. It should also be noted that these calculations imply a severe geometry change upon a $\sigma \rightarrow \sigma^*$ excitation of a cyclopropane, opening up a CC bond to give this excited trimethylene.

The general shape of the ground- and excited-state potential surfaces is not difficult to understand. As mentioned previously[9] the extended Hückel calculations give as the highest occupied and lowest unoccupied orbitals in cyclopropane precisely the linear combinations suggested by Walsh[10] some time ago. In Figure 2 these are illustrated; one specific form of the degenerate pair χ_1, χ_2 is chosen, namely that adapted to the ensuing stretching of the C_2–C_3 bond. The motion of the energy levels as one bond is slowly broken is apparent—those levels which are bonding in the region of bond cleavage are destabilized, while those antibonding in the same region are stabilized. Since some 2–3 bonding is

(9) R. Hoffmann, *J. Chem. Phys.*, **40**, 2480 (1964); *Tetrahedron Letters*, 3819 (1965).
(10) A. D. Walsh, *Trans. Faraday Soc.*, **45**, 179 (1949).

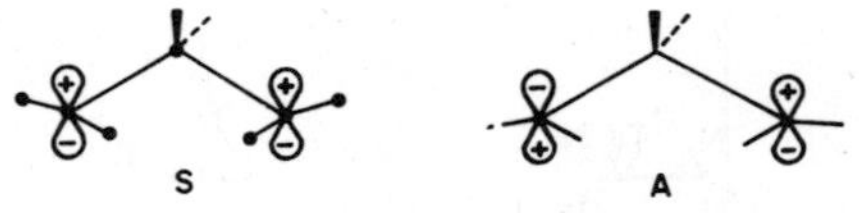

Figure 3. The symmetric (S) and antisymmetric (A) π-type molecular orbitals in trimethylene.

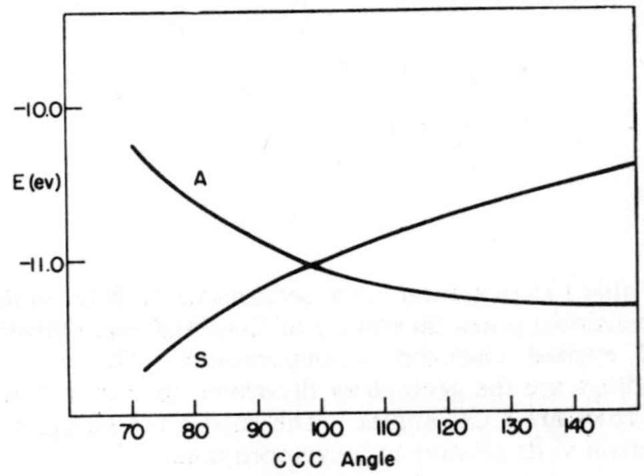

Figure 4. Energies of nonbonding orbitals of trimethylene as a function of CCC angle.

retained even when the CCC angle is quite large there remains a ground configuration barrier to rotating a methylene group. But the excited configuration has an electron promoted from χ_2 to χ_3. This not only weakens C_2–C_3 bonding but also C_1–C_2 and C_1–C_3 as well, leading to a very flexible excited molecule.

Returning to the ground configuration, the electronic structure of the trimethylene intermediate is extremely interesting. *A priori* one would have expected that if any 1,3 interaction were present, the molecular orbital which is the symmetric combination of terminal $2p_z$ orbitals (Figure 3, S) would be stabilized over the antisymmetric combination (Figure 3, A). Figure 4 shows the energies of the S and A levels as a function of CCC angle. While the expected order is observed at small angles, at angles above ~100° the A level becomes increasingly stabilized over the S, and in particular for the equilibrium calculated geometry of the trimethylene intermediate the A level is below the S level by approximately 0.55 eV. This unexpected behavior can be understood as the result of two competing factors: (1) the direct 1,3 interaction which favors the positive 1,3 overlap S combination disfavors the A molecular orbital; (2) an indirect coupling in which the central methylene group far from being a mere insulator participates in molecular orbital formation (in a manner of mixing which is nothing else but hyperconjugation) and, while leaving the A level unaffected, destabilizes the S level. The clue to the second factor is provided by the form of the S molecular orbital at a CCC angle of 120°. It is

$$0.6860[C_1 2p_z + C_3 2p_z] - 0.1401 C_2 2p_z - 0.2629(H_3 - H_4)$$

Note that the hydrogens on the central carbon contribute significantly to this MO and that there are antibonding interactions as a result of this mixing (C_1 and C_3 with C_2 and $(H_3 - H_4)$). The mixing pattern can be understood by examining a little interaction diagram (Figure 5) in which are sketched the S and A π levels

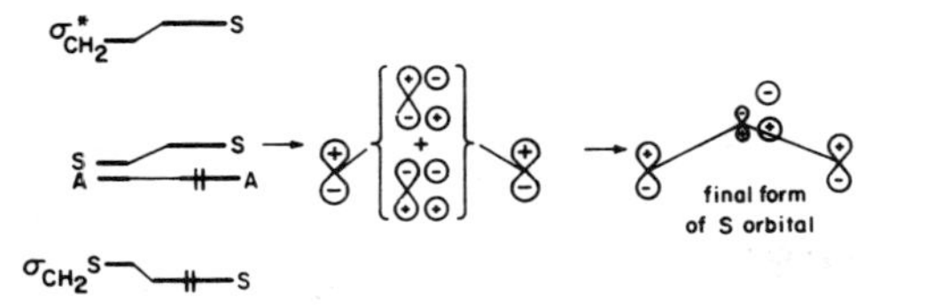

Figure 5. The interaction of the central CH_2 group with the nonbonding S orbital. The center of the figure illustrates the mixing which results in the final composition of the orbital shown at right.

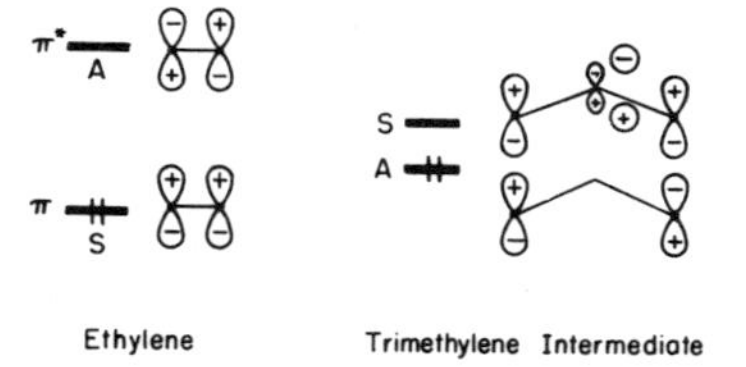

Figure 6. The electronic structure of trimethylene compared to that of ethylene.

and the σ and σ^* combinations of π symmetry arising from the C–H bonds of the central methylene group before and after interaction. The A level by symmetry cannot mix with σ and σ^*. The S level mixes and the decisive factor in its energy movement seems to be that the interaction with σ is stronger than that with σ^* as a result of (1) being closer in energy to σ than to σ^*, (2) more efficient overlap with σ than σ^*. The form of the interacted S level may be obtained from the general rule that if two orbitals of different energy mix, then the lower (energy) one of the two will mix into itself the higher one in a bonding way but the upper orbital will mix into itself the lower one in an antibonding way, so as to create a node between the two. This principle is a general consequence of the correlation of an increasing number of nodes in a wave function with higher energy. This kind of mixing is illustrated in the middle of Figure 5 with the net result, reinforcement at $H_{3,4}$, cancellation at C_2, shown in the figure at right and indicated by the form of the S orbital specified above.

Thus at large CCC angles the mixing of the central methylene group, destabilizing the S level, dominates, while at small angles the direct interaction stabilizing the S level wins out. The trimethylene intermediate lies in the former region and this has two immediate and important consequences for its reactions. First, using either simple orbital overlap considerations or constructing correlation diagrams it is clear that a species with two electrons in an A level (Figure 6) should close to a cyclopropane in a conrotatory manner. Second, such a species being essentially in electronic structure an "antiethylene," *i.e.*, having its order of levels precisely reversed from that in ethylene (Figure 6), should have selection rules for concerted cycloaddition precisely the opposite of those of an ethylene.[6c] This is illustrated by the correlation diagrams in Figure 7 for the cycloaddition of trimethylene to ethylene and butadiene. The trimethylene intermediate lowest singlet (two electrons in A) should add stereospecifically to an olefin and not a diene. The conrotatory closing is what is needed to rationalize the observed preferences for inversion of stereochemistry in the pyrazoline to cyclopropane pyrolysis studied by Crawford and by McGreer.[3] Because of the weak splitting between S and A trimethylene should in fact exhibit the least stereospecificity in this regard when compared to isoelectronic species. The direct cycloaddition of trimethylenes has not been observed despite attempts in that direction,[11] but that of some isoelectronic species is very well known and will be discussed below.

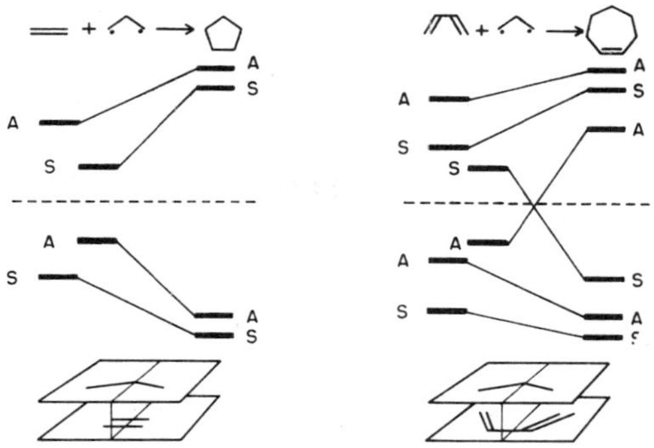

Figure 7. Level correlation diagrams for the addition of trimethylene to ethylene and butadiene.

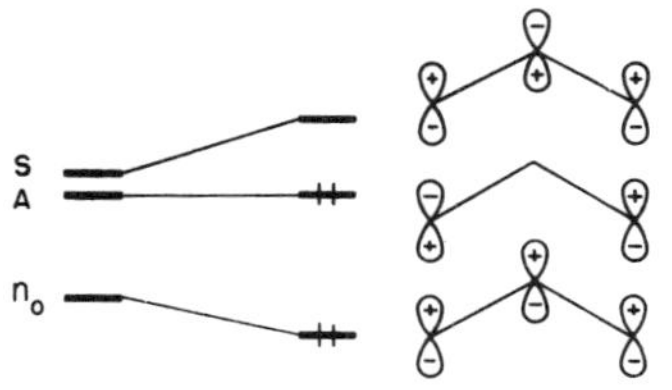

Figure 8. An interaction diagram showing the formation of the level structure of CH_2OCH_2 (ring-opened ethylene oxide) and the resemblance of the levels to those of allyl anion.

Since the one-electron energy difference between the "ground-state" $(A)^2$ configuration and the "excited-state" $(A)^1(S)^1$ configuration is only about 0.55 eV in the trimethylene intermediate geometry, it is anticipated that the triplet $(A)^1(S)^1$ state, a molecule as mentioned before quite free in geometry, would in fact be at lower energy than the singlet $(A)^2$ trimethylene intermediate (but of course at higher energy than the singlet cyclopropane). Some preliminary calculations by Simmons indicate that this is certainly so for the geometry of the trimethylene intermediate.[12]

We can now proceed to justify some of the theoretical license we took in simplifying the problem to trimethylene. Consider a carbonyl ylide—the ring-opened 1,3 dipole derived from ethylene oxide. Figure 8 shows an interaction diagram mixing the central oxygen lone pair with the S combination of terminal CH_2 orbitals. Clearly the resultant 4π-electron system with a highest occupied antisymmetric level resembles the allyl anion. The same would clearly be true of an azomethine ylide—the ring-opened 1,3-dipolar structure of an ethylenimine or aziridine. The resemblance to allyl anion is even more obvious in the parent compound to this isoelectronic series of 1,3 dipoles "without octet stabilization," *i.e.*, ozone. Figure 9 shows a simplified molecular orbital view of the electronic

(11) M. C. Flowers and H. M. Frey, *J. Chem. Soc.*, 2758 (1960).
(12) H. E. Simmons, private communication.

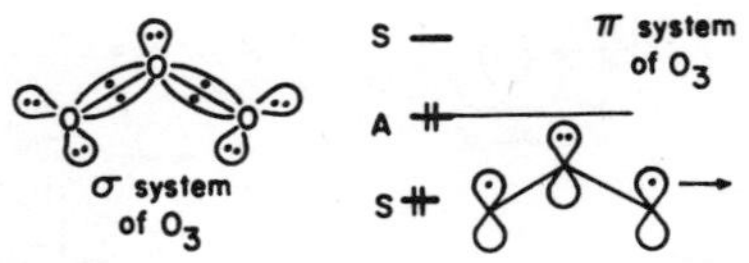

Figure 9. The electronic structure of ozone.

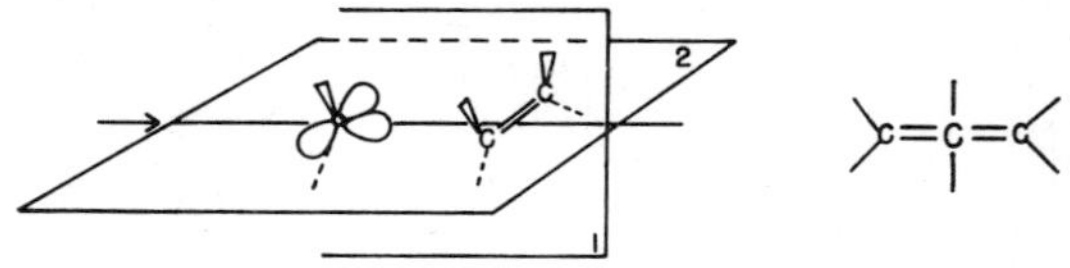

Figure 10. The most symmetrical approach of a methylene to an ethylene.

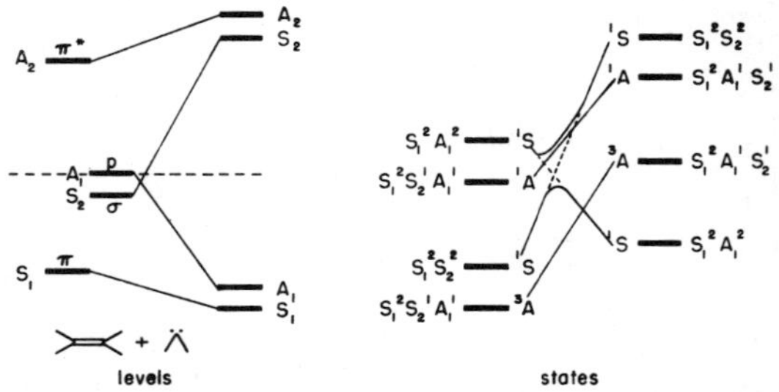

Figure 11. Level and state correlation diagrams for the approach of Figure 10. Since all the levels considered are symmetric with respect to plane 2, the symmetry designations are noted only for plane 1.

structure of ozone. Superimposed on a σ system of normal single bonds and lone pairs is an allylic π system into which must be placed four electrons.[13] For all of these molecules a consequence of the resemblance to allyl anion is that they should reclose to their three-membered minima in a conrotatory fashion and that they should add in a concerted manner to π-electron systems with $4q + 2$ electrons. A further discussion of the selection rules for 1,3-dipolar addition will be given elsewhere,[14] and an account of the electronic structure of cyclopropanone may be found at the end of this paper.

The existence of a trimethylene intermediate or an isoelectronic species has been seriously implicated in at least three areas. First this has come up in the broad area of 1,3-dipolar additions which formed the stimulus for this study. It seems to me that the conrotatory reclosure and the concerted addition to ethylenes make the identification of the experimentally inferred intermediate with the calculated geometry fairly clear. In the second instance, a trimethylene diradical has been the keystone of a consistent kinetic scheme for the structural and geometrical isomerization of cyclopropane.[15] The suggestion of a diradical intermediate goes back to the first studies of this reaction, but the primary experimental facts are due to Rabinovitch and coworkers[16] and the kinetic scheme was suggested by Benson.[15] Is the trimethylene intermediate computed here identical with Benson's trimethylene? The geometry of Benson's trimethylene is not specified, but it is implied that the molecule has small barriers to internal rotation and that an activation energy of approximately 10 kcal is needed to effect the reclosure to cyclopropane. Its estimated heat of formation puts it 54 kcal/mol above cyclopropane. The trimethylene intermediate calculated here is 44 kcal/mol less stable than cyclopropane and has a calculated activation energy to reclosure of 1 kcal and a barrier to internal rotation of a single terminal methylene group of about 10 kcal. In view of the poor quality of the calculations it seems to me that the differences are not drastic, and I would like to identify the potential minimum calculated here with Benson's suggested trimethylene intermediate with the reservation that some confidence is felt in the calculated existence of sizable rotation barriers in the molecule.

In the third instance it has been suggested that a trimethylene intervenes in the stereospecific addition of singlet methylene to ethylene.[17] The next section describes our results on this reaction.

(13) See A. D. Walsh, *J. Chem. Soc.*, 2266 (1953), for a discussion of the electronic structure of ozone.
(14) R. Hoffmann and R. B. Woodward, to be published.
(15) S. W. Benson, *J. Chem. Phys.*, **34**, 521 (1961).
(16) B. S. Rabinovitch, E. W. Schlag, and K. B. Wiberg, *ibid.*, **28**, 504 (1958).
(17) W. B. De More and S. W. Benson, *Advan. Photochem.*, **2**, 219 (1964).

The Addition of Methylene to Ethylene

Figure 10 shows the most symmetrical transition state in the approach of a bent methylene to an ethylene. This transition state possesses C_{2v} symmetry and is closest in geometry to the addition product, cyclopropane. A correlation diagram for levels and states in the C_{2v} approach is drawn in Figure 11. Since all the orbitals involved in the reaction are symmetric with respect to reflection in plane 2, only the symmetry with respect to reflection in plane 1 is used in classifying states. The methylene p orbital has been placed at the nonbonding level and the methylene σ orbital below it, but above the ethylene π level, since we estimate that in ethylene the π level lies perhaps 2.5 eV below nonbonding, while the methylene σ is no more than 1.5 eV in one electron energy below the p. Here and in what follows below I am assuming the reader is familiar with the problems associated with the electronic structure of methylene[18,19] —a molecule whose ground state is a linear or nearly linear triplet, $^3B_1$ in C_{2v} or $^3\Sigma_g^-$ in $D_{\infty h}$, which arises from what we would call an excited configuration σ^1p^1 (see Figure 12). The lowest singlet, thought to lie considerably less than 1 eV above the ground-state triplet, is that of the ground electronic configuration σ^2, $^1A_1$. The singlet component of the excited configuration, $^1B_1$, σ^1p^1, lies about 0.88 eV above the lowest singlet.

Following our previous interpretations of such correlation diagrams as that of Figure 11 we would conclude that the most symmetrical addition is a forbidden process for the σ^2 configuration but an allowed one for σ^1p^1. In other words if the singlet addition is an observed process, as it certainly is, it should not occur in this highly symmetrical manner but in some unsymmetrical manner. This prompted a detailed exploration of various geometry approaches of methylene to ethylene.

(18) G. Herzberg, *Proc. Roy. Soc., Ser. A*, **262**, 291 (1961).
(19) P. P. Gaspar and G. S. Hammond in "Carbene Chemistry," W. Kirmse, Ed., Academic Press Inc., New York, N. Y., 1964, p 235.

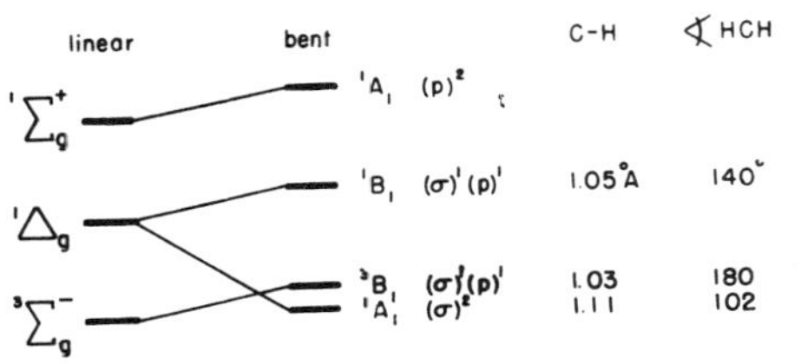

Figure 12. Some experimental molecular parameters for the lower methylene states. The parent configurations are also indicated. For uncertainties in experimental parameters see the original references.

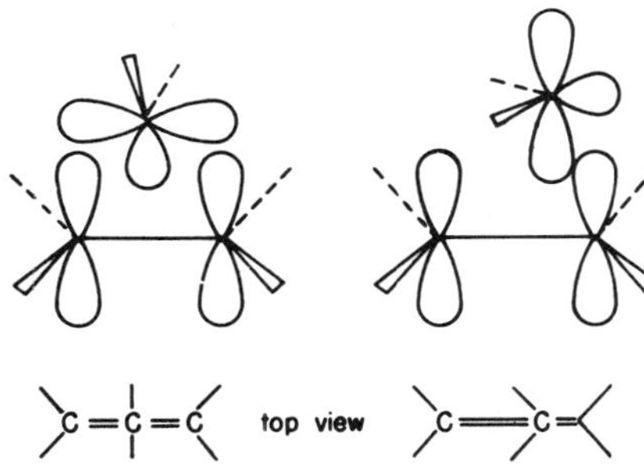

Figure 13. Two geometries of approach of a methylene to an ethylene. At left σ; at right π.

Two extreme geometries of approach are illustrated in Figure 13. In the σ approach the σ orbital of the methylene impinges on the ethylene π system and the methylene hydrogens are in a position already close to that of the hydrogens in the product cyclopropane. In the π approach the approaching methylene lies in a plane parallel to the ethylene plane, *i.e.*, the p orbital of the methylene impinges on the π system of the ethylene. The correlation diagram of Figure 11 implies that the most symmetrical σ approach is not likely, and that an approach with less symmetry than C_{2v} should be favored. But it says nothing about the relative merits of various less symmetrical reaction paths, *e.g.*, an unsymmetrical σ approach with the axis of the CH_2 initially over a carbon of the ethylene *vs.* a π approach, both of C_s symmetry. The detailed exploration of the potential surface assumed a fixed ethylene and methylene geometry. Some preliminary calculations showed that the methylene carbon moved in the plane perpendicular to the ethylene and containing the ethylene carbons. This left five degrees of freedom in specifying the approach: a distance R and angle θ to locate the methylene carbon with respect to some reference point in the ethylene, and three Euler angles ϕ', θ', ψ' to specify the orientation in space of the methylene. These are shown in Figure 14. What we have called a σ approach would be one with $\theta' = 0$, a π approach one with $\theta' = 90°$. We do not have the resources to study the complete potential surface so we will present only the reaction path—the line of minimum energy. Even this is not uniquely defined, for the reaction path varies according to the choice made for the origin from which R is measured. If the location of the ethylene is given by Cartesian coordinates, $(0, \pm 0.67, 0)$ for the carbons, $(\pm 0.952627, \pm 1.22, 0)$ for the hydrogens, then two logical origins for R are (a) the center of the ethylene (0,0,0) and (b) a point near where the methylene group has to end up, *e.g.*, (0,0,1.33368). The calculations show that along each of these reaction paths $\phi' = \psi' = 0$. Thus the path is completely specified by R, θ, θ'. Figure 15 shows the interesting segment of the reaction path, specifically the projection of all atoms on the *yz* plane for reaction paths a and b. Note the difference between a and b.

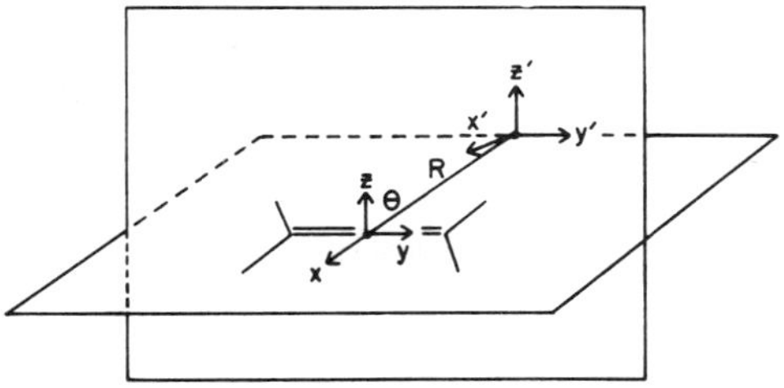

Figure 14. Definition of variables in search of methylene + ethylene potential surface. The origin is taken here at the midpoint of the ethylene (case a) and spherical coordinates R and θ are used to locate the CH_2 carbon (ϕ is not needed since the CH_2 carbon always remains in the perpendicular plane.) Euler angles ϕ', θ', ψ' are defined in the x', y', z' system using the convention of H. Goldstein, "Classical Mechanics," Addison-Wesley, Inc., Reading, Pa., 1950, p 106.

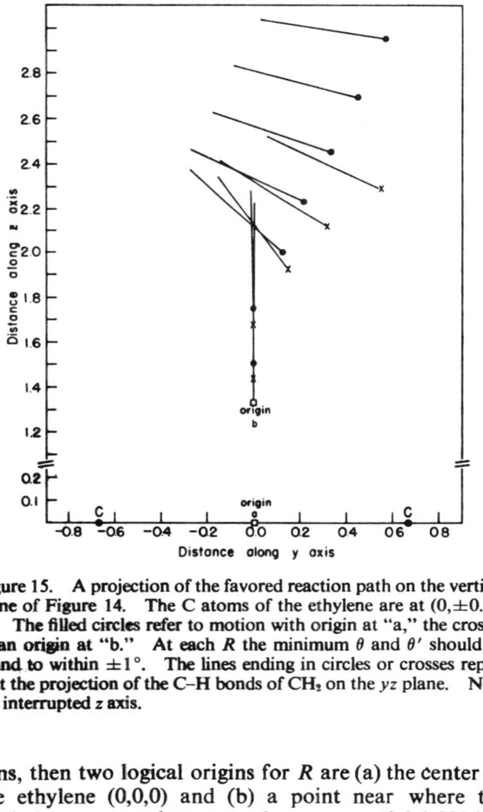

Figure 15. A projection of the favored reaction path on the vertical plane of Figure 14. The C atoms of the ethylene are at $(0, \pm 0.67, 0)$. The filled circles refer to motion with origin at "a," the crosses to an origin at "b." At each R the minimum θ and θ' should be found to within ±1°. The lines ending in circles or crosses represent the projection of the C–H bonds of CH_2 on the *yz* plane. Note the interrupted *z* axis.

In words the reaction pathway computed can be described as follows. At large separation a π approach is favored, slightly off-center presumably as a result of steric interactions of the hydrogens. At short separation the above calculations are worthless since they do not allow the geometry changes necessary in the ethylene component to convert it into a part of the cyclopropane. But close in the geometry *must* be that of a ground-state cyclopropane, *i.e.*, a σ approach with $\theta = 0$. At intermediate distances the two extremes are connected by a decrease in θ, and a transition from a π to a σ approach, *i.e.*, from $\theta' = 90$ to $0°$. What makes a π approach favorable at large separation can be attributed to the virtues of a three-orbital, two-electron transition state over a four-electron one—until the point is reached where there is so much to be gained from having the two new σ bonds formed at the same time that a transition to a σ approach is made. Or to put it in other words, there is a lot of energy to be gained by forming two bonds at the same time; eventually, therefore the optimum θ is never very great. But if θ is close to zero the σ approach is a forbidden one, meaning that there is a prohibitive electronic energy hill to traverse if one forces the reaction that way, and it is better to initiate a π approach. This is strikingly illustrated in Figure 16 where the σ approach is compared to the optimum calculated reaction path.

At this point it is worthwhile to remind the reader of the limitations of the numerical experiment that is being performed here. A potential surface for a chemical reaction of methylene and ethylene is being constructed using a semiempirical method which appears to be trustworthy for qualitative predictions but which is hardly infallible. Not only is there a question of the trustworthiness of the method, but even accepting its simulation of reality the exploration of all the degrees of freedom in this potential surface is out of the question. Each point on the surface consumed about 10 sec on a CDC 1604 computer, so since the focus of interest was the *approach* of the methylene to ethylene, the number of degrees of freedom was reduced to five or six. Even so the potential surface is not being explored, only the reaction path—the minimum of the energy valley for the reaction. Any pair of molecules may choose to effect the reaction in any peculiar way desired—what is being computed here is not a molecular collision but the potential surface on which such collisions may take their course. Perhaps the quantitative features of Figure 15 are not to be trusted, but experience with extended Hückel calculations leads one to accept the qualitative approach geometry obtained.

It should also be pointed out that the energy differences between the reaction path and deviations from it are sometimes small and naturally smaller at larger R. At $R = 3.00$ Å, path a, the minimum energy is -315.947 eV for a point with $\theta = 11.0°$, $\theta' = 79.6°$. For the same R some other computed points have the energies listed in Table II.

Table II

θ	ϕ'	θ'	ψ'	E, eV
0	0	0	0	−315.593
0	0	0	90	−315.593
0	0	90	0	−315.942
0	90	90	0	−315.929

For the excited configuration of methylene ($^{1,3}B_1$) approaching a ground-state ethylene the situation is somewhat more complex. At large separation a symmetrical σ approach is preferred ($\theta = \phi' = \theta' = \psi' = 0$),

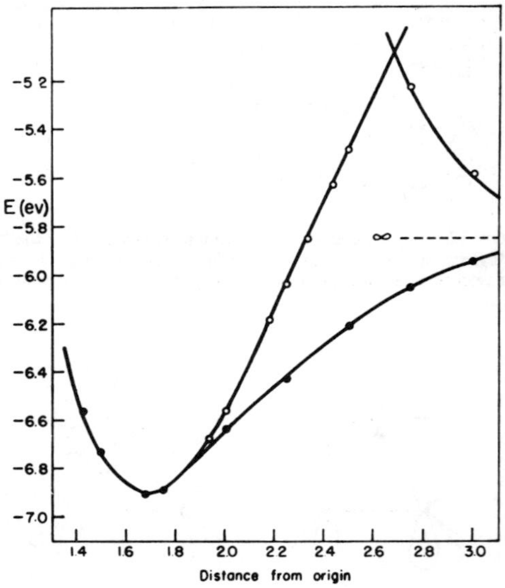

Figure 16. Energy along reaction coordinate measured from origin a (●) and along the symmetrical σ approach (O). The energy scale is with respect to an arbitrary zero.

as would have been anticipated from the correlation diagram. However this complex of an excited configuration of a methylene and a ground-state ethylene *must* correlate with an excited configuration of trimethylene or cyclopropane. As was shown in Table I our calculations give for that configuration a loose, floppy molecule with an open CCC angle, so that while the excited methylene begins its approach at $\theta = 0$ it proceeds by increasing θ until it is in the single loose minimum of the excited trimethylene. These calculations thus lead to the conclusion that insofar as the preservation of stereochemistry at the original ethylene fragment is concerned, the addition of the lowest $^1A_1$ of methylene should be stereospecific (but unsymmetrical) while the addition of triplet *or* singlet B_1 methylene should be nonstereospecific. Spin has not entered this argument and we will return to that momentarily.

As for the geometry of the $^1A_1$ methylene approach, the conclusion reached here is in essence that originally made by Skell and coworkers in their important and original series of papers on carbene reactions.[20] They drew an initial π approach guided by the similarity in the selectivity of carbenes and carbonium ions. The same transition state has been discussed by Moore and coworkers.[21] It remains difficult today to find unambiguous evidence arguing for or against these approach geometries.

As for the degree of specificity in methylene reactions, Skell argued that while it was possible for singlet ($^1A_1$) methylene to form two bonds in a single, concerted, stereospecific, addition process, a triplet ($^3B_1$) methylene attacking a singlet, ground-state ethylene could not produce in a concerted manner a ground-state cyclopropane but must give a triplet intermediate tri-

(20) P. S. Skell and A. Y. Garner, *J. Amer. Chem. Soc.*, **78**, 5430 (1956).

(21) W. R. Moore, W. R. Moser, and J. E. La Prade, *J. Org. Chem.*, **28**, 2200 (1963).

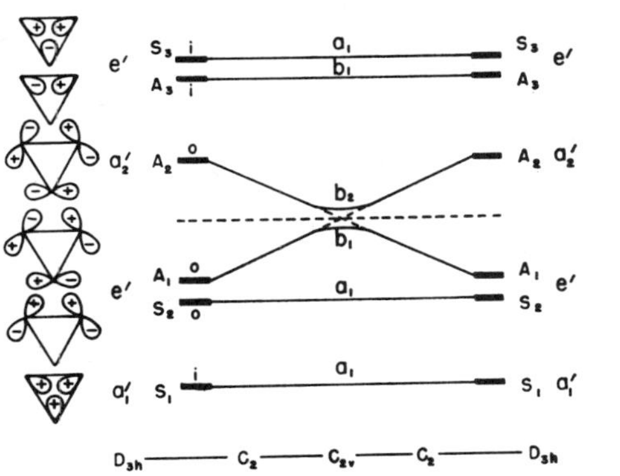

Figure 17. A correlation diagram for a twisting of a methylene group in cyclopropane. See text for notation.

Figure 18. The b_2(AA) and b_1(AS) levels of a C_{2v} twisted transition state.

methylene. In this diradical rotation about what would now be formal single bonds should be rapid and certainly competitive with the radiationless decay to a singlet. Thus one would account for the lack of stereospecificity in triplet addition.[20] This argument by Skell has proven most fruitful; it has also survived a good deal of criticism.[19,22,23]

Though the conclusions regarding the stereospecificity of the observed methylene additions are the same here and in Skell's argument, the reasons for these conclusions are quite distinct. In Skell's explanation one focuses on the spin state of the methylene, singlet or triplet. In the argument presented here one emphasizes the spatial part of the wave function. The $^1A_1$ adds stereospecifically not because it is a singlet but because it can correlate with the lowest singlet configuration of a trimethylene and thus with the ground state of a cyclopropane. The $^3B_1$ methylene adds nonstereospecifically not because it is a triplet but because the complex of it and a ground-state ethylene must correlate with a triplet state of an excited configuration of the trimethylene, one in which there are no barriers to rotation around terminal bonds. The $^1B_1$ methylene must also correlate with a singlet state of the same excited configuration, and should also add without retention of stereochemistry, even though it is a singlet. The prediction of nonstereospecific addition of $^1B_1$ methylene is perhaps the only different conclusion which one would draw from these calculations, though it is not clear that Skell's hypothesis would in fact necessitate stereospecific addition of an excited singlet. At any rate this prediction is hardly a readily verifiable one; even if we had a source of $^1B_1$ methylene separated from $^1A_1$ and $^3B_1$ the resultant trimethylene would be an extremely "hot" species and any nonstereospecificity could probably be equally well attributed to the high excitation.

Clearly these calculations agree with the basic conclusions of Skell that the addition of $^1A_1$ methylene while initially unsymmetrical is a concerted process which does not proceed through an intermediate, and that the addition of $^3B_1$ methylene proceeds through an intermediate. While there is an intermediate on the ground trimethylene potential surface, I do not think it is involved in the addition reaction because it is too far removed in structure from either reactants or products and because all stereochemical information would be lost if in fact the addition went through this intermediate. Moreover the reaction path traced in the calculated potential energy surface has no activation energy and yet easily bypasses the trimethylene intermediate potential valley. These conclusions are thus in contrast with the opinion of De More and Benson[17] who feel that in the addition of singlet methylene to ethylene there must be involved the same trimethylene intermediate that intervenes in the geometrical and structural isomerizations. The kinetic parameters they adduce explain the stereospecificity by having the rate of reclosure of the diradical to cyclopropane faster than the rate of rotation around the trimethylene C–C bonds.

A suggestion has also been made that in the additions of $^3B_1$ methylene to ethylene the lowest triplet of cyclopropane is initially formed.[24] This is in accord with our calculations which say that the lowest triplet state of cyclopropane does not retain the cyclopropane geometry but is in fact the lowest triplet (and probable ground state) of trimethylene.

One Mechanism for the Isomerization of Cyclopropanes

At one time a simple mechanism for *cis–trans* isomerization of cyclopropane had been suggested in the literature, in which a single methylene group rotated in place proceeding from D_{3h} through C_2 to a C_{2v} transition state in which the methylene group was symmetrically located in the plane of the three carbons of the ring.[25] It was also suggested that an unsymmetrical (C_s) motion from the C_{2v} geometry could lead to the propylene isomerization. This transition-state proposal was criticized[15] and did not gain support; it is nevertheless of interest to reexamine part of it with the methods at disposal here. Figure 17 is a correlation diagram which traces what happens to the six Walsh orbitals in cyclopropane under such a motion. The proper linear combinations for the reduction of symmetry to C_2 are shown; their initial ordering and their disposition in the transition state are taken from an extended Hückel calculation. The sp^2 hybrids pointing into the center of the cyclopropane ring are labeled "i"; the peripheral p-orbital combinations "o." The following considerations make the construction of the correlation diagram understandable: (1) all the i orbitals (S_1, A_3, S_3) are not affected by the rotation in the first approximation; (2) S_2 is also not much affected since it initially has no contribution at C_1; (3) A_1 is 1–2, 1–3 bonding, so it is destabilized by the rotation; A_2 is 1–2, 1–3 antibonding

(22) H. M. Frey, "Progress in Reaction Kinetics," Vol. 2, G. Porter, Ed., Pergamon Press Inc., New York, N. Y., 1964, p 131.
(23) B. J. Herold and P. P. Gaspar, *Fortschr. Chem. Forsch.*, **5**, 89 (1965).
(24) J. A. Bell, *J. Amer. Chem. Soc.*, **87**, 4996 (1965); R. F. W. Bader and J. I. Generosa, *Can. J. Chem.*, **43**, 1631 (1965); see also R. J. Cvetanovic, H. E. Avery, and R. S. Irwin, *J. Chem. Phys.*, **46**, 1993 (1967).
(25) F. T. Smith, *ibid.*, **29**, 235 (1958).

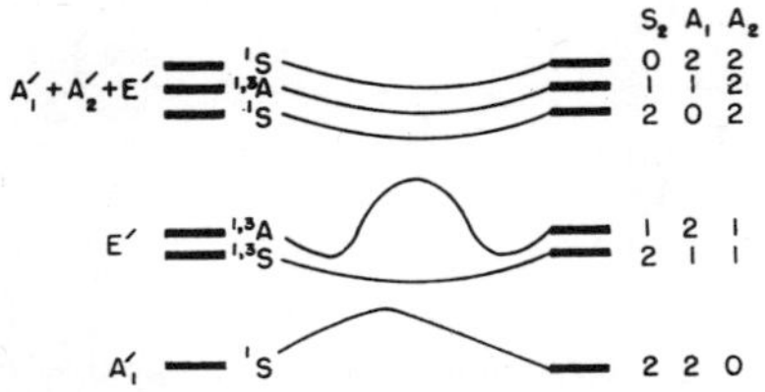

Figure 19. The state correlation diagram for the twisting of a methylene group.

and is stabilized by the motion; (4) formally, a rotation takes A_1 into A_2 and *vice versa*, and the C_{2v} transition-state symmetry allows a crossing there. In fact in the C_{2v} transition state the AS and AA levels have widely different compositions and energies. They are shown in Figure 18. AA should be at higher energy because it is 2–3 antibonding. The actual energies for C–C distances kept as in cyclopropane are AS at −10.906 eV, AA at −8.154 eV. The calculated energy of this transition state is 4.19 eV above cyclopropane. Energy may be gained by pulling the rotating CH_2 group away from the rest of the molecule but it is difficult to see how this would be consistent with a possible transition state leading to propylene isomerization where in fact either the C_1–C_2 or C_1–C_3 bond must become a double bond. Also pulling a CH_2 group away would seem to be heading down the road to a methylene plus ethylene rather than back to cyclopropane. Figure 19 shows the state correlation diagram for this motion. The excited states of cyclopropane nearly all prefer the twisting motion but in fact still better turn out to be a simple bond stretching in one bond. One amusing feature is that one level, $^{1,3}A$ in C_2, has a double minimum potential curve with a preferred equilibrium geometry of C_2 rather than D_{3h} or C_{2v}.

According to these calculations this would also then appear to be an energetically unfavorable thermal isomerization mechanism. It is however worthwhile to point out that such a rotation does not do such violence to the electronic structure of the molecule as might have been naively expected. The peculiarity of the electronic structure of cyclopropane, so nicely described by the Walsh model, leads to the equivalent of only one bond being broken in the rotation, not two; notice how only A_1 and not S_1 or S_2 are affected by the motion. We hope to return to a more detailed examination of the isomerization to propylene soon.

Cyclopropanone

The preceding discussion noted that whereas 1,3 dipoles isoelectronic with trimethylene underwent 1,2 cycloadditions, cyclopropanone, in the few cases studied, preferred a 1,4 cycloaddition.[4,26–28] This implied a fundamental difference between trimethylene and the ring-opened cyclopropanone. Since recent work has lead to clearer isolation of this molecule[28] it became of interest to examine its electronic structure.

Assuming a C=O distance of 1.20 Å and a C—C distance of 1.50 Å some sections through a potential surface of cyclopropanone very similar to those studied for trimethylene were taken. The results are shown in Table III; the column headings are references to geometrical configurations of the terminal methylenes that were defined in Figure 1.

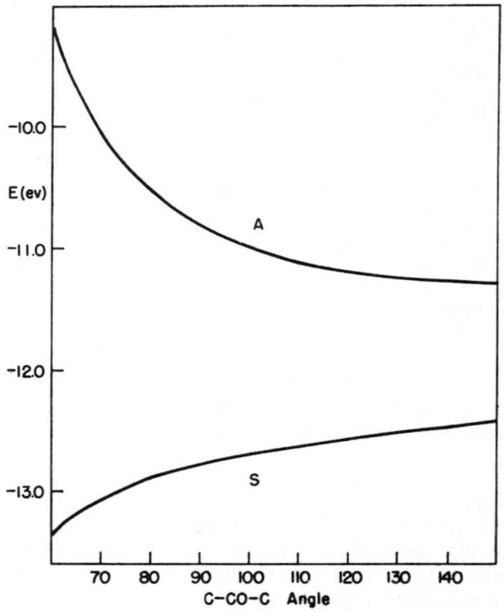

Figure 20. Energy of S and A levels of a 0,0 trigonal cyclopropanone as a function of bending.

Table III. Sections through a CH_2COCH_2 Potential Surface[a]

Angle CCC, deg	Ground configuration 0,0	0,90	90,90	Excited configuration 0,0	0,90	90,90
60	11.840	4.801	1.827	13.193	8.195	4.979
70	9.449	2.918	0.811	9.681	6.625	4.269
80	4.300	1.891	0.586	5.643	4.676	3.777
90	1.763	1.336	0.723	3.461	3.179	3.606
100	0.633	1.005	0.998	2.306	2.330	3.079
110	0.145	0.809	1.317	1.636	1.873	2.560
120	0.001	0.737	1.663	1.365	1.684	2.349
130	0.092	0.796	2.050	1.360	1.710	2.368
140	0.383	0.989	2.498	1.571	1.928	2.575
150	0.872	1.317	2.884	1.987	2.333	2.956

[a] The geometries are defined in Figure 1 and the entries are energies in electron volts measured from an arbitrary zero of energy at the most stable point in the system, (0,0) at a CCC angle of 121.1°.

The general features of the surface are similar to those of trimethylene. There is now a much deeper minimum in the ground configuration, for a 0,0 geometry and a CCC angle of around 121°. The excited configuration is conformationally fairly stable in contrast to the floppy trimethylene.

The interesting and anticipated difference here is that in the ring-opened intermediate (which we will call *oxyallyl*) the symmetric S level is lower than A. The splitting is quite pronounced, and Figure 20 shows that the S level is lower at all CCC angles. The immediate consequence of this electronic structure is that oxyallyl

(26) R. C. Cookson, M. J. Nye, and G. Subrahmanyam, *J. Chem. Soc., Sect. C*, 473 (1967).
(27) A. W. Fort, *J. Amer. Chem. Soc.*, **84**, 2620, 2625, 4979 (1962).
(28) N. J. Turro and W. B. Hammond, *ibid.*, **87**, 3258 (1965); **88**, 3672 (1966); N. B. Hammond and N. J. Turro, *ibid.*, **88**, 2880 (1966).

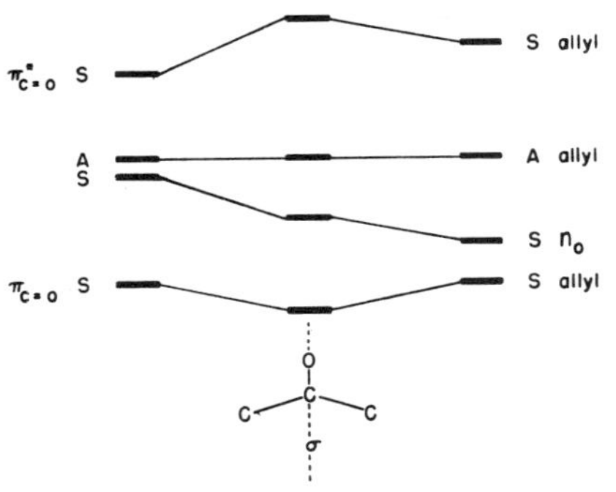

Figure 21. Two ways of rationalizing the ordering of levels in oxyallyl. On the left is the interaction of a carbonyl π and π^* with two nonbonding orbitals. On the right is the mixing of an allyl system with an oxygen p orbital.

should collapse to cyclopropanone in a disrotatory manner and add in a concerted fashion 1,4 (or 1,8, etc.).[29] The preference for the disrotatory closing has been checked directly by an extended Hückel calculation.

The simplest way to explain this behavior contrasted to that of trimethylene is just to say that a good valence bond representation of this species is given by the resonance structure V below, and that the characteristic modes of reaction that ensue are just those of an allyl

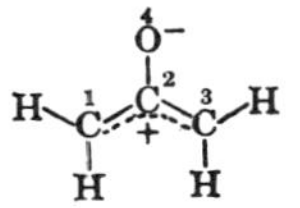

V

cation. A more direct way to see this is to draw an interaction diagram (Figure 21) for the mixing of carbonyl π and π^* levels with our S and A combinations. The carbonyl levels are both of S symmetry, but it is important to note that the energy center of gravity of the C=O double bond is below the energy of a 2p electron on carbon, *i.e.*, the C=O π^* is relatively low lying and therefore interacts most strongly with the S combination, pushing it down in energy. Another way to obtain the same level order is to allow the interaction of an allyl system with an oxygen 2p orbital, also shown in Figure 21.

The Pariser–Parr–Pople SCF procedure to be described later on in this paper gives the molecular orbitals of Table IV. The correlation diagrams for the closures

Table IV. SCF Orbitals and One-Electron Energies for Oxyallyl[a]

	1	2	3	4
E_i (eV)	−14.913	−9.540	−3.379	+2.604
C_1	0.3194	0.5387	0.7071	0.3283
C_2	0.6185	0.1078	0	0.7784
C_3	0.3194	0.5387	−0.7071	−0.3283
C_4	0.6431	−0.6387	0	−0.4225

[a] The AO ordering is given in structure V.

and cycloadditions which lead to the above predictions are then simple to construct. It should of course be mentioned that there is nothing original about oxyallyl, the "dipolar" or "zwitterion" form of cyclopropanone. It has a long history as a possible intermediate in the Favorskii rearrangement[30,31] and has been discussed theoretically before.[32] It would appear in general that for the molecule VI S would come below A if X=Y had a low-lying π^* level, A below S if X=Y had a high lying π level. The case when X=Y becomes $C{=}CH_2$ is the trimethylenemethyl radical VII which for a perfectly symmetrical structure has a degenerate central level pair. A two-electron system like that of allyl cation can also result if the central atom contributes no electrons to the π system; *e.g.*, the hypothetical borane VIII.

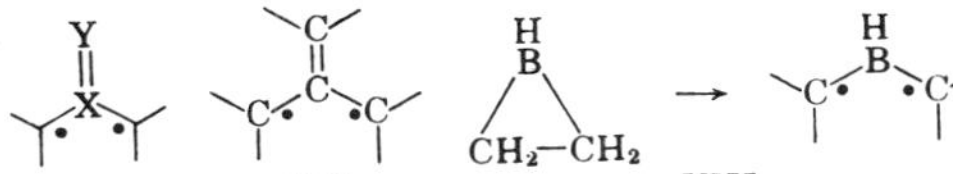

VI VII VIII

Whereas trimethylene is a small valley on the hillside leading to the much deeper cyclopropane minimum, Table III shows that oxyallyl lies in a most deep potential well. Assuming model geometries for classical cyclopropanone and allene epoxide, the extended Hückel calculations result in a cyclopropanone 1.0 eV less stable than oxyallyl, and an allene epoxide structure 0.9 eV less stable than oxyallyl. A careful examination of the potential surface connecting these molecules shows no activation energy for the conversion of cyclopropanone to oxyallyl.

These are of course approximate calculations, sometimes unreliable. Can they be trusted in this instance? The physical evidence for cyclopropanones appears unambiguous.[28,33] It may be that the result that oxyallyl is more stable than cyclopropanone is reliable, but that the calculations are incorrect in exhibiting no activation energy for the interconversion. We are engaged in some further, more rigorous, calculations on this problem. We also intend to examine the possible stability of ring-opened forms IX and X.

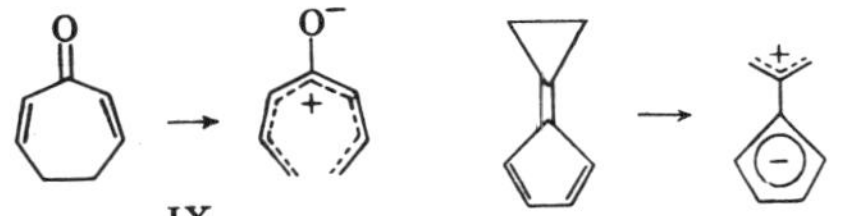

IX X

Since the splitting between S and A levels in the optimum geometry of oxyallyl is quite large, it was considered important to inquire if the ground state of oxyallyl could possibly be a singlet and not the anticipated triplet.

To do this a Pariser–Parr–Pople SCF–CI calculation was carried out on the π electrons of this molecule, at a CCC angle of 120°, C–O 1.20, C–C 1.50 Å. The details of the method are described elsewhere.[34] We in-

(29) R. B. Woodward and R. Hoffmann, *J. Amer. Chem. Soc.*, **87**, 395 (1965); R. Hoffmann and R. B. Woodward, *ibid.*, **87**, 2046 (1965).

(30) References to the earlier literature may be found in a review by A. S. Kende, *Org. Reactions*, **11**, 261 (1960).

(31) (a) G. Stork and I. Borowitz, *J. Amer. Chem. Soc.*, **82**, 4307 (1960); (b) H. O. House and W. Gilmore, *ibid.*, **83**, 3972, 3980 (1961); H. O. House and W. Thompson, *J. Org. Chem.*, **28**, 164 (1963); (c) E. Smissman, E. Lemke, and O. Kristiansen, *J. Amer. Chem. Soc.*, **88**, 334 (1966).

(32) J. G. Burr, Jr., and M. J. S. Dewar, *J. Chem. Soc.*, 1201 (1954).

(33) J. F. Pazos and F. D. Greene, *J. Amer. Chem. Soc.*, **89**, 1030 (1967).

(34) R. Hoffmann, A. Imamura, and G. D. Zeiss, *ibid.*, **89**, 5215 (1967).

cluded all configurations in the configuration interaction treatment and obtained the ordering of levels in Table V. It should of course be kept in mind that these are

Table V. SCF–CI Results on Oxyallyl[a]

$^1A_1$	4.136[b]	$^1A_1$	0.112[b]
$^1B_1$	1.967	$^3B_1$	[0.0]

[a] The states are labeled in C_{2v} symmetry, with the *xz* plane being that of the molecule. [b] In electron volts.

π-electron states only. In particular (n,π^*) excited states may arise at low energy in this molecule, but we have no reliable way to estimate their position. The $^1A_1$ states are strong mixtures of the $(S_1)^2(S_2)^2$ and $(S_1)^2$-$(A_1)^2$ configurations of Figure 21 while $^1B_1$ and $^3B_1$ represent predominantly the $(S_1)^2(S_2)^1(A_1)^1$ configuration. $^3B_1$ is the molecular ground state, but the singlet is not far above it in energy.

Acknowledgment. I am grateful to H. E. Simmons, R. J. Crawford, and R. B. Woodward for discussions concerning this work, and to A. Imamura, P. Clark, W. J. Hehre, and G. D. Zeiss for assistance with the calculations. The minimum searching methods used in exploring the approach of methylene to ethylene are due to K. D. Gibson. Motivation for a good part of this work was found in the interesting studies of R. C. Cookson. Generous support was provided by the Chevron Research Company, the Sloan Foundation, the National Institutes of Health, and the National Science Foundation.

IV-5a

Some Organic Chemical Applications of Sigma Molecular Orbital Theory

KENNETH B. WIBERG

The π-electron MO methods have been of considerable interest to many organic chemists and have led to a large number of experimental studies.[1] Although the π-electron methods continue to be useful, they are severely limited in that they can be used with conjugated unsaturated systems. Attempts have been made to apply the methods to homoconjugated systems (those in which a saturated unit separates two unsaturated centers), but these attempts generally violate the basic premise of the π-electron methods, the separability of π- and σ-electron systems. Such separation is appropriate only with planar conjugated systems in which the σ bonds lie in the nodal plane of the π-electron system.

The recent development of several σ molecular orbital methods promises to provide the organic chemist with a tool which may be applied to a wide variety of chemical problems. Here one hopes to be able to calculate the heats of formation (or of atomization) and the equilibrium geometry of molecules, transient species, and activated complexes.

The Pople-Santry-Segal SCF procedure (CNDO) was chosen as a starting point for an investigation of the possibility of obtaining the above quantities from a σ-MO calculation. The CNDO method seemed preferable to the extended Hückel method because electron repulsion will certainly be of importance in determining the heat of formation. In an article which has been included in Chapter III, an attempt was made to find a set of semiempirical parameters which would give correct equilibrium geometries, approximately correct C–H dipole moments (i.e., ~0), and calculated heats of atomization which were proportional to the observed values. For a series of hydrocarbons, a set of parameters could be derived which were satisfactory for all three criteria stated above. The equilibrium bond lengths agreed with the experimental values to ±0.03 Å, and the bond angles agreed to ±1°. The calculated heats of atomization were found to correlate with the experimental values provided the calculated values were considered in arbitrary units (rather than atomic units) with a conversion factor of 218 kcal/mole/unit.

It is not surprising that the slope was less than the expected value (628 kcal/mole/a.u.), since the CNDO approach involves major approximations and neglects electron correlation. The fact that a linear relationship was found would appear significant and suggests that further improvements could be made. It should be noted that the experimental values of the heats of atomization are corrected to 0°K and for zero-point energies. This has not always been done and makes other correlations less convincing.

The failings of an approximate theory are as important as its successes. First, a calculation of the torsional barrier in ethane using the parameters and scaling factor given above leads to a barrier of 0.7 kcal/mole, and about one fourth that which is observed. Second, a calculation of the geometry of cyclobutane leads to a planar structure as that of lowest energy. This appears to be a reflection of the low calculated torsional barrier. The calculated energy of cyclobutane increases only very slowly as the molecule is puckered. Thus, if the torsional barrier were estimated correctly, the lowest energy structure would be found to be puckered. Third, strain energy appears to be underestimated both for cyclopropane and cyclobutane.

Attempts were made to extend the reparametrization to oxygen- and nitrogen-containing compounds. The experimental data for nitrogen-containing compounds were not sufficient to permit a meaningful test to be made. With oxygen-containing compounds, the difference in bond length between single and double bonds could not be reproduced. The change in bond length with bond character is underestimated with carbon compounds, and the deviation becomes even larger with the oxygen-containing compounds. It appears not unlikely that the situation could be improved if the repulsion integrals between different types of orbitals were calculated and used directly rather than using only an average repulsion integral which does not distinguish between s and p orbitals.

Having found that it is possible to obtain meaningful results concerning the geometry and energy for hydrocarbons, it was of interest to extend the calculations to intermediates and activated complexes. It was found that reasonable minimum-energy geometries could be calculated for cations and radicals and that the calculated energies were related to the observed values using the same scaling factor as found above. The only change was in the intercept, which appeared to be related to the charge type. The results were sufficiently encouraging to suggest further extensions.

One of the problems of recent interest to organic chemists is concerned with the nature of the species which are formed when cyclopropylcarbinyl and cyclobutyl compounds ionize. With the cyclopropylcarbinyl compounds, considerable experimental information is available concerning the structure, and the evidence favors a cation in which the carbinyl p-orbital achieves maximum overlap with the cyclopropyl ring orbitals.

In the paper which follows it was found that the calculations agreed well with the experimental observations. Further, the different effects of fusing a second ring onto the cyclopropane ring fit in well with the changes in bond character for the 1,2 and 2,3 ring bonds. Possibly of more importance, the calculations suggest that the effect should be much smaller with the cyclobutylcarbinyl cation and should be negligible with the isobutyl cation. This indicates one of the more significant uses of these calculations—as a guide to experiment. If a significant barrier to rotation were calculated for the isobutyl cation, it would be interesting to attempt to create molecules in which this effect could be observed. In this case, the calculations suggest that such experimental work would be fruitless.

The cyclobutyl cation is a particularly interesting species. Both extended Hückel calculations using the parameters given by Hoffmann and CNDO calculations using the parameters given by Pople suggest that the lowest energy form for the cyclobutyl cation is the planar form. However, later calculations using the CNDO parameters which were optimized for hydrocarbons have led to a puckered ion as the lowest energy form. Experimental data agree with the latter result. The calculations suggest that the important factor in stabilizing the cyclobutyl cation is a 1,3-cross ring interaction.

In considering bonding in compounds such as the above, as well as considering hybridization, we have found it useful to define a new quantity, the bond index. The bond index is the sum of the squares of the bond orders between the atoms in question. It can be shown to be related to the covalent bond character of a given bond, and it agrees very well with the bond orders which are derived from valence-bond treatments. The bond index between a pair of orbitals is a measure of hybridization, and, in the particular case of 1s-2s, it is a direct measure of the per cent s character. These values correlate well with the C^{13}—H coupling constants.

The general subject of electron distribution in cations is of interest. For example, attempts have been made to determine the electron distribution in the conjugate acid of benzene via proton NMR spectroscopy. The results were[2]

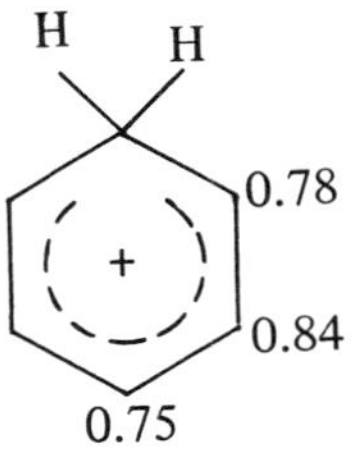

Observed

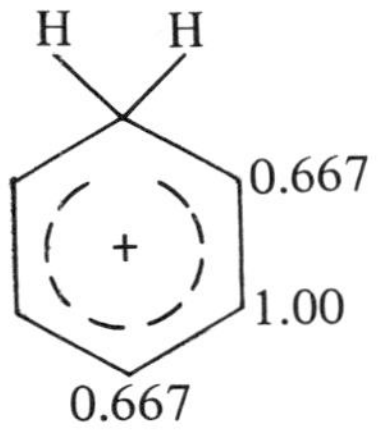

Calculated
(π-electron SCF)

It was argued that inductive effects would lead to a leveling of the electron distribution, and that the π-electron SCF results were incorrect because they did not include polarization of the σ core.

We have calculated the electron distribution using the CNDO method which does include the σ core and find

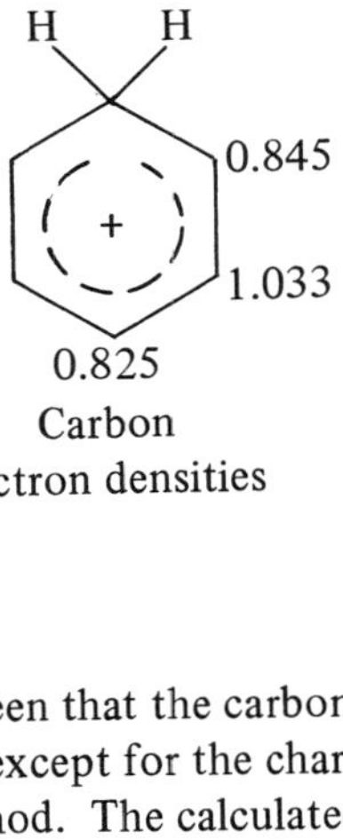

Carbon
electron densities

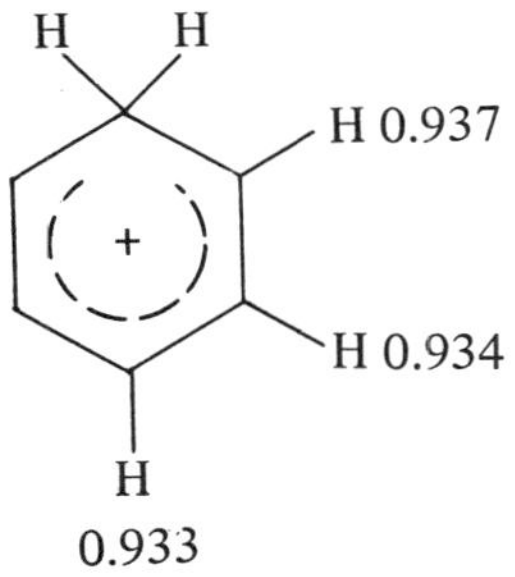

Hydrogen
electron densities

It can be seen that the carbon electron densities agree well with the results of the π-electron calculations, except for the charge transfer from hydrogen to carbon, which is included only in the σ-MO method. The calculated hydrogen electron densities are essentially the same, which agrees reasonably well with the *proton* chemical shifts. It seems not unlikely that the proton NMR chemical shifts may be a poor indication of the electron density at the carbon to which they are attached, except in the case of species with high symmetry (cyclopentadienyl anion, cycloheptatrienyl cation, etc.)

The difference between the carbon and hydrogen electron density distributions results from the alternation of charge normally predicted by MO methods. Although this alternation is well accepted for the π electrons, it has only recently been considered for the σ system. Yet, there is no reason for one group of electrons to behave differently from another. It would appear important to obtain further experimental evidence concerning the extent of charge alternation in cations.

Finally, the paper considers the thermal transformation of bicyclobutane into butadiene.

Here an attempt was made to follow the bonding changes during the course of the reaction. Using the bond index as a criterion for bonding, several of the possible processes could be shown to be electronically forbidden. Only one of the possible stereochemical results was found to be allowed, and subsequently, this has been shown to be the correct result.[3]

References

1. See A. Streitwieser, Jr., *Molecular Orbital Theory for Organic Chemists*, Wiley, New York, 1961.
2. J. P. Colpa, C. Maclean, and E. L. Mackor, *Tetrahedron* **19** (Suppl. 2), 65 (1963).
3. K. B. Wiberg and G. Szeimies, *Tetrahedron Letters* **1968**, 1235; G. L. Closs and P. E. Pfeffer, *J. Am. Chem. Soc.* **90**, 2452 (1968).

IV-5b Reprinted from *Tetrahedron* **24,** 1083 (1968)

APPLICATION OF THE POPLE-SANTRY-SEGAL CNDO METHOD TO THE CYCLOPROPYLCARBINYL AND CYCLOBUTYL CATION AND TO BICYCLOBUTANE[1]

K. B. Wiberg

Department of Chemistry, Yale University, New Haven, Connecticut

(*Received in USA* 2 *May* 1967; *accepted for publication* 21 *June* 1967)

Abstract—The CNDO method has been applied to the cyclopropylcarbinyl and cyclobutyl cations, and has given results which are in very good accord with experimental data. A cross-ring interaction is calculated to be of importance with cyclobutyl derivatives, and agrees with the large difference in rate observed with equatorial and axial leaving groups. Some properties of bicyclobutane as well as the relative energies for some models of the activated complex for the thermal rearrangement of bicyclobutane have also been calculated and compared with experimental data. The CNDO method appears to have considerable promise in the investigation of organic chemical phenomena.

Much of the recent progress in the study of unsaturated organic molecules has its origin in the MO calculations based on the π-electron approximation. A similar treatment of saturated compounds would allow an examination of a wider range of phenomena. A number of approaches to such a treatment have been suggested.[2–5] One of them, the extended Hückel method, has been very successfully applied by Hoffmann to a variety of organic chemical problems,[6] and provides the theoretical basis for the Woodward–Hoffmann rules for electrocyclic and related reactions.[7]

The extended Hückel method has a major drawback in that it is a one electron approximation. Recently, Pople *et al.*[4,5] have developed an approximate SCF treatment (CNDO-complete neglect of differential overlap) which has the virtue of being independent of the coordinate system used. This paper will explore the application of the CNDO method to the cyclopropylcarbinyl-cyclobutyl cation problem, and to the properties and thermal rearrangement of bicyclobutane.

The CNDO treatment begins with extended Hückel molecular orbitals derived from the $2s$ and $2p$ atomic orbitals on carbon and the $1s$ atomic orbital on hydrogen. The proper linear combinations of these orbitals are obtained by constructing and diagonalizing the Hückel secular determinant in which the on-diagonal elements are the average ionization potentials and the off-diagonal elements are made proportional to the overlap integral between the atomic orbitals. The bond orders and electron densities are calculated using the occupied molecular orbitals, and from this the approximate Hartree–Fock matrix is constructed in a fashion similar to that previously used by Pople for π-electronic systems.[8] Diagonalization of the latter leads to a new set of molecular orbitals. The procedure is repeated until there is no longer a significant change in the coefficients.

The hybridization at a given carbon is not specified at the beginning of the calculation; only the coordinate of the atoms, the average ionization potential of each type of atomic orbital used, the core charge, the β proportionality constant, and the exponent for the Slater type atomic orbital are required. The values used are summarized in Table 1. The final hybridization is determined by the geometry initially supplied and by the difference in the above parameters between 2*s* and 2*p* orbitals.

TABLE 1. PARAMETERS USED IN CNDO CALCULATIONS

Orbital	Average ionization potential	Core charge	Slater exponent	β proportionality constant
C, 2*s*	14·051	4	1·625	21.
C, 2*p*	5·572	4	1·625	21.
H, 1*s*	7·1761	1	1·2	9.

A. *Cyclopropylmethyl and cyclobutyl cations*

The structure and energy of the cyclopropylmethyl cation has received much attention in the recent literature,[9,10] and has been considered by Hoffmann using the extended Hückel theory.[6] All of these results are in accord with the "bisected" structure as the more stable, the stabilization being derived from overlap between the empty *p* orbital and the C—C bond orbitals of the cyclopropane ring:

A fairly large barrier to rotation of the carbinyl carbon is suggested by the NMR data of Olah *et al.*[9]

The energies required for the removal of a hydrogen atom and an electron from a series of hydrocarbons have been calculated by the CNDO method, giving the results in Table 2.* The energies for forming the methyl, ethyl, 2-propyl and t-butyl cations from the hydrocarbons fall in a reasonable order and provide a guide for interpreting the following entries. With the isobutyl, cyclopropylmethyl and cyclobutyl cations, two conformations are possible. One corresponds to the conformation of the cyclopropylcarbinyl cation described above (out of plane or bisected) and the other has the plane of the CH_2 group roughly parallel to that of the rest of the molecule (referred to as "in plane").

The calculated difference in energy between the two conformations is large for cyclopropylmethyl, much smaller for cyclobutylmethyl and quite small for the isobutyl cation. The *in plane* conformation for the cyclopropylmethyl cation is

* The parameters used by Pople *et al.*[4,5] lead to somewhat short equilibrium C—C bond lengths (1·44 Å). In the calculations reported herein, the experimental bond lengths and angles were used whenever possible; analogous values were chosen when experimental data were not available. This approach has been used by Hoffmann in his extended Hückel calculations. For the cyclopropane ring, r(C—C) was 1·53 Å and r(C—H) was 1·09 Å, for the bond to a methyl group, r(C—C) was 1·54 Å, and in a methyl group, r(C—H) was 1·11 Å. The HCH bond angle in cyclopropane was taken as 116°, and for cyclobutane it was taken as 114°. The C—C bond length in cyclobutane was 1·55 Å, and the other C—C lengths were taken as 1·54 Å.

TABLE 2. ENERGIES OF FORMATION OF SOME CATIONS[a]

Reaction			ΔE(a.u.)
CH_4	→	$CH_3^+ + H^{\bullet} + e$	0·9190
C_2H_6	→	$CH_3CH_2^+ + H^{\bullet} + e^-$	0·8230
C_3H_8	→	$CH_3\overset{+}{C}HCH_3 + H^{\bullet} + e^-$	0·7559
$i\text{-}C_4H_{10}$	→	$t\text{-}C_4H_9^+ + H^{\bullet} + e^-$	0·7004
$i\text{-}C_4H_{10}$	→	$i\text{-}C_4H_9^+ + H^{\bullet} + e^-$	
		a. in plane	0·8045 } ΔΔE = 0·0061
		b. out of plane	0·7984
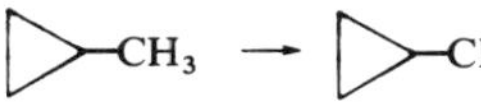	→	$CH_2^+ + H^{\bullet} + e^-$	
		a. in plane	0·8178 } ΔΔE = 0·0400
		b. out of plane	0·7778
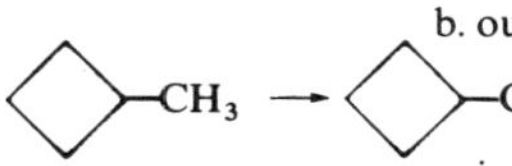	→	$CH_2^+ + H^{\bullet} + e^-$	
		a. in plane	0·8016 } ΔΔE = 0·0118
		b. out of plane	0·7896
	→	(planar) $+H^{\bullet} + e^-$	0·8124 (tetrahedral) 0·7651 (trigonal)
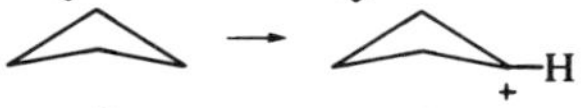	→	(40°) $+H^{\bullet} + e^-$	0·8539
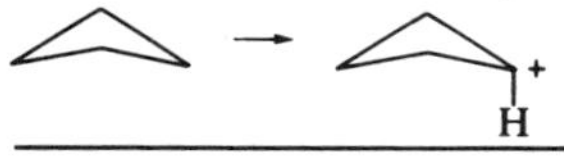	→	(40°) $+H^{\bullet} + e^-$	0·7764

[a] Although the energies are calculated in a.u., the values obtained using the CNDO method are invariably too large by a factor of about 3. A detailed comparison of calculated and observed energies will be made elsewhere.

predicted to be destabilized in comparison to the isobutyl cation, whereas the *out of plane* conformation is predicted to have a considerably lower energy than the isobutyl cation. The latter result is in good agreement with experiment and with previous calculations.

It is interesting that whereas the *in plane* conformation of the cyclobutylmethyl cation is calculated to have about the same energy as the isobutyl cation, the *out of plane* conformation is predicted to have a significantly lower energy. It is, however, difficult to test this prediction since rearrangement to a cyclopentyl cation would give a secondary ion and relieve 20 kcal/mole of ring strain.

A comparison of the results obtained by the extended Hückel method and the CNDO method is shown in Fig. 1. It is valuable first to consider the results for methylcyclopropane. The extended Hückel method generally gives exaggerated charge distributions, in contrast to the CNDO method which predicts relatively small charge separation.* The same is true in the present case.

* For example, the carbon charges for n-pentane by the two methods are:

Extended Hückel[6]

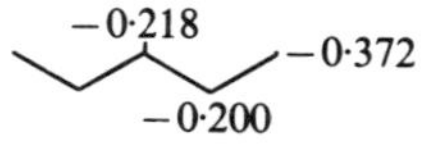
−0·218
−0·372
−0·200

CNDO

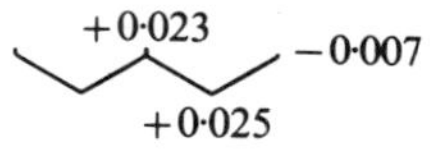
+0·023
−0·007
+0·025

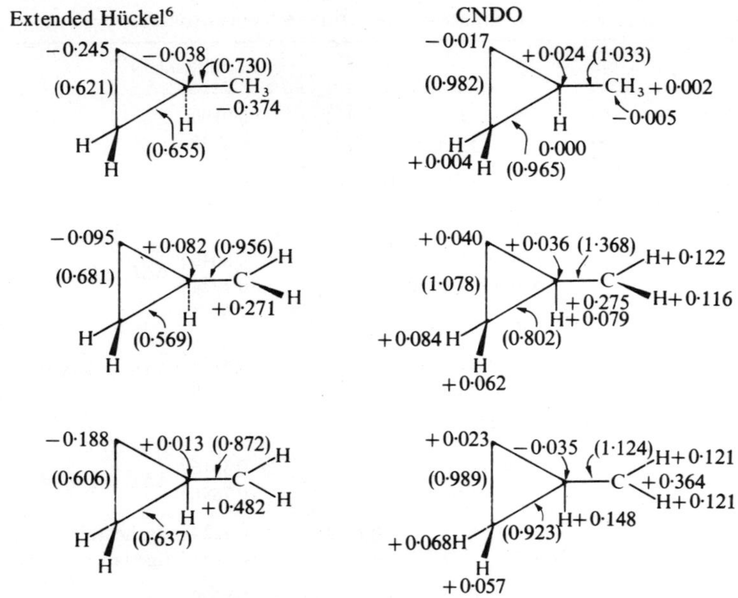

FIG. 1. Comparison of the results obtained using the extended Hückel and CNDO methods. The charge densities are given as signed numbers; the overlap populations for the extended Hückel method and the bond indices for the CNDO method are given in parentheses.

The bond populations given for the extended Hückel calculation are those derived by the method of Mulliken.[11] In the case of the CNDO results, the values given are the sum of the squares of the bond orders between the atoms in question. This sum of squares, which we should like to call "bond indices" appear to be closely related to the bond character.*

If the change in parameters on going from the hydrocarbon to the ion is compared for the two calculations, it is seen that the results are in good agreement. The *in plane* conformation has a relatively small change in bond index to the carbinyl carbon, and a relatively small transfer of charge into the cyclopropane ring as compared to the *out of plane* conformation. For the latter, the C_1—C_2 bond index is markedly reduced whereas the C_2—C_3 bond index increases.*

The results are in accord not only with the structural data on the cyclopropyl-

* The sum of the squares of the bond orders (p_{jk}) between any one atomic orbital and all other orbitals in a molecule can be shown by a derivation similar to that given by L. Salem, *Molecular Orbital Theory of Conjugated Systems*, p. 39. W. A. Benjamin, New York (1966), to be two times the charge density in that orbital (p_{jj}) less the square of the charge density:

$$\sum_k p_{jk}^2 = 2p_{jj} - p_{jj}^2.$$

For a unit charge density, the value is 1 whereas it goes to zero for $p_{jj} = 2$ (a non-bonded pair) or for $p_{jj} = 0$ (an empty orbital). Correspondingly, the sum of the squares of the bond orders to an atom corresponds to the number of covalent bonds formed by that atom, corrected for the ionic character in each bond.

carbinyl ion, but also with the effect of bridging the cyclopropane ring with a second ring. Some significant data are:

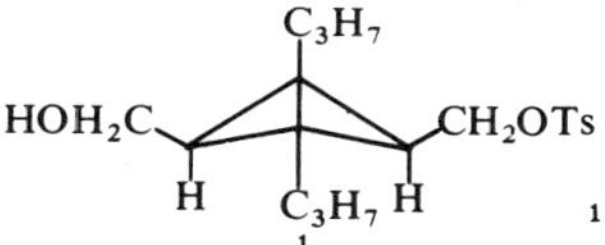
12

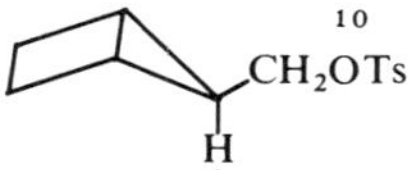
10

k_{rel} (HOAc): $\frac{1}{80}$; $\frac{1}{40}$

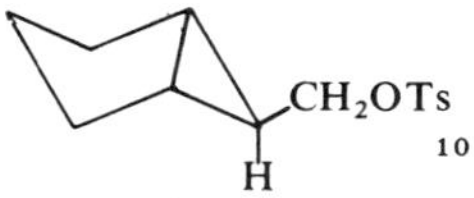
10

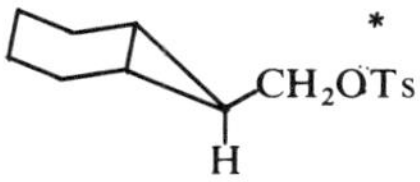
*

k_{rel} (HOAc): 1 ; 5

14

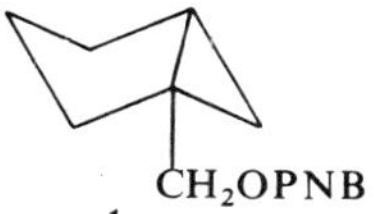
15

k_{rel} (60% aq. acetone): 550 ; 1

With the 2,3-bridged compounds, the normal trend is toward slightly lower rates when the bridging ring is decreased in size. The calculations predict an increase in the 2,3 bond index, which should result in a decrease in the 2,3-bond length. If such a change in bond length occurred, the strain energy would increase for the smaller bridging ring and should lead to a small decrease in the rate of reaction.

With the 1,2-bridged compounds, a decrease in ring size leads to a marked increase in rate. The calculations predict a considerable decrease in the 1,2 bond order on going to the cation, which should lead to an increase in bond length and a relief of strain for the smaller ring. This may, of course, not be the only effect contributing to the changes in rate constant, but it is in the correct direction.

The energy of the cyclobutyl cation also was calculated. The planar conformation with a trigonal cationic center was calculated to be slightly destabilized with respect to the 2-propyl cation as would be expected. However, the trigonal cation may not be a good model for the activated complex formed in the solvolysis of a cyclobutyl derivative. Bonding to the leaving group will still be important in the activated complex, and so, an ion with a tetrahedral arrangement about the cationic center might be a better model.

The energy of a cyclobutyl cation with a puckered conformation (as found in cyclobutane itself) and a tetrahedral cationic center was found to be dependent on the orientation of the empty orbital. The structure derived from an equatorial leaving

* Estimated from data on 3,5-dinitrobenzoates.[13]

group was calculated to have considerably lower energy than one derived from an axial leaving group.* This is in good accord with our experimental results:[17]

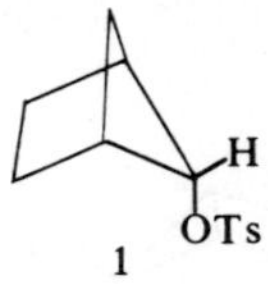

1

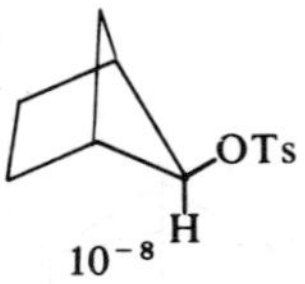

10^{-8}

k_{rel} (HOAc)

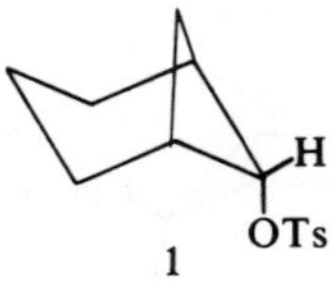

1

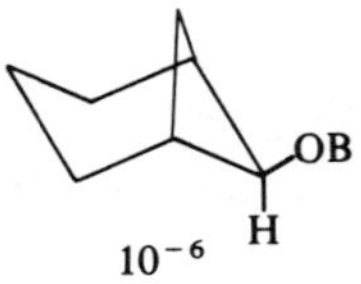

10^{-6}

A marked difference in reactivity is noted and the isomer with the equatorial tosylate group is in each case the more reactive one.

The calculated charge densities and bond indices for the two conformations are summarized below:

Charge densities

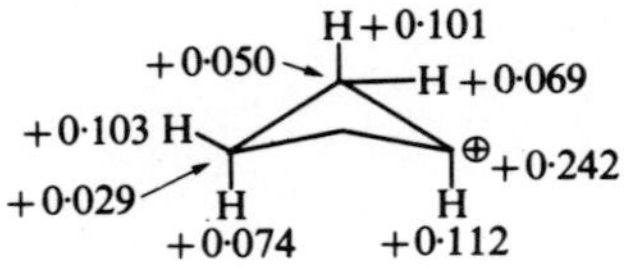

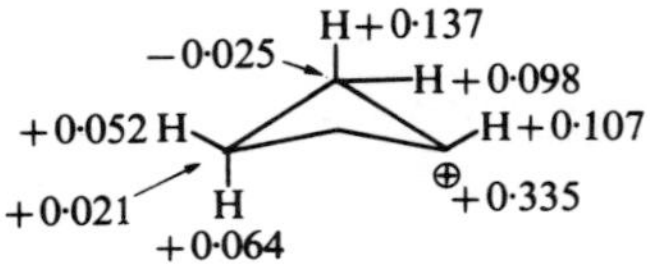

Bond indices

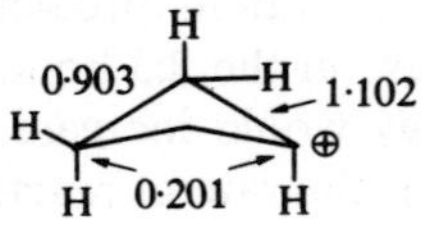

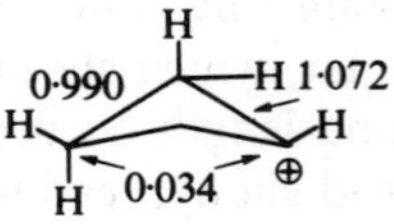

It can be seen that the equatorial leaving group leads to a cation which has a large cross-ring bond index, and a better charge distribution than the other cation. The cross-ring bond index results largely from a 2*p*–2*p* interaction, and thus involves the cationic orbital and the *p*-component of the cross ring C—H bond orbital to the equatorial hydrogen. Correspondingly, considerable charge is transferred to this hydrogen.

The interaction referred to above leads to what might be described as a symmetrical bicyclobutonium ion (as contrasted to the unsymmetrical ion proposed by Roberts and his co-workers).[18] Such an ion (or at least an activated complex) leads to a ready explanation of a variety of observations concerning the solvolytic reactivity of cyclobutyl derivatives. For example, Wilcox and Nealy[19] have shown that *trans*-3-hydroxy-2,2,4,4-tetramethylcyclobutyl tosylate is 800 times as reactive as the *cis*-isomer, and is about as reactive as 2,2,4,4-tetramethylcyclobutyl tosylate. The decreased reactivity of the *cis*-isomer may result from the introduction of the electron

* The same conclusions have been reached using the extended Hückel method.[16]

withdrawing hydroxy group into the cross ring equatorial position which interacts strongly with the developing cationic center.

TABLE 3. BOND INDICES AND CHARGES

Compound	Hydrogen	Charge	Bond indices 1s–2s ⟶	1s–2p	$J_{C_{13}-H}$
Propane	CH_2	0·998	0·220	0·739	123
Cyclobutane	equatorial	1·024	0·237	0·724	
	axial	0·995	0·265	0·700	134[a]
Cyclopropane		0·995	0·288	0·680	161
Bicyclobutane	exo	1·009	0·283	0·686	152
	endo	1·014	0·301	0·663	170
	bridgehead	0·979	0·366	0·600	203

Compound	C—C bond	Bond indices 2s–2s	2s–2p	2p–2p
Propane		0·067	0·422	0·552
Cyclobutane		0·046	0·389	0·557
Cyclopropane		0·049	0·308	0·629
Bicyclobutane	side	0·050	0·327	0·596
	central	0·037	0·168	0·688

[a] The coupling is averaged over the two hydrogen positions because of the rapid conformation inversion. The 1s–2s bond index for the planar conformation is 0·244.

We should like to be able to make a direct comparison of the calculated energies of the planar and non-planar cyclobutyl cations and the cyclopropylcarbinyl cation. However, in order to do this, it would first be necessary to minimize the energy with respect to structure. This, in turn, requires that the Pople parameters used be adjusted to give the best fit for the ground state structures. The problem of obtaining such parameters is currently under investigation.

B. *Bicyclobutane and its thermal rearrangements*

Bicyclobutane is an unusually interesting hydrocarbon. It has been found to have a dipole moment of 0·7 D[20] which is remarkably high for a saturated hydrocarbon. The bridgehead hydrogens are found to be remarkably acidic, and to have an unusually large C^{13}-H NMR coupling constant.[21] The electronic structure of bicyclobutane has been discussed using two models differing in hybridization,[22,*] but no calculations have been reported.

Using the bond lengths suggested by the microwave data, the energy was calculated as a function of the dihedral angle between the two cyclopropane planes. A 60° angle

* It is unfortunate that the Coulson–Moffitt representation of cyclopropane[23] has been considered as basically different than that of Walsh.[24] Coulson and Moffitt started with 2s and 2p atomic orbitals as the basis functions and calculated the coefficients of the basis functions which led to the lowest energy bond orbitals. Walsh developed a model which used sp^2 hybrid orbitals and *p* orbitals. However, Walsh did not minimize the energy with respect to hybridization (in fact, he did not calculate the energy). If he did, and both calculations were done correctly using the same method, the final results should be the same.

gave a significantly lower energy than either 50° or 70°, and this is in good accord with the available data.[29] The dipole moment was calculated to be 1·1 D, which again is in good accord with the experimental value.

The calculated data for the C—H bonds of bicyclobutane and other hydrocarbons are summarized in Table 3. The 1*s*–2*s* bond index is higher for the bridgehead protons of bicyclobutane than for any of the other compounds.* In order to see if the 1*s*–2*s* bond indices are correlated with the C^{13}–H NMR coupling constants, a plot of the data was prepared (Fig. 2). It is seen that a good linear relationship is found, and this is in accord with the postulated relationship between the coupling constant and the *s*-character of the bond orbitals.[26]

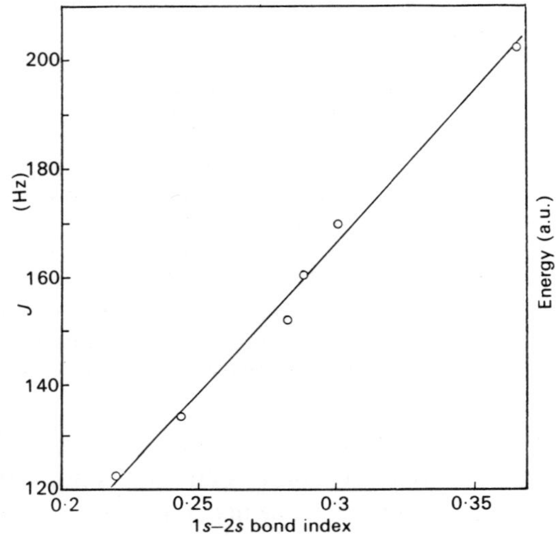

FIG. 2. Correlation of C^{13}–H NMR coupling constants with 1*s*–2*s* bond indices.

The calculations also suggest that the central bond of bicyclobutane has considerably more *p*-character than the C—C bonds of cyclopropane. Consequently, the facile reaction with electrophiles and the reaction with carbenes is not surprising.[27] The large 1*s*–2*s* bond index for the bridgehead protons, coupled with the calculated low electron density, provides a rationale for the acidity of these protons. It can be seen that the CNDO method gives results which are in good accord with the properties of bicyclobutane.

It was then of interest to extend the calculation to the activated complex formed in the thermal decomposition of bicyclobutane to butadiene. The reaction leading directly to butadiene may proceed in any one of six ways. First, the reaction may lead either to *cis*- or *trans*-butadiene as the initial product.

* Just as the sum of the squares of the bond orders to a given orbital gives the bond index, the square of the bond order between two orbitals is related to hybridization. For example, with methane, the 1*s*–2*s* bond order is 0·50 and the 1*s*–2*p* bond order is 0·866, whereas the squares are 0·25 and 0·75 and correspond to the hybridization (sp^3).

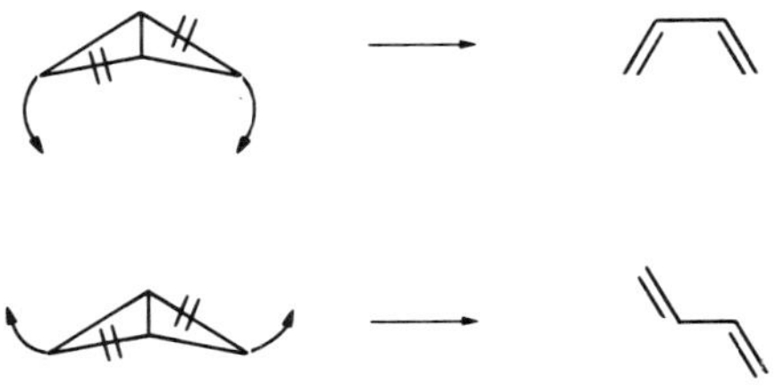

For each of these modes of reaction, the two CH_2 groups may rotate so that the *exo*-hydrogens move toward the bonds being broken (hereafter referred to as $+\,+$); they may rotate so that the *exo*-hydrogens move away from the bonds being broken $(-\,-)$; and they may move so that one hydrogen moves toward and the other away from the bonds being broken $(+\,-)$. The geometrical consequences are:

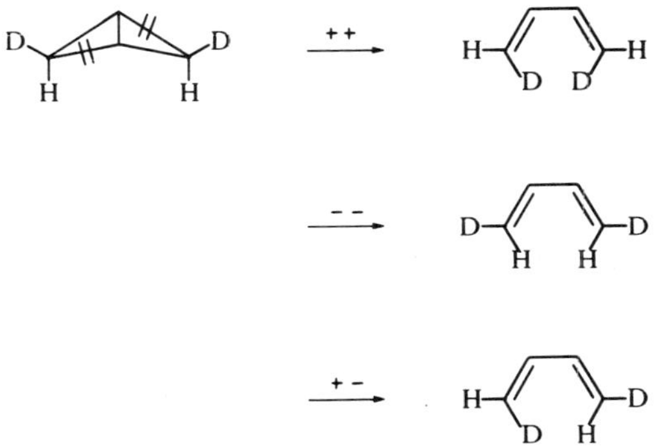

Corresponding results are found for opening to *trans*-butadiene.

Models of the possible activated complexes were constructed as follows. First, the bonds to be broken were stretched from 1·54 Å to 2·15 Å:

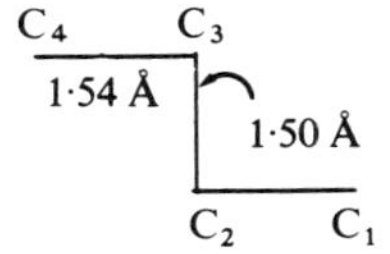

Then, the energy was calculated as a function of the C_1–C_2–C_3–C_4 dihedral angle. The change in energy with dihedral angle is shown in Fig. 3, and the corresponding changes in bond indices are shown in Fig. 4. It can be seen that the *trans*-conformation (180°) represents a maximum in energy, and that when the dihedral angle is reduced to a value below that in bicyclobutane (120°), the energy drops rapidly and reaches a minimum with the *cis*-conformation (0°). The product of the reaction, as described above, is, of course, cyclobutene.

The bond index curve shows a smooth transition from a C_2—C_3 single bond to a double bond, with the C_1—C_3 bond index essentially vanishing. Besides indicating the nature of the bonding changes which occur in the reactions, the fact that the bond indices are continuous functions of the dihedral angle indicates that the process is electronically allowed. If there were an interchange of bonding and antibonding orbitals at some point during the change in angle, the curves would be discontinuous. The above result suggests that a possible mechanism for the reaction is one in which

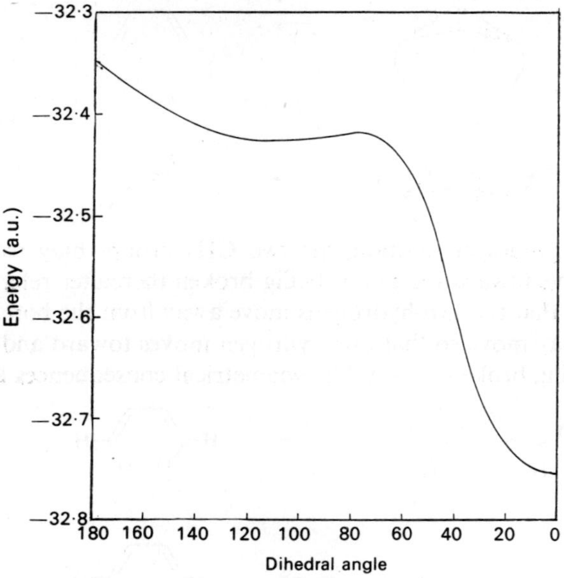

FIG. 3. Effect of dihedral angle on the energy of the model of the activated complex for bicyclobutane isomerization

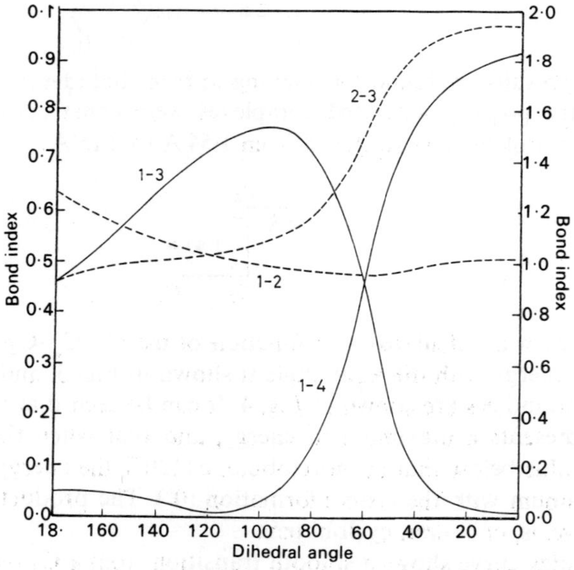

FIG. 4. Effect of dihedral angle on the bond indices for the model of the activated complex for bicyclobutane isomerization. The solid lines refers to the lefthand axis; the dashed lines to the righthand axis.

a vibrationally excited cyclobutene is the initial product, and then rapidly rearranges to butadiene. The geometrical consequences of such a process are:

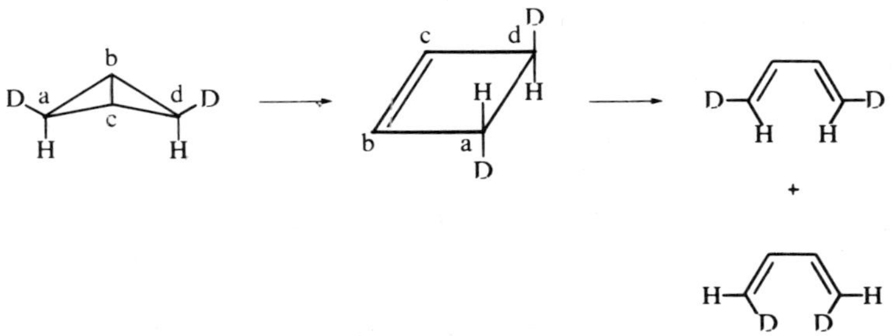

The rearrangement of the cyclobutene would be expected to be a conrotatory process,[28] and since there is no steric difference between hydrogen and deuterium, the product should be an equal mixture of *cis*- and *trans*-dideuterobutadienes. This would be in accord with our experimental result[29] in which bicyclobutane-2-*exo*-d_1 gave a 1:1 mixture of butadiene-*cis*-1-d_1 and butadiene-*trans*-1-d_1.

The effect of twisting the CH_2 groups was then examined. When a dihedral angle of 150° was used, any direction of twist led to an increase in the calculated energy. Thus, the calculations suggest that the formation of *trans*-butadiene is not a favored process.* A dihedral angle of 90° was then chosen and the effect of the three possible modes of CH_2 twist was examined. The $+-$ mode led to a decrease in energy, but the $++$ and $--$ modes initially led to an increase in energy. The C_1–C_2 bond index is plotted as a function of the angle of twist in Fig. 5. The $+-$ mode leads to a continuous function of the angle but the $++$ and $--$ modes lead to a discontunity indicating that these modes are not allowed.† ‡ Thus, a direct reaction leading

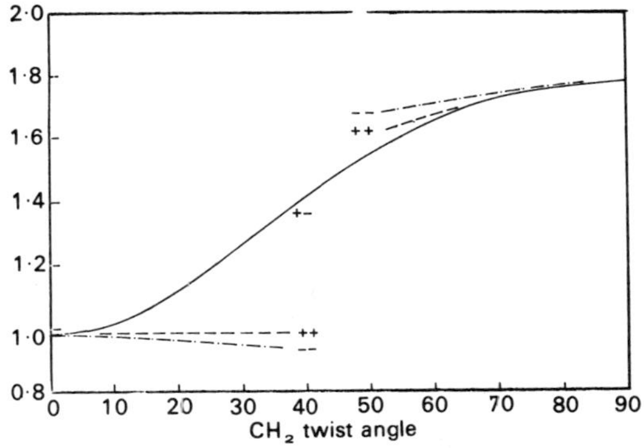

FIG. 5. Effect of CH_2 twist on the C_1–C_2 bond index for a model of the activated complex for bicyclobutane isomerization. The dihedral angle is 90°.

* Since only a limited number of models were examined, it is possible that a geometry exists which would favor opening to *trans*-butadiene. However, of the three possible twisting modes, only the $+-$ mode is allowed. The $++$ and $--$ modes were shown not to be allowed in the same fashion as later described for opening to *cis*-butadiene.

† The same conclusion may be reached in a qualitative way starting with the two bonding π-orbitals of *cis*-butadiene. In the $++$ and $--$ modes, one of the π-orbitals will give a bonding orbital whereas the other will give an antibonding orbital. In the $+-$ mode, each of the π-orbitals will give one bonding C—C bond orbital. The simple orbital representation we have earlier[29] should be disregarded for it, among other things, would lead to a *trans*-fused bicyclobutane!

‡ Corresponding results were obtained when a shorter C_1—C_3 distance of 2·0 Å was used in the calculations.

to butadiene would occur as follows:

This, again, is in accord with our experimental findings.[29] It is clear that a study of the rearrangement of bicyclobutane-2,4-d_2 will be required in order to establish the course of the reaction.

Because of the large number of geometrical variables, it is not practical to search out the minimum energy path between bicyclobutane and butadiene. Since the calculations indicate that a decrease in energy may occur with less dihedral angle distortion if the CH_2 group is twisted during the reaction, it appears likely that the lowest energy path is the one which leads directly to *cis*-butadiene.* However, even if this is the case, suitable structural modification may lead to a cyclobutene as an intermediate.

The above conclusions permit a reexamination of the interesting results reported by D'yakonov *et al.*[30] in their study of the thermal rearrangement of a highly substituted bicyclobutane:

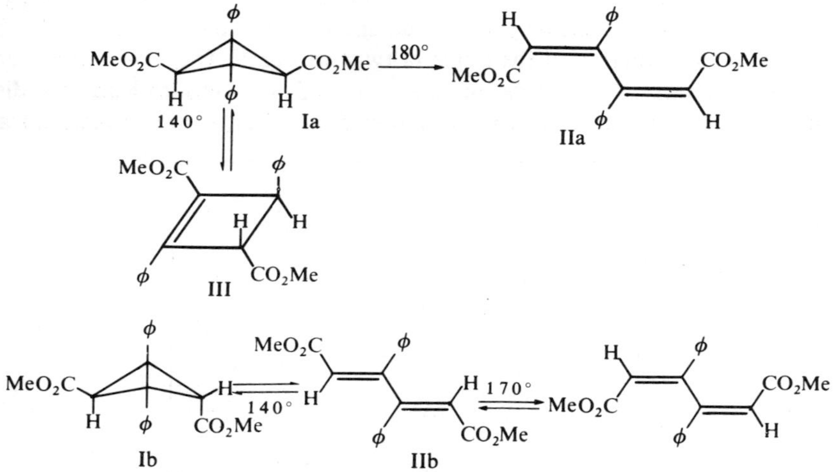

Whereas bicyclobutane and 1,3-dimethylbicyclobutane[31] have been shown not to form a cyclobutene during thermal rearrangement, the diphenyl substituted bicyclobutane, Ia, not only gives the cyclobutene III, but the latter is in thermal equilibrium with Ia. This indicates that phenyl substitution has markedly stabilized the bicyclobutane since the rearrangement of bicyclobutane to cyclobutene would be exothermic by 17 kcal/mole.

The rearrangement of Ia to III would require a mechanism in which the central bond of Ia is opened to a diradical, which then undergoes a hydrogen migration to give III. It might be reasoned that phenyl substitution would stabilize the intermediate diradical and permit a mode of reaction which is not realized with bicyclobutane.

* The conversion of bicyclobutane to cyclobutene has $\Delta H = -16$ kcal/mole whereas the reaction to butadiene has $\Delta H = -27$ kcal/mole. Thus, the energetics of the reaction also favor the direct process.

However, it should be noted that the NMR spectrum reported for III is unusual and corresponds to a more symmetrical structure. It is tempting to postulate that the product is actually IV, formed by the process suggested by the CNDO calculations:

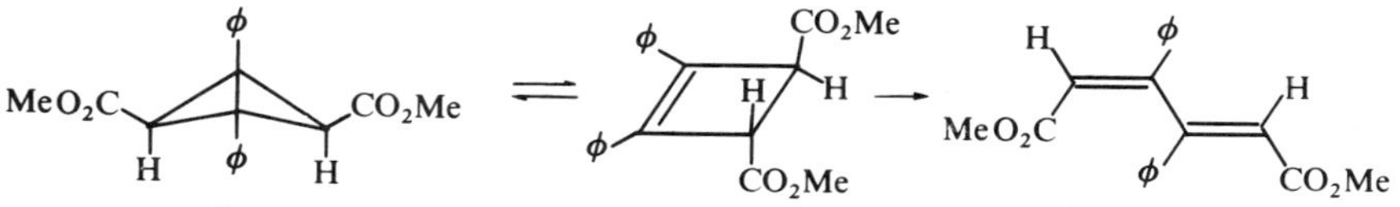

Ia IV IIa

The product obtained at a higher temperature, IIa, is the one which would be expected from the normal conrotatory ring opening of IV. The reaction could lead either to the *cis-cis* or *trans-trans* isomer (IIb or IIa). The latter is the more stable and presumably is therefore formed.

The rearrangement of Ib to IIb does not appear to involve a cyclobutene intermediate (at least none was found under the conditions employed for Ia). If the reaction proceeded directly to a butadiene, it should involve the $+-$ mode of methylene rotation. The observed product, IIb, corresponds to this prediction. A further electronic effect must also be operative, since the product is the less stable of the two isomers which could be formed via the $+-$ mode of reaction.

EXPERIMENTAL

The calculations described above were carried out using the program written by G. A. Segal[5] which was made available through the Quantum Chemistry Program Exchange at Indiana University. The matrix diagonalization subroutine in this program was found not to be satisfactory with some of the matrices involved in these calculations, and was replaced by the M.I.T. subroutine HDIAG. Also, the program was modified so that the bond indices were calculated.

REFERENCES

1 This investigation was supported by the U.S. Army Research Office, Durham.
2 R. S. Mulliken, *J. Chim. Phys.* **46**, 497, 675 (1949); M. Wolfsberg and L. Helmholtz, *J. Chem. Phys.* **20**, 837 (1952); R. Hoffmann, *Ibid.* **39**, 1397 (1963); J. A. Pople and D. P. Santry, *Mol. Phys.* **7**, 269 (1964).
3 G. Klopman, *J. Am. Chem. Soc.* **86**, 4450 (1964); H. A. Pohl, R. Appel, and K. Appel, *J. Chem. Phys.* **41**, 3385 (1964); J. J. Kaufman, *Ibid.* **43**, 5152 (1965).
4 J. A. Pople, D. P. Santry and G. A. Segal, *Ibid.* **43**, 5129 (1965).
5 J. A. Pople and G. A. Segal, *Ibid.*, **43**, 5236 (1965); *Ibid.* **44**, 3289 (1966).
6 R. Hoffmann, *Ibid.*, **40**, 2474, 2480, 2745 (1964); *J. Am. Chem. Soc.* **86**, 1259 (1964); *Tetrahedron* **22**, 521, 539 (1966).
7 R. B. Woodward and R. Hoffmann, *J. Am. Chem. Soc.* **87**, 395, 2511 (1965); R. Hoffmann and R. B. Woodward, *Ibid.* **87**, 2046, 4388, 4389 (1965).
8 J. A. Pople, *J. Phys. Chem.* **61**, 6 (1957).
9 G. A. Olah, M. B. Comisaro, C. A. Cupas and C. U. Pittman, Jr., *J. Am. Chem. Soc.* **87**, 2997 (1965); C. U. Pittman, Jr., and G. A. Olah, *Ibid.* **87**, 2998, 5123 (1965); N. C. Deno, J. S. Liu, J. O. Turner, D. N. Lincoln and R. E. Fruit, *Ibid.* **87**, 3000 (1965); N. C. Deno, H. G. Richey, Jr., J. S. Liu, D. N. Lincoln and J. O. Turner, *Ibid.* **87**, 4533 (1965); G. L. Closs and H. B. Klinger, *Ibid.* **87**, 3265 (1965); H. C. Brown and J. D. Cleveland, *Ibid.* **88**, 2051 (1966).
10 K. B. Wiberg and A. J. Ashe, III, *Tetrahedron Letters* 1553, 4245 (1965); P. v. R. Schleyer and G. W. Van Dine, *J. Am. Chem. Soc.* **88**, 2321 (1966).
11 R. S. Mulliken, *J. Chem. Phys.* **23**, 1833, 1841, 2338, 2343 (1955).
12 P. Wolff, Ph.D. Thesis, Columbia University 1965.
13 P. v. R. Schleyer and G. W. Van Dine, *J. Am. Chem. Soc.* **88**, 2321 (1966).

[14] W. G. Dauben and J. Wiseman, *J. Am. Chem. Soc.* **89**, 3545 (1967).
[15] W. D. Closson and G. T. Kwiatkowski, *Tetrahedron* **21**, 2779 (1965).
[16] R. Hoffmann, private communication.
[17] K. B. Wiberg and R. Fenoglio, *Tetrahedron Letters* 1273 (1963); K. B. Wiberg and B. A. Hess, Jr., *J. Am. Chem. Soc.* **89**, 3015 (1967).
[18] J. D. Roberts and R. H. Mazur, *J. Am. Chem. Soc.* **73**, 2509 (1951); R. H. Mazur, W. N. White, D. A. Semenow, C. C. Lee, M. S. Silver and J. D. Roberts, *Ibid.* **81**, 4390 (1959).
[19] C. F. Wilcox, Jr., and D. L. Nealy, *J. Org. Chem.* **28**, 3450 (1963).
[20] M. D. Harmony and K. Cox, *J. Am. Chem. Soc.* **88**, 5049 (1966).
[21] G. L. Closs and L. E. Closs, *Ibid.* **83**, 1003 (1961), *Ibid.* **85**, 2022 (1963).
[22] M. Pomerantz and E. W. Abrahamson, *Ibid.* **88**, 3970 (1966).
[23] C. A. Coulson and W. Moffitt, *J. Chem. Phys.* **15**, 151 (1947); *Phil. Mag.* **40**, 1 (1949); C. A. Coulson and T. H. Goodwin, *J. Chem. Soc.* 2851 (1962).
[24] A. D. Walsh, *Nature, Lond.* **159**, 165, 712 (1947).
[25] I. Haller and R. Srinivasan, *J. Chem. Phys.* **41**, 2745 (1964).
[26] N. Muller and D. E. Pritchard, *J. Chem. Phys.* **31**, 768, 1471 (1959).
[27] K. B. Wiberg, G. M. Lampman, R. P. Ciula, D. S. Connor, P. Schertler and J. Lavanish, *Tetrahedron* **21**. 2749 (1965).
[28] R. B. Woodward and R. Hoffmann, *J. Am. Chem. Soc.* **87**, 396 (1965).
[29] K. B. Wiberg and J. Lavanish, *J. Am. Chem. Soc.* **88**, 5272 (1966).
[30] I. A. D'yakonov, V. V. Razin and M. I. Komendantov, *Tetrahedron Letters* 1127, 1135 (1966).
[31] W. v. E. Doering and M. Pomerantz, *Ibid.* 961 (1964).

IV-6

A CNDO Treatment of the Arylmethyl Cations

A. STREITWIESER, JR., AND R. G. JESAITIS

1. Introduction

In the application of MO methods to the study of reactivities of organic compounds, three types of techniques may be distinguished. One is based on the general form, symmetry, and nodal characteristics of molecular orbitals and has led to such important generalizations as the Hückel 4n + 2 rule and the Woodward-Hoffmann stereoselectivity rules. Another method depends on the energetics of individual molecular orbitals. Applications involving the highest occupied or lowest vacant orbitals and perturbation methods come under this heading. Both of these general methods have been notably successful in interpreting a wide range of organic chemistry, both qualitatively and quantitatively. The third general method makes use of estimates of the total electronic energy. In the simple HMO method $E = \sum^{occ} n_j \epsilon_j$, the sum of occupied orbital energies. If such a calculation is made for both a reactant and a product or model of a transition state, the corresponding ΔE is compared with an experimental energy change for the respective equilibrium or rate constant. Such correlations with the HMO method, particularly for reactions involving ions, are often mediocre.[1] It is quite clear that this poor performance results primarily from the inherent neglect of electron repulsion terms in the HMO method. Corresponding treatments by SCF methods are generally more successful.[2]

Nevertheless, these successful applications have been restricted entirely to π-electronic systems. In such systems, the stabilization of ions is presumably by one mechanism only—charge delocalization. A second important effect is the interaction of charge with a substituent dipole —the field effect. Only limited treatments of this interaction mechanism have been attempted within a molecular orbital framework.[3] The Pople-Santry-Segal CNDO-SCF method[4] is not limited to π systems and should be applicable to many kinds of organic ions. In this paper we describe the application of the CNDO method (actually, the CNDO/2 method) to a reaction involving arylmethyl cations.

Our procedure was to compare the CNDO method with π methods for unsubstituted polycyclic arylmethyl cations in which charge delocalization is expected to be an important stabilizing mechanism. The CNDO procedure was then applied to the corresponding reactions involving m- and p-substituted benzyl cations in which the substituents consist of first-row elements and for which charge-dipole interactions are expected to be important. Appropriate data was available in our laboratory for the acetolysis of arylmethyl tosylates. In both, the reaction mechanism has been amply studied and reaction-rate constants are known not only for a wide range of polycyclic benzenoid systems[5] but also for sufficient substituents on the parent benzyl compound to establish an appropriate $\rho\sigma^+$ correlation.[6]

2. Pi and Sigma Calculations

For comparisons with π-electron calculations, several studies have been made of solvolyses of polycyclic arylmethyl halides and sulfonates.[1] The results of Dewar and Sampson[7] on formolysis of the chlorides and the studies in our laboratory[8] on acetolysis of the tosylates both represent systems that are sufficiently limiting in character and involve transition states that are represented adequately for the present purpose as the corresponding arylmethyl carbonium ions. However, only in the acetolysis system are data also available for a range of substituted benzyl compounds.[6] Hence, only this cationic reaction system is treated in this paper.

For our purposes, the mechanism for the acetolysis of benzyl tosylate (and other arylmethyl tosylates) can be represented by

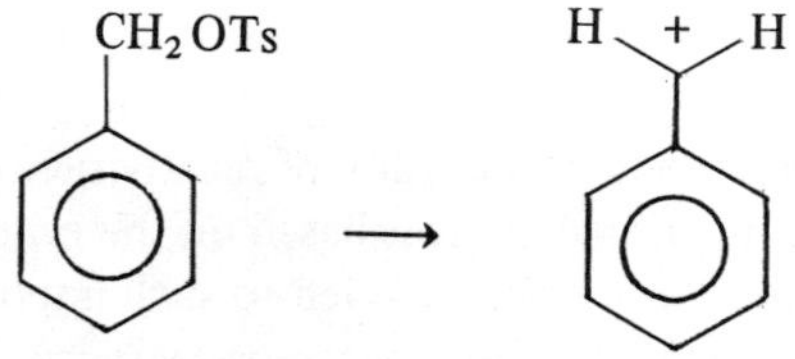

The energy changes calculated for the above process by an MO method applied to a series of related compounds should be proportional to the logarithms of the rate constants of the corresponding solvolysis reactions, given the usual assumption that a linear free-energy relation applied. Further approximations concerning the species actually calculated are made to simplify the various calculations.

The structures which are calculated by HMO and SCF-π methods are the π systems:

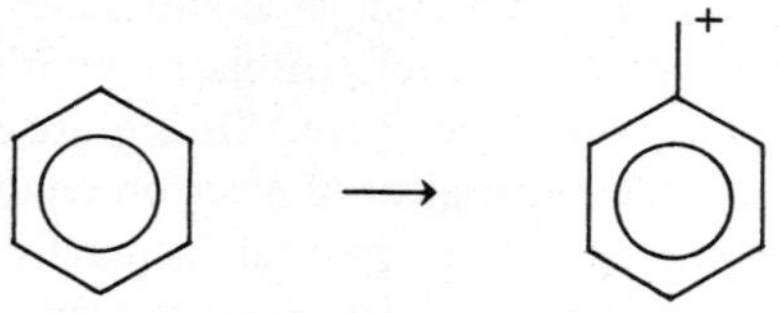

The CH_2OTs group is assumed to make only a minor perturbation to the π system and is thus usually neglected. In the HMO method, no further specification of geometry need be made. The HMO energies were taken from Ref. 9. In the SCF π calculations, with which comparisons will be made, those of Hammond[8,10] and Wright[3] in our laboratory parallel the calculations of Dewar and Thompson except for the use of different parameter values. The various sets of values used are listed in Table 1; the behavior of the different sets of the repulsion integrals as a

Table 1. Parameter Values for SCF-π Calculations

Calculation	γ_{11}	γ_{12}	β (1.40 Å)
PPP-Hammond[10]	11.08	5.32	-1.75
PPP-Dewar[2]	10.53	7.30	-1.75
PPP-Dewar-SCB[2]	10.53	7.30	-1.75[a]
Mataga[3,10]	11.08	5.32[b]	-2.231[a]

[a]Calculated from thermochemical formula; see A. L. H. Chung and M. T. S. Dewar, *J. Chem. Phys.* **42**, 756 (1965).

[b]$\gamma_{\mu\nu} = e^2/(r_{\mu\nu} + 1.229)$.

function of internuclear distance is illustrated in Fig. 1. The ring systems were treated as regular hexagons wherever possible, with all C–C bond lengths set at 1.40 Å.

In the CNDO/2 method all atoms and valence electrons are included. The process we calcu-

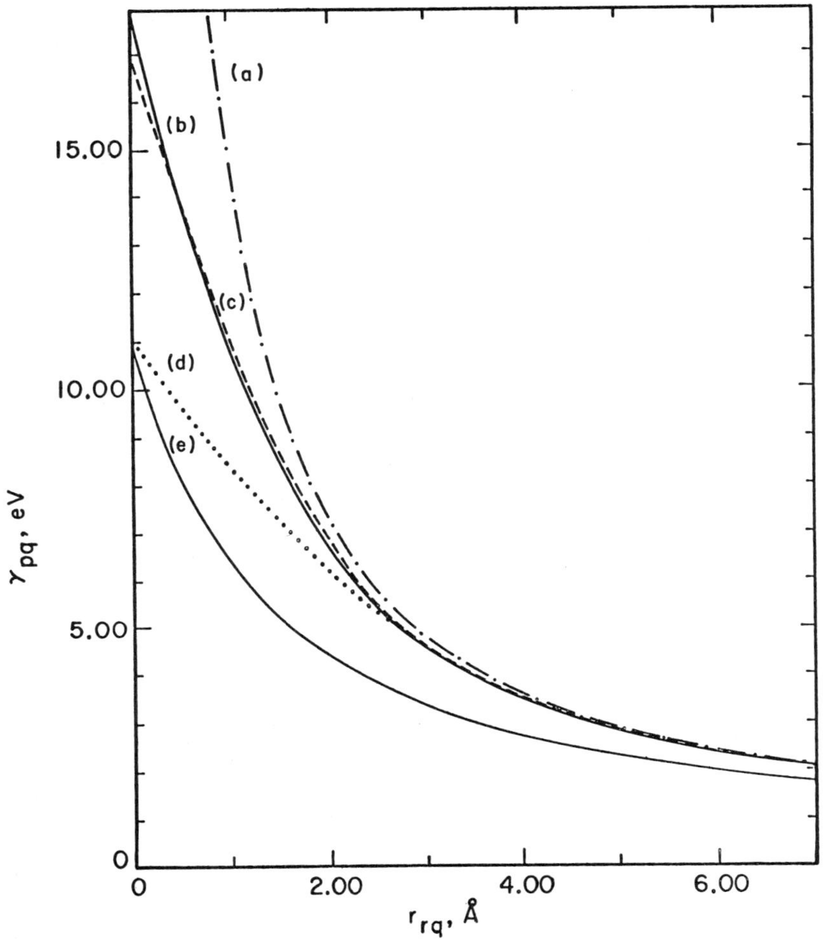

Fig. 1. Approximations to repulsion integrals. γ_{pq},eV, versus r_{pq}, Å: (a) e^2/r_{pq}; (b) uniformly charged spheres (Z = 3.31); (c) 2p Slater orbitals (Z = 3.18); (d) Pariser-Parr; (e) Mataga (a = 1.229). (From Ref. 10.)

late is of the type

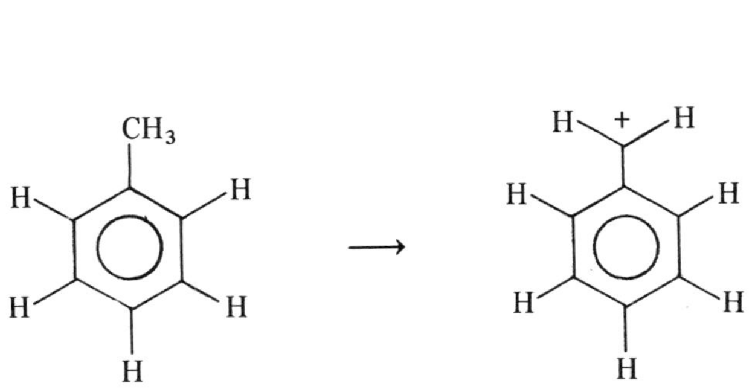

A methyl group was allowed to represent a $-CH_2OTs$ group to simplify the calculations. The reasonableness of this assumption was demonstrated by calculation of some of the energy differences using a $-CH_2OH$ group in place of a methyl group; no significant differences resulted. All C–C bond lengths were set at 1.39 Å except for the $C-CH_3$ bond (1.54 Å) and the long bond of biphenylene (1.50 Å). The C–H distances were all set at 1.09 Å. Little geometric variation was performed except as noted below. The CNDO calculations were made with a modification of QCPE Program No. 100.

3. Discussion

The HMO correlations of solvolytic reactivities have been amply discussed previously[1,7] and will be examined here only briefly. The correlation of HMO energy differences with the present acetolysis data (summarized in Table 3) is shown in Fig. 2 and is clearly rather poor.

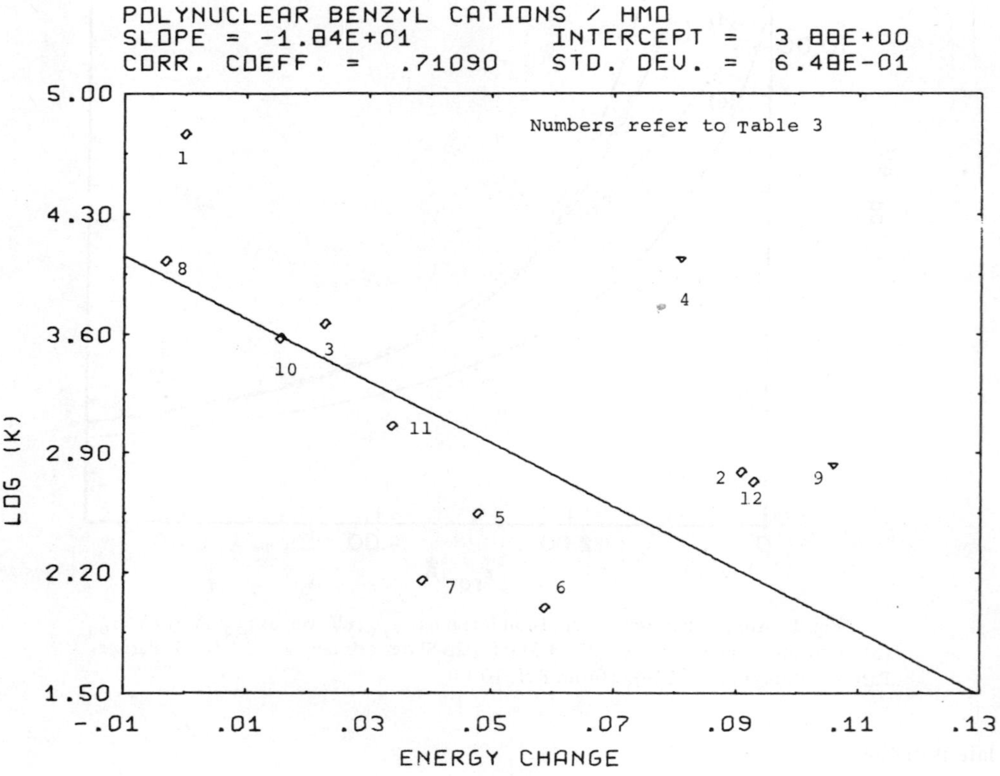

Fig. 2

The line shown is a least-squares line through all the points. The data actually fall into two classes, the benzyl or β-naphthyl type and the α-naphthyl type, in which the reaction center is bonded to a carbon next to a ring juncture. Earlier discussion of this dispersion was in terms of steric hindrance effects,[1,7] but more recent SCF π calculations show that differences in the neglected electron repulsion effects are responsible.

The Pople-Pariser-Parr calculations of Dewar and Thompson[2] give a satisfactory correlation with the Dewar and Sampson formolysis data.[5] The parallel calculations of Hammond[10] with our acetolysis results are comparable and the π calculations with Mataga repulsion integrals give a correlation of similar quality. Only the latter correlation ("Mataga SCF-π") is illustrated in Fig. 3, in which it is apparent that a fair correlation is obtained if one omits the two points farthest from the least-squares line, 2-biphenylenyl and 4-pyrenyl (triangles in Fig. 3). Among the variations attempted to improve the results are the use of a Ar-CH_2^+ bond length of 1.44

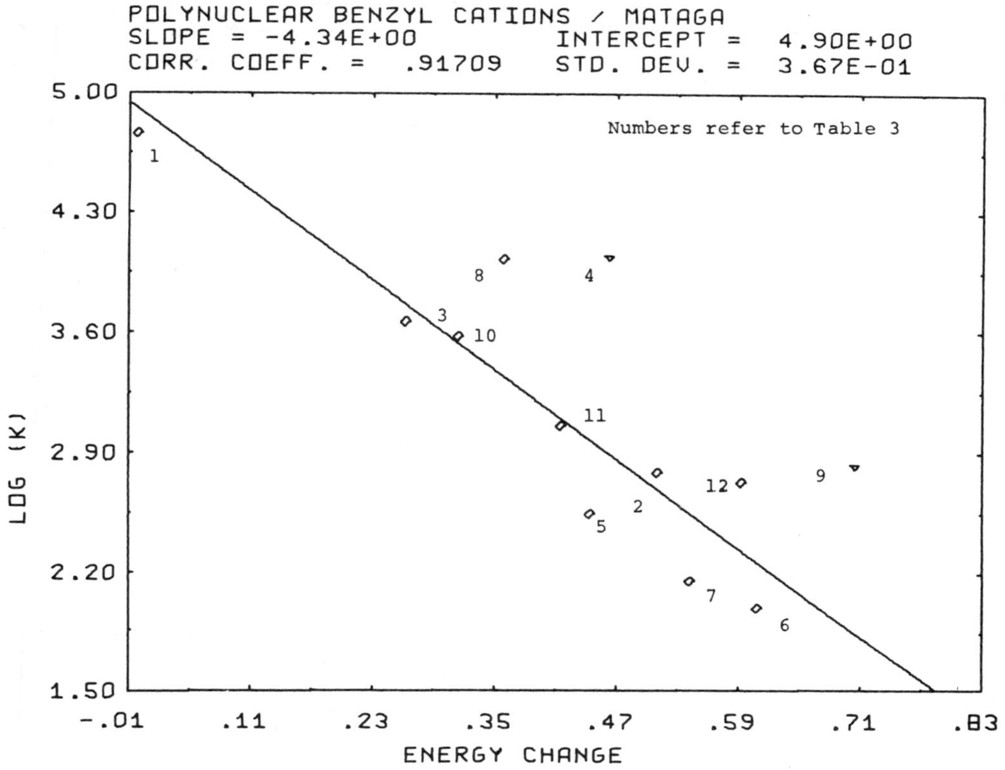

Fig. 3

Table 2. Correlation Results–Benzyl Cations

Energy relations	Points omitted	Slope[a]	Standard deviation	Corr. coeff.
CNDO/substituted	None	5.97	0.34	0.987
CNDO/polynuclear	2-biph, 4-pyr	2.49	0.22	0.971
CNDO/polynuclear	2-biph	2.25	0.28	0.938
HMO	2-biph, 4-pyr	18.36[b]	6.42	0.711
HMO	2-biph	14.60[b]	5.50	0.661
PPP	2-biph, 4-pyr	3.69	0.50	0.931
PPP	2-biph	3.24	0.55	0.890
PPP-Dewar	2-biph, 4-pyr	3.82	0.64	0.905
PPP-Dewar	2-biph	3.13	0.67	0.842
PPP-Dewar-SCβ	2-biph, 4-pyr	4.23	0.62	0.923
PPP-Dewar-SCβ	2-biph	4.03	0.60	0.912
Mataga	2-biph, 4-pyr	4.33	0.66	0.917
Mataga	2-biph	3.64	0.71	0.862
Mataga-1.40 real	2-biph, 4-pyr	5.34	0.80	0.920
Mataga-1.40 real	2-biph	3.83	0.92	0.809
Mataga-1.44 real	2-biph, 4-pyr	5.25	0.67	0.940
Mataga-1.44 real	2-biph	3.60	0.87	0.810
Charge relations				
CNDO/substituted	None	127.50[c]	27.60	0.853
CNDO/polynuclear	2-biph, 4-pyr	20.30[c]	2.40	0.949
CNDO/polynuclear	2-biph	17.10[c]	2.90	0.890

[a] Units of $\log(k/k_o)$ per eV except where indicated.
[b] $\log(k/k_o)$ per β.
[c] $\log(k/k_o)$ per unit charge at methylene position.

Å, the use of "real" structures as given by X-ray crystallographic measurements on the parent hydrocarbons rather than regular hexagon structures, and Dewar's variation of β to self-consistent bond lengths (SCβ). The correlations with these methods are summarized in Table 2, in which the slopes, standard deviations, and correlations of the least-squares regression lines are tabulated. The improvements that result by the different procedures are generally small and depend on the specific data used. None of the correlation coefficients of the π methods is better than fair. The universal deviation of the 2-biphenylenyl point is undoubtedly associated with the inability of these methods to cope with the four-membered ring, but we cannot account for the generally poor representation of the relative reactivity of the 4-pyrenyl system.

The energy changes for the process $ArCH_3 \longrightarrow ArCH_2^+$ were calculated by the CNDO/2 method using the Pople-Santry parameters. The results are summarized in Table 3 and the correlation of the experimental reactivities with the calculated energy changes is shown in Fig. 4 and summarized in Table 2. The constant energy of the other product in the model

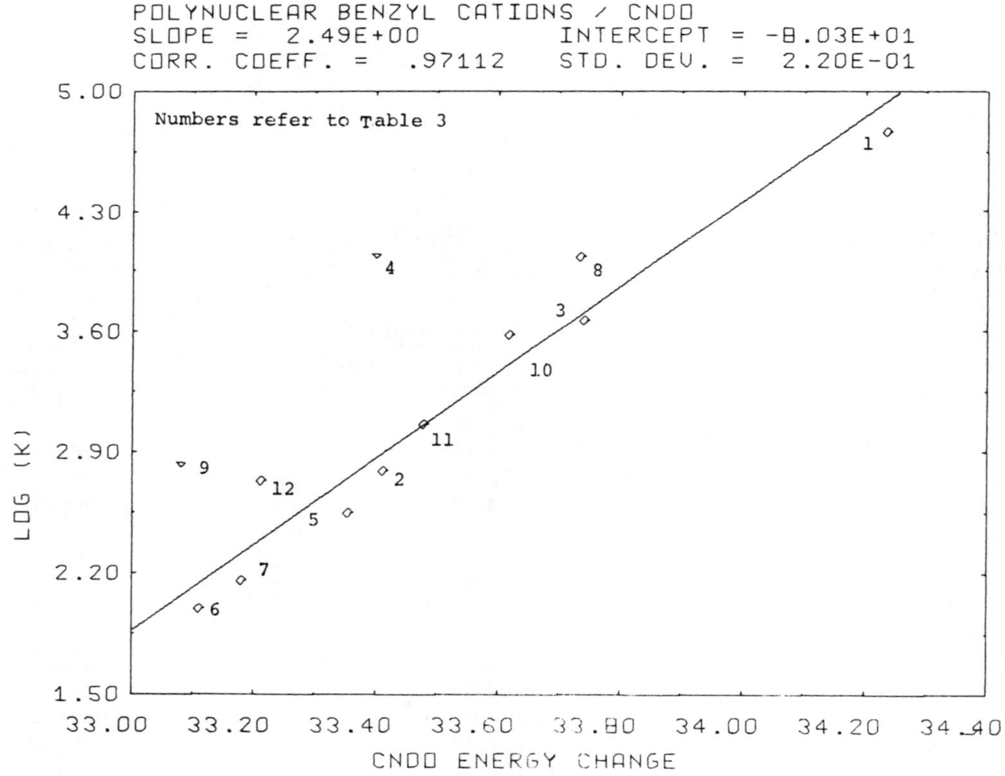

Fig. 4

reaction, H^-, is not included in the results shown. The 2-biphenylenyl and 4-pyrenyl systems still show substantial deviations compared to the other systems; when these points are omitted a rather satisfactory correlation (correlation coefficient = 0.97) is obtained. A plot of the CNDO all-valence-electron results against the Mataga SCF-π results (Fig. 5) shows that the net changes are well represented by the π calculations alone. The deviations from linearity that do

Table 3. Results of CNDO Calculations on Benzyl Cations

Compound	-Energy $ArCH_3$	-Energy $ArCH_2^+$	ΔE	Δg (CH_2)	-log k	No.
Polynuclear						
Phenyl	1518.028	1483.791	34.237	+0.4101	4.77	1
1-Naphthyl	2333.406	2299.990	33.416	+0.3103	2.80	2
2-Naphthyl	2333.450	2299.707	33.743	+0.3556	3.67	3
2-Biphenylenyl	2718.283	2684.882	33.401	+0.3073	4.03	4
2-Anthryl	3148.499	3115.142	33.357	+0.3000	2.55	5
3-Fluoranthyl	3536.131	3503.019	33.112	+0.2753	2.01	6
8-Fluoranthyl	3536.158	3502.976	33.182	+0.3176	2.17	7
2-Pyrenyl	3537.723	3503.986	33.737	+0.3681	4.04	8
4-Pyrenyl	3537.687	3504.605	33.082	+0.2644	2.82	9
2-Phenanthryl	3149.205	3115.584	33.621	+0.3495	3.59	10
3-Phenanthryl	3149.208	3115.727	33.481	+0.3274	3.07	11
9-Phenanthryl	3149.163	3115.948	33.215	+0.2919	2.74	12
Substituted benzyl						
m-Me	1754.467	1720.283	34.184	+0.4051	4.35	
p-Me	1754.458	1720.611	33.847	+0.3801	3.02	
p-F	2252.072	2217.951	34.121	+0.3892	4.38	
m-F	2252.100	2217.499	34.601	+0.4147	(6.78)[a]	
m-CF_3	3957.595	3922.923	34.672	+0.4124	6.35(7.74)[a]	
p-CF_3	3957.608	3922.880	34.728	+0.4091	6.56(8.27)	
m-OCH_3	2255.837	2221.506	34.331	+0.4060	(5.04)[a,b]	
p-OCH_3	2255.811	2222.016	33.795	+0.3701	(0.34)[a,b]	
p-OCH_3	2256.004	2222.412	33.592	+0.3596	c	
p-NO_2	2819.079	2784.011	35.086	+0.4114	(9.29)[a]	

[a]Calculated from $\rho\sigma^+$ correlation. See the text σ^+ values are: $(NO_2)_p = 0.790$, $(CF_3)_p = 0.612$, $(CF_3)_m = 0.52$, $(F)_m = 0.352$, $(OCH_3)_m = 0.047$, $(OCH_3)_p = 0.778$.
[b]C—O = 1.43 Å.
[c]C—O = 1.36Å.

occur in Fig. 5 are generally in the direction to improve the fit to the experimental results by the CNDO method.

Because of the linear relation between SCF-π and CNDO energy changes, it is instructive to examine the charge distributions given by these methods. The charge distributions shown for benzyl cation in Fig. 6 show some comparatively small differences between the HMO and Mataga SCF-π results; the general distribution is remarkably similar in both methods, but the SCF-π procedure puts more charge in the m and p positions at the expense of the α position. Noteworthy also is the almost neutral condition of the branch carbon 1 with the Mataga integrals; other parameter sets generally put substantial negative charge at this position. The CNDO result looks superficially rather different from the π results, but further analysis shows the differences to result primarily from the dissection of a given charge between C and its attached H's. Pople et al.[4] have already pointed out that such differences are sensitive to the relative parameter values adapted for C and H. We define a "regional charge" as that of a carbon plus its attached hydrogens. Such regional charges for benzyl cation are almost identical with the Mataga SCF-π results.

Also instructive is the plot in Fig. 7 of log k against the change in the regional charge of the methyl group of the parent $ArCH_3$ and the methylene group of the product, $ArCH_2^+$. The methyl regional charge is actually small and almost constant. The linearity of this plot shows convincingly that the principal stabilizing mechanism is that of charge delocalization. Once again, the 2-biphenylenyl and 4-pyrenyl systems show the strongest deviations from the linearity displayed by the other systems.

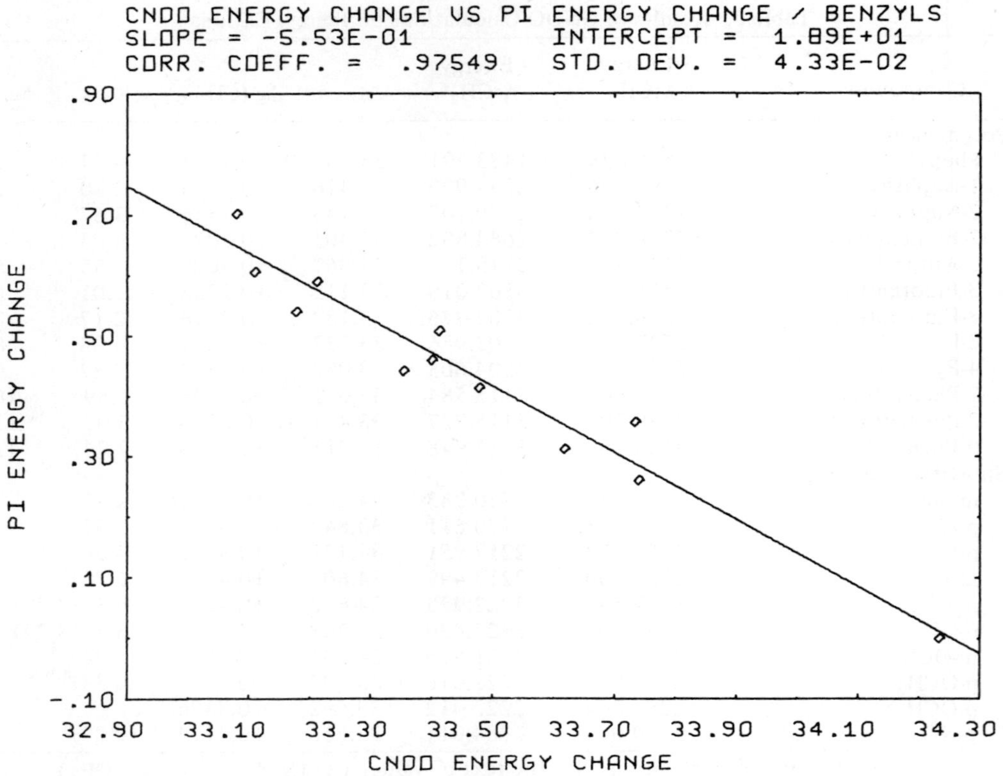

Fig. 5

We conclude that the CNDO method provides a satisfactory account of energy changes for delocalized cations but is not markedly better than π-electron theory. However, having established a successful correlation with reactivities involving delocalized arylmethyl cations, we can now examine the effect of polar substituents.

A $\rho\sigma^+$ treatment of the experimental acetolysis rates of some substituted benzyl tosylates shows dispersion into two linear correlations.[6] m-Chloro and m- and p-trifluoromethyl substituents give a linear σ^+ correlation with $\rho = -2.33$, a rather low value indicative of substantial direct displacement character by solvent. p-Fluoro, m- and p-methyl, and hydrogen give a linear σ^+ correlation with $\rho = -5.71$, a rather high value suggestive of more limiting solvolysis character, a conclusion reinforced by other evidence.[9] Using this value of ρ, we can put all substituents, including the polycyclic aryl systems, on a common basis; that is, in comparing the effect of strongly electron-withdrawing substituents (e.g., CF_3) with the theoretical calculations, we use not the experimental reactivities but the values calculated from the known σ^+ values[10] and $\rho = -5.71$. The resulting relative rates are those expected for solvolysis by the more limiting solvolysis mechanism of other compounds compared in the calculations. Both sets of reactivities are given in Table 3.

CNDO calculations were made for p-NO_2, m-F, and m- and p-OCH_3 groups for comparisons with rates also estimated by the $\rho\sigma^+$ treatment. Although experimental values were available for m- and p-chloro substituents, our CNDO program is not yet capable of treating second-row elements.

A plot of log k as the CNDO energy changes is shown in Fig. 8. An excellent linear corre-

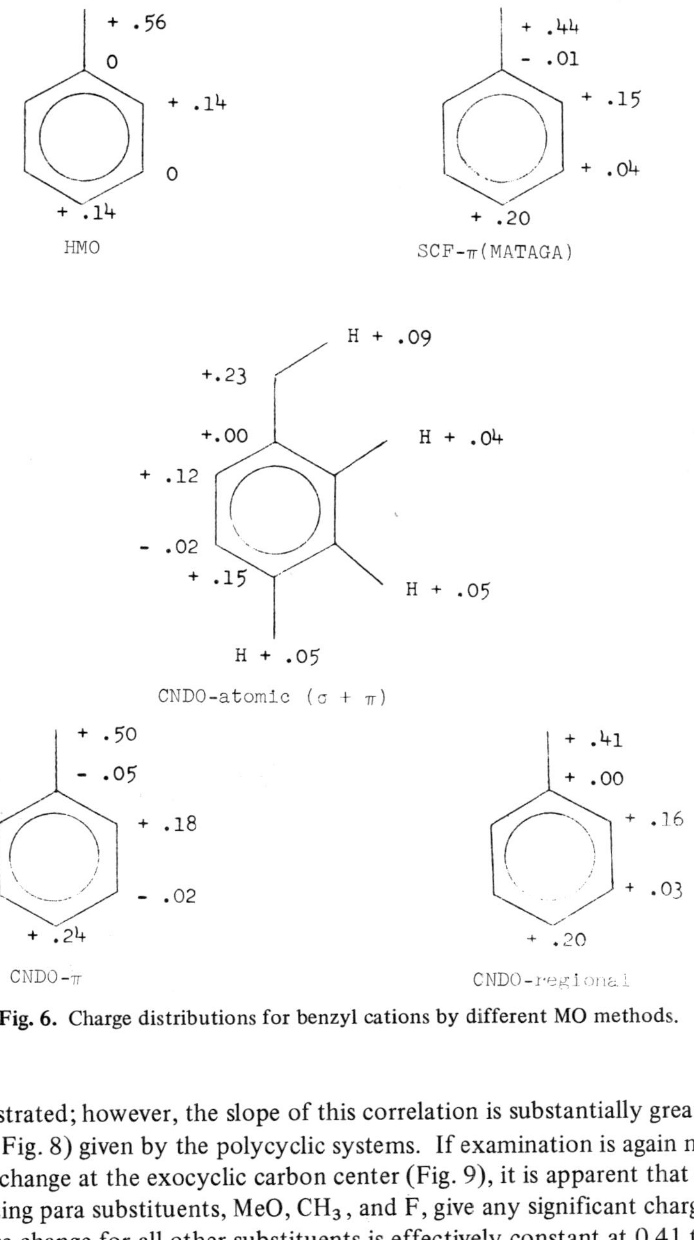

Fig. 6. Charge distributions for benzyl cations by different MO methods.

lation is demonstrated; however, the slope of this correlation is substantially greater than that (dashed line on Fig. 8) given by the polycyclic systems. If examination is again made of the regional charge change at the exocyclic carbon center (Fig. 9), it is apparent that only the strongly stabilizing para substituents, MeO, CH_3, and F, give any significant charge delocalization. The charge change for all other substituents is effectively constant at 0.41 to 0.42 unit. Even for the delocalizing substituents the slope of the line in Fig. 9 is higher than that (dashed line in Fig. 9) given by the polycyclic systems. For substituted benzyl cations, charge delocalization is clearly less important than charge-dipole interactions. Furthermore, the present results show that *the CNDO method does not scale these two interaction mechanisms on the same basis*. Other examples of such unequal scaling of different "effects" in the CNDO method have been documented; for example, binding energies are greatly overestimated, yet the barrier to rotation in ethane is underestimated.[11]

To evaluate the magnitude of the effect of geometric distortion in going to the transition

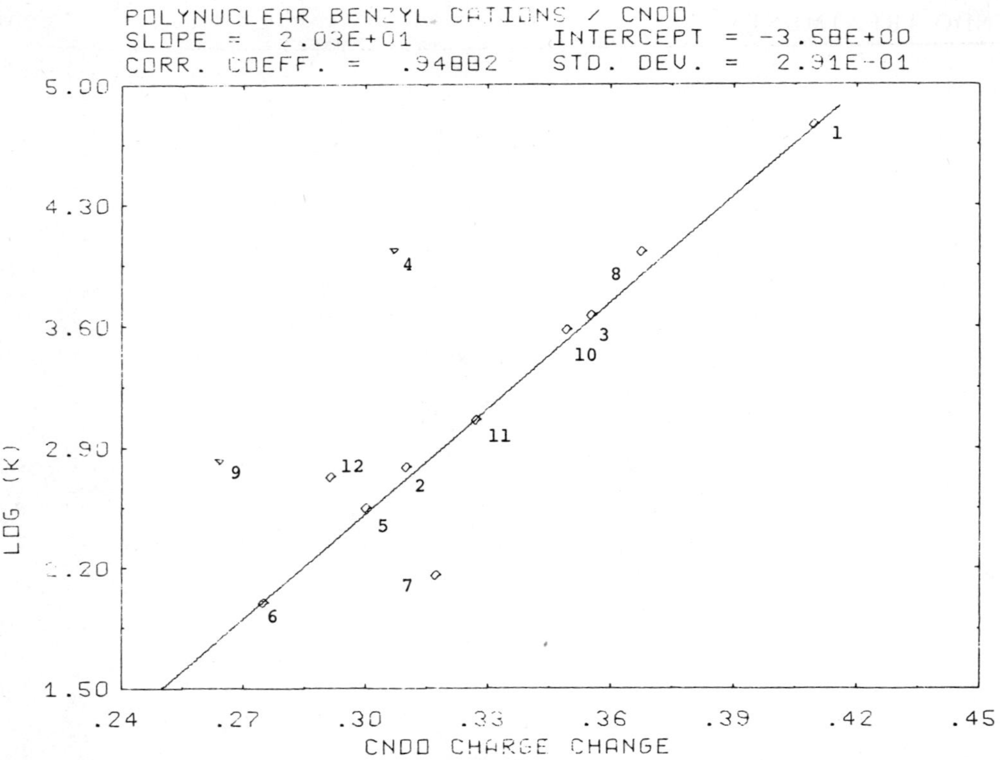
POLYNUCLEAR BENZYL CATIONS / CNDO
SLOPE = 2.03E+01
INTERCEPT = -3.58E+00
CORR. COEFF. = .94882
STD. DEV. = 2.91E-01
LOG (K)
CNDO CHARGE CHANGE

Fig. 7

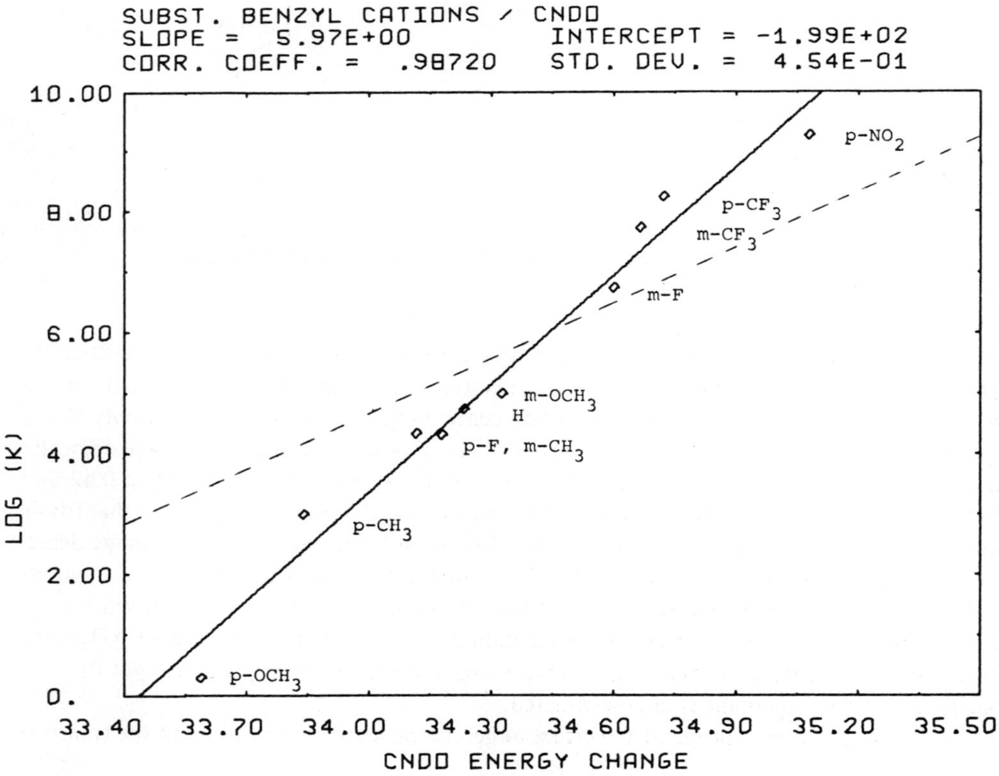
SUBST. BENZYL CATIONS / CNDO
SLOPE = 5.97E+00
INTERCEPT = -1.99E+02
CORR. COEFF. = .98720
STD. DEV. = 4.54E-01
LOG (K)
CNDO ENERGY CHANGE
p-NO2
p-CF3
m-CF3
m-F
m-OCH3
H
p-F, m-CH3
p-CH3
p-OCH3

Fig. 8

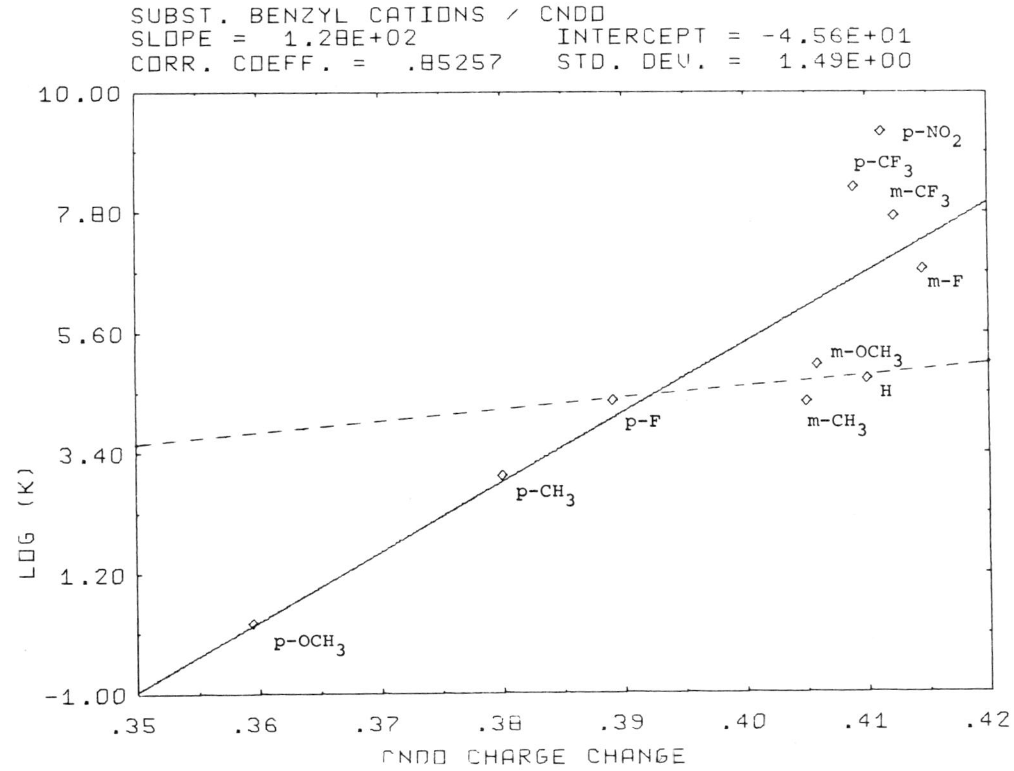

Fig. 9

state, some additional calculations were made. The p-methoxy group was chosen for examination because of the bond-shortening effect of a structure such as

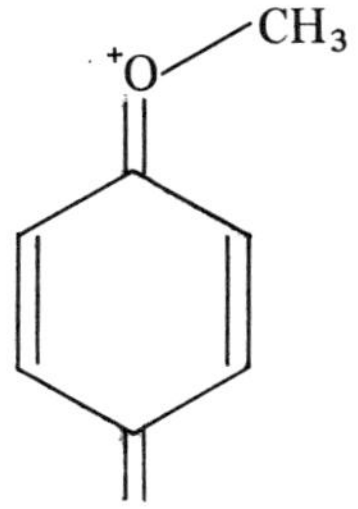

This ion was recalculated at ring-oxygen bond lengths of 1.43, 1.30, and 1.20 Å, and the results were extrapolated to give the ion of minimum energy at 1.36 Å. It was found that substantial stabilization resulted from the use of this shorter bond length. However, the parent molecule, p-methoxytoluene, was also somewhat more stable at this distance, which shows, incidentally, that proper bond distances are not usually given by CNDO. However, extra stabilization that occurs at this distance can be ascribed to the ability of the methoxy group to conjugate in the cation more than in the ground state. If anything, this extra stabilization only makes the fit of the substituted benzyl data even better.. We could certainly get even more stabilization by varying all the bond lengths, but then, of course, we would have to do the same for all the compounds. Such extensive variation is impractical and is unlikely to affect the present results significantly.

We conclude that the CNDO method is valuable for comparing closely related types of systems within which the interaction mechanisms are substantially the same; however, when more than one interaction mechanism undergoes significant variation along a series, CNDO results should be handled with due circumspection.

Acknowledgments

This research was supported in part by U.S. Air Force Office of Scientific Research Grant 68-1364. Many of the calculations were run on a CDC 6400 with time donated by the Computer Center of the University of California, Berkeley.

References and Notes

1. A. Streitwieser, Jr., *Molecular Orbital Theory for Organic Chemists*, Wiley, New York, 1961.
2. M. J. S. Dewar and C. C. Thompson, *J. Am. Chem. Soc.* 87, 4414 (1965).
3. J. Wright, Dissertation, University of California, Berkeley, Calif., 1968.
4. J. A. Pople, D. P. Santry, and G. A. Segal, *J. Chem. Phys.* **43**, S129 (1965); J. A. Pople and G. A. Segal, *ibid.* **43**, S136 (1965), **44**, 3289 (1966).
5. A. Streitwieser, Jr., R. M. Williams, R. H. Jagow, and H. H. Hammond, to be published.
6. A. Streitwieser, Jr., H. A. Hammond, R. H. Jagow, and C. J. Chang, to be published.
7. M. J. S. Dewar and R. J. Sampson, *J. Chem. Soc.* **1956**, 2789; **1957**, 2946, 2952.
8. Preliminary results for the polycyclic arylmethyl systems were presented in Ref. 1; full details are currently in couse of publication elsewhere (Ref. 5).
9. A. Streitwieser, Jr., and J. I. Brauman, *Supplemental Tables of Molecular Orbital Calculations*, Pergamon Press, Oxford 1965.
10. H. A. Hammond, Dissertation, University of California, Berkeley, Calif., 1966.
11. K. B. Wiberg, *J. Am. Chem. Soc.* **90**, 59 (1968).

Chapter V

Local Orbitals and Hybridization

V-1

Semiempirical Orbital Localization and Its Chemical Applications

CARL TRINDLE AND OKTAY SINANOĞLU

1. Introduction

The idea that local bonds connect adjacent atoms in molecules is a very useful concept in chemistry. Definition of properties associated with these bonds often provides a surprisingly quantitative description of molecules of imposing size. Molecular calculations would stand to gain considerable intuitive appeal if they contained some direct recognition of localized bonds. Herein we summarize our semiempirical studies of orbital localization in organic molecules of moderate size: The topics include hybridization, hyperconjugation, intermolecular transfer of local orbitals (LO's), the LO guide to nuclear motion in molecules, and electron correlation in the localized basis.

Local bond functions were naturally employed in the valence-bond theory of molecular electronic structure.[1] But as useful as the qualitative form of the VB theory was, its mathematical drawbacks and its difficulties with magnetic properties and spectra[2] led to an emphasis on the method of molecular orbitals.[3] The latter approach produced orbitals delocalized throughout a molecule, sacrificing the intuitive advantages of local bonds for the conveniences of symmetry.

The definition of equivalent orbitals by Hall and Lennard-Jones[4] showed the possibility of an MO description of bonds. The equivalent orbitals share the permutation properties of bonds, being interconverted by the symmetry operations of the molecular point group, and are highly localized in the bond regions. Equivalent orbitals ϕ_k are obtained from the delocalized symmetry-classed molecular orbitals ψ_ℓ by a unitary transform $T\psi = \phi$. In a few cases symmetry completely determines the matrix T,[4] but in general auxiliary conditions are necessary. Lennard-Jones and Pople[5] suggested that since the ϕ_k are highly localized, an electron occupying one of the ϕ_k would have maximal Coulomb interaction with the electron sharing that orbital. Edmiston and Reudenberg[6] devised an iterative numerical method for implementing this orbital self-energy localization criterion and applied it to a number of small molecules. A steepest-ascents method, which promises an improvement in efficiency, has also been suggested for this localization measure.[7]

Overlap criteria for localization are of interest, since their definition need not involve the troublesome two-electron integrals. Peters[8] demanded that bond orbitals contain no contribution from basis functions on atoms not participating in the bond. Del Re et al.[9] required that hybrids in a local bond orbital have minimum overlap with the hybrids associated with other bonds. Magnasco and Perico[10] defined as a measure of the localization the population of orbitals in local sets of basic functions, appropriate to individual bonds.

Foster and Boys[11] have formed localized orbitals which maximize the sum of separations between the orbitals' centroids of charge. This is the only localization analysis which has been extended to multiconfigurational wave functions.[12]

The promise that bond orbitals may be transferable among molecules, implied by empirical additivity relations for experimental quantities, has been exploited by Adams,[13] who sets conditions for optimizing the transfer.

The direct calculation of local orbitals has been discussed by Edmiston and Reudenberg and in more detail by Gilbert.[14] Peters[15] has defined eigenvalue equations for individual bond orbitals, and a technique for the direct computation of two-center orbitals has been presented by Letcher and Dunning.[16]

2. Semiempirical Localization

A semiempirical exploration of localization is suggested for several reasons. For example, a good deal of experience with localization can be gained without investment of a great amount of computer time. Furthermore, molecules of considerable size are accessible, permitting the study of chemical problems. We will see that few of the conclusions on localization based on semiempirical work would be altered by more rigorous calculation.

The method used in the localization studies discussed here[17] is semiempirical in two senses. First, the original delocalized MO's are determined by the CNDO-2 approximation of Pople and Segal[18] to the Roothaan SCF-MO equations.[19] This method is discussed in Chapter II; but of particular relevance in this work is the invariance of the CNDO expectation values to hybridization of the basis and rotation of the local atomic Cartesian axes. Second, the equations for the localization transform derived by Edmiston and Reudenberg are simplified by approximations to the two-electron integrals parallel to the INDO scheme of Pople et al.[20]

It is of interest to see what errors arise in the semiempirical localization relative to more rigorous calculations. Direct comparison is possible for two molecules, water and ammonia. Edmiston and Reudenberg have applied their method without approximation to minimal-basis-set LCAO-MO-SCF wave functions[21]: Their T matrix for water is shown in Table 1. Set beside the rigorously determined matrix is the semiempirical counterpart.

Table 1. Localization Transforms for H_2O^a

	Core(O)	1p1(O)	1p2(O)	OH′ bond	OH″ bond
$1a_1$	−0.9899	−0.1268	−0.0629	0.0000	−0.0000
$2a_1$	0.0898	−0.4242	−0.5587	0.0000	−0.7071
$3a_1$	0.0898	0.4242	0.5584	0.0001	−0.7071
$1b_2$	0.0442	−0.5588	0.4316	−0.7070	−0.0002
$1b_1$	0.0441	−0.5583	0.4311	0.7075	−0.0001
$1a_1$	−1.0000	0.0000	0.0000	0.0000	0.0000
$2a_1$	0.0000	−0.4044	−0.5800	0.0000	−0.7071
$3a_1$	0.0000	0.4044	0.5800	0.0000	−0.7071
$1b_2$	0.0000	−0.5801	0.4045	−0.7071	0.0000
$1b_1$	0.0000	−0.5801	0.4045	0.7071	0.0000

[a]The unitary matrices **T** which transform the canonical orbitals to localized orbitals are shown. The column at the left contains the symmetry labels of the canonical orbitals, and the top row describes the local orbitals. The top matrix is that of Edmiston and Reudenberg; the lower matrix was determined semiempirically in this work.

Note first that since the CNDO method treats only the valence electrons, the semiempirical matrix is factored by assumption into valence and core blocks. The factoring is only approximately observed in the rigorously determined matrix.

Considerable symmetry obtains in this molecule; in fact, the last two columns in each matrix are fixed by symmetry. The meaningful comparison of the matrices is between the outlined blocks. Errors in the elements of the semiempirical T matrix are less than 0.03, comparable to the errors in the CNDO coefficients. The result gives some indication that errors in the wave function are not badly exaggerated by the approximate localization transform.

3. Localization of Systems Containing Pi Electrons

Edmiston and Reudenberg[6] found in their localization of diatomic molecules that (for example) N_2 contained three equivalent "banana" bonds rather than the σ and π bonds to which we have become accustomed. The question arose whether the CC bonds in ethylene would be localized to banana bonds: If so, the LO description of alkenes may be inconsistent (but not necessarily) with the very fruitful and widely used assumption of σ-π separability.[22] We found that in ethylene the CC bond orbitals which produce a maximum localization measure are of the separated σ and π type. The separation is apparently due to the σ plane definitely established in ethylene but absent in diatomics. An exactly analogous separation occurs in butadiene.

Localization of systems such as the butenes or cyclopropene, which contain an isolated double bond in a nonplanar environment, also produced a local σ-π separation. Therefore, discussion of distinct (and literal) σ and π bonds, even in nonplanar molecules, is consistent with the present localization method.

4. Hybridization and Localization

The definition of hybridization is intimately related to the feasibility of a localized description of a molecule. The idea of hybridization arose in the context of valence-bond theory: Hybridization of s and p orbitals of one atom toward another atom is relevant to a single (localized) valence-bond structure. If that single structure is sufficient for a description of that molecule, the hybridization in the structure is a relevant property of the wave function. To the extent that other valence-bond structures contribute to the wave function, the value of a definition of hybridization is reduced.

Localized orbitals are an ideal means for studying hybridization: If localization is complete, each LO will contain the hybrids appropriate to the bond it describes. But if delocalization is sufficient to affect the hybridization, its presence is immediately disclosed by the LO coefficients.

A LO study of hybridization in highly localizable alkanes, alkenes, amines, alcohols, esters, carbonyl compounds, and carboxylic acids showed considerable deviations from the commonly presumed sp^2 and sp^3 values of the p character, X_p.[23] The deviations may be understood when we consider that hybridization is a response of AO's to a local field: sp^2 and sp^3 values of X_p would be expected only in strictly trigonal or tetrahedral fields. Consider a carbon atom in ethylene: The neighboring carbon may be considered to exert a strong axial perturbation on an otherwise trigonal field. Thus the p character of the CC σ hybrids is shifted from the trigonal value (66.7 per cent) toward the linear value (50.0 per cent), giving a final $X_p = 62.5$ per cent.

A succinct expression of the argument above known as the rule of Bent[24] accounts for the hybridization of C atoms in molecules containing hetero atoms. The rule states that s

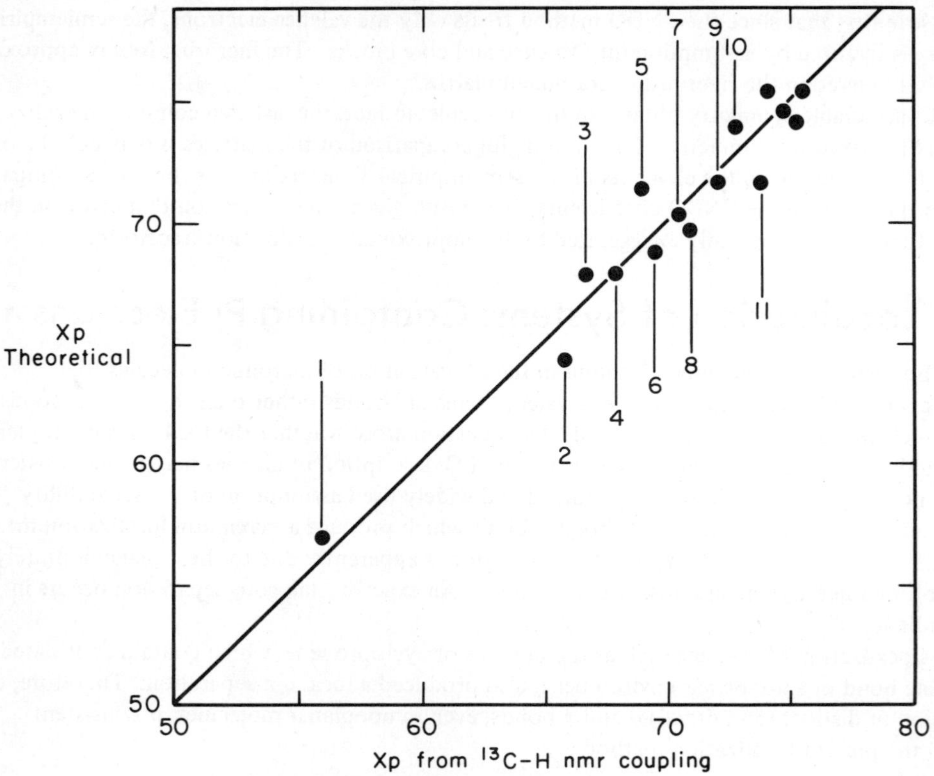

Fig. 1. Comparison of experimental and theoretical predictions of the p character X_p of the carbon hybrid in CH bonds. Experimental X_p is deduced from the NMR coupling constants $J(^{13}CH)$, Ref. 14a, b, and c. (See the text). Circles represent local orbital values of X_p. The straight line would result from perfect agreement of the experimental and theoretical estimates of X_p. Numerical labels correspond to: 1, HCOOH; 2, CH_3CHO; 3, ethyleneimine; 4, cyclopropane; 5, isobutene (methylene); 6, ethylene; 7, trimethylene oxide (α); 8, CH_3OH (methyl); 9, CH_3NH_2 (methyl); 10, CH_3COOH (methyl); 11, trimethylene oxide (β).

character is concentrated in the C hybrid directed toward the most electropositive substituent. The LO analysis of CNDO wave functions and Bent's rule are entirely consistent.[23]

The linear dependence of the constant $J(^{13}CH)$, a measure of the magnetic coupling between a C_{13} nucleus and a proton directly bound to it, on the p character allows a comparison of the LO predictions of X_p with an empirical estimate.[25] Figure 1 illustrates the comparison for CH bonds in a variety of molecules and indicates a substantial agreement between experimental and theoretical values of X_p.

5. Departures from Complete Localization

A localized description of molecules is suggested by the fact that many molecular properties are well represented by additive functions of the atoms and bonds contained in the system. The most striking exceptions to additivity rules are "nonclassical" molecules, for which a single valence-bond structure is a misleading representation. The familiar nonclassical aromatics have highly mobile charge distributions. Since the major departures from additivity are therefore associated with delocalization, it is tempting to ascribe the smaller deviations from additivity observed in classical molecules to "hyperconjugative" delocalization.[26]

The semiempirical localization analysis is a useful way to study hyperconjugation, since departures from complete localization are made apparent. No restrictions on hybridization are imposed, so that the possibility that hybridization changes can account for all effects attributed to hyperconjugation is given full consideration. Moreover, the self-consistent account of electron repulsion avoids the objection that calculations neglecting electron repulsion must overestimate delocalization.

Delocalization becomes significant when "excited" valence-bond structures approach the "ground" valence-bond structure in energy. In benzene the two Kekulé structures are degenerate, which is an extreme example. In butadiene the π orbitals are largely localized into ethylenic CC π-bond orbitals, but the ionic structures are sufficiently important that the off-bond carbon p orbitals have coefficients of ca. 0.10. Delocalization of about the same order is observed in carboxylic acids and amides.

These examples involve electrons of relatively polarizable π bonds and lone pairs. However, hyperconjugation is often invoked where electrons in σ CC and CH bonds must forsake these more stable situations.[27] For example, hyperconjugation is found to be considerable in carbonium ions, according to several Hückel and approximate SCF treatments of charged species,[28] as well as the LO analysis of the $C_4H_7^+$ system.[29]

Another situation where hyperconjugation is often assumed is in the description of coupled π and σ systems, such as the alkylbenzenes.[30] Examination of the LO's for the isomeric butenes shows that hyperconjugation is smaller in this type of molecule than in either carbonium ions or carboxylic acids. However, the hyperconjugation in cyclopropene is comparable with that of the latter systems.

Both hyperconjugation and changes in hybridization are attempts to describe the effects on local bond orbitals of the neighboring bonds. We expect these effects to be exaggerated in strained systems, such as cyclopropene. The expectation is borne out by localization analyses of some highly strained polycyclic hydrocarbons[31]: Cubane, prismane, tetrahedrane, and bicyclo(1,1,1)pentane display quite unorthodox hybridization, which has the effect of billowing out the bond orbitals to reduce their interaction. While hybridization changes are large, hyperconjugation is by no means negligible. Delocalization in these molecules ranges from 3 to 6 per cent of the total charge, while unstrained linear alkanes are less than 1 per cent delocalized. The delocalization provides an explanation for the rather large long-range nuclear magnetic coupling observed in polycyclic molecules, most notably in bicyclo(1,1,1)pentane.[32]

6. Synthesis of Wave Functions for Large Molecules

An attractive application of localized orbitals is the construction of wave functions for large molecules by assembly of LO's representing the bonds, cores, and lone pairs of the molecule. The prerequisite for any success of this technique is, of course, a high degree of transferability of the local orbitals. There are several cases in which transferability cannot be expected. The highly polarizable lone pairs are hardly ever transferable per se, although sets of LO's representing functional groups can sometimes be transferred. Essentially delocalized π systems do not admit bond by bond transfer. Molecules with unusual strain, leading to unorthodox hybridization and hyperconjugative delocalization, cannot be well represented by transferred orbitals. However, there remains a large number of classical molecules which might be expected to be composed of transferable local orbitals.

An exploratory study of LO transfer followed this procedure[33]: LO's suitable to bonds (or groups, for regions of the molecule containing atoms with lone pairs) were oriented in directions consistent with the geometry of a molecule to be synthesized. The LO's were then orthogonalized by the Löwdin method,[34] which preserves symmetry-dictated equivalences among the orbitals. The SCF Hamiltonian matrix was constructed, using the orthonormal transferred LO's to define the Coulomb and exchange operator contributions, with the CNDO approximations determining the remainder. A single diagonalization of this Hamiltonian matrix produced the "synthesized" MO energies and coefficients, bond-order matrix, atomic charges, dipole moment, and total energy. Data for *cis*-but-2-ene and isopropanol are presented in Table 2. The results are compared with the direct CNDO-SCF treatment. If transfer were

Table 2. Transfer of Local Orbitals: Construction of Wave Function

	CNDO[a]	Transfer[b]
Isopropanol:		
$(C_{(2)}H''')_2C_{(1)}H''OH'$		
Dipole moment, D	2.14	2.11
Atomic charges (electrons) on atom:		
0	6.25	6.244
$C_{(1)}$	3.852	3.823
$C_{(2)}$	4.025	4.036
H'''	0.991	0.996
H''	1.016	1.006
H'	0.885	0.877
E_{el}, a.u.	-126.4329	-126.2595
$\epsilon_{orbital}$, a.u.	-1.7709	-1.7703
	-1.3984	-1.3875
	-1.2761	-1.2674
	-1.0657	-1.0507
	-0.9012	-0.8875
	-0.8690	-0.8592
	-0.8083	-0.7948
	-0.7012	-0.7045
	-0.7004	-0.7029
	-0.6290	-0.6266
	-0.5827	-0.5811
	-0.5465	-0.5430
	-0.5264	-0.5215
cis-Z-Butene:		
$H_3CHC = C^{(1)}H^{(1)}C^{(2)}H_3{}^{(2)}$		
Dipole moment, D	0.06	0.07
Atomic charges (electrons) on atom:		
$C^{(1)}$	4.006	4.023
$C^{(2)}$	4.020	4.020
$H^{(1)}$	1.001	0.991
$H^{(2)}$	0.990	0.989
E_{el}, a.u.	-101.4139	-101.5506
$\epsilon_{orbital}$, a.u.	-1.6440	-1.6481
	-1.3097	-1.3065
	-1.2068	-1.2032
	-0.9557	-0.9511
	-0.8973	-0.8905
	-0.8468	-0.8397
	-0.7720	-0.7688
	-0.7430	-0.7374
	-0.6239	-0.6249
	-0.6009	-0.6076
	-0.5443	-0.5575
	-0.4891	-0.5046

[a]Values determined by a direct CNDO calculation. [b]Values for isopropanol obtained from the wave function, assembled from ethane CH LO's and CC bond LO's and methanol carbon-oxygen and hydroxyl LO's.

exact, the synthesized values would agree with the SCF results. Although this goal was not quite achieved, the synthesized MO's are a considerable improvement over the extended Hückel MO's and require only slightly more effort (i.e., computer time) to obtain. Despite this measure of success, our view is one of guarded pessimism, in view of the many exceptions to successful transfer.

7. Local Orbital Guide to Nuclear Motion

The topics treated to this point have only indirect application to the problems which confront the practicing chemist. In this section we discuss a use of the local orbital analysis which can provide insights into many chemical reactions.

The Hoffmann-Woodward guide to the low-energy paths for nuclear motion in molecules provide an invaluable means of predicting the course of chemical reactions.[35] Their methods

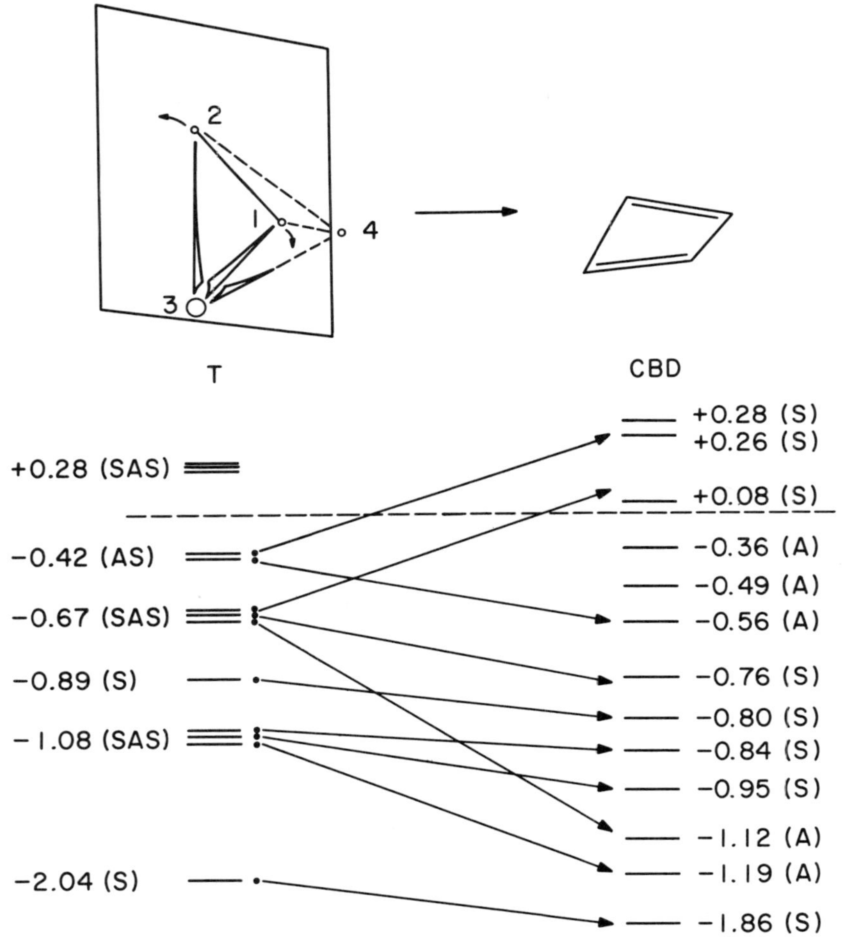

Fig. 2. Hoffmann-Woodward analysis of the rearrangement of tetrahedrane (T) to cyclobutadiene (CBD). The motion of C_1 and C_2 in one of the planes of symmetry, as shown, causes two occupied orbitals of T to correlate with two virtual orbitals of CBD. For that reason, the process is thermally forbidden. The orbital energies shown are obtained from a CNDO-MO computation. The labels S and A associated with the orbitals refer to the symmetry of the orbital relative to reflection in the plane: S, symmetric, A, antisymmetric.

have been discussed extensively elsewhere,[36] but briefly, their predictions are based on the correlation of the orbitals occupied in some geometry A with the orbitals of another geometry B. If each occupied orbital of A passes into an occupied orbital of B, the motion A → B requires little energy and can occur under ordinary thermal conditions. If no correspondence between the sets of occupied orbitals of A and B exists, A can form B only by free-radical or photochemical pathways.

Application of the Hoffmann-Woodward approach and arguments analogous to it becomes more difficult in large three-dimensional systems of low symmetry. A formulation of the symmetry arguments in the language of localized molecular orbitals evades this difficulty.[29] We rely on the fact that, if a set of MO's ϕ_a corresponds to another set ϕ_b, one may form arbitrary linear combinations within each set without destroying the correspondence of the sets. The energy-localized orbitals are particularly suitable linear combinations in this context because of the degree of transferability of many local orbitals. The transferability makes the correspondence of individual orbitals very easy to see; for example, a CH bond orbital in the original geometry A usually has an obvious counterpart in other geometries, B_1, B_2, etc.

Signs are associated with each local orbital by the unitary localization transform **T**. These signs are meaningful if a convention is employed to fix the signs of the canonical molecular orbitals. (For example, we could demand that the first LCAO coefficient not zero by

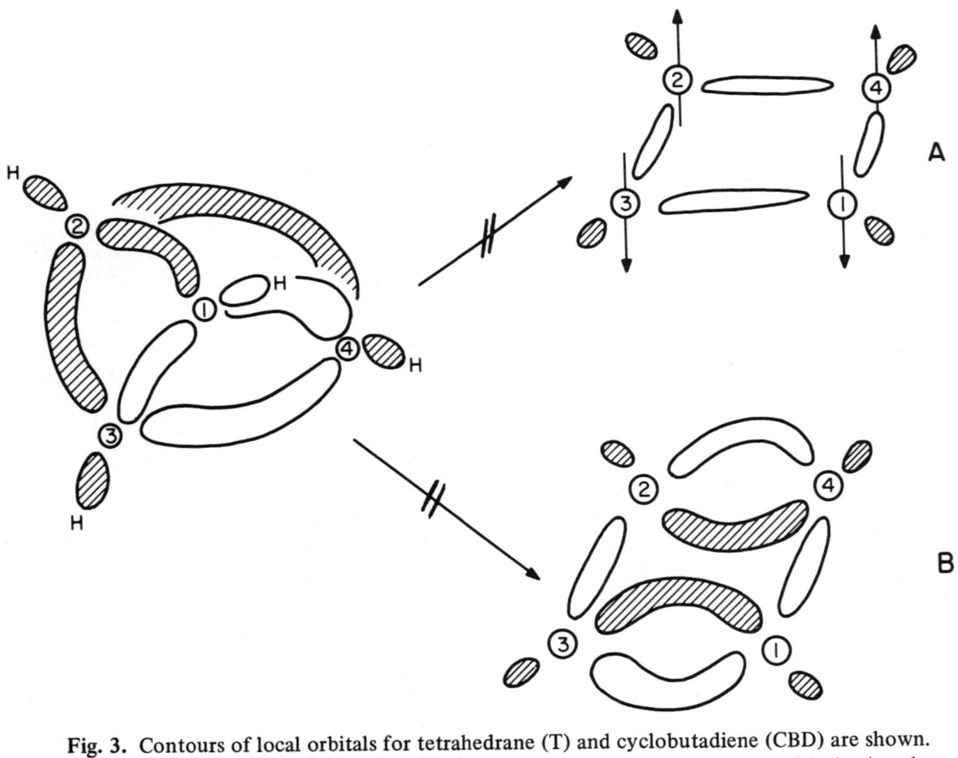

Fig. 3. Contours of local orbitals for tetrahedrane (T) and cyclobutadiene (CBD) are shown. Shaded contours imply negatively signed orbitals, while open contours indicate positively signed orbitals. A and B are nearly equivalently localized representations of CBD. In A the σ and π orbitals were localized separately; in B the σ and π CC bond orbitals were combined. B indicates more clearly that CBD has six negative and four positive orbitals. T, however, has seven negative orbitals, three positive orbitals. In order for T to pass to CBD, a bond orbital must change sign, producing a node in a region of strong bonding. Therefore, the process is not thermally allowed, requiring free radical at photochemical pathways.

symmetry be positive in each MO, for some consistent numbering of the atomic orbitals.[37]) The use of the signs is illustrated by tetrahedrane, which can be analyzed with equal ease by the Hoffmann-Woodward MO classification or by the LO method. The rearrangement of tetrahedrane to cyclobutadiene is presumed to proceed by a motion of two carbons in one of the planes of symmetry. The correlation diagram between MO's of the two molecules may be constructed with the aid of the symmetry properties of the MO's relative to that plane (Fig. 2) and shows that the process is thermally forbidden. The sketches of the LO's, identified by their signs (Fig. 3), show that the transformation tetrahedrane ⟶ cyclobutadiene must change the sign of one of the CC bond orbitals. Since in order to change the sign a node must be formed in a region of strong bonding, free-radical or photochemical pathways are indicated. The conclusion agrees with the Hoffmann-Woodward analysis, but does not rest on any assumption on the nuclear motion or depend on an explicit recognition of symmetry.

The advantages of the LO method are illustrated by a reaction without helpful symmetry, the rotation of *anti*-allylcarbinyl carbonium ion to *syn*-allylcarbinyl ion and the closure of the latter to the cyclobutyl carbonium ion. As the sketches of the LO's in Fig. 4 show, the orbitals of the *anti* ion bear a strong resemblance to the respective orbitals of the *syn* ion. The only major difference, the orientation of the LO's, follows the nuclear motion closely.

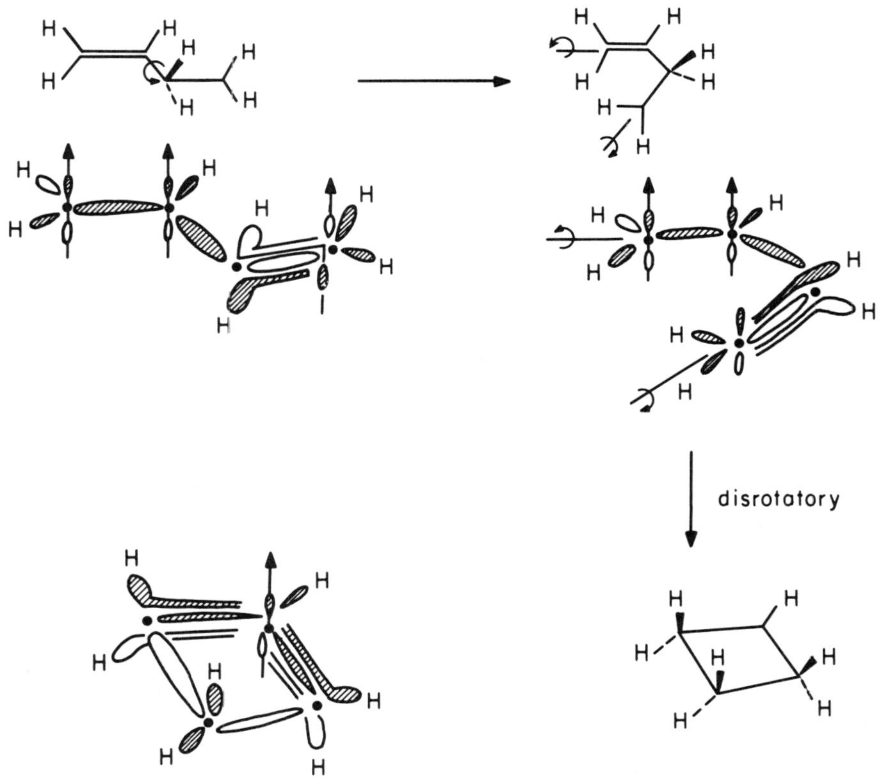

Fig. 4. LO guide to allowed nuclear motion. Diagrams of the local orbitals in *anti*-allylcarbinyl cation (a), *syn*-allylcarbinyl cation (b), and cyclobutyl cation (c) are shown. Open contours denote LO's of positive sign, shaded contours indicate negatively signed LO's. The rotation of (a) about the central CC bond produces (b) with no change in sign of the LO's, indicating that the process is allowed. In order for (b) to close to (c) with no change in sign of the LO's, the motion of the terminal CH_2 groups must be disrotatory.

Given the *syn*-allylcarbinyl ion LO's, it is evident that the ring closure must be disrotatory, in order that the cyclobutyl LO's be achieved.[38]

Parallel analyses have been employed in a study of nuclear motion and other configurations in the $C_4H_7^+$ system. On the basis of this experience, we believe that the LO method would be applicable wherever the Hoffmann-Woodward approach is appropriate, and would be easier to use in asymmetric systems.

8. Electron Correlation and Localized Molecular Orbitals

An important reason for going from the MO description to the LO description is to make the connection with the additive and transferable bond energies of thermochemistry. Clearly, to recover the bond-energies picture one needs the localized orbitals. However, it is also necessary to see how correlation energy transforms to the LO description, as it contributes very significantly to binding energy. We shall restrict ourselves to only closed-shell saturated molecules.

In the Hartree-Fock MO description the correlation energy of a ground state has been shown by Sinanoğlu, to a good approximation, to be the sum of all MO pair correlations. The theory (MET) in the MO description is discussed in Chapter VII-6. It has been shown[39] that when the molecular orbitals are transformed into LO's, the sum of MO pair correlations becomes a sum over all LO-LO′ pair correlations:

$$E_{corr} = \sum_{\rho,\nu}^{N} \epsilon_{\rho\nu}$$

where ρ and ν are the occupied LO spin orbitals, of which there are N, and there are $N(N-1)/2$ terms in the sum. This expression contains both the correlations in doubly occupied LO's and the ones between different LO's, on an equal footing. It is expected that the doubly occupied pairs should be insensitive to the details of environmental charge distributions and thus be transferable from one saturated molecule to another. The next question is, however, how significant are the interorbital, LO-LO′ correlations. It has been customary to assume that, especially with the local orbital description, only the correlations within bonds would be important. The theory (MET) has shown, however, that the interbond correlations can contribute more than the sum of bond correlations (see Chapter V-2). Thus the earlier intuitive pictures and methods based on them using the products of only the bond wave functions (geminals) are invalid. The article (Chapter V-2) obtains from MET the LO pair correlations for tetrahedral hybrids in the neon atom, showing how large the correlations between separately localized orbitals can be. Similar results are found from molecules such as CH_4 using semi-empirical methods based on MET.

An extensive semiempirical molecular correlation energy method has been developed over the past few years by Pamuk and Sinanoğlu. The method depends on expanding the MO or LO pair correlations in the LCAO form.

$$\epsilon_{\kappa\lambda} = \left\langle B(\kappa_1\lambda_2) \left| \frac{1}{r_{12}} \right| \hat{u}_{\kappa\lambda}(12) \right\rangle$$

$$= \sum_{pqrs} C^{*}_{\kappa p} C^{*}_{\lambda q} C_{\kappa r} C_{\lambda s} \left\langle B(pq) \left| \frac{1}{r_{12}} \right| \hat{u}_{rs} \right\rangle$$

$$= \sum_{pqrs} C^{+}_{\kappa p} C^{+}_{rq} C_{\kappa r} C_{\lambda s} \, \epsilon^{ABCD}_{pqrs}$$

In this equation B is the antisymmetrizer and $\hat{\mu}_{\kappa\lambda}$ is the correlation function associated with orbitals κ and λ. The transformation properties of $\hat{\mu}_{\kappa\lambda}$ are the same as those of the antisymmetrized orbital product $B(\kappa\lambda)$. The p, q, r, and s refer to AO's centered on atoms A, B, C, and D, respectively.

To use the above expression it is necessary to introduce further approximations. A convenient form is obtained by assuming the atomic orbitals to be an orthogonalized AO basis. The coefficients are obtained from an approximate MO theory such as CNDO-MO. All the one-centered atomic pairs are retained and are identified with pair energies obtained by MET from first-row atoms. In the Pamuk-Sinanoğlu version of this method, ZDO-like assumptions are made on the correlation terms, and all terms except one-center and two-center correlations are neglected. The two-center pair terms are related by semiempirical formulas to overlap integrals and one-center atomic pairs. We shall not discuss this method further here.

In a different version of the semiempirical theory, Trindle reduces the two-center pairs to one-center terms by the Reudenberg[40] expansion of orbitals on a center A in terms of orbitals on another center B:

$$U^A = \sum_{W} S_{UW} W^B$$

Here W^B extends over the valence shell of atom B, assuring that the approximation is rotationally invariant. In the evaluation of $< B(p^A q^B) \mid 1/r_{12} \mid \hat{\mu}(r^A s^B) > = \epsilon^{ABAB}_{pqrs}$, the view is taken that B(pq) and $\hat{\mu}_{rs}$ may be independently expanded on either centers A or B. The four terms resulting are averaged. In the light of the work of Cook et al.[41] on the effects of orthogonalizing the basis on the magnitudes of two-election integrals, one-center integrals are increased by 10 per cent while two-center integrals are reduced by 10 per cent.[42]

Preliminary results indicate a high degree of transferability of the intra-LO correlations. The CH bond correlation is identical in CH_4 and C_2H_6 and is only very slightly less in C_2H_4. Furthermore, the intra-lone-pair correlation in the series NH_3, H_2O, HF is constant at −0.58 eV.

The expectation of transferability of local orbital correlations seems to be borne out. Also, the *inter*orbital correlations again make an important contribution. The fraction interbond versus total valence-shell correlation increases in the AH_n series until in HF the interorbital correlation accounts for 70 per cent of the valence-shell correlation. More rigorous calculations on tetrahedrally hybridized orbitals in Ne yield a value of 60 per cent interorbital correlation, 40 per cent intraorbital correlation (Chapter V-2). In favorable cases, transfer is still possible. The CH–CH′ interactions in CH_4 and C_2H_6 are identical, while the interaction between CH bonds on different carbons of C_2H_6 is very small (<0.005 eV in magnitude).

One should note that the discussion above has used the total pair correlations of ground states. It is known in nonclosed shells, more generally, that there are three types of correlation energy[43]: "internal," "semi-internal," and "all-external" correlations. A closed-shell atom such as Ne contains only the "all-external" correlations. The calculations above use the untested assumption that only the "all-external" effect is predominant in saturated molecules. Trindle's calculations suggest that this assumption is accurate to roughly 0.3 eV for AH_n molecules and the series C_2H_2, C_2H_4, and C_2H_6. But for the series N_2, CO, and BF, the error in the prediction of the correlation energy is +1.7, +1.2, and +0.6 eV, respectively. If this discrepancy is to be ascribed to nondynamical contributions to the correlation, the questions remain: (a) can it be incorporated into an approximate calculation in a simple way, and (b) can it be represented by local terms? These problems need further study.

References and Notes

1. L. Pauling, *The Nature of the Chemical Bond*, Cornell University Press, Ithaca, N.Y., 3rd ed., 1960.
2. C. Sandorfy, *Electronic Spectra and Quantum Chemistry*, Prentice-Hall, Englewood Cliffs, N.J., 1964.
3. See J. C. Slater, *Quantum Theory of Molecules and Solids*, Vol. 1, McGraw-Hill, New York, 1963.
4. G. G. Hall and J. E. Lennard-Jones, *Proc. Roy. Soc. (London)* **A202**, 155 (1950); *ibid.* **A205**, 357 (1951); G. G. Hall, *Proc. Roy. Soc. (London)* **A202**, 336 (1950), *ibid.* **A205**, 541 (1951).
5. J. E. Lennard-Jones and J. A. Pople, *Proc. Roy. Soc. (London)* **A202**, 446 (1950); *ibid.* **A210**, 190 (1951).
6. C. Edmiston and K. Reudenberg, *Rev. Mod. Phys.* **35**, 457 (1963); *J. Chem. Phys.* **43**, 597 (1965).
7. W. J. Taylor, *J. Chem. Phys.* **48**, 2385 (1968).
8. D. Peters, *J. Chem. Soc.* **1963**, 2003, 2015, 4017.
9. G. Del Re, U. Esposito, and M. Carpentieri, *Theoret. Chim. Acta* **6**, 36 (1968).
10. V. Magnasco and A. Perico, *J. Chem. Phys.* **47**, 971 (1967); *ibid.* **48**, 800 (1968).
11. J. M. Foster and S. F. Boys, *Rev. Mod. Phys.* **32**, 300 (1960).
12. S. Boys, in "*Quantum Theory of Atoms, Molecules and the Solid State*, P.-O. Löwdin, ed., Academic Press, New York, 1966, p. 253.
13. W. Adams, *J. Chem. Phys.* **37**, 2009 (1962).
14. T. Gilbert, in *Molecular Orbitals in Chemistry, Physics and Biology: A Tribute to R. S. Mulliken*, B. Pullman and P. -O. Löwdin, eds., Academic Press, New York, 1964, p. 405.
15. D. Peters, *J. Chem. Phys.* **46**, 4427 (1967).
16. J. H. Letcher and T. H. Dunning, *J. Chem. Phys.* **48**, 4538 (1968).
17. C. Trindle and O. Sinanoğlu, *J. Chem. Phys.* **49**, 65 (1968).
18. J. A. Pople and G. A. Segal, *J. Chem. Phys.* **43**, S136 (1965); *ibid.* **44**, 3289 (1966): Quantum Chemistry Program Exchange No. 91, Department of Chemistry, Indiana University, Bloomington, Ind.
19. C. C. J. Roothaan, *Rev. Mod. Phys.* **23**, 69 (1951).
20. J. A. Pople, D. L. Beveridge, and P. A. Dobosh, *J. Chem. Phys.* **47**, 2026 (1967).
21. C. Edmiston and K. Reudenberg, *Molecular Orbital Theory of Atoms, Molecules and the Solid State*, P. -O. Lowdin, ed., Academic Press, New York, 1966, p. 263.
22. P. G. Lykos and R. G. Parr, *J. Chem. Phys.* **24**, 1166 (1956).
23. C. Trindle and O. Sinanoğlu, Localization and Hybridization in Molecular Orbitals, *J. Am. Chem. Soc.* **91**, 853 (1969).
24. H. A. Bent, *Chem. Rev.* **61**, 275 (1961).
25. C. Juan and H. S. Gutowsky, *J. Chem. Phys.* **37**, 2198 (1962).
26. M. J. S. Dewar, *Hyperconjugation*, Ronald Press, New York, 1962.
27. Ref. 26, Chaps. 5 and 6 and references therein.
28. R. Hoffmann, *J. Chem. Phys.* **40**, 2480 (1964).
29. C. Trindle and O. Sinanoğlu, Local Orbital Guide to Interconversions of $C_4H_7^+$ Ions, *J. Am. Chem. Soc.*, July 1969.
30. J. W. Baker, *Hyperconjugation*, Oxford University Press, New York, 1952.
31. C. Trindle and O. Sinanoğlu, Hybridization and Hyperconjugation in Polycyclic Hydrocarbons (to be published).
32. K. B. Wiberg, G. M. Lampman, R. P. Ciula, D. S. Connor, P. Schertler, and J. Low; *Tetrahedron* **21**, 2749 (1965); K. Wiberg and D. S. Connor, *J. Am. Chem. Soc.* 88, 4437 (1966).
33. C. Trindle, unpublished work.
34. P. -O. Löwdin, *J. Chem. Phys.* **18**, 365 (1950).
35. R. B. Woodward and R. Hoffmann, *J. Am. Chem. Soc.* **87**, 395, 2511 (1965); R. Hoffmann and R. B. Woodward, *J. Am. Chem. Soc.* 87, 2046, 4388, 4389 (1965).
36. R. Hoffmann and R. B. Woodward, *Accounts Chem. Res.* 1, 17 (1968); J. J. Vollmer and K. L. Servis, *J. Chem. Ed.* **45**, 214 (1968).
37. Ref. 29 provides a more detailed discussion of the problem.
38. Consistent with an extensive CNDO study of cyclobutyl ion opening, by K. B. Wiberg and G. Szeimies (private communications). Extensive bibliographies on orbital localization prior to 1965 are given by Gilbert (Ref. 14) and O. Sinanoğlu and D. T. Tuan, *Ann. Rev. Phys. Chem.* **15**, 251 (1964).
39. O. Sinanoğlu, *J. Chem. Phys.* **37**, 191 (1962).
40. K. Reudenberg *J. Chem. Phys.* **19**, 1433 (1951). [See P. J. A. Ruttink for discussions of rotational invariance of this approximation. *Theoret. Chim. Acta* **6**, 83 (1966).]
41. D. B. Cook, P. C. Hollis, and R. McWeeny, *Mol. Phys.* **13**, 454 (1967).
42. The analogy of the ϵ_{pqrs} with the Coulomb and exchange integrals studied in Ref. 41 is not firmly established, although intuitively appealing.
43. O. Sinanoğlu and I. Öksüz, *Phys. Rev. Letters* **21**, 507 (1968).

V-2 Reprinted from *Chemical Physics Letters* **1**, 699 (1968)

CORRELATIONS BETWEEN TETRAHEDRALLY LOCALIZED ORBITALS

Oktay SINANOĞLU and Bolesh SKUTNIK *
Sterling Chemistry Laboratory, Yale University, New Haven, Connecticut, USA

Received 11 March 1968

It is shown that correlations between neighboring tetrahedral orbitals contribute about twice as much to the valence shell correlation energies of neon and methane, as the correlations within the four doubly occupied ("bonds") tetrahedral pairs. "Non-bonded attractions" are thus not negligible but may be an important part of the correlation energies of saturated molecules.

The empirical bond additivity of thermochemistry leads one to expect that the valence shell correlation energies of molecules like CH_4, C_2H_6, ... would consist mainly of bond correlations. This assumption is built into "geminal" or "separated-pair" type approaches [1]. To consider that correlation energy resides mainly in doubly occupied orbitals is an assumption also in some recent empirical discussions of correlation [2]. On this basis an N-electron atom or molecule contains (N 2)-"geminal" correlations with additional effects as small corrections.

In the rigorous wave function and energy of an N-electron ground state on the other hand, written out in a cluster form [3], intra- and inter-orbital correlations occur on pretty much the same footing. It has been found first by perturbation theory [4], next by a more general, non-perturbative treatment [5,3a] (referred to as "M.E.T.", "many electron theory of atoms and molecules") that the correlation energy of a ground state is given to a good approximation as the sum of $\frac{1}{2}N(N-1)$ spin-orbital pair correlations each of whoch may be evaluated separately non-empirically or by a combination of computational and semi-empirical ways [6]. In refs. [3], [4] and [5], three types of pair correlations, related to each other via unitary transformations, were introduced: (i) Reducible ("B(ij)-type"), i.e. Hartree-Fock spin-orbital pairs, ϵ_{ij}, (ii) "Irreducible-pairs" $\epsilon_K^{\mathrm{irr}}$, and (iii) localized orbital (LO)-pairs, $\epsilon_{\rho\nu}$. Thus

$$E_{\mathrm{CORR}}(N\text{-electronic ground state}) \cong \sum_{i>j}^{N} \epsilon_{ij} = \sum_{K}^{\frac{1}{2}N(N-1)} \epsilon_K^{\mathrm{irr.}} = \sum_{\rho>\nu}^{N} \epsilon_{\rho\nu} . \qquad (1)$$

The significant magnitudes of inter-orbital correlations were first noted in [6] where sizable 2s-2p correlations, ϵ(2s-2p), were found in first row atoms, also in He-He with $\epsilon(1\sigma_g^2) \approx \epsilon(1\sigma_u^2) \approx \epsilon(1\sigma_g^\alpha 1\sigma_u^\beta)$ [7], and in the computations of Kelly [8], Grimaldi [9], Davidson [10] and others. The sizable role of inter-orbital correlations remains even after transformation to the LO-description, $\epsilon_{\rho\nu}$'s [11].

In the present note, the correlations between tetrahedral orbitals are evaluated. In M.E.T. the valence shell E_{CORR} of neon atom and of CH_4 may be writen rigorously as the sum of four "bond" correlations ("intra-bond") plus the correlations between the bonds ("inter-bod") (cf. also [4a]). It is shown below that the sum of "inter-bond" correlations contribute roughly two times the four (sp^3 or C-H) "bond" correlations in Ne and CH_4.

Let η_1, η_3, η_5, η_7 be the tetrahedral, (sp^3)-hybrid spin-LO's of neon with α-spins. The η_2, η_4, η_6, η_8 have β-spins. The tetrahedron is oriented such that η_1 and η_2 are along the Z-axis and η_3 and η_4 (the same spatial LO) are in the XZ-plane. The LO-transformation matrix t is

* Present address: Department of Chemistry, Brandeis University, Waltham, Massachusetts, USA.

April 1968

$$\begin{pmatrix}\eta_1\\ \eta_3\\ \eta_5\\ \eta_7\end{pmatrix} = \frac{1}{2}\begin{pmatrix}1 & \sqrt{3} & 0 & 0\\ 1 & -\frac{1}{\sqrt{3}} & 2\sqrt{\frac{2}{3}} & 0\\ 1 & -\frac{1}{\sqrt{3}} & -\sqrt{\frac{2}{3}} & \sqrt{2}\\ 1 & -\frac{1}{\sqrt{3}} & -\sqrt{\frac{2}{3}} & -\sqrt{2}\end{pmatrix}\begin{pmatrix}2s\alpha\\ 2p_z\alpha\\ 2p_x\alpha\\ 2p_y\alpha\end{pmatrix} \tag{2}$$

The pair functions $\hat{u}_{ij}$ are then transformed by the direct product matrix $T = t \times t$ (cf. [5b,4a]). Each LO-pair correlation $\epsilon_{\eta_i\eta_j}$ may then be expressed in terms of the "reducible" pairs ϵ_{kl} and the "cross pairs" $\epsilon_{kl;mn}$ [12], where

$$\epsilon_{kl;mn} = \langle B(kl) \frac{1}{r_{12}} \hat{u}_{mn} \; . \tag{3}$$

We have for an $(LO)^2$-correlation,

$$\epsilon_{\eta_1\eta_2} = \frac{1}{16}\Big[\epsilon(2s^2) + 6\epsilon(2s\overline{2p}_z) + 9\epsilon(2p_z^2) + 6\epsilon(2s^2;2p_z^2) + 6\epsilon(2s\overline{2p}_z;2p_z\overline{2s})\Big] \tag{4}$$

where ($\overline{2p}_z \equiv 2p_z\beta$; $2p_z \equiv 2p_z\alpha$). For an inter-orbital correlation $(LO)_\alpha$-$(LO')_\beta$:

$$\epsilon_{\eta_1\eta_4} = \frac{1}{16}\Big[\epsilon(2s^2) + 6\epsilon(2s\overline{2p}_z) + \epsilon(2p_z^2) + 8\epsilon(2p_z\overline{2p}_x) - 2\epsilon(2s^2;2p_z^2) - 2\epsilon(2s\overline{2p}_z;2p_z\overline{2s})\Big] \tag{5}$$

and for a $(LO)_\alpha$-$(LO')_\alpha$:

$$\epsilon_{\eta_1\eta_3} = \frac{1}{2}\Big[\epsilon(\overline{2s}\overline{2p}_z) + \epsilon(\overline{2p}_z\overline{2p}_x)\Big] . \tag{6}$$

To get just the total E_{CORR}, we would not need the "cross-pairs". However here we ask for the individual $\epsilon_{\eta_i\eta_j}$, hence we must know the values of the $\epsilon_{kl;mn}$ as well as the ϵ_{ij}. All "cross-pairs", except $\epsilon(2s^2;2p_z^2)$ are evaluated by group theory in terms of "irreducible" (symmetry) pairs ϵ_K^{irr}. One obtains

$$\epsilon_{\eta_1\eta_2} = \frac{1}{16}\Big[\epsilon(2s^2) + 6\epsilon(2s2p;{}^1P) + 3\epsilon(2p^2;{}^1S) + 6\epsilon(2p^2;{}^1D) + \frac{6}{\sqrt{3}}\epsilon(2s^2\,{}^1S;2p^2\,{}^1S)\Big] \tag{7a}$$

$$\epsilon_{\eta_1\eta_4} = \frac{1}{16}\Big[\epsilon(2s^2) - 2\epsilon(2s2p;{}^1P) + 4\epsilon(2s2p;{}^3P) + \frac{1}{3}\epsilon(2p^2;{}^1S) + \frac{14}{3}\epsilon(2p^2;{}^1D) + 4\epsilon(2p^2;{}^3P) - \frac{2}{\sqrt{3}}\epsilon(2s^2\,{}^1S;2p^2\,{}^1S)\Big] \tag{7b}$$

$$\epsilon_{\eta_2\eta_3} = \frac{1}{2}\Big[\epsilon(2s2p;{}^3P) + \epsilon(2p^2;{}^3P)\Big] . \tag{7c}$$

The total $(2s^22p^6)$ L-shell correlation energy of Ne in the three different descriptions [5,4] is:
In "reducible pairs" (s,x,y,z form):

$$E_{CORR}(\text{L-shell;Ne}) = \epsilon(2s^2) + 6\epsilon(\overline{2s}\overline{2p}_z) + 6\epsilon(2s\overline{2p}_z) + 3\epsilon(2p_x^2) + 6\epsilon(2p_x2p_y) + 6\epsilon(p_x\overline{p}_y) . \tag{8}$$

In "irreducible pairs":

$$E_{CORR}(\text{L-shell;Ne}) = \epsilon(2s^2) + 3\epsilon(2s2p;{}^1P) + 9\epsilon(2s2p;{}^3P) + 9\epsilon(2p^2;{}^3P) + 5\epsilon(2p^2;{}^1D) + \epsilon(2p^2;{}^1S) . \tag{9}$$

In tetrahedral "LO-pairs":

$$E_{CORR}(\text{L-shell;Ne}) = 4\epsilon_{\eta_1\eta_2} + 12\epsilon_{\eta_1\eta_4} + 12\epsilon_{\eta_1\eta_3} . \tag{10}$$

Also

$$E_{CORR}(\text{Ne}) = \epsilon(1s^2) + 2\epsilon(1s^2 \to 2s) + 6\epsilon(1s^2 \to 2p) + E_{CORR}(\text{L-shell}) . \tag{11}$$

Very similar equations apply to CH_4 with s, p_x, p_y, p_z replaced by a, t_x, t_y, t_z MO's, and irreducible pairs defined with respect to the point group T_d instead of the rotation group O(3).

We have for Ne and CH_4:

$$E_{CORR}(\text{L-shell}) = 4\epsilon_{\eta_1\eta_2} \times (1 + \text{Rt}). \tag{12}$$

The $4\epsilon_{\eta_1\eta_4}$ would be just the "bonds only" correlation. Rt is the ratio indicating the deviation from this concept, the ratio of the total inter-"bond" correlations to the "bonds" part:

$$\text{Rt} \equiv \frac{E_{CORR}^{inter}}{E_{CORR}^{intra}} = \frac{3(\epsilon_{\eta_1\eta_4} + \epsilon_{\eta_1\eta_3})}{\epsilon_{\eta_1\eta_2}}. \tag{13}$$

The ϵ_{ij} and ϵ_K^{irr} for first row atoms have been evaluated using a non-closed shell version of M.E.T. by a combination of computation and experimental E_{CORR}'s on $1s^2 2s^n 2p^m$ configurations [13]. For Ne the values are (in eV): $\epsilon(2s^2) = -0.275$; $\epsilon(2s2p,{}^3P) = -0.123$; $\epsilon(2s2p;{}^1P) = -0.585$; $\epsilon(2p^2;{}^3P) = -0.174$; $\epsilon(2p^2;{}^1D) = -0.545$; $\epsilon(2p^2;{}^1S) = -1.202$. Also, $\epsilon(1s^2 \to 2p) = -0.102$; $\epsilon(1s^2 \to 2s) = -0.075$, and $\epsilon(1s^2) = -1.272$ eV. These values not only reproduce the ground state E_{CORR} of Ne, but also those of many Ne-ions and excited states after the addition of non-closed shell effects [12,13].

In eqs. (7), we also need the $\epsilon(2s^2;2p_z^2)$ cross term. The individual LO-pair $\epsilon_{\eta_i\eta_j}$ values are shown in table 1 for two estimates of $\epsilon(2s^2;2p_z)$. To second order in energy one has $\epsilon^{(2)}(2s^2;2p_z^2) = \epsilon^{(2)}(2p^2;2s_z^2)$. Taking this as valid for the "exact pairs", and neglecting the small dynamical $\hat{u}_{2s^2}$ [6], we have

$$\text{(A)} \quad \epsilon(2s^2;2p_z^2) \approx \epsilon(2p_z^2;2s^2)_{\text{non-dyn.}} = 0. \tag{14}$$

On the other hand, a rough estimate, (likely to be on the large side) is obtained as

$$\text{(B)} \quad \frac{1}{\sqrt{3}}\epsilon(2s^2;2p^2\,{}^1S) = \epsilon(2s^2;2p_z^2) \approx -0.313 \text{ eV}. \tag{15}$$

This estimate is based on applying the Cauchy-Schwartz inequality twice to $\langle B(2s^2)\, g_{12} | \hat{u}_{2p_z^2}\rangle$. Using also the rough rule $\langle \hat{u}_{ij} | \hat{u}_{ij}\rangle \approx \epsilon_{ij}/E_{H.F.}$, one gets

$$\langle B(2s^2)\, g_{12}\, \hat{u}_{2p_z^2}\rangle \approx J_{2s^2} \sqrt{|\epsilon(2p_z^2)|/E_{H.F.}} \tag{16}$$

where J_{2s^2} is the 2s-2s Coulomb integral. Three different equations of this type are obtained yielding 0.411, 0.298 and 0.250 eV with the geometric mean 0.313 eV in eq. (15).

Table 1 indicates that the $\epsilon_{\eta_i\eta_j}$ and the Rt ("inter-bonds"/"intra-bonds" ratio) are not affected in their general features by (A) or (B). The inter-LO correlations contribute to the valence shell E_{CORR} of Ne, as much as 1.8 to 2.2 times the 4-tetrahedrally localized pairs.

Does the same drastic picture hold true in CH_4? How large are the non-bonded (C-H)--(C-H) attractions within the same molecule [14], compared to the four (C-H) bond correlations?

Table 1

Correlations between and within tetrahedral orbitals in the valence shell of neon[a] (in eV). Even with such localized orbitals, correlations between "bonds" are found to contribute more than the four "bonds" (Rt > 1).

	A[b]	B[b]
$\epsilon_{\eta_1\eta_2}$ (a "bond" or "lone pair" type)	-0.666 eV	-0.784 eV
$\epsilon_{\eta_1\eta_4}$ ("inter-bond" opposite spin)	-0.349 eV	-0.310 eV
$\epsilon_{\eta_1\eta_3}$ ("inter-bond" same spin)	-0.149 eV	-0.149 eV
$\sum_{i>j}^{N} \epsilon_{\eta\,\eta} = E_{CORR}^{(Ne;L\text{-}shell)}$	-8.64 eV	-8.64 eV
$\text{Rt} \equiv \dfrac{E_{CORR}^{inter}}{E_{CORR}^{intra}} =$	2.24	1.75

[a] The irreducible pair correlation values of İ. Öksüz and O. Sinanoğlu [13] for neon are used.
[b] Calculation A uses eq. (14); B uses eq. (15) in text.

The CH_4. LO-pair correlations have been obtained using a semi-empirical MO-pairs (or LO-pairs) correlation theory [15] by Pamuk. These CH_4 values are: $\epsilon_{\eta_1\eta_2}$ (C-H bond) = -0.64 eV, $(\epsilon_{\eta_1\eta_4}+\epsilon_{\eta_1\eta_3})$ = = -0.40 eV, and Rt = 1.9. Thus the tentative CH_4 results are in line with those in table 1 for Ne. The tetrahedral (C-H) bond correlation value as well as the Ne values in table 1 may be compared with the bonding MO-pair ϵ_{ij} = -0.87 eV in the 2π C-H radical, calculated non-empirically by Davidson [10]. The latter value however is for a different hybridization as well as being for a non-closed shell system in which total pair correlation (including the "semi-internal" correlation effect which is zero in CH_4 and Ne (cf. refs. [12,13])) are expected to be larger.

In conclusion: the inter-bond correlations above and in other molecules studied in this laboratory are found to contribute at least as much to valence shell correlation energies as the "bond" correlations. Both effects however naturally occur and are predicted in the perturbation theory of correlation [4] and in M.E.T. [5]. A semi-empirical MO-pairs version of M.E.T. [15] is being applied to classes of molecules whose LO's have been obtained by a recently developed localization method [16].

This work was supported by grants from the U.S. National Science Foundation and the Alfred P. Sloan Foundation. It is a pleasure to thank Mr. İskender Öksüz and Mr. Ö. Pamuk for helpful discussions.

References

[1] a) A.C. Hurley, J.E. Lennard-Jones and J.A. Pople, Proc. Roy. Soc. (London) A220 (1953) 446.
b) R. McWeeny, Rev. Mod. Phys. 32 (1960) 335.
c) C. Edmiston and Ruedenberg, Rev. Mod. Phys. 35 (1963) 457.
d) R.K. Nesbet, Advs. Chem. Phys. 9 (1965) 321, cf. p. 355.
e) cf. also remarks on this point by E. Kapey, J. Chem. Phys. 44 (1966) 956; and by C. Edmiston, J. Chem. Phys. 39 (1963) 2394.

[2] a) D.C. Pan and L.C. Allen, J. Chem. Phys. 46 (1967) 1797; cf. also p. 49 of L.C. Allen, in: Quantum Theory of Atoms, Molecules and the Solid State, ed. P.O. Löwdin (Acad. Press, New York, 1966).
b) E. Clementi, J. Chem. Phys. 38 (1963) 2248; and L.C. Allen, E. Clementi and H.M. Gladney, Rev. Mod. Phys. 35 (1963) 465.

[3] a) O. Sinanoğlu, Proc. Nat'l. Acad. Sci. U.S. 47 (1961) 1217;
b) Rev. Mod. Phys. 35 (1963) 517.

[4] a) O. Sinanoğlu, J. Chem. Phys. 33 (1960) 1212;
Proc. Roy. Soc. (London) A260 (1961) 379.

[5] a) O. Sinanoğlu, J. Chem. Phys. 36 (1962) 706 [M.E.T. 1]
b) Advs. Chem. Phys. 6 (1964) and later papers of M.E.T. series.

[6] V. McKoy and O. Sinanoğlu, J. Chem. Phys. 41 (1964) 2689.

[7] O. Sinanoğlu, in: Modern Quantum Chemistry, 1964 Istanbul Lectures, Vol. II (Acad. Press, New York, 1965) p. 231.

[8] H.P. Kelly, Phys. Rev. 144 (1966) 39.

[9] F. Grimaldi, J. Chem. Phys. 43 (1965) S59.

[10] C.F. Bender and E.R. Davidson, J. Phys. Chem. 70 (1966) 2675; and other papers. E.R. Davidson, to be published. That the correlations between different 2p-orbitals are also important is pointed out by C.F. Bender and E.R. Davidson, J. Chem. Phys. 47 (1967) 360.

[11] As presented and emphasized by one of us (O.S.), in a number of recent symposia (e.g. Canadian Qu. Chem. Symp., Montreal 1967; Qu. Chem. Symp., Kutna Hora, Czechoslovakia 1967; Nato Frascati Summer School on Electron Correlation 1967; Yale Seminar on Sigma MO Theories 1967).

[12] See e.g. ref. [7], also O. Sinanoğlu, in: Advances in Chemical Physics (1968), (Frascati volume; R. Lefebvre and C. Moser editors). In closed shells, the sum of all cross-pairs is zero. They occur only in transforming individual ϵ_{ij}'s. They do not enter into the separate variational calculation of each ϵ_{ij}. The "cross-pairs" are particularly significant in non-closed shell states where they do not cancel out.

[13] İ. Öksüz and O. Sinanoğlu, to be published (cf. also ref. [12]). These values use many excited states of different N's and Z's to extract ϵ_{ij}'s, by a generalization of the methods used in ref. [6]. Ref. [6] has used only ground states, hence less detail and accuracy were attainable. Some preliminary calculations of the above detailed type have also been made by B. Skutnik (Ph.D. Thesis, 1967: Yale University).

[14] Intra-molecular C/R^6 London dispersion forces were invoked by K.S. Pitzer (Advs. Chem. Phys. 2 (1959) 59), in estimating heats of isomerization. Inter-bond LO-correlations are a more fundamental formulation of the intra-molecular attraction concept [4a].

[15] Ö. Pamuk and O. Sinanoğlu, to be published.

[16] C. Trindle and O. Sinanoğlu, J. Chem. Phys., in press.

Chapter VI

Tests and Comparisons of Sigma Molecular Orbital Theories and Their Approximations

VI-1

In this chapter the different semiempirical methods are compared with each other and with nonempirical calculations so that their successes and limitations may be better understood.

L. C. Allen (Chapter VI-2) has made a thorough study of the geometry predictions of extended Hückel theory (EHT), through nonempirical calculations. For various molecules, he compares the sum of one-electron orbital energies with the total energy, both plotted versus angle variables. As the orbital energy sum used here is also nonempirical, a direct test of one of the basic aspects of the method itself without questions of parameters is thereby provided. It is found that the geometry predictions of extended Hückel theory get worse for ionic systems, the method starting to break down when the electronegativity difference between two atoms exceeds about 1.3 Pauling. The study of the geometry of excited states by the extended Hückel method gives insight into Walsh's rules, showing that both Walsh's rules and the EHT give the correct excited-state geometry when this is determined by a "dominant orbital" rather than by the total N-electron wave function.

The question of which AO basis, the Hartree-Fock AO's or single Slater orbitals (STO), should be used to form the molecular orbitals is studied by Allen and taken up in detail by Cusachs and Corrington (Chapter VI-4) for heavier atoms as well. The actual Hartree-Fock SCF-MO's may be considered made up rigorously of "distorted" Hartree-Fock AO's" as shown by Gilbert in Chapter VI-3. The question therefore is how to approximate these with, for example, STO's, by properly choosing their parameters (Chapter VI-4). In the analysis of Gilbert, a bridge between the actual Hartree-Fock MO-SCF and extended Hückel theory is attempted by the expansion of the rigorous matrix elements and distorted Hartree-Fock AO's in terms of overlap integrals. The reader is referred on this topic also to a series of papers by Lipscomb and co-workers [*J. Am. Chem. Soc.* 88, 2353, 2361, 2367 (1967)].

The "zero-differential-overlap" (ZDO) approximation, familiar from π-electron theory, is basic also to the σ methods. The basis of various approximations involved in π-electron theories, including the analytic derivation of the ZDO, will be found in *Modern Quantum Chemistry*, Vol. I, Orbitals, O. Sinanoğlu, ed., Academic Press, New York, 1965, with some of the results pertinent or extendable to σ theory as well. The additional problem that comes up with σ systems, the local rotational invariance problem, and the number of ways it may be overcome at the expense of different approximations was studied in the important paper by Pople, Santry, and Segal (Chapter VI-5).

Semiempirical MO methods are already proving useful, not only in organic chemistry, but for the molecules of biochemistry as well. In this regard, the heterocycles provide important test cases, owing to the presence of heteroatoms, for the charge distributions obtainable by various σ-MO methods. An excellent critical study on these systems is undertaken by A. Pullman on purines and pyrimidines (Chapter VI-6). This article outlines the extended Hückel, the iterative extended Hückel, and the CNDO/2 SCF methods, and compares the orbital energies, charge distributions, and dipole moments obtained on ground states. Calculations on the π electrons alone are also compared with σ and π together. The CNDO/2 gives inner MO's which are too deep in energy. The interesting result is obtained that the orbital energies are not bunched separately as σ and π sets, but in the ground states their energies are intermingled. In the different semiempirical methods and in nonempirical calculations analogous to the one on pyrolle included in Chapter VII-4, the highest occupied MO is found to come out sometimes σ, sometimes π, depending on the method used. The σ distribution affects the π charges importantly, whereas the π-electron distribution is found (within the ground state) not to have much effect on the σ part. The CNDO/2 is found to give good dipole moments when both σ and π electrons are included. The gross AO populations are also satisfactory as given by the semiempirical methods.

McGlynn and Bertus (Chapter VI-7) study also the excited states, the spectra, and the ordering of levels in pyridine by an iterative extended Hückel method. An open-shell MO method is used, and different molecular orbitals are calculated for each overall electronic state. Good agreement with the experimental order of states is found. The authors caution the reader that present nonempirical calculations on molecules such as pyridine (for example, Chapter VII-4), are still at a rudimentary stage and far from actual Hartree-Fock SCF-MO, so that they are no more valid than the semiempirical methods at the present stage, yielding in this case an order of orbital energies in disagreement with experiment.

Chapter VII contains additional material on σ-π separation questions and nonempirical computation methods.

VI-2

Why Three-Dimensional Hückel Theory Works and Where It Breaks Down

LELAND C. ALLEN

1. Introduction

During the last few years three-dimensional Hückel theory (i.e., σ electrons as well as π's) has made a significant impact upon organic chemistry. Concomitant to this has been a general heightening of interest in theory among organic chemists and perhaps a beginning toward closing the "credibility gap" that has long separated electronic structure theory from the real world of organic chemistry. The propitious timing and organization of this σ-electron book by Sinanoğlu (theory) and Wiberg (organic + theory) and the recent Yale seminar on the subject have constituted an important move in this direction.

At present, strictly *ab initio* methods for constructing polyatomic electronic wave functions are still too time consuming and size limited to meet everyday requirements for investigating organic mechanisms. This is unfortunate, because greater overall unity can be brought to a chemical area when the theory has its direct origin in the basic laws of physics. It is also true that among those chemists with a physical turn of mind, *ab initio* methods are more instinctively believable. Since we are denied direct use of *ab initio* theory, the best we can do is attempt an *ab initio* underpinning for the largely *ad hoc* three-dimensional Hückel model. The understanding so gained leads directly to criteria for the validity of Hückel theory and expectations as to the range of chemical systems and properties for which one can expect reasonable results.

The general way in which we approach the connection between *ab initio* theory and model theory is *inductive* rather than *deductive.* Our connection is established by an extensive series of molecular case studies—much in the manner of experimental chemistry—instead of through a formal mathematical apparatus. There exist in the literature a number of formal mathematical "derivations" of Hückel theory. But we do not believe they tell us a great deal more about chemical systems than does a statement of Schrödinger's equation itself. The trouble lies in an information mismatch. Formal mathematical deductions from Schrödinger's equation, or any *ab initio* formulation for that matter, contain about a billion times more bits of information than does the almost trivially simple three-dimensional Hückel theory. On the other hand, the contemporary success of the Hückel model demonstrates that it is capable of providing at least a moderate amount of important chemical information and this in turn implies the need to select a detailed and very specific aspect of an *ab initio* formulation for comparison with the simple model.

The first step in making the *ab initio* model theory connection is choice of a particular *ab initio* formulation. This choice is obvious. The Hückel MO model is to be matched against an LCAO-MO approximation to the Hartree-Fock equations carried out via the Roothaan scheme. Unique to this choice is that the input AO basis set, the output one-electron energies, and the definition of a one-electron MO can be made identical for both levels of theory.

The second step is a still more specific choice as to what properties of the one-electron energies and MO's are to be compared. The answer is angular properties and the two levels of theory are to be related by use of Walsh diagrams (molecular orbital energies versus bond angle). These diagrams have a long history of systematizing and elucidating many aspects of spectroscopy and inorganic chemistry. Their development and widespread application has firmly established the existence of generic bond combinations—AH_2, AH_3, AB_3, H_2AB, HAAH, etc.; of a relationship between bond angles and the number of valence electrons; and of a scheme for predicting the shape of molecules in excited states. In a rough way one can say that these observations demonstrate that there exist a set of reasonably well codified rules governing hybridization. We may anticipate one of the general conclusions of this paper by stating that the information content required of a model to satisfy these rules is relatively small. In its usual form, three-dimensional Hückel theory is a scheme for weighting average atomic orbital energies with orbital overlaps, and this generates just sufficient information to reproduce the Walsh diagrams and adequately specify hybridization. The third step in our *ab initio* model matching is concerned with criteria for the breakdown of the Walsh rules and limitations on the usefulness of Hückel theory. Again anticipating our conclusions qualitatively, we find that Hückel results are meaningful only when the charge distribution is relatively uniform.

2. Methods for Construction and Analysis of the Wave Functions

Ab initio wave functions and their resultant energies provide the fundamental data in terms of which we analyze the Hückel model. For the quantities of interest to us here (geometry and first ionization potentials mostly) our *ab initio* solutions give values so close to the experimental numbers that they may be considered as absolute reference points. The Hartree-Fock equations rigorously define the molecular orbitals, and the LCAO approximation to these equations formulated by Roothaan has been well developed and extensively studied in numerous laboratories. The input atomic basis orbitals are quite close to atomic Hartree-Fock solutions, and the output molecular orbitals are similarly close to molecular Hartree-Fock solutions.

$$\phi_i = \sum_j C_{ij}\,\eta_j$$

ϕ_i = one-electron MO's

η_j = AO basis set

C_{ij} = linear variation coefficients

All electrons in the molecule are treated equally and included simultaneously. A rigid nuclear framework is assumed for each geometry. The molecular orbitals are distributed over the whole nuclear framework and they may be classified according to the irreducible representations of the molecular point group, thus taking on all aspects of the symmetry inherent to the molecule. The latter is a very important point because the same clamped nuclei and orbital expansion given above is employed by Hückel theory and this (along with the symmetry properties of the Hückel matrix elements) assures that the Hückel MO's will have the same symmetry characteristics as the *ab initio* solutions. A great deal of the success of three-dimensional Hückel theory may be attributed to *orbital* symmetry considerations. Since we are dealing with wave amplitude functions, sign information provides an additional aspect to symmetry arguments not realizable from the macroscopic symmetry of a molecular system.

The output of our *ab initio* calculations is the collection of MO expansion coefficients, C_{ij}, their corresponding one-electron energies, ϵ_i, and the total energy, E_T.

$$E_T = \sum_{i}^{\text{val.}} \epsilon_i + \left(\sum_{i}^{\text{core}} \epsilon_i - V_{ee} + V_{nn} \right)$$

V_{nn} = nuclear framework potential

$$V_{ee} = \frac{1}{2} \sum_{i,j} \left\{ (ii|jj) - (ij|ji) \right\}$$ = screening potential produced by molecular Coulomb and exchange integrals

The electronic charge distribution in a molecule is determined by a rather delicate balance between the one-electron attraction effects represented in the ϵ_i working against the nuclear framework and electron screening repulsion of V_{nn} and V_{ee}.[1] The *ab initio* Hartree-Fock equations are solved by an iterative procedure which converges to a solution when the proper screening potential is chosen so that all terms in the energy expression remain unchanged under small changes in the C_{ij}. A large amount of information is contained in V_{ee}. For molecules with 5 to 50 electrons the number of six-dimensional integrals which constitute V_{ee} (each required to about eight significant figures) ranges from a few hundred to a few millions, and the number of numbers involved is many times that for all the other terms considered. Further, there is no cancellation between V_{ee} and V_{nn}—they always differ by many tens of electron volts and they differ by different amounts for different geometries. It is these matters which are at the heart of our discussion, because Hückel theory contains neither V_{nn} nor V_{ee}.

The device by which we make connection between the two levels of theory is the Walsh diagram.[2] The Walsh diagram for type-AH_2 molecules (e.g., BH_2^+, BH_2^-, H_2O, BeH_2) is shown in Fig. 1. Figure 2 is a schematic for AH_2 showing how the two hydrogen 1s orbitals and the A

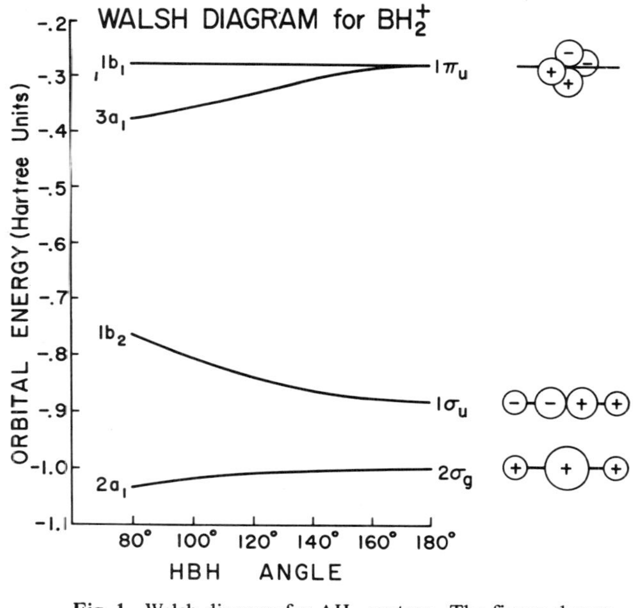

Fig. 1. Walsh diagram for AH_2 system. The figure shown is that computed for the BH_2^+ ion.

1s, 2s, $2p_x$, $2p_y$, $2p_z$ mix to form the various MO's for linear and bent conformations. The Walsh diagrams are basically hybridization graphs. It is worthwhile cataloguing an example in detail because, when valid, Hückel theory's chief attribute is its ability to generate reasonably good Walsh diagrams. Thus a description of the Walsh diagram is in effect a description of the workings of Hückel theory. The explanation of the orbital trends given here resulted

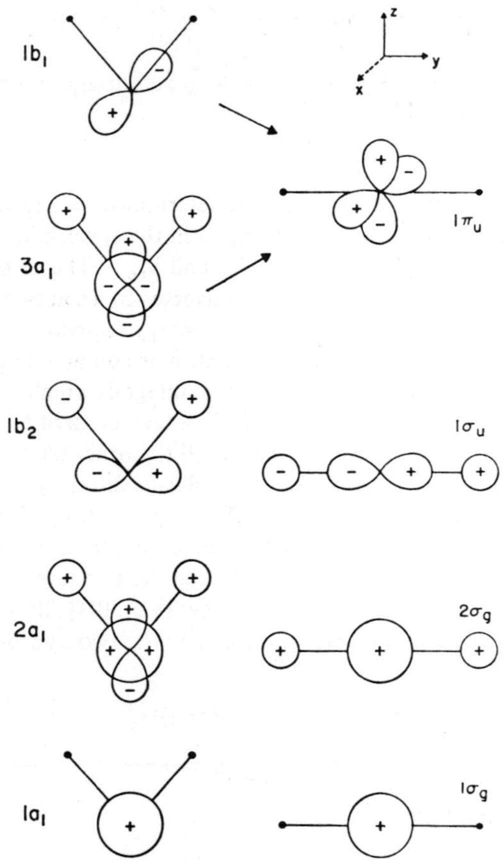

Fig. 2. Orbital schematic for AH_2 system in bent and linear conformations. Energy of the molecular orbitals increases from bottom to top. Atomic basis set is restricted to H 1s and to the 2s and $2p_x$, $2p_y$, and $2p_z$ on the central atom.

from an analysis of *ab initio* calculations for numerous molecules. The $1\sigma_g \longrightarrow 1a_1$ MO is almost entirely A 1s in both linear and bent conformations. It is chemically uninteresting and thus not displayed in Fig. 1. The $2\sigma_g$ is the bonding combination of the two H 1s orbitals and the A 2s (only the 2s because of the transformation properties of the $D_{\infty h}$ group). As bending occurs $D_{\infty h} \longrightarrow C_{2v}$ and A $2p_z$ mixes in as well, increasing its contribution as the angle decreases. The orbital energy decreases as angle decreases because the p_x throws an increasing amount of charge into the triangle. The A 2s contribution decreases with decreasing angle. The two H 1s contributions are nearly independent of angle. To keep competing effects clearly in mind we now describe the higher energy molecular orbital of the same symmetry in C_{2v}, $1\sigma_u \longrightarrow 3a_1$. At 180° this is a pure $2p_z$ orbital (in contrast to the $2\sigma_g \longrightarrow 2a_1$, which was pure 2s at 180°). For smaller angles the 2s mixes increasingly with the $2p_z$, and this lowers the molecular orbital energy for two reasons: first, because the 2s atomic orbital is lower in energy than the 2p and the mixture lowers the average and, second, because 2s, 2 p mixing lowers the electron repulsion in V_{ee}. In $3a_1$ the A 2s comes in with a sign opposite to that in the $2a_1$ and so the 2s,

2p mixture produces a lone pair with charge thrown out of the triangle. $1\sigma_u \longrightarrow 1b_2$ forms a $p\sigma$ bond from the H 1 s orbitals and the A p_y at 180°. As bending occurs the overlap between the p lobes and the H 1 s orbitals is not so great; thus the $1b_2$ increases in energy with decreasing angle. $1\pi_u \longrightarrow 1b_1$ is $2p_x$, a 2p orbital on A perpendicular to the molecular plane. It therefore does not mix with other orbitals and its energy is independent of angle.

The Walsh diagrams were originally constructed empirically from experimental spectroscopic data on ionization potential and excitation energies.[2] The curve shapes and their orbital mixing rationale were more qualitative and cruder than given above. Formerly, it was not certain precisely what physical quantity was implied by "orbital energy." Likewise, the question of just how invariant the diagrams were for a specific molecular type, AH_2, AH_3, AB_2, etc., was open. *Ab initio* computations for many molecules[3] have now shown unequivocally that the ordinate in the diagram is indeed the one-electron molecular orbital energy which appears as the eigenvalue in the canonical form on the Hartree-Fock equation,

$$\mathcal{H}(1,2)\,\phi_i(1) = \epsilon_i \phi_i(1)$$

and which is identified with first ionization potentials through Koopman's theorem. At the same time the *ab initio* investigations have led to considerable confidence as to the universality of diagrams for a particular molecular type. A typical set of diagrams is shown for three AH_3 species, BeH_3^-, BH_3, and CH_3^+, in Fig. 3. These diagrams all lead unambigously to the same hy-

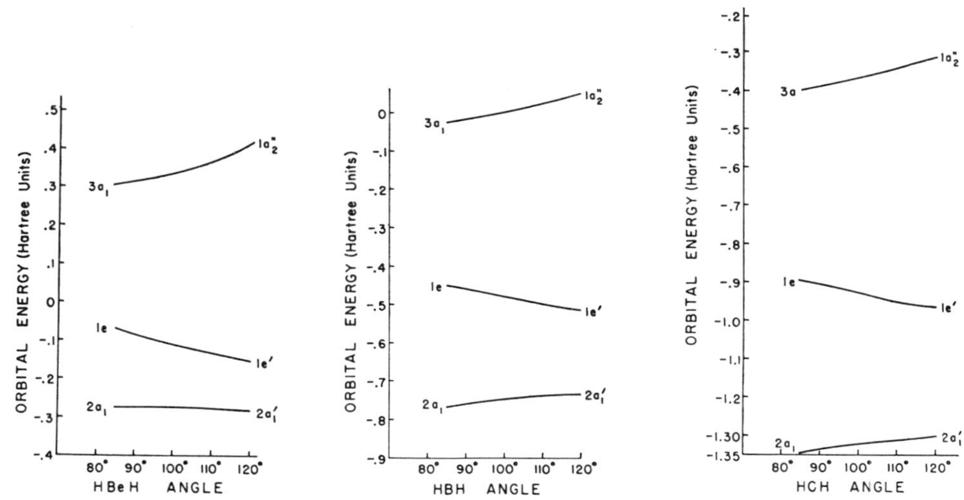

Fig. 3. Walsh diagrams for AH_3 system. Z of central atom increasing from left to right.

bridization predictions, but one can observe the trend toward larger splittings as the effective field of A becomes greater. The principal reason which called into question the "orbital energy" definition was Walsh's use of a simple sum of orbital energies as a criteria for molecular shape. It has been known for a long time that $\Sigma\epsilon_i \neq E_T$ (as can be seen immediately from the E_T relation given above), and thus it was thought that some other quantity than ϵ_i should be the Walsh diagram ordinate. This dilemma has been resolved[4] and the answer proves to be very simple: All that is required is for $\partial \Sigma_i^{\text{val.}} \epsilon_i/\partial\theta$ to have the same sign as a function of angle as $\partial E_T/\partial\theta$ (it is, of course, $\partial E_T/\partial\theta = 0$, which sets the equilibrium molecular angle). This is true when either of the following inequalities are satisfied:

$$\frac{\partial}{\partial\theta}\left(\sum_{i}^{\text{val.}} \epsilon_i\right) \bigg/ \frac{\partial}{\partial\theta}\left(\sum_{i}^{\text{core}} \epsilon_i - V_{ee} + V_{nn}\right) > 0$$

$$\left|\frac{\partial}{\partial\theta}\left(\sum_{i}^{\text{val.}} \epsilon_i\right)\right| > \left|\frac{\partial}{\partial\theta}\left(\sum_{i}^{\text{core}} \epsilon_i - V_{ee} + V_{nn}\right)\right|$$

These inequalities are found numerically to be satisfied for hydrocarbons and, in fact, a majority of all molecules. Physically, it requires that there be sufficiently little charge transfer from one atom to its neighbors to prevent the buildup of sizable atomic charges, which then gives rise to a large change in electrostatic potential with angle. Roughly speaking, the molecular charge distribution must be smooth enough to avoid large local charge separations. When this is true, $\Sigma^{\text{val.}}\,\epsilon$ will predict the same equilibrium angle as E_T even if the $\Sigma^{\text{val.}}\,\epsilon$ curve has a

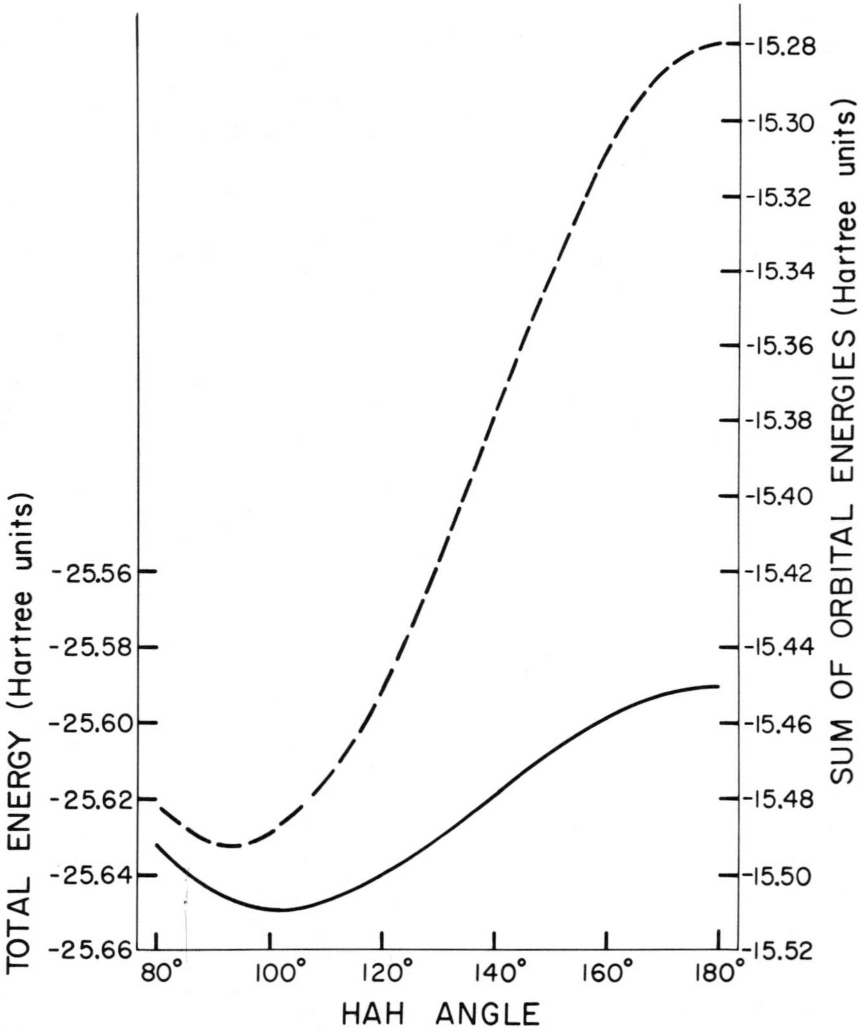

Fig. 4. Total energy (solid line) and $\Sigma^{\text{val.}}\,\epsilon$ (dashed lines) for covalent AH_2 system. The figure shown is that computed for the BH_2^+ ion.

markedly different shape than E_T. Thus one can conclude that molecular shape is a property corresponding to a relatively small amount of information. Figure 4 shows a typical plot of the two quantities. If adjacent atoms in a molecule differ greatly in electronegativity, we may expect the $\Sigma\,\epsilon$ criteria of molecular shape to be invalid. Figure 5 displays $\Sigma^{\text{val.}}\,\epsilon$ and E_T for the highly ionic molecule Li_2O and the breakdown of the criteria is apparent.

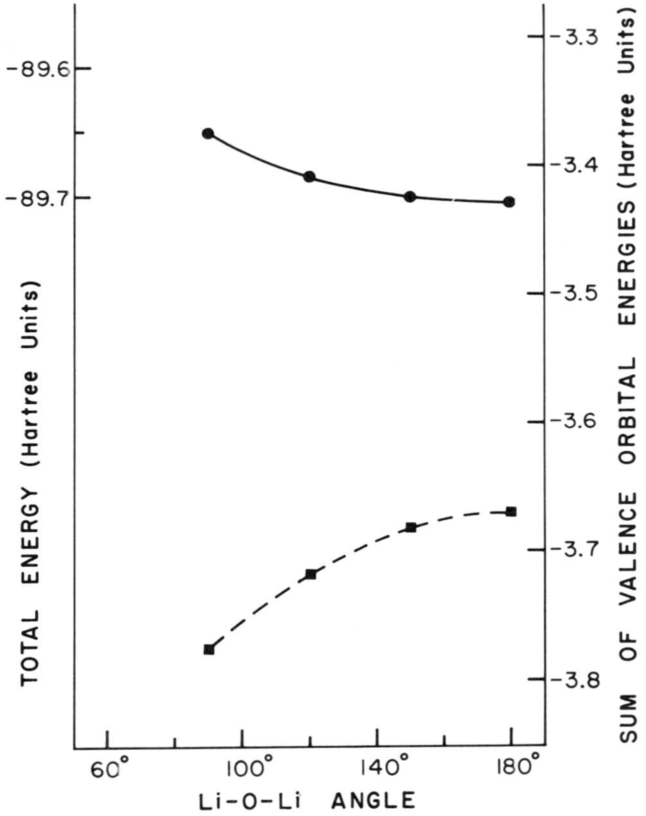

Fig. 5. Total energy (solid line) and $\Sigma^{\text{val.}}\,\epsilon$ for ionic AB_2 system. The figure shown is that computed for Li_2O.

The concept of a low information content inherent to equilibrium angle predictions immediately suggests the possibility that three-dimensional Hückel theory may be almost as quantitatively valid for angular energetics as *ab initio* results themselves. To study this we chose a widely applied and simple formulation of three-dimensional Hückel theory: the Ballhausen-Gray form[4] of the Mulliken proposal[5] (the indices refer to nuclear centers; the orbital index has been suppressed):

$$H_{ii} = \alpha_i$$

$$H_{ij} = KS_{ij}\sqrt{\alpha_i\alpha_j}$$

α_i = valence-state ionization potential of an orbital on center i

S_{ij} = overlap integral between an orbital on center i and another on center j

These matrix elements are used to form a secular equation, $|H_{ij} - S_{ij}\,\epsilon| = 0$, and the output is a set of $C_{k\ell}$'s and ϵ_k's. As noted previously, the expansion of the one-electron molecular orbitals is identical in form to that of the *ab initio* formulation except that valence orbitals only are included in the expansion. Also, most other workers have employed a function of the form $r^n \exp(-ar)$ for each atomic orbital of the basis set, η_ℓ, instead of Hartree-Fock atomic orbitals. We have carried out computations with both basis sets. For the single exponential basis sets we used free-atom optimized exponents[6] and explored a range of values for the splitting parameter, K. The K value which comes closest to matching *ab initio* results for equilibrium bond angles and for the curve shapes in Walsh diagrams is $K = -1.5$. Our other set of Hückel computations used the *ab initio* basis set, and $K = -1.75$ was found to be the best match. It is an important and satisfying result that for all the many molecules investigated, a better fit to the *ab initio* results was achieved when the identical (and physically superior) basis set was utilized.

We may note that the Hückel theory discussed above does not employ an iterative self-consistent procedure for determining the orbitals. When one compares the coefficients, $C_{k\ell}$, computed by this model with *ab initio* results, it is immediately apparent that the swings in magnitude from one coefficient to another in a given MO, or in corresponding coefficients from one MO to another, are much too great and this may in part be attributed to lack of an iterative scheme. Several iterative methods have been proposed (generally relating the coefficients identified with a given orbital to the α for that orbital), and there is no doubt that they accomplish the desirable numerical effect of smoothing out the overly large changes in the coefficients. On the other hand, the direct physical concept of self-consistency which is manifest in the Hartree-Fock equations involves readjustment of V_{ee}, and we have already noted that this term is specifically avoided in Hückel theory. This raises doubts about the basic meaning of iterative methods and at least focuses attention on the necessarily *ad hoc* criteria for parameter choice in such a scheme.

At this point we present a rationalization for the model-theory matrix elements and α that we believe to be as tightly reasoned as any yet offered. This is an instructive exercise because it couples the model to chemical concepts and clearly indicates that the supportive arguments are open ended, thus demonstrating that the model must unavoidably provide only limited chemical information. It seems intuitively obvious that α should be identified with the electronegativity of the corresponding orbital, and this is, in fact, the general feeling that has arisen among many scholars over the years. Many years ago Mulliken[7] proposed the most straightforward physical definition of electronegativity: $\frac{1}{2}(I + A)$, where I is the ionization potential and A is the electron affinity. Even then he applied this definition not only to atoms and ions as a whole, but to the orbitals themselves (particularly to various orbital occupancies representing different valence states). An extensive knowledge of these so-called valence-state ionization potentials has been built up from the analysis of experimental spectra,[8] and they are the values most frequently used for α by other workers and also in our studies here. In some unpublished work, the present author has sought to make up a set of purely theoretical values for valence-shell atomic orbitals (atoms Li through F) that would also reflect the Mulliken concept. Data from Hartree-Fock solutions of neutral and negative ions were examined and for a given shell the one-electron energies of the two species were averaged. The center of gravity of the atomic multiplet was used where necessary. This theoretical set of electronegativities was compared with the high-valence-state experimentally derived values for 2s and 2p orbitals and they agreed to within a few per cent in all cases.[9] Having convinced ourselves that it is possible to obtain a reasonably consistent and complete set of values for the α's, we can ask if there is any way to relate the analytical form of the orbital interaction (H_{ij}) to these ideas. This may be accomplished through some long-standing work on atomic and molecular electronegativity by Sanderson.[10] His atomic electronegativities are directly proportional to a normalized atomic charge density. The atomic charge density is Z divided by the atomic volume determined from the atomic radius. This density is normalized by dividing by the density of the noble gas atom in the same row of the periodic table. The electronegativities so defined are found to display a dependence on Z very similar to that of Mulliken, Pauling, and several other of the common electronegativity scales. When two atoms join these electronegativities will equalize (match their Fermi levels). Sanderson hypothesized that each atom, and thus the bond, would achieve a resultant electronegativity whose value was the geometric mean of the initial electronegativities. We might term this a "self-consistent" electronegativity for two interacting atoms, and Sanderson was able to successfully relate it to the length, energy, and polarity of bonds in many molecules. Just as in the case of Mulliken's electronegativity concept, we may translate this combining rule into orbital form, yielding $H_{ij} \sim \sqrt{\alpha_i \alpha_j}$. S_{ij} is often taken as a measure of bond strength and therefore comes in as a multiplicative factor for $\sqrt{\alpha_i \alpha_j}$.

Now, there exist a number of interesting papers on the orbital electronegativity concept that are far more sophisticated than our treatment here,[11] and a similar set concerning the dependence of H_{ij} on S_{ij}.[12] It is apparent that the language and descriptive mode for the simple model level is decidedly different than that appropriate to the *ab initio* level, and it is quite a trick to find a quantitatively useful meeting ground between the two. Section 3 presents numerical and pictorial results pertinent to this problem.

3. Results

Ab initio wave functions have been computed as a function of geometry in this laboratory by the method described in Section 2 for the following species: BH_2^+, BH_2^-, NH_2^+, BeH_2, BH_3, CH_3^+, BeH_3^-, Li_2O, LiOH, F_2O, FOH, H_2O, CH_4, HCN, CH_3^-, NH_3, H_3O^+, CH_5^+, B_2H_6, CH_3CH_3, CH_3CH_2F, CH_3OH, H_2O_2, CH_3NH_2, N_2H_4, NH_2OH, N_3^-, O_3, NO_2^+, NO_2, NO_2^-, CO_2, CH_3F, CH_2F_2, CHF_3, BH_2F, BHF_2, BF_3, and about 20 others.[3] For those listed above, three-dimensional Hückel calculations also have been carried out as a function of angle (either completely or sufficiently to uncover any new features) for both basis sets. The species studied were chosen to include most of the Walsh types; representatives of covalent, ionic, and intermediate bonding; positive and negative ions; electron-deficient species; several isoelectronic sequences; successively fluorinated molecules; species existing in several states of ionization, such as NO_2^+, NO_2, NO_2^-; and certain other pivotal series, such as F_2O, H_2O, and Li_2O. Part of the comparison between *ab initio* results and single-exponential-basis-set Hückel calculations has been reported before.[13]

We have previously seen (Fig. 4) that the *ab initio* $\Sigma\, \epsilon$ gave an equilibrium-angle prediction remarkably near to that of E_T for BH_2^+. Another pair of typical covalent species, BH_2^- and F_2O, is shown in Fig. 6, and we again observe the similarity in angle predictions between *ab initio* $\Sigma\, \epsilon$ (dashed lines) and E_T (solid lines). However, the question we wish to address is the adequacy of Hückel theory, and the staggered lines in Fig. 6 displays $\Sigma\, \epsilon$ from single-exponential-basis orbital Hückel results. It is gratifying that even such a relatively crude reproduction of the *ab initio* MO's as this yields $\Sigma\, \epsilon$ predictions not appreciably different than the *ab initio* $\Sigma\, \epsilon$. Errors must show up in the equilibrium values, however, because it is just in the equilibrium region that $\partial \Sigma \epsilon / \partial \theta$ is small and that the differences between *ab initio* $\Sigma \epsilon$ and Hückel $\Sigma \epsilon$ will be most noticeable. *Ab initio* $\Sigma \epsilon$ often has a more intricate shape over the angular domain than does Hückel $\Sigma\, \epsilon$. This is once again a manifestation of information content: There are many more physical forces contributing to the shape of the *ab initio* $\Sigma \epsilon$ and it is capable of finer-grained chemical distinctions than Hückel $\Sigma\, \epsilon$. The pair of graphs in Fig. 7 show the striking parallelism that is sometimes present between E_T and $\Sigma\, \epsilon$ curves in both ground and excited states. The excited state illustrated is a Σ state corresponding to the promotion of two electrons from the highest occupied ground-state orbital to the next available orbital, $(6a_1)^2 \rightarrow (2b_2)^2$. Although the $\Sigma\, \epsilon$ curves displayed are *ab initio*, Hückel theory with a

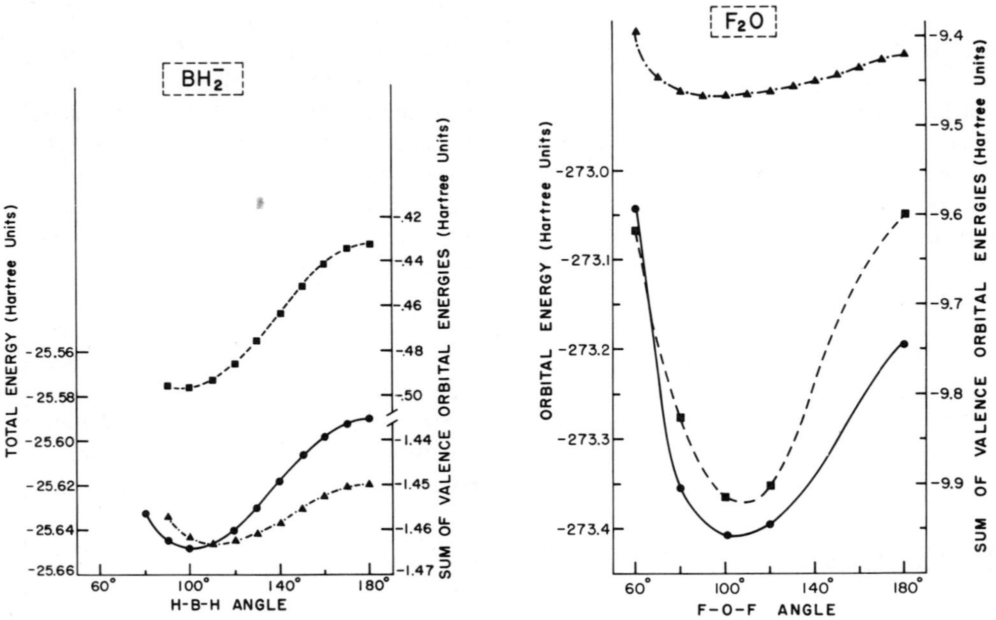

Fig. 6. Total energy and $\Sigma^{\text{val.}}\ \epsilon$ for covalent molecules. Total energy (solid line), *ab initio* $\Sigma\ \epsilon$ (dashed line), single-exponetial Hückel $\Sigma\ \epsilon$ (staggered line). BH_2^- in figure on left, F_2O in figure on right.

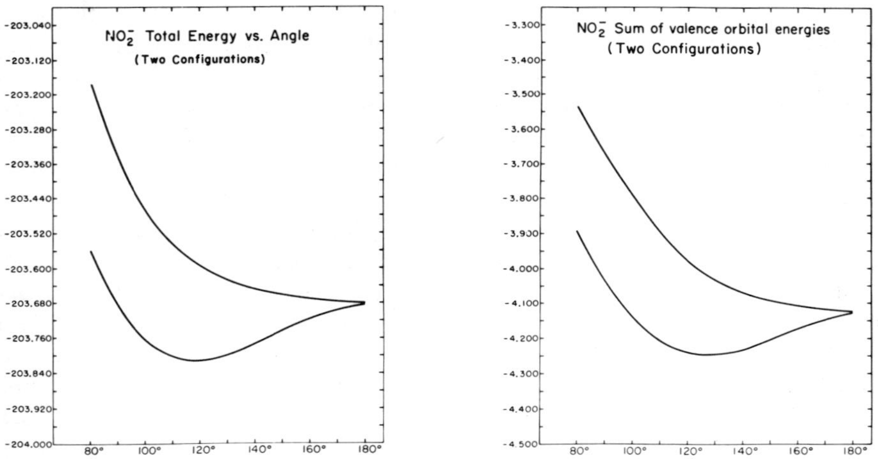

Fig. 7. Comparison of total energy and $\Sigma^{\text{val.}}\ \epsilon$ potential-energy curves for ground and excited states of NO_2^-. Total energy versus O–N–O angle on left graph, $\Sigma\ \epsilon$ versus O–N–O angle on right graph.

Hartree-Fock AO basis set does almost as well. In Fig. 8 the pair of graphs illustrates all levels of our calculations for the partially ionic molecule, H_2O. The dashed line of large excursion in the graph on the left is the *ab initio* $\Sigma\ \epsilon$ and we see a partial breakdown of the Walsh summation criteria. Small changes in the Hückel parameters would change the equilibrium predicted by the single-exponential Hückel result (staggered line, left graph), but both basis sets (Hartree-

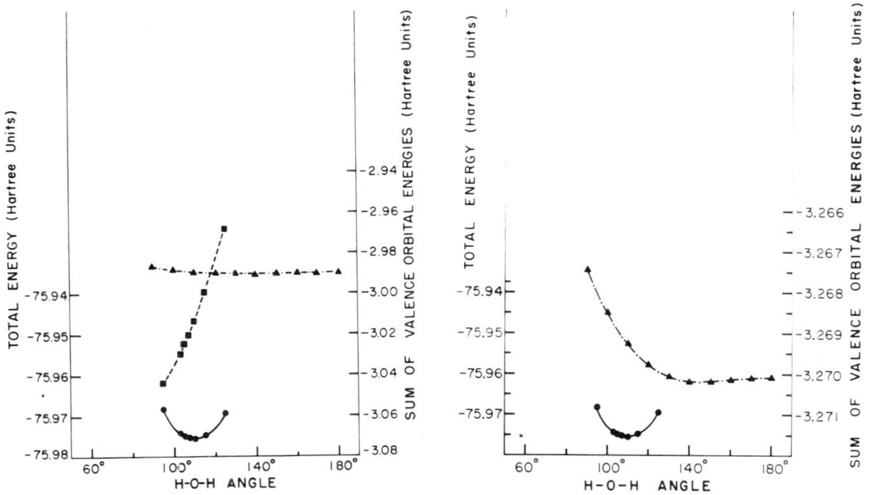

Fig. 8. Total energy and $\Sigma^{val.}\ \epsilon$ for partially ionic molecule (H_2O). Left figure: total energy (solid line), *ab initio* $\Sigma\ \epsilon$ (dashed line), single-exponential-basis Hückel $\Sigma\ \epsilon$ (staggered line). Right figure: total energy (solid line), Hartree-Fock-basis Hückel $\Sigma\ \epsilon$ (staggered line).

Fock AO's generate the staggered curve on the right) lead to Hückel $\Sigma\ \epsilon$ quantitatively at variance with *ab initio* $\Sigma\ \epsilon$ and E_T. Nevertheless, significant chemical information is being conveyed simply by stating that the molecule is bent instead of linear, and Hückel theory with a good basis set (Hartree-Fock AO's) is capable of bringing this out. No utility may be obtained by noticing that there exists for this case a set of Hückel parameters that will give a better $\Sigma\ \epsilon$ result than *ab initio* $\Sigma\ \epsilon$ because examination of other partially ionic molecules (e.g., FOH) shows this not to be a universal situation. When the electronegativity difference between one pair of adjacent atoms exceeds about 1.3 on Pauling's scale or about 1.0 on Sanderson's scale (the O–H difference is 1.4 and 1.15, respectively), we can expect the beginning of a breakdown in the $\Sigma\ \epsilon$ criterion. Strongly ionic Li_2O (Fig. 9) unambiguously violates the summation rule for both basis sets in the Hückel calculations.

Here again, even for ionic species, the better basis orbital set produces a Hückel result that is closer to the *ab initio* result. The origin and details of angle prediction failure for the partially and strongly ionic test cases studied here provides the basis for understanding the long recognized uncertainties and inadequacies found when Hückel theory (two dimensional as well as three dimensional) is applied to organic systems containing heteroatoms. As a final example of the use of $\Sigma\ \epsilon$ we consider barriers to internal rotation. $\Sigma\ \epsilon$ from three-dimensional Hückel was shown some time ago[14] to yield the correct energy dependence with dihedral angle and a reasonable barrier magnitude for ethane. Our *ab initio* $\Sigma\ \epsilon$ computations confirm this Hückel theory result and it is thus quite interesting to look into barriers for other molecules. Barriers heights are only a few kilocalories–an order of magnitude smaller than other energy changes we have been considering–so that such factors as K and the assumed form of the matrix elements can greatly modify predictions. Thus it is only useful to report *ab initio* $\Sigma\ \epsilon$ values, and these will tell us the potential that Hückel theory might realize. For ethane and ethyl fluoride values are ΔE_T = 2.58 and 2.59 kcal, $\Delta(\Sigma\ \epsilon)$ = 4.02 and 4.77 kcal, respectively.[15] These numbers are quite encouraging and symmetry alone guarantees that ethane, ethyl fluoride, methylamine, and methanol will all be properly threefold. ΔE_T for ethane, methylamine, and methanol yields a reasonable approximation to the well-known 3:2:1 rule for these barriers but, unfortunately, $\Delta(\Sigma\ \epsilon)$ gives methylamine 60 per cent higher than ethane and

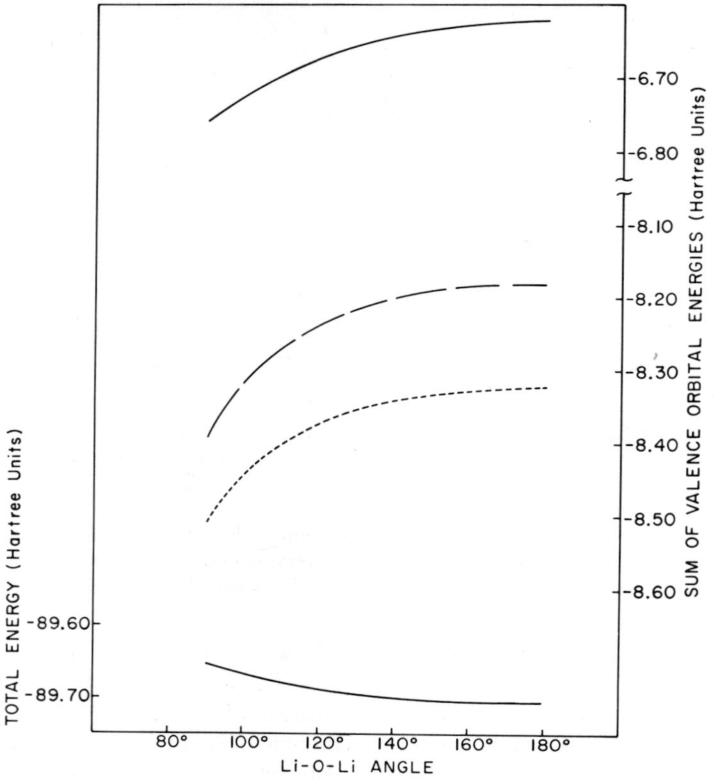

Fig. 9. Total energy and $\Sigma^{\text{val.}}\epsilon$ for strongly ionic molecule (Li_2O). Total energy (solid line at bottom of graph), *ab initio* $\Sigma\epsilon$ (solid line at top of graph), Hartree-Fock-basis Hückel (long dashes), single-exponential-basis Hückel (short dashes).

methanol 20 per cent higher than ethane. In hydrogen peroxide Fig. 10 shows $\Sigma\epsilon$ to yield an angular dependence entirely opposite to both experimental and *ab initio* E_T curves. Thus $\Sigma\epsilon$ does not produce useful information on rotational barriers.

The second major method for establishing the *ab initio*-model theory connection is comparison of Walsh diagrams. Figure 11 is the Walsh diagram for covalent BH_3. *Ab initio*, Hückel single-exponential AO basis, and Hückel Hartree-Fock AO basis are all essentially the same, with Hartree-Fock AO's again being slightly superior. For partially ionic H_2O there was a breakdown in the $\Sigma\epsilon$ criteria, and it is particularly interesting to see if a similar type of deviation separates *ab initio* from model-theory Walsh diagrams. We find the deterioration to be less immediately apparent. The direction of slopes and level ordering are well preserved (Fig. 12). Although omitted for clarity, the Hartree-Fock AO Hückel curves are almost exactly midway between *ab initio* and single-exponential AO Hückel results in both slope magnitudes and level ordering. Figure 13 for partially ionic FOH well illustrates the comparison between our three levels of approximation. On the left the full Walsh diagram for the valence orbitals is shown for the *ab initio* (dashed lines) and single-exponential AO Hückel calculations (solid lines). The *ab initio* level pattern is well reproduced except for an incorrect level inversion between 1b and 6a levels of the Hückel calculations. On the right-hand graph an enlargement of the $1\pi \longrightarrow 1b$, $6a$, $5\sigma \longrightarrow 5a$ region is given and Hartree-Fock AO Hückel results are added. The superior basis

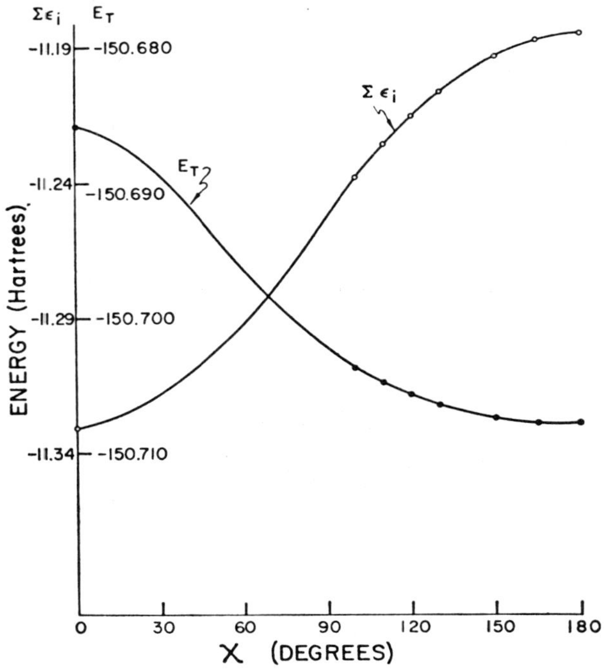

Fig. 10. Rotational barrier in hydrogen peroxide. Total energy (decreasing with increasing dihedral angle), *ab initio* $\Sigma \epsilon$ (increasing with increasing dihedral angle).

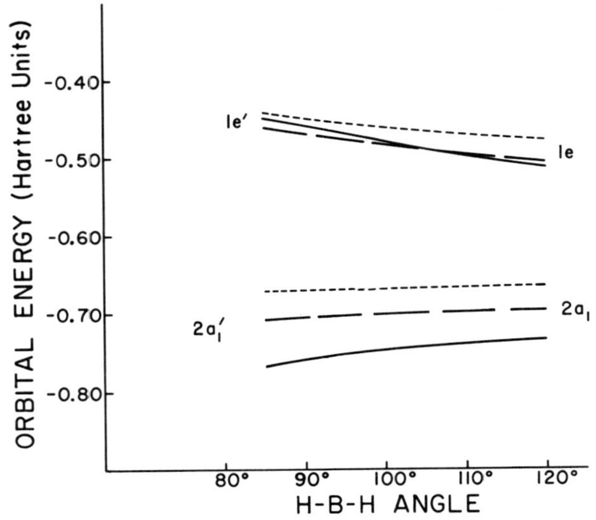

Fig. 11. Walsh diagram for covalent AH_3 system (BH_3). *Ab initio* (solid lines), Hartree-Fock basis Hückel (long dashes), single-exponential-basis Hückel (short dashes).

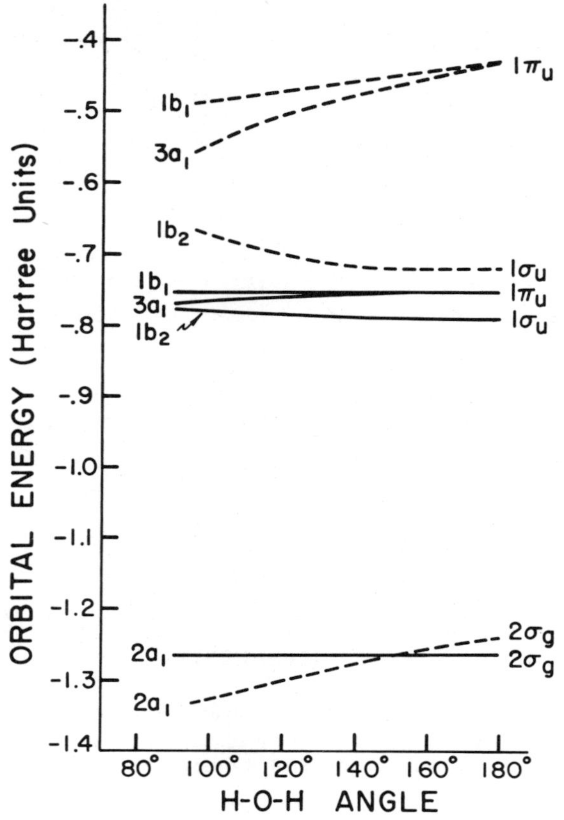

Fig. 12. Walsh diagram for partially ionic AH_2 system (H_2O). *Ab initio* (dashed lines), single-exponential-basis Hückel (solid lines).

set (long dashes) corrects the level inversion and significantly improves the agreement of Hückel theory with the *ab initio* curves (solid lines). (*N. B.*: A notation inconsistency is present in Figs. 12 and 14 and the left graph of Fig. 13. In these, *ab initio* curves are shown by dashed lines and single exponential AO Hückel curves by solid lines.) Figure 14 is the Walsh diagram for Li_2O with single-exponential AO Hückel and *ab initio* results displayed. Again there is a level reversal for the close lying $2\sigma_u$, 1π MO's, and again this is removed by a Hartree-Fock AO basis. In both this and the FOH case the error arises from the generally poor representation of the p-orbital radial distribution afforded by single exponential AO's. Overall examination of the figures given here and many others representing all bonding types reveals Hückel generated Walsh diagrams to be relatively flat and featureless compared to *ab initio* curves. This is due to the omission of two-electron matrix elements as described in Section 2.[16] Very crudely, these terms are $\sim 1/R$, and when this dependence is added it spreads out the curves more in line with *ab initio* results at small angles.

The reason Walsh diagrams preserve their generic form over a wide range of bonding situations is due to the fact that, to first order, hybridization is an atomic effect and this atomic character is maintained in the molecular wavefunction because an LCAO representation (with Hartree-Fock AO's) of the MO's is a good approximation to the true *ab initio* molecular Hartree-Fock solution for any molecule. With a Hartree-Fock AO basis Hückel theory reproduces a good part of the hybridization effects. However, as the molecular environment becomes more

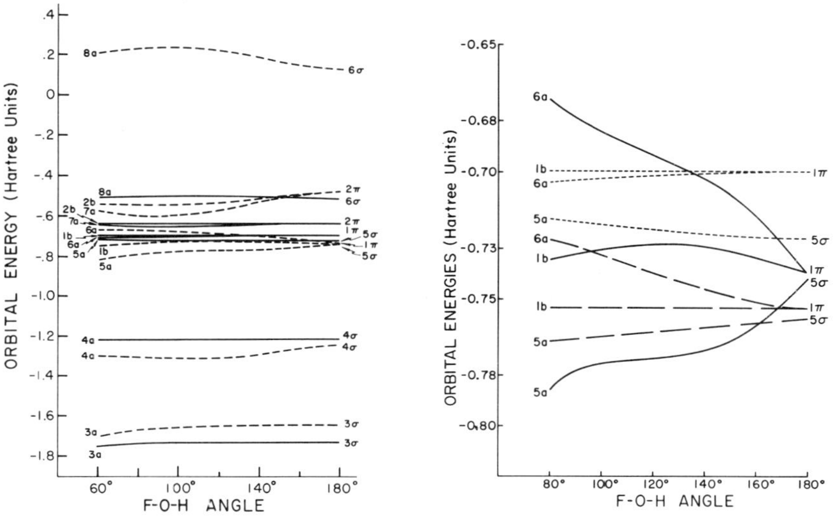

Fig. 13. Walsh diagram for partially ionic HAB system (FOH). Left figure shows the complete set of valence levels: *ab initio* (dashed lines), single-exponential-basis Hückel (solid lines). Right figure shows selected levels: *ab initio* (solid lines), Hartree-Fock-basis Hückel (long dashes), single-exponential-basis Hückel (short dashes). (*N.B.*: There is a change of convention between the left and right graphs.)

ionic, occupancy of the atomic orbitals is strongly modified and the Walsh diagrams supply less information about the actual molecular charge distribution.

The purpose of Figs. 15 and 16 is to demonstrate the use of Walsh diagrams in comparing two related species that may be connected by an angular variable. It is use of this sort which underlies much of the potential efficacy of three-dimensional Hückel theory for elucidating organic mechanisms.[17] Using these diagrams we can understand the geometry and relative stabilities of NO_2^+ and NO_2^- (Fig. 15) and of diborane compared to ethane (Fig. 16).[18] Figure 15 is a Walsh diagram for NO_2^- as a function of angle for all its occupied orbitals. Similar *ab initio* calculations for NO_2^+ yield a Walsh diagram for the occupied orbitals which is essentially identical for Fig. 15 except that the $6a_1$ orbital is missing. (This is simply a confirmation of the generic nature of Walsh diagrams.) Thus the linear conformation of NO_2^+ and the bent conformation (115°) of NO_2^- can be understood almost entirely in terms of the dramatic change with angle of the single $6a_1$ orbital. Figure 16 is a Walsh diagram for diborane as a function of the BBH_1 angle, where H_1 is moving from a bridged hydrogen position in D_{2h} geometry through C_{2h} to a terminal hydrogen position coming together with two other equivalent hydrogens as in ethane (D_{3d}).[19] Calculations starting with ethane produce a Walsh diagram of almost identical appearance to that of Fig. 16.[20] The difference between diborane and ethane is double occupancy of the dramatically changing $3a_{1g}$ orbital. Unoccupied in diborane, this orbital has a large $p\sigma$ contribution in ethane, putting charge between the heavy centers where none exists in diborane. We can gain further insight into use of Walsh diagrams by returning to the rotational barrier problem. An additional feature of Hoffman's original calculation on ethane was his observation that the highest (doubly degenerate) occupied MO's were responsible for the major portion of the barrier energy. *Ab initio* calculations show the highest occupied MO's to be dominant in methylamine and methanol as in ethane, but this is not at all the case for ethyl fluoride, hydrogen peroxide, hydrazine, or hydroxylamine. In these molecules several of the molecular orbitals are changing by large and roughly equal, but unrelated, amounts. We may conclude from our study (illustrated here by the diborane, NO_2^-, and rotational barrier exam-

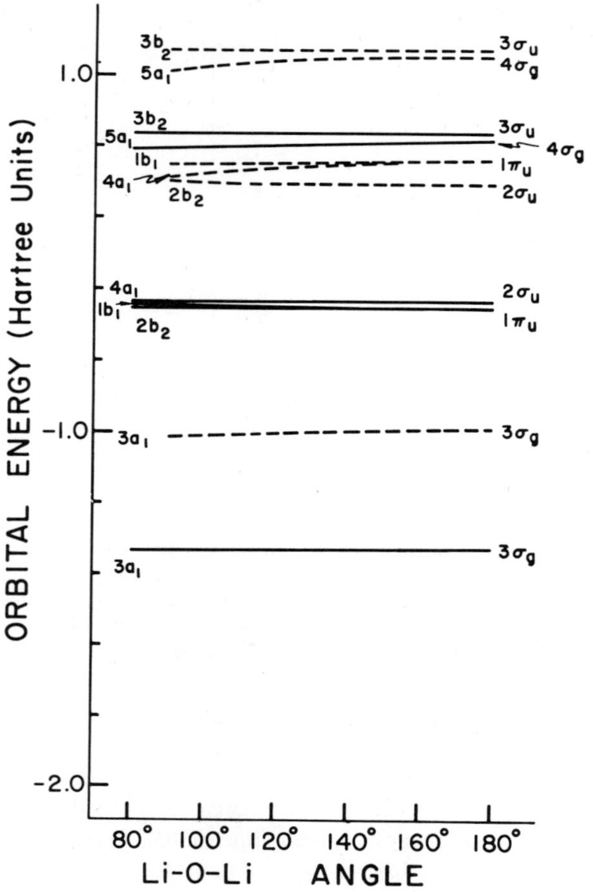

Fig. 14. Walsh diagram for strongly ionic AB_2 system (Li_2O). *Ab initio* (dashed lines), single-exponential-basis Hückel (solid lines).

ples) that the type of molecular transformation suitable to a simple orbital analysis (and thus, in principle, accessible to three-dimensional Hückel theory) is characterized by a "dominant orbital" in an appropriate Walsh diagram for the combined system. The dominant orbital(s) must change quite sharply with the independent variable (while the other orbitals are more or less flat) because Walsh diagrams for the two systems to be compared are not perfectly the same and because otherwise competing changes in other orbitals may greatly complicate the documentation of changes and make a straightforward analysis impossible.

Also, the systems must not be highly ionic (marked change in the degree of ionic bonding between the two systems is particularly to be avoided) because the occupancy of the atomic orbitals may not be properly reflected in a Walsh diagram and modification in the orbital environment, owing to the effective electron repulsion screening, is not correctly represented in this case. Not only do Walsh diagrams themselves contain less information for ionic bonding, but Hückel-generated Walsh diagrams tend to deviate increasingly from *ab initio*-generated diagrams with increasing electronegativity difference between adjacent atoms.

There are three more aspects of Hückel theory that are worth discussing. The first concerns Hückel theory's domain of capability; in particular we offer notes on properties for

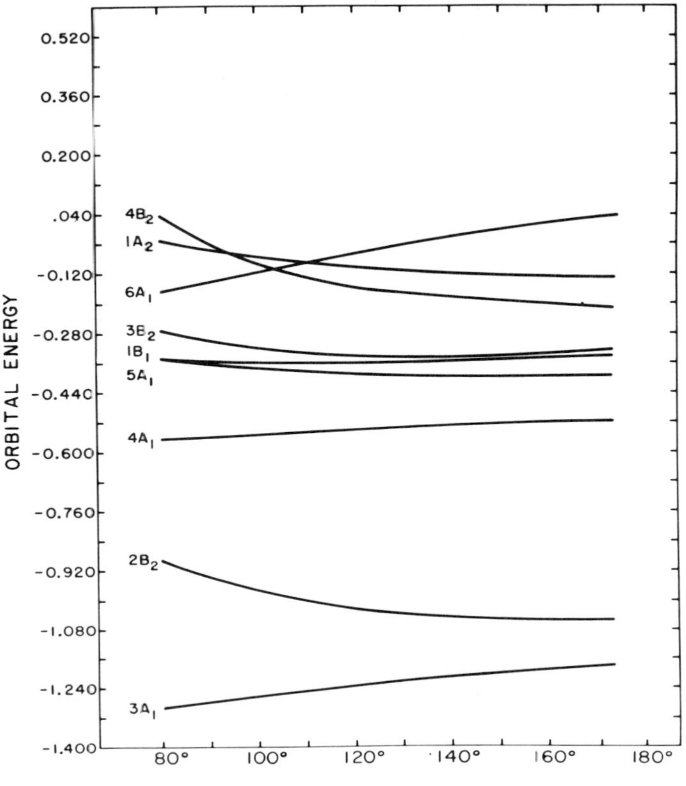

Fig. 15. Walsh diagram for NO_2^- Orbital energy in Hartree units versus O–N–O angle. (*N.B.*: The levels marked $3A_1$, $2B_2$, $4A_1$, etc., are usually designated $3a_1$, $2b_2$, $4a_1$, etc.)

which Hückel theory is unlikely to make useful predictions. Bond length, in contrast to bond angles, is one of these. Bond length involves a balance between one-electron and two-electron (V_{ee})effects, and examination of *ab initio* results shows that $\partial \Sigma^{\text{val.}} \epsilon/\partial R$ does not dominate the corresponding radial derivatives of V_{nn} and V_{ee}. In fact, it is easy to see that three-dimensional Hückel theory achieves $\Sigma \epsilon$ versus R minima from changes in sign of the overlap integrals involving p-orbital lobes–an effect not related to the physical forces causing equilibrium distances in real molecules and leading to no minima for molecules with only s electrons. Mulliken[21] and Coulson[22] have qualitatively related shorter bond lengths to greater occupancy of bonding orbitals (ϵ decreasing with decreasing bond length). There is also the well-known bond order–bond length relations. Use of these ideas is about as well as one can do with the data produced by Hückel theory. To obtain a good dipole moment it is necessary to include d orbitals (this is true for *ab initio* calculations as well), and it is hard to guess the appropriate Hückel α. In addition, dipole moments are sensitive to orbital coefficients and Hückel theory's relatively crude representation of *ab initio* MO coefficients does not provide a basis for much confidence in the predictions. The adequacy of molecular ionization potentials is also mixed. Using Koopman's theorem, *ab initio* results themselves do quite well in bringing out chemical trends and have an absolute error of $\sim$10 per cent. Comparative data for *ab initio* and single-exponential Hückel computations are given in Ref. 13. The Hückel theory is subject to errors on top of *ab initio* results and their magnitude follows the same patterns we have come to expect from our foregoing analysis: Small deviations from *ab initio* results are associated with covalent molecules

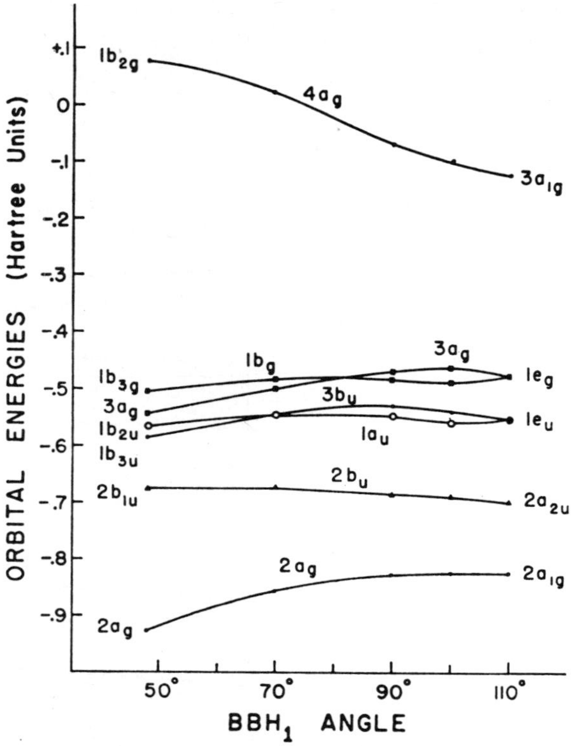

Fig. 16. Walsh diagram of diborane for the angular position of a bridge hydrogen. Positions progress from D_{2h} at small angles (normal diborane) through C_{2h} to D_{3d} at large angles (normal ethane conformation).

(∿5 per cent for BH_3). They become greater the more ionic the species (∿40 per cent for H_2O, a factor of $2\frac{1}{2}$ for Li_2O). When Hartree-Fock AO's are introduced the deviations are cut in half. The rather violent fluctuations in Hückel theory coefficients relative to *ab initio* solutions make it more difficult to follow chemical trends with Hückel theory. Use of an iterative scheme with Hückel theory specifically calibrated for molecular ionization potentials is probably the most reasonable procedure from a physical viewpoint, as well as the most practically useful from a chemistry viewpoint.[23] However, the molecular ionization potentials observed for organic compounds generally correspond to only a few per cent deviation from a relatively fixed number (about 10 to 12 eV); thus even for the case where there exists a natural parallelism between the *ab initio* and model theory, it is still not easy to produce quantitatively useful numbers from three-dimensional Hückel theory.

Another aspect concerns excited states. In Fig. 7 we have already pointed out one encouraging, but rather specialized, case of an excited-state potential curve quantitatively described by $\Sigma\ \epsilon$. However, theoretical calculations and theorems are rare for excited states, and this is unfortunate in view of the long-standing interest of organic chemists in photochemistry. Coulson[24] has given the best existing survey of theory and specific molecular systems. The problem is obviously vastly more complicated than the ground state because there will be many possible levels, often rather closely spaced, with potential-energy surface crossings and degeneracies (giving rise to Jahn-Teller distortions). Quantitative level ordering predictions for many closely spaced states is very difficult and the configuration interaction treatment necessary for handling degeneracies and near degeneracies is not within the framework of three-dimensional Hückel theory. A good fraction of observed excited states are Rydberg levels, which involve transitions to atomic-like levels of higher quantum numbers than the atomic valence orbitals represented in the LCAO-MO expansion. To make Hückel theory produce quantitative Rydberg levels is not so much a question of enlarging the basis set but (like the d functions for dipole moments) rather the severe problem of choosing α's. Singlet-triplet separations are an important aspect of excited-state studies, but since their separation is governed by the two-electron Coulomb and exchange terms, they are also outside the domain of the Hückel model. In spite of this list of troubles, we can still obtain useful information on excitations—again by application of Walsh's diagrams. In his empirical studies Walsh found that he could catalogue the excited states of some molecules by the following rule: The shape of an n-valence electron molecule in its first excited state resembles that of the ground state of a similar molecule containing n + 1 or n + 2 valence electrons. For example, consider the molecule BeH_2 and refer to the AH_2 Walsh diagram of Fig. 1. BeH_2 is linear in its ground state, $(2\sigma_g)^2(1\sigma_u)^2$, and bent in its first excited state, $(2a_1)^2(1b_2)(3a_1)$. This state is achieved by a one-electron excitation from $1b_2 - 1\sigma_u$ to the $3a_1 - 1\pi_u$ and $\Sigma\,\epsilon$ shows this to be bent, just as one would predict from an n + 1 valence electron molecule with the valence configuration $(2a_1)^2(1b_2)^2(3a_1)$. (In Walsh's original hypothesized diagrams, levels such as $1b_2 - 1\sigma_u$ were flatter and the $3a_1$ was pictured as undergoing a very large change with decreasing angle, thus making the n + 1 rule seem obvious. It is still largely true for our *ab initio* derived diagrams but the balance is more delicate.) Examination of Walsh diagrams for his various molecular types show that every other or the next succeeding energy levels have a slope reversal some place in the diagram (there are many cases of level crossings). For the n + 1 or n + 2 rule to work well there must be "dominant orbitals" such as those we described above. An extreme charge redistribution can accompany the excitation process and this may take a covalent ground state into a strongly ionic excited state (or vice versa), giving rise to the same problem we have already considered for ground states. There is another novel feature of three-dimensional Hückel theory that makes it attractive for the sort of excitation we are discussing here: The energy levels resulting from the Hückel secular equation are not affected by orbital occupancy and thus they nicely match the conceptual framework of the Walsh diagram. On the other hand, it is well recognized that because of the manner in which the two-electron effective potential is constructed in Hartree-Fock theory, only the energy positions of occupied levels have quantitative validity for *ab initio* results.[25]

Our final item to discuss is the matter of "improvements" to the structure of the three-dimensional Hückel model. There are four categories of physically motivated effects that are frequently mentioned as methods to enhance the general accuracy of results but which appear to retain Hückel theory's basic simplicity: (1) addition of kinetic-energy terms to diagonal and/or off-diagonal matrix elements, (2) addition of ionic (Madelung) terms to diagonal and/or off-diagonal matrix elements, (3) introduction of an interative procedure, and (4) determination of singlet-triplet separations. All of these modifications represent practical and quite modest computational efforts. Even the multicenter two-electron integrals required for singlet-triplet computations are readily available from a number of existing digital computer programs. However, each of these effects must be *ad hoc* in the sense that they do not enter the model theory with a direct mathematical and physical parallel in *ab initio* theory. It must also be recognized that insofar as addition of these effects are meant to upgrade the general information content of the model theory, there are equally important two-electron physical effects that are being omitted and it is doubtful that the addition of one or all of the four categories can significantly enhance the ability of Hückel theory to more nearly match Schrödinger equation solutions in a general way. The net of all this is that to get more quantitative information out of the Hückel theory beyond the simple form discussed here, we are forced to focus on one or perhaps two specific chemical quantities at a time; i.e., a modified Hückel theory may be developed to raise prediction accuracy for a particular expectation value, but we cannot expect the new model to simultaneously raise the accuracy for a whole collection of expectation values. To improve the overall quality one has to go to the next level of sophistication, and this always involves some approximate way of bringing in two-electron terms (the most popular of these during the last few years has been the various approximate forms of Hartree-Fock theory introduced by Pople et al.[26]). Lest one be left with the feeling that this higher level of theory ends the matter, we may note that there remains in all current theories the unresolved physical and conceptual question of three- and four-center, two-electron integral omission, and this gives rise to its own inadequacies for chemical problems.

4. Summary and Conclusions

The approach adopted in this article is to acknowledge that three-dimensional Hückel theory is capable of producing significant chemical information—particularly for organic chemistry—and then to ask why and how this comes about.

In the introduction we have pointed out the desirability of creating an *ab initio* foundation for Hückel theory. To establish a common meeting ground we have made the *ab initio* model theory comparison with Walsh diagrams (molecular orbital energies versus angle) and with plots of total energy and $\Sigma^{\text{val.}}\,\epsilon$.

In Section 2 are stated the principal physical and mathematical aspects of *ab initio* molecular orbital theory and three-dimensional Hückel molecular orbital theory. The origin, generic features, and range of validity of Walsh diagrams are established and related to the corresponding properties of Hückel theory. A macroscopic, heuristic "derivation" of the Hückel model is presented and its limitations are noted.

Results in the form of graphs are displayed. The independent variable for every figure is angular displacement. This coordinate can be either an internal angle in one molecule or an angular variable connecting the conformations of two separate molecular species. The graphs show total energies, *ab initio* $\Sigma\,\epsilon$, and Hückel $\Sigma\,\epsilon$ for a variety of covalent, partially ionic, and strongly ionic molecules and also sets of Walsh diagrams for the same bonding types. From these results we conclude that three-dimensional Hückel theory gives a good account of AO hybridization and its attendant energetics and nodal structure, molecular shape in ground and some excited states, and the ordering of MO energy levels.

The quality of results is significantly better when a Hartree-Fock AO basis set is employed instead of the more usual single-exponential set. In all cases the ability of Hückel theory to adequately represent the true molecular charge distribution decreases as the electronegativity difference between adjacent atoms increases.

An important application of Hückel-generated Walsh diagrams is the study of molecular transformation mechanisms in organic chemistry. MO energetics and hybridization effects may often be followed from one species to another by choice of an appropriate angular coordinate. In the paper we have given an example of the transformation of NO_2^+ to NO_2^- and of diborane to ethane. Both of these examples correspond to an increase of two electrons, but this scheme of representation has quite general utility whether or not the number of electrons changes. The simple isoelectronic example discussed was the conformation change in threefold rotational barriers, but examples more pertinent to organic reaction mechanisms are apparent. In these applications it was shown that successful use of the Walsh diagram technique is generally dependent upon the existence of a "dominant" or "controlling" orbital(s) whose sharp change in energy with angle is the predominating characteristic of the transformation.

The final part of the paper concerns chemical quantities for which Hückel theory has little chance of making useful predictions. These include bond length, dipole moment (and other inner and outer moments of the charge distribution), binding energies, and even for molecular ionization potentials special care must be exercised. The possibility of improvements to Hückel theory, while still retaining its simplicity, was also discussed, and we conclude that this may be accomplished for one property at a time but that no modification can notably raise the accuracy for many expectation-values simultaneously.

Note. After this manuscript was completed we received Vol. 10, Fasc. 4, of *Theoretica Chimica Acta* containing Blyholder and Coulson's article, Basis of Extended Hückel Formalism [G. Blyholder and C. A. Coulson, *Theoret. Chim. Acta* **10**, 316 (1968)]. It is satisfying to find that they have pointed out the uniform charge-density requirement for Hückel theory validity. Their approach is a deductive one in an attempt to derive Hückel theory from the same Hartree-Fock-Roothaan equations that we have employed as our *ab initio* reference point. They have not, however, had the aid of rigorous polyatomic wave-function calculations, and they have made only a limited attempt to specify what quantum mechanical quantities or what chemical properties Hückel theory is capable of predicting.

Acknowledgments

This research was supported by the Directorate of Chemical Science of the U.S. Air Force Office of Scientific Research, Contract AF 49(638)-1625.

I would like to acknowledge a stimulating conversation with Prof. John D. Roberts which encouraged me to write this article.

References and Notes

1. The total energy is written in this form because of the central role ϵ plays in Hückel theory. Other equivalent forms of the total energy expression which separate electron nuclear attraction from kinetic energy are often useful in explaining bonding effects in other problems. The identification of ϵ with ionization potential and thus with the binding energy of a given type of electron has a heuristic appeal for simple model theory but ϵ contains both the electron nuclear attraction term and the kinetic energy which acts like a repulsive potential. Thus the specific behavior of the kinetic energy term tends to be conceptually hidden in this formalism.
2. A. D. Walsh, *J. Chem. Soc.* **1953**, 2260. R. S. Mulliken, *Rev. Mod. Phys.* **14**, 204 (1942). Orbital energy versus angle graphs should really be called Walsh-Mulliken diagrams.
3. *Ab initio* calculations carried out in our laboratory at Princeton University include BH_2^+, BH_2^-, NH_2^+, BeH_2, BH_3, CH_3^+, and BeH_3^-, *J. Chem. Phys.* **45**, 734 (1966). B_2H_6 and C_2H_6, *J. Chem. Phys.* **45**, 2835 (1966). F_2O, Li_2O, FOH, LiOH, *J. Chem. Phys.* **45**, 3682 (1966). O_3 and N_3^-, *J. Chem. Phys.* **47**, 1953 (1967). $HCOO^-$, *J. Chem. Phys.* **1967**, 349. FNO, *Theoret. Chim. Acta* **9**, 103 (1967). C_2H_2, C_2H_4, *J. Chem. Phys.* **46**, 2029 (1967). CO_2 and BeF_2, *J. Chem. Phys.* **46**, 1797 (1967). CH_3OH and H_2O_2, *J. Chem. Phys.* **46**, 2261 (1967). CH_3NH_2, *J. Chem. Phys.* **46**, 2276 (1967). C_2H_5F (submitted to *J. Chem. Phys.*), NO_2^+ and NO_2^- (submitted to *J. Chem. Phys.*). N_2H_4 and NH_2OH, *J. Chem. Phys.* **47**, 895 (1967). CH_4, H_2O, NH_3, CH_5^+, NH_4^+, H_3O^+, H_2F^+, and BH_4^- (submitted to *J. Chem. Phys.*). CH_3F, CH_2F_2, CHF_3 (to be submitted to *J. Am. Chem. Soc.*). BH_3, BH_2F, BHF_2, BF_3 (to be submitted to *J. Chem. Phys.*). CH_3CHO (to be submitted to *J. Chem. Phys.*).
4. C. J. Ballhausen and H. B. Gray, *Inorg. Chem.* **1**, 111 (1962).
5. R. S. Mulliken, *J. Chim. Phys*, **46**, 497 (1949).
6. E. Clementi and D. L. Raimondi, *J. Chem. Phys.* **38**, 2685 (1963).
7. R. S. Mulliken, *J. Chem. Phys.* **2**, 782 (1934).
8. G. Pilcher and H. A. Skinner, *J. Inorg. Nucl. Chem.* **24**, 937 (1962). H. O. Pritchard and H. A. Skinner, *Chem. Rev.* **55**, 745 (1955).
9. Many of the atoms differ in the p-shell occupancy but not in the 2s shell; thus it is perhaps a little sur-

prising to find that ϵ_{2s} changes in magnitude from neutral atom to negative ion almost as much as the ϵ_{2p} values. It is also surprising that the multiplet averaged one-electron average, $(\epsilon + \epsilon^-)/2$, correlates so well with the highest valence values.

10. R. T. Sanderson, *Chemical Periodicity*, Reinhold, New York, 1960.
11. H. O. Pritchard and H. A. Skinner, *Chem. Rev.* **55,** 745 (1955). H. O. Pritchard and F. H. Sumner, *Proc. Roy. Soc. London* **A235,** 136 (1956). J. Hinze and H. H. Jaffé, *J. Am. Chem. Soc.* **84,** 540 (1962). J. Hinze, M. A. Whitehead, and H. H. Jaffé, *J. Am. Chem. Soc.* **85,** 148 (1963). G. Klopman, *J. Am. Chem. Soc.* **86,** 1463, 4550 (1964), **87,** 3300 (1965); *J. Chem. Phys.* **43,** S124 (1965). Among other things, these works show that Coulomb repulsions, bond type, bond length, and charges on adjacent atoms can all significantly modify electronegativity predictions.
12. One can demonstrate qualitatively that the potential-energy part of the matrix elements of *ab initio* calculations are roughly proportional to S, while it can be easily demonstrated that the kinetic-energy matrix elements are approximately proportional to S^2. Thus one can argue that for the off-diagonal Hückel element, one should either calculate the kinetic-energy terms directly or have a term in the element $\sim S^2$ [M. D. Newton, F. P. Boer, and W. N. Lipscomb, *Proc. Natl. Acad. Sci.* **53,** 1089 (1965); *J. Am. Chem. Soc.* **88,** 2353 2361, 2367 (1967); L. C. Cusachs, *J. Chem. Phys.* **43,** S157 (1965)]. One can also argue for an S^2 term in the diagonal Hückel matrix element due to neighbor-atom effects. On the other hand, the higher level approximate theory devised by Pople, which includes two-electron terms and yields results superior to Hückel theory, does not include an explicit dependence on kinetic-energy terms and its off-diagonal matrix element is linearly dependent on S [J. A. Pople, D. L. Beveridge, and P. A. Dobash, *J. Chem. Phys.* **47,** 2026 (1967)]. Further discussion of improvements to Hückel theory is given at the end of Section 3. These points illustrate the incompleteness of our discussion. However, our general approach here is a macroscopic one, in that we are not trying to establish a detailed one-to-one relation with *ab initio* theory. On an overall, superficial level, the intuitive interpretation of ϵ is consistent with our argument. Our attitude is to acknowledge that three-dimensional Hückel theory in its simplest form has made a significant contribution to chemistry and to ask why and how it does so.
13. L. C. Allen and J. D. Russell, *J. Chem. Phys.* **46,** 1029 (1967). Similar data for Hückel theory with the Hartree-Fock AO basis set is to be published by J. E. Williams, J. D. Russell, and L. C. Allen.
14. R. Hoffmann, *J. Chem. Phys.* **39,** 1397 (1963).
15. L. C. Allen and H. Basch, The Electronic Structure and Rotational Barrier in Ethyl Fluoride and Ethane (submitted to the *J. Chem. Phys.*).
16. As noted in Ref. 12, kinetic-energy effects will also contribute—approximately as S^2—and this tends to give more structure in the curves at large angles.
17. Current applications of molecular orbital energy diagrams have often achieved a connection between two species by orbital symmetry arguments alone. In our discussion we mean to employ the most general interpretation; thus "connected by an angular variable" implies full use of orbital symmetry. In cases where symmetry completely specifies the desired connection, the problem may be reducible to a two-point correlation diagram with no bond-coordinate displacement at all. Examples of recent work in this area are R. Hoffmann and R. B. Woodward, *J. Am. Chem. Soc.* **87,** 395, 2511, 2046, 4388 (1965). Hoffmann has been instrumental in popularizing three-dimensional Hückel theory (using the off-diagonal Hückel matrix elements in the arithmetic rather than a geometric mean form) starting with his article in *J. Chem. Phys.* **39,** 1397 (1963).
18. The results in Figs. 15 and 16 are *ab initio*, but similar diagrams are produced by Hückel theory.
19. Both bond lengths and some other angles change slightly in the transition from diborane to ethane, and these have been set at their equilibrium values for the various points on our graph, but the potential surface is so flat for these other lengths and angles that it would not make any difference qualitatively if they had not been adjusted.
20. The $3a_{1g}$ is not occupied in diborane, and so its position in this *ab initio* produced diagram is too high. In the similar calculation for ethane all the curves have the same shape but the position of the $3a_{1g}$ falls slightly below the $1e_g$. This effect is not present in three-dimensional Hückel theory.
21. R. S. Mulliken, *Rev. Mod. Phys.* **14,** 204 (1942).
22. C. A. Coulson, *Valence,* Oxford University Press, New York, 2nd ed., 1961.
23. Rein, Fukuda, Win, Clarke, and Harris, *J. Chem. Phys.* **1966,** 4743.
24. C. A. Coulson, chapter titled Theoretical Aspects of Electronic Excitation in Molecules, in *Reactivity of the Photoexcited Organic Molecule* (Solvay Institute of Chemistry, Proceedings of the 13th Conference, Brussells, 1964), Wiley-Interscience, New York, 1967.
25. We have already noted this effect in Fig. 16 (see Ref. 20).
26. J. A. Pople, D. L. Beveridge, and P. A. Dobash, *J. Chem. Phys.* **47,** 2026 (1967). Also many other articles stretching back to Pople's original introduction of his method, *Trans. Faraday Soc.* **49,** 1375 (1953).

VI-3

A Derivation of the Extended Hückel Theory from an Overlap Expansion of the Hartree-Fock Hamiltonian Matrix for Distorted-Atom Orbitals

T. L. GILBERT

1. Introduction

The extended Hückel (EH) theory may be regarded as a method of simulating Hartree-Fock (HF) calculations by guessing elements of the HF Hamiltonian.[1] A number of prescriptions for evaluating the matrix elements have been given.[2,3] In all the prescriptions, the dominant part of the contribution has the form

$$F_{ij} = K_{ij}S_{ij} \tag{1}$$

where F_{ij} are matrix elements of the one-electron Hamiltonian,[4] S_{ij} are the orbital overlap integrals, and the factor K_{ij} is obtained by a variety of prescriptions. Equation (1) suggests that one should be able to obtain the prescription in a formal manner by expanding the matrix elements of the Fock operator in powers of the overlap integrals and keeping only the first-order terms.

An analysis of this kind was initially undertaken in connection with an investigation of the properties of color centers.[5] Although the method is quite straightforward, and is doubtless known to many of workers in the field, it does not appear to be available in the published literature. The analysis is relevant to the theory of σ-MO methods.

The analysis will be limited to systems of closed-shell atoms or ions. The extension to systems of open-shell atoms or ions is immediate if the wave function for the entire system can be approximated by a single determinant and one restricts oneself to orbitals which can be obtained from the occupied orbitals by a linear transformation. An extension to the most general case in which one wishes to use atomic orbitals for a system composed of open-shell atoms or ions is nontrivial because a multideterminant representation of the wave function for the entire system must be used. Although it appears feasible to extend the analysis to this general case, it has not yet been carried through.

Extensive numerical tests of the method have not yet been attempted. It has been used to calculate the valence-band structure of the alkali chlorides, with reasonable results, and has been found to give good results for the LCAO matrix elements for He_2 at moderate values of the overlap.[6] In other cases, e.g., for the LCAO matrix elements of O_2^{4-}(Ref. 7) and for He_2 at very large internuclear distances,[6] preliminary results indicate that the agreement may be less satisfactory. Further studies are needed to establish the quantitative validity of the expansion. In this contribution only the formal analysis will be presented.

The formal analysis leads to the result that, by expanding the Fock matrix elements for a polyatomic system in powers of the overlap integrals, one obtains an expression in which the only molecular integrals appearing are the two-center kinetic energy and overlap integrals. The approximation obtained by neglecting terms of second and higher order in the overlap may be referred to as the "KO approximation." This designation seemed appropriate for an approximation which eliminates all molecular integrals except the kinetic-energy and overlap integrals.

2. Analysis

The Fock operator for an arbitrary polyatomic system consisting of closed-shell atoms may be written[8]

$$F = T + V^Z + V^C[\rho] + V^E[\rho] \tag{2}$$

where $T = -\frac{1}{2}\nabla^2$ is the kinetic-energy operator,[9] $V^Z = \Sigma V_a^Z = \Sigma - Z_a/|\mathbf{r} - \mathbf{R}_a|$ is the Coulomb field of the nuclei,

$$V^C[\rho] = 2\int |\mathbf{r} - \mathbf{r}'|^{-1}\rho(\mathbf{r}',\mathbf{r}')\,d\mathbf{r}' \tag{3}$$

is the electronic Coulomb potential [a local potential, $V^C(\mathbf{r})$],

$$V^E[\rho] = -\rho(\mathbf{r},\mathbf{r}')/|\mathbf{r} - \mathbf{r}'| \tag{4}$$

is the exchange potential [a nonlocal potential, $V^E(\mathbf{r},\mathbf{r}')$], and $\rho(\mathbf{r},\mathbf{r}')$ is the spin-independent part of the kernel of the density operator.

When ρ is expressed in terms of atomic orbitals, we have

$$\rho(\mathbf{r},\mathbf{r}') = \Sigma_{ai,bj}\phi_{ai}(\mathbf{r})S^{-1}_{ai,bj}\phi^\dagger_{bj}(\mathbf{r}') \tag{5}$$

where $\mathbf{S}^{-1} = [S^{-1}_{ai,bj}]$ is the inverse of the overlap matrix $\mathbf{S} = [\mathbf{S}_{ai,bj}] = [\langle\phi_{ai}|\phi_{bj}\rangle]$.[10] For our purposes it is convenient to write ρ as the sum

$$\rho = \Sigma_a\rho_a + \Sigma_a\delta\rho_a + \Sigma'_{ab}\rho_{ab} \tag{6}$$

where

$$\rho_a = \Sigma_i\phi_{ai}\phi^\dagger_{ai} \tag{7}$$

is the atomic density operator, which is identical with the density operator for a free atom in the LCAO approximation,

$$\delta\rho_a = \Sigma_{ij}\phi_{ai}(S^{-1}_{ai,aj} - \delta_{ij})\phi^\dagger_{aj} \tag{8}$$

is the intraatomic overlap density operator, which corresponds to an added electronic charge distribution within the atom, and

$$\rho_{ab} = \Sigma_{ij}\phi_{ai}S^{-1}_{ai,bj}\phi^\dagger_{bj} \tag{9}$$

is a component of the interatomic overlap density operator $\sigma_a = \Sigma'_b\rho_{ab}$, which, on the average, corresponds to a positive charge density between the atoms.[11] The prime on the sum indicates that the term a = b is omitted.

The expressions for the Coulomb and exchange potentials are linear in the density operator, so that we may write

$$\begin{aligned} V^C[\rho] &= \Sigma_a(V^C[\rho_a] + V^C[\delta\rho_a]) + \Sigma'_{ab}V^C[\rho_{ab}] \\ V^E[\rho] &= \Sigma_a(V^E[\rho_a] + V^E[\delta\rho_a]) + \Sigma'_{ab}V^E[\rho_{ab}] \end{aligned} \tag{10}$$

In order to simplify the notation in what follows, we shall set $V^C[\rho_a] = V_a^C$, $V^C[\delta\rho_a] = \delta V_a^C$, $V^C[\rho_{ab}] = V_{ab}^C$, and similarly for the components of V^E. These expansions may be used to separate those terms which give large contributions to the matrix elements between atomic orbitals from those terms which give small contributions.

The appropriate separation for evaluating one-center matrix elements is

$$F = F_a + U_a \tag{11}$$

where

$$F_a = T + V_a^Z + V_a^C + V_a^E \tag{12}$$

is the Fock operator for the ath atom or ion and

$$U_a = \Sigma_b'(V_b^Z + V_b^C + V_b^E) + \Sigma_b(\delta V_b^C + \delta V_b^E) + \Sigma_{bc}'(V_{bc}^C + V_{bc}^E) \tag{13}$$

is the potential field produced by the environment of the ath atom or ion in the polyatomic system (the atomic environment potential).

The appropriate separation for evaluating two-center matrix elements is

$$F = F_a + F_b + U_{ab} - T \tag{14}$$

where F_a and F_b are the atomic Fock operators defined by equation (12) and

$$U_{ab} = \Sigma_c''(V_c^Z + V_c^C + V_c^E) + \Sigma_c(\delta V_c^C + \delta V_c^E) + \Sigma_{cd}'(V_{cd}^C + V_{cd}^E) \tag{15}$$

is the nonlocal potential field produced by the atoms or ions which comprise the environment of the ath and bth atoms. The double prime on the first sum indicates omission of the terms for which c = a and c = b. The interaction $V_{ab}^C + V_{ab}^E$ between the two atoms is included in U_{ab}, so that it is not the true diatomic environment potential.[12]

We must now choose the orbitals with respect to which the matrix elements of the Fock operator are to be evaluated. One could choose atomic orbitals for the separated atoms; however, this would immediately restrict us to an LCAO approximation. We would like to show that the extended Hückel theory can be derived systematically from an SCF approximation without using the LCAO approximation as an intermediate step. This can be done by using screened atomic orbitals (also referred to as "Adam's orbitals"[13] or "distorted atomic orbitals"[14]).

The distorted atomic orbitals may be defined by the pseudoeigenvalue equation

$$[F_a + U_a - \rho U_a \rho]\phi_{ai} = \epsilon_{ai}\phi_{ai} \tag{16}$$

where $\rho U_a \rho$ designates a triple operator product, $\int \rho(\mathbf{r},\mathbf{r}'')\,U_a(\mathbf{r},\mathbf{r}''')\,\rho(\mathbf{r},\mathbf{r}')\,d\mathbf{r}''d\mathbf{r}'''$. It is the projection of the atomic environment potential U_a on the Hartree-Fock manifold for the entire system. This implies that $\rho U_a \rho$ acts as a null operator for any orbital which is orthogonal to all occupied Hartree-Fock orbitals, i.e., $\rho U_a \rho \chi = 0$ if $\langle \chi | \phi_{ai} \rangle = 0$, but gives identically the same matrix elements as U_a for any two occupied Hartree-Fock orbitals, i.e., $\langle \phi_{ai} | \rho U_a \rho | \phi_{bj} \rangle = \langle \phi_{ai} | U_a | \phi_{bj} \rangle$. There is a separate equation for each atom, a = 1, 2, . . . , N, and only the localized orbitals, i = 1, 2, . . . , n_a, corresponding to the n_a lowest eigenvalues are used. The total number of orbitals, $\sum_{a=1}^{N} n_a$, will equal the number of occupied orbitals. The number of independent equations which must be solved will correspond to the number of atoms at inequivalent sites.

Let us define the screened environment potential,

$$U_a' = U_a - \rho U_a \rho \tag{17}$$

We note that the matrix elements $\langle \phi_{ai} | U_a' | \phi_{bj} \rangle$ between any two occupied Hartree-Fock orbitals will vanish. If one introduces the approximation of setting the screened environment potential identically equal to zero, then equation (16) reduces to the Hartree-Fock equations for separated atoms,

$$F_a^0 \phi_a^0 = \epsilon_{ai}^0 \phi_{ai}^0 \tag{18}$$

It should be noted that F_a and F_a^0 are not identical because equations (16) and (18) are pseudoeigenvalue equations which must be solved by means of the SCF procedure. F_a^0 will, therefore, be a function of the density operator ρ^0 constructed from the undistorted orbitals ϕ_{ai}^0, while F_a will be the same function of the density operator ρ constructed from the distorted orbitals ϕ_{ai} [see equation (5)].

The LCAO-MO approximation consists in constructing a set of molecular orbitals from linear combinations of undistorted atomic orbitals. In the SCF-MO approximation, one may either solve the standard Hartree-Fock equations $F\phi_{i\lambda\alpha} = \epsilon_{i\lambda\alpha}\phi_{i\lambda\alpha}$ to obtain the molecular orbitals $\phi_{i\lambda\alpha}$ directly, or one may construct the molecular orbitals from linear combinations of the distorted atomic orbitals. If the distorted atomic orbitals are chosen to be self-consistent solutions of equation (16), then one can always find linear combinations which will be identical with the occupied Hartree-Fock MO's. *One may, therefore, regard the difference between an LCAO and an SCF wave function as a difference which is due to the distortion, $\delta\phi_{ai} = \phi_{ai} - \phi_{ai}^0$, of the atomic orbitals.* This distortion may, in turn, be regarded as an effect produced by the screened environment potential U_a'. The important consideration for what follows is that the derivation which leads to the KO approximation is the same for distorted or undistorted orbitals; hence the formal expressions are valid for both the SCF and LCAO approximations.[15]

Introducing the notation $|ai\rangle$ for $|\phi_{ai}\rangle$, the relations which we shall need for deriving the KO approximation are

$$\begin{aligned} \langle ai|F_a|bj\rangle &= \langle ai|F_a + U_a'|bj\rangle = \epsilon_{ai}\langle ai|bj\rangle \\ \langle ai|F_b|bj\rangle &= \langle ai|F_b + U_b'|bj\rangle = \epsilon_{bj}\langle ai|bj\rangle \end{aligned} \tag{19}$$

which hold for distorted and, mutatis mutandis, undistorted orbitals.

Using equations (11) and (19), the one-center matrix elements of the Fock operator between atomic orbitals may be written

$$\langle ai|F|bj\rangle = \epsilon_{ai}\delta_{ij} + \langle ai|U_a|bj\rangle \tag{20}$$

where ϵ_{ai} are the atomic orbital energies obtained from equation (16).

The difference $\delta\epsilon_{ai} = \epsilon_{ai} - \epsilon_{ai}^0$ between the eigenvalues for distorted and undistorted atoms can be expected to be of second order in the overlap, so that it is reasonable to replace ϵ_{ai} by ϵ_{ai}^0, which can easily be obtained from equation (18), for actual calculations.

Using equations (14) and (19), the two-center matrix elements of the Fock operator become

$$\langle ai|F|bj\rangle = \langle ai|bj\rangle(\epsilon_{ai} + \epsilon_{bj}) + \langle ai|U_{ab}|bj\rangle - \langle ai|T|bj\rangle \tag{21}$$

The KO approximation consists in keeping only those parts of the matrix elements $\langle ai|U_a|bj\rangle$ and $\langle ai|U_{ab}|bj\rangle$ which are of first order in the overlap.

To separate out terms of various order in the overlap we use the familiar Löwdin expansion,

$$S_{ai,bj}^{-1} = \delta_{ai,bj} - (S_{ai,bj} - \delta_{ai,bj}) + \sum_{ck} (1 - \delta_{ca})(1 - \delta_{cb}) S_{ai,ck} S_{ck,bj} + \cdots \tag{22}$$

where $S_{ai,bj} = \langle ai|bj\rangle$. When this expansion is introduced into the intraatomic overlap density operator we obtain, omitting terms of second order in the overlap,

$$\delta\rho_a \approx -\sum_{ij}{}' \phi_{ai} S_{ai,aj} \phi^\dagger_{aj} = 0 \tag{23}$$

The second equality is a consequence of the fact that the distorted atomic orbitals on a given center form an orthonormal set ($S_{ai,aj} = \delta_{ij}$).

The components of the interatomic overlap density operator, up to terms of second order in the overlap, are

$$\rho_{ab} = -\sum_{ij} \phi_{ai} S_{ai,bj} \phi^\dagger_{bj} \tag{24}$$

The overlap integrals $S_{ai,bj}$ do not vanish for $a \neq b$.

We now use these various expansions to calculate the matrix elements $\langle ai|F|bj\rangle$ up to terms of second order in the overlap.

One-Center Matrix Elements

Referring to equation (13), one can readily establish that the terms of $\langle ai|U_a|bj\rangle$ coming from the last two sums and from $\langle ai|V^E_b|bj\rangle$ are of second order or higher in the overlap. Hence, for the purpose of evaluating the matrix elements, we may write

$$\begin{aligned} U_a &\approx \sum_b{}' (V^Z_b + V^C_b) \\ &= \sum_b{}' \frac{N_b - Z_b}{|\mathbf{r} - \mathbf{r}_b|} + \sum_b{}' \int \frac{2\rho(\mathbf{r}', \mathbf{r}') - N_b \delta(\mathbf{r}' - \mathbf{R}_b)}{|\mathbf{r} - \mathbf{r}'|} d\mathbf{r}' \\ &= V^{PI}_a + \sum_b{}' V^0_b \end{aligned} \tag{25}$$

where V^{PI}_a is the field of the point-ion lattice (with the ath ion removed) and V^0_b is the total (electronic and nuclear) Coulomb potential of the bth atom or neutral ion (i.e., an ion in which the true nuclear charge Z_a has been replaced by an effective nuclear charge N_a to give a neutral system). The matrix elements $\langle ai|V^0_b|aj\rangle$ will be negligible when the overlap integrals $\langle ai|bj\rangle$ all vanish, because the nuclear charge N_a will then be completely shielded.[16]

The question now arises: Of what order in the overlap is $\langle ai|V^0_b|ai\rangle$? At this point we argue that $\langle ai|V^0_b|ai\rangle$ must always be negative for small overlap, so that, if we expanded it in powers of the overlap integrals, the leading term would have to be of second order in the overlap. This same argument can be used for the off-diagonal matrix elements $\langle ai|V^0_b|aj\rangle$, which will, in general, be much smaller than diagonal elements $\langle ai|V^0_b|ai\rangle$. Hence, up to terms of second order in the overlap, the one-center matrix elements of the atomic environment potential reduce to

$$\langle ai|U_a|aj\rangle \approx \langle ai|V^{PI}_a|aj\rangle \tag{26}$$

and the one-center matrix elements of the Fock operator become

$$\langle ai|F|aj\rangle = \delta_{ij}\epsilon_{ai} + \langle ai|V^{PI}_a|aj\rangle \tag{27}$$

If the atom or ion is located at a point of high symmetry, then the additional simplification

$$\langle ai|V^{PI}_a|aj\rangle \approx \delta_{ij} V^{PI}_a \tag{28}$$

can be used, where $V_a^{PI} = V_a^{PI}(\mathbf{R}_a)$ is the point-ion field evaluated at the position, $\mathbf{R}_a$, of the missing atom or ion.

Two-Center Matrix Elements

The only contribution to the matrix elements of $\langle ai|U_{ab}|bj\rangle$ which can be of first order in the overlap must come from terms in U_{ab} which are, in effect, of zero order. One may readily establish that there are only two terms which can do this.[17] One is the potential field V_{ab}^{PI} of a point-ion lattice with the ath and bth ions removed. If the midpoint between the ions is a point of high symmetry, one may be able to use the approximation

$$\langle ai|V_{ab}^{PI}|bj\rangle \approx S_{ai,bj}V_{ab}^{M} \tag{29}$$

where V_{ab}^{PI} is the point-ion potential $V_{ab}^{PI}(\mathbf{r})$ evaluated at the center of the overlap charge distribution $\rho_{ai,bj}(\mathbf{r}) = \phi_{ai}(\mathbf{r})\phi_{bj}^{\dagger}(\mathbf{r})$.[18]

The other term which cannot be neglected is V_{ab}^{E}. For this term we have [using Equation (24)]

$$\begin{aligned}\langle ai|V_{ab}^{E}|bj\rangle &= -\iint \frac{\phi_{ai}^{\dagger}(\mathbf{r})\rho_{ab}(\mathbf{r},\mathbf{r}')\phi_{bj}(\mathbf{r}')}{|\mathbf{r}-\mathbf{r}'|}\,d\mathbf{r}\,d\mathbf{r}' \\ &= +\sum_{kh}\iint \frac{\phi_{ai}^{\dagger}(\mathbf{r})\phi_{ak}(\mathbf{r})S_{ak,bh}\phi_{bh}^{\dagger}(\mathbf{r}')\phi_{bj}(\mathbf{r}')}{|\mathbf{r}-\mathbf{r}'|}\,d\mathbf{r}\,d\mathbf{r}' \\ &\approx S_{ai,bj}\iint \frac{|\phi_{ai}(\mathbf{r})|^2\,|\phi_{bj}(\mathbf{r}')|^2}{|\mathbf{r}-\mathbf{r}'|}\,d\mathbf{r}\,d\mathbf{r}' \\ &\approx S_{ai,bj}/|\mathbf{R}_a-\mathbf{R}_b|\end{aligned} \tag{30}$$

The error introduced by the approximation of replacing the integral by $|\mathbf{R}_a - \mathbf{R}_b|^{-1}$ is of second order in the overlap; hence it can be used for obtaining the first-order contribution to the matrix element.

The two-center matrix elements for U_{ab} may, therefore, be reduced to

$$\langle ai|U_{ab}|bj\rangle = \frac{S_{ai,bj}}{|\mathbf{R}_a-\mathbf{R}_b|} + \langle ai|V_{ab}^{PI}|bj\rangle \tag{31}$$

and the two-center matrix elements of the Fock operator become [from equations (21) and (31)]

$$\langle ai|F|bj\rangle = S_{ai,bj}\left(\epsilon_{ai} + \epsilon_{bj} + \frac{1}{|\mathbf{R}_a-\mathbf{R}_b|}\right) - T_{ai,bj} + \langle ai|V_{ab}^{PI}|bj\rangle \tag{32}$$

where $T_{ai,bj} = \langle ai|-\frac{1}{2}\nabla^2|bj\rangle$. Equation (29) may sometimes be used for further simplification of this expression.

Equations (27) and (32), with (28) and (29) when applicable, constitute the KO approximation. Equation (32) is similar, but not identical, to some of the prescriptions which have been introduced for extended Hückel calculations.[19] Numerical studies and further work to include modification for open-shell systems will be needed to establish whether it offers any advantages. At this stage it has only the esthetic asset of being derivable in a systematic manner from first principles.

Acknowledgment

Based on work performed under the auspices of the U.S. Atomic Energy Commission.

References and Notes

1. F. P. Boer, M. D. Newton, and W. N. Lipscomb, *Proc. Natl. Acad. Sci. U.S.* **52**, 890 (1964).
2. K. Fukui, in *Modern Quantum Chemistry*, Part 1, Orbitals, O. Sinanoğlu, ed., Academic Press, New York, 1965, pp. 49–85. This review article includes references to earlier work.
3. M. D. Newton, F. P. Boer, W. E. Palke, and W. N. Lipscomb, *Proc. Natl. Acad. Sci. U.S.* **53**, 1089 (1965).
4. The symbol F is used for the one-electron Hamiltonian (the "Fock operator") to emphasize the distinction between this operator and the true many-body Hamiltonian for the system.
5. T. L. Gilbert, Hole Centers in Alkali Halides, unpublished notes for lectures given at the NATO Summer School in Solid State Physics, Ghent, Belgium, Sept. 5–16, 1966.
6. T. L. Gilbert and A. C. Wahl, unpublished calculations.
7. F. Farate, unpublished calculations.
8. The word atom should be interpreted throughout as atom or ion.
9. Atomic units are used throughout, so that $e = m = \hbar = 1$.
10. S^{-1} is also called the charge and bond-order matrix.
11. If we write $\Sigma'_{ab}\rho_{ab} = \Sigma_a \sigma_a$, where we can either set $\sigma_a = \Sigma'_b \rho_{ab}$ (unsymmetric form) or $\sigma_a = \frac{1}{2}\Sigma'_b (\rho_{ab} + \rho_{ba})$ (symmetric form), then we have $\Sigma_a \operatorname{tr}(\delta\rho_a + \sigma_a) = 0$, and the negative charge distributions, $\delta\rho_a(\mathbf{r},\mathbf{r})$, can be paired off with the surrounding positive charge distributions, $\sigma_a(\mathbf{r},\mathbf{r})$. This is important for infinite systems to avoid divergent sums.
12. The diatomic environment potential is obtained by writing $F = F_{ab} + U_{ab}$, where F_{ab} is the Fock operator for a diatomic molecule. The true diatomic environment potential, $F - F_{ab}$, includes $V_{ab}^C + V_{ab}^E$, some small terms associated with the contribution of atom b to $\delta\rho_a$, etc.
13. W. H. Adams, *J. Chem. Phys.* **34**, 89 (1961); **37**, 2009 (1962).
14. There are a variety of criteria for defining distorted atomic orbitals. Adam's criterion consists in minimizing the energy of each atom in the polyatomic system which, in effect, minimizes the difference between the energy of the separated atom and the self-energy of the same atom within the polyatomic system. Other criteria, such as that of minimizing the overlap of the distorted and undistorted orbitals, could also be used. Adam's criteria has a number of conceptual advantages. For example, it minimizes the changes in the self-energy, thereby placing most of the information concerning the potential interaction into the explicit interaction terms; it leads to an intuitively appealing form for the environment potential in the equations for the localized orbitals [see equation (16)]; and it permits one to write the interaction energy for closed-shell systems in a simple form which contains no kinetic-energy terms (see Ref. 13). For a discussion of various localization criteria, see T. L. Gilbert, in *Molecular Orbitals in Chemistry, Physics and Biology* P. O. Löwdin, ed., Academic Press, New York, 1964, p. 405.
15. One is not constrained to use atomic orbitals. An alternative would be to divide the system up into groups of atoms (constituent molecules). The analysis would be the same. The only difference would be the reinterpretation of the index as a label for constituent molecules rather than constituent atoms.
16. The multipole contributions due to distortion are assumed to be negligible for closed-shell atoms or ions. For open-shell systems there would be appreciable multipole contributions, which would have to be separated out and added to the point-ion field.
17. We are here ignoring matrix elements between orbitals which are separated far enough so that another atom can lie between them. If the overlap between such a pair of centers were nonnegligible, the overlap between nearest and next-nearest neighbors would be so large that the KO approximation could not be used.
18. In simple ionic lattices, such as the alkali halides, the center of the overlap charge distribution is at the midpoint between like ions but is slightly away from the midpoint (toward the positive ion) between unlike ions. In the former case $V_{ab}^M = (2Z_a/R_{ab}) + (2Z_b/R_{ab})$ and is large. In the latter case, $V_{ab}^{PI} = 0$ at the midpoint, so that V_{ab}^M will be small. This approximation is not an essential part of the KO approximation. In environments with low symmetry, it may often be necessary to evaluate $\langle\phi_{ai}|V_{ab}^{PI}|\phi_{bj}\rangle$ exactly using three-center nuclear attraction integrals.
19. See Ref. 2, p. 55.

Atomic Orbitals for Semiempirical Molecular Orbital Calculations

LOUIS CHOPIN CUSACHS and JOYCE H. CORRINGTON

1. Simplified Representation of Atomic Orbitals

Numerous researchers have employed the Hartree-Fock technique to formulate numerical wave functions for all the neutral atoms (and their ions) of the periodic table. Such numerical functions are quite difficult to employ in semiempirical molecular orbital calculations (so-called extended Hückel calculations which require explicit atomic orbitals for the evaluation of overlap integrals). Clementi[1] employed Roothaan's technique to tabulate "analytic" atomic wave functions, expressing each orbital in terms of a basis set of from 4 to 11 Slater-type orbitals (STO) of the form

$$\psi(n,\ell,m,z) = r^{n-1} \exp(-zr) Y_1^m(\theta,\phi)$$

where Y_1^m are ordinary spherical harmonics, z is the orbital exponent equal to Z_e/n, and Z_e is the effective nuclear charge. The multi-z functions are necessary not only to create nodes (a single STO is nodeless) but also to approximate the fact that the Z_e for an orbital should really be expressed as a function of r, having the value of the nuclear charge near the nucleus and decreasing to a value of 1 (for a neutral atom) far from the nucleus. Hence a single STO function will always be inadequate to represent even the outer lobe of an orbital accurately, but the present computational speed for large molecules demands, even so, that simplified representations be employed in molecular orbital calculations.

Several procedures for selecting a single STO for each valence orbital have been suggested. Notably, Slater[2] proposed screening constant formulas to estimate the effective nuclear charge. Although these are widely used, Brown and Fitzpatrick[3] have extensively investigated d-orbital representations and have concluded that the Slater functions inadequately describe any orbital property. Clementi and Raimondi[4] have presented a minimum basis set of STO's (one for each n,1 combination) for atoms through krypton which were calculated by minimizing the atomic energy. Brown and Fitzpatrick found that they do reproduce d-orbital kinetic and nuclear potential energies but do not reproduce the overlap region well. Burns[5] has presented functions which best match value of $\langle r^m \rangle$ calculated from the SCF functions (m varying from −4 to +3). Brown and Fitzpatrick found that they reproduce kinetic and nuclear potential energies poorly but approximate, although not too accurately, the overlap region. Cusachs et al.[6] proposed, for a limited number of atoms, functions of empirical z and n selected to represent as accurately as possible by a single STO the overlap region as defined by the SCF functions.

This latter work suggests what can be done accurately by a single STO representation. Granting that it is impossible to adequately represent all the properties of an SCF function by a single STO representation, it still may be possible to represent any single property accurately by a single STO representation.

2. Orbital Exponents for Energy

As a first example, the kinetic-energy operator for a STO has been shown to be

$$T = -\tfrac{1}{2}z^2\left[1 - \frac{2n}{zr} + \frac{(n+\ell)(n-\ell-1)}{z^2 r^2}\right]$$

Applying this operator to an STO orbital function, ψ_1, to determine the kinetic energy becomes simply a matter of determining the average value of r^{-1} and of r^{-2} for that orbital.

$$\int_0^\infty \psi_i^* T\psi_i dv = -\tfrac{1}{2} z_i^2 \left[1 - \frac{2n_i}{z_i} \langle r^{-1}\rangle + (n_i + \ell_i)(n_i - \ell_i - 1)\frac{\langle r^{-2}\rangle}{z_i^2}\right]$$

where

$$\langle r^{-m}\rangle = \frac{(2z)^{2n+1}}{2n!} \int_0^\infty r^{2n} r^m e^{-2zr}\, dr = \frac{(2n+m)!}{2n!(2z)^m}$$

For a nodeless orbital (1s, 2p, 3d, 4f), where $\ell = n - 1$, the $(n + \ell)(n - \ell - 1)\langle r^{-2}\rangle$ factor disappears and the kinetic energy is simply a function of $\langle r^{-1}\rangle$. Hence, if one determines the value of $\langle r^{-1}\rangle$ for the exact SCF wave function, a single STO can be constructed which reproduces this value by using the relation

$$\langle r^{-1}\rangle = 2z/2n$$

to define the z (assuming the nominal quantum number value of n). This function is characteristic of the energy of the orbital and thus should correspond to the Clementi-Raimondi function for the same orbital which was determined by energy minimization. Good agreement is found (Table 1).

Table 1. Values of $z = n\langle r^{-1}\rangle$ Calculated from SCF Atomic Functions[1] Compared to Energy-Optimized Values

Element	Orbital	$1/\langle r^{-1}\rangle$, a.u.	$z = n\langle r^{-1}\rangle$	z energy[4]
B	2p	1.653	1.21	1.21
C	2p	1.276	1.57	1.57
N	2p	1.044	1.92	1.92
O	2p	0.900	2.22	2.23
F	2p	0.786	2.54	2.55
Sc	3d	1.252	2.40	2.37
Ti	3d	1.102	2.72	2.71
V	3d	1.002	2.99	2.99
Cr	3d	0.923	3.25	3.25
Mn	3d	0.857	3.50	3.51
Fe	3d	0.809	3.71	3.73
Co	3d	0.763	3.93	3.95
Ni	3d	0.723	4.15	4.18
Cu	3d	0.687	4.37	4.40
Zn	3d	0.653	4.74	4.63

In the case of functions with nodes ($\ell \neq n - 1$), it can readily be seen that for a single STO to represent the kinetic energy it must reproduce the $\langle r^{-1}\rangle$ and the $\langle r^{-2}\rangle$ values of the SCF function by satisfying the relations

$$\langle r^{-1}\rangle = 2z/2n$$

$$\langle r^{-2}\rangle = (2z)^2/(2n)(2n - 1)$$

Thus n can no longer be arbitrarily chosen to be the principal quantum number but must be determined by these relations. Since Clementi and Raimondi did not allow for this, further close agreement between their orbital coefficients and those estimated by this procedure would not be expected. It does appear, though, that as the Clementi-Raimondi z's for the nodeless functions almost exactly predict the orbital kinetic energy (note, not the orbital energy) by

$$T_i = -\frac{1}{2} z_i^2 \left[1 - \frac{2n_i}{z_i} (z_i/n_i)\right] = \frac{1}{2} z_{CRI}^2$$

so they continue to approximate it for the functions with nodes by

$$T_j = -\frac{1}{2} z_j^2 \left[1 - \frac{2n_j}{z_j} \langle r^{-1} \rangle + \frac{(n_j + \ell_j)(n_j - \ell_j - 1)\langle r^{-2} \rangle}{z_j^2}\right] \approx \frac{1}{2} z_{CRj}^2$$

One observes by applying the virial theorem

$$E_{atom} = T + V = T - 2T = -T = \sum_i T_i \approx \frac{1}{2} \sum_i z_{CRi}^2$$

that the atomic energies are approximately equal to half the sum of the squares of the Clementi-Raimondi z's (see Table 2).

Table 2. Atomic Energies Calculated from Clementi-Raimondi[4] STO Orbital Coefficients

Element	$\frac{1}{2} \Sigma_i z_i^2$, a.u.	−Atomic energy, a.u.
He	2.85	2.86
Li	7.44	7.43
Be	14.49	14.57
B	24.29	24.53
C	37.12	37.69
N	53.64	54.40
O	73.60	74.81
F	97.65	99.41
Ne	126.13	128.55

3. Orbital Exponents for Overlap

A second example of great interest is representing the overlap region accurately by a single STO. It is not apparent, as in the case of kinetic energy, what if any average value can be obtained from the SCF function to fix the single STO which would best accomplish this. However, Cusachs et al.[6] have provided empirical values of n and z optimized to reproduce SCF overlap against which any model can be checked. Values of z calculated from $\langle r^{-1} \rangle$, $\langle r \rangle$, and $\langle r^2 \rangle$ are compared to the overlap optimized values (Table 3). It is found that the values

Table 3. Values of z Calculated from $< r^m >$ Values of SCF[1] Functions Compared to Overlap Optimized z

Element	Orbital	$z = \frac{n}{\langle r^{-1} \rangle}$	$z = \frac{n + \frac{1}{2}}{\langle r \rangle}$	$z = \sqrt{\frac{(n + \frac{1}{2})(n + 1)}{\langle r^2 \rangle}}$	z overlap[6]
C	2s	1.79	1.58	1.57	1.60
N	2s	2.14	1.88	1.87	1.90
O	2s	2.52	2.19	2.18	2.20
F	2s	2.88	2.50	2.48	2.50
C	2p	1.56	1.46	1.41	1.43
N	2p	1.92	1.77	1.72	1.69
O	2p	2.22	2.03	1.95	1.95
F	2p	2.54	2.30	2.21	2.21

of z calculated from $\langle r^{-1} \rangle$ are not good representations for overlap, but that the values calculated using the relations

$$\langle r \rangle = \frac{2n + 1}{2z}$$

$$\langle r^2 \rangle = \frac{(2n + 1)(2n + 2)}{(2z)^2}$$

do closely reproduce the overlap optimized values.

Note that Cusachs et al.'s selected values for n equal to the principal quantum number and also equal to one less than the principal quantum number and got a substantially better overlap match with the second value. If the overlap STO is required to reproduce both $\langle r \rangle$ and $\langle r^2 \rangle$, then n can no longer be arbitrarily chosen but is determined by these relations. Table 4

Table 4. $\langle r \rangle$ and $\langle r^2 \rangle$ Values for Ground-State Configurations of Atoms and Ions Calculated from SCF Atomic Functions[1,8]

		$\langle r \rangle$, a.u.			$\langle r^2 \rangle$, a.u.^{2}		
Element	Orbital	+ 1 ion	Atom	- 1 ion	+1 ion	Atom	- 1 ion
Li	2s		3.87			17.74	
Be	2s	1.98	2.65		6.10	8.42	
B	2s	1.80	1.98	2.12	3.83	4.71	5.44
C	2s	1.47	1.59	1.69	2.59	3.05	3.52
N	2s	1.25	1.33	1.39	1.87	2.15	2.37
O	2s	1.09	1.14	1.18	1.42	1.58	1.71
F	2s	0.96	1.00	1.03	1.10	1.22	1.32
B	2p		2.20	3.17		6.14	14.22
C	2p	1.47	1.71	2.15	2.69	3.75	6.35
N	2p	1.24	1.41	1.76	1.93	2.54	4.35
O	2p	1.08	1.23	1.47	1.46	1.97	3.04
F	2p	0.97	1.09	1.26	1.20	1.54	2.01
Na	3s		4.21			20.70	
Mg	3s	2.85	3.25		9.39	12.41	
Al	3s	2.44	2.60	2.73	6.88	7.91	8.87
Si	3s	2.09	2.21	2.31	5.06	5.67	6.34
P	3s	1.85	1.93	2.00	3.94	4.35	4.72
S	3s	1.66	1.72	1.78	3.18	3.44	3.71
Cl	3s	1.51	1.56	1.60	2.62	2.81	3.01
Aℓ	3p		3.43	4.53		13.95	26.38
Si	3p	2.48	2.75	3.23	7.19	8.98	12.95
P	3p	2.14	2.32	2.71	5.33	6.39	9.21
S	3p	1.88	2.06	2.32	4.15	5.07	6.71
Cℓ	3p	1.71	1.84	2.03	3.44	4.06	5.09
K	4s		5.24			31.52	
Ca	4s		4.22			20.46	
Sc	4s		3.96	5.84		18.06	42.41
Ti	4s		3.78	5.78		16.51	42.29
V	4s		3.63	5.62		15.24	40.38
Cr	4s		3.49	5.49		14.17	38.60
Mn	4s		3.38	5.24		13.30	35.30
Fe	4s		3.26	5.10		12.37	33.62
Co	4s		3.15	4.92		11.62	31.22
Ni	4s		3.06	4.66		10.97	27.16
Cu	4s		3.42	4.68		14.20	28.35
Zn	4s		2.90			9.85	

Table 4 (*continued*)

		⟨r⟩, a.u.			⟨r²⟩, a.u.²		
Element	Orbital	+ 1 ion	Atom	− 1 ion	+1 ion	Atom	− 1 ion
Ga	4s		2.49	2.59	6.31	7.21	7.97
Ge	4s		2.22	2.32	5.15	5.72	6.33
As	4s		2.03	2.09	4.34	4.74	5.12
Se	4s		1.87	1.92	3.73	4.02	4.30
Br	4s		1.74	1.78	3.25	3.47	3.69
Sc	3d	1.87	1.68	2.12	4.56	3.66	6.37
Ti	3d	1.62	1.46	1.73	3.42	2.74	4.08
V	3d	1.45	1.32	1.52	2.77	2.26	3.13
Cr	3d	1.32	1.22	1.37	2.28	1.92	2.52
Mn	3d	1.24	1.31	1.28	2.03	1.65	2.20
Fe	3d	1.16	1.07	1.19	1.79	1.49	1.91
Co	3d	1.09	1.02	1.12	1.58	1.35	1.71
Ni	3d	1.03	0.97	1.05	1.43	1.22	1.52
Cu	3d	0.98	0.99	0.99	1.28	1.33	1.34
Zn	3d	0.87	0.88		0.98	1.00	
Ga	4p		3.42	4.61		13.83	27.61
Ge	4p	2.59	2.87	3.34	7.75	9.65	13.74
As	4p	2.32	2.51	2.91	6.20	7.37	10.39
Se	4p	2.11	2.30	2.57	5.13	6.18	8.05
Br	4p	1.97	2.11	2.30	4.48	5.22	6.40
Rb	5s		5.63			36.18	
Sr	5s		4.63			24.50	
Y	5s		4.30			21.14	
Zr	5s		4.08			19.05	
Nb	5s		3.91			17.52	
Mo	5s		3.77			16.32	
Tc	5s		3.65			15.34	
Ru	5s		3.55			14.52	
Rh	5s		3.46			13.82	
Pd	5s		3.38			13.20	
Ag	5s		3.30			12.66	
Cd	5s		3.24			12.17	
In	5s		2.84			9.30	
Sn	5s		2.58			7.63	
Sb	5s		2.39			6.48	
Te	5s		2.23			5.63	
I	5s		2.09			4.97	
Xe	5s		1.98			4.44	
Y	4d		2.44			7.27	
Zr	4d		2.14			5.55	
Nb	4d		1.93			4.52	
Mo	4d		1.78			3.81	
Tc	4d		1.65			3.29	
Ru	4d		1.55			2.88	
Rh	4d		1.46			2.56	
Pd	4d		1.38			2.29	
Ag	4d		1.31			2.07	
Cd	4d		1.25			1.88	
In	5p		3.78			16.65	
Sn	5p		3.29			12.53	
Sb	5p		2.95			10.07	
Te	5p		2.70			8.41	
I	5p		2.50			7.20	
Xe	5p		2.34			6.28	

Table 4 (*continued*)

Element	Orbital	⟨r⟩, a.u.			⟨r²⟩, a.u.²		
		+ 1 ion	Atom	- 1 ion	+1 ion	Atom	- 1 ion
Cs	6s		6.31			44.99	
Ba	6s		5.26			31.27	
La[a]	6s		5.18			20.36	
Hf	6s		4.07			18.92	
Ta	6s		3.92			17.59	
W	6s		3.80			16.54	
Re	6s		3.69			15.67	
Os	6s		3.60			14.93	
Ir	6s		3.52			14.29	
Pt	6s		3.45			13.73	
Au	6s		3.39			13.24	
Hg	6s		3.33			12.80	
Tℓ	6s		2.97			10.07	
Pb	6s		2.72			8.44	
Bi	6s		2.54			7.31	
Po	6s		2.39			6.45	
At	6s		2.26			5.78	
Rn	6s		2.16			5.23	
La[a]	4f		1.07			1.50	
Hf	5d		2.23			5.98	
Ta	5d		2.05			5.03	
W	5d		1.91			4.35	
Re	5d		1.80			3.85	
Os	5d		1.71			3.44	
Ir	5d		1.62			3.11	
Pt	5d		1.55			2.84	
Au	5d		1.49			2.61	
Hg	5d		1.43			2.41	
Tℓ	6p		3.93			17.88	
Pb	6p		3.46			13.78	
Bi	6p		3.14			11.30	
Po	6p		2.89			9.59	
At	6p		2.70			8.34	
Rn	6p		2.54			7.37	

[a] See Ref. 8 for rare earth elements.

presents SCF values of ⟨r⟩ and ⟨r²⟩. Table 5 presents values of z and n satisfying these relations and shows why a value of n less than the quantum number is desirable to reproduce SCF overlap values. Figure 1 indicates the importance of a reduced n representation. Since many common overlap computation techniques cannot use noninteger n values, a good approximation to the optimum values can be found by selecting the integer value closest to the noninteger value of n in Table 5 and calculating z so that the STO reproduces the SCF ⟨r⟩ (see Table 6).

These functions are often similar to Burns' functions (see Fig. 2), but the distinctions are as follows. It is noted that energy is primarily a function of $\langle r^{-1}\rangle$ and $\langle r^{-2}\rangle$ and that overlap is primarily a function of ⟨r⟩ and ⟨r²⟩. Although a function can be constructed approximating these moments (Burns), Brown and Fitzpatrick have shown such a single function only roughly approximates the orbital kinetic energy and overlap. Much better results are obtained by requiring a single STO to reproduce only a single orbital property. The STO

Table 5. Single STO Functions to Reproduce SCF ⟨r⟩ and ⟨r²⟩ Values

		STO's to reproduce ⟨r⟩ and ⟨r²⟩ for overlap					
		+1 ion		Atom		−1 ion	
Element	Orbital	n	z	n	z	n	z
Li	2s			2.23	0.70		
Be	2s	2.36	1.26	1.97	0.93		
B	2s	2.18	1.49	1.92	1.23	1.64	1.02
C	2s	2.11	1.77	1.88	1.50	1.64	1.27
N	2s	2.05	2.04	1.85	1.76	1.66	1.56
O	2s	2.00	2.30	1.83	2.04	1.71	1.87
F	2s	1.98	2.59	1.81	2.31	1.66	2.09
B	2p			1.39	0.86	0.71	0.38
C	2p	1.58	1.41	1.32	1.06	0.84	0.62
N	2p	1.51	1.62	1.28	1.26	0.75	0.71
O	2p	1.45	1.81	1.16	1.35	0.74	0.84
F	2p	1.35	1.91	1.10	1.48	0.75	1.00
Na	3s			2.46	0.70	1.39	0.31
Mg	3s	2.76	1.14	2.38	0.89		
Aℓ	3s	2.69	1.31	2.46	1.14	2.15	0.97
Si	2s	2.74	1.55	2.53	1.37	2.22	1.18
P	3s	2.73	1.75	2.53	1.57	2.30	1.40
S	3s	2.72	1.94	2.57	1.79	2.35	1.60
Cℓ	3s	2.73	2.14	2.58	1.98	2.37	1.79
Aℓ	3p			2.19	0.78	1.26	0.39
Si	3p	2.46	1.20	2.20	0.98	1.57	0.64
P	3p	2.46	1.39	2.21	1.17	1.49	0.74
S	Sp	2.45	1.57	2.09	1.26	1.54	0.88
Cℓ	3p	2.34	1.66	2.04	1.38	1.59	1.03
K	4s			2.92	0.65		
Ca	4s	3.41	1.05	2.84	0.79		
Sc	4s			2.76	0.83	1.56	0.35
Ti	4s			2.68	0.84	1.37	0.32
V	4s			2.65	0.87	1.30	0.32
Cr	4s			2.59	0.89	1.27	0.32
Mn	4s			2.55	0.90	1.26	0.34
Fe	4s			2.51	0.93	1.21	0.33
Co	4s			2.46	0.94	1.22	0.35
Ni	4s			2.41	0.95	1.47	0.42
Cu	4s			1.86	0.69	1.21	0.36
Zn	4s			2.37	0.99		
Ga	4s	2.88	1.44	2.54	1.22	2.22	1.05
Se	4s	2.94	1.62	2.68	1.43	2.35	1.23
As	4s	2.98	1.79	2.76	1.61	2.50	1.43
Se	4s	3.02	1.95	2.80	1.77	2.58	1.60
Br	4s	3.04	2.10	2.86	1.93	2.64	1.76
Sc	3d	1.11	0.86	1.17	1.00	0.71	0.57
Ti	3d	1.12	1.00	1.25	1.20	0.87	0.79
V	3d	1.12	1.12	1.25	1.32	0.91	0.93
Cr	3d	1.13	1.23	1.22	1.41	0.93	1.05
Mn	3d	1.06	1.26	1.24	1.54	0.88	1.08
Fe	3d	1.03	1.32	1.18	1.57	0.92	1.19
Co	3d	1.03	1.40	1.16	1.63	0.86	1.22
Ni	3d	0.99	1.44	1.12	1.68	0.86	1.29
Cu	3d	0.98	1.51	0.91	1.43	0.89	1.40
Zn	3d	1.14	1.89	1.10	1.83		
Ga	4p			2.23	0.80	1.17	0.36
Ge	4p	2.76	1.26	2.37	1.00	1.68	0.65
As	4p	2.82	1.43	2.47	1.18	1.68	0.75
Se	4p	2.87	1.60	2.42	1.27	1.75	0.88
Br	4p	2.77	1.66	2.40	1.37	1.89	1.04

Table 5 (*continued*)

		STO's to reproduce ⟨r⟩ and ⟨r^2⟩ for overlap					
		+1 ion		Atom		−1 ion	
Element	Orbital	n	z	n	z	n	z
Rb	5s			3.05	0.63		
Sr	5s			3.00	0.76		
Y	5s			3.00	0.81		
Zr	5s			2.97	0.85		
Nb	5s			2.92	0.88		
Mo	5s			2.88	0.90		
Tc	5s			2.79	0.90		
Ru	5s			2.79	0.93		
Rh	5s			2.73	0.93		
Pd	5s			2.71	0.98		
Ag	5s			2.57	0.93		
Cd	5s			2.64	0.97		
In	5s			2.79	1.16		
Sn	5s			2.92	1.33		
Sb	5s			3.20	1.55		
Te	5s			3.26	1.69		
I	5s			3.15	1.75		
Xe	5s			3.26	1.90		
Y	4d			1.75	0.92		
Zr	4d			1.86	1.10		
Nb	4d			1.83	1.21		
Mo	4d			1.98	1.39		
Tc	4d			1.88	1.44		
Ru	4d			2.00	1.68		
Rh	4d			1.98	1.70		
Pd	4d			1.94	1.77		
Ag	4d			1.96	1.88		
Cd	4d			1.94	1.95		
In	5p			2.53	0.80		
Sn	5p			2.66	0.96		
Sb	5p			2.68	1.08		
Te	5p			2.75	1.20		
I	5p			2.79	1.32		
Xe	5p			2.92	1.46		
Cs	6s			3.35	0.61		
Ba	6s			3.35	0.73		
La	6s			3.29	0.73		
Hf	6s			2.00	0.86		
Ta	6s			2.97	0.89		
W	6s			2.95	0.91		
Re	6s			2.81	0.90		
Os	6s			2.79	0.91		
Ir	6s			2.77	0.93		
Pt	6s			2.75	0.94		
Au	6s			2.79	0.97		
Hg	6s			2.75	0.98		
Tℓ	6s			3.02	1.19		
Pb	6s			3.05	1.31		
Bi	6s			3.26	1.48		
Po	6s			3.35	1.61		
At	6s			3.32	1.69		
Rn	6s			3.67	1.93		
La	4f			1.08	1.48		
Hf	5d			1.96	1.10		
Ta	5d			2.03	1.23		
W	5d			2.10	1.36		

Table 5 (*continued*)

Element	Orbital	STO's to reproduce ⟨r⟩ and ⟨r²⟩ for overlap					
		+1 ion		Atom		−1 ion	
		n	z	n	z	n	z
Re	5d			2.16	1.48		
Os	5d			2.31	1.64		
Ir	5d			2.17	1.65		
Pt	5d			2.23	1.76		
Au	5d			2.34	1.91		
Hg	5d			2.26	1.93		
Tℓ	6p			2.66	0.80		
Pb	6p			2.81	0.96		
Bi	6p			2.92	1.09		
Po	6p			2.86	1.16		
At	6p			2.97	1.29		
Rn	6p			3.00	1.38		

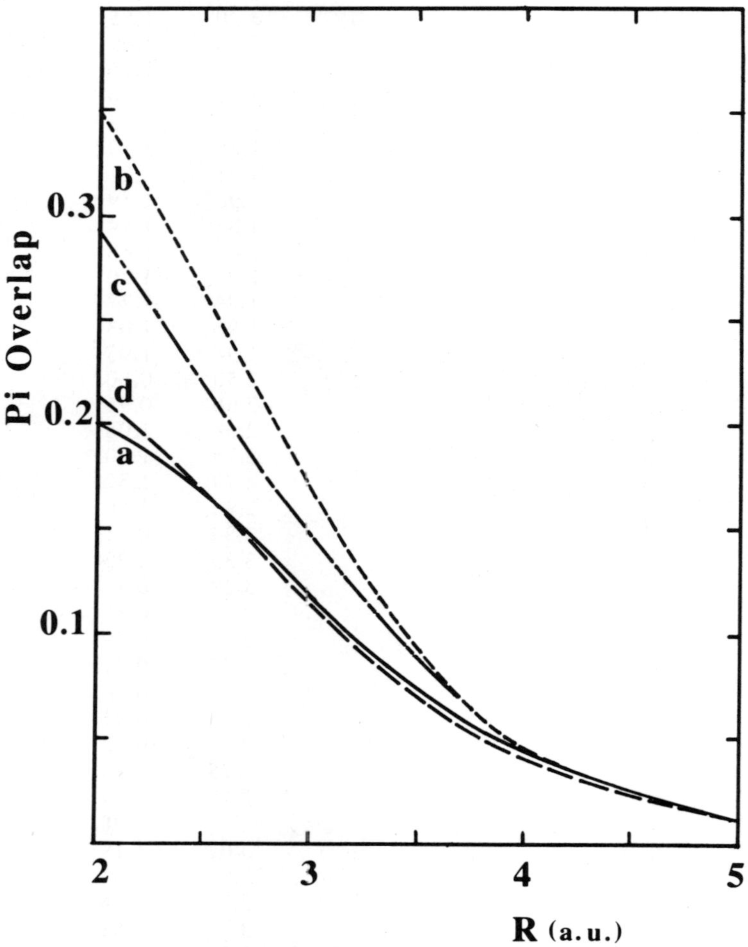

Fig. 1. π overlap integral between Ni^+ 3d orbitals as a function of distance: curve a (solid), SCF function (Ref. 9); curve b (dotted), STO, n = 3, z = 2.65, overlap matched at R = 4 a.u.; curve c (dot-dash), STO, n = 2, z = 2.06, overlap matched at R = 4 a.u.; curve d (dashed) STO, n = 0.99, z = 1.44, from Table 5.

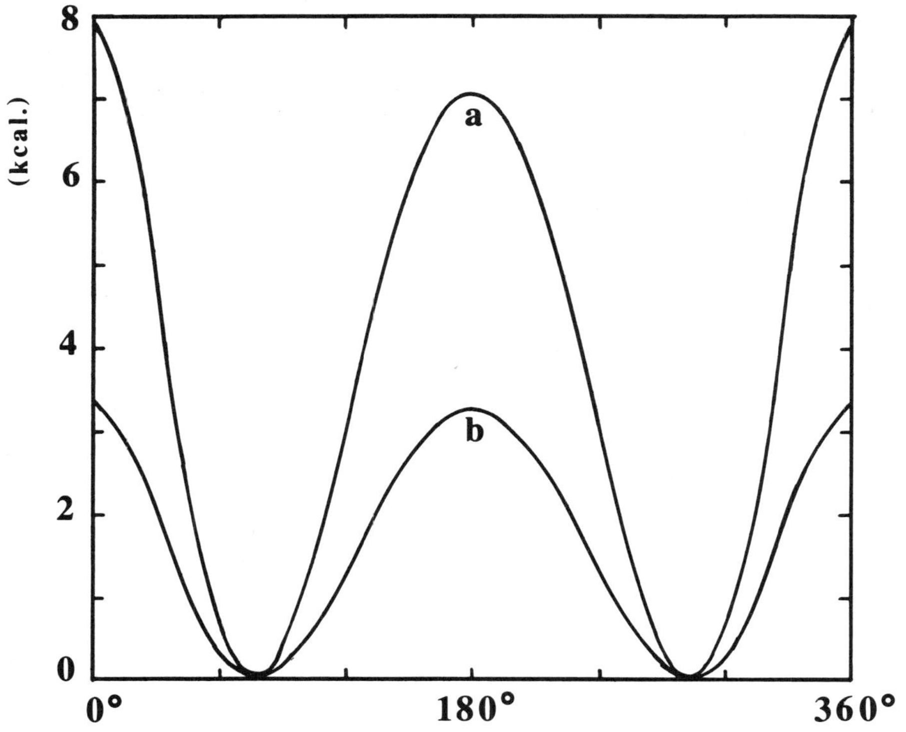

Fig. 2. Barrier to internal rotation in H_2S_2: (a) Burns' functions, (Ref. 5) and (b) Clementi-Raimondi orbitals (Ref. 4). Calculations by Trus and Cusachs (Ref. 19) using method described by Cusachs and Cusachs (Ref. 20). Calculations are identical except for choice of orbital exponents.

to best do this is determined by applying the property operator to the SCF functions to determine average values of the principal variables, then constructing a STO to reproduce these average values. Also, if noninteger n's are allowable in the molecular orbital calculation procedure used, then n should be treated as a variable[7] and the optimum value selected to reproduce the SCF average values of importance.

Since SCF functions are available for most ions, it is possible to determine single STO's for the ionic orbital functions and hence to interpolate a value for the atomic orbital in a molecular environment. However, care should be exercised in using such interpolated values for z. First, because it is still debatable how overlap charge density should be assigned to the atoms. Second, because where charge transfer does occur, the charged neighbor atoms exert electrostatic potentials which can significantly alter the effective nuclear potential, hence the z. Charge transfer and neighbor-atom effects tend to cancel, and the molecular z would not generally be expected to vary a large amount from the neutral atom value.

Table 6. Single STO Functions to Reproduce SCF ⟨r⟩ Values Using Integer n Values Optimized for Overlap

		STO's to reproduce ⟨r⟩				STO's to reproduce ⟨r⟩	
Element	Orbital	n overlap	$z = \frac{n + \frac{1}{2}}{\langle r \rangle}$	Element	Orbital	n overlap	$z = \frac{n + \frac{1}{2}}{\langle r \rangle}$
Li	2s	2	0.65	Ga	4s	3	1.41
Be	2s	2	0.94	Ge	4s	3	1.58
B	2s	2	1.26	As	4s	3	1.72
C	2s	2	1.57	Se	4s	3	1.87
N	2s	2	1.88	Br	4s	3	2.01
O	2s	2	2.19				
F	2s	2	2.50	Sc	3d	1	0.89
				Ti	3d	1	1.03
B	2p	1	0.68	V	3d	1	1.14
C	2p	1	0.88	Cr	3d	1	1.23
N	2p	1	1.06	Mn	3d	1	1.15
O	2p	1	1.22	Fe	3d	1	1.40
F	2p	1	1.38	Co	3d	1	1.47
				Ni	3d	1	1.66
Na	3s	2	0.59	Cu	3d	1	1.52
Mg	3s	2	0.77	Zn	3d	1	1.71
Aℓ	3s	2	0.96				
Si	3s	3	1.58	Ga	4p	2	0.73
P	3s	3	1.81	Ge	4p	2	0.87
S	3s	3	2.03	As	4p	2	1.00
Cℓ	3s	3	2.24	Se	4p	2	1.09
				Br	4p	2	1.18
Aℓ	3p	2	0.73				
Si	3p	2	0.91	Rb	5s	3	0.62
P	3p	2	1.08	Sr	5s	3	0.76
S	3p	2	1.21	Y	5s	3	0.81
Cℓ	3p	2	1.36	Zr	5s	3	0.86
				Nb	5s	3	0.90
K	4s	3	0.67	Mo	5s	3	0.93
Ca	4s	3	0.83	Tc	5s	3	0.99
Sc	4s	3	0.91	Ru	5s	3	0.99
Ti	4s	3	0.93	Rh	5s	3	1.01
V	4s	3	0.96	Pd	5s	3	1.04
Cr	4s	3	1.00	Ag	5s	3	1.06
Mn	4s	3	1.04	Cd	5s	3	1.08
Fe	4s	3	1.07	In	5s	3	1.23
Co	4s	2	0.79	Sn	5s	3	1.36
Ni	4s	2	0.82	Sb	5s	3	1.46
Cu	4s	2	0.84	Te	5s	3	1.57
Zn	4s	2	0.86	I	5s	3	1.67
				Xe	5s	3	1.77

Table 6 (*continued*)

		STO's to reproduce ⟨r⟩				STO's to reproduce ⟨r⟩	
Element	Orbital	n overlap	$z = \frac{n + \frac{1}{2}}{\langle r \rangle}$	Element	Orbital	n overlap	$z = \frac{n + \frac{1}{2}}{\langle r \rangle}$
Y	4d	2	1.02	Pt	6s	3	1.01
Zr	4d	2	1.17	Au	6s	3	1.03
Nb	4d	2	1.30	Hg	6s	3	1.05
Mo	4d	2	1.40	Tℓ	6s	3	1.18
Tc	4d	2	1.52	Pb	6s	3	1.29
Ru	4d	2	1.61	Bi	6s	3	1.38
Rh	4d	2	1.71	Po	6s	3	1.46
Pd	4d	2	1.81	At	6s	3	1.55
Ag	4d	2	1.91	Rn	6s	3	1.62
Cd	4d	2	2.08	La	4f	1	1.40
In	5p	3	0.93	Hf	5d	2	1.12
Sn	5p	3	1.06	Ta	5d	2	1.22
Sb	5p	3	1.19	W	5d	2	1.31
Re	5p	3	1.30	Re	5d	2	1.39
I	5p	3	1.40	Os	5d	2	1.46
Xe	5p	3	1.50	Ir	5d	2	1.54
Cs	6s	3	0.55	Pt	5d	2	1.61
Ba	6s	3	0.67	Au	5d	2	1.68
La	6s	3	0.68	Hg	5d	2	1.75
Hf	6s	3	0.86	Tℓ	6p	3	0.89
Ta	6s	3	0.89	Pb	6p	3	1.01
W	6s	3	0.92	Bi	6p	3	1.11
Re	6s	3	0.95	Po	6p	3	1.21
Os	6s	3	0.97	At	6p	3	1.30
Ir	6s	3	0.99	Rn	6p	3	1.38

4. Orbital-Energy Parameters

Semiempirical LCAO-MO calculations require matrix elements of an effective Hamiltonian, **H** or **F**. Some appropriate operator is defined, and the resulting integrals evaluated or identified with atomic parameters. Direct computation is usually the least satisfactory for routine chemical applications, since minimum-basis SCF-MO calculations for polyatomic molecules require a large investment in labor and in computer time for modest returns in quality of agreement with experiment. However, when approximations are introduced, and terms neglected, to reduce the computational effort, the quantities retained can be adjusted to go beyond the

SCF limitations as well as compensating for the terms neglected. Insight into this process may be obtained by analysis of atomic calculations.

Table 7 displays empirical atomic VSIP data along with the negatives of orbital energies from atomic calculations. From the data presented, it can be seen that the empirical valence-

Table 7. VSIP from Atomic Orbital-Energy Parameters

Atom		VSIP[a]	HFS[a]	CR[a]	C[a]	DZ[a]
Li	2s	5.39	5.493	5.301	5.340	
Be	2s	9.32	8.176	8.395	8.412	
B	2s	14.04	12.565	13.162	13.455	
	2p	9.09	6.661	8.170	8.428	
C	2s	19.54	17.537	18.428	19.193	
	2p	11.20	8.980	10.924	11.787	
N	2s	25.51	23.064	24.277	25.710	
	2p	14.02	11.485	13.691	15.437	
O	2s	33.21	29.158	31.304	33.844	
	2p	16.05	14.156	13.689	17.186	
F	2s	40.16	35.835	38.913	42.773	
	2p	19.85	17.003	14.318	19.856	
Na	3s	5.14	5.137	4.767	4.953	4.927
Mg	3s	7.64	6.869	6.554	6.883	6.862
Aℓ	3s	11.32	10.125	9.434	10.703	10.681
	3p	5.98	4.871	4.742	5.717	5.698
Si	3s	14.76	13.566	13.391	14.686	14.634
	3p	7.90	6.531	7.338	8.081	8.048
P	3s	18.17	17.120	17.500	18.943	18.911
	3p	10.74	8.349	9.751	10.654	10.660
S	3s	21.28	20.808	22.077	23.926	23.912
	3p	11.53	10.284	10.530	11.897	11.889
Cℓ	3s	–	24.649	26.888	29.188	28.982
	3p	14.55	12.331	11.945	13.777	13.598
Sc	3d		7.219	5.168	9.346	
	4s		5.860	5.173	5.715	
Ti	3d		8.539	6.622	11.976	
	4s		6.226	5.334	6.006	
V	3d		9.769	6.982	13.856	
	4s		6.554	5.481	6.274	
Cr	3d		6.513	6.804	15.473	
	4s		5.864	5.614	6.258	
Mn	3d		12.048	6.846	17.386	
	4s		7.144	5.905	6.746	
Fe	3d		13.090	4.261	17.601	
	4s		7.413	5.850	7.026	
Co	3d		14.164	2.188	18.375	
	4s		77.680	5.957	7.274	
Ni	3d		15.165	0.004	19.227	
	4s		7.930	6.058	7.514	
Cu	3d		10.106	b	20.138	
	4s		6.924	6.152	7.746	
Zn	3d		17.112	b	21.288	
	4s		8.412	6.241	7.955	

Table 7 (*continued*)

Atom	VSIP[a]	HFS[a]	CR[a]	C[a]	DZ[a]
Ga 4s		11.393	9.043	11.546	
4p		4.933	4.602	5.671	
Ge 4s		14.376	12.408	15.056	
4p		6.369	6.527	7.815	
As 4s		17.345	16.059	18.657	
4p		7.923	6.419	10.056	
Se 4s		20.336	19.569	22.774	
4p		9.540	8.957	10.956	
Br 4s		23.372	23.411	27.006	
4p		11.216	10.005	12.431	

[a] VSIP, valence-state ionization-potential data, averaged over occupation of p orbitals[11]; HFS, Hartree-Fock-Slater approximation, data from Herman and Skillman[10]; CR, Clementi and Raimondi minimum-basis-set atomic SCF functions[4]; C, Clementi extended-basis-set atomic SCF functions[1]; DZ, double-ζ atomic SCF functions of Clementi.[1]

[b] Orbital energy positive; unacceptable value for this use.

state parameters are closely approximated by the more accurate SCF calculations. Further comparison of the Clementi[1] analytic function orbital energies with those from Froese's numerical SCF calculations,[8] not shown, reveals only small differences, owing to the fact that Clementi did his calculations for spectroscopic states, whereas Froese determined her orbitals for the average of the ground configuration. The differences between the double-ζ quantities and those from the extended basis calculations are insignificant for approximate molecular calculations, while the minimum-basis functions, particularly for the 3d series, are unacceptable, even qualitatively. The Hartree-Fock-Slater functions of Herman and Skillman[10] could be used with small corrections proportional to the number of valence electrons or the electron-interaction terms slightly modified to yield valence-state data directly. An attraction of the HFS model is the possibility of determining reasonable 3d and 4s functions for S, P, etc., in the same calculations as the occupied orbitals, which is beyond any methods using the regular Hartree-Fock exchange potential.

5. Generality of Koopmans' Theorem

The widespread practice of identifying orbital energies in atoms and molecules with ionization energies is based on the assumption that Koopmans' theorem[12] applies. The traditional development of the theorem shows only that the IP calculated by difference of SCF total energies is equal to the negative of the orbital energy provided that the orbitals of the initial state are also used for the ion, and that both states may be described by single-determinant wave functions.

Slater[12] has extended the theorem to open-shell SCF theory, when the two states compared are configuration averages. This is very similar to the assumptions of valence-state theory.

It can be shown further that Koopmans' theorem is not limited to the SCF approximation but holds for more general electron-interaction models if the interaction terms are defined in the same way for both states, and if the initial state orbitals are again used for both states. In particular, it applies to methods employing averaged or effective exchange potentials, such as the charge self-consistent one-electron models, to the same level of approximation as their resemblance to SCF orbital energies. All that this theorem requires is that the orbital energy of the electron removed include the full value of its effective interaction with all the other electrons. It may fail for some models in which the parameters are independent of the number of electrons present. It should be recalled that IP values by Koopmans' theorem are usually much closer to the experimental values (usually within 1 eV) than the differences between optimized SCF total energies for both states of the system, which are seriously affected by correlation energy. Mulliken [13] has indicated how the cancellation of errors producing this situation may be generally expected, so that it is reasonable to define the orbital energies of a simplified model as the negatives of ionization potentials. Only atomic and molecular orbitals with negative orbital energies can be regarded as physically significant; computation schemes leading to positive orbital energies for basis functions should be regarded with great suspicion. This happy situation does not extend to electron affinities.

6. Generality of Brillouin's Theorem

Brillouin's theorem[14] states that there will be no nonvanishing configuration interaction matrix elements between the ground-state determinental wave function and those determinants resulting from the excitation of one electron to an empty (virtual) orbital of the initial SCF calculation. The demonstration follows from the distinction between the filled and the virtual orbitals in SCF theory. It is permissible to make arbitrary unitary transformations among the filled orbitals, and among the virtual orbitals, respectively, without changing predictions of total energy or charge density. Thus non-SCF methods, insofar as they lead to the same charge-density function, or one-electron density matrix, as would the LCAO-SCF in the same basis, lead to orbitals which are at most such transformations of the SCF orbitals of the filled and empty sets separately. This is a convenience in the simplified treatment of electronic spectra, for a one-electron model calculation may be followed by the introduction of explicit electron-interaction functions to lift space or spin degeneracies by first-order perturbation. All that is required for such a procedure to incorporate Brillouin's theorem is that the computed ground-state charge distribution closely approximate that which would have been obtained by the analogous SCF procedure. Not only is this easily achieved in practice, but the computation further benefits from the variation theorem with respect to the relation between errors in the wave function and errors in the energy.

7. Transition Energies in Non-SCF Models

In the SCF approximation, the transition energy between the ground state and the excited one produced by excitation of an electron from filled orbital a to virtual orbital b is[13,15]

$$^{2\mp1}E_{ab} - E_N = e_b - e_a - (J_{ab} - K_{ab}) \pm K_{ab}$$

where e_a and e_b are the orbital energies and J_{ab} and K_{ab} the two-electron Coulomb and exchange integrals. The term $(J_{ab} - K_{ab})$ arises from the difference between the exchange potential for occupied and that for virtual orbits. Non-SCF methods using a uniform exchange potential that does not distinguish between occupied and empty orbitals require the simpler expression

$$^{2\mp1}E_{ab} - E_N = e_b - e_a \pm K_{ab}$$

This is a case where the assumption that one-electron models are approximations to SCF theory can be seriously misleading. For both SCF and non-SCF methods, the transition-energy expressions become more complicated if both orbitals a and b are degenerate, but the remaining computation runs parallel. It is important to observe that in both SCF and non-SCF approximations it is implicitly assumed that the charge density does not change significantly between the ground and exited states. When charge redistribution is significant, it is necessary either to iterate in the excited configuration or to employ extensive configuration interaction. This problem arises primarily in calculations involving explicit consideration of σ molecular orbitals. Present limited experience suggests that reiteration in the excited configuration will be more economical of effort for larger molecules.

8. Incompatibility of Approximations

Despite the vast number of possible choices of input data and computational approximations, it is possible to show that some simplifications imply others. Whatever the details of the effective Hamiltonian selected, the off-diagonal elements, H_{ij}, may be reduced to the sum of a kinetic term, T_{ij}, and potentials due to the atom on which the first orbital, i, is centered, V^a, that due to the second atom, V^b, and the remainder of the molecule, V^m. Thus

$$H_{ij} = T_{ij} + V^a_{ij} + V^b_{ij} + V^m_{ij}$$

Available evidence[16] shows that the Mulliken approximation, $V_{ij} \simeq S_{ij}(V_{ii} + V_{jj})/2$, holds well for typical potentials in molecules. Applying it to the expression for H_{ij} above reduces it to

$$H_{ij} \simeq T_{ij} + C_i S_{ij}$$

with the constant, C_1, containing terms varying slowly compared to S_{ij}. The Goeppert-Mayer and Sklar approximation,[17] in the form

$$T_{ij} + V^a_{ij} \simeq e_i S_{ij} \qquad T_{ij} + V^b_{ij} \simeq e_j S_{ij}$$

is frequently adopted to eliminate the potential terms in favor of the e_i. The Mulliken approximation can be applied to reduce V^m_{ij}, and the combination becomes

$$H_{ij} \simeq -T_{ij} + C_2 S_{ij}$$

If the two forms are equated, it is found that

$$T_{ij} = S_{ij}(C_2 - C_1)/2$$

or the two-center kinetic term should be proportional to the overlap. Beyond the aesthetic satisfaction of an alternative derivation of the Mulliken-Wolfsberg-Helmholz approximation from more mysterious expressions, this implies that it is basically inconsistent to mix the Mulliken approximation, the Goeppert-Mayer and Sklar approximation, and explicit calculation of T_{ij}. The least reliable of the three relations is the GMS approximation, which was not even originally proposed as accurate, but adopted discretely to avoid integrals that could not be evaluated at the prevailing state of the art.[18] Although there must be considerable latitude in assigning properties to atomic orbitals in semiempirical calculations, the sacrifice of internal consistency to such combinations of approximations in several procedures recently proposed does not inspire confidence in their reliability.

Acknowledgments

The present book and the recent Yale Seminar on sigma molecular orbital theory which preceeded it has afforded a unique perspective of the current state of *ab initio* and semiempirical molecular orbital theory. We wish to thank Professors Sinanoğlu and Wiberg for affording us this stimulus. One of us (J.H.C.) has benefited from a National Science Foundation Traineeship, and the research described is based on work supported in part by American Cancer Society Grant E-3478B and by the Tulane Computer Laboratory.

References

1. E. Clementi, *J. Chem. Phys.* **38**, 996, 1001 (1963); E. Clementi, *J. Chem. Phys.* **41**, 295, 303 (1964); E. Clementi, *IBM J. Res. Devel.* 9, 2 (1965), and supplement.
2. J. C. Slater, *Phys. Rev.* **36**, 57 (1930).
3. D. A. Brown and N. J. Fitzpatrick, *J. Chem. Soc.* **A1966**, 941; D. A. Brown and N. J. Fitzpatrick, *J. Chem. Soc.* **A1967**, 316.
4. E. Clementi and D. L. Raimondi, *J. Chem. Phys.* **38**, 2686 (1963) and supplement.
5. G. Burns, *J. Chem. Phys.* **41**, 1521 (1964).
6. L. C. Cusachs, D. G. Carroll, B. L. Trus, and S. P. McGlynn, *Intern. J. Quantum Chem.* **1** (Slater Symposium Issue), 159 (1967).
7. R. G. Parr, private communication.
8. C. Froese, *J. Chem. Phys.* **45**, 1417 (1966), and supplement, Hartree-Fock Parameters for the Atoms Helium to Radon, Department of Mathematics, University of British Columbia, Vancouver, B. C., 1966.
9. J. W. Richardson, W. C. Nieuwpoort, R. R. Powell, and W. F. Edgell, *J. Chem. Phys.* **36**, 1057 (1962); **38**, 796 (1963).
10. F. Herman and S. Skillman, *Atomic Structure Calculations*, Prentice-Hall, Englewood Cliffs, N. J., 1960.
11. L. C. Cusachs and J. W. Reynolds, *J. Chem. Phys.*, **43**, S160 (1965); L. C. Cusachs, J. W. Reynolds, and D. Barnard, *J. Chem. Phys.*, **44**, 835 (1966).
12. J. C. Slater, *Quantum Theory of Atomic Structure*, Vol. II, McGraw-Hill, New York, 1960, Chap. 17.
13. R. S. Mulliken, *J. Chim. Phys.* **46**, 497ff. (1949).
14. L. Brillouin, *Acta Sci. Ind. No. 159,* Hermann, Paris, 1934.
15. C. C. J. Roothaan, *Rev. Mod. Phys.* **23**, 69 (1951).
16. M. D. Newton, F. P. Boer, and W. N. Lipscomb, *J. Am. Chem. Soc.* **88**, 2353 (1966), and following papers.
17. M. Goeppert-Mayer and A. L. Sklar, *J. Chem. Phys.* **6**, 645 (1938).
18. K. Ruedenberg, *J. Chem. Phys.* **34**, 1861 (1961).
19. B. L. Trus and L. C. Cusachs, to be published.
20. L. C. Cusachs and B. B. Cusachs, *J. Phys. Chem.* **71**, 1060 (1967).

VI-5 Reprinted from *The Journal of Chemical Physics* **43**, S129 (1965)

Approximate Self-Consistent Molecular Orbital Theory. I. Invariant Procedures

J. A. Pople
Carnegie Institute of Technology and *Mellon Institute, Pittsburgh, Pennsylvania*
AND
D. P. Santry and G. A. Segal
Carnegie Institute of Technology, Pittsburgh, Pennsylvania
(Received 13 April 1965)

A general discussion of approximate methods for obtaining self-consistent molecular orbitals for all valence electrons of large molecules is presented. It is shown that the procedure of neglecting differential overlap in electron-interaction integrals (familiar in π-electron theory) without further adjustment may lead to results which are not invariant to simple transformations of the atomic orbital basis set such as rotation of axes or replacement of s, p orbitals by hybrids. The behavior of approximate methods in this context is examined in detail and two schemes are found which are invariant to transformations among atomic orbitals on a given atom. One of these (the simpler but more approximate) involves the complete neglect of differential overlap (CNDO) in all basis sets connected by such transformations. The other involves the neglect of diatomic differential overlap (NDDO) only, that is only products of orbitals on different atoms being neglected in the electron-repulsion integrals.

1. INTRODUCTION

THE self-consistent-field theory attempts to calculate atomic or molecular orbitals using the full many-electron Hamiltonian and a single determinantal wavefunction (or the smallest number of determinants required to produce a function with the appropriate symmetry). Optimization of the orbitals using the variational principle leads to a set of differential equations[1,2] which are normally intractable, so most applications of the theory have used linear combinations of atomic orbitals (LCAO). If the coefficients in the LCAO orbitals are chosen to minimize the total energy, one obtains LCAOSCF orbitals, the best LCAO approximations to the self-consistent functions, for which

[1] J. E. Lennard-Jones, Proc. Roy. Soc. (London) **A198**, 1 (1949).
[2] C. C. J. Roothaan, Rev. Mod. Phys. **23**, 69 (1951).

equations were first given by Hall[3] and by Roothaan.[2] Accurate calculations using this technique have since been made for many small systems. However, the use of the LCAOSCF method without further approximation is limited by computational difficulties and there remains a need for simpler methods which retain the principal features determining the electron distribution but which are sufficiently tractable to be applied to large molecules.

Considerable progress in a more approximate MO theory has been made for the π electrons of aromatic systems where the Roothaan LCAOSCF equations have been simplified by the "neglect of differential overlap" approximation.[4,5] This involves the neglect of the product of pairs of different atomic orbitals in certain electron-interaction integrals. By combining this approximation with a semiempirical approach to the determination of the remaining parameters, a theory correlating many physical properties of aromatic molecules has been developed.[6] This kind of treatment is intermediate in complexity between full LCAOSCF calculations for π electrons and the very simple Hückel approach which does not handle electron interaction in any explicit manner.

The value of molecular orbital calculations would be greatly increased if these more approximate treatments could be extended to *all* valence electrons, rather than just π electrons. Such an advance would permit a full treatment of σ and π electrons in planar molecules and application of the theory to the great range of molecules where σ–π separation is not possible. An approach to problems of this sort has already been made with "extended Hückel methods" treating electrons independently[7–10] but these have disadvantages similar to those possessed by the π-electron version when applied to nonuniform electron distributions. It is clearly desirable to extend the intermediate type of theory to take at least some account of electron repulsion. Some treatments of this kind for two-electron systems have been developed by Klopman[11] and by Pohl *et al.*[12]

In this and following papers we shall develop self-consistent methods based on neglect of differential overlap for all valence orbitals. This procedure leads to simplified LCAOSCF equations which become tractable for quite large systems. In this first paper, we discuss various levels at which such approximations can be made systematically and develop the general method to be used for closed shell molecules.

---

[3] G. G. Hall, Proc. Roy. Soc. (London) **A205**, 541 (1951).
[4] R. Pariser and R. G. Parr, J. Chem. Phys. **21**, 466 (1953).
[5] J. A. Pople, Trans. Faraday Soc. **49**, 1375 (1953).
[6] J. A. Pople, J. Phys. Chem. **61**, 6 (1957).
[7] R. S. Mulliken, J. Chem. Phys. **46**, 497, 675 (1949).
[8] M. Wolfsberg and L. Helmholtz, J. Chem. Phys. **20**, 837 (1952).
[9] R. Hoffman, J. Chem. Phys. **39**, 1397 (1963).
[10] J. A. Pople and D. P. Santry, Mol. Phys. **7**, 269 (1964).
[11] G. Klopman, J. Am. Chem. Soc. **86**, 4550 (1964).
[12] H. A. Pohl, R. Appel, and K. Appel, J. Chem. Phys. **41**, 3385 (1964).

2. SCF EQUATIONS AND TRANSFORMATION PROPERTIES

We begin this section with a discussion of Roothaan's LCAOSCF equations[2] as a basis for calculating the molecular orbitals of a molecule of arbitrary symmetry, paying particular attention to certain of their transformation properties which limit the manner in which they may be approximated. Only those electrons in the valence shell are considered explicitly, all inner shells being treated as part of an unpolarizable core. To begin with, we only deal with atoms up to fluorine, for which the core consists of the bare nucleus for hydrogen and the nucleus screened by two 1*s* inner-shell electrons for Li to F. However, generalization to other atoms presents little difficulty. The valence electrons are then assigned to LCAO molecular orbitals

$$\psi_i=\sum_\nu \phi_\nu c_{\nu i}, \qquad (2.1)$$

where ϕ_ν are valence atomic orbitals (1*s* for hydrogen; 2*s*, 2*px*, 2*py*, 2*pz* for lithium to fluorine). Greek suffixes will be used to denote atomic orbitals in the basis set. For molecules, with a closed-shell configuration, variational treatment of the orbital coefficients $c_{\nu i}$ leads to the Roothaan equations[2]

$$\sum_\nu F_{\mu\nu}c_{\nu i}=\sum_\nu S_{\mu\nu}c_{\nu i}\epsilon_i, \qquad (2.2)$$

where (using atomic units)

$$F_{\mu\nu}=H_{\mu\nu}+G_{\mu\nu}, \qquad (2.3)$$

$$H_{\mu\nu}=\int\bar{\phi}_\mu[-\tfrac{1}{2}\nabla^2-\sum_A V_A(\mathbf{r})]\phi_\nu d\tau, \qquad (2.4)$$

$$G_{\mu\nu}=\sum_{\lambda,\sigma}P_{\lambda\sigma}[(\mu\nu\mid\lambda\sigma)-\tfrac{1}{2}(\mu\sigma\mid\nu\lambda)], \qquad (2.5)$$

$$(\mu\nu\mid\lambda\sigma)=\iint\bar{\phi}_\mu(1)\phi_\nu(1)(r_{12})^{-1}\bar{\phi}_\lambda(2)\phi_\sigma(2)d\tau_1d\tau_2, \qquad (2.6)$$

$$S_{\mu\nu}=\int\bar{\phi}_\mu\phi_\nu d\tau, \qquad (2.7)$$

$$P_{\lambda\sigma}=2\sum_i^{\mathrm{occ}}\bar{c}_{i\lambda}c_{i\sigma}. \qquad (2.8)$$

In these equations ϵ_i is the orbital energy for the molecular orbital ψ_i (giving a theoretical estimate of the corresponding ionization potential if the orbital is occupied). $H_{\mu\nu}$ is the matrix element of the one-electron Hamiltonian including the kinetic energy and the potential energy in the electrostatic field of the core, this being written as a sum of potentials $V_A(\mathbf{r})$ for various atoms A in the molecule. $G_{\mu\nu}$ is the matrix element of the potential due to other valence electrons and depends on the molecular orbitals via the population matrix $P_{\lambda\sigma}$. The summation

$$\sum_i^{occ}$$

in (2.8) is over occupied molecular orbitals only.

The orbital energies ϵ_i are roots of the secular equation

$$|F_{\mu\nu}-\epsilon S_{\mu\nu}|=0, \qquad (2.9)$$

and the total electronic energy of the valence electrons is

$$E_{\text{electronic}}=\tfrac{1}{2}\sum_{\mu,\nu}P_{\mu\nu}(H_{\mu\nu}+F_{\mu\nu}). \qquad (2.10)$$

The total energy of the molecule (relative to separated valence electrons and isolated cores) is obtained by adding the repulsion energy between cores. This latter can be well approximated by a point-charge model so that

$$E_{\text{total}}=E_{\text{electronic}}+\sum\sum_{A<B}Z_AZ_B/R_{AB}, \qquad (2.11)$$

where Z_A is the *core* charge of Atom A and R_{AB} is the A–B internuclear distance. Later, we shall find it convenient to discuss Eqs. (2.2) in the matrix form

$$\mathbf{F}\mathbf{c}_i=(\mathbf{H}+\mathbf{G})\mathbf{c}_i=\mathbf{S}\mathbf{c}_i\epsilon_i, \qquad (2.12)$$

where **F** and **S** are the matrices $F_{\mu\nu}$ and $S_{\mu\nu}$, respectively, and $\mathbf{c}_i$ is the column matrix $c_{\nu i}$ and ϵ_i the orbital energy, a scalar quantity.

We now wish to simplify these equations by approximations comparable to those already used in π-electron theory. There are, however, certain restrictions on the form of these approximations which are imposed by invariance properties of the SCF equations but which are not apparent in the corresponding theory of π electrons. It is well known[1] that molecular wavefunctions, formed from antisymmetrized products of one-electron functions, are invariant with respect to an orthogonal (or unitary) transformation of the *occupied molecular orbitals* among themselves. The molecular wavefunction formed from an antisymmetrized product of LCAOMO's, and the corresponding orbital energies have the further property of being invariant with respect to an orthogonal (or unitary) transformation between the *atomic orbital basis functions*. An example of such a transformation would be the replacement of carbon $2s$, $2px$, $2py$, $2pz$ atomic orbitals in the basis set by four sp^3 hybrid functions. A *full* LCAOSCF calculation based on hybrid orbitals would give the same energies as one based on $2s$ and $2p$ functions. To investigate the effect of approximations on this invariance, we need to study this transformation in some detail.

Suppose a new basis set of functions t_m is chosen, related to the original set ϕ_μ by the matrix equation

$$\mathbf{t}=\mathbf{O}\boldsymbol{\phi}, \qquad (2.13)$$

where **O** is an orthogonal transformation matrix. Under this transformation, the square matrices **M** in (2.12) acquire new values

$$\mathbf{M}'=\mathbf{O}\mathbf{M}\mathbf{O}^{-1}, \qquad (2.14)$$

and the Roothaan equation can be rewritten

$$(\mathbf{O}^{-1}\mathbf{H}'\mathbf{O}+\mathbf{O}^{-1}\mathbf{G}'\mathbf{O})\mathbf{c}_i=\mathbf{O}^{-1}\mathbf{S}'\mathbf{O}\mathbf{c}_i\epsilon_i, \qquad (2.15)$$

where **H′** and **G′** and **S′** are referred to the new basis functions **t**. The elements of the electron-repulsion matrix **G′** are given by

$$\begin{aligned}G_{mn}'&=\sum_{\mu,\nu}O_{m\mu}O_{n\nu}G_{\mu\nu}\\&=\sum_{\lambda\sigma}P_{\lambda\sigma}[(mn\,|\,\lambda\sigma)-\tfrac{1}{2}(m\sigma\,|\,n\lambda)] \qquad (2.16)\\&=\sum_{l,s}P_{ls}'[(mn\,|\,ls)-\tfrac{1}{2}(ms\,|\,nl)],\end{aligned}$$

where Greek letters are used for the old basis functions and

$$P_{ls}'=\sum_{\lambda,\sigma}O_{l\lambda}O_{s\sigma}P_{\lambda\sigma}. \qquad (2.17)$$

Multiplication of (2.15) on the left by O gives

$$[\mathbf{H}'+\mathbf{G}'](\mathbf{O}\mathbf{c}_i)=\mathbf{S}'(\mathbf{O}\mathbf{c})\epsilon_i \qquad (2.18)$$

indicating that $(\mathbf{O}\mathbf{c}_i)$ are the orbital coefficients for the solution of the Roothaan equations using the new basis set. The orbital energies ϵ_i are the same as in the equations using the basis $\boldsymbol{\phi}$. It then follows that P'_{ls} is the new population matrix and the equations are therefore invariant.

The total electronic energy (2.10) is also invariant under the transformation for

$$E_{\text{electronic}}=\sum_i^{occ}\epsilon_i+\tfrac{1}{2}\sum_{\mu\nu}P_{\mu\nu}H_{\mu\nu}=\sum_i^{occ}\epsilon_i+\tfrac{1}{2}\sum_{m,n}P'_{mn}H'_{mn}. \qquad (2.19)$$

Clearly, the essential feature of this invariance proof is that the one- and two-electron integrals transform according to

$$H'_{mn}=\sum_{\mu\nu}O_{m\mu}O_{n\nu}H_{\mu\nu}, \qquad (2.20)$$

$$(mn\,|\,ls)=\sum_\mu\sum_\nu\sum_\lambda\sum_\sigma O_{m\mu}O_{n\nu}O_{l\lambda}O_{s\sigma}(\mu\nu\,|\,\lambda\sigma). \qquad (2.21)$$

Approximate LCAOSCF calculations will only be invariant insofar as they satisfy these two transformation conditions.

Possible transformations O may be classified in successive degrees of complexity as follows:

(i) transformations which only mix $2px$, $2py$, and $2pz$ atomic orbitals centered on the same atom;

(ii) transformations which only mix $2s$ and $2p$ orbitals on the same atom;

(iii) general transformations which mix atomic orbitals on different atoms (leading to a *nonatomic* basis set).

The first type simply corresponds to a rotation of local atomic axes. Invariance to this transformation must be an *essential* feature of an approximate LCAOSCF theory, especially for molecules of low symmetry when there is no unique choice for the axes.

The second transformation corresponds to the hybridization of the orbitals on the various atoms. Although the choice of a particular hybrid form does not strictly make any difference to the calculation of the electronic structure of a molecule, the concept of hybridization is very useful in theoretical chemistry, even within the framework of molecular orbital theory. It is thus *desirable* that the approximate SCF equations should retain the invariance to the hybridization of the atomic orbital basis function.

The retention of invariance under the first two types of transformations places some restrictions on simplifications of Roothaan equations. We shall now consider "neglect of differential overlap" approximations (analogous to those already proposed in π-electron theory) from this point of view. Within this framework, it transpires that there are two degrees of approximation which retain this type of invariance. These are described in the next two sections.

3. THEORY WITH COMPLETE NEGLECT OF DIFFERENTIAL OVERLAP (CNDO)

We first consider the simplest version of the theory, which involves the complete neglect of differential overlap between atomic orbitals on the same atom. The approximations used are given below.

Approximation 1

The ϕ_μ are treated as if they form an orthonormal set; that is, the overlap integrals $S_{\mu\nu}$ are put equal to zero unless $\mu=\nu$, in which case they are unity. The coefficients $c_{i\mu}$ then form an orthogonal matrix and the orthonormality condition for ψ_i becomes

$$\sum_\mu c_{\mu i}c_{\mu j}=\delta_{ij}, \tag{3.1}$$

where δ_{ij} is unity if $i=j$ and zero otherwise. For different orbitals on the same atom, the overlap integral $S_{\mu\nu}$ is already zero; for orbitals on different atoms, the approximation is not quantitatively accurate, but it does simplify much of the subsequent analysis and permits extensive application of the method.

With this approximation, the diagonal matrix elements $P_{\mu\mu}$ correspond to the electron populations of the atomic orbitals ϕ_μ and

$$\sum_\mu P_{\mu\mu}=2N, \tag{3.2}$$

where $2N$ is the number of valence electrons. The off-diagonal elements $P_{\mu\nu}$ are usually referred to as molecular orbital bond orders although they only correlate with bond energies if μ and ν are on neighboring atoms. For more distant atoms, the term long-range bond order is sometimes used.

Approximation 2

All two-electron integrals (2.5) which depend on the overlapping of charge densities of different basis orbitals are neglected. This means that $(\mu\nu\mid\lambda\sigma)$ is zero unless $\mu=\nu$ and $\lambda=\sigma$. The nonzero values will sometimes be written $\gamma_{\lambda\mu}$ where

$$\gamma_{\lambda\mu}=(\lambda\lambda\mid\mu\mu). \tag{3.3}$$

As in the π-electron theory, this approximation can be partially justified by noting that it is consistent with Approximation 1, for $S_{\mu\nu}$ is the magnitude of the "overlap distribution" $\phi_\mu\phi_\nu$ which is ignored in calculating electron interactions. It should be noted, however, that certain one-center exchange integrals such as $(2s, 2px\mid 2s, 2px)$ are omitted in this form of the theory, whereas in the π-electron theory, only one atomic orbital per atom is used and the neglected integrals are only two-center hybrid, two-center exchange, and three- and four-center integrals.

At this point, the theory is not invariant under a rotation of local axes or under hybridization. For example, it is easily shown that rotation of the local axes on Atom A by 45° about the z axis leads to the two-electron integral transformation.

$$2(2px_A', 2py_A'\mid 2s_B, 2s_B)=(2py_A, 2py_A\mid 2s_B, 2s_B) - (2px_A, 2px_A\mid 2s_B,2s_B). \tag{3.4}$$

With Approximations 1 and 2, the left-hand side of (3.4) is neglected but the right-hand side is not. Further approximations must therefore be made to restore invariance.

Approximation 3

The electron-interaction integrals $\gamma_{\mu\nu}$ are assumed to depend only on the atoms to which the orbitals ϕ_μ and ϕ_ν belong and not to the actual type of orbital. This means that there remains only a set of atomic electronic-interaction integrals γ_{AB} measuring an average repulsion between an electron in a valence atomic orbital on A and another in a valence orbital on Atom B.

As mentioned above this further approximation is necessitated by the adoption of Approximation 2. To elaborate further, suppose that the set of atomic orbitals on *one* particular Atom A is replaced by an alternative set t_m, where

$$t_m=\sum_\mu O_{m\mu}\phi_\mu, \quad (3.5)$$

the elements $O_{m\mu}$ forming an orthogonal matrix corresponding to transformations of Types (i) or (ii) listed in Sec. 2. A general electron-repulsion integral in this new basis set is given by

$$(t_m t_n \mid \phi_\lambda\phi_\sigma)=\sum_{\mu\nu} O_{m\mu}O_{n\nu}(\phi_\mu\phi_\nu \mid \phi_\lambda\phi_\sigma), \quad (3.6)$$

ϕ_λ, ϕ_σ being atomic orbitals on another Atom B. The notation in (3.6) is an obvious elaboration of (2.6).

Now if differential overlap is neglected between the ϕ orbitals, (Approximation 2) integrals on the right of (3.6) vanish unless $\mu=\nu$ and $\lambda=\sigma$. Hence we find

$$(t_m t_n \mid \phi_\lambda\phi_\lambda)=\sum_\mu O_{m\mu}O_{n\mu}(\phi_\mu\phi_\mu \mid \phi_\lambda\phi_\lambda). \quad (3.7)$$

This expression does not generally vanish even if $m\neq n$, so that differential overlap is not neglected in the transformed system. However, under the application of Approximation 3, Eq. (3.7) becomes

$$(t_m t_n \mid \phi_\lambda\phi_\lambda)=\gamma_{AB}\sum_\mu O_{m\mu}O_{n\mu}=\gamma_{AB}\delta_{mn}$$

and the invariance required by (2.21) is restored. Hence although Approximation 3 appears severe, it is imposed on us by the nature of the equations once we have adopted Approximation 2.

Using Approximations 1 and 2, the matrix elements $F_{\mu\nu}$ become

$$F_{\mu\mu}=H_{\mu\mu}+\tfrac{1}{2}P_{\mu\mu}\gamma_{\mu\mu}+\sum_{\sigma(\neq\mu)}P_{\sigma\sigma}\gamma_{\mu\sigma}, \quad (3.8)$$

$$F_{\mu\nu}=H_{\mu\nu}-\tfrac{1}{2}P_{\mu\nu}\gamma_{\mu\nu}, \quad (\mu\neq\nu). \quad (3.9)$$

With the addition of Approximation 3, $F_{\mu\mu}$ may be rewritten

$$F_{\mu\mu}=H_{\mu\mu}-\tfrac{1}{2}P_{\mu\mu}\gamma_{AA}+P_{AA}\gamma_{AA}+\sum_{(B\neq A)}P_{BB}\gamma_{AB}, \quad (3.10)$$

where μ belongs to Atom A and P_{BB} is the total valence electron density on Atom B

$$P_{BB}=\sum_\nu{}^{B}P_{\nu\nu}. \quad (3.11)$$

Some care must be exercised in developing the diagonal core matrix elements $H_{\mu\mu}$. These, according to (2.4), include the interaction of an electron in the atomic orbital ϕ_μ with the cores of other atoms. These are conveniently separated, so that we write

$$H_{\mu\mu}=(\mu \mid -\tfrac{1}{2}\nabla^2-V_A \mid \mu)-\sum_{B(\neq A)}(\mu \mid V_B \mid \mu)$$
$$=U_{\mu\mu}-\sum_{B(\neq A)}(\mu \mid V_B \mid \mu), \quad (3.12)$$

where $U_{\mu\mu}$ is the diagonal matrix element of ϕ_μ with respect to the one-electron Hamiltonian containing only the core of its own atom.

$U_{\mu\mu}$ is an essentially atomic quantity measuring the energy of the atomic orbital. It may either be evaluated from approximate atomic orbitals or chosen semi-empirically from experimental data on atomic energy levels or possibly even used as a purely empirical parameter in molecular calculations. Discussion of actual numerical values used will be postponed to the following paper.

The remaining terms in (3.12) give the interaction of an electron in ϕ_μ with the cores of other Atoms B. A comparable separation was made in the corresponding π-electron theory, the principal difference being that the cores now consist only of the nucleus and inner-shell electron, whereas in the previous treatment they included all σ electrons.

We also have to consider the off-diagonal core matrix elements $H_{\mu\nu}$. Here it is convenient to distinguish cases where ϕ_μ and ϕ_ν are on the same or different atoms. If both belong to the same atom (A, say), $H_{\mu\nu}$ may be written analogously to (3.12)

$$H_{\mu\nu}=U_{\mu\nu}-\sum_{B(\neq A)}(\mu \mid V_B \mid \nu), \quad (3.13)$$

where again $U_{\mu\nu}$ is the one-electron matrix element using the local core Hamiltonian. If ϕ_μ, ϕ_ν are s, p, d functions, this is zero by symmetry. (However, this would not be so if the basic functions were hybrids.) The remaining terms in (3.13) represent the interaction of the distribution $\phi_\mu\phi_\nu$ with cores of other atoms. Since we are neglecting differential overlap in corresponding electron-interaction integrals, it is consistent to neglect these contributions. Hence we have Approximation 4.

Approximation 4

Integrals $(\mu \mid V_B \mid \nu)$ where ϕ_μ and ϕ_ν belong to Atom A are put equal to zero if $\mu\neq\nu$. Further, if $\mu=\nu$, the integral is taken to be the same for all valence atomic orbitals on Atom A and we write

$$(\mu \mid V_B \mid \mu)=V_{AB}. \quad (3.14)$$

It should be noted that the matrix V_{AB} need not be symmetric. The second part of this approximation is necessary to ensure that $H_{\mu\nu}$ transforms correctly [Eq. (2.20)] for reasons comparable to those already given for replacing $\gamma_{\mu\nu}$ by γ_{AB}. In consequence, we may write (for $s, p, d,\cdots$ basis functions)

$$H_{\mu\mu}=U_{\mu\mu}-\sum_{B(\neq A)}V_{AB} \quad (\mu \text{ on Atom A}), \quad (3.15)$$

$$H_{\mu\nu}=0 \quad (\mu\neq\nu, \text{ but both on the same atom}). \quad (3.16)$$

To complete the specification of the calculation, we

need to discuss the matrix elements $H_{\mu\nu}$ where ϕ_μ and ϕ_ν are on different atoms. We shall again neglect the interaction of the distribution $\phi_\mu\phi_\nu$ with distant cores and suppose that $H_{\mu\nu}$ depends on the local environment between the two atoms. It is then a measure of the possible lowering of energy levels by being in the electrostatic field of two atoms simultaneously. Following common usage, it will then be referred to as a "resonance integral" and denoted by the symbol $\beta_{\mu\nu}$. To estimate it we adopt Approximation 5.

Approximation 5

Off-diagonal core matrix elements between atomic orbitals on different atoms are estimated by a formula

$$H_{\mu\nu}=\beta_{\mu\nu}=\beta_{AB}^{0}S_{\mu\nu}, \tag{3.17}$$

where $S_{\mu\nu}$ is the overlap integral and β_{AB}^{0} is a parameter depending only on the nature of the atoms A and B.

This type of approximation is comparable to some made in independent-electron calculations. However, the proportionality factor between $H_{\mu\nu}$ and $S_{\mu\nu}$ must be taken the same for all atomic orbitals on two given atoms if the calculations are to be invariant under transformation of the atomic basis sets. The only flexibility permitted is a possible dependence of β_{AB}^{0} on the AB distance. We shall not attempt to calculate this parameter; rather, empirical values will be chosen either to fit experimental data or to reproduce results already obtained by complete a priori calculations.

The $F_{\mu\nu}$ matrix elements now reduce to the form (ϕ_μ belonging to Atom A and ϕ_ν to Atom B)

$$F_{\mu\mu}=U_{\mu\mu}+(P_{AA}-\tfrac{1}{2}P_{\mu\mu})\gamma_{AA}+\sum_{B(\neq A)}(P_{BB}\gamma_{AB}-V_{AB}), \tag{3.18}$$

$$F_{\mu\nu}=\beta_{AB}^{0}S_{\mu\nu}-\tfrac{1}{2}P_{\mu\nu}\gamma_{AB}, \quad (\mu\neq\nu). \tag{3.19}$$

Equation (3.19) applies even if ϕ_μ and ϕ_ν are on the same atom A (when $S_{\mu\nu}=0$ and γ_{AB} is replaced by γ_{AA}).

Using the same approximations, we may derive an expression for the total energy given by (2.11). The terms are conveniently collected into one- and two-atom types, giving

$$E_{\text{total}}=\sum_{A}E_A+\sum\sum_{A<B}E_{AB}, \tag{3.20}$$

where

$$E_A=\sum_{\mu}^{A}P_{\mu\mu}U_{\mu\mu}+\tfrac{1}{2}\sum_{\mu}^{A}\sum_{\nu}^{A}(P_{\mu\mu}P_{\nu\nu}-\tfrac{1}{2}P_{\mu\nu}^{2})\gamma_{AA} \tag{3.21}$$

and

$$E_{AB}=\sum_{\mu}^{A}\sum_{\nu}^{B}(2P_{\mu\nu}\beta_{\mu\nu}-\tfrac{1}{2}P_{\mu\nu}^{2}\gamma_{AB})$$
$$+(Z_AZ_BR_{AB}^{-1}-P_{AA}V_{AB}-P_{BB}V_{BA}+P_{AA}P_{BB}\gamma_{AB}). \tag{3.22}$$

For atoms with large separation, the quantities γ_{AB}, $Z_B^{-1}V_{AB}$, $Z_A^{-1}V_{BA}$ approximate to R_{AB}^{-1}, so that the last group of terms in (3.22) become $Q_AQ_BR_{AB}^{-1}$ where Q_A is the net charge on Atom A

$$Q_A=Z_A-P_{AA}, \tag{3.23}$$

showing that the theory takes proper account of the electrostatic interaction between charged atoms in a molecule.

4. THEORY WITH NEGLECT OF DIATOMIC DIFFERENTIAL OVERLAP (NDDO)

In this section we consider a less approximate form of the theory in which differential overlap in two-electron integrals is neglected only for atomic orbitals on different atoms. To make the scheme internally consistent, *all* products $\phi_\mu\phi_\nu$ for different orbitals on the same atom are retained.

Starting with the accurate LCAOSCF scheme (Eqs. 2.2 to 2.11), Approximation 1 remains unchanged. Approximation 2 is replaced by 2A as follows.

Approximation 2A

All two-electron integrals (2.6) which depend on the overlapping charge densities of basis orbitals on different atoms are neglected. Thus $(\mu\nu|\lambda\sigma)$ is zero unless μ, ν belong to the same atom A and λ, σ are centered on a common atom B (which could, of course, be the same atom as A). All remaining two-electron integrals are of the two-center type with one electron associated with each atom. These may either be calculated from given atomic orbitals, or, if chosen empirically, must satisfy the conditions (2.21) if the transformation is between orbitals on the same atom. Approximation 3 of the previous section is then no longer necessary.

With these approximations, the matrix elements $F_{\mu\nu}$ may be written

$$F_{\mu\nu}=H_{\mu\nu}+\sum_{B}\sum_{\lambda,\sigma}^{B}P_{\lambda\sigma}(\mu\nu|\lambda\sigma)-\tfrac{1}{2}\sum_{\lambda,\sigma}P_{\lambda\sigma}(\mu\sigma|\nu\lambda)$$
$$(\mu,\ \nu \text{ both on A}), \tag{4.1}$$

$$F_{\mu\nu}=H_{\mu\nu}-\tfrac{1}{2}\sum_{\sigma}^{A}\sum_{\lambda}^{B}P_{\lambda\sigma}(\mu\sigma|\nu\lambda)$$
$$(\mu \text{ on A},\ \nu \text{ on B}). \tag{4.2}$$

As before, the core Hamiltonian is split into atomic parts, and we write for the matrix elements between orbitals on the same atom,

$$H_{\mu\nu}=U_{\mu\nu}-\sum_{B(\neq A)}(\mu|V_B|\nu). \tag{4.3}$$

Since the product $\phi_\mu\phi_\nu$ is no longer neglected, all terms in (4.3) are retained so that Approximation 4 of the previous section is no longer needed. However, each term in (4.3) must transform correctly according to (2.20). The remaining matrix elements $H_{\mu\nu}$, where μ and ν are on different atoms are calculated using Approximation 5 of the previous section.

5. CONCLUSIONS

Roothaan's SCF equations and their transformation properties have been discussed from the point of view of an approximate scheme for the inclusion of electron repulsion in the calculation of the LCAOMO's for a molecule of any size or symmetry. It is desirable that some of these transformation properties carry over into a more approximate theory, particularly for molecules of low symmetry where no unique choice for the local atomic axes exists. It is clearly *very important* that any proposed approximate theory be independent of any choice of these axes, as is the full SCF theory. The full equations also possess the property of being invariant with respect to the hybridization of the atomic orbital basis functions, and it seemed *reasonable* that our approximate theory should also be so invariant. This is because it is convenient to be able to translate the calculated molecular orbitals or bond orders of one basis set to another, such a procedure often illuminating a physical significance not apparent when the MO's are referred to atomic orbital basis functions. These transformation conditions were found to impose restrictions as to the manner in which we approximated to the full SCF equations. Within the framework of a theory based on the neglect of differential overlap these restrictions took the following form.

(i) There are two internally consistent approximations to the full SCF equations. These two approximations have been developed in the preceding sections and termed the complete neglect of differential overlap (CNDO) and the neglect of diatomic differential overlap (NDDO) approximations, respectively.

(ii) The adoption of the complete neglect of differential overlap approximation necessitates the simplification of all Coulomb integrals to a few which measure an average repulsion between electrons in the various valence orbitals.

(iii) The adoption of a more refined theory requires the inclusion of *all* exchange integrals of the form

$$(\phi_\mu\phi_\nu \mid \phi_\sigma\phi_\lambda),$$

where ϕ_μ and ϕ_ν have a common center and ϕ_σ and ϕ_λ have a common center (which could of course be the same for both pairs).

(iv) If the resonance integral $\beta_{\mu\nu}$ is chosen to be proportional to the corresponding overlap integral $S_{\mu\nu}$, then the constant of proportionality must be independent of the type of orbital ϕ_μ or ϕ_ν (i.e., as to whether they are s, p, $d\cdots$) and only depend on the nature of the two participating atoms.

Finally, we may note that this last criterion is not satisfied by some approximations that have already been used in independent-electron theories. An example is the Wolfsberg–Helmholtz approximation to the resonance integral,[8] used by Hoffman[9] in his extended Hückel theory. According to this

$$\beta_{\mu\nu} = (L/2)(I_{\mu\mu} + I_{\nu\nu})\, S_{\mu\nu}, \tag{5.1}$$

where L is a constant and $I_{\mu\mu}$, $I_{\nu\nu}$ are valence-state ionization potentials for the atomic orbitals ϕ_μ and ϕ_ν. As the two sides of (5.1) transform in different ways, the energies and orbitals obtained using this approximation will not be invariant to hybridization of the atomic orbital basis set.

ACKNOWLEDGMENT

The research described in this paper was supported by a grant from the National Science Foundation.

VI-6

Sigma Electrons in Conjugated Heterocycles, with Special Emphasis on Biological Purines and Pyrimidines

ALBERTE PULLMAN

1. Introduction

Our laboratory has been involved in the last decade in the study of the electronic structure of a number of large heterocyclic molecules of biological interest.[1] In the early calculations, the simple Hückel procedure in the π-electron approximation was used and the general image of the π-electronic structure so obtained has been largely confirmed by calculations using the Pariser-Parr-Pople method.[2]

It was soon recognized, however, that the influence of the σ framework was to be taken into account both implicitly in the parametrization of the self-consistent procedure[3] for π electrons and also explicitly for the accurate evaluation of the dipole moments. A simple localized-bond approximation[4] has been adapted to conjugated molecules[5,6] and used to evaluate the σ-electron distribution in the nucleic bases. Meanwhile a careful optimization of the Pariser-Parr-Pople set of integrals was performed on a series of reference compounds and used to calculate the π-electron structure which was to be associated with the σ distribution.[7,8] Although the σ and the π electrons were still treated separately, their mutual influence was, however, implicitly taken into account by the fitting of the calculated properties (e.g., dipole moments, ionization potentials) on experimental data for the reference compounds. The agreement with the known properties in a large number of molecules[7–10] gave confidence in the image of the overall electron distribution obtained in this fashion.

Nevertheless, the simultaneous treatment of the σ and π electrons of these compounds was highly desirable to gain further insight into their actual electronic properties and to assess the validity of the picture of the σ-electron distribution utilized.

The recent development of new methods within the MO formalism, together with the availability of high-speed computers, today permits such simultaneous treatment of σ and π electrons in large conjugated molecules, and three of these new procedures have been recently used in our laboratory for all-valence-electron computation of the main bases of the nucleic acids.[11–13] This gives us the possibility to present a *comparative* study of the results obtained. We feel that, since the available all-valence-electron methods are still quite semiempirical in nature and are rooted in different theoretical concepts, the comparison of their results for the same compounds is highly desirable.

The comparison will be carried out for adenine, guanine, uracil, and cytosine (A, G, U, and C) and supplemented by a study of pyrrole and pyridine, which often serve as model compounds in the study of larger heterocycles. In the last two cases, nonempirical SCF calculations in a contracted Gaussian basis set have recently become available,[14,15] and comparisons will be made with these computations wherever possible.

2. Analysis of the Methods

For the all-valence-electron calculations three methods were selected: (1) the extended Hückel theory (EHT) developed by Hoffmann[16,17]; (2) one of its iterative refinements, which includes both Cusachs' modification[18] and iteration on the charges[19] (the version utilized has been called IEHC[13]); and (3) the CNDO/2 procedure.[20]

The details of the procedures can be found in the original papers. Only a discussion of what seems to us relevant for a comparison will be given here. In the three procedures the molecular orbitals are obtained as linear combinations of all valence atomic orbitals:

$$\phi_i = \sum_{\mu} c_{i\mu} \chi_\mu$$

where χ_μ are Slater-type orbitals including 2s, $2p_x$, $2p_y$, and $2p_z$ orbitals for C, N, and O and the 1s orbital for hydrogen. The exponents for the χ's differ however in the three procedures, as summarized in Table 1.

Table 1. Outstanding Features of EHT, IEHC, and CNDO

	EHT	IEHC	CNDO/2
Exponents in χ_μ	Slater exponents	Clementi-Raimondi's "best single Slater exponents" for atoms	Slater exponents except $\zeta = 1.2$ for the 1s orbital of hydrogen
Hamiltonian	Hückel-type nonexplicit		Hartree-Fock $H_{\mu\nu} = I_{\mu\nu} + G_{\mu\nu}$
$H_{\mu\nu}$	Wolfsberg-Helmholtz	Cusachs	$(\beta_A^0 + \beta_B^0)\frac{S_{\mu\nu}}{2} - \frac{1}{2} P_{\mu\nu}\gamma_{AB}$
$H_{\mu\mu}$	$-I_\mu$	$-I_\mu + \lambda q_A$	$X_\mu + \gamma_{AA}(q_A - \frac{1}{2}q_\mu) + \sum_B \gamma_{AB} q_B$
Process	No iteration	Iteration on charges	Iteration on charges

The general equation of the linear variation method is used in every case for finding the coefficients $c_{i\mu}$ and the orbital energies by

$$\sum_{\nu} c_{i\nu} (H_{\mu\nu} - ES_{\mu\nu}) = 0$$

in which all bonded and nonbonded elements are included.

The Hamiltonian used in EHT and IECH is an effective nonexplicited Hamiltonian of the Hückel type. In CNDO/2 it is a Hartree-Fock Hamiltonian expressed as usual as a sum of core + interactions. From there on, the philosophies of the methods diverge.

In the two Hückel procedures all overlaps are retained and the calculations are made after choosing empirical values for the diagonal elements $H_{\mu\mu}$ and empirical formulas for the nondiagonal elements $H_{\mu\nu}$. In EHT, $H_{\mu\mu}$ is equated to the negative of the valence-state ionization potential of orbital μ and the Wolfsberg-Helmholtz approximation is chosen for calculating the $H_{\mu\nu}$'s:

$$H_{\mu\nu} = 1.75 S_{\mu\nu} \frac{H_{\mu\mu} + H_{\nu\nu}}{2}$$

In EHT the calculations are terminated after solving the equations once.

IEHC differs from EHT in two aspects:

a. The Cusachs formula is used for the nondiagonal matrix elements:

$$H_{\mu\nu} = (2 - |S_{\mu\nu}|)\, S_{\mu\nu} \frac{H_{\mu\mu} + H_{\nu\nu}}{2}$$

b. An iteration on the charges is carried out until self-consistency is obtained: At each iteration the total net charge on atom A serves to recalculate every $H_{\mu\mu}$ of A by

$$H_{\mu\mu} = - I_\mu + \lambda q_A$$

In the CNDO method, the fundamental hypothesis of zero differential overlap and the averaging of all Coulomb interactions as γ_{2s2s}, associated with the assimilation of the core attractions $\langle \chi_\mu^A V_B \chi_\mu^A \rangle$ to $- Z_B \gamma_{AB}$, yield a simple expression of the diagonal elements:

$$H_{\mu\mu} = X_\mu + (q_A - \tfrac{1}{2} q_\mu)\gamma_{AA} + \sum_B q_B \gamma_{AB}$$

in terms of the Mulliken orbital electronegativity X_μ and of the net charges q. On the other hand, the assumption of proportionality of the *core* integrals $I_{\mu\nu}$ to the overlap $S_{\mu\nu}$ gives for the matrix elements of the total Hamiltonian

$$H_{\mu\nu} = \frac{\beta_A^0 + \beta_B^0}{2} S_{\mu\nu} - \tfrac{1}{2} P_{\mu\nu}\gamma_{AB}$$

where $P_{\mu\nu}$ is the bond order for the couple $\mu\nu$ and where β_A^0 and β_B^0 are empirical parameters fitted so as to reproduce accurate minimal-basis-set SCF calculations.

This parallel description shows clearly both the similarities and the differences among the three methods. In particular, as far as the iteration procedures are concerned, two fundamental differences appear: On the one hand, IEHC iterates on the *local charge* of the atom to which μ belongs, whereas CNDO takes into account *all* the atomic charges—a feature which may very well eliminate in some cases the effect of the local correction. On the other hand, the starting point of the iterative scheme is fundamentally different in the two procedures, insofar as IEHC starts from the orbital *ionization potential*, while CNDO starts from the mean value of this ionization potential and of the electron affinity, thus from much smaller values of the $H_{\mu\mu}$'s. This can be seen in Table 2, which gives the numerical values of the diagonal core parameters used in EHT and of the starting values of the same quantities in the two iterative procedures.

Table 2. Values of the Diagonal Parameters Used in EHT (a) and as a Starting Point in IEHC (b) and CNDO (c) (in eV)

	s			p		
	(a)	(b)	(c)	(a)	(b)	(c)
H	13.6	13.8	7.176			
C	21.4	21.2	14.051	11.4	11.4	5.572
N	26.0	25.5[a]	19.316	13.4	14.4[b]	7.275
O	35.3	32	25.390	17.76	16.45	9.111

[a] 27.4 for pyrrolic nitrogens.
[b] 12.5 for p_z of pyrrolic nitrogens.

The parametrizations indicated in Table 2 are

a. The values used by Hoffmann for H, C, and N in his original papers.[16,17] For consistency the corresponding valence-state ionization potentials of Pritchard and Skinner[21] have been used for oxygen.

b. The "smoothed" valence-state ionization potentials of Cusachs and Raynolds[22] for C, N, and O. [An average of A(p,1) and A(p,2) has been taken for the p orbitals of oxygen to avoid the a priori particularization of one lone pair.] The values for pyridine- and pyrrole-like nitrogens correspond to s^2xyz and $sxyz^2$ of Ref. 22, respectively. In fact, as will be seen later, this last particularization of the p_z orbital of NH is probably not appropriate. An example of a calculation of pyrrole itself with the same initial $H_{\mu\mu}$ values for p_x, p_y, and p_z of NH has been performed and seems more satisfactory.

c. The values given by Pople and Segal[20] with no modification.

3. Comparison of the Results

From a comparative point of view, the most interesting quantities which can be calculated by the three aforementioned methods are the individual orbital energies, the atomic orbital populations, and the general charge distribution, with the resulting dipole moments, as well as its separation into the σ and π components. Each of these aspects will be studied briefly.

Figure 1 recalls the chemical structure and the numbering of the atoms in the molecules studied, as well as the directions choosen for the x and y axes.

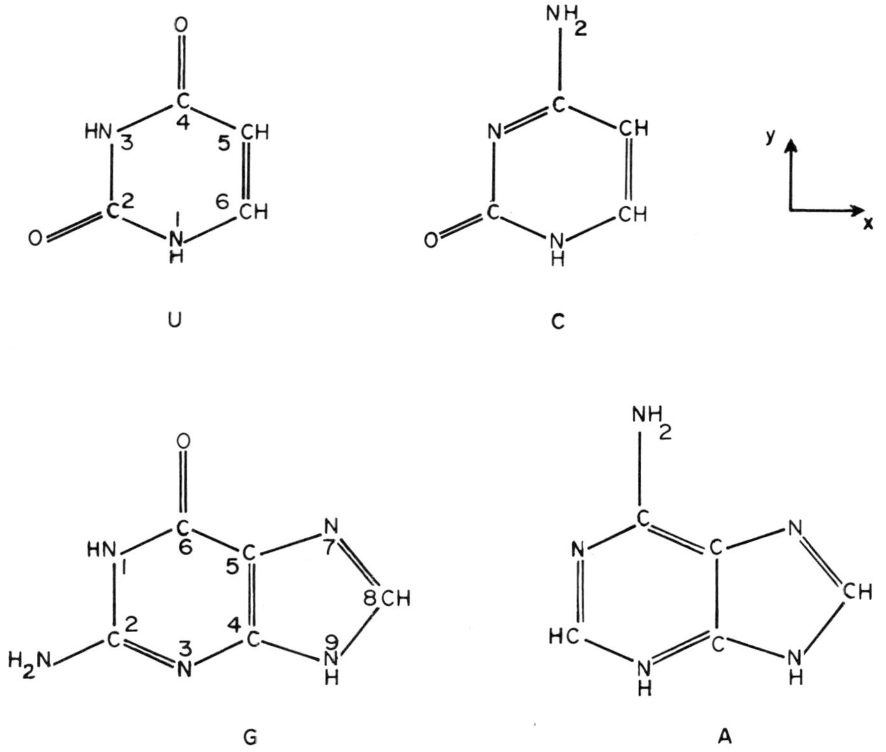

Fig. 1. Chemical formulas for U, C, G, and A and directions of the x and y axes.

Energies and Characteristics of Molecular Orbitals

Table 3 gives the individual orbital energies obtained by the three procedures for uracil, cytosine, guanine, and adenine. Corresponding values for pyrrole and pyridine are gathered in Table 4 together with the orbital energies obtained in the nonempirical calculations by Clementi. (Note that two parametrizations have been used in IEHC for pyrrole.)

It is seen that EHT and IEHC give a similar spread of 20 to 25 eV in energy levels, whereas the CNDO levels extend deeper, features which are respectively characteristic of the procedures.

Table 3. Energies of the Occupied Molecular Orbitals (in eV) in the Four Nucleic Bases U, C, G, A[a]

U			C		
(a)	(b)	(c)	(a)	(b)	(c)
37.19	33.09	52.87	37.03	31.35	52.36
36.93	29.53	45.09	31.39	28.56	44.91
30.39	27.96	43.64	29.51	26.76	41.69
28.18	25.98	40.22	27.29	23.87	38.85
25.45	23.63	36.88	25.42	22.89	36.61
20.56	20.03	32.25	20.25	18.94	31.66
20.04	19.14	27.41	19.87	18.28	27.75
18.41	17.70	π 26.38	π 18.14	16.92	25.74
π 18.18	16.56	26.36	17.90	16.23	π 25.53
π 18.14	16.33	25.71	17.67	15.63	25.45
18.12	π 16.29	21.39	16.78	π 15.25	20.96
17.59	π 14.53	π 20.07	16.16	14.28	20.01
17.18	14.29	19.83	15.44	13.72	π 19.91
16.52	13.85	19.12	π 15.27	π 13.68	18.40
15.58	π 13.70	π 18.65	14.86	13.11	17.91
π 14.92	12.19	17.72	14.40	π 12.99	π 17.20
14.74	11.46	16.42	π 14.40	11.17	15.56
14.07	π 11.31	13.83	13.83	10.86	13.68
π 13.90	π 11.08	π 13.18	π 13.51	π 10.54	π 13.21
13.37	9.89	12.90	12.58	π 10.39	11.81
π 12.71	9.43	π 11.88	π 12.50	9.07	π 10.78

G			A		
(a)	(b)	(c)	(a)	(b)	(c)
37.04	32.36	55.49	32.74	31.56	55.33
32.56	30.60	51.27	30.99	29.33	50.39
31.22	28.73	48.23	29.74	26.97	44.13
28.45	25.77	42.73	27.50	24.94	38.86
27.66	25.13	39.50	26.82	23.67	37.34
26.97	24.84	38.98	23.97	22.07	33.94
23.52	21.54	32.08	20.03	18.85	30.41
20.05	19.12	30.57	19.33	18.49	29.94
19.22	18.67	28.50	18.55	17.98	27.73
18.78	18.24	27.82	16.67	16.81	π 27.07
π 18.15	17.02	π 27.60	16.28	15.99	25.64
17.86	16.50	26.48	15.79	π 15.86	22.62
17.48	π 16.10	π 22.82	π 15.78	15.34	π 21.15
16.45	15.54	22.46	15.35	π 14.28	20.51
16.19	15.40	21.61	π 15.05	14.20	19.82
π 15.74	π 14.74	20.60	14.85	14.13	18.07
15.65	14.68	π 19.66	π 14.54	13.59	π 17.89
15.32	π 14.14	19.55	14.32	π 13.44	17.31
π 15.12	13.71	17.38	14.08	13.01	15.76
14.69	12.87	16.75	π 13.59	π 12.15	15.11
14.23	π 12.46	15.65	13.58	11.84	π 14.14
π 14.10	11.97	π 15.34	π 13.35	π 11.67	13.11
13.73	π 11.73	15.08	12.92	11.07	π 12.57
π 13.52	11.52	π 14.21	12.08	π 10.43	11.46
13.51	π 10.93	π 13.11	π 11.95	10.26	π 10.08
π 13.39	10.43	12.77			
12.38	π 10.31	11.55			
π 11.84	9.21	π 9.05			

[a](a) EHT, (b) IEHC, (c) CNDO. Unlabeled orbitals are σ levels. All signs are reserved.

Table 4. Energies of Occupied Levels (in eV, sign reversed) in Pyrrole and Pyridine by Different Methods

Pyrrole					Pyridine			
EHT	IEHC[a]	IEHC[b]	CNDO	Ref. 14	EHC	IEHC	CNDO	Ref. 15
31.1	29.9	29.6	52.9	35.9	30.7	29.0	51.4	36.1
25.9	23.5	23.5	37.1	29.9	27.0	24.7	39.0	31.5
24.5	22.1	22.1	34.5	28.2	25.8	23.6	37.4	30.2
19.2	17.4	17.4	27.9	22.4	20.5	18.9	30.9	25.1
18.6	16.7	16.6	27.1	21.6	20.1	18.2	30.1	24.6
16.9	16.6	16.5	26.5	21.1	16.8	16.1	28.4	21.2
π 15.1	π 14.0	π 14.6	π 24.1	17.6	16.3	π 15.0	π 23.5	19.7
14.6	13.9	13.8	17.5	π 17.2	15.1	14.7	22.1	19.1
14.4	13.7	13.7	17.4	17.0	π 14.9	14.7	20.0	18.2
13.8	12.6	12.7	16.0	16.4	16.6	14.0	18.6	17.4
13.5	12.4	12.5	15.9	15.6	14.4	13.4	18.1	π 16.9
πS 12.8	πAS 11.2_4	πS 11.8	πS 13.3	πS 11.6	π 13.4	π 12.6	π 14.4	15.2
πAS 12.1	πS 11.2_2	πAS 11.4	πAS 11.9	πAS 10.5	13.2	π 12.3	14.3	12.7
					π 12.8	12.2	π 13.9	π 12.5
					12.3	10.7	13.0	π 12.2

[a] $H_{\mu\mu}$ values of Table 2.
[b] The same as footnote a, except p_z of NH equivalent to other p.

Although the numerical values of the orbital energies should not be taken as absolute in any of the methods, one can make a few important observations:

a. Everywhere one observes a very large intermingling of the σ and π levels with no appearance of a superficial "π shell," as was sometimes implicitly admitted in π-electron calculations. Although the details may differ, the interspersing of the σ levels among the π orbitals seems to be a constant feature in all calculations. The same mixing appears also in the nonempirical SCF calculations made in a Gaussian basis set. Such a constancy in the outcome of procedures which are quite different in their principles is probably a strong indication in favor of this picture. It may be added that this interspersion of σ and π levels occurs much less in the virtual orbitals, in which the π^* levels are often grouped below the σ^* levels.

b. The danger of taking for granted the results of one single procedure appears particularly clearly when examining the nature of the highest occupied orbital (*homo*); thus

(1) For pyrrole, all methods, including Clementi's, yield a π homo, followed by another π orbital. But, while this homo is generally antisymmetrical (AS) with respect to the molecular symmetry axis in all procedures, the usual diagonal parametrization of IEHC gives a symmetrical *homo* closely followed by the antisymmetrical one; however, when the same $H_{\mu\mu}$ is used for all the p orbitals of nitrogen, the usual order is restored without changing much the other molecular orbitals.

(2) In pyridine, the agreement on the character of the homo is far from being obtained: The three semiempirical methods yield a σ homo, whereas Clementi obtains even two π homos.

(3) In the fundamental bases of the nucleic acids, the homo is π in EHT and CNDO, but in IEHC it is a σ orbital, the first π level coming below. (In uracil there are even two such σ orbitals above the first π level.) A discussion of the π homos for the nucleic bases has been given elsewhere.[23] From a comparative point of view it is interesting to mention that all procedures predict guanine to be the best π-electron donor and uracil the worst one among the four bases, in agreement with previous Pariser-Parr-Pople π calculations[7] and also with recently measured values of the ionization potentials for some of the bases.[24]

As to the σ character of the homo predicted in some cases, it may be remarked that in U, C, G, and A, this σ homo is obtained by IEHC only, but this procedure has been shown on the example of pyrrole to be sensitive to relatively small parameter changes. On the other hand, the available experimental data both on the fundamental heterocycles[25] and on the nucleic bases[24] do not seem to favor a σ-first ionization potential. The direct comparison of theory with experiment must, however, be done with caution until the effect of introducing electron correlation both in the molecule and in the ion will have been properly assessed.

An important question which arises in such heteromolecules is the fate of the lone-pair atomic orbitals. Thus it is interesting to analyze the eigenvectors corresponding to the individual molecular energy levels. By such inspection of the IEHC orbital wave functions, one discovers that in all the carbonyl compounds (U, C, G) the σ homo(s) are essentially made of the oxygen p_x and p_y orbitals. Moreover, looking at the same compounds in the CNDO methods, one observes that, here, the σ homo (that is, the second homo) is also made principally of the p_x and p_y orbitals of the oxygens. In EHT, similarly weighted molecular orbitals appear also but they are buried rather deep toward the middle of the energy range.

In contrast to these results, all procedures show a more appreciable mixing of the nitrogen atomic orbitals with the other atomic orbitals in the molecular eigenfunctions. At any rate it seems more inappropriate to speak of pure nitrogen lone pairs than of oxygen lone pairs (on carbonyls) in these molecules both energy-wise and population-wise (*vide infra*).

Populations on Atomic Orbitals

Tables 5, 6, and 7 give the Mulliken orbital populations for the s, x, and y orbitals for each atom of the four compounds studied, together with the corresponding values for pyrrole, py-

Table 5. σ Populations on Oxygen Atomic Orbitals

			EHT				IEHC				CNDO			
Type	Atom	Mol.	s	p	x	y	s	p	x	y	s	p	x	y
	$O(C_2)$	U	1.77	3.73	1.81	1.92	1.65	3.42	1.60	1.82	1.73	3.18	1.42	1.76
	$O(C_4)$	U	1.77	3.74	1.98	1.76	1.70	3.43	1.96	1.47	1.73	3.21	1.93	1.28
C=O	O	C	1.78	3.74	1.82	1.92	1.67	3.42	1.62	1.80	1.74	3.16	1.42	1.74
	O	G	1.77	3.74	1.98	1.76	1.70	3.42	1.96	1.46	1.73	3.19	1.92	1.27
	O	a	1.77	3.74	1.98	1.76	1.74	3.49	1.96	1.53	1.73	3.29	1.92	1.37

[a]Formaldehyde.

Table 6. σ Populations on Nitrogen Atomic Orbital

			EHT				IEHC				CNDO/2			
Type	Atom	Mol.	s	p	x	y	s	p	x	y	s	p	x	y
HN	N	a	1.40	2.34	1.16	1.18	1.46	2.29	1.19	1.10	1.19	2.25	1.19	1.06
	N_1	U	1.39	2.37	1.15	1.22	1.39	2.28	1.13	1.15	1.18	2.28	1.17	1.11
	N_3	U	1.39	2.40	1.22	1.18	1.39	2.27	1.14	1.13	1.21	2.29	1.15	1.14
	N_1	C	1.39	2.39	1.16	1.23	1.41	2.30	1.14	1.16	1.20	2.30	1.19	1.11
	N_1	G	1.39	2.39	1.22	1.17	1.39	2.27	1.13	1.14	1.21	2.30	1.15	1.15
	N_9	G	1.38	2.37	1.20	1.17	1.42	2.30	1.19	1.11	1.19	2.32	1.22	1.10
	N_9	A	1.39	2.38	1.20	1.18	1.42	2.30	1.20	1.10	1.19	2.29	1.20	1.09
H_2N	N	C	1.37	2.55	1.36	1.19	1.47	2.23	1.16	1.07	1.21	2.20	1.07	1.13
	N	G	1.40	2.47	1.20	1.27	1.42	2.17	1.06	1.11	1.21	2.20	1.11	1.09
	N	A	1.41	2.47	1.30	1.17	1.45	2.20	1.14	1.06	1.21	2.17	1.05	1.12
N	N	b	1.52	2.97	1.14	1.83	1.40	2.83	1.08	1.75	1.45	2.64	1.11	1.53
	N_3	C	1.49	2.99	1.65	1.34	1.34	2.74	1.55	1.19	1.45	2.52	1.35	1.17
	N_3	G	1.48	2.98	1.17	1.81	1.32	2.70	1.02	1.68	1.43	2.53	1.05	1.48
	N_1	A	1.48	2.98	1.68	1.30	1.37	2.79	1.59	1.20	1.44	2.60	1.41	1.19
	N_3	A	1.47	2.98	1.16	1.82	1.36	2.76	1.04	1.72	1.42	2.61	1.07	1.54
	N_7	G	1.51	3.00	1.28	1.72	1.42	2.75	1.14	1.61	1.49	2.53	1.15	1.38
	N_7	A	1.52	2.99	1.26	1.73	1.41	2.73	1.12	1.61	1.50	2.51	1.13	1.38

[a]Pyrrole.
[b]Pyridine.

Table 7. σ Populations on Carbon Atomic Orbitals

		EHT				IEHC				CNDO			
Atom	Mol.	s	p	x	y	s	p	x	y	s	p	x	y
C_2	U	0.92	1.12	0.52	0.60	1.06	1.51	0.75	0.76	1.00	1.76	0.89	- 0.87
C_4	U	0.97	1.26	0.76	0.50	1.10	1.64	0.88	0.76	0.99	1.82	0.89	0.93
C_5	U	1.18	1.95	0.96	1.15	1.10	1.86	0.94	0.92	1.01	1.97	0.99	0.98
C_6	U	1.15	1.73	0.83	0.90	1.13	1.76	0.83	0.93	0.99	1.82	0.89	0.93
C_2	C	0.91	1.10	0.52	0.58	1.13	1.56	0.74	0.82	1.02	1.77	0.86	0.91
C_4	C	1.07	1.45	0.75	0.70	1.11	1.66	0.90	0.75	1.00	1.93	0.92	1.01
C_5	C	1.16	2.04	1.04	1.00	1.09	1.90	0.94	0.96	1.00	1.97	0.97	1.00
C_6	C	1.14	1.75	0.84	0.91	1.13	1.77	0.83	0.94	1.01	1.90	1.01	0.89
C_2	G	1.02	1.31	0.65	0.66	1.07	1.56	0.81	0.75	1.00	1.80	0.92	0.88
C_4	G	1.06	1.47	0.62	0.85	1.09	1.65	0.69	0.96	0.98	1.84	0.83	1.01
C_5	G	1.10	1.66	0.73	0.93	1.08	1.77	0.81	0.96	0.97	1.91	0.92	0.99
C_6	G	0.96	1.28	0.78	0.50	1.10	1.64	0.89	0.75	1.00	1.82	0.89	0.93
C_8	G	1.10	1.50	0.83	0.67	1.14	1.61	0.85	0.76	1.03	1.77	0.88	0.89
C_2	A	1.08	1.57	0.84	0.73	1.19	1.70	0.87	0.83	1.02	1.87	0.93	0.94
C_4	A	1.06	1.48	0.62	0.86	1.09	1.66	0.69	0.97	0.99	1.85	0.84	1.01
C_5	A	1.10	1.66	0.73	0.93	1.08	1.78	0.81	0.97	0.97	1.94	0.93	1.01
C_6	A	1.06	1.47	0.78	0.69	1.08	1.67	0.91	0.76	0.98	1.88	1.00	0.88
C_8	A	1.10	1.51	0.84	0.67	1.14	1.62	0.85	0.76	1.04	1.81	0.90	0.91
C_α	a	1.15	1.71	0.82	0.89	1.18	1.78	0.87	0.91	1.02	1.91	0.93	0.98
C_β	a	1.18	1.93	0.98	0.95	1.14	1.89	0.94	0.95	1.01	2.00	0.98	1.02
C_γ	a	1.18	1.91	0.92	0.99	1.16	1.93	0.96	0.97	1.01	1.99	1.03	0.96
C_α	b	1.16	1.72	0.84	0.88	1.14	1.70	0.79	0.91	1.01	1.86	0.87	0.98
C_β	b	1.19	1.92	0.95	0.97	1.16	1.83	0.93	0.90	1.00	1.95	0.99	0.96

[a] Pyrrole.
[b] Pyridine.

ridine, and formaldehyde. Examination of the table allows very important conclusions concerning the σ-electron distributions. For convenience, the atoms have been grouped according to their nature: oxygens, nitrogens, and carbons.

Oxygen atoms. Oxygens in all the compounds studied are carbonyl oxygens. In each procedure they appear with a very constant structure:

$$s^{1.8}p^{3.7}; \quad s^{1.7}p^{3.4}; \quad s^{1.7}p^{3.2}$$

in EHT, IEHC, and CNDO, respectively. Moreover, the decomposition of p into its x and y components shows that in each method there is *very nearly a pure lone pair in the direction perpendicular to the C=O bond.* The lone pair on a carbonyl oxygen does not depart much from the classical hybridization di^2 di $x^2\pi$, ($s^{1.5}y^{1.5}x^2\pi$), but the s character in the σ orbitals is larger than in a pure digonal orbital, the y orbital being slightly "depopulated" in CNDO.

Nitrogen atoms. Two kinds of nitrogens exist in the molecules studied, the pyrrole or aniline-like NH (or NH_2) nitrogens and the pyridine-like nitrogens. In the classical representation, the first kind derives from the valence-state s xy π^2, the second kind from s^2 xy π; furthermore, trigonal hybridization is generally assumed to take place in the molecular plane, yielding sp^2 and $s^{1.33}p^{2.66}$ hybridization ratios, respectively. The data of Table 6 show, here again, a striking constancy for each category of nitrogens in each method; thus in EHT, IEHC, and CNDO/2, respectively,

$$\text{NH:} \quad s^{1.4}p^{2.4}; s^{1.4}p^{2.3}; s^{1.2}p^{2.3}$$

$$\text{NH}_2\text{:} \quad s^{1.4}p^{2.5}; s^{1.5}p^{2.2}; s^{1.2}p^{2.2}$$

$$\text{N:} \quad s^{1.5}p^{3.0}; s^{1.4}p^{2.7/2.8}; s^{1.4/1.5}p^{2.5/2.6}$$

The two kinds of nitrogens belong to different series, the imino or amino nitrogens being farther from the trigonal representation than the pyridine-like ones. Moreover, both NH and N populations are very similar in all compounds to the corresponding values in pyrrole and pyridine, respectively. In all methods the s character is larger than in the fundamental sp^2 hybrids.

Carbon atoms. It is generally assumed that in aromatic molecules, carbon atoms exhibit trigonal sp^2 hybridization (sxy). The numerous carbon atoms included in Table 7 display ratios going from

$s^{0.9}p^{1.1}$	to	$s^{1.2}p^{2.0}$	in EHT
$s^{1.1}p^{1.5}$	to	$s^{1.2}p^{1.9}$	in IEHC
$s^{1.0}p^{1.8}$	to	$s^{1.0}p^{2.0}$	in CNDO

Clearly, EHT emphasizes strongly the displacements of electrons, particularly when the atom is in a very electronegative environment (between two nitrogens and one oxygen, for instance). The ω-technique iteration process smoothes out the phenomenon, which is still less pronounced in CNDO. The differences between the three procedures are obviously a feature of the parametrization, but a very important observation concerns the fact that, in the three methods, the maximum population (very close to sp^2) is observed for carbons environed by three other carbons or two carbons and one hydrogen (C_γ of pyridine, C_β of pyrrole, C_5 of U and C); then a decrease in population is observed for environments including one or two nitrogens, the largest withdrawal of p_σ electrons with respect to the trigonal valence state occurring with NNN, NNO, or NCO neighborhoods.

It may thus clearly be concluded from this survey that *the σ population for a given atom in a given neighborhood is fairly constant, that it is very little dependent upon nonbonded atoms and, moreover, that it is very little affected by the π-electron displacements* (whereas the reverse is not true). This last conclusion is shown also when one follows the convergency in the iteration processes: The convergency is very quickly reached for the σ orbitals, whereas the π convergency is obtained much later.

The implication of this conclusion is important in that it suggests the possibility of saving a fair amount of computer time by choosing a priori a set of σ-electron populations for a given molecule and iterating only on the π system. This choice could be done rather easily by inspection of large tables of already existing data similar to those of Tables 5 to 7. Or one could run the program so as to obtain the σ, π, and total energies, but again starting from a near-final σ distribution instead of the usual pure atomic distribution. These different possibilities are being investigated in our laboratory and will be reported elsewhere.

It can also be remarked that these conclusions concerning the σ populations lend support to the fundamental idea used in Del Re's approximate procedure for treating the σ bonds, in which each bond is represented by a two-by-two Hamiltonian, the matrix elements of which depend, however, on all the *adjacent* bonds.

Dipole Moments and Charge Distributions

One of the incentives for explicitly introducing the σ electrons in calculations on heterocycles has been the need for an accurate evaluation of dipole moments. This is particularly important in biomolecules for which very few values of the dipole moments have been measured (often on account of solubility problems). Thus, for example, while values of dipole moments have been obtained for some simple derivatives of adenine and uracil,[26] no experimental information is available about the moments of guanine or cytosine. As recalled in Section 1, we have, in the last few years, calculated the σ contribution to the dipole moments in heterocycles by a carefully parametrized localized-bond approach. The values of the moments obtained in

Table 8. Total Dipole Moment (in D)[a]

	σ/π_1^b		σ/π_2^c		EHT[d]		IEHC[d]		CNDO[d]		Experimental value[e]
A	3.2	77	2.3	73	6.1	65	4.4	104	2.9	64	3.0
U	3.6	39	4.0	35	12.2	41	7.6	36	4.6	36	3.9
G	6.8	-36	7.2	-36	17.2	−16,5	13.6	-32	7.5	-31	–
C	7.2	97	7.1	110	17.3	106	12.9	125	7.6	102	–

[a]The first number is the moment, the second its angle $\theta°$ with the vertical axis (measured counterclockwise).

[b]π moment obtained from the best Hückel parameters and added to the σ moments evaluated as in Ref. 6.

[c]π moment obtained by the Pariser-Parr-Pople computations[7] and added to the same σ moment.

[d]Hybrid moments included.

[e]For 9-methyladenine and 1,3-dimethyluracil.[24]

this fashion for U, C, G, and A are recalled in Table 8 as well as their directions and their σ and π components. It must be pointed out, however, that Del Re's procedure does not permit a decomposition of the σ dipole into different contributions and that, in particular, although the hybrid or lone-pair moments are not explicitly introduced, they must, in fact, be considered as hidden in each bond contribution insofar as the parametrization is made so as to reproduce dipole moments for reference compounds.

It is thus very instructive to examine comparatively these old results together with those obtained now in the all-valence-electron calculations. Table 8 summarizes the results for the

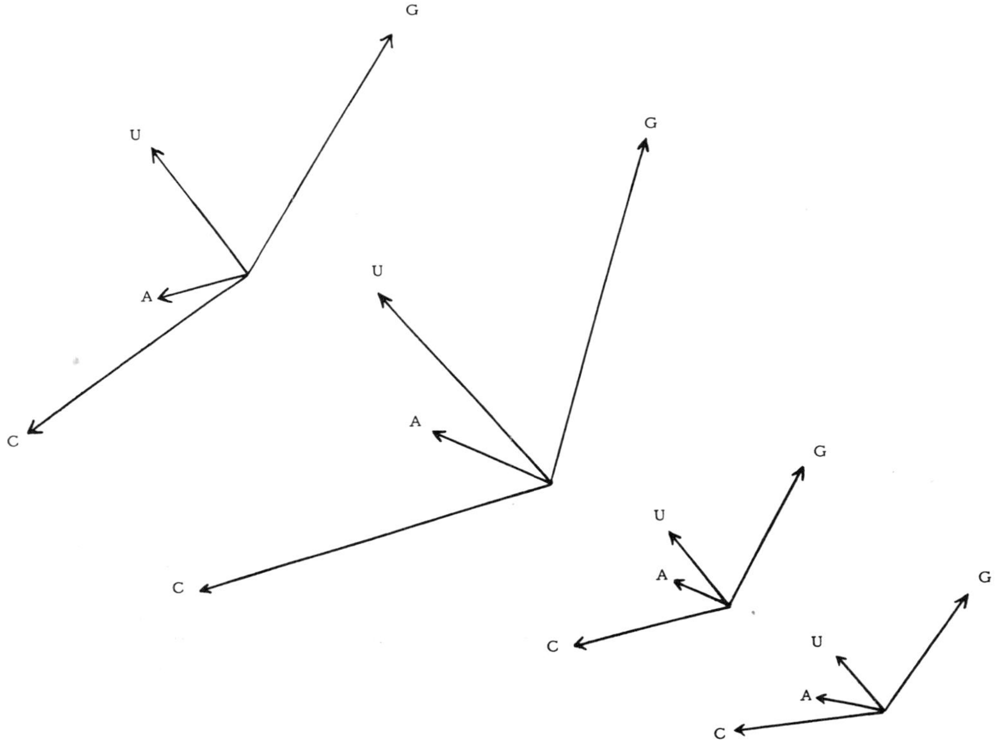

Fig. 2. Total dipole moments drawn on the same scale (y axis vertical). From right to left: simple method of Ref. 6; CNDO; EHT; IEHC.

total dipole moments. In both EHT, IEHC, and CNDO, these moments are obtained in the point-charge approximation, with addition of a hybrid moment μ_{sp} calculated as suggested by Pople and Segal.[27]

It may be observed that the EHT method yields relatively large dipole moments, a feature which was predictable in view of the large values of the $H_{\mu\mu}$ elements and the absence of iteration. The CNDO procedure gives moderate values of the dipole moments, very close to those obtained by combining the Berthod-Pullman σ distribution[6] with either a Hückel or the Pariser-Parr-Pople π distribution. The IEHC method yields intermediate values.

But what is very striking is that all methods predict greater dipole moments for guanine and cytosine than for uracil, which in turn should have a greater moment than adenine. The comparison of the theoretical values with the experimental ones available only for simple derivatives of adenine and uracil indicates that the CNDO and the $\sigma + \pi$ (Hückel or PPP) calculations are in very satisfactory agreement with experiment. It may therefore reasonably be expected that the values predicted for G and C in these last two procedures are the reliable ones.

What is still more striking is the similarity in the directions of the moments predicted for the bases by all methods (Fig. 2). Obviously in spite of the differences in charge distributions (*vide infra*), the overall polarity of the molecules is well reproduced by all procedures, although somewhat exaggerated by some of them. It may be observed that the IEHC dipoles show the largest departures from the general trends, particularly in the direction of the adenine dipole moment. This is most probably due to the parametrization of the NH and NH_2 nitrogens (see Section 2), which overemphasizes the delocalization of the π lone pairs of these nitrogen atoms.[23]

The σ and π contributions to the total moments in the three procedures are compared in Fig. 3(a) and 3(b), respectively, with the values of the simple representation. In the all-valence-electron calculation, the σ components include the hybrid moments. It is seen that in these molecules, the general trends observed on the total moments are essentially present in their

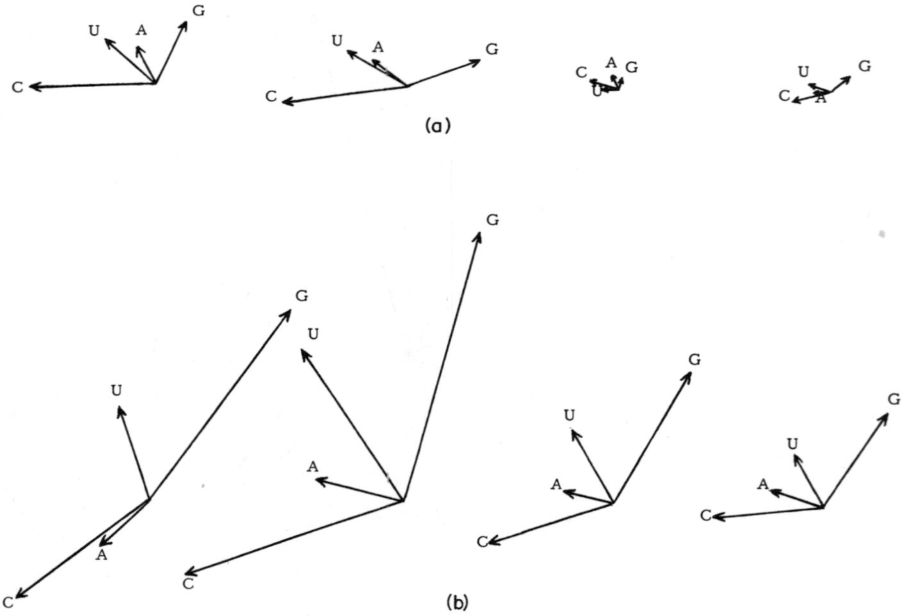

Fig. 3. Decomposition of the total dipole moments into their σ (upper row) and π (lower row) components. Same order and same scale as in Fig. 2.

π components, both in direction and in relative order. The σ contribution is appreciably smaller than the π contribution, the largest μ_σ occurring in cytosine. The differences observed in the σ components, although small enough (as μ_σ itself) to have only a small effect on the final moments, invite further investigation. Calculations using the correct dipole-moment integrals will be reported elsewhere.

There remains to study comparatively and carefully the distribution of the electronic charges themselves as obtained by the different methods. We shall carry out this huge work in detail in a separate publication. In this article we would simply like to illustrate the general aspects of the comparison of the σ-charge distribution. This is done in Fig. 4, which shows the

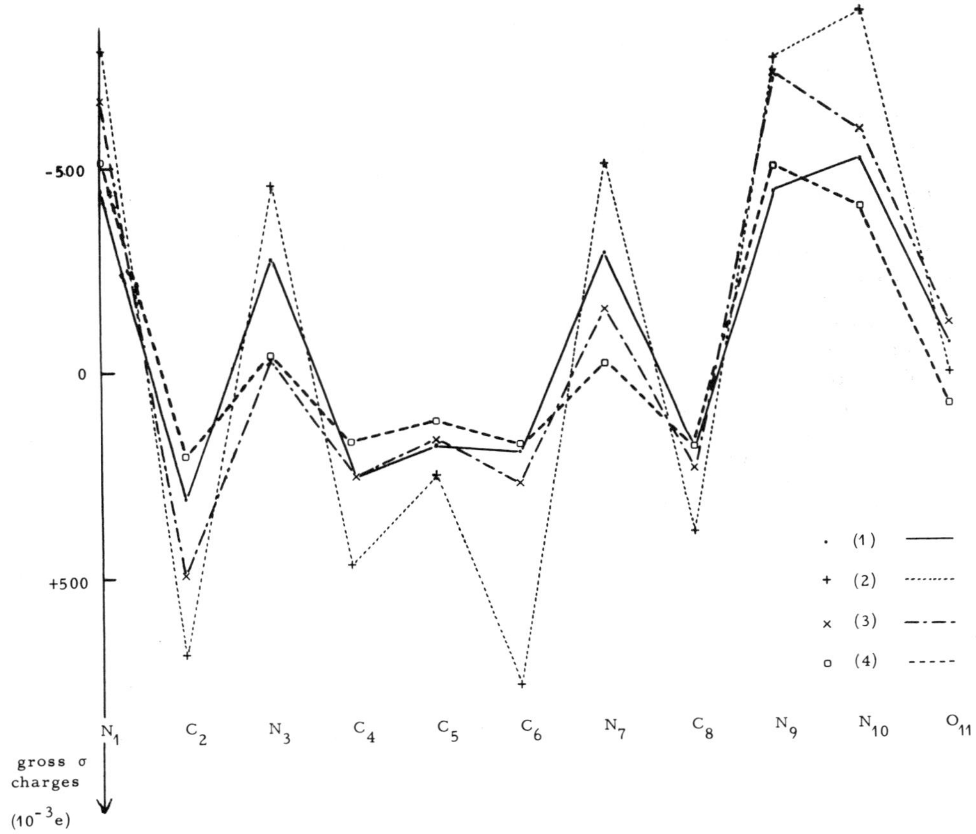

Fig. 4. Gross σ atomic populations in guanine. (1) as in Ref. 7; (2) EHT; (3) IEHC; (4) CNDO.

gross σ charges on the different atoms of guanine as obtained by the four procedures. It can be seen that, generally speaking, all the methods agree in their appreciation of the relative values of the charges as a function of the atomic positions. They differ of course in their evaluation of the numerical values of the charges, the greatest spread occurring in the EHT method, the smallest in the CNDO procedure.

Acknowledgment

This work was supported by Grant 67-00-532 of the Délégation Générale à la Recherche Scientifique et Technique (Comité de Biologie Moléculaire).

References

1. A. Pullman and B. Pullman, *Quantum Biochemistry*, Wiley-Interscience, New York, 1963.
2. B. Pullman, *Ann. N. Y. Acad. Sci.* (Proceedings of an International Symposium on Electronic Aspects of Biochemistry, Dec. 1967), **158**, 1 (1969).
3. A. Pullman, and M. Rossi, *Biochim. Biophys. Acta* **88**, 211 (1964).
4. G. Del Re, *J. Chem. Soc.* **1958**, 4031.
5. H. Berthod and A. Pullman, *Biopolymers* **2**, 483 (1964).
6. H. Berthod and A. Pullman, *J. Chim. Phys.* **55**, 942 (1966).
7. H. Berthod, C. Giessner-Prettre, and A. Pullman, *Theoret. Chim. Acta* **5**, 53 (1966).
8. H. Berthod, C. Giessner-Prettre, and A. Pullman, *Intern. J. Quantum Chem.*, **1**, 123 (1967).
9. A. Denis and A. Pullman, *Theoret. Chim. Acta* **7**, 110 (1967).
10. A. Pullman and B. Pullman, *Advances in Quantum Chemistry* **4**, 267 (1968).
11. F. Jordan and B. Pullman, *Theoret. Chim. Acta* **9**, 242 (1968), and references therein.
12. C. Giessner-Prettre and A. Pullman, *Theoret. Chim. Acta* **9**, 279 (1968).
13. A. Pullman, E. Kochanski, M. Gilbert, and A. Denis, *Theoret. Chim. Acta* **10**, 231 (1968).
14. E. Clementi, H. Clementi, and D. R. Davis, *J. Chem. Phys.* **46**, 4725 (1967).
15. E. Clementi, *J. Chem. Phys.* **46**, 4731 (1967).
16. R. Hoffmann, *J. Chem. Phys.* **39**, 1397 (1963).
17. R. Hoffmann, *J. Chem. Phys.* **40**, 2745 (1964).
18. L. C. Cusachs, Semi-empirical Molecular Orbitals for General Polyatomic Molecules I, Report 50 of the Quantum Theory Project, University of Florida, Gainesville, Fla., 1964, p. 36.
19. D. G. Carroll, A. T. Armstrong, S. P. McGlynn, *J. Chem. Phys.* **44**, 1865 (1966).
20. J. A. Pople and G. A. Segal, *J. Chem. Phys.* **44**, 3289 (1966).
21. H. O. Pritchard and H. A. Skinner, *Chem. Rev.* **55**, 745 (1955).
22. L. C. Cusachs and J. W. Raynolds, *J. Chem. Phys.* **43**, S160 (1965).
23. A. Pullman, *Ann. N. Y. Acad. Sci.* (Proceedings of an International Symposium on Electronic Aspects of Biochemistry. Dec. 1967), **158**, 65 (1969).
24. C. Lifschitz, E. D. Bergmann, and B. Pullman, *Tetrahedron Letters* **46**, 4583 (1967).
25. M. A. El Sayed, M. Kasha, and Y. Tanaka, *J. Chem. Phys.* **34**, 334 (1961).
26. H. De Voe and I. Tinoco, Jr., *J. Mol. Biol.* **4**, 500 (1962).
27. J. A. Pople and G. A. Segal, *J. Chem. Phys.* **43**, S136 (1965).

VI-7

Semiempirical Molecular Orbital Calculations: Open-Shell Computations on Pyridine

B. J. BERTUS AND S. P. McGLYNN

1. Introduction

The pyridine molecule is of considerable theoretical importance: The lowest-energy ionization potential refers to removal of a π electron, whereas the lowest-energy electronic excitation is of $\pi^* \leftarrow$ n orbital type. Such behavior is inconsistent with any simple MO attitude. It has been suggested[1] that the release of electron repulsion generated in the $\pi^* \leftarrow$ n transition is much greater than that associated with the $\pi^* \leftarrow \pi$ excitation and that it is this difference which causes reversal of the expected order of energy levels. Apart from the fact that this latter rationalization is quantitatively unsatisfactory, the work which will be reported here indicates that quite different sets of MO's must be used for the ground state and for the $\Gamma_{n\pi*}$ configuration; furthermore, it will appear that the energy reversal $E_{n\pi*} < E_{\pi\pi*}$ is best associated with σ,π polarization effects. In other words, σ,π separability is not a valid assumption for discussions of the spectroscopy of simple azines.

Finally, the results obtained in this work reflect on the computational methodologies employed in their generation. Thus, to the extent that open-shell charge-self-consistent Mulliken-Wolfsberg-Helmholtz (M-W-H) methodologies provide agreement with experiment, or a reasonable interpretation of experiment, these methodologies must be considered valuable spectroscopic tools. It is our opinion that the results of this work, concerned as they are with a molecule of crucial theoretical importance, do much to enhance the validity of the M-W-H schematization which is used. Given the validity of this last assertion, it is equally clear that some effort must also be made to provide a basic understanding of just what is being done quantum mechanically in the M-W-H formulation.

2. Method

A Mulliken-Wolfsberg-Helmholtz computation was performed on the ground closed-shell configuration and several excited open-shell configurations of pyridine. A single configuration approximation was adopted for all pseudo-one-electron computations, whereas a multiconfiguration mixing was used in all computations restricted to π electrons solely. The electron configurations which were subjected to M-W-H computation are shown in Fig. 1.

The M-W-H computations were iterated until self-consistency of the charge density matrix was obtained; convergence was achieved computationally using heavily damped iterative procedures.[2] Considerable care and much intuition was required to prevent oscillatory divergence during those periods of the convergence process in which accidental near-degeneracies of MO's occurred.

The Coulomb integrals (H_{ii}) were approximated by the corresponding valence-state ionization energies (VSIE's) and were written as functions of atomic charge (q) and orbital population (P). The Coulomb integral expressions which were used are given in Table 1. The reso-

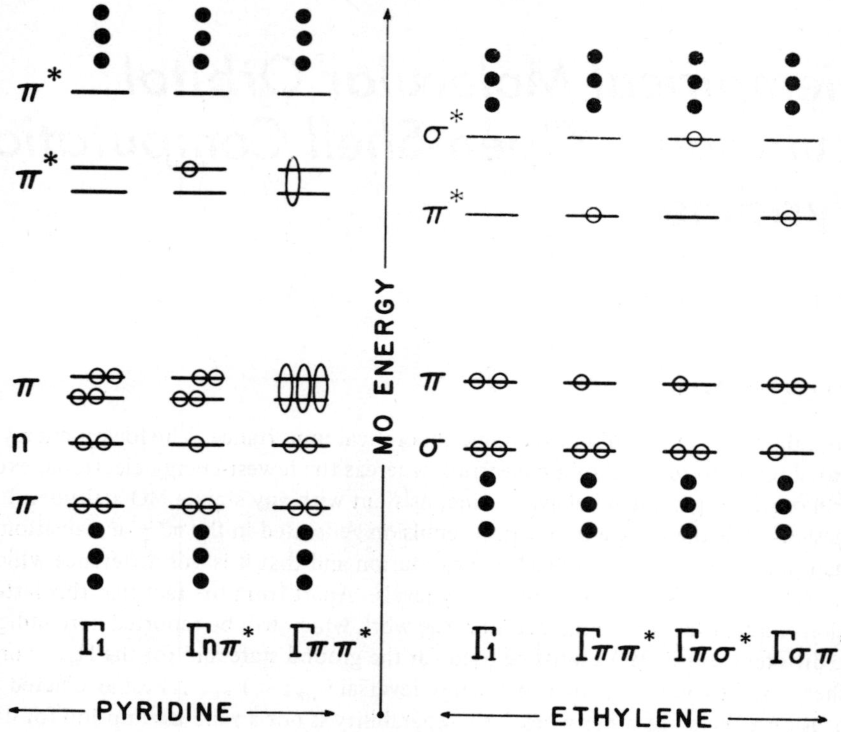

Fig. 1. Schematic diagram of the various configurations of pyridine and ethylene which were subjected to M-W-H computation. The $\Gamma_{\pi\pi*}$ configuration of pyridine is drawn somewhat differently to the others to emphasize the fact that four different configurations are involved in the $\pi^* \leftarrow \pi$ excitation. Two of the π-MO sets in pyridine are almost degenerate.

Table 1. Coulomb Integrals[a]

Carbon	$H_{2s,2s} = 1.7P_{2s} - 22.9 - 11.9q$
	$H_{2p,2p} = 1.5P_{2p} - 12.9 - 11.9q$
Nitrogen	$H_{2s,2s} = 1.9P_{2s} - 29.3 - 13.7q$
	$H_{2p,2p} = 1.9P_{2p} - 16.3 - 13.7q$
Hydrogen	$H_{1s,1s} = 0.121q^3 - 13.97q^2 - 26.93q - 13.6$

[a] P_i is the orbital population of atomic orbital i; q is the atom charge; energies are in electron volts.

nance integrals (H_{ij}) were approximated by the Cusachs expression[3]

$$H_{ij} = (2 - |S_{ij}|)(H_{ii} + (H_{ij})S_{ij}/2$$

where S_{ij} is the overlap integral $<\chi_i/\chi_j>$.

The atomic orbitals used were of single Slater type and were engineered so as to mimic the overlap properties of the atomic SCF functions over a range of representative bond distances.[4]

The energy of a given molecular state was assumed to be identical to that of the corresponding MO configuration and given by

$$E = \sum_{j=1}^{i} n_j \epsilon_j$$

where ϵ_j is the MO eigenvalue and n_j is the number of electrons in the jth MO.

An additional empirical factor was introduced in most of the M-W-H calculations. The AO Coulomb integrals were adjusted in such a way as to split the input $2p\pi$,$2p\sigma$-AO degeneracies; this adjustment was an attempt to empiricize the "atoms-in-molecule" effect.[5] The VSIE for the $2p\pi$-AO in pyridine was usually weighted by a factor of 0.74 and in ethylene was usually weighted by a factor of 0.80, whereas the $2p\sigma$-AO integrals were retained, in most instances, as in Table 1. Calculations were also performed without such correction factors. The general effect of these correction factors was to render the $2p\pi$-AO less bonding than the $2p\sigma$-AO's.

The planarity and detailed geometry of pyridine was retained invariant in all configurations. Axes and atom-numbering conventions are given in Fig. 2. Structural parameters for pyridine were taken from Bak et al.[6]

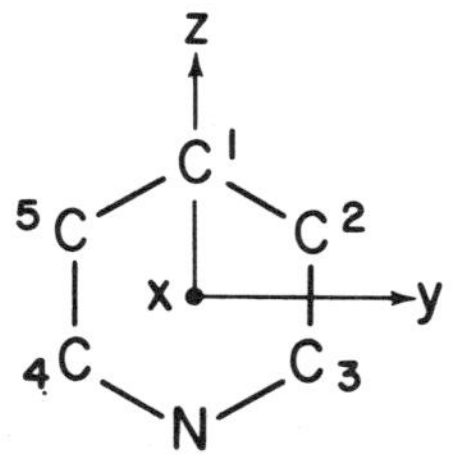

Fig. 2. Axes and atom-numbering conventions.

3. Results and Discussion of Pyridine

Ground Configuration

The order of highest energy MO filling for the closed-shell configuration of pyridine (Γ_1) is π (− 9.35 eV, a_2), π(−9.37 eV, b_1), n (− 10.44 eV, a_1), and π (− 10.89 eV, b_1). This sequence appears to coincide with the available experimental evidence (see Table 2) with respect to both the energy and nature of each MO. The energy ordering of the MO eigenvalues from this M-W-H computation does not agree with recent work reported by Clementi,[7] which indicated

Table 2. Orbital Energies Associated with the Ground Configuration of Pyridine[a]

Calculated	Symmetry	Experimental		Type of orbital
9.35	a_2	9.28[b]	9.26[c]	π
9.37	b_1			
10.44	a_1	10.54	10.30	n
10.89	b_1	12.22	11.56	π
12.22	b_2	14.44	—	↑
13.28	a_1	15.59	—	
13.89	b_2	16.94	—	
14.27	b_2	—	—	
14.64	a_1	—	—	
16.37	a_1	—	—	σ
18.06	b_2	—	—	
18.49	a_1	—	—	
23.46	b_2	—	—	
24.54	a_1	—	—	
29.00	a_1	—	—	↓

[a]All eigenvalues in this table are negative and in electron volts.
[b]Values in this column are from M. I. Al-Joboury and D. W. Turner, *J. Chem. Soc.* **1964**, 4434.
[c]Values in this column are from M. A. El-Sayed, M. Kasha, and Y. Tanaka, *J. Chem. Phys.* **34**, 334 (1961).

that the order of filling should be $\pi(a_2)$, $\pi(b_1)$, $n(a_1)$, $\sigma(b_2)$, $\pi(b_1)$. Thus, in the Clementi results, two σ MO's separate the highest-energy and lowest-energy filled π MO's. Experimental evidence does not justify such an ordering. It is our opinion that this misordering of the σ-MO energy arises from the assumption made in Clementi's work relative to the identity, apart from angular factors, of $2p_x$, $2p_y$, and $2p_z$ atomic orbitals on any one center throughout the iterative procedure. Although a similar assumption is made in the present work, it is not quite so crucial as in the Clementi case: The AO's in M-W-H theory are used only for overlap considerations; furthermore, the use of different Coulomb terms for the various 2p-AO's is equivalent to the use of different AO's not only for the diagonal H_{ii} terms but also, via the Cusachs relation, for H_{ij} terms as well. Thus, because of the minimal usage of the wave functions in M-W-H computations, the identity of the $2p\pi$,$2p\sigma$ radial wave functions is effectively removed in the present instance.

The MO at -10.43 eV (a_1) is a bonding σ orbital. This orbital corresponds to the classical[8] n-MO but has significant charge density throughout the σ skeleton of the molecule (see Table 3). Quantitative considerations of the form of the n-MO have become possible only recently. It has been postulated that the n-electrons are not entirely located in a particular 2p nitrogen orbital but are involved in σ bonding. In this work the $2p_z$-AO on nitrogen was found to

Table 3. Some MO Eigenvectors and Eigenvalues of Closed-Shell Pyridine (Γ_1) Configuration[a]

MO designation:	$\pi(b_1)$	$n(a_1)$	$\pi(b_1)$	$\pi(a_2)$	$\pi^*(a_2)$	$\pi^*(b_1)$	$\pi^*(b_1)$
MO energy:	-10.869	-10.436	-9.367	-9.350	-6.403	-6.085	-3.904
Carbon 1 2s	0.000	0.002	0.000	0.000	0.000	0.000	0.000
$2p_z$	0.000	0.030	0.000	0.000	0.000	0.000	0.000
$2p_x$	0.230	0.000	0.582	0.000	0.000	-0.714	-0.446
$2p_y$	0.000	0.000	0.000	0.000	0.000	0.000	0.000
Carbon 2 2s	0.000	0.095	0.000	0.000	0.000	0.000	0.000
$2p_z$	0.000	-0.252	0.000	0.000	0.000	0.000	0.000
$2p_x$	0.266	0.000	0.340	-0.444	0.583	0.405	0.461
$2p_y$	0.000	-0.014	0.000	0.000	0.000	0.000	0.000
Carbon 5 2s	0.000	0.095	0.000	0.000	0.000	0.000	0.000
$2p_z$	0.000	-0.252	0.000	0.000	0.000	0.000	0.000
$2p_x$	0.266	0.000	0.340	0.444	-0.583	0.405	0.461
$2p_y$	0.000	0.014	0.000	0.000	0.000	0.000	0.000
Carbon 3 2s	0.000	-0.027	0.000	0.000	0.000	0.000	0.000
$2p_z$	0.000	0.286	0.000	0.000	0.000	0.000	0.000
$2p_x$	0.352	0.000	-0.185	-0.477	-0.556	0.293	-0.593
$2p_y$	0.000	0.076	0.000	0.000	0.000	0.000	0.000
Carbon 4 2s	0.000	-0.027	0.000	0.000	0.000	0.000	0.000
$2p_z$	0.000	0.286	0.000	0.000	0.000	0.000	0.000
$2p_x$	0.352	0.000	-0.185	0.477	0.556	0.293	-0.593
$2p_y$	0.000	-0.076	0.000	0.000	0.000	0.000	0.000
Nitrogen 2s	0.000	0.266	0.000	0.000	0.000	0.000	0.000
$2p_z$	0.000	-0.827	0.000	0.000	0.000	0.000	0.000
$2p_x$	0.454	0.000	-0.447	0.000	0.000	0.000	0.000
$2p_y$	0.000	0.000	0.000	0.000	0.000	0.000	0.000
$H-C_1$ 1s	0.000	0.119	0.000	0.000	0.000	0.000	0.000
$H-C_2$ 1s	0.000	-0.092	0.000	0.000	0.000	0.000	0.000
$H-C_5$ 1s	0.000	-0.092	0.000	0.000	0.000	0.000	0.000
$H-C_3$ 1s	0.000	-0.135	0.000	0.000	0.000	0.000	0.000
$H-C_4$ 1s	0.000	-0.135	0.000	0.000	0.000	0.000	0.000

[a]Energies are in electron volts; MO's are displayed vertically.

be the largest contributor to the n-MO for all configurations (Γ_1, $\Gamma_{\pi\pi^*}$, $\Gamma_{n\pi^*}$) studied: However, the general results do accord with postulates as to the semidiffuse nature of the MO. Anno and Sado[9] estimated the 2s nitrogen AO character of the n-MO to be less than 10 per cent that of the 2p nitrogen AO character in this MO. The present computations are more or less in agreement with this result; in addition, the 2p-AO's on carbons 2, 3, 4, and 5 also contribute to the n-MO. The molecular extent and nitrogen AO character of the σ orbital is in total accord with the computations of Robin et al.[10] and Clementi.[7]

Excited Configurations

The experimentally determined symmetries[11,13] of the first two singlet excited states of pyridine are ${}^1\Gamma_{n\pi^*} = {}^1B_1$ and ${}^1\Gamma_{\pi\pi^*} = {}^1B_2$. The $\Gamma_{n\pi^*}$ configuration has only one plausible origin: An n electron (- 10.46 eV, a_1) is excited to a π^* MO (- 6.09 eV, b_1). The two other possible n → π^* excitations are readily eliminated: The transition to the π^* MO (- 6.04 eV, a_2) generates a ${}^1A_2$ configuration, whereas the energy gap to the π^* (- 3.90 eV, b_1) is prohibitively large. The $\Gamma_{\pi\pi^*}$ configuration may arise from excitation of a π electron (- 9.35 eV, a_2) to a π^* MO (- 6.09 eV, b_1) and/or excitation of a π electron (- 9.36 eV, b_1) to a π^* MO (- 6.40 eV, a_2). Both of these $\pi^* \leftarrow \pi$ excitations were considered here.

The results of the open-shell computations are shown in Fig. 3. The $\Gamma_{n\pi^*}$ (B_1) configura-

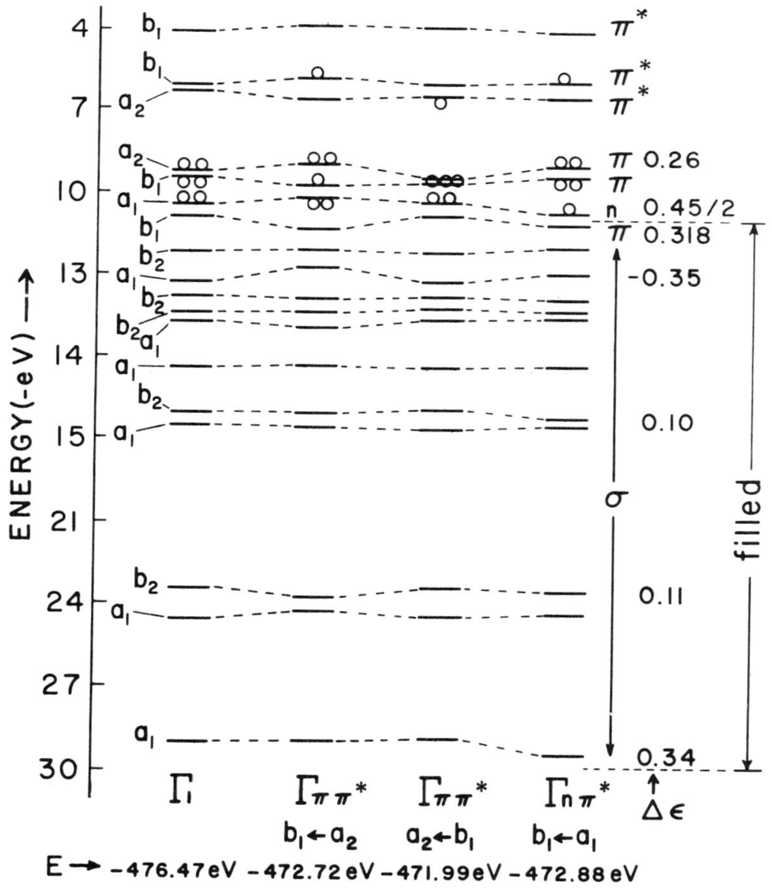

Fig. 3. MO's of pyridine in the closed-and various open-shell configurations. $\Delta\epsilon$ is the stabilization energy of an MO in the Γ_1 (A_1) configuration.

tion is lower in energy than the $\Gamma_{n\pi^*}$ (B_2) configuration. The calculated energy difference is $0 \leqslant \Delta E \leqslant 1.0$ eV. This difference agrees with the energy spacings found in the electronic absorption spectra of pyridine. The M-W-H computations predict the $\Gamma_{n\pi^*}$ (B_1) configuration to be 3.59 eV above the closed-shell configuration and the $\Gamma_{\pi\pi^*}$ (B_2) configuration to be between 3.75 eV ($b_1 \longleftarrow a_2$) and 4.48 eV ($a_2 \longleftarrow b_1$) higher than the filled-shell configuration.

The n-MO in the closed-shell configuration is lower in energy (~1 eV) than the two highest-energy filled π-MO's. The order of configuration energies, however, is reversed in the sense that the $\Gamma_{n\pi^*}$ (B_1) (-472.88 eV) configuration is of lower energy than the $\Gamma_{\pi\pi^*}$ (B_2) (-472.72 eV and -471.99) configurations. Individually small but additively significant changes in several of the lower energy (particularly σ-) MO's facilitate this apparent anomaly. The energy difference of corresponding MO levels of the $\Gamma_{n\pi^*}$ and Γ_1 (A_1) configurations are denoted by $\Delta\epsilon$ in Fig. 3. It is readily seen that the lowering of the $\Gamma_{n\pi^*}$ configuration below the $\Gamma_{\pi\pi^*}$ configurations is caused by effects which cannot derive from a simple electron repulsion model[1] based on the same set of MO's for different configurations; indeed the effects observed are best described as σ,π polarization effects. The relationship of MO energy differences ($\Delta\epsilon$) to charge shifts can be perceived by comparison of the $\Delta\epsilon$'s to the population analysis of Table 4 and/or the eigen-

Table 4. Population Analysis of Pyridine for Closed- and Open-Shell Configurations

Configuration MO excitation		Γ_1	$\Gamma_{\pi\pi^*}$ $b_1 \leftarrow a_2$	$\Gamma_{\pi\pi^*}$ $a_2 \leftarrow b_1$	$\Gamma_{n\pi^*}$ $b_1 \leftarrow a_1$
Carbon 1	2s	1.140	1.100	1.148	1.112
	$2p_z$	0.946	0.901	0.958	0.886
	$2p_x$	0.998	1.228	0.957	1.215
	$2p_y$	0.943	0.869	0.964	0.887
Carbon 2	2s	1.141	1.158	1.138	1.146
	$2p_z$	0.955	0.965	0.954	0.864
	$2p_x$	0.975	0.903	0.984	1.067
	$2p_y$	0.947	0.978	0.941	0.956
Carbon 3	2s	1.121	1.145	1.117	1.104
	$2p_z$	0.944	0.972	0.938	0.854
	$2p_x$	0.996	0.878	1.013	1.103
	$2p_y$	0.939	0.969	0.933	0.920
Nitrogen	2s	1.499	1.468	1.507	1.505
	$2p_z$	1.700	1.683	1.710	1.228
	$2p_x$	1.060	1.210	1.050	1.427
	$2p_y$	0.893	0.830	0.894	0.931
$H\text{-}C_1$	1s	0.962	0.986	0.960	0.980
$H\text{-}C_2$	1s	0.958	0.953	0.958	0.960
$H\text{-}C_3$	1s	0.953	0.942	0.954	0.941

vectors of Table 3. A considerable redistribution of charge is associated with the $\pi^* \longleftarrow n$ transition, but no such redistribution is associated with the $\pi^* \longleftarrow \pi$ excitation. It is this gross change in the electron density matrix caused by the $\pi^* \longleftarrow n$ transition, but not by the $\pi^* \longleftarrow \pi$ transition, which makes feasible the energy inversion of $\Gamma_{n\pi^*}$ and $\Gamma_{\pi\pi^*}$ configurations. Therefore, we conclude that density changes and the concomitant σ,π polarization effects must be considered to properly evaluate configuration energies. In other words, open-shell considerations are a prerequisite to the understanding of electronic spectra which are of an intramolecular charge-transfer nature.

Electron Repulsion Considerations

The σ potential from the M-W-H calculation of the pyridine closed-shell configuration was used in a variable electronegativity self-consistent-field (VESCF) computation.[14] The computed ionization potential was - 10.00 eV, which is similar to the general run of numbers generated for pyridine by VESCF methods.[14] Configuration interaction (CI) was added and the first excited singlet state ($^1B_2$) was found to lie at 5.21 eV and to consist of ~50 per cent $\pi^* \leftarrow \pi$ ($b_1 \leftarrow a_2$) and ~50 per cent $\pi^* \leftarrow \pi$ ($a_2 \leftarrow b_1$) configurations. The lowest triplet state ($^3A_1$) was located at 3.74 eV.

The electronic exchange integral for the $\pi^* \leftarrow n$ excitation was evaluated in a one-center approximation limited solely to the nitrogen center—a valid procedure[15] because of the high contribution of 2p-nitrogen in the n-MO. The exchange integral $K_{n\pi^*}$ (nitrogen) was found to be 0.294 eV, which corresponds to a singlet-triplet split of 0.589 eV. Several experimental singlet-triplet splittings for $\pi^* \leftarrow n$ transitions of azines taken from spectral data [13,16] indicate an experimental energy difference of approximately 0.5 eV.

The order of state energies which is predicted is $E(^3\Gamma_{\pi\pi^*}) < E(^3\Gamma_{n\pi^*}) < E(^1\Gamma_{n\pi^*}) < E(^1\Gamma_{\pi\pi^*})$. This ordering results from either M-W-H computations with exchange energy added or from considerations of the latter type conjoined with VESCF π-electron computations. Furthermore, the correct state symmetries and energy ranges are predicted. Thus the agreement with experiment must be considered complete.

Spin-Orbital Coupling

A spin-orbital-coupling calculation[15,17] using closed-shell pyridine eigenvectors was performed. The lifetime of the dipole emission from the lowest triplet state ($^3\Gamma_{\pi\pi^*}$; $^3A_1$) to the ground state ($^1\Gamma_1$; $^1A_1$) was computed to be $\sim 1 \times 10^5$ sec. This lifetime is probably too long by a factor of approximately 10^4. The mixings which were considered to take place under the influence of the spin-orbit coupling operator were $^1\Gamma_{n\pi^*}$ states with $^3\Gamma_{\pi\pi^*}$ and $^3\Gamma_{n\pi^*}$ states with $^1\Gamma_1$. Although the correct polarization of phosphorescence results from consideration of such mixing, it is equally clear that insufficient allowedness is conferred. Thus further allowedness must be sought in mixing of higher-energy states: $^{1,3}\Gamma_{\sigma\pi^*}$ into $^3\Gamma_{\pi\pi^*}$ and $^1\Gamma_1$, respectively. Certainly, vibronic origins of the allowedness may not be blithely inferred until this possibility is investigated and eliminated. Thus σ,π mixing considerations also become relevant to spin-orbit coupling in azines. In other words, the $^{1,3}\Gamma_{\pi\sigma^*}$ and $^{1,3}\Gamma_{\sigma\pi^*}$ states are not of such high energies as has been usually supposed and first-order perturbation considerations based on supposedly large perturbation denominators lose much of their relevance.

4. Conclusions

The electronic states of pyridine, in order of increasing energy, are $^1\Gamma_1$, $^3\Gamma_{\pi\pi^*}$, $^3\Gamma_{n\pi^*}$, $^3\Gamma_{n\pi^*}$, $^1\Gamma_{\pi\pi^*}$, ... This ordering coincides with the best rationalization of experiment presently available. The S-T split in the $\Gamma_{n\pi^*}$ configuration is predicted to be small, as it must be to conform with experiment; this result derives from the small amount of $2s_N$ character in the n-MO. The $^3\Gamma_{\pi\pi^*} \leftarrow {}^1\Gamma_1$ transition derives little allowedness from spin-orbit mixing with states derived from $\pi^* \leftarrow n$ excitations; consequently, the nonemissive nature of the $^3\Gamma_{\pi\pi^*}$ state is rationalized, and it is suggested that states derived from $\sigma^* \leftarrow \pi$ and $\pi^* \leftarrow \sigma$ excitations may dominate the routes by which allowedness is fed to the $^3\Gamma_{\pi\pi^*} \rightarrow {}^1\Gamma_1$ transition. The lowest energy ionization is predicted to be concerned with π-electron removal.

The artificial separation of σ- and π-electron problems is generally inappropriate. In particular, the use of this supposed separability in computations of the energy of transitions possessing intramolecular charge-transfer character leads to grossly wrong predictions. The results of the present work very clearly point out the mutual interdependence of σ and π charge density distributions and the absolute necessity of iteration, whether of a priori or empirical nature, on open-shell configurations resulting from $\pi^* \leftarrow \sigma$, $\sigma^* \leftarrow \pi$, or $\pi^* \leftarrow n$ excitations. This requirement can be eased somewhat for configurations resulting from $\pi^* \leftarrow \pi$ excitations.

In sum, we believe that the present computations possess three significant facets not present in previous work. First, most a priori computations are based on Hartree-Fock SCF functions using $2p_x$, $2p_y$, and $2p_z$ carbon AO's whose radial parts are retained identical throughout the iterative process; although this might be computationally convenient, it is certainly not physically correct. Second, the Hartree-Fock MO's in question usually pertain to the ground-state configuration, whereas our considerations indicate that such MO's possess little or no relevance when discussing open-shell configurations and excitation energies involving electron promotions of charge-transfer type (i.e., $\sigma^* \leftarrow \pi$, $\pi^* \leftarrow \sigma$, $\pi^* \leftarrow n$, $\sigma^* \leftarrow n$, etc.). Finally, the very process of converging on the ground-state energy using a Hartree-Fock operator is particularly deleterious to the form of the unfilled MO's. In contrast, the method used here–involving, as it does, such a gross input of empirical information and based, as it appears to be, on a Hamiltonian which is probably more nearly Hartree-Fock-Slater than Hartree-Fock–seems to provide a satisfactory correlation with experiment. Indeed the results obtained here, as well as others which we have available on the ethylene configurations shown in Fig. 1, indicate that electron repulsions are rather fully accounted for and that no gross "self-interactions," such as characterize Hartree-Fock open-shell considerations, exist.

It is these three facets, we believe, plus the semiempirical nature of the computations and their recurring normalization to experiment, which provide any successes evident in the reported computational work on pyridine.

Acknowledgments

Supported by a contract between the U.S. Atomic Energy Commission–Biology Branch and the Louisiana State University. Mr. McGlynn is a Fellow of the Sloan Foundation.

References

1. J. W. Sidman, *J. Chem. Phys.* **27**, 429 (1957).
2. D. G. Carroll, A. T. Armstrong, and S. P. McGlynn, *J. Chem. Phys.* **44**, 1865 (1966).
3. L. C. Cusachs, *J. Chem. Phys.* **43**, 157s (1965).
4. E. Clementi, Table of Atomic Functions, supplement to a paper in *IBM J. Res. Develop.* **9**, 2 (1965).
5. A. T. Armstrong, B. Bertus, and S. P. McGlynn, *Spectroscopy Letters* **1**, 43 (1968).
6. B. Bak, L. Hansen-Nygaard, and J. Rastrup-Anderson, *J. Mol. Spectry.* **2**, 54 (1956).
7. E. Clementi, *J. Chem. Phys.* **46**, 4731 (1967).
8. M. Kasha, *Discussions Faraday Soc.* **9**, 14 (1950).
9. T. Anno and A. Sado, *J. Chem. Phys.* **29**, 1171 (1958).
10. M. R. Robin, R. H. Hart, and N. A. Kuebler, *J. Am. Chem. Soc.* **89**, 1564 (1967).
11. H. Sponer and J. Rush, *J. Chem. Phys.* **20**, 1847 (1952).
12. H. Stephenson, *J. Chem. Phys.* **22**, 1077 (1954).
13. K. K. Innes, J. P. Byrne, and I. G. Ross, *J. Mol. Spectry.* **22**, 125 (1967).
14. R. D. Brown and M. L. Heffernan, *Australian J. Chem.* **12**, 554 (1959).
15. S. P. McGlynn, *Molecular Spectroscopy of the Triplet State*, Prentice-Hall, Englewood Cliffs, N.J., 1969.
16. F. J. Smith and S. P. McGlynn, *Photochem. Photobiol.* **3**, 369 (1964).
17. D. G. Carroll, L. G. Vanquickenborne, and S. P. McGlynn, *J. Chem. Phys.* **45**, 2777 (1966).

Chapter VII

Nonempirical Methods and Theory

VII-1

Nowadays the computer is used for almost any type of calculation. With time-sharing consoles, it has become practically a household tool. In its most powerful form, however, the computer becomes quite a different instrument, a probe into the fine details of the electronic structure of an atom or molecule, an instrument which from the most basic data would produce any desired physical, chemical, thermodynamic, or kinetic property. This point of view is brought out in a most stimulating manner by Wahl (Chapter VII-2). The goal of an ideal input-output machine is still far away; however, occasionally it is remarkably approached, as in the evaluation of the thermodynamic properties of alkali vapors in the work of Wahl.

As mentioned earlier, totally nonempirical Hartree-Fock MO-SCF calculations provide the norm for approximate MO methods. Such total Hartree-Fock calculations have been carried out so far (other than atoms) only on diatomics by Wahl and by Roothaan and his co-workers. The Hartree-Fock method yields close to exact charge distributions, even near dissociations, provided it is properly modified so as to involve several determinants. The electronic charge distributions obtained from this method on the computer directly on film as a molecule dissociates should be of great educational value (Chapter VII-2).

Even point-group theory may be automated, as shown in the remarkable work of Bouman, Chung, and Goodman (Chapter VII-3). Usually in extensive computer calculations, group theory has been more of a hindrance than help, its algebraic operations not being the simplest to program. For this reason, one has often bypassed group-theoretic simplifications and simply let the computer work some more. With the new advance, however, it becomes possible to feed in essentially just the positions of all the atoms, from which the automatic programs figure out the finite group involved, its representations, symmetry basis, and coupling coefficients. These automatic programs which are being made available, combined with other programs, bring the goal of fully automated computations on larger molecules closer.

For molecules of common interest in organic or inorganic chemistry, beyond the smallest, no actual Hartree-Fock MO-SCF results are available. Calculations on sizable molecules such as pyridine, although fully nonempirical, retaining all the integrals, are still restricted to some form of limited-basis LCAO-MO-SCF. In calculations which thus approximate the actual SCF orbitals, it has been customary to use either Slater-type orbitals (STO), or Gaussian basis sets. Gaussians require many more terms to represent the fewer STO's, but this disadvantage is made up for by the simplicity of the integral evaluations. Several ways of using Gaussian bases have been developed, as in the works of Allen (Chapter VI-2), Moscowitz (Chapter VIII-2), and Clementi (Chapter VII-4). With contracted Gaussian bases, the charge distributions in the ground state are thought to be quite accurate, but the energies may differ from the actual Hartree-Fock energies by 10 to 20 eV, as judged from the 1 to 2 eV accuracy with which the contracted Gaussians reproduce the Hartree-Fock orbitals of each free atom. In large-scale machine computations, a crucial factor is, of course, the computer time required and the flexibility of the programs written with regard to applications to other molecules. It would be of interest to have this information in papers reporting such calculations.

In the work of Frost (Chapter VII-5), Gaussians are used in quite a different way. For each bond in a saturated molecule a single Gaussian is taken whose location, however, is variable and determined so as to minimize the total energy. As was mentioned in Chapter VI, these calculations may be contrasted with those in Chapter VI-2. There, with nonempirical extended Hückel-type calculations, the wave functions used were more detailed, but only part of the total Hamiltonian was used. Here the wave function is as simple as it can be, yet the full expectation value of the total Hamiltonian is obtained. Both methods, although much simplified in different directions, are surprisingly successful, particularly on geometries.

Electron Correlation and How It Supplements MO Theory

As discussed in Chapter III, in a nonempirical MO method, to obtain energy properties it is necessary to add correlation to a Hartree-Fock SCF-MO result. The conventional way to go beyond SCF-MO has been by the straightforward use of the configuration-interaction (CI) method, which, however, is usually slowly convergent and difficult to apply systematically. In a study entitled "Many-Electron Theory of Atoms and Molecules (MET)," Sinanoğlu and co-workers (1959 to date) have found that the N-electron problem, that is, electron correlation, may be reduced mathematically and physically into a number of simple principles which also provide convenient estimation methods. The basic notions arrived at are:

a. In a ground state the exact wave function Ψ is made up of clusters of electrons involving correlations in one, two, three, . . . , N electrons at a time, and the independent products of all these cluster functions. Similarly, the energy is also rigorously given as a finite sum of terms involving successively larger numbers of electrons. These pieces of a many-electron atom or molecule are calculable separately by "subvariational" principles [see, for example, Sinanoğlu, *Advances in Chemical Physics,* Vol. 6 (esp. pp. 342 ff.), Wiley-Interscience, New York, 1964]. These essentially reduce the N-electron problem into a set of, say, two-body, three-body, etc., problems, which are easier to handle than the total problem itself.

b. Again, in a ground state, it is found that the exact wave function is mainly made up of all possible pair correlations between different molecular orbitals occupied in the initial Hartree-Fock MO-SCF determinant. The one-electron corrections beyond Hartree-Fock are small, thus justifying the good charge distributions that may be expected from MO calculations by themselves. It has been customary to assume in the previous literature that correlations only in doubly occupied and localized orbitals would be significant. However, MET shows that both inter- and intra-MO pair correlations are equally significant, as indicated for the molecular orbitals of benzene (Chapter VII-6b). Even after the molecular orbitals of a closed-shell determinant have been transformed without changing the overall wave function, into localized orbitals (see Chapter V), the interbond, LO-LO′ correlations contribute more than the bond correlations in a system such as methane (Chapter V-2). Thus the intuitive notion of a correlation energy made up of Lewis-Langmuir type electron pairs ("geminals") and wave functions based on this idea are inadequate and difficult to relate to the correct form.

Although in atoms various methods such as the ones on helium atom have been shown to be useful for the calculation of individual pairs, in molecules so far the only practicable way seems to be the evaluation of each pair correlation by the individual "pair-CI" method (Chapter VII-6a), or more difficultly, by total CI. With any form of CI, however, the crucial practical problem that remains is on the efficient choice of virtual orbitals. This problem has been successfully handled in two different ways by Wahl and by Davidson (Chapter VII-6c). Both authors have calculated some of the correlation effects on diatomic molecules.

c. As shown in Chapter VII-6a, the correlation effects that remain after taking into account all the pairs may be estimated by a series of successive CI's. The magnitude of these higher electron correlations is crucial to the role of the pairs alone. Few such calculations have as yet been carried out. Especially in a molecule such as benzene, where delocalization may make the electrons behave almost like metallic electrons, exploratory calculations would be of great interest.

d. The notion of inner cores unaffected by the number and state of the outer electrons is basic in π-electron theory. It is also implied in the calculation of ionization potentials and excitation energies by MO methods. MET shows, however, that the correlation energy of, for example, the $1s^2 2s^2$ cores of first-row atoms and of the σ cores of conjugated systems decreases in absolute magnitude as p or π electrons are added, an effect called "exclusion effect" in the theory (Chapter VI-6a). An analogous effect exists in diatomic molecules and gives a direct contribution to binding energy (see Chapter III-2d). In Chapter III-2d, the effect has been crudely estimated semiempirically. Nonempirical values for it, however, have been obtained recently by Wahl et al. and by Davidson (Chapter VII-6c).

e. In non-closed-shell excited states, additional and novel correlation effects arise. The total correlation separates into three distinct types: "internal," "semi-internal," and "all-external" correlations. The last one is just like those in closed shells, made up mainly of pairs. The first, internal, is what is usually calculated in the finite CI [for example, the π-electron CI involving only $2p_z$ orbitals in benzene (Chapter VII-6b)]. The three effects have been evaluated in a great many atomic excited states with applications also to properties such as spectral transitions and electron affinities [Sinanoğlu and Öksüz, *Phys. Rev. Letters* **21**, 507 (1968)]. All three types of correlations are equally important. Little is known, however, about their actual magnitudes in molecules. They are expected to contribute significantly to electronic spectra. [For more applications cf. papers by Sinanoğlu, Öksüz, and Westhaus, *Phys. Rev.* **181**, 42, 54(1969); **189**, 56(1969).]

Sigma-Pi Separation

We have mentioned that the σ core correlations may change significantly as the outer π electrons are ionized or excited. Also, the correlations between the σ and π electrons are large. The resulting noncancellations affect the ionization potentials and excitation energies but are omitted in either π or in σ and π MO theory. The work in Chapter VI-6 and 7 showed that the σ and π orbital energies too are intermingled, which is another source of ambiguity in σ-π separation. Aside from these effects, the MO-SCF calculations have to be carried out anew on each excited state or ion, as shown in the calculations of Griffith and Goodman (Chapter VII-8). It would not be satisfactory to calculate expectation values for excited states using the orbitals obtained from ground states, as the overall MO-SCF reveals significant σ-π charge reorganization.

Recently, Goodman and Pamuk have supplemented the nonempirical σ-π calculations on acetylene by adding correlation corrections to the different ions and states of this molecule, using approximate methods similar to those in Chapter III-2d.

We conclude this section by remarking on a way of doing the semiempirical MO calculations which would make them much more directly comparable with the nonempirical Hartree-Fock MO-SCF. Assuming that the H-F MO SCF would be sufficiently closely approximated by the linear combinations of Hartree-Fock atomic orbitals, the integrals of approximate MO-SCF may be obtained from Hartree-Fock AO's. Such a procedure was carried out to evaluate the Coulomb repulsion integrals in an a priori way for π-electron theory (Chapter VII-7). If the parameters can be evaluated in an analogous way for the σ methods, one would get a direct test of one crucial aspect of the σ-MO methods, the neglect of various integrals by various forms of ZDO. If, with the Hartree-Fock AO-type parameters, one obtained approximate MO results close to the actual Hartree-Fock MO-SCF energies, one could then correct for correlation for binding properties by simply adding these on, as discussed in Chapter III. At present, the ambiguous nature of the semiempirical parameters in σ-MO theory with orbital and correlation effects mixed prevents a clearer comparison of semiempirical methods with the results of nonempirical calculations.

VII-2

Chemistry from Computers: A New Instrument for the Experimentalist

ARNOLD C. WAHL

1. Introduction

Some time ago, being fresh from the enthusiasm of computing Hartree-Fock (best molecular orbital) wave functions for nontrivial molecules, I gave a talk[1] entitled "Hartree-Fock Is Here—What Next?" and I received, from an experimentalist, a reprint request for the article "Hartree-Fock Is Here—Who Cares?" This misstating of the title of my optimistic talk, in addition to being humorous, contained a very substantial bit of truth: Computers have brought us a great deal— vast numerical tables of molecular properties, pretty pictures, detailed wave functions from many small molecules (in many cases, so precise that they are unusable), and perhaps more "theoretical chemists" than ever before. But how much chemistry have they really given to us? This question certainly needs to be answered, and if we are to make chemists happy we must agree to answer it on their terms. Thus, we need a precision in potential-energy surfaces of about $\frac{1}{10}$ eV; we need *better than* 5 per cent precision in ionization potentials, binding energies, vibrational frequencies, term values, and spectroscopic constants. Further, we must go beyond isolated calculations and into their comprehensive coupling with the traditional tools of the chemist to allow us to obtain macroscopic properties.

In this article I would like to explore how close we, at Argonne, are to achieving such results from computers. It will become apparent that we must be cautious but that in *certain* cases we are able to obtain truly reliable chemical information from our a priori computing systems and that our research is most properly viewed as the development of a *new instrument* for the chemist by which he can obtain detailed answers often not accessible experimentally. An intriguing and very important aspect of this new apparatus is that it permits us to "look" in unprecedented detail with arbitrary magnification of time scale (when quantum mechanically legal) at a chemical process under study, be it molecular electronic excitation, vibration, collision, or the entire path of a chemical reaction.[1a] (See the figures.[1b])

2. Molecular Orbital Model

As a first step in tracing the development of our new *ab initio* instrument for exploring molecular structures, let us look at a very popular model of electronic structure: the Hartree-Fock MO model.[2] This model currently is being applied widely in the name of chemistry to all kinds of systems. However, some typical results obtained by the Hartree-Fock model show that although it is adequate for some molecular properties, the Hartree-Fock model has very well-known and well-substantiated deficiencies which make it difficult to *really* do a priori chemistry from Hartree-Fock calculations.

First, let us briefly review what the Hartree-Fock model[2] is. The Hartree-Fock model is the *best* orbital model. In the Hartree-Fock model[20] we place electrons in individual three-dimensional functions ϕ_i, include spin functions, and then form a properly antisymmetrized product of these spin orbitals.

$$\psi = A\pi_i\phi_i \qquad (1)$$

to obtain the total atomic or molecular wave function ψ. Mathematically, the orbitals ϕ_i are solutions of integrodifferential equations of the form

$$F\phi_i = \epsilon_i\phi_i$$

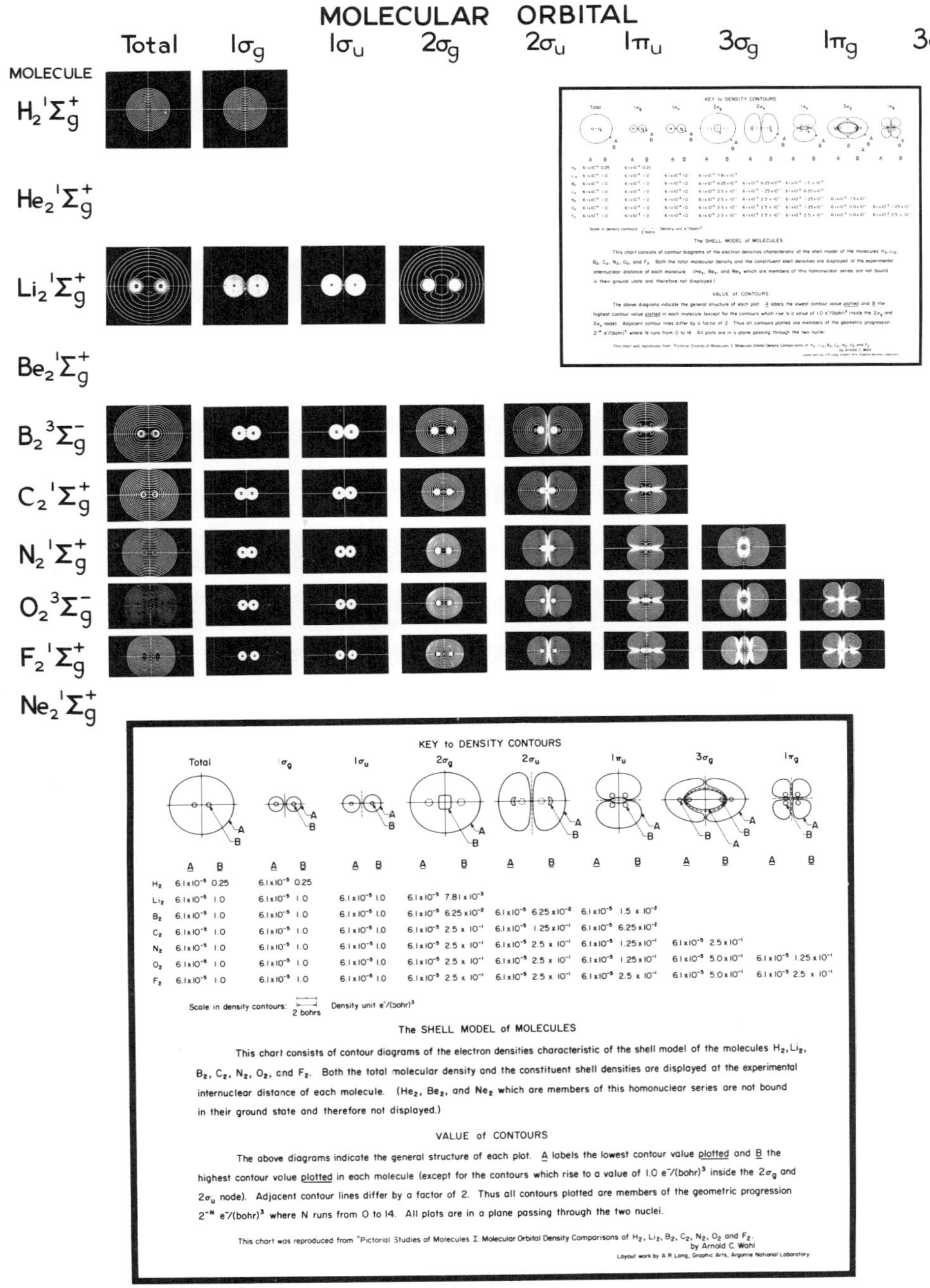

Fig. 1a. Orbital model of homonuclear diatomic molecules (Ref. 30). All such diagrams in this article are produced automatically by digital computers linked to a cathode-ray tube. Large wall charts similar to Figs. 1a and 2a are available (Ref. 31).

where F is an operator depending upon all electrons and nuclei of the systems and arising from the variation of its total electronic energy. These equations must be solved iteratively,[28] since the orbitals ϕ_i determine the operator F and vice versa. Convergence on the "best" set of orbitals is achieved when the ϕ_i's from two successive iterations agree within some permissible numerical threshold. In Figs. 1 through 3 some typical pictures of best MO and total electron densities for a variety of diatomic molecules are shown. From these pictures it immediately becomes apparent that the Hartree-Fock model forms a very appealing, conceptual, and, in fact, symbolically beautiful framework for thinking about molecules.[30,31]

The homopolar cases (Fig. 1) have symmetry and covalency; in the ionic systems (Fig. 2) the molecular orbitals are really localized on the individual ions and are very much like the isolated ionic orbitals. In looking at "pictures" of molecular processes–ionization[32] and excitation[32,33] (Fig. 3)–it appears that in the orbital framework we get a qualitative feeling for what is happening. An orbital shrinks when we remove an electron, it gets a little smaller, a little tighter, because there is less electron repulsion. Further, the nonactive orbitals are rela-

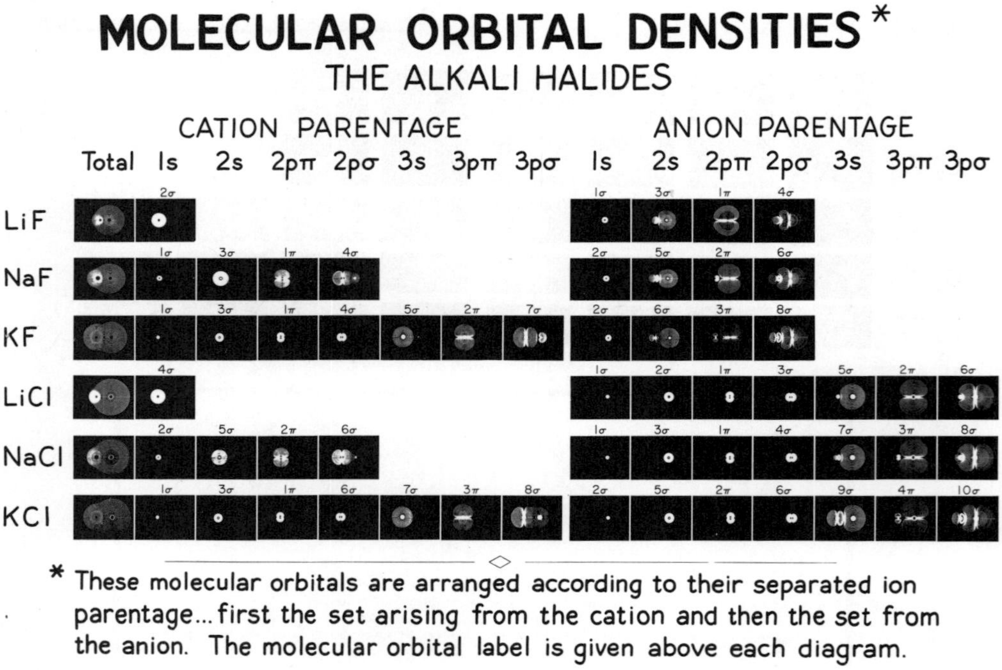

Fig. 2a. Orbital model of some alkali halides conventions explained in Refs. 30 and 31.

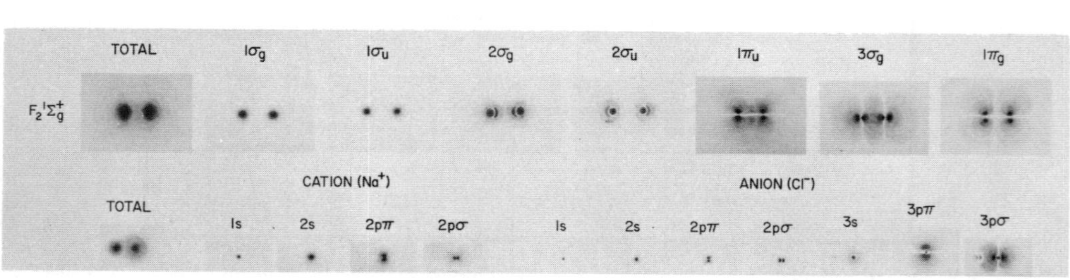

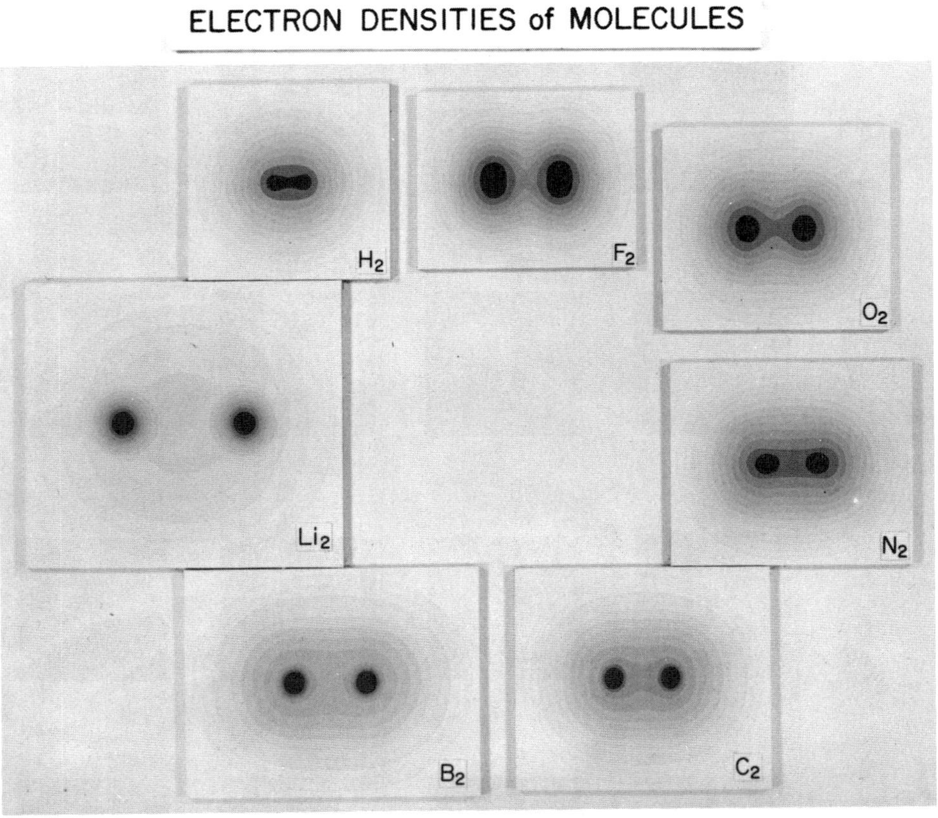

Fig. 1c. Total electron probability clouds for some homonuclear diatomic molecules which can be mapped by contour lines as in Fig. 1a.

tively insensitive to the ionization process, Fig. 3 (a). In Fig. 3 (b) we "see" quite dramatically a σ to π excitation[33] in the hydrogen molecule followed by ionization.

Now let us turn to some typical quantitative results obtained from the model. In Tables 1 and 2 recent results of Hartree-Fock calculations on diatomic systems[34-63] are shown. We are first struck by the rather *good* geometry predictions—we can predict internuclear distances, often within less than 1 per cent. Also, a typical one-electron property such as the dipole moment is well predicted. But, looking further at the tables, we find, from a chemist's viewpoint, that the binding energies are terrible. We can, of course, correct them semiempirically, but this, after all, cannot be called a priori chemistry from computers. We see that the dissociation energies are often off by 100 per cent; take F_2 as an outstanding example of this deficiency. Also, vibrational frequencies really do not allow us to distinguish between excited states of the same molecule; the precision is not sufficient. We find, for instance, in the excited states of the nitrogen molecule ion[42] (Fig. 4) that there is an inversion of molecular energy levels that does not allow us to interpret spectroscopy. Now, the reasons for these defects are well known. The orbital picture has an inherent error built into it, the correlation error. The orbital picture does not allow electrons to get out of each other's way explicitly and instantaneously. It happens that this error for the *atoms* is *not* the same as the error for the molecule built from these atoms. Therefore, binding energies are not good, and further, the error differs for different states of the same system. Thus, term values are not accurate.

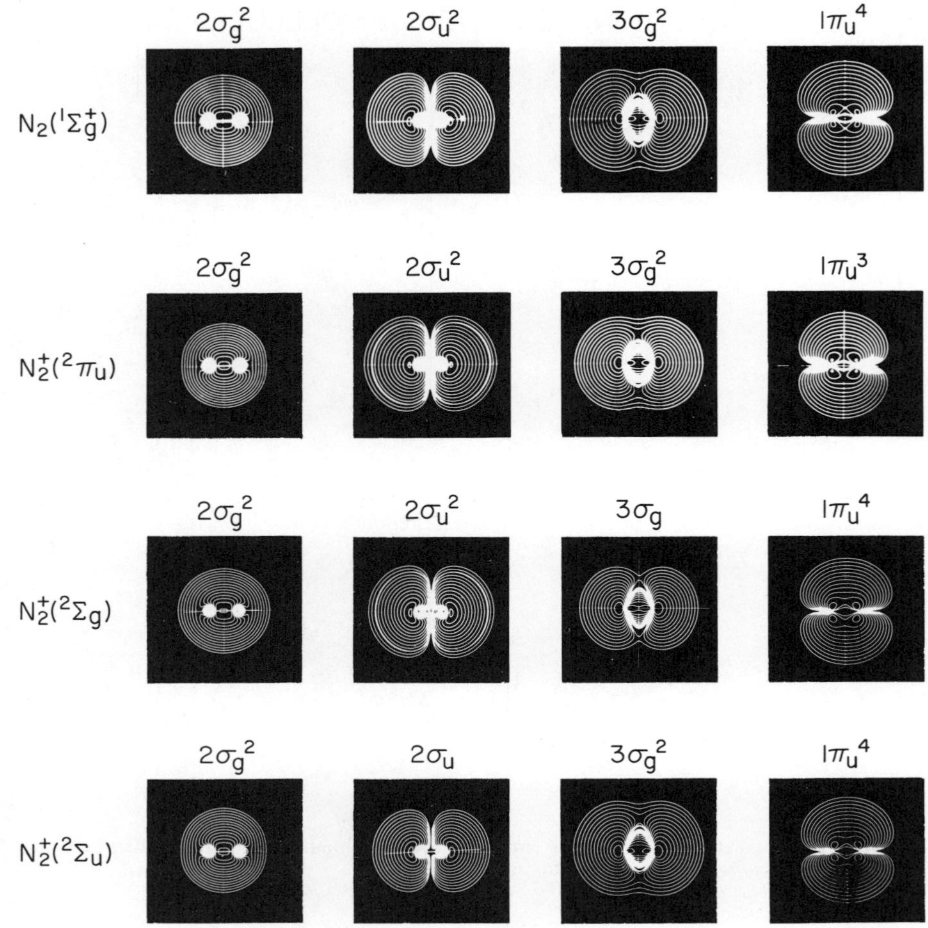

Fig. 3a. In the orbital picture of molecular ionization for the nitrogen molecule only valence orbitals have been plotted. Note that it is primarily the orbital from which an electron has been removed which changes. See Ref. 30 for conventions and Ref. 42 for wave functions.

Second, a simple MO picture, although relatively good at the equilibrium configuration of most molecules, deteriorates rapidly as you try to pull the molecule apart. Thus, the shape of the potential curve is distorted by the constrained form of the MO picture. These two defects of the Hartree-Fock model account for its most serious shortcomings: improperly shaped energy surfaces, bad binding energies, and badly computed term values for transitions between electronic states.

Now, there are notable cases where the Hartree-Fock model *does* provide some chemistry, but certainly not a "chemist's" chemistry. For instance, in Figs. 5, 6, and 7 the potential curves obtained from Hartree-Fock calculations[47] on the rare gases He, Ne, and Ar are plotted. Here, we are obtaining from our calculations a precision just about as high as currently is available from experiment. This is due to the fact that the correlation error for these closed-shell systems remains relatively constant as the atoms are forced together. There are no new electron pairs formed and thus no new strong electron correlations associated with molecular formation.

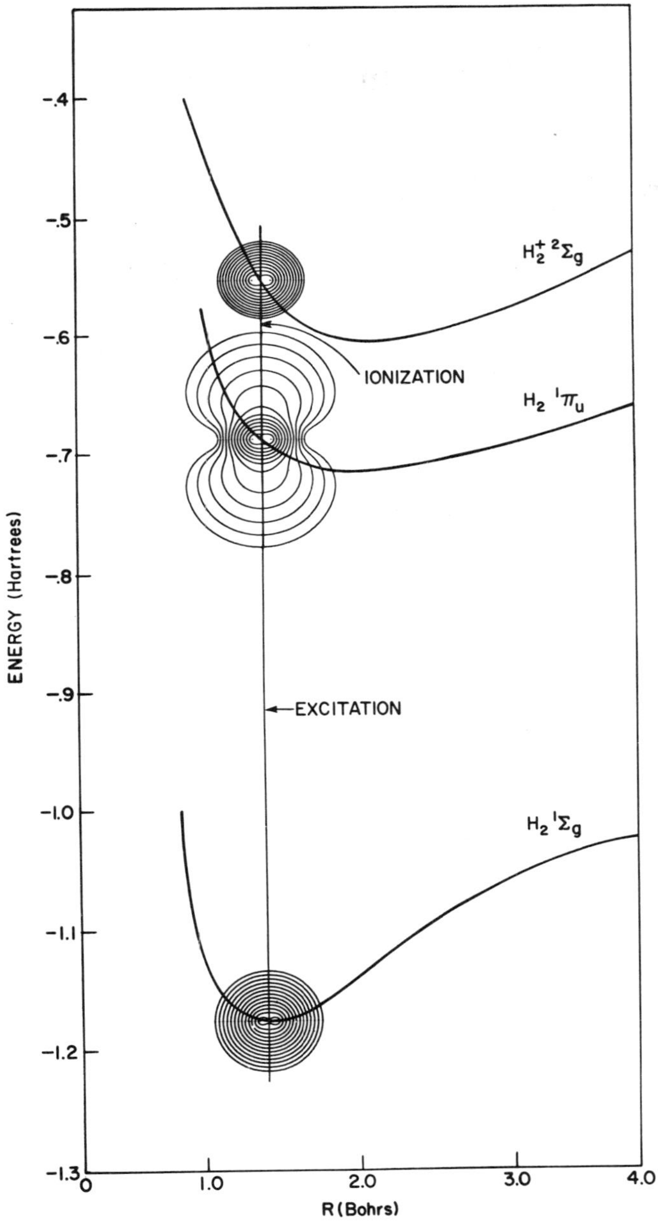

Fig. 3b. The diagrams represent the two fundamental processes in chemistry: excitation and ionization. The diagram representing the excited and ionized states are contour plots of the total electron densities of the three systems: H_2 in its ground state, H_2 in an excited π state, and H_2^+ in its ground state. In all three diagrams the outermost contour has a value of 6.1×10^{-5} electron/Bohr^3, and each successive inner contour increases in value by a factor of 2. Note the disappearance of the inner contour value of 0.25 in H_2 after it has undergone excitation or ionization. These diagrams were drawn automatically by electronic computers and are based on accurate *ab initio* calculations of the wave functions for the H_2 system (Refs. 33 and 143).

Table 1. Progress from Minimal LCAO-MO-SCF Wave Functions (1960) to Recent Extended-Basis-Set SCF Wave Functions (1960–1967) for Li_2, N_2, F_2, CO, BF, LiF, HF, and LiH

	Total energy, Hartrees			Dissociation energy, eV		
Molecule	Minimal[a] STO set	Extended STO set	Exptl.	Minimal[a] STO set	Extended STO set	Exptl.
LiH	-7.9667	-7.9860[c]	-8.0703	1.39	1.44	2.52
Li_2	-14.84075	-14.8715[e]	-14.9944	0.15	0.17	1.05
N_2	-108.57362	-108.9928[f]	-109.586	1.19	5.31	9.90
F_2	-197.85686	-198.7683[g]	-199.670	-0.30	-1.37	1.68
HG	-99.4785	-100.0571[h]	-100.527	1.21	4.11[i]	6.08
LiF	-106.3652	-106.9885[j]	-107.502	0.37	4.03	5.99
CO	-112.34357	-112.7860[k]	-113.377	5.38	7.84	11.24
BF	-123.61550	-124.1659[l]	-124.777	5.24	6.18	8.58

	Ionization potential (Koopmans'), eV		
Molecule	Minimal[a] STO set	Extended[b] STO set	Exptl.
LiH	8.24	8.54[d]	6.5 ± 0.5
Li_2	4.86	4.93	4.96
N_2	14.82	17.36	15.77
F_2	11.75	18.04	15.7
HG	12.65	18.16[i]	15.77
LiF	9.26	13.02	
CO	13.08	14.978	14.01
BF	9.61	10.999	10.969

	Dipole moments		
	Minimal[a] STO set	Extended STO set	Exptl.
CO	0.730 $C{-}O^+$	0.274[k] C^+O^-	0.118 $C{-}O^+$
BF	2.16 $B{-}F^+$	0.945[l] $B{-}F^+$	—
LiF	2.94 Li^+F^-	6.297[l] Li^+F^-	6.284 Li^+F^-
HF	0.878 H^+F^-	1.827 H^+F^-	1.818 H^+F^-
LiH	6.41 Li^+H^-	5.888[c] Li^+H^-	5.882 Li^+H^-

[a] B. J. Ransil, *Rev. Mod. Phys.* **32**, 245 (1960).

[b] Better values of ionization potentials can sometimes be obtained by direct computation: $IP = E_{ion}^{SCF} - E_{neutral}^{SCF}$.

[c] Ref. 36. Slightly improved values have been obtained recently (Ref. 43).

[d] Ref. 45.

[e] The first-row homonuclear diatomics, F_2 and N_2, have been published (Refs. 40 and 42); others will appear shortly.

[f] Ref. 42.

[g] Ref. 40 and 103.

[h] Ref. 35. Slightly improved values have been recently obtained (Ref. 45).

[i] Ref. 34.

[j] Ref. 37.

[k] Ref. 41.

[l] Ref. 41.

3. Beyond the Molecular Orbital Model[65-136]

The Hartree-Fock model is good for predicting geometry and many one-electron properties (this is also true for similar calculations on polyatomic systems). The Hartree-Fock model also yields adequate results for a variety of molecular systems: those arising from closed-shell interactions, for instance the noble gases,[47] systems in which there is only a one-electron bond, as typified by He_2^+,[48,49] LiHe, NaHe,[64] NeH^+, HeH^+[56] (some rather bizarre systems from the chemist's viewpoint), and highly ionic systems for which the shape of the potential curve is rather good, as typified by the alkali halides,[52,50] which are, in fact, analogous to the rare gas systems held together by a Coulomb force.

Table 2. Some Typical Calculated (Extended Basis Set) SCF–MO and Experimental Values of R_e, ω_e, and D_e for Diatomic Molecules

	R_e, Bohrs		ω_e, cm^{-1}		D_e, eV		
Molecule	Calc.	Exptl.	Calc.	Exptl.	Calc.	Exptl.	δD_e
$H_2(X^1\Sigma_g^+)$[a]	1.385	1.400	4561	4400	3.64	4.75	1.11
$Li_2(X^1\Sigma_g^+)$[b]	5.25	5.051	326	351	0.17	1.05	0.88
$N_2^+(X^2\Sigma_g^+)$[c]	2.041	2.113	2571	2207	3.24	8.85	5.61
$N_2(X^1\Sigma_g^+)$[c]	2.013	2.075	2730	2358	5.27	9.91	4.64
$F_2(X^1\Sigma_g^+)$[d]	2.51	2.725	1257	892	-1.63	1.35	3.28
$LiF(X^1\Sigma_g^+)$[e]	2.941	2.955	1033	964	4.08	5.95	1.87
$BF(X^1\Sigma+)$[f]	2.355	2.391	1496	1402	6.18	8.58	2.40
$CO(X^1\Sigma+)$[f]	2.08	2.132	2431	2170	7.84	11.23	3.39
$Cl_2(X^1\Sigma_g^+)$[g]	3.78	3.76	577	559	0.87	2.51	1.64
$BeO(A^1\pi)$[h]	2.74	2.76	1208	1144	4.93	–	–
$He_2^+(X^2\Sigma_g^+)$[i]	2.0	–	1790	–	2.7	–	–
$F_2^-(X^2\Sigma_g^+)$[j]	3.6	–	510	–	1.66	–	–
$Ne_2^+(X^2\Sigma_g^+)$[i]	3.2	–	660	–	1.65	–	–
$Cl_2^-(X^2\Sigma_g^+)$[j]	5.0	–	260	–	1.28	–	–
$Ar_2^+(X^2\Sigma_g^+)$[i]	4.6	–	300	–	1.25	–	–
$NaH(X^1\Sigma+)$[k]	3.62	3.57	1187	1172	0.932	(2.3	1.4
$SiH(X^2II_r)$[k]	2.86	2.87	2144	2042	2.23	3.32 ± 0.25	1.1 ± 0.25
$HCl(X^1\Sigma+)$[k]	2.39	2.41	3181	2989	3.48	4.616	1.14
$NaF(X^1\Sigma+)$[l]	3.65	3.64	590	536	3.08	4.49	1.41

[a]Ref. 103.
[b]Ref. 51.
[c]Ref. 42.
[d]Ref. 40.
[e]Calculated from results in Ref. 50.
[f]Ref. 41.
[g]Ref. 49.
[h]Ref. 44.
[i]Ref. 48.
[j]Ref. 49.
[k]Ref. 46.
[l]Ref. 52.

How can we improve this model without losing its valuable features? I would like to describe what we have been doing to go beyond the Hartree-Fock model with the goal in mind of obtaining results on diatomic and, eventually, larger systems. A condition is that these results be of genuine *quantitative* use to the experimental chemist and serve to complement his efforts, particularly where the experiment is difficult to perform. Such situations might involve high temperatures, highly corrosive materials, or very short-lived transients. The essence of our scheme,[6,103-106] which I am now going to describe, involves a direct pursuit of the two major defects of the Hartree-Fock model: (a) The molecule is allowed to dissociate properly (in many cases, this is a trivial extension of the Hartree-Fock picture), and (b) the correlation error in the molecule is made the same as the correlation error in the atoms. The latter is more difficult to achieve—especially in the general case. However, for systems in which the chemical bond is isolated by being one quantum number *higher* than the rest of the molecular core, the correlation energy changes associated with molecular formation are isolated by being placed in this bonding region and we can do a rather good job of making the correlation error constant, as a function of internuclear distance.

We see some typical results for the systems H_2, Li_2, and NaF in Figs. 8, 9, and 10. We see in Figs. 8b and 9b the orbitals necessary, in addition to the Hartree-Fock model, for the bonding electrons to avoid each other as the molecule forms, and to give us a correct continuous

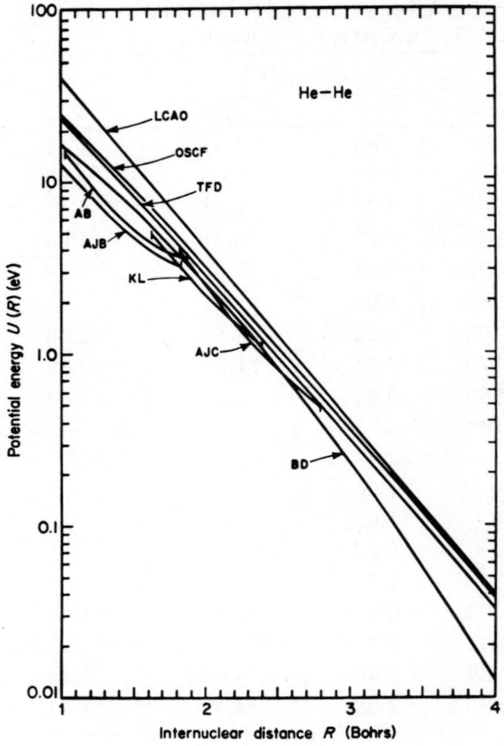

Fig. 4. He-He potential curves. AJC, semiempirical curve from scattering data [I. Amdur, J. D. Jordan, and S. O. Colgate, *J. Chem. Phys.* **34**, 1525 (1961)]. AB, semiempirical curve from scattering data [I. Amdur and R. R. Bertrand, *J. Chem. Phys.* **36**, 1078 (1962)]. BD, semiempirical curve from composite data [R. A. Buckingham and D. M. Duparc, *Progress in International Research on Thermodynamic and Transport Properties* (Symposium on Thermophysical Properties, Princeton University, 1962; American Society of Mechanical Engineers), Academic Press, New York, 1962, p. 378]. TFD, Thomas-Fermi-Dirac method [A. A. Abrahamson, *Phys. Rev.* **130**, 693 (1963)]. AJB, semiempirical curve from scattering data [I. Amdur, J. E. Jordan, and R. R. Bertrand, *Atomic Collision Processes*, M. R. C. McDowell, ed. (Proceedings of the Third International Conference on the Physics of Electronic and Atomic Collisions, London, 1963), North-Holland, Amsterdam, 1964, p. 934]. KL, semiempirical curve from scattering data [A. B. Kamney and V. B. Leonas, *Soviet Phys. Doklady* **10**, 529 (1965)]. LCAO, linear-combination-of-atomic-orbitals method.[21] OSCF, optimized self-consistent-field (Hartree-Fock) method.[21]

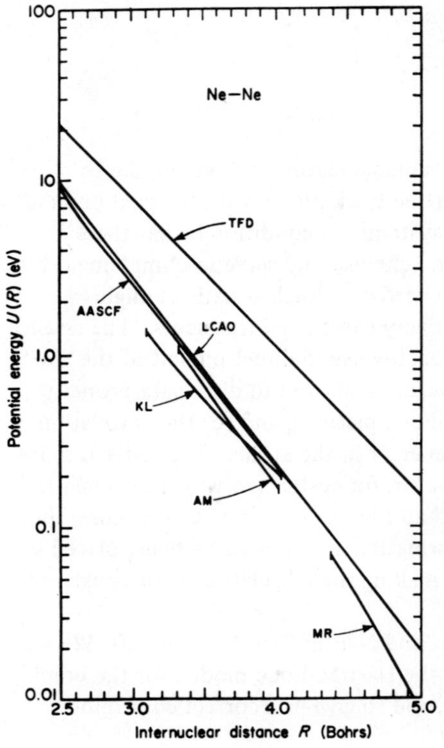

Fig. 5. Ne-Ne potential curves. MR, semiempirical curve from scattering data [E. A. Mason and W. R. Rice, *J. Chem. Phys.* **22**, 843 (1954)]. AM, semiempirical curve from scattering data [I. Amdur and E. A. Mason, *J. Chem. Phys.* **23**, 415 (1955)]. TFD, Thomas-Fermi-Dirac method [A. A. Abrahamson, *Phys. Rev.* **130**, 693 (1963)]. KL, semiempirical curve from scattering data [A. B. Kamnev and V. B. Leonas, *Soviet Phys. Doklady* **10**, 529 (1965)]. LCAO, linear-combination-of-atomic-orbitals method.[21] AASCF, augmented asymptotic self-consistent-field method.[21]

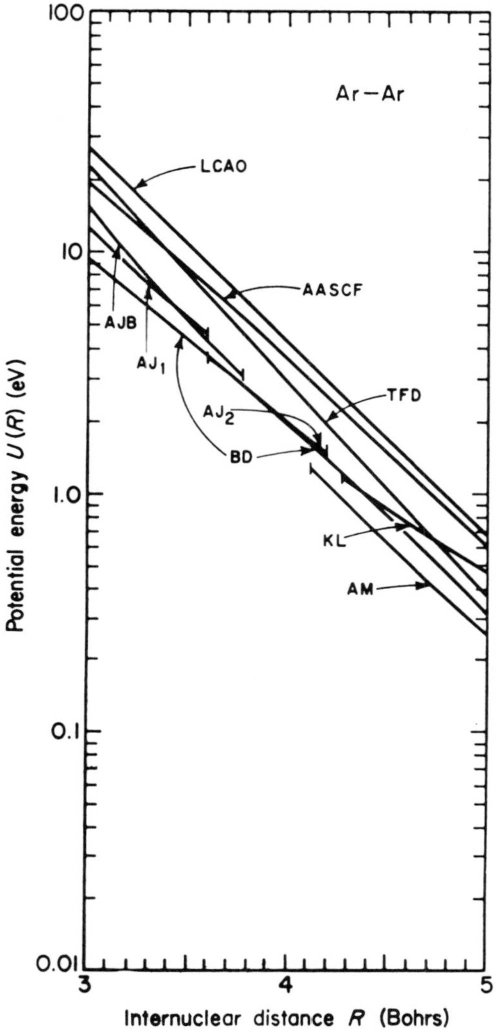

Fig. 6. Ar-Ar potential curves. LCAO, linear-combination-of-atomic-orbitals method.[21] AASCF, augmented asymptotic self-consistent-field method.[21] TFD, Thomas-Fermi-Dirac method [A. A. Abrahamson, *Phys. Rev.* **130,** 693 (1963)]. BP, semiempirical curve from composite data (J. A. Barker and A. Pompe, to be published). AJP, semiempirical curve from scattering data [I. Amdur, J. E. Jordan, and R. R. Bertrand, *Atomic Collision Processes*, M. R. C. McDowell, ed. (Proceedings of the Third International Conference on the Physics of Electronic and Atomic Collisions, London, 1963), North-Holland, Amsterdam, 1964, p. 934]. AJ_1 and AJ_2 semiempirical curves from scattering data (I. Amdur and J. E. Jordan, quoted in D. D. Konowalow and S. Carra, Morse Potential Functions for Nonpolar Gases, *Rept. WIS-TC1-74* from the Theoretical Chemistry Institute of the University of Wisconsin, Dec. 1964). KL, semiempirical curve from scattering data [A. B. Kamnev and V. B. Leonas, *Soviet Phys. Doklady* **10,** 529 (1965)]. AM, semiempirical curve from scattering data [I. Amdur and E. A. Mason, *J. Chem. Phys.* **22,** 670 (1954)].

picture of chemical bonding (Figs. 8c and 9c). These correlation terms may be conveniently stated or categorized as in-out, left-right, and angular. When added or subtracted to the ground-state molecular orbital they move the electrons away from each other in an in-out, left-right, or angular direction. For systems such as H_2, in which there is a single bond—no other electrons, and Li_2, in which the bond is quantum level 2 and the core in quantum level 1, the quantitative results are truly gratifying and, in fact, are of sufficient quality to be useful to the experimentalist. For F_2 we have obtained significantly better results than those obtainable from the model, but by only correlating the 36g electrons did not do well enough to really aid the experimentalist (Table 3 and Fig. 11). The reason for this is that although F_2 is considered to be a single-bonded molecule, the two electrons forming the bond are in the same radial quantum

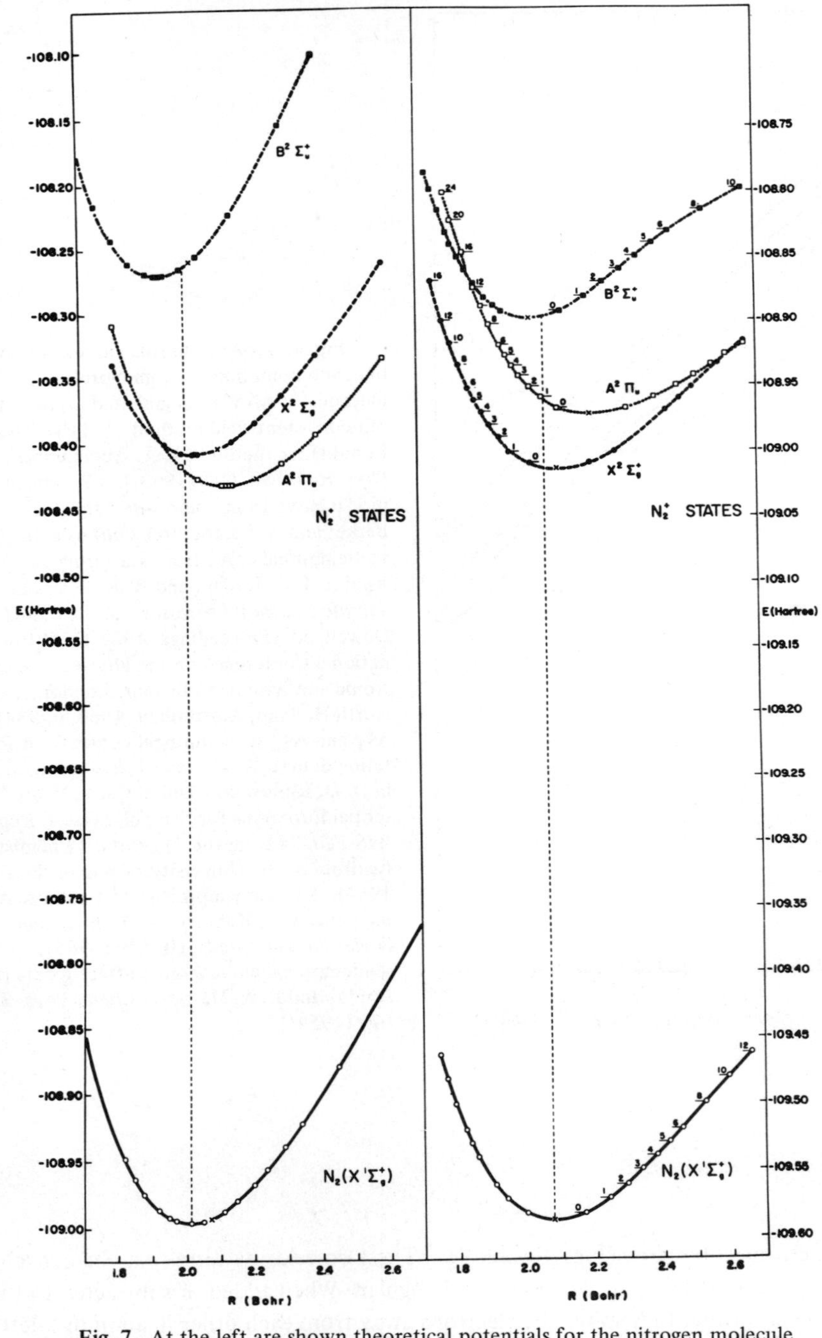

Fig. 7. At the left are shown theoretical potentials for the nitrogen molecule and its ions; at the right experimental curves. Note the inversion of the $^2\Sigma_g^+$ and $^2\Pi_u$ state—a defect of the orbital picture (see Ref. 42). See Fig. 3a for the changes taking place in the orbital picture of these ionizations.

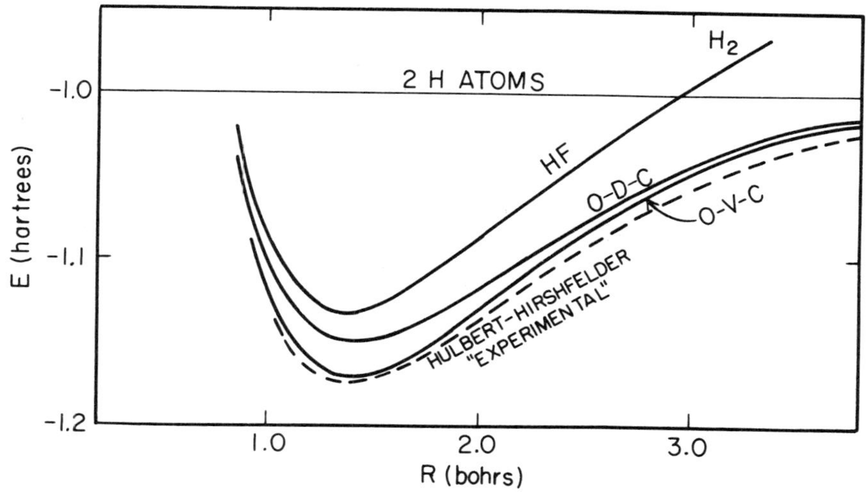

Fig. 8a. Comparison of Hartree-Fock optimal double configuration (ODC) and optimal valence configuration (OVC) potential curves with the experimental one for H_2.

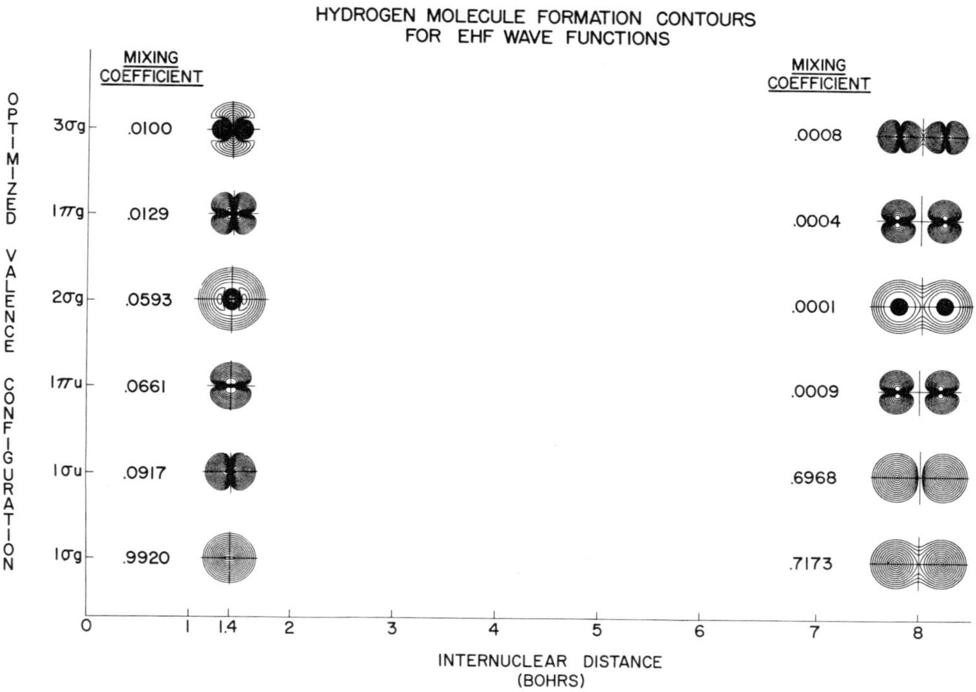

Fig. 8b. Orbitals making up the H_2 OVC wave function at R_e and near dissociation. In Figs. 8b, 8c, 9b, and 9c, the outermost contour in all cases corresponds to a density of 6.1×10^{-5} $e^-/Bohr^3$. Each successive inner contour then increases by a factor of 2.

shell and thus occupy the same physical space as twelve other electrons. Therefore, there are subtle *intershell* correlation effects, which must be taken into account. We have recently included such intershell effects and obtain a preliminary binding energy of 1.4 eV for F_2.

Let me now review the essential ideas of the optimized valence configuration (OVC) method and describe its recent application to an experimentally uncharacterized system.

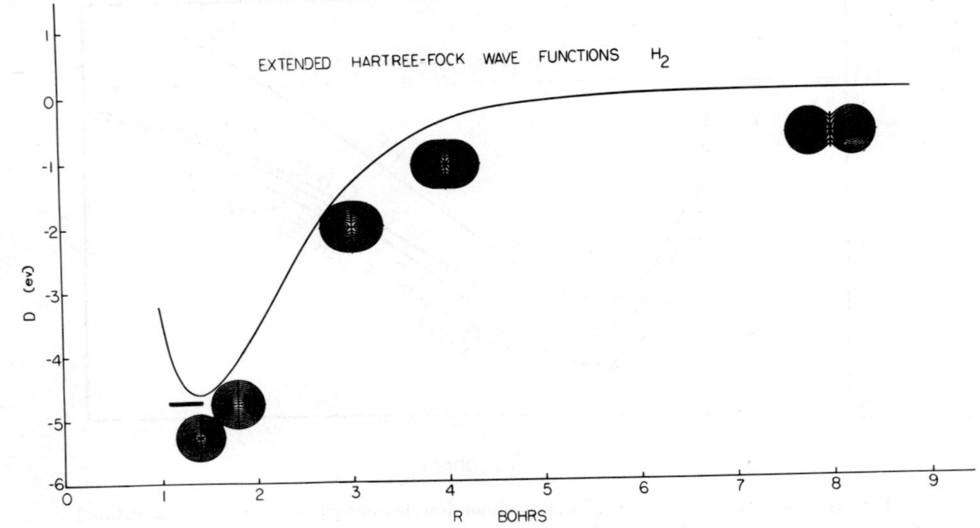

Fig. 8c. Total charge-density contours of the H_2 OVC wave function[1] as the molecule forms. The horizontal line is the level of the H_2 experimental energy.

Table 3. Comparison of Hartree-Fock, Optimized Double Configuration (ODC), and Optimized Valence Configuration (OVC) Results for H_2, Li_2, NaLi, F_2, and NaF with Experiment

	ω_e, cm^{-1}	R_e, Bohrs	D_e, eV
H_2			
HF[a]	4561	1.39	3.64
ODC[a]	4214	1.42	4.13
OVC[a]	4398	1.40	4.63
Exptl.	4400	1.40	4.75
Li_2			
HF[a,b,c]	326	5.26	0.17
ODC[a]	344	5.43	0.46
OVC[a]	345	5.09	0.93
OVC[b]	345	5.09	0.99
Exptl.	351	5.05	1.05
NaLi			
HF[d]	264	5.56	0.05
OVC[a]	256	5.48	0.85
Exptl.	–	–	–
F_2			
HF[a,f]	1257	2.50	-1.37
ODC[a]	678	2.74	0.54
OVC ($3\sigma_g$ only)[g]	704	2.72	0.81
OVC (all promotions into $3\sigma_u$)[g]	N.C.[j]	2.70	0.95
OVC ($p_g - p_\pi$ correlation included)[k]	N.C.	2.71	1.54
Exptl.	892	2.68	1.60[l]
NaF			
HF[h]	570	3.65	3.08
OVC[g]	570	3.66	3.70
Exptl.[i]	536	3.64	4.94

[a] See Ref. 103.
[b] See Ref. 104.
[c] See Ref. 51.
[d] See Ref. 54.
[e] See Ref. 138.
[f] See Ref. 40.
[g] See Ref. 137.
[h] See Ref. 52.
[i] See Ref. 52.
[j] Not computed.
[k] Calculations in progress.
[l] Most recent revised experimental value [Dibeler, Walker, and McCulloh, *J. Chem. Phys.* (1969)].

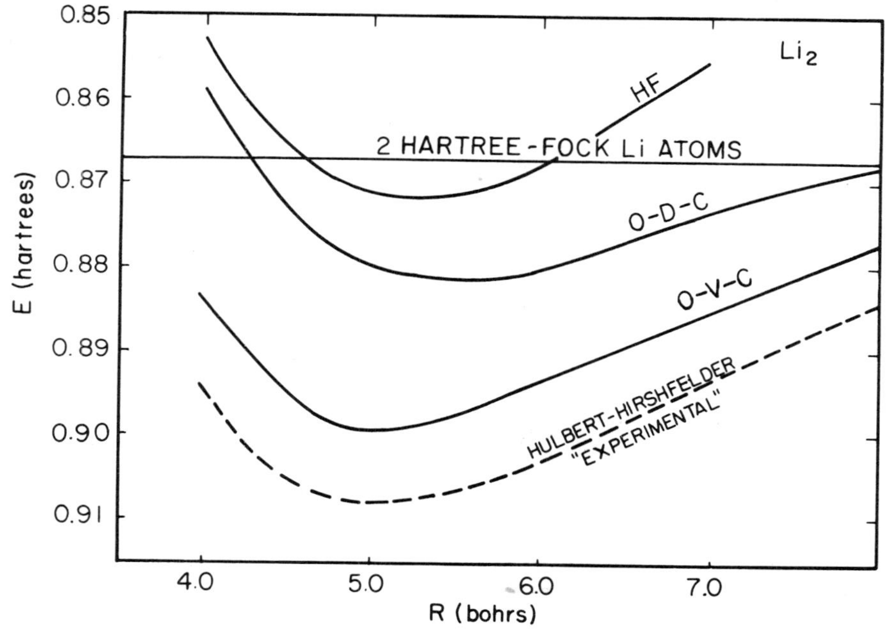

Fig. 9a. Comparsion of Hartree-Fock and optimal valence configuration potential curves with experimental one for Li_2 (Ref. 103).

Configurations were added to the Hartree-Fock picture to allow proper *dissociation* of the molecule. (In H_2 this is merely one extra configuration which subtracts the ionic terms as the atoms separate. Inclusion of such a configuration allows us to "look" continuously at molecular formation. We could not do this with the simple model of the molecular orbitals.) Additional configurations allowed the bonding electrons to avoid each other as the molecule formed, since they were, of course, avoiding each other completely when the constituent atoms were at infinite distances. Included were angular and in-out terms in addition to the left-right term, which brought about proper dissociation.

The optimized valence configuration wave function thus has the form

$$\psi = A_0 \psi_0 + \Sigma_k A_k \psi_k$$

where the ψ_k are antisymmetrized products of orbitals ϕ_i, similar to equation (1). However, the ϕ_i are now in the general solution of *individual* integrodifferential equations of the form[103–106,121]

$$F_i \phi_i = \epsilon_i \phi_i$$

where the F_i are more complicated operators depending upon all orbitals of the system and all mixing coefficients A_k. Convergence to best orbitals is achieved in this model when the energy of the system is stationary with respect to variation of both the coefficients A_k and the orbitals ϕ_i. We "see" some of these necessary additional orbitals again in Figs. 8c and 9c. Thus we evaluate only the increase in correlation energy associated with molecular formation, and we, in addition, allow the core to distort and polarize, but we do not try to correlate the core. We assume that the correlation error in the core is constant. This assumption is supported by calculations on the rare gas systems (Figs. 4 to 6) and also by our calculations on Li_2 (Table 3). We now feel that we understand this model very well, and, as we have stated earlier, for the type of bond which is isolated we can do a very good job in obtaining the proper molecular energy surface. For an ionic bond, the alkali halide system was handled quite well with the Hartree-

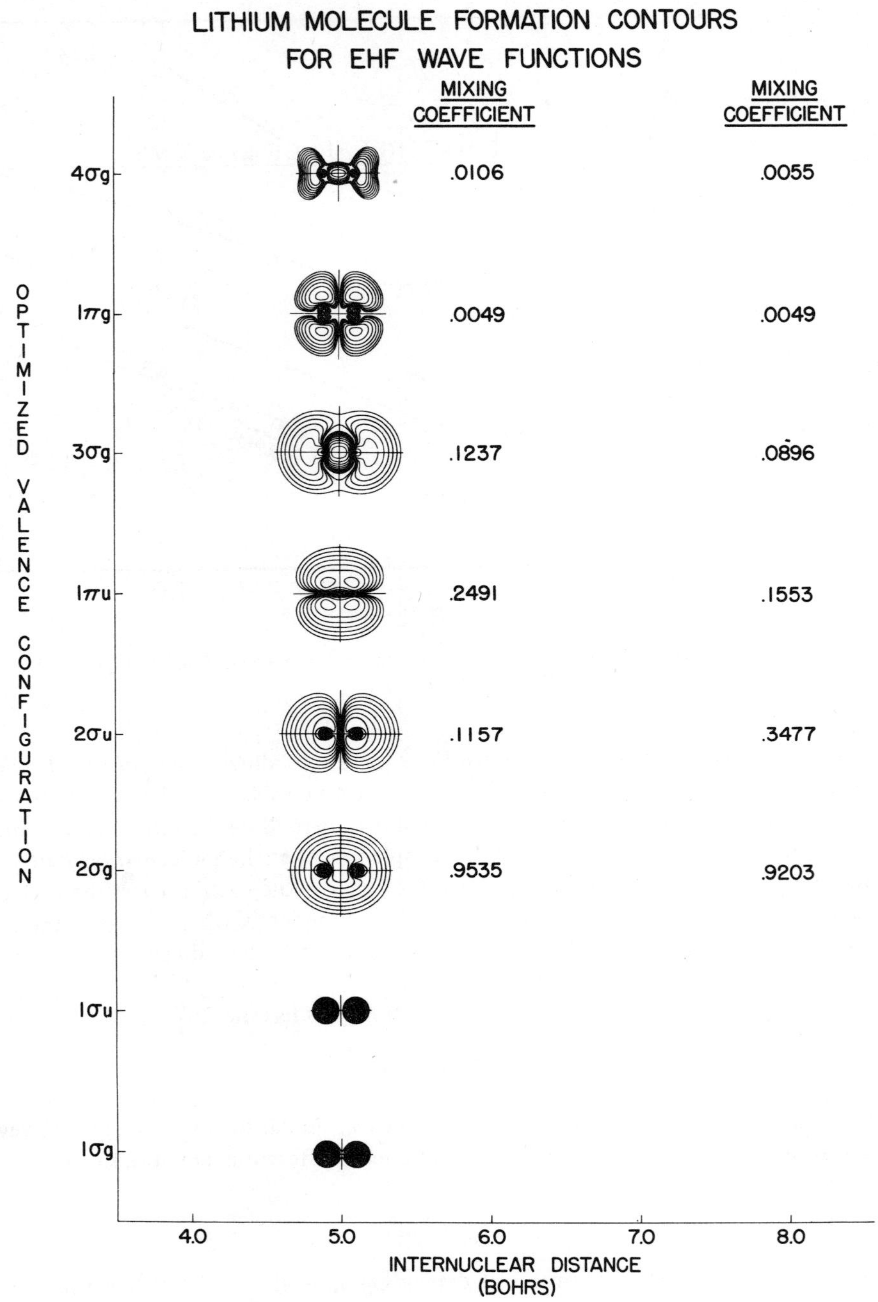

Fig. 9b. Orbitals making up the Li_2 OVC wave function at R_e (Refs. 103 and 105).

Fock approximation and we needed only one additional configuration to allow for proper dissociation. Because of intershell effects, we did not calculate the correct binding energy. However, techniques now developed appear to provide the necessary improvement. (See the NaF results in Table 3 and Fig. 10.)

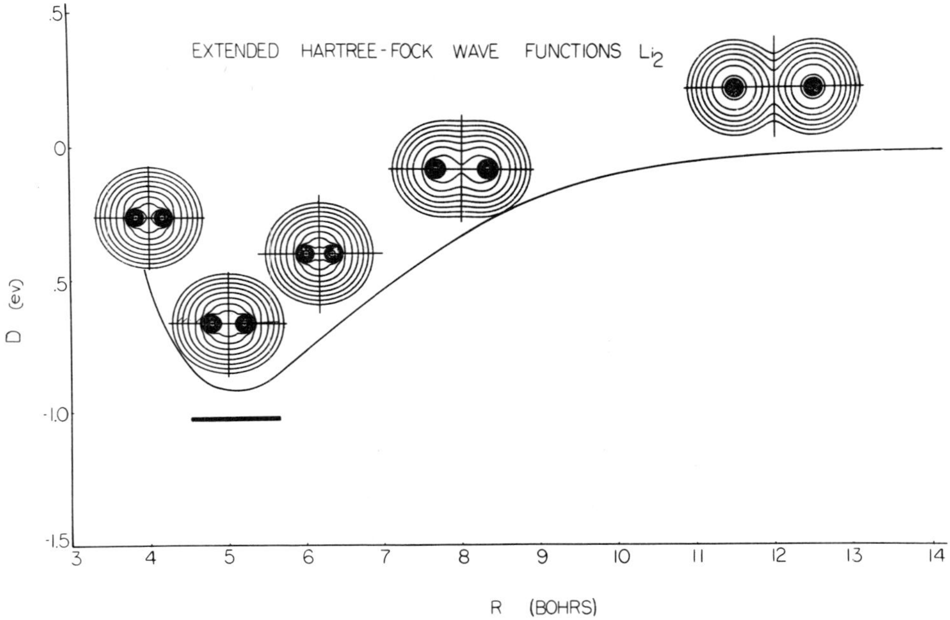

Fig. 9c. Total charge-density contours of the Li_2 OVC wave function[1] as the molecule forms. The horizontal line is the level of the Li_2 experimental binding energy (Refs. 103 and 105).

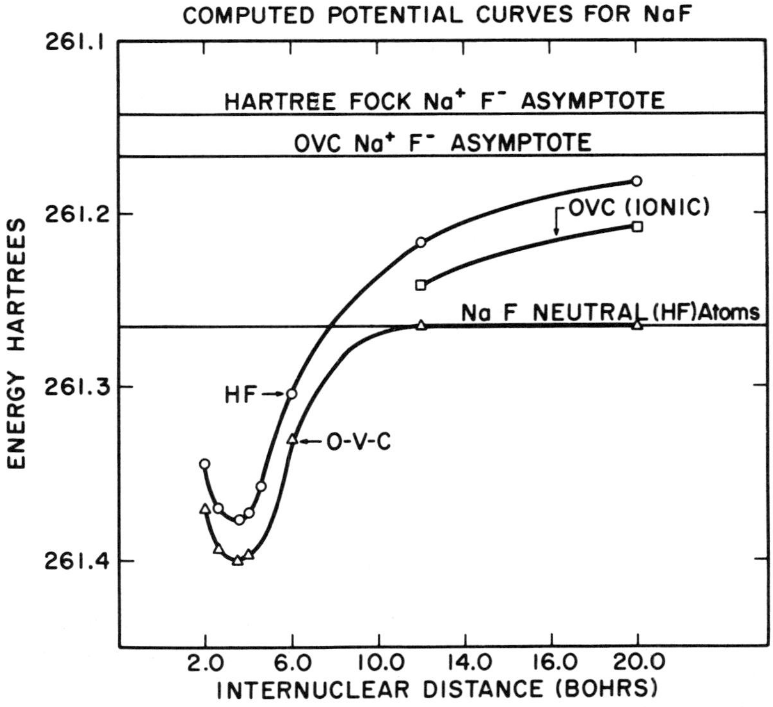

Fig. 10. Comparison of Hartree-Fock and optimized valence configuration potential curves for NaF (Ref. 6).

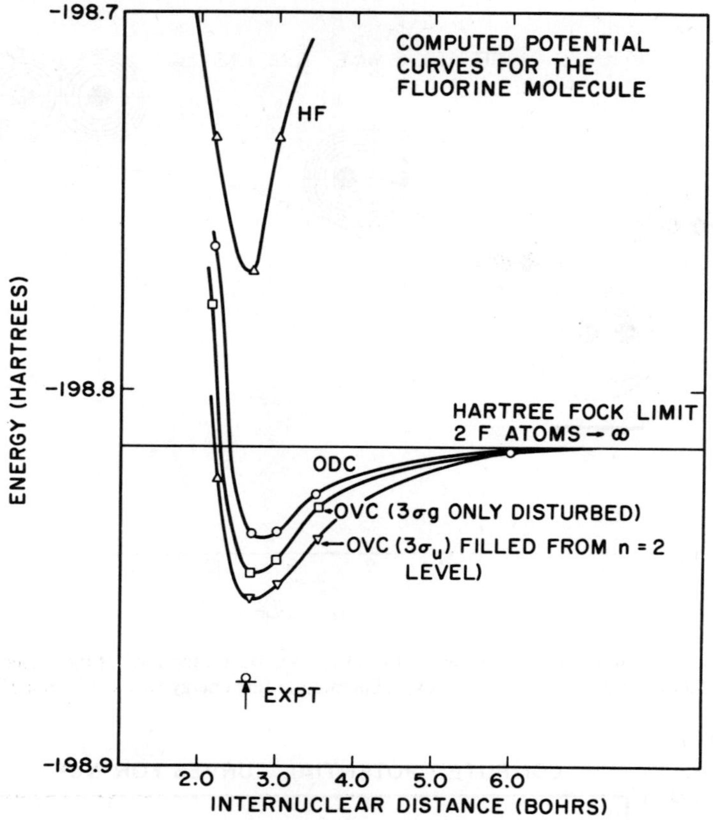

Fig. 11. Comparison of Hartree-Fock ODC and OVC potential curves for F_2 (Ref. 103).

For F_2 we see that our previous results were not satisfactory (Table 3 and Fig. 11), again because of intershell effects. When the correlations between p_ϵ and p_π electrons are taken into account, however, we obtain the very encouraging binding energy of 1.4 eV (calculations in progress).

However, bolstered by our confidence that we can do very well on the alkali-type systems, we decided to try to do a *predictive* calculation concerning a system of which little is known experimentally. Applying our method, we obtained the results[138] for NaLi presented in Table 3 and Figs. 12 and 13. A binding energy of about 0.85 eV (quite substantial) and the calculated vibrational frequency, etc., are predicted. We feel we can attach a precision to these results of a few per cent, and we are hoping that these results will be used by the experimentalist in the near future.

4. New "Instrument" for Chemists

The next question is: Since we now have some *ab initio* results on the energy surface of the simple but experimentally uncharacterized system, NaLi, what can be done with such results to enhance their chemical utility? As one example of such an application we can turn to thermodynamics. We know that we could describe a vapor mixture of Na and Li thermodynamically if we could obtain its equation of state.[139] We must therefore evaluate the virial coefficients. In the case of NaLi we are fortunate because, as indicated by *calculated* equilibrium constants

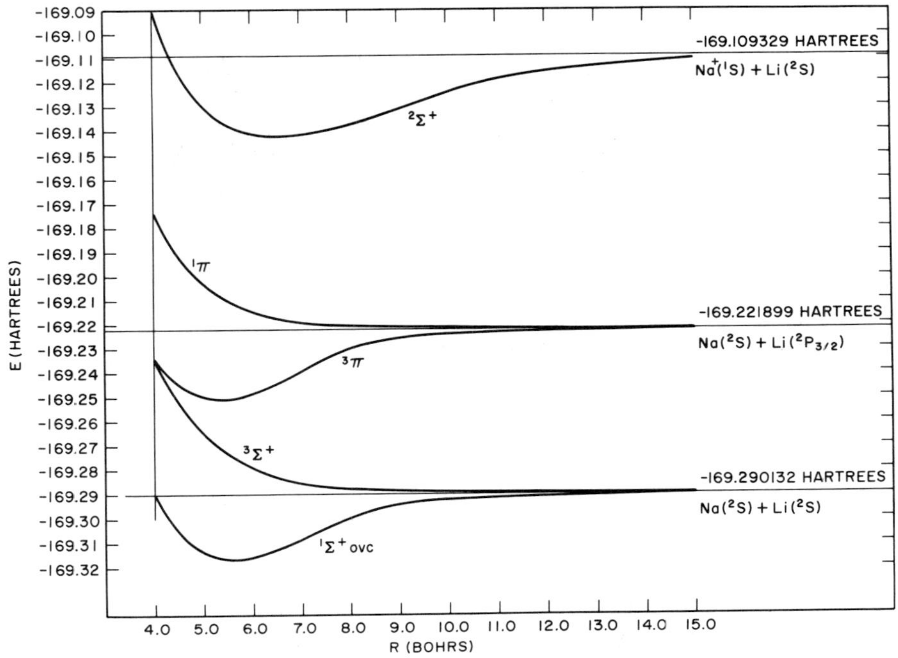

Fig. 12a. Calculated potential curves for Na and Li (Ref. 138).

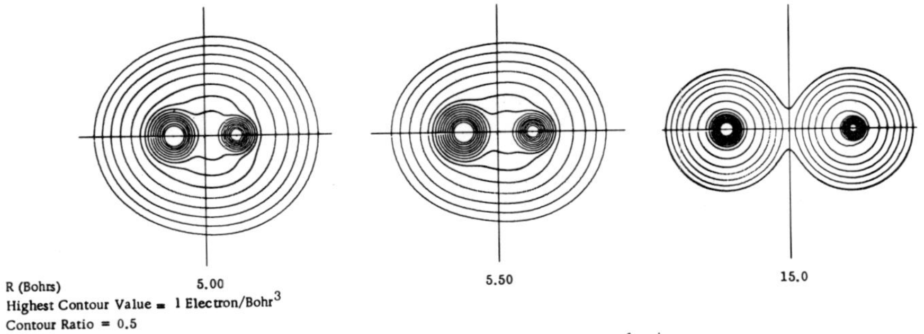

Fig. 12b. OVC and total charge-density contours for the $^1\Sigma^+$state of NaLi at 15, 5.5, and 5.0 Bohrs. Picture at 15 Bohrs is two thirds the scale of the other two pictures (Ref. 138).

for a NaLi mixture (Tables 4 and 5), at the temperatures and pressure of interest, we may assume that the system consists overwhelmingly of isolated atoms. Therefore, we may calculate only the second virial coefficient, assuming the concentrations to be the initial concentrations of the atoms. The second virial coefficient, B(T), in the virial equation of state may be obtained for a NaLi mixture from the relation

$$B(T) = B_{Na2}(T)\, X_{Na}^2 + B_{Li2}(T)\, X_{Li}^2 + B_{NaLi}(T)\, X_{Li} X_{Na}$$

where X_i represents the mole fraction of component i.

In order to calculate the three $B_{AB}(T)$ we of course needed the AB interaction potential V_{AB} for both the $^1\Sigma$ ground state and $^3\Sigma$ excited repulsive state arising from the three types of collision of atoms in their ground state: NaLi, Li-Li, Na-Na. It is here that the *ab initio* calcu-

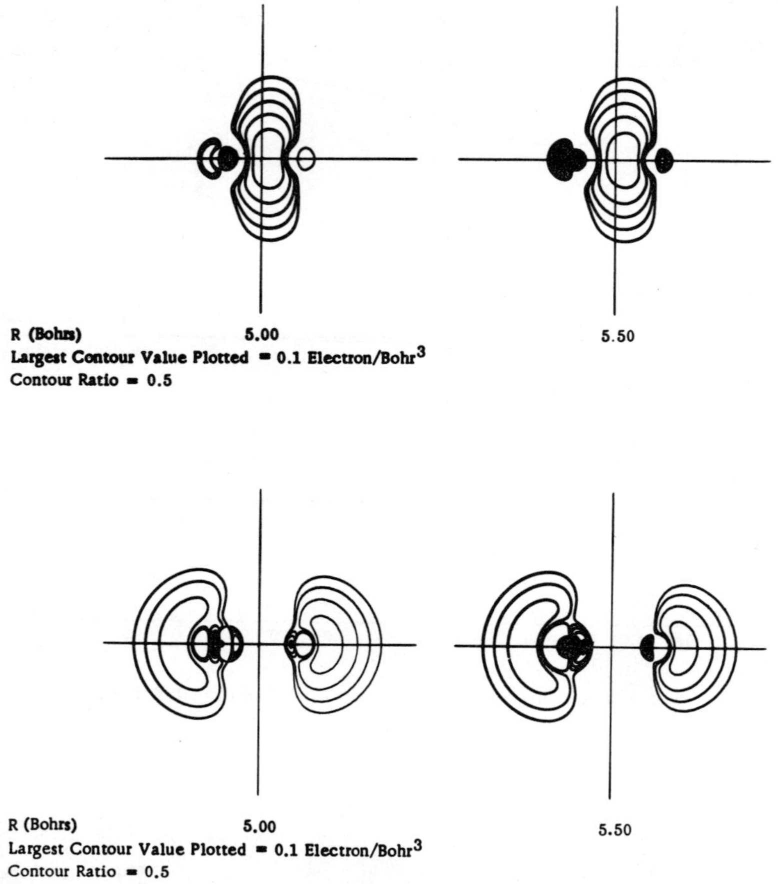

Fig. 13. Contours of charge-density difference associated with the formation of the chemical bond in the $^1\Sigma$ state of NaLi. Upper diagram shows the region where charge has increased relative to dissociated atoms. Lower diagram shows where charge density has decreased relative to dissociated atoms (Ref. 138).

lation plays an important role. The $^1\Sigma$ potentials are available experimentally for Li_2 or Na_2 and we used our calculated one for NaLi. A variety of ways for evaluating $V_{AB}(R)$ for the $^3\Sigma$ state are used consisting mostly of semiempirical forms or "scaled" or "reduced" potentials:

Table 4. Calculated and Measured Values of r_e, ω_e, and D_e for Na_2, Li_2, and NaLi

	$^{23}Na_2$	$^7Li_2$	$^{23}Na^7Li$
r_e, Å	3.078	2.673[a]	Not avail.
	(in prog.)	(2.691)	(2.939)
ω_e, cm^{-1}	159.2	351.4	Not avail.
	(in prog.)	(345.3)	(249.9)
D_e, eV	0.75	1.03	Not avail.
	(in prog.)	(0.99)	(0.85)

[a]The upper number is the experimental one.

Table 5. Equilibrium Constant K_p Computed from r_e, ω_e, and D_e as a Function of Temperature for the Reaction Na + Li ⇌ NaLi

T, °K	$K_p^{Na_2}$	K_p^{NaLi}	$K_p^{Li_2}$
1123	5.88016	1.27468	0.3504
1000	2.13976	0.4104	0.0850
973	1.65798	0.3083	0.0565
873	5.63966×10^{-1}	0.0979	0.0139
773	1.46223×10^{-1}	0.0202	0.0019
673	2.56338×10^{-1}	0.0029	0.0002

namely, the Rydberg, Morse, anti-Morse, and power series.[140] We have used our *ab initio* potentials (in which we have considerable confidence) and compared them with various popular semiempirical forms.

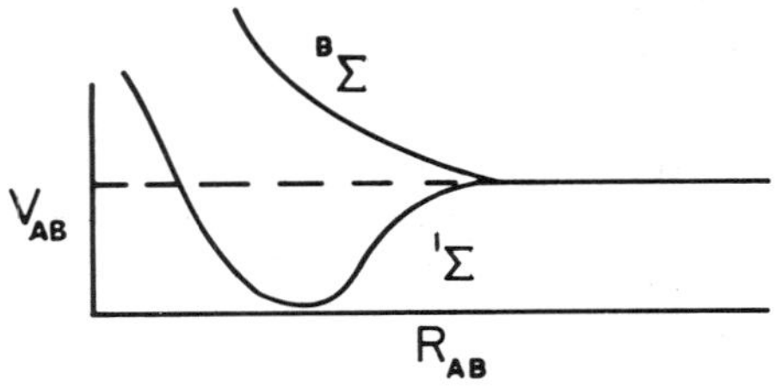

Through the relations[140]

$$B = \frac{1}{4}B^1 + \frac{3}{4}B^3$$

and

$$B_{AB}^{\frac{1}{3}} = 2\pi Na \int_0^\infty \left\{ 1 - \exp\left[- \frac{V^{\frac{1}{3}}(R)}{kt} \right] \right\} R^2 \, dR$$

we can obtain from these potentials the three required virial coefficients.

Now, what have we done with all these results? We have predicted the existence of and computed molecular properties for NaLi that were not available experimentally; we have gone further and used our energy surfaces to calculate the second virial coefficients and thus the degree of ideality of alkali vapors (alkali systems are difficult systems to look at experimentally and, in fact, NaLi, as noted previously, has not been observed), and are now calculating some transport properties for the vapor mixture.[141] We also have been able to criticize the assumed form on the $^3\Sigma$ potential curves customarily used and are suggesting a better form based on our calculations. This study of the alkali vapor presents a good example of how our *ab initio* results can be dovetailed into traditional chemistry.

By the further development of the techniques outlined above to handle cross-shell correlations and other types of molecules, we feel that comparable energy surfaces can be obtained for other diatomic molecules. In fact, this is currently our main mission, in addition to extending these methods to open-shell excited-state systems so that we may obtain accurate theoretical term values.

It should be emphasized that there is no reason to believe that what we have learned in diatomic systems will be any simpler in polyatomic systems. In fact, it is important to develop

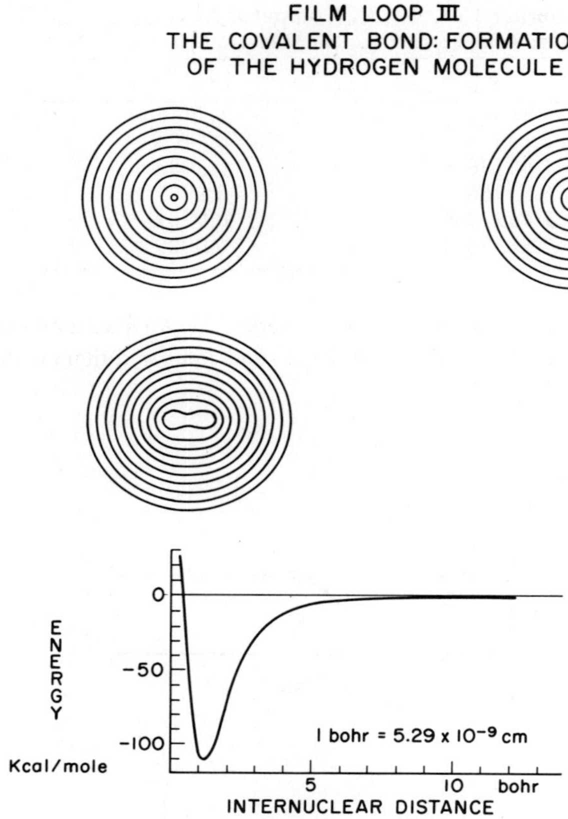

Fig. 14. Displays the beginning and end diagrams in the formation of the H_2 molecule. The innermost contour corresponds to a probable electron density of $0.25\ e^-/\text{Bohr}^3$, and each successive outer contour decreases by a factor of 2 down to $4.9 \times 10^{-4} e^-/\text{Bohr}^3$. [Figures 14–17 summarize part of a series of films and charts on atomic and molecular structure (Ref. 31).]

diatomic theory and diatomic calculations to a point where they are of use to the experimentalist before jumping into complicated polyatomic systems using old methods with their well-known deficiencies. This is the compelling reason for refining our tools on diatomic molecules, which, although perhaps of less immediate interest to the traditional chemist, provide a much more economical medium for the evolution of new techniques.

We feel that the calculated results on the alkali dimers do represent an example of chemistry from computers. The isolated energy point or a wave function did not represent chemistry, but a potential-energy surface then used to predict binding energy, and ultimately transport properties must be considered genuine chemistry. More important, we are now thinking more comprehensively in terms of complete chemical processes, potential-energy surfaces, ground and excited states, state functions, and virial coefficients—all of which for selected simple systems can now be obtained from computers and from computer calculations. It is more important now, therefore, than previously that theoreticians talk to experimental chemists, because I think that many chemists do not know how far the field of computational chemistry has come. Likewise, theoreticians do not know how intimately new experimental techniques are probing molecules and how much interesting chemistry there is in the world of small molecules. It is

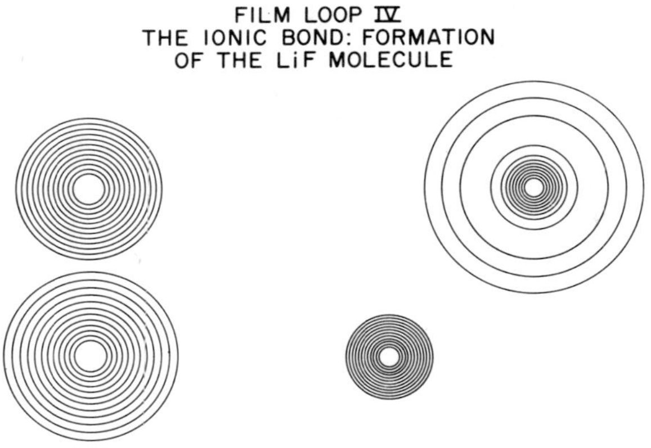

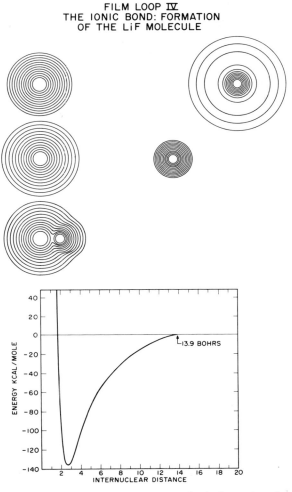

Fig. 15. Displays three sequences in the formation of the highly ionic system LiF from the Li and fluorine atoms. Note change from Li and F atoms to Li^+ and F^- atoms at a distance of 13.9 Bohrs, where the ionic configuration becomes more stable. In all diagrams the innermost contour corresponds to a probable electron density of $1.0e^-/Bohr^3$, and each successive outer contour decreases by a factor of 2 down to 4.9×10^{-4} $e^-/Bohr^3$.

also important that theoreticians *and* the experimentalists feed these *ab initio* numbers into existing semiempirical theories, and that through this process the semiempirical rules be upgraded by what might be called semi-*ab initio* rules into which new relationships and concepts of molecular interactions obtained from accurate calculations are embodied. This is an important added bonus available from the calculations due to the "adiabatic" nature of the computing process. We can "freeze" our numeric model and "look" at any stage of a process. For example, we can now watch a molecule form or atoms collide in terms of their changing electronic charge density continuously being displayed on a cathode-ray tube controlled by digital computers (Fig. 5) during the chemical process numerically under way[31] (Figs. 8, 9, 12–17).

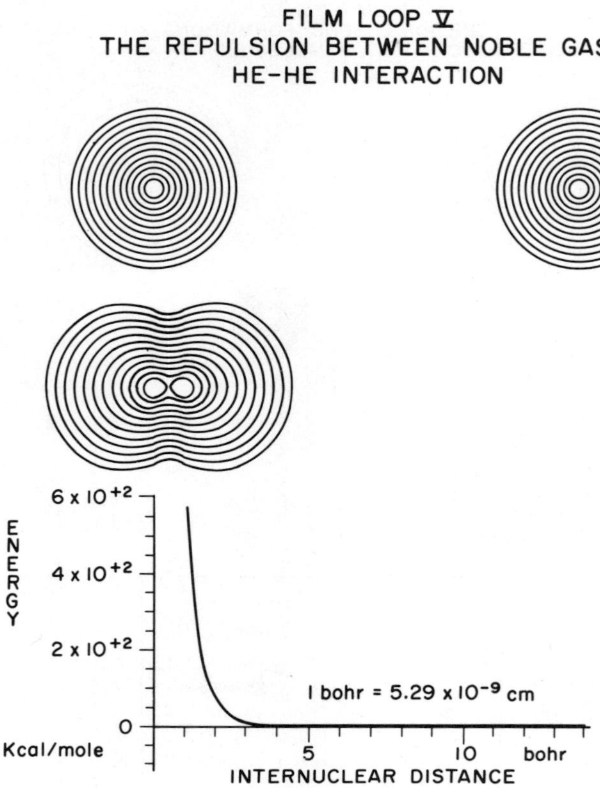

Fig. 16. Displays the beginning and end diagrams in the repulsive interaction of two helium atoms. As atoms move together, note how the electron charge is squeezed from between the nuclei to the ends of the molecule (a consequence of the Pauli exclusion principle). In all diagrams the innermost contour corresponds to a probable electron density of 1.0 $e^-/Bohr^3$, and each successive outer contour decreases by a factor of 2 down to 4.9×10^{-4} $e^-/Bohr^3$.

In conclusion, we should say that the chemist is not going to be replaced by the computer—at least, certainly not by the computer used in a purely theoretical manner. However, I think that times have changed in the sense that the chemist can obtain more sophisticated, relevant, and accurate answers from computers, particularly where difficult experimental conditions are required, where the calculation is feasible but the experiment is not.[142] (Computers don't burn up or melt when the temperature of your theoretical model is increased; computers do not corrode if you perform a calculation on fluorine; and the time scale of your numeric experiment can be expanded or contracted at will.)

The new capabilities we have with electronic computers should be interwoven into the experimentalist's thinking and he really should consider the alternatives when he seeks a particular physical property of a simple system: Should I measure it or should I compute it? We shall find that the limits of error in the computation often are comparable to limits of error obtainable experimentally, and sometimes that the derivation is more tractable and economical.

Toward this view we "computer quantum chemists" are developing our theory and procedures into self-contained "packages" or systems which should be treated as a new piece of experimental apparatus requiring little or no intimate knowledge of its detailed structure.[143-146]

FILM LOOP VI
The Formation of the Water Molecule from H and OH

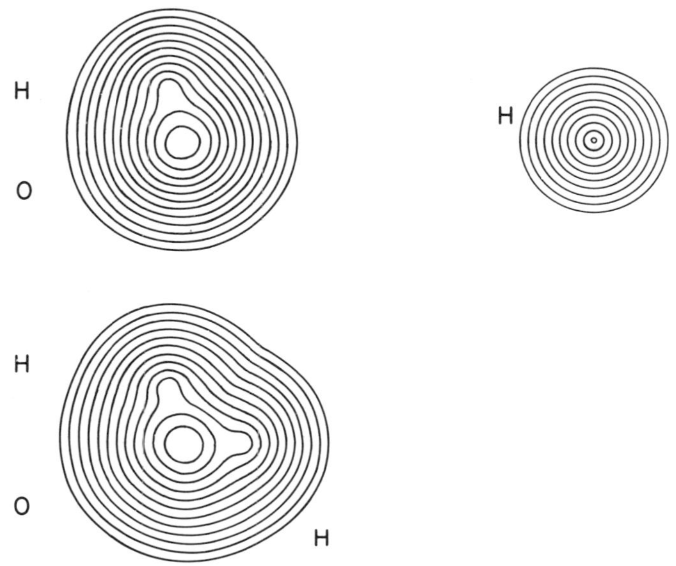

Fig. 17. Displays electron-density contour diagrams for the OH molecule and the H atom and the final diagrams for the H_2O molecule in its equilibrium configuration. In the OH and H_2O diagrams the innermost contour corresponds to a probable electron density of 1.0 $e^-/Bohr^3$ and each successive outer contour decreases by a factor of 2 down to 4.9×10^{-4} $e^-/Bohr^3$; in the H diagram the innermost contour has a value of 0.25 $e^-/Bohr^3$.

Our ultimate goal is to make these chemical computing systems so self-contained that the experimental chemist can ask of them truly *chemical* questions[143] such as: What is the dipole moment of CaO? What are the vibrational levels of $VC\ell_2$? What are the transport properties of alkali vapors? This requires, of course, that our computing systems embody not only mathematical analysis but our *procedures* and experience in utilizing these theoretical techniques and that such systems are capable of continuous growth, revision, and eventually learning from the system's own accummulating experience. It is further mandatory if these systems are to be useful to the nonspecialist that a guarantee of *reliability* and *precision* relative to *external reality* be made. This has formulated our philosophy that only methods capable of routinely producing chemically useful precision be incorporated into the final system concept. The nonspecialist should not be seduced into believing numbers simply because they are produced in an officious manner. Owing to the meteoric progress being made in computer design and capacity and the increasing sophistication[143-149] of "computer chemists," I have confidence that these and more provocative chemical questions shall be posed to, answered by, and eventually formulated by our new numeric apparatus.

Acknowledgments

The author is indebted to his collaborators, P. J. Bertoncini, G. Das, F. Janiszewski, K. Kaiser, R. H. Land, P. Maldonado, and D. Miller, for their cooperation in preparing as yet unpublished results on many of the systems mentioned. Also, gratitude is extended to Max Matheson and Oliver Simpson for their generous support of the development of our computational tools to the point where we can begin to think comprehensively about chemistry from computers, and to P. J. Bertoncini, F. Janiszewski, T. L. Gilbert, G. Goodman, K. Kaiser, A. Chung, R. H. Land, T. Bowman, and C. Trindle for stimulating discussions concerning these concepts.

References and Notes

1. A. C. Wahl, "Hartree-Fock Is Here–What Next?," a talk delivered at the Alberta Symposium of Quantum Chemistry, University of Alberta, Edmonton, Canada, June 1965.

1a. Although this article is concerned with the amalgamation of theory and computing techniques based primarily upon first principles, its conceptual implications stimulate us to think about the construction of semiempirical and empirical numeric instruments, such as a "synthesizer" for the organic chemist. This numeric synthesizer would be able to test a variety of possible preparation steps and, on the basis of programmed empirical yields and reaction conditions, decide upon the optimum one. It would seem that such an instrument could certainly remove much tedium and, if well conceived, could be expected to find new preparations and, when taxed to its limit, new compounds.

Another tool we have thought about is a "chemistry data-reduction system." This would link, on an interactive basis, mathematical steps and techniques commonly used by the experimental chemist, such as least-squares fitting, spectral analysis, and error analysis, into a fluidly manipulatable system. A "kinetics data analyzer" is also suggested, which would permit the efficient determination of rate constants from experimental data by interactively exploring and fitting the kinetics data with different parameters defining the possibly relevant differential equations. The area of such application is myriad and, of course, we finally seek the orchestration of our theoretical semiempirical and empirical numerical apparatus into a complete chemistry system.

1b. Larger copies of all illustrations are available from the author upon request.

2. Recently much has been written elsewhere about the development and present state of the MO method; therefore we shall confine ourselves to a brief statement of crucial steps in its history.

The MO method as introduced by Mulliken[8,10] and Hund[11] was used extensively in the semiempirical interpretation of band spectra. However, mathematically and computationally the concept matured rather slowly. Its early development (and the search for the "best" MO's) may be traced from the recognition by Lennard-Jones[12] of its relationship to Hartree's[13] SCF work on atoms followed by the introduction of the determinantal form for the wave function[14,15] with the application of the variational principle (Fock[16] and Slater[17,18]) to yield the now familiar pseudo-eigenvalue equations of the form

$$F\phi_i = \epsilon_i\phi_i$$

known as the Hartree-Fock equations, which provide a rigorous mathematical definition of best orbitals. Lennard-Jones[12] presented the equations for an arbitrary system; Coulson foreshadowed their solution by the expansion method; and Roothaan[20,21] developed and perfected the extensively used matrix formulation of the expansion method. Important also are the proofs by Delbrük,[22] Löwdin,[23] and Roothaan[21] that the Hartree-Fock functions are always self-consistent, symmetry-adapted, and correspond to a specific minimum of the total energy. Extremely relevant to the potency and appeal of Hartree-Fock wave functions was the work of Brillouin,[24,25] Moller and Plesset,[26] Sinanoğlu,[102] and others, on corrections to the Hartree-Fock approximation. They showed that one-electron properties computed from Hartree-Fock wave functions have first-order corrections in perturbation theory which vanish provided degeneracy is not present. Koopman[27] developed similar theorems for ionization potentials.

Having clearly defined the Hartree-Fock model of a molecular system it still remained a formidable practical problem to obtain the MO's ϕ_i. In 1951, Roothaan had cast the Hartree-Fock equations into a solid computational framework remarkably suitable for the then-embryonic digital computers. In what is now referred to as the Roothaan[20,21] method, the orbital ϕ_i is expanded in terms of some suitable truncated basis set χ_p,

$$\phi_i = \sum_p c_{ip}\chi_p$$

The expansion coefficients c_{ip} are optimized through the iterative SCF process.[28] In the full numerology of the process the best truncated set of basis functions is also hunted down, usually by brute-force methods. In practice a very close approximation to the Hartree-Fock molecular orbitals can be obtained in this way. Calculations of this type, utilizing analysis and computer programs developed recently,[29,30] have resulted in the determination of the molecular orbitals for a large number of diatomic molecules. These functions, in which the basis set χ_p consists of many Slater-type orbitals (STO's), are very close to the Hartree-Fock result. Such wave functions form the basis for the SCF results presented in this article.

3. P.-O. Löwdin and B. Pullman, *Molecular Orbitals in Chemistry, Physics, and Biology*, Academic Press, New York, 1964.
4. J. C. Slater, *J. Chem. Phys.* **43,** S11 (1965).
5. R. K. Nesbit, *J. Chem. Phys.* **43,** 530 (1965).
6. A. C. Wahl, P. J. Bertoncini, G. Das, and T. L. Gilbert, *Intern. J. Quantum Chem.* **1S,** 123 (1967).
7. R. K. Nesbet, in *Advances in Quantum Chemistry* Vol. 3, P.-O. Löwdin, ed., Academic Press, New York, 1967.
8. R. S. Mulliken, *Phys. Rev.* **32,** 186, 388, 761 (1928).
9. R. S. Mulliken, *Phys. Rev.* **33,** 730 (1929).
10. R. S. Mulliken, *Phys. Rev.* **40,** 55; **41,** 49, 751 (1932).
11. F. Hund, *Z. Physik* **51,** 759 (1928).
12. J. E. Lennard-Jones, *Trans. Faraday Soc.* **25,** 668 (1929).
13. D. R. Hartree, *Proc. Cambridge Phil. Soc.* **24,** 89 (1928).
14. J. C. Slater, *Phys. Rev.* **35,** 509 (1930).
15. J. C. Slater, *Phys. Rev.* **41,** 255 (1932).
16. V. Fock, *Z. Physik* **61,** 126 (1930).
17. J. C. Slater, *Phys. Rev.* **34,** 1929 (1929).
18. J. C. Slater, *Phys. Rev.* **35,** 509 (1930).
19. C. A. Coulson, *Proc. Cambridge Phil. Soc.* **34,** 204 (1938).
20. C. C. J. Roothaan, *Rev. Mod. Phys.* **23,** 69 (1951).
21. C. C. J. Roothaan, *Rev. Mod. Phys.* **32,** 179 (1960).
22. M. Delbruck, *Proc Roy. Soc. (London)* **A129,** 686 (1930).
23. P.-O. Löwdin, *J. Appl. Phys. (Suppl.)* **33,** 270 (1962).
24. L. Brillouin, *Actualities Sci. Ind.* **71,** (1933).
25. L. Brillouin, *Actualities Sci. Ind.* **159** (1934).
26. C. Moller and M. S. Plesset, (1934). *Phys. Rev.* **46,** 618 (1934).
27. T. Koopman, *Physica* **1,** 104 (1933).
28. C. C. J. Roothaan and P. Bagus, in, *Methods in Computational Physics*, Vol. 2, B. Alder, S. Fernbach, and M. Rotenberg, eds., Academic Press, New York, 1963, p. 47.
29. A. C. Wahl, *J. Chem. Phys.* **43,** 2600 (1964); A. C. Wahl, P. E. Cade, and C. C. J. Roothaan, *J. Chem. Phys.* **41,** 2578 (1964); A. C. Wahl, P. J. Bertocini, K. Kaiser, and R. H. Land, *Argonne National Laboratory Rept. 7271*, 1968.
30. A. C. Wahl, *Argonne National Laboratory Rept. 7076*; also *Science* **151,** 961 (1966), and F. Janiszewski and A. C. Wahl (in preparation).
31. A. C. Wahl and U. Blukis, *Atoms to Molecules* (film series), McGraw-Hill, New York, 1969; A. C. Wahl and M. T. Wahl, *Four Wall Charts of Atomic and Molecular Structure*, McGraw-Hill, New York, 1969; *J. Chem. Ed.* **45,** 787 (1968).
32. A. C. Wahl, P. E. Cade, and K. S. Sales, unpublished.
33. W. Zemke, P. Lykos, and A. C. Wahl, Ph.D. Thesis, Joint IIT-ANL and ANL Report.
34. E. Clementi, *J. Chem. Phys.* **36,** 33 (1962).
35. R. K. Nesbet, *J. Chem Phys.* **36,** 1518 (1962).
36. S. L. Kahalas and R. K. Nesbet, *J. Chem. Phys.* **39,** 259 (1963).
37. A. D. McLean, *J. Chem. Phys.* **39,** 2653, (1963).
38. M. Yoshimine, *J. Chem. Phys.* **40,** 2970 (1964).
39. R. K. Nesbet, *J. Chem. Phys.* **41,** 100 (1964).
40. A. C. Wahl, *J. Chem. Phys.* **41,** 2600 (1964).
41. W. Huo, *J. Chem. Phys.* **43,** 624 (1965); *J. Chem. Phys.* **45,** 1554 (1966).
42. P. E. Cade, K. S. Sales, and A. C. Wahl, *J. Chem. Phys.* **44,** 1973 (1966).
43. W. Huo, to appear in *J. Chem. Phys.*
44. W. Huo, to appear in *J. Chem. Phys.*
45. P. E. Cade and W. Huo, *J. Chem. Phys.* **47,** 614 (1967).
46. P. E. Cade and W. Huo, *J. Chem. Phys.* **47,** 619 (1967).
47. T. L. Gilbert and A. C. Wahl, *J. Chem. Phys,* **47,** 3425 (1967).
48. T. L. Gilbert and A. C. Wahl, to be submitted to *J. Chem. Phys.*
49. T. L. Gilbert and A. C. Wahl, in preparation.
50. A. D. McLean and M. Toshimine, *Tables of Linear Molecule Wave Functions* (IBM Research Publication, San Jose, Calif., 1967); *Intern. J. Quantum Chem.* **1S,** 313 (1967).

51. P. E. Cade, J. B. Greenshields, G. Malli, K. D. Sales, and A. C. Wahl, unpublished work.
52. P. J. Bertoncini, T. L. Gilbert, and A. C. Wahl, unpublished work.
53. P. E. Cade, unpublished work.
54. P. J. Bertoncini and A. C. Wahl, unpublished work.
55. R. L. Matcha and R. K. Nesbet, *IBM Research Paper RJ-429,* to be published.
56. S. Peyerimhoff, *J. Chem. Phys.* **43,** 998 (1965).
57. K. D. Carlson and C. M. Moser, *J. Chem. Phys.* **67,** 2644 (1963).
58. K. D. Carlson and C. M. Moser, *J. Chem. Phys.* **44,** 3295 (1966).
59. W. G. Richards, G. Verhagen, and C. M. Moser, *J. Chem. Phys.* **45,** 3226 (1966).
60. G. Verhaegen, W. G. Richards, and C. M. Moser, *J. Chem. Phys.* **46,** 160; (1967).
61. K. D. Carlson, E. Ludena, and C. M. Moser, *J. Chem. Phys.* **43,** 2408 (1965).
62. K. D. Carlson and R. K. Nesbet, *J. Chem. Phys.* **41,** 1051 (1964).
63. K. D. Carlson, C. M. Moser, K. Kaiser, and A. C. Wahl, CaO, *J. Chem. Phys.* (in press).
64. P. Maldonado and A. C. Wahl, A Priori Calculations on LiHe, LiNe, NaHe, NaNe (in progress).
65. Substantial progress beyond the Hartree-Fock scheme for molecules has been achieved by a variety of approaches. Direct inclusion of correlation in the wave function is illustrated by the classic work of Kolos and Roothaan[41] and later Kolos and Wolniewicz,[67,69] where the ground and several excited states of H_2 have been computed to an accuracy which matches experiment. However, it is unlikely that this approach can be extended to larger molecules.

Other more practical methods for going beyond the quantitative limits of the Hartree-Fock approximation include a variety of configuration interaction methods, exemplified by the work of Matsen and co-workers, Browne, Boys and co-workers, Nesbet, Mosser, Harris, Taylor, Michels, Davidson, McLean, Allen, and others.[70-100] One of the most useful extensions, which has been developed to the point where reliable predictions[101] can be made on a wide variety of molecules, is the many-electron theory of Sinanoğlu.[102] We regard these various methods, and the optimized valence configuration method described in this article, as complementary rather than competitive. Further calculations are needed to ascertain whether any one scheme will turn out to be so simple, general, and powerful that it supersedes all others.

There are at least three general approaches to the problem of going beyond the Hartree-Fock approximation: perturbation theory, correlated wave functions, and configuration interaction. If one chooses the third approach, as we have done, then one is immediately faced with the problem of the slow convergence of the multiconfiguration expansion of a many-electron wave function. At this point one must make a choice between trying to cope with very large numbers of configurations and improving the convergence. Although the former choice is now feasible with the large computers available (and one can always extract simple quantities from a complicated wave function and leave everything else out of sight in the machine), it seemed to us that one would be more likely to be able to comprehend the steps of the calculation and find unexpected relations and regularities by working directly with the most convergent expansion obtainable. If one constructs all configurations from a common set of orthonormal orbitals, then this reasoning leads immediately to variational equations for a finite set of orbitals which must be solved simultaneously with the secular equation for the configuration expansion coefficients.
These equations are sometimes referred to as the extended Hartree-Fock equations,[103-105,107-121] although this name is also used for a more restricted set of equations obtained by projecting a wave function from a single determinant before varying the orbitals.[117] In the general case the orbitals which satisfy the extended Hartree-Fock equations are similar to, but not identical with, the natural spin orbitals[117,122-124] of the system. This is essentially what we would like to evaluate in an a priori way, for the natural-orbital analyses of atoms and molecules have shown that natural orbitals provide a very compact and informative way of characterizing various wave functions.[125-136]

The extended Hartree-Fock method in its most general form is still rather complicated for the first step beyond the Hartree-Fock approximation, and perhaps more complicated than is needed. Hence, a simpler scheme, the optimized valence configuration (OVC) method, has been proposed by Das and Wahl and used successfully for practical calculations on a number of diatomic molecules.[6,103-105,137,138]
66. W. Kolos and C. C. J. Roothaan, *Rev. Mod. Phys.* **32,** 205 (1960).
67. W. Kolos and L. Wolniewicz, *Rev. Mod. Phys.* **35,** 473 (1963).
68. W. Kolos and L. Wolniewicz, *J. Chem. Phys.* **45,** 509 (1966).
69. W. Kolos and L. Wolniewicz, *J. Chem. Phys.* **41,** 3663 (1964).
70. S. Rothenberg and E. R. Davidson, *J. Chem. Phys.* **44,** 730 (1966).
71. W. M. Wright and E. R. Davidson, *J. Chem. Phys.* **43,** 840 (1965).
72. J. C. Browne, *J. Chem. Phys.* **40,** 43 (1964).

73. J. C. Browne, *J. Chem. Phys.* **41,** 1583 (1964).
74. E. R. Davidson, *J. Chem. Phys.* **35,** 1189 (1961).
75. E. R. Davidson, *J. Chem. Phys.* **33,** 1577 (1960).
76. W. Kolos and C. C. J. Roothaan, *Rev. Mod. Phys.* **32,** 219 (1960).
77. A. D. McLean, A. Weiss, and M. Yoshimine, *Rev. Mod. Phys.* **32,** 211 (1960).
78. J. C. Browne, *J. Chem. Phys.* **42,** 1428 (1965).
79. S. Fraga and B. J. Ransil, *J. Chem. Phys.* **37,** 1112, (1962).
80. B. G. Anex, *J. Chem. Phys.* **38,** 1651 (1963).
81. F. E. Harris, *J. Chem. Phys.* **44,** 3636, (1966).
82. P. N. Reagen, J. C. Browne, and F. A. Matsen, *Phys. Rev.* **132,** A304 (1963).
83. J. F. Browne, *J. Chem. Phys.* **45,** 2709 (1966).
84. H. H. Michels and F. E. Harris, *J. Chem. Phys.* **39,** 1464 (1963).
85. C. F. Bender and E. R. Davidson, *J. Phys. Chem.* **70,** 2675 (1966).
86. H. S. Taylor and F. E. Harris, *Mol. Phys.* **7,** 287 (1964).
87. J. C. Browne, *J. Chem. Phys.* **41,** 3495 (1964).
88. R. D. Poshusta and F. A. Matsen, *Phys. Rev.* **132,** A307 (1953).
89. P. E. Phillipson, *Phys. Rev.* **125,** 1981 (1962).
90. H. S. Taylor and F. E. Harris, *Mol. Phys.* **1,** 287 (1964).
91. J. C. Browne, *J. Chem. Phys.* **42,** 2826 (1965).
92. D. R. Scott, E. M. Greenwalt, J. C. Browne, and F. A. Matsen, *J. Chem. Phys.* **44,** 2981 (1966).
93. J. C. Browne, *Phys. Rev.* **138,** A9 (1965).
94. J. C. Browne and F. A. Matsen, *Phys. Rev.* **135,** A1227 (1964).
95. D. D. Ebbing, *J. Chem. Phys.* **36,** 1361 (1962).
96. J. C. Browne, *J. Chem. Phys.* **36,** 1814 (1962).
97. F. E. Harris and H. S. Taylor, *Physica* **30,** 105 (1964).
98. P. L. Moore, J. C. Browne, and F. A. Matsen, *J. Chem. Phys.* **43,** 903 (1965).
99. R. F. Fourgere and R. K. Nesbet, *J. Chem. Phys.* **44,** 285, (1966).
100. L. C. Allen, *Quantum Theory of Atoms, Molecules and the Solid State,* Academic Press, New York, 1966.
101. C. Hollister and O. Sinanoğlu, *J. Am. Chem. Soc.* **88,** 13 (1966).
102. O. Sinanoğlu, *Advan. Chem. Phys.* **2** (1964); see also *Proc Roy. Soc. (London)* **A260,** 379 (1961) and *Ann. Rev. Phys. Chem.* **15,** 251 (1964).
103. G. Das and Arnold C. Wahl, *J. Chem. Phys.* **44,** 87 (1966).
104. G. Das, *J. Chem. Phys.* **46,** 1568 (1967).
105. G. Das and Arnold C. Wahl, *J. Chem. Phys.* **47,** 2934 (1967).
106. Refs. 107 through 117 present the historical development of the formation of best orbitals for a multi-configurational wave function.
107. J. Frenkel, *Wave Mechanics, Advanced General Theory,* Oxford University Press, New York, 1934; reprinted by Dover, New York, 1950, pp. 460–462.
108. J. C. Slater, *Phys. Rev.* **91,** 528 (1953).
109. P.-O. Löwdin, *Phys. Rev.* **97,** 1497 (1955), Sec. 5.
110. (a) R. McWeeny, *Proc. Roy. Soc. (London),* **A232,** 114 (1955); (b) R. McWeeny and R. Steiner, *Advan. Quantum Chem.* **2,** 93 (1965).
111. D. R. Hartree, W. Hartree, and B. Swirles, *Phil. Trans. Roy. Soc. (London),* **A238,** 292 (1939).
112. A. P. Yutsis, *Zh. Eksperim. i Tenet. Fiz.* **23,** 129 (1952).
113. A. P. Yutsis, V. V. Kibartas, and I. I. Glembotskiy, *Zh. Eksperim. i Theo. Fiz.* **24,** 425 (1954).
114. V. V. Kibartas, V. I. Kauetskis, and A. P. Yutsis, *Soviet Phys. JETP* **2g,** G23 (1955).
115. R. E. Watson, *Ann. Phys.* **13,** 250 (1961).
116. T. L. Gilbert, *J. Chem. Phys.* **43,** S248 (1965).
117. P.-O. Löwdin, *Phys. Rev.* **97,** 1509 (1955).
118. A. Veillard, *Theoret. Chim. Acta* **4,** 22 (1966).
119. E. Clementi and A. Veillard, *J. Chem. Phys.* **44,** 3050 (1966); *Theoret. Chim. Acta* **7,** 133 (1967).
120. E. Clementi, *J. Chem. Phys.* **46,** 3842 (1967).
121. J. Hinze and Clemens C. J. Roothaan, to be published.
122. W. Kutzelnigg, *J. Chem. Phys.* **40,** 3640 (1964).
123. P.-O. Löwdin, *Advan. Chem. Phys.* **2** (1959).
124. H. Shull and P.-O. Löwdin, *J. Chem. Phys.* **30,** 617 (1959).
125. S. Rothenberg and E. R. Davidson, *J. Chem. Phys.* **45,** 2560 (1966).
126. C. F. Bender and E. R. Davidson, *J. Chem. Phys.* **70,** 2675 (1966).
127. E. R. Davidson and L. L. Jones, *J. Chem. Phys.* **37,** 1918 (1962).

128. E. R. Davidson and L. L. Jones, *J. Chem. Phys.* **37,** 2966 (1962).
129. M. A. Eliason and J. O. Hirschfelder, *J. Chem. Phys.* **30,** 1397 (1959).
130. H. Shull, *J. Chem. Phys.* **30,** 1405 (1959).
131. S. Hagstrom and H. Shull, *Rev. Mod. Phys.* **35,** 624 (1963).
132. H. Shull and F. Prosser, *J. Chem. Phys.* **40,** 233 (1964).
133. C. Edmiston and M. Krauss, *J. Chem. Phys.* **45,** 1833 (1966).
134. C. F. Bender and E. R. Davidson, *J. Chem. Phys.* **70,** 2675 (1966).
135. D. D. Ebbing and R. C. Henderson, *J. Chem. Phys.* **42,** 2225 (1965).
136. W. Lyon and J. O. Hirschfelder, *J. Chem. Phys.* **46,** 1788 (1967).
137. G. Das and A. C. Wahl, Optimized Valence Configuration Wavefunctions for NaF and F_2, in preparation.
138. P. Bertoncini, G. Das, and A. C. Wahl, A Theoretical Study of the NaLi System, *J. Chem. Phys.* (in press).
139. See, for example, J. O. Hirschfelder, C. F. Curtiss, and R. B. Bird, *Molecular Theory of Gases and Liquids*, Wiley, New York, 1954.
140. O. Sinanoğlu and K. S. Pitzer, *J. Chem. Phys.* **31,** 960 (1959); R. H. Davis, E. A. Mason, and R. J. Munn, *Phys. Fluids* **8,** 444 (1965); F. Jenc, *J. Mol. Spec.* **24,** 284 (1967); D. Miller, *Argonne National Laboratory Rept. 7210*, 280, 1966.
141. D. Miller, P. J. Bertoncini, and A. C. Wahl, *Comparison of Ab Initio and Semiempirical Interatomic Potentials in the Calculation of the Transport Properties of Alkali Vapors*, in preparation.
142. R. S. Mulliken, *Science* **157,** 13 (1967) (Nobel Prize lecture), and *Physics Today* **21,** 52 (1968).
143. A. C. Wahl, P. J. Bertoncini, K. Kaiser, and R. H. Land, BISON: A New Instrument for the Experimentalist, *Intern. J. Quantum Chem.* (Sanibel Symposium Issue, 1969).
144. A. C. Wahl, P. J. Bertoncini, K. Kaiser, and R. H. Land, BISON: A Fortran Computing System for the Calculation of Wavefunctions, Properties and Charge Densities for Diatomic Molecules, *Argonne National Laboratory Rept. 7271*, 1968.
145. S. Rottenberg, *MOLE: A System for Quantum Chemistry* (available from the author).
146. L. M. Sachs and M. Geller, *Intern. J. Quantum Chem.* **1S,** 445 (1967).
147. A. Clementi and A. Veillard, *IBMOL: Computation of Wave-function for Molecules of General Geometry,* IBM Research Laboratories, San Jose, Calif.
148. T. D. Bouman, A. L. H. Chung, and G. L. Goodman, Chap. VII-3, this volume.
149. We chemists are also becoming aware of parallel developments in other disciplines. An outstanding example of this is the work of Stanley Cohen, of the Argonne Physics Division, on interactive computing and its application to nuclear structure problems.

VII-3

Automation of Molecular Point-Group Theory

T. D. BOUMAN, A. L. H. CHUNG, AND G. L. GOODMAN

1. Introduction

The theorems and techniques of group theory as applied to molecules with symmetry have long been recognized as a powerful tool for the theoretical chemist.[1-3] Use of symmetry-based selection rules, for example, has shed much light on the interpretation of atomic and molecular spectra[4,5] and has led to many simplifications in quantum mechanical calculations.[6] Methods developed by Racah[7,8] for atomic spectroscopy and extended to molecules by Tanabe and coworkers[9,10] and by Griffith[11,12] have made possible the determination of quantitative as well as qualitative relationships from the symmetry properties of a problem.

Until fairly recently, however, molecular symmetry properties have been generally ignored in organizing problems in molecular quantum mechanics for solution on large, high-speed digital computers. The resulting type of procedure, in which one allows the computer to calculate and manipulate all quantities in the same way, even those which group theory would say must vanish, may be termed a "brute-force" approach. The tendency to use this type of approach arises because of difficulties in translating the nonnumeric operations and manipulations of molecular point-group theory to a form that can be conveniently handled on a digital computer. Nevertheless, quantum chemists have recently come increasingly to appreciate again that an investigator who ignores symmetry not only gives up an important means of alleviating computational bottlenecks but foregoes insights which symmetry can give to the results of a large calculation. Thus increased attention is currently being paid to techniques for implementing symmetry aspects for machine calculations.

Several different lines of attack have been followed in efforts to take account of symmetry in designing machine programs for quantum chemistry; a few examples, which are by no means exhaustive, will serve to illustrate them. One method is to design a computer program around the specific symmetry properties of a particular point group and thus take full advantage of the qualitative simplifications occurring in a restricted class of molecules. Examples of this approach are found in the work of Joshi[13] on hydrides with C_{3v} symmetry, that of Wahl[14] on homonuclear diatomic molecules (symmetry $D_{\infty h}$), and that of McLean and Yoshimine[15] on linear molecules in general.

A second category includes computer programs which are designed to accommodate the simplifications arising in a wider class of point groups but which require the specification of various amounts of group-theoretical information as input. The IBMOL program system of Clementi and Davis,[16] for example, makes use of selection rules in eliminating the handling of zeros; however, the user must provide the program with a symmetry-adapted basis set and certain indices characterizing the point group. The POLYATOM system[17] generates lists of nonvanishing integrals and useful identities among them. Here the user must specify the effects of certain symmetry operations in terms of permutations of the atomic basis functions. Symmetry- and spin-adapted Slater determinantal configuration functions for this same class of molecules may be generated automatically by the projection-operator method of Gershgorn and Shavitt,[18] if a symmetry-adapted atomic basis set is provided. A semiempirical calculation organized for symmetry has been programmed by Glarum,[19] who requires the user to supply the irreducible representation matrices for the point group under consideration. Again, no degeneracies higher than 2 are permitted. A final example in this category is the automation by several investigators of problems in molecular vibration and rotation.[20] The methods have in common the requirement that the user provide, among other things, the transformation from internal to symmetry coördinates.

Recently, several attempts have been made to generate the required group-theoretical information on the computer. Possibly the first such method to be implemented was developed by Flodmark and Blokker.[21] These authors concentrate on the detailed properties of the group itself, such as the actual irreducible representation matrices, and generate symmetry-adapted functions with an automated version of the same techniques and formalism one employs in discussing the symmetry aspects of such a problem in a group-theory textbook. The user must supply the program with the multiplication table of the group. Gabriel[22,23] has developed and is implementing a computer-oriented method for finding symmetry-adapted functions which avoids explicit use of the representation matrices but makes use of the generators for the group and their commutator algebra.

An approach which is much closer in spirit to the ones we have implemented has been proposed by Moccia.[24] Rather than working with the symmetry-group operations and irreducible representations themselves, he uses group-theoretical theorems which basis functions and operators must satisfy if they are to be symmetry adapted. Standard techniques of calculating quantum mechanical integrals and diagonalizing matrices are employed to generate a symmetrized basis set in a numerical way well suited to the capabilities of a digital computer. Moccia's method, however, is dependent on the assumption that the unsymmetrized atomic basis functions comprise sets of "equivalent orbitals"[25] which are permuted among themselves by the symmetry operations of the point group. Thus a pretransformation of the atomic basis functions into spatially equivalent orbitals must be performed before the method can be used; this may be a nontrivial task, especially if extended basis sets are employed. Furthermore, the construction of spatially equivalent orbitals in some cases requires hybridization of the atomic basis functions; this involves assumptions about the nature of the basis orbitals which go beyond those strictly determined by symmetry.

To obtain symmetry coordinates for their work on vibrational spectra of polymers and other molecules, Gussoni and Zerbi[26] have implemented a method similar in spirit to our method 1 described below. By setting up the G matrix for a set of internal nuclear displacement coordinates and diagonalizing it, a transformation is produced that takes the original internal coordinates into symmetry coordinates.

Our present work is motivated in part by the desire to develop an efficient and systematic way of introducing symmetry into our planned large-scale computing efforts, and also by the wish to explore new algorithms for solving standard problems—algorithms suggested by the numerical capabilities of digital computers and their associated scientific programming languages. If such a method is to be widely used, it should fulfill the requirements of speed, generality, simplicity of input, and perhaps also transferability from one type of problem to another.

In the following sections, we describe two methods—similar in some respects—of generating group-theoretic information on a machine in a form which is readily usable for subsequent computation. As we show, it is, in fact, possible to obtain virtually all relevant symmetry information, both qualitative and quantitative, for a quantum mechanical problem without recourse to any standard textbook representation for, or explicit consideration of, the elements forming the point group. We adopt the point of view that what really interests us is not rotations, reflections, etc., or group multiplication tables and irreducible representation matrices, or even how functions transform under the symmetry operations. We are interested rather only in numerical relationships, such as whether a matrix element is zero or nonzero, and what ratios exist between nonzero elements. Hence we work with matrix elements from the start.

In the methods to be described, we rely substantially on the following theorem in group theory, which we state without proof:

Theorem 1. The eigenvectors of a matrix, which has a structure embodying the full symmetry of the particular nuclear geometry, form bases for irreducible representations of the point group of the molecule, provided that there are no accidental degeneracies in the eigenvalues.

Two ways in which a matrix having the structure referred to in the theorem may be formed are as follows. First, it may be the matrix of a rotationally invariant operator evaluated over a set of functions that includes at least one function centered on each atom in the molecule and involves the same type of rotationally complete basis set on each symmetry-equivalent atom. Second, it may be the matrix of an operator that itself contains the symmetry of the nuclear framework, in which case the former of the two restrictions on the basis set given above is relaxed, and only the latter must hold.

We have implemented two distinct programs following, respectively, the approaches just outlined. In the first, the irreducible representations spanned by the given atomic basis set are characterized by studying the matrices for two spherically symmetric operators, the overlap and kinetic energy, which occur in most MO calculations. A symmetry-adapted basis set is generated in this way. In the second method, two artificial operators totally symmetric with respect to the nuclear framework are constructed, and a set of functions is chosen to span *all* the symmetry species occurring for the point group.[27] These functions then comprise a standard set against which the symmetry aspects of any other functions can be aligned, and with which other group-theoretic information can be obtained.

The first method is described in detail in Section 2. Section 3 describes the construction of the "standard set" required in the second method, and its application to symmetry adapting an AO basis set is shown in Section 4. In Section 5 the standard set is used to obtain quantitative group-theoretical information, and the two approaches are summarized and discussed in Section 6.

2. Method 1. Symmetry Adaptation of an Atomic Orbital Basis Set through Spherically Symmetric Operators

We begin this section with a brief review of representation theory. Let $\boldsymbol{\phi}$ be a set of functions $\phi_1, \phi_2, \ldots$ associated in some way with a molecule belonging to a point group G. We assume that $\boldsymbol{\phi}$ is sufficiently complete so that any symmetry operation $\underline{R}$ of G transforms each of the functions, ϕ_j say, into a linear combination of the other functions in the set:

$$\phi_j' \equiv \underline{R}\phi_j = \sum_k \phi_k A_{kj}(\underline{R}) \tag{1}$$

or, in matrix-vector notation,

$$\boldsymbol{\phi}' = \boldsymbol{\phi}\mathbf{A}(\underline{R}) \tag{2}$$

Thus, each such operation $\underline{R}$ has a matrix $\mathbf{A}(\underline{R})$ associated with it. One can easily show (see, e.g., Ref. 6) that the set of matrices $\mathbf{A}(\underline{R})$ forms a *representation* of the group G. Although this representation is in general not the simplest possible, we can find a reduction matrix $\mathbf{U}$, corresponding to a unitary transformation of ϕ that generates a new set of representation matrices $\mathbf{B}(\underline{R})$, so these factor into the simplest possible or *irreducible representations* of G:

$$\mathbf{B}(\underline{R}) = \mathbf{U}^{-1}\mathbf{A}(\underline{R})\mathbf{U} \tag{3}$$

This same transformation applied to the functions ϕ converts them to a set of functions ψ which form bases for irreducible representations of G.

$$\psi = \phi\mathbf{U} \tag{4}$$

This new set of functions is said to be *symmetry adapted* to the group G, and satisfies all the usual symmetry selection rules and other relationships.

Our interest therefore centers on the determination of the reduction matrix **U** for a given set of functions, in a computer-oriented way. We identify three major stages in our procedures. First, the basis set is transformed to a set decomposed according to definite symmetry types; second, particularly simple functions are extracted from these initial symmetrized functions; and, third, a standard phase choice is imposed to yield canonical symmetrized functions.

Let **X** be the set of all AO basis functions to be used in a subsequent calculation; a typical member is $\chi^{A}_{n\ell m}(r_A, \theta_A, \phi_A)$, whose quantum numbers are n, ℓ, and m and whose origin is at atom A. From **X** we may extract a smaller set χ, which contains all values of the angular indices, ℓ and m, represented in **X**, but with only one representative radial part for each ℓ value within a class of symmetry-related atoms, or *atom type.*[28] This extraction involves no loss of information, because no point-group operation changes the radial part of a function (except for a change of origin) and therefore functions which differ only in their radial parts have the same transformation coefficients associated with them. The final set of symmetry-adapted functions from χ, which are to be used in a quantum mechanical calculation, are denoted σ:

$$\sigma = \chi\mathbf{U} \tag{5}$$

By Theorem 1 of Section 1, a unitary transformation $\mathbf{W}'$ which diagonalizes a totally symmetric matrix **E**,

$$\mathbf{W}'^{-1}\mathbf{E}\mathbf{W}' = \mathbf{E}_D \tag{6}$$

produces eigenvectors which belong to definite symmetry types. In our method 1 we form a totally symmetric matrix satisfying Theorem 1 by computing the overlaps between all pairs of atomic functions in χ:

$$S_{\mu\nu} = \langle\chi_\mu | \chi_\nu\rangle \tag{7}$$

The rotationally invariant operator used here is just the unit, or identity, operator. We diagonalize the matrix **S** by means of a suitable transformation $\mathbf{W}'$, and form a new set of functions,

$$\phi = \chi\mathbf{W}' \tag{8}$$

These new functions ϕ do already transform irreducibly, but they are not yet in a sufficiently well-defined form to be suitable for subsequent applications.

In particular, we must first find out how many different *symmetry species* are spanned by the set ϕ, and identify which functions belong to which species. To do this, we sort the ϕ first according to the degeneracies of their associated eigenvalues. We still must distinguish among species with the same degeneracy but different symmetry behavior. For this purpose we choose a second spherically symmetric operator, the kinetic energy $-\frac{1}{2}\nabla^2$, and calculate the matrix **K**,

$$K_{\mu\nu} = \langle\chi_\mu | -\tfrac{1}{2}\nabla^2 | \chi_\nu\rangle \tag{9}$$

In general, the matrix **W**′ which diagonalized **S** does not completely diagonalize **K**, but produces a matrix,

$$\mathbf{W}'^{-1}\mathbf{K}\mathbf{W}' = \mathbf{M} \tag{10}$$

having the structure shown in Fig. 1. The nonzero off-diagonal elements of **M** connect functions ϕ_i belonging to the same symmetry species, so we can now assign a unique symmetry label to each function.

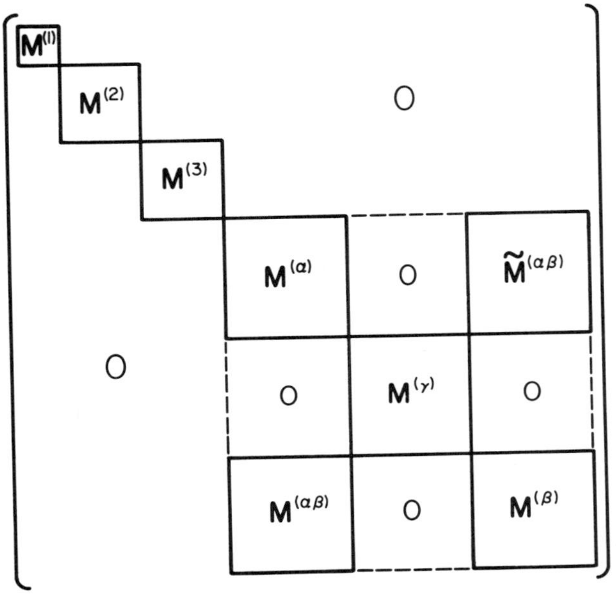

Fig. 1. Structure of matrix **M** from equation (10), showing off-diagonal blocks connecting sets of functions belonging to the same symmetry species. The blocks on the diagonal are themselves diagonal, with equal diagonal elements within a block.

At this stage, however, a given symmetrized function ϕ_i contains, in general, contributions from each of the original atomic orbitals in χ. A moment's thought shows that, as ϕ_i is subjected to the various symmetry operations of G, certain subsets of the constituent functions χ transform only among themselves and never into members of another subset. In particular, no rotation or reflection changes the quantum numbers n or ℓ of a function, nor can it place the function on a new nucleus which is not related by symmetry to the initial nucleus. We denote a subset with a given ℓ value and atom-type τ by $\chi^{\ell\tau}$; the subset contains $(2\ell + 1) \times n_\tau$ basis orbitals, where n_τ is the number of atoms in τ. The symmetry-adapted functions we desire contain components only within a given subset; we therefore dissect the functions ϕ into separate subsets and renormalize each fragment to obtain new functions ψ. Formally, this corresponds to a unitary transformation **A**:

$$\psi = \phi\mathbf{A} = \chi\mathbf{W}'\mathbf{A} = \chi\mathbf{W} \tag{11}$$

but it is not necessary to evaluate **A** explicitly. Schmidt orthogonalization may be necessary in some cases to generate enough linearly independent functions ψ for a given symmetry species.

Finally, we want all occurrences of a given degenerate symmetry species in ψ to have the same phase and axis choice. We may ensure that all such sets have parallel axis systems by means of a unitary transformation **T** on ψ. The determination of **T** is described in the Appendix.

The resulting functions σ',

$$\sigma' = \psi \mathbf{T} = \chi \mathbf{W} \mathbf{T} = \chi \mathbf{U}' \tag{12}$$

have all the symmetry properties necessary to utilize selection rules and to induce the maximum factorization of matrices in which they appear. However, the phases of the degenerate symmetry species, while internally consistent, bear no simple relationship to the master coordinate system or principal axis system of the molecule. Furthermore, two or more representatives of a given symmetry species derived from the same orbital subset $\chi^{\ell\tau}$ mix together in a way determined by the operators we have used and hence in a way that bears no interpretive significance.

The way these problems are solved is discussed in full detail elsewhere.[29] In brief, the former problem–that of reorienting the phases of the degenerate representations–is handled by constructing artificial operators which strongly emphasize the master coordinate axes, and whose eigenvectors determine the transformation which rotates phases into a simple relationship to the master axes. The second task, "undoing" an arbitrary mixing of functions which have the same symmetry properties, is accomplished by a final transformation which maximizes the number of zeros in **U** while preserving the symmetry properties. These two steps are combined formally in the matrix **B**, giving

$$\sigma = \sigma' \mathbf{B} = \chi \mathbf{U}' \mathbf{B} = \chi \mathbf{U} \tag{13}$$

The set σ constitutes the set of symmetry-adapted basis functions on which computations may be performed.

3. Method 2. Generation of Standard Representatives for the Symmetry Species of the Finite Point Group

Method 2 shares many of the same concepts and algorithms as the approach just described, and in fact was developed as an outgrowth of it. These two approaches differ, not only in the practical ways outlined in Section 1 but also in some degree in their philosophy. In method 1 we accepted almost all the features of the given basis set, such as the particular radial dependences, as absolute and fixed; in method 2 we abstract even more from the detailed forms of any operators and functions those properties which are sufficient to determine their symmetry behavior.

We address ourselves first to the choice of an operator $\mathcal{P}$ which embodies the symmetry of any given nuclear geometry. The nuclear attraction and the full Hamiltonian for a molecule are examples of such operators which normally occur in a quantum mechanical calculation. We want to construct, however, an artificial one for which the matrix elements are almost trivial to compute, and which has no more and no less invariance under rotations and reflections than do these naturally occurring ones.

A very simple operator having these symmetry properties, while also avoiding all symmetry-dictated nodal surfaces of the various functions, is one which annihilates a function everywhere except at a small set of points S:

$$\mathcal{P}(\mathbf{r}) = \sum_{\nu \in S} \delta(\mathbf{r} - \mathbf{r}_\nu) \tag{14}$$

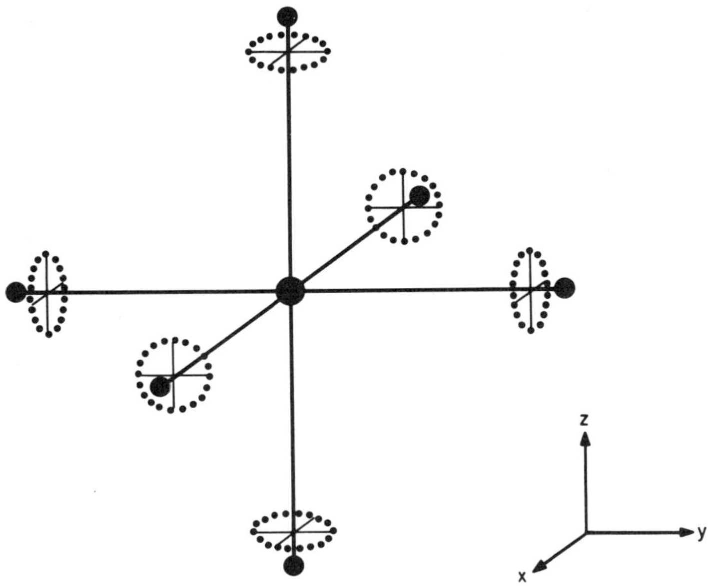

Fig. 2. Disposition of the set of points S at which the operator $\mathcal{P}$ of equation (14) is defined, shown for an octahedral MX_6 molecule.

the points being arranged in a series of circles about the lines joining each nucleus to the center of nuclear charge of the molecule as shown in Fig. 2. Evaluation of the matrix elements of $\mathcal{P}$ then reduces to summing the values of the integrand at the points in S.

This set of points defining $\mathcal{P}$ may be viewed either as one that goes identically into itself under all the operations of the point group G or as an approximation to a set of symmetrically disposed circles [where the summation in equation (14) is replaced by a sum of line integrals]. In practice, it appears that the two interpretations are equivalent, in that no accuracy is lost if the points do not exactly transform into one another. Transformation coefficients good to 11 significant figures are generally obtained with 24 equally spaced points on each circle, even in cases that correspond to a fivefold symmetry axis. Our choice of operator has the advantage not only of making the computations simple, but also of freeing as much as possible the determination of symmetry properties from other details of any problem in which they are to be used.

In choosing a "standard set" of functions to symmetry adapt, we require that each symmetry species occur at least once in the representation spanned by these functions. A particularly convenient choice is the set of 2L + 1 orbital angular momentum eigenfunctions, $Y_{LM}(\theta, \phi)$, of appropriately high L value. In a spherically symmetric environment these functions span a (2L + 1)-fold degenerate irreducible representation of the three-dimensional full rotation group SO(3). As the symmetry of the environment is lowered, this erstwhile irreducible representation splits into the various irreducible representations appropriate to the finite point group. By choosing L high enough, therefore, we are assured that each of the possible new representations will be spanned at least once. If a center of inversion is present, one, in fact, needs two distinct L values: an even L, L_g, for the representations which are even with respect to inversion, and an odd L, L_u, for those of odd parity. (The selection of the proper set of functions for any particular case of chemical significance, i.e., for any point group having no higher than an eightfold rotation axis, is built into the program. It can be shown that L = 16 is the highest one needs to span all the symmetry species which occur in these cases.)

For convenience of computation and ease of visualization we choose the combinations of the Y_{LM} which are real and for which the ϕ dependence goes as $\cos|M|\phi$ and $\sin|M|\phi$; this entails no loss of generality. We index the functions sequentially by the integer K, K = 1, 2, . . . , 2L + 1. Let this set be denoted $S_K^{(L)}$, or simply $\mathbf{S}^{(L)}$; a simple radial dependence, $R(r) = r^{-1}$, is assumed a part of the functions $\mathbf{S}^{(L)}$ to avoid accidental degeneracies. The functions are centered at the origin of the master coordinate system, which is chosen at the center of nuclear charge.

We compute the matrix of $\mathcal{P}$ over the set $\mathbf{S}^{(L)}$ and diagonalize it as in Section 2, by means of a unitary matrix **W**.[30] This transformation defines a symmetry-adapted set of functions $\mathbf{F}^{(L)}$ through

$$\mathbf{F}^{(L)} = \mathbf{S}^{(L)}\mathbf{W} \tag{15}$$

As in Section 2, we identify functions belonging to the same symmetry species by constructing a second totally symmetric operator $\mathcal{P}'$, of the type of equation (14) but defined on a different set of points S′. The matrix **W** does not completely diagonalize **P′**, formed from the set $\mathbf{S}^{(L)}$, but produces a matrix **M**,

$$\mathbf{W}^{-1}\mathbf{P}'\mathbf{W} = \mathbf{M} \tag{16}$$

whose nonzero off-diagonal elements connect functions of the same symmetry type (Fig. 1). If all sets of functions have the same phase, the individual off-diagonal blocks $\mathbf{M}^{(\alpha\beta)}$ are also diagonal, and we have achieved the maximum factorization in the matrix. The different occurrences of the same degenerate representation are aligned by a unitary transformation **T**, described in the Appendix. The functions $\psi^{(L)}$,

$$\psi^{(L)} = \mathbf{F}^{(L)}\mathbf{T} = \mathbf{S}^{(L)}\mathbf{W}\mathbf{T} = \mathbf{S}^{(L)}\mathbf{V} \tag{17}$$

comprise the standard set, called *prototype functions*, representing each possible type of symmetry behavior under the group G. A further phase transformation may be applied to the prototype functions to force a correspondence with particular sign conventions found in the literature, but this is incidental to our argument and will not be discussed further here.

As an example of the prototype functions and the standard phase choices we have implemented, we consider the octahedral group O_h. The irreducible representations of O_h and their degeneracies are indicated in Table 1. The lowest L values which span all the even and

Table 1. Symmetry Species of the Point Group O_h

Symbol		
Even parity	Odd parity	Degeneracy
A_{1g}	A_{1u}	1
A_{2g}	A_{2u}	1
E_g	E_u	2
T_{1g}	T_{1u}	3
T_{2g}	T_{2u}	3

odd parity representations are $L_g = 6$ and $L_u = 9$, respectively. If $\Gamma(L = 6)$ and $\Gamma(L = 9)$ are the corresponding representations, then under O_h symmetry they reduce as follows:

$$\begin{aligned} \Gamma(L = 6) &\longrightarrow A_{1g} + A_{2g} + E_g + T_{1g} + 2T_{2g} \\ \Gamma(L = 9) &\longrightarrow A_{1u} + A_{2u} + E_u + 3T_{1u} + 2T_{2u} \end{aligned} \tag{18}$$

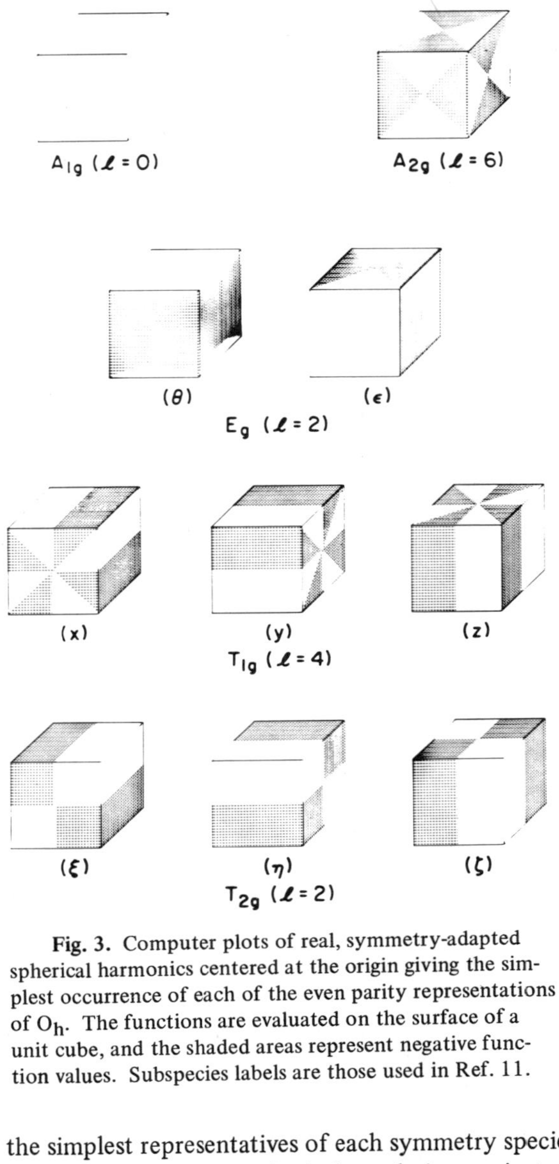

Fig. 3. Computer plots of real, symmetry-adapted spherical harmonics centered at the origin giving the simplest occurrence of each of the even parity representations of O_h. The functions are evaluated on the surface of a unit cube, and the shaded areas represent negative function values. Subspecies labels are those used in Ref. 11.

Figures 3 and 4 show the simplest representatives of each symmetry species in O_h, derived from real spherical harmonics, and indicate the standard phase choices against which the prototype functions are aligned. The behavior of the symmetry-adapted prototype functions for $L = L_g$ and $L = L_u$ is shown in Figs. 5 and 6. These figures were generated and plotted by the computer as a program option and serve to illustrate a type of incidental application for which the program might be useful. The transformation coefficients making up the matrix **V** in equation (17) are real combinations of those given in Appendix 2 of Ref. 11.[31]

4. Method 2. Symmetry Adaptation of an Arbitrary Closed Set of Functions

We consider now the task of symmetry adapting an arbitrary set of functions associated with the molecule, that is, of finding the reduction matrix **U** of equation (4). For this to be

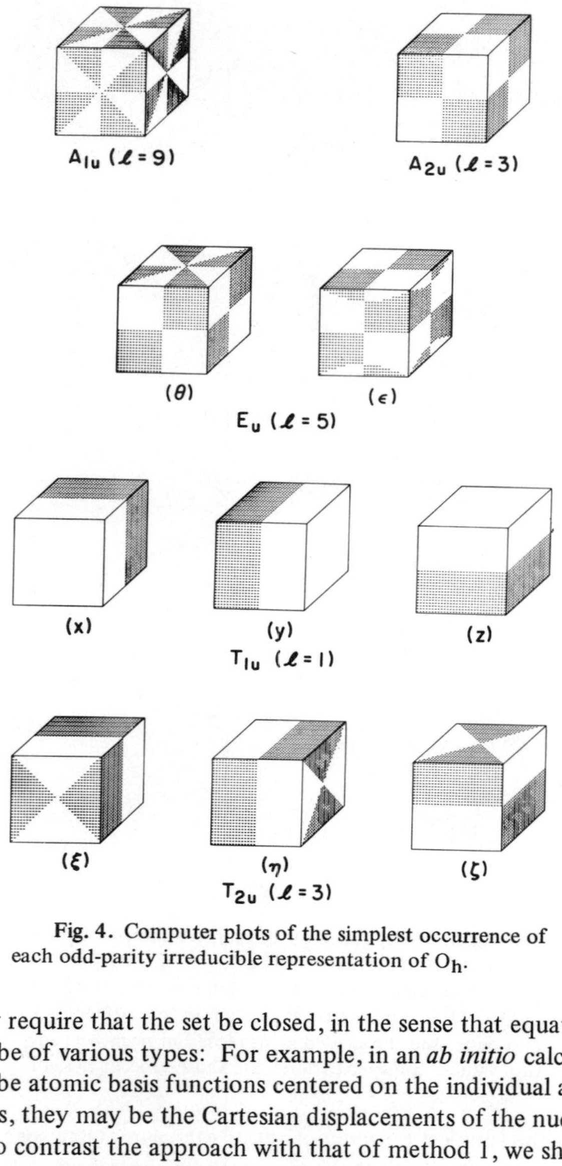

Fig. 4. Computer plots of the simplest occurrence of each odd-parity irreducible representation of O_h.

possible we need only require that the set be closed, in the sense that equation (1) is valid. These functions may be of various types: For example, in an *ab initio* calculation of electronic properties, they may be atomic basis functions centered on the individual atoms. For problems in molecular dynamics, they may be the Cartesian displacements of the nuclei. To illustrate an important case, and to contrast the approach with that of method 1, we shall carry through the example of determining a symmetry-adapted atomic basis set.

Let χ be a set of real atomic basis functions $\chi^{A}_{n\ell m}$ (r_A, θ_A, ϕ_A) such as that described in Section 2. Since the operator $\mathcal{P}$ has the symmetry of the nuclear framework, the previous restriction—that the basis set contain functions centered on all the atoms in the molecule—no longer applies here, and we may carry out the division of χ into subsets $\chi^{\ell\tau}$ at the outset. We recognize further that the transformation matrix **U** also factors into submatrices $\mathbf{U}^{\ell\tau}$, permitting each subset to be symmetry adapted independently:

$$\sigma^{\ell\tau} = \chi^{\ell\tau}\mathbf{U}^{\ell\tau} \qquad (19)$$

Each subset in χ is treated in the same way; in the remainder of the section, therefore, the superscripts will be omitted for brevity. Further, since the actual radial dependence factors out of the symmetry transformation, we ignore it and replace it by $R(r) = r^{-1}$ for ease of computa-

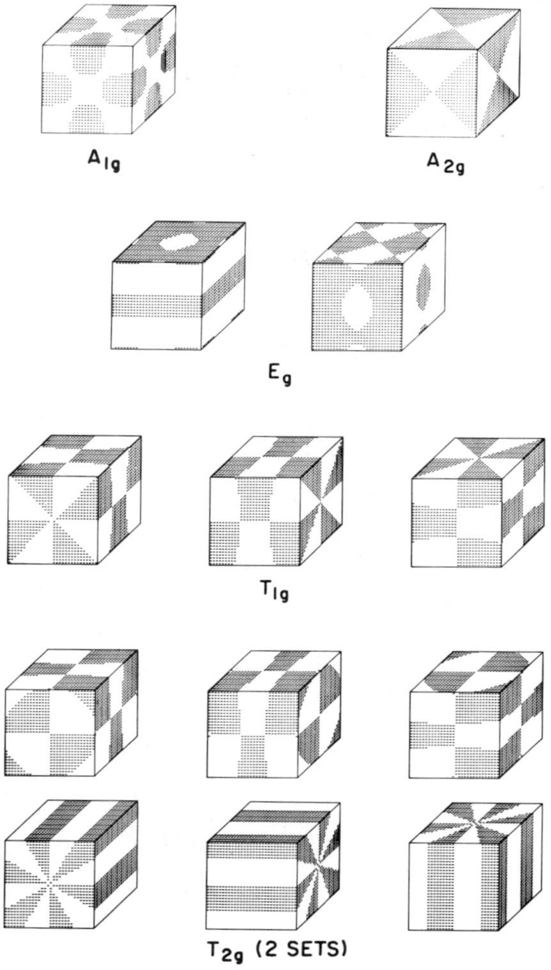

(ℓ = 6)

Fig. 5. Symmetry-adapted prototype functions (L = 6) spanning all the even-parity representations of O_h.

tion; the only requirement on R(r) is that it be sufficiently distinctive to avoid accidental degeneracies.

The initial stages of symmetry adaptation proceed as outlined previously: The matrix of $\mathcal{P}$ over the subset χ is calculated and diagonalized by a transformation **W**. A new set of functions ρ is formed from χ,

$$\rho = \chi W \tag{20}$$

To determine what representations are contained in ρ and to line up phases, we reintroduce the real harmonics $\mathbf{S}^{(L)}$ and calculate a connection matrix **Q**,

$$Q_{ij} = \langle S_i^{(L)} | \mathcal{P} | \chi_j \rangle \tag{21}$$

Then, letting $\mathbf{V}^{(\alpha)}$ be the set of prototype transformation vectors associated with the αth prototype eigenvalue, of degeneracy n_α, and $\mathbf{W}^{(\beta)}$ the set of atomic basis function eigenvectors from

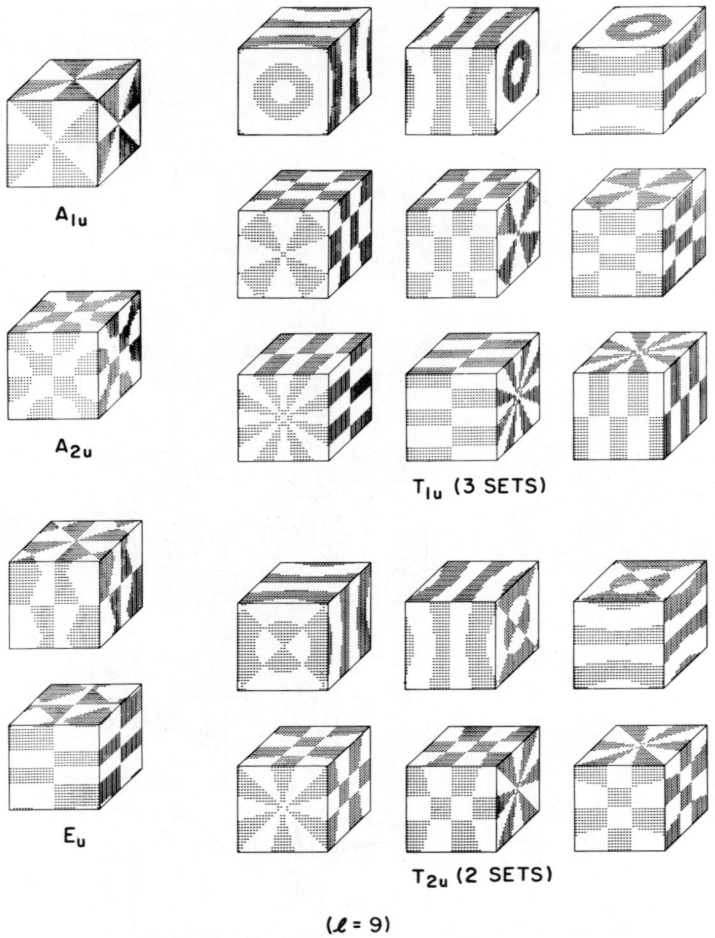

Fig. 6. Symmetry-adapted prototype functions (L = 9) spanning all the odd-parity representations of O_h.

eigenvalue β, degeneracy $n_\beta = n_\alpha$, we compute the $n_\alpha \times n_\alpha$ matrix,

$$\mathbf{M}^{(\alpha\beta)} = \mathbf{V}^{-1(\alpha)}\mathbf{Q}\mathbf{W}^{(\beta)} \tag{22}$$

If $\mathbf{M}^{(\alpha\beta)}$ contains any nonzero elements, then $\mathbf{W}^{(\beta)}$ carries the same symmetry species as $\mathbf{V}^{(\alpha)}$. The phase of $\mathbf{W}^{(\beta)}$ is made to correspond to that of $\mathbf{V}^{(\alpha)}$ by the method of the Appendix.

The new basis functions σ',

$$\sigma' = \rho\mathbf{T} = \chi\mathbf{W}\mathbf{T} = \chi\mathbf{U}' \tag{23}$$

have the well-defined transformation properties necessary to induce maximum symmetry factorization in a calculation. However, if the same irreducible representation is spanned more than once in the set of functions σ', the different sets still mix together in a way determined by the operator $\mathcal{P}$. The method described in the last part of Section 2 and Ref. 29 determines a final transformation $\mathbf{B}'$, which produces a simple set of coefficients in $\mathbf{U}$:

$$\sigma = \sigma'\mathbf{B}' = \chi\mathbf{U}'\mathbf{B}' = \chi\mathbf{U} \tag{24}$$

The set σ corresponds to that generated by method 1, for each subset of atomic basis functions.

5. Method 2. Finite-Point-Group Vector-Coupling Coefficients

Let $\mathbf{F}_\Gamma$ and $\mathbf{G}_{\Gamma}'$ be two sets of functions forming bases, respectively, for the irreducible representations Γ and Γ' of the point group. We can form a new set **H** with $n_\Gamma \times n_{\Gamma'}$ members by taking all possible products of pairs $F_{\Gamma\gamma} \times G_{\Gamma'\gamma'}$, where γ and γ' label the different subspecies in Γ and Γ'. If the representation Γ'' is contained in the direct product $\Gamma \times \Gamma'$, we can construct symmetry-adapted functions $H_{\Gamma''\gamma''}$ from **H** by means of a suitable unitary transformation,[12]

$$H_{\Gamma''\gamma''} = \sum_{\gamma} \sum_{\gamma'} (\Gamma''\gamma'' | \Gamma\gamma\Gamma'\gamma') F_{\Gamma\gamma} G_{\Gamma'\gamma'} \tag{25}$$

The coefficients $(\Gamma''\gamma'' | \Gamma\gamma\Gamma'\gamma')$ in equation (25) are called vector-coupling coefficients. In the full rotation group, these coefficients occur in the coupling of two angular momenta to form a third:

$$Y_{\ell''m''} = \sum_{m} \sum_{m'} (\ell''m'' | \ell m \ell' m') Y_{\ell m} Y_{\ell' m'} \tag{26}$$

In this case the coefficients are also called Wigner or Clebsh-Gordan coefficients.[8] The vector-coupling coefficients are orthonormal in the sense

$$\begin{aligned} \sum_{\gamma} \sum_{\gamma'} (\Gamma_1\gamma_1 | \Gamma\gamma\Gamma'\gamma')(\Gamma\gamma\Gamma'\gamma' | \Gamma_2\gamma_2) &= \delta_{\Gamma_1\Gamma_2}\delta_{\gamma_1\gamma_2}\delta(\Gamma, \Gamma', \Gamma_1) \\ \sum_{\Gamma''} \sum_{\gamma''} (\Gamma\gamma_1\Gamma'\gamma_1' | \Gamma''\gamma'')(\Gamma''\gamma'' | \Gamma\gamma_2\Gamma'\gamma_2') &= \delta_{\gamma_1\gamma_2}\delta_{\gamma_1'\gamma_2'} \end{aligned} \tag{27}$$

where $\delta(\Gamma, \Gamma', \Gamma_1)$ is unity if $\Gamma \times \Gamma'$ contains Γ_1, and is zero otherwise. An important use of vector-coupling coefficients occurs in factoring a matrix element into a part determined by symmetry and one which depends on the particular functions and operator:

$$\langle F\alpha''\Gamma''\gamma'' | A_{\Gamma\gamma} | G\alpha'\Gamma'\gamma' \rangle = (\Gamma''\gamma'' | \Gamma\gamma\Gamma'\gamma') \langle F\alpha''\Gamma'' \| A_\Gamma \| G\alpha'\Gamma' \rangle \tag{28}$$

The second factor on the right is called the *reduced matrix* element.[12] If Γ, Γ', and Γ'' are degenerate irreducible representations, considerable labor may be saved in calculating matrix elements among the members of the degenerate sets.

To compute the vector-coupling coefficients for finite point groups, we make use of the fact that our real spherical harmonics are symmetry adapted to the full rotation group. In showing the derivation, it is convenient to revert to the complex functions Y_{LM}, which are then related to the prototype functions $\psi^{(L)}$ of equation (17) by a complex transformation $\mathbf{C}^{(L)}$,

$$\psi^{(L)} = \mathbf{Y}^{(L)}\mathbf{C}^{(L)} \tag{29}$$

or

$$\psi^{(L)}_{\alpha\Gamma\gamma} = \sum_{M=-L}^{L} Y_{LM} C^{\alpha\Gamma\gamma}_{LM} \tag{30}$$

Substituting in equation (25), with the obvious change of notation, and making use of equations (26) and (28), we obtain finally

$$(\Gamma''\gamma'' | \Gamma\gamma\Gamma'\gamma') = \mathfrak{N} \sum_{M''} \sum_{M} \sum_{M'} C^{*\alpha''\Gamma''\gamma''}_{L''M''} C^{\alpha\Gamma\gamma}_{LM} C^{\alpha'\Gamma'\gamma'}_{L'M'} (L''M'' | LML'M') \tag{31}$$

The $(L''M''|LML'M')$ are just the Clebsh-Gordan coefficients, which are easily computed with standard analytical expressions. The coefficients **C** are all known from the matrix **V** of Section 3, and we can choose the constant $\mathfrak{N}$ to normalize the $(\Gamma''\gamma''|\Gamma\gamma\Gamma'\gamma')$. Since these vector-coupling coefficients are not dependent upon the inversion parity of the representations involved, we may take $L = L' = L'' = L_g$, the value which spans all the even representations of the group.

Equation (31) fails for some combinations $(\Gamma, \Gamma', \Gamma'')$ if all three sets of spherical harmonics have even L values, because of the behavior of the Clebsh-Gordan coefficients under permutation of L values. This condition arises when two of the Γ's are the same and the third belongs to the set of antisymmetric squares (see, e.g., Ref. 11). The summation may be made not to vanish if one of the sets of Y_{LM} is antisymmetric. Let the functions $Y_{L''M''}$, Y_{LM}, $Y_{L'M'}$ of equation (26) be chosen so that $L = L' = L_g$ and $L'' = L_u$. The functions $Y_{L''M''}$ are then not usual spherical harmonics, because they are antisymmetric with respect to interchange of L and L′ and have *even* inversion parity, as the product of two even-parity sets. The coefficients $C^{\alpha''\Gamma''\gamma''}_{L''M''}$ which symmetry adapt the $Y_{L''M''}$ can be inserted in equation (31) to obtain the desired antisymmetric vector-coupling coefficients.

Vector-coupling coefficients are produced in this way for any finite point group, even for one such as the icosahedral group, which has two linearly independent sets of coefficients for certain combinations of irreducible representations.[11] With proper attention to phase choices, it should be possible also to generate the V, W, and X coefficients of Griffith,[12] which are the respective analogues of the Wigner 3-j, 6-j, and 9-j symbols.[8]

6. Discussion

We have presented two computer-oriented approaches for generating symmetry information for subsequent machine manipulations. Methods 1 and 2 have been coded in Fortran IV and are in operation on the CDC 3600 and the IBM 360/75, respectively. In both cases, the time taken to compute the symmetry information—of the order of 1 to 2 minutes—is completely insignificant in comparison to the time required for a moderately elaborate quantum mechanical calculation. Both programs require only the specification of the Cartesian coordinates of the nuclei and the indices characterizing the basis set. The two programs arrange their results in block COMMON storage in a form convenient for interfacing to applications programs. Table 2 summarizes the input and output for the two programs

Table 2. Summary of Input and Output for the Computer Programs as Currently Implemented

- Input
  - Chemical symbols and Cartesian coordinates of atoms
  - Parameters describing basis set (either read in or selected from the tables stored in program)
  - Control parameters (or use of default options)
- Output
  - Method 1
    - Transformation coefficients for converting atomic basis set to symmetry-adapted set
    - Labels and degeneracies for symmetry species spanned by basis set
    - Internal bookkeeping information
  - Method 2, in addition
    - Characterization of all symmetry species in point group
    - Identification of point group
    - Set of symmetry-adapted prototype functions
    - Selection rules for direct products of representations
    - Finite-point-group vector-coupling coefficients

Methods 1 and 2 do differ, however, in their scope and in the extent to which they might be applied to other classes of problems; this has already been touched on in the foregoing sections. Method 1 finds its chief utility in molecular orbital calculations of electronic structure in which selection rules for matrix elements of totally symmetric operators are sufficient for the contemplated application, in other words, for the type of computation discussed by Roothaan.[6] As such, many characteristics peculiar to an atomic orbital basis set, for example, the handling of the parameters n, ℓ, m, and ζ, are embedded in the structure of the method. The present implementation is for Slater-type basis functions, but the computer program is sufficiently modular to allow replacement of these by functions having different radial dependences.

Method 2, as we have shown, is also capable of symmetry adapting an atomic orbital basis set. However, this is done as an application germane to a particular type of quantum mechanical problem rather than as the core of the approach. As has already been intimated in Section 4, one may hope to symmetry adapt classes of functions used in other kinds of problems; a case in point is the set of Cartesian nuclear displacements which arise in molecular dynamics calculations. The capability of providing finite-point-group vector-coupling coefficients not only achieves the automation of selection rules appropriate for nontotally symmetric operators, such as those arising in the calculation of transition probabilities but also provides a means of facilitating machine computations using irreducible tensor methods for molecules.[12]

The applications just discussed deal with finite point groups and functions of one set of variables. It appears feasible, within the framework of Sections 3 and 4, to investigate and automate the determination of symmetry properties, first, of other groups associated with the molecule, such as the double groups arising from the inclusion of spin; and second, of functions of more than one set of variables. In the former case, the key lies in choosing a set of prototype functions (in a generalized sense) to span the irreducible representations of the group in question. For double groups, an appropriate choice appears to be the complex functions ("spinors") which form bases for "half-integral" irreducible representations of the full rotation group.[1]

In the second case mentioned, one generally requires that the function be properly antisymmetrized with respect to particle interchange. Unless there is some way of distinguishing among different sets of variables, however, an operator like $\mathcal{P}$, defined in ordinary three-dimensional space, causes all matrix elements to vanish identically, since $F(1)G(2) - G(1)F(2)$ and functions like it vanish at every point. This problem is solved by allowing $\mathcal{P}$ to be defined on separate, topologically equivalent sets of points for each set of variables involved. If the association of a particular set of variables with its own distinct set of points is preserved, then quantities like $F(1)G(2) - G(1)F(2)$ do not vanish. Although the magnitudes have little significance, the resulting matrix elements display the same symmetry transformation properties as if we had performed a full, 3N-dimensional integration over all space. In fact, we have used this method in Section 5, in constructing the set of antisymmetric prototype functions. Extension to many-particle functions appears straightforward and also timely in view of recent investigations into their symmetry properties.[18,32-34]

A further possible extension of method 2 is the computation of intermolecular selection rules which involve the state of orbital symmetries of the individual interacting species. Examples of this type occur in scattering studies,[35] and in the Woodward-Hoffman symmetry rules for concerted organic reactions.[36] If such an extension were implemented on an interactive time-sharing computer system, an investigator could hope to obtain, in a short time, yes-no answers to various possible reaction intermediates or processes by providing a small amount of input to the program.

The calculations described in Sections 2 through 5 reflect the current state of implementation of the two methods. We plan, of course, to make the programs and their documentation available through the usual channels. We hope to investigate further the extensions of method 2 just outlined, and, finally, we hope that the present work will encourage full use of symmetry in the design and coding of future large-scale computer programs in many areas of quantum chemistry.

Appendix: Phase Alignment of Sets of Functions Carrying the Same Symmetry Species

Let $\mathbf{V}^{(\alpha)}$ and $\mathbf{W}^{(\beta)}$ be the sets of transformation vectors associated, respectively, with eigenvalues α, β of degeneracy $n_\alpha = n_\beta$. $\mathbf{V}^{(\alpha)}$ and $\mathbf{W}^{(\beta)}$ need not arise from the same set of functions, but $\mathbf{V}^{(\alpha)}$ is assumed to have the standard phase choice. We seek a unitary transformation $\mathbf{T}^{(\alpha\beta)}$,

$$\mathbf{W}'^{(\beta)} = \mathbf{W}^{(\beta)}\mathbf{T}^{(\alpha\beta)} \tag{A-1}$$

such that

$$\mathbf{M}^{(\alpha\beta)}\mathbf{T}^{(\alpha\beta)} = \mathbf{M}_D^{(\alpha\beta)} \tag{A-2}$$

where $\mathbf{M}^{(\alpha\beta)}$ is the matrix connecting $\mathbf{V}^{(\alpha)}$ and $\mathbf{W}^{(\beta)}$, and $\mathbf{M}_D^{(\alpha\beta)}$ is diagonal, with equal diagonal elements λ. $\mathbf{M}^{(\alpha\beta)}$ is defined by

$$\mathbf{M}^{(\alpha\beta)} = \mathbf{V}^{-1\,(\alpha)}\mathbf{P}\mathbf{W}^{(\beta)} \tag{A-3}$$

where $\mathbf{P}$ is the matrix of a totally symmetric operator over the basis functions represented in $\mathbf{V}^{(\alpha)}$ and $\mathbf{W}^{(\beta)}$

Suppressing the superscripts and rewriting,

$$\mathbf{M} = \mathbf{M}_D\mathbf{T}^{-1} \tag{A-4}$$

or

$$m_{ij} = \sum_k \lambda_{ik}\delta_{ik}t_{kj}^{-1} = \lambda t_{ij}^{-1} \tag{A-5}$$

Since $\mathbf{T}$ is unitary, we may write

$$\sum_j |m_{ij}|^2 = \lambda^2 \sum_j |t_{ij}^{-1}|^2 = \lambda^2 \cdot 1 \tag{A-6}$$

Then $\lambda = \sqrt{\sum_j |m_{ij}|^2}$, and $t_{ij}^{-1} = t_{ji}^* = m_{ij}/\lambda$. These small transformations $\mathbf{T}^{(\alpha\beta)}$ are all that need be calculated, although formally they may be collected in one unitary transformation $\mathbf{T}$ applied to the entire matrix $\mathbf{W}$.

Acknowledgments

Work performed under the auspices of the U.S. Atomic Energy Commission.

We are grateful to Dr. Arnold C. Wahl and Dr. Robert H. Land of this Laboratory for many helpful discussions, and for making available to us their one-electron, diatomic-integrals package, OETCSUT, which is used in the computer program for method 1.

References and Notes

1. E. P. Wigner, *Group Theory and Its Applications to the Quantum Mechanics of Atomic Spectra*, Academic Press, New York, 1959.
2. H. Bethe, *Ann. Physik* **3** [5], 133 (1929).
3. R. S. Mulliken, *Phys. Rev.* **40**, 55; **41**, 49,751 (1932); **43**, 279 (1933).
4. E. U. Condon and G. H. Shortley, *The Theory of Atomic Spectra*, Cambridge University Press, New York, 1935.
5. C. J. Ballhausen, *Introduction to Ligand Field Theory*, McGraw-Hill, New York, 1962.
6. C. C. J. Roothaan, *Rev. Mod. Phys.* **23**, 69 (1951).
7. G. Racah, *Phys. Rev.* **62**, 438 (1942); **63**, 367 (1943); **76**, 1352 (1949).
8. A. R. Edmonds, *Angular Momentum in Quantum Mechanics*, Princeton University Press, Princeton, N.J., 1957.
9. Y. Tanabe and S. Sugano, *J. Phys. Soc. Japan* **9**, 753 (1954).
10. Y. Tanabe and H. Kamimura, *J. Phys. Soc. Japan* **13**, 394 (1958).
11. J. S. Griffith, *The Theory of Transition Metal Ions*, Cambridge University Press, New York, 1961.
12. J. S. Griffith, *The Irreducible Tensor Method for Molecular Symmetry Groups*, Prentice-Hall, Englewood wood Cliffs, N.J., 1962.
13. B. D. Joshi, *J. Chem. Phys.* **43**, S40 (1965).
14. A. C. Wahl, *J. Chem. Phys.* **41**, 2600 (1964).
15. A. D. McLean and M. Yoshimine, *IBM J. Res. Develop.* **12**, 206 (1968).
16. E. Clementi and D. R. Davis, *J. Comp. Phys.* **2**, 223 (1967).
17. I. G. Csizmadia, M. C. Harrison, J. W. Moscowitz, and B. T. Sutcliffe, *Theoret. Chim. Acta* **6**, 191 (1966).
18. Z. Gershgorn and I. Shavitt, *Intern. J. Quantum Chem.* **1S**, 403 (1967).
19. S. H. Glarum, private communication.
20. P. Pulay, G. Borossay, and F. Török, *J. Mol. Struc.* **2**, 336 (1968), and references therein.
21. S. Flodmark and E. Blokker, *Intern. J. Quantum Chem.* **1S**, 703 (1967).
22. J. R. Gabriel, *J. Math. Phys.* **5**, 494 (1964); **9**, 973 (1968).
23. J. R. Gabriel, *J. Math. Phys.*, in press.
24. R. Moccia, *Theoret. Chim. Acta* **7**, 85 (1967).
25. R. McWeeny, *Symmetry–An Introduction to Group Theory*, Pergamon Press, Oxford, 1963, p. 195.
26. M. Gussoni and G. Zerbi, *J. Mol. Spectry.* **26**, 185 (1968).
27. It is clear at this point that linear molecules should be excluded from this method, because their possible symmetry species are not finite in number. However, this class of molecules is already well endowed with computer-oriented techniques taking advantage of symmetry.
28. Operationally, we define an atom type to include all atoms of a given chemical element which are "equidistant" from the center of nuclear charge. According to the looseness of the criterion applied to determine "equidistance," a user may optionally include atoms that are not strictly related by a symmetry operation. In studies of small distortions from an equilibrium geometry, this option allows one to avoid abrupt changes of symmetry adaptation as the nuclear geometry changes continuously.
29. A. L. H. Chung and G. L. Goodman, to be published.
30. The choice of algorithm for diagonalization in this method is not a matter of indifference. If degeneracies are present, the Givens-Householder technique introduces mixing in the eigenvectors where there is none in the matrix, corresponding to an arbitrary rotation of the coordinate axes. The Jacobi method gives eigenvectors which retain the simple structure of the original matrix, and is therefore preferable. Indeed, for the set $S^{(L)}$, the axis systems evinced by sets of degenerate eigenvectors already bear the simple relationship to the master axis system that one hopes to obtain.
31. If a symmetry species occurs more than once in a set of functions, e.g., T_{2g} in the $L = 6$ functions, the transformation vectors **V** are determined only to within a unitary transformation among the like sets. Simple sets of coefficients can be obtained by the method described in Section 2.
32. H. F. King, *J. Chem. Phys.* **46**, 705 (1967).
33. H. J. Silverstone and O. Sinanoğlu, *J. Chem. Phys.* **46**, 854 (1967).
34. R. McWeeny and W. Kutzelnigg, *Intern. J. Quantum Chem.* **2**, 187 (1968).
35. J. J. Kaufman, private communication.
36. R. B. Woodward and R. Hoffman, *J. Am. Chem. Soc.* **87**, 395, 2511 (1965); R. Hoffman and R. B. Woodward, *ibid.* **87**, 2046, 4388, 4389 (1965).

VII-4

Reprinted from *The Journal of Chemical Physics* **46**, 4725 (1967)

Study of the Electronic Structure of Molecules. III. Pyrrole Ground-State Wavefunction

E. Clementi, H. Clementi, and D. R. Davis
IBM San Jose Research Laboratory, San Jose, California

(Received 17 January 1967)

An SCF LCAO MO ground-state wavefunction is presented for the pyrrole molecule. All electrons, σ and π, are considered, and all the necessary many-center integrals are included in this work. The analysis of the wavefunction reveals that there is a strong two-way charge transfer on the nitrogen atom. (This reopens the problem of the validity of π-electron computations, as discussed in the paper.) The lowest π-electron MO is below the highest σ electron. The hybridization of the σ electrons in the carbon and nitrogen atoms is not s^1p^2, but deviates strongly for the nitrogen atom and slightly for the carbon atoms.

I. INTRODUCTION

In this work we report an all-electron computation for the wavefunction of pyrrole (C_4H_5N). This computation does not give a Hartree–Fock energy since the basis set we have used is not sufficiently large. The basis set is built with special linear combinations of Gaussian functions—the "contracted Gaussian function"—as previously done for the ethane molecule (C_2H_6) [1] or for the NH_3–HCl reaction study.[2] Ad-

[1] E. Clementi and D. R. Davis, J. Chem. Phys. **45**, 2595 (1966).
[2] E. Clementi, J. Chem. Phys. **46**, 3851 (1967).

TABLE I. Molecular geometry for the pyrrole molecule.[a]

Center	N	C(1)	C(2)	C(3)	C(4)	H(1)	H(2)	H(3)	H(4)	H(5)
y	0.0	−2.128939	2.128939	−1.360632	1.360632	−4.093449	4.093449	−2.571379	2.571379	0.0
z	1.633593	0.0	0.0	−2.433109	−2.433109	0.619408	0.619408	−4.099559	−4.099559	3.598950

[a] Distances are given in atomic units; the value of the x coordinate is 0, since the molecule is in the yz plane.

TABLE II. Uncontracted Gaussian set for pyrrole.

Centers[a]	Type[b]	Indices[c]	Orbital exponent
N	s	1	636.101
N	s	2	105.386
N	s	3	27.5167
N	s	4	9.02708
N	s	5	3.33086
N	s	6	0.828625
N	s	7	0.243109
C(1), C(2), C(3), C(4)	s	8, 15, 22, 29	391.445
C(1), C(2), C(3), C(4)	s	9, 16, 23, 30	64.7358
C(1), C(2), C(3), C(4)	s	10, 17, 24, 31	16.2247
C(1), C(2), C(3), C(4)	s	11, 18, 25, 32	5.33460
C(1), C(2), C(3), C(4)	s	12, 19, 26, 33	2.00995
C(1), C(2), C(3), C(4)	s	13, 20, 27, 34	0.502323
C(1), C(2), C(3), C(4)	s	14, 21, 28, 35	0.155139
H(1), H(2), H(3), H(4), H(5)	s	36, 39, 42, 45, 48	0.151374
H(1), H(2), H(3), H(4), H(5)	s	37, 40, 43, 46, 49	0.681277
H(1), H(2), H(3), H(4), H(5)	s	38, 41, 44, 47, 50	4.50037
N	x, y, z	51, 66, 81	5.19829
N	x, y, z	52, 67, 82	1.10716
N	x, y, z	53, 68, 83	0.261750
C(1), C(2), C(3), C(4)	x, y, z	54, 57, 60, 63, 69, 72, 75, 78, 84, 87, 90, 93	4.31613
C(1), C(2), C(3), C(4)	x, y, z	55, 58, 61, 64, 70, 73, 76, 79, 85, 88, 91, 94	0.873682
C(1), C(2), C(3), C(4)	x, y, z	56, 59, 62, 65, 71, 74, 77, 80, 86, 89, 92, 95	0.20286

[a] The value in parentheses (1), (2), ··· in the "Centers" column, refers to the first, second, ··· equivalent atom.

[b] s, x, y, z indicates Gaussian function of $1s$, $2p_x$, $2p_y$, $2p_z$ type.

[c] The indices label a given Gaussian function specified on its origin (center), n and l values (type), and orbital exponent (last column).

TABLE III. Contracted Gaussian set for pyrrole.

Contracted Gaussian	Type	Center
$\gamma_1 = 0.01817\beta_1 + 0.10776\beta_2 + 0.32320\beta_3 + 0.47673\beta_4 + 0.22046\beta_5$	s; $1s$	N
$\gamma_2 = 0.50027\beta_6 + 0.63917\beta_7$	s; $2s$	N
$\gamma_3 = 0.02220\beta_8 + 0.13285\beta_9 + 0.38435\beta_{10} + 0.45798\beta_{11} + 0.15441\beta_{12}$	s; $1s$	C(1)
$\gamma_4 = 0.56673\beta_{13} + 0.55692\beta_{14}$[a]	s; $2s$	C(1)
$\gamma_{11} = 0.64767\beta_{36} + 0.40789\beta_{37} + 0.07048\beta_{38}$[b]	s; $1s$	H(1)
$\gamma_{16} = 0.13843\beta_{51} + 0.49760\beta_{52} + 0.57505\beta_{53}$	x; $2p_x$	N
$\gamma_{17} = 0.10845\beta_{54} + 0.46116\beta_{55} + 0.63043\beta_{56}$[c]	x; $2p_x$	C(1)
$\gamma_{21} = 0.13843\beta_{66} + 0.49760\beta_{67} + 0.57505\beta_{68}$	y; $2p_y$	N
$\gamma_{22} = 0.10845\beta_{69} + 0.46116\beta_{70} + 0.63043\beta_{71}$[d]	y; $2p_y$	C(1)
$\gamma_{26} = 0.13843\beta_{81} + 0.49760\beta_{82} + 0.57505\beta_{83}$	z; $2p_z$	N
$\gamma_{27} = 0.10845\beta_{84} + 0.46116\beta_{85} + 0.63043\beta_{86}$[e]	z; $2p_z$	C(1)

[a] γ_5, γ_6 [centered on C(2)], γ_7, γ_8 [centered on C(3)], γ_9, γ_{10} [centered on C(4)] are not reported since there is a one-to-one correspondence between these and γ_3 and γ_4.

[b] γ_{12}, γ_{13}, γ_{14}, γ_{15} [centered on H(2), H(3), H(4), H(5), respectively] are not reported since there is a one-to-one correspondence between these and γ_{11}.

[c] γ_{18}, γ_{19}, γ_{20} [centered on C(2), C(3), C(4), respectively] are not reported since there is a one-to-one correspondence between these and γ_{17}.

[d] γ_{23}, γ_{24}, γ_{25} [centered on C(2), C(3), C(4) respectively] are not reported since there is a one-to-one correspondence between these and γ_{22}.

[e] γ_{28}, γ_{29}, γ_{30} [centered on C(2), C(3), C(4), respectively] are not reported since there is a one-to-one correspondence between these and γ_{27}

ditional information on the contracted Gaussian representation is available elsewhere.[3]

The basis set we have used gives nearly Hartree–Fock energy for the separated atom. The C(^3P) and the N(^4S) energies obtained with this set are −37.6296 and −54.310 a.u., respectively, to be compared with the Hartree–Fock energy of −37.6886 and −54.4009 a.u., respectively (see for instance our work on Hartree–Fock atomic functions[4]); the energy for H(^2S) is 0.4998 a.u., to be compared with the exact energy of 0.50000 a.u. One of the reasons for the use of a truncated basis set is that we wish to compute a number of aromatic molecules of larger size than pyrrole, with the same basis set, so as to make possible comparison among functions for different molecules. A rather extended study of this nature is of importance if we wish to assess the reliability of semiempirical computations. As known, molecules of the size of pyrrole have not been previously computed on *ab initio* basis, and only semiempirical work is available for such molecules. In order to assess semiempirical studies it is quite useful to have *ab initio* computations for a variety of molecules with the same type of basis set. Pyrrole is the first example we report, and in subsequent papers we shall present pyridine and pyrazine, benzene, and a few other molecules.

TABLE IV. Symmetry-adapted functions for pyrrole (unnormalized).

$\chi_1=\gamma_1$	$\chi_{16}=\gamma_3-\gamma_5$
$\chi_2=\gamma_2$	$\chi_{17}=\gamma_4-\gamma_6$
$\chi_3=\gamma_3+\gamma_5$	$\chi_{18}=\gamma_7-\gamma_9$
$\chi_4=\gamma_4+\gamma_6$	$\chi_{19}=\gamma_8-\gamma_{10}$
$\chi_5=\gamma_7+\gamma_9$	$\chi_{20}=\gamma_{22}+\gamma_{23}$
$\chi_6=\gamma_8+\gamma_{10}$	$\chi_{21}=\gamma_{24}+\gamma_{25}$
$\chi_7=\gamma_{22}-\gamma_{23}$	$\chi_{22}=\gamma_{11}-\gamma_{12}$
$\chi_8=\gamma_{24}-\gamma_{25}$	$\chi_{23}=\gamma_{13}-\gamma_{14}$
$\chi_9=\gamma_{26}$	$\chi_{24}=\gamma_{27}-\gamma_{28}$
$\chi_{10}=\gamma_{27}+\gamma_{28}$	$\chi_{25}=\gamma_{29}-\gamma_{30}$
$\chi_{11}=\gamma_{29}+\gamma_{30}$	$\chi_{26}=\gamma_{16}$
$\chi_{12}=\gamma_{11}+\gamma_{12}$	$\chi_{27}=\gamma_{17}+\gamma_{18}$
$\chi_{13}=\gamma_{13}+\gamma_{14}$	$\chi_{28}=\gamma_{19}+\gamma_{20}$
$\chi_{14}=\gamma_{15}$	$\chi_{29}=\gamma_{17}-\gamma_{18}$
$\chi_{15}=\gamma_{21}$	$\chi_{30}=\gamma_{19}-\gamma_{20}$

II. SCF WAVEFUNCTION FOR PYRROLE

The geometry for pyrrole is given in Table I, and follows the values reported by Sutton.[5] The uncontracted Gaussians are given in Table II. This set consists of 95 Gaussian functions and the orbital exponents have been optimized for the carbon atom using a program written by S. Huzinaga. Seven s-type Gaussians are used for the 1s and 2s orbitals of the C (or N) atom; three Gaussian functions are used for the $2p_x$ (or $2p_y$ or $2p_z$) orbitals of the C (or N) atom. The 1s orbital of the hydrogen is described by three Gaussian functions. In Table II the orbital exponents of the uncontracted Gaussian functions are given. In Table III the above basis set is contracted into 30 functions of "the contracted Gaussian set," and in Table IV the 30 contracted Gaussians are combined in symmetry-adapted functions which transform as the C_{2v} irreducible representations. The electronic configuration

TABLE V. Total energy and orbital energies for pyrrole (in atomic units).[a]

$A_1(\sigma)$	$B_2(\sigma)$	$B_1(\pi)$	$A_2(\pi)$
−15.71000	−11.42526	−0.63133	−0.38794
−11.42520	−11.37850	−0.42529	
−11.37931	− 1.03448		
− 1.32387	− 0.79702		
− 1.09548	− 0.62429		
− 0.82508	− 0.60219		
− 0.77787			
− 0.64759			
− 0.57659			

[a] Total energy is −207.93135 a.u.

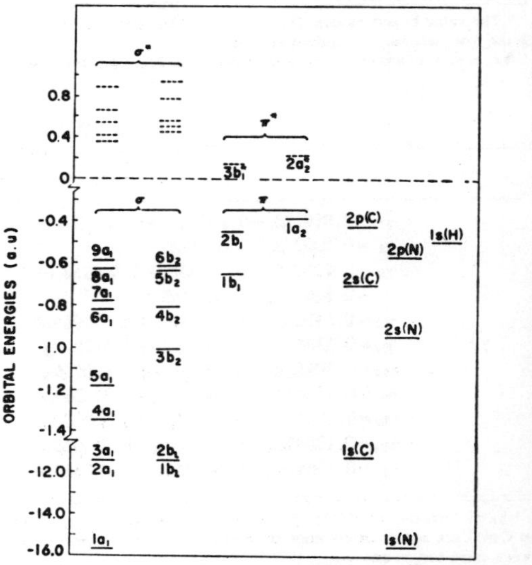

FIG. 1. Orbital energies for the ground state of the pyrrole molecule. In this figure we have added the orbital energies of the separated atoms as well as a few of the virtual orbital energies.

[3] E. Clementi and D. R. Davis, J. Comput. Phys. 1, 223 (1966).
[4] E. Clementi, "Tables of Atomic Functions," Supplement to IBM J. Res. Devel. 9, 1 (1965).
[5] L. E. Sutton, *Tables of Interatomic Distances* (The Chemical Society, London, 1958).

TABLE VI. Expansion coefficients for MO's of A_1 symmetry.

	$1a_1$	$2a_1$	$3a_1$	$4a_1$	$5a_1$	$6a_1$	$7a_1$	$8a_1$	$9a_1$
χ_1	−0.99681	−0.00022	0.00004	−0.21170	0.16600	−0.07051	−0.02170	0.02452	0.03073
χ_2	−0.01113	−0.00026	−0.00008	0.54834	−0.46622	0.21406	0.04118	−0.07535	−0.10202
χ_3	−0.00029	0.99861	−0.01438	−0.15640	−0.04691	0.13820	0.10518	0.00516	−0.01266
χ_4	0.00223	0.00446	−0.00078	0.33719	0.09557	−0.35898	−0.34560	−0.05113	0.05245
χ_5	−0.00001	0.01414	0.99838	−0.09934	−0.21905	−0.12508	0.00992	−0.00207	−0.01015
χ_6	−0.00001	−0.00029	0.00536	0.23217	0.58308	0.40143	−0.07608	−0.04916	0.03584
χ_7	0.00154	0.00036	−0.00028	0.15776	−0.04328	0.01635	0.47721	−0.02826	−0.31127
χ_8	−0.00009	−0.00023	0.00025	0.03469	0.14583	0.19685	0.29119	0.15802	0.81598
χ_9	0.00056	−0.00006	0.00020	−0.09629	−0.06862	0.40158	−0.29502	0.45521	0.00478
χ_{10}	0.00109	−0.00008	0.00009	0.01281	−0.25127	0.00428	0.00642	−0.39954	0.34069
χ_{11}	0.00004	0.00002	0.00386	0.09951	0.10661	−0.30185	0.05412	0.57438	−0.05607
χ_{12}	0.00009	−0.00012	−0.00017	0.04406	0.01410	−0.14654	−0.29337	−0.07737	0.30619
χ_{13}	0.00004	−0.00013	−0.00015	0.02619	0.09712	0.17394	−0.09155	−0.29562	−0.25265
χ_{14}	0.00154	−0.00022	−0.00002	0.07120	−0.11902	0.26163	−0.11960	0.27915	−0.06025

TABLE VII. Expansion coefficients for MO's of B_2 symmetry.

	$1b_2$	$2b_2$	$3b_2$	$4b_2$	$5b_2$	$6b_2$
χ_{15}	0.0	0.0	−0.22333	0.38696	−0.36812	0.28682
χ_{16}	0.99850	−0.01507	−0.24160	0.07856	−0.00412	−0.05868
χ_{17}	0.00430	−0.00048	0.55424	−0.19403	−0.00319	0.14445
χ_{18}	0.01471	0.99852	−0.14584	−0.16124	−0.04761	0.04789
χ_{19}	−0.00003	0.00415	0.27304	0.31300	0.09896	−0.12662
χ_{20}	0.00082	0.00019	0.05221	−0.00094	0.58491	−0.02998
χ_{21}	0.00006	0.00019	−0.10685	−0.16637	0.01212	0.12350
χ_{22}	0.00004	−0.00003	0.13621	−0.14561	−0.42817	0.03762
χ_{23}	0.0	0.00002	0.07219	0.21633	−0.02298	−0.43184
χ_{24}	−0.00009	−0.00006	−0.10260	−0.38867	−0.10511	−0.39039
χ_{25}	0.00032	0.00073	0.15678	−0.01661	0.16626	0.49002

for pyrrole in the ground state is as follows:

σ type: $1a_1^2 2a_1^2 3a_1^2 4a_1^2 5a_1^2 6a_1^2 7a_1^2 8a_1^2 9a_1^2$,

σ type: $1b_2^2 2b_2^2 3b_2^2 4b_2^2 5b_2^2, 6b_2^2$,

π type: $1b_1^2 2b_1^2 1a_2^2$.

The total energy and the orbital energies of our computation are given in Table V and the expansion coefficients of the SCF function are given in Tables VI–VIII.

III. DISCUSSION

Let us analyze the MO's. This is done both by considering the expansion coefficients of Tables VI–VIII, and by making use of the gross population analyses given in Table IX (for the a_1 orbitals), Table X (for the b_2 orbitals), and Table XI (for the b_1 and a_2 orbitals).

TABLE VIII. Expansion coefficients for MO's of B_1 and A_2 symmetry.

	$1b_1$	$2b_1$		$1a_2$
χ_{26}	0.58590	0.66165	χ_{29}	0.77260
χ_{27}	0.46059	−0.09656	χ_{30}	0.43901
χ_{28}	0.35074	−0.78297		

The $1a_1$, $2a_1$, and $3a_1$ MO's are clearly the $1s$ atomic orbitals for the nitrogen and carbon atoms, respectively. The remaining two $1s$ electrons for the carbon atoms are the $1b_2$ and $2b_2$ MO's. This can be seen either from the expansion coefficients, or by the population analysis results, or from Fig. 1, where the orbital energies of the MO are given together with the orbital energies of the separated atoms (^3P for C, ^4S for N, and ^2S for H). The $1a_1$, $2a_1$, $3a_1$, $1b_2$, and $2b_2$ are the inner-shell MO's and to a large extent they can be considered undistorted atomic orbitals of the separated atoms in the molecular field.

The second group of σ electrons are responsible for the C–C, C–N, C–H, and N–H bond formation. The $4a_1$ is mainly constructed from the $2s$ atomic orbital on C and N. This orbital flows over the entire molecular skeleton with maximum density at the nitrogen atom [the gross charge on $2s$(N) is 0.85], lesser density at the C_1 and C_2 positions (gross charges on C_1 and C_2 are 0.34 partially polarized), and again lesser density at the C_3 and C_4 positions (gross charges on C_3 and C_4 are 0.12, again partially polarized). This MO therefore envelopes the molecule, has maximum density at nitrogen, and is polarized. It has an analog in the lowest π MO, the $1b_1$, which has very similar charge distribution. This makes us question the long-standing idea in chemistry that the π electrons are much more delocalized than the σ electrons.[6] As far as we can see, the $4a_1$ MO and the $1b_1$ are very similar in character (of course the former is much more bound than the latter).

The $5a_1$ MO has maximum charge at the C_3 and C_4 positions, lesser charge at the C_1 and C_2 positions, and intermediate charges at the N position. This orbital in part tends to reverse the charge distribution given by the $4a_1$ by concentrating charges on C_3 and C_4. In addition, whereas the $4a_1$ orbital is mainly $2p_y$ polarized, the $5a_1$ is mainly $2p_z$ polarized. Note that the H_5 contributes to both the $4a_1$ and the $5a_1$, by 0.06 and 0.10 fractional electrons, respectively.

The remaining MO's of a_1 symmetry have the charge distribution on all the 10 atoms of the molecule. The $5a_1$, the $6a_1$, and the $8a_1$ are responsible for the N–H bond. The set $6a_1$ to $9a_1$ as well as the set $3b_2$ to $6b_2$ are responsible for the C_1–H_1, C_2–H_2, C_3–H_3, and C_4–H_4 bonds. There is no single MO which can be identified with a given bond: this is the nature of the MO theory.

The π occupied MO's are the $1b_1$, $2b_1$, and $1a_2$. The $1b_1$ flows over the full molecular skeleton, with a density maximum at the nitrogen atom. The $2b_1$ has a node between the nitrogen atom and the rest of the molecule; it has a density maximum both at the C_3 and the C_4 positions and at the nitrogen atom; a density minimum is present at the C_1 and C_2 positions (the C_3, C_4 maximum is higher than the nitrogen maximum). The a_2 has no charge on the nitrogen (by symmetry); it has high density at the C_1 and C_2 positions, and low density at the C_3 and C_4 positions; for symmetry considerations, it has a node in the C_2 symmetry axis and, being the lowest of that symmetry, has the same phase on C_1 and C_2 and on C_3 and C_4. Therefore, there is extended conjugation. Pyrrole is an aromatic compound, because of the six π electrons; however, there is one π MO deep in the σ MO's (see Fig. 1).

The charge distribution in pyrrole is summarized in Table XII. The nitrogen atom has the following charges:

$$1s^2 2s^{1.87} 2p_\sigma^{2.38} 2p_\pi^{1.66},$$

which could be compared with the original (separated-

TABLE IX. Pyrrole—gross population analysis (A_1).

	$1a_1$	$2a_1$	$3a_1$	$4a_1$	$5a_1$	$6a_1$	$7a_1$	$8a_1$	$9a_1$	Total
$N(1s)$	1.99356	0.0	0.0	0.00236	0.00132	0.00020	0.00002	0.00002	0.00004	1.99752
$N(2s)$	0.00718	−0.00002	0.0	0.85534	0.42728	0.07166	0.00094	0.00150	0.00702	1.37090
$C_1(1s)+C_2(1s)$	0.0	1.99700	0.00042	0.00096	0.00008	0.00050	0.00024	0.0	0.0	1.99920
$C_1(2s)+C_2(2s)$	−0.00020	0.00266	−0.00008	0.51234	0.03734	0.22734	0.23380	0.00140	0.00752	1.02212
$C_3(1s)+C_4(1s)$	0.0	0.00040	1.99654	0.00040	0.00166	0.00040	0.0	0.0	0.0	1.99940
$C_3(2s)+C_4(2s)$	0.0	−0.00002	0.00316	0.22062	0.85246	0.26344	0.01096	0.00900	−0.00076	1.35386
$C_1(2p_y)-C_2(2p_y)$	−0.00020	0.00002	−0.00002	0.17066	0.00174	0.00280	0.69752	−0.00228	0.13268	1.00292
$C_3(2p_y)-C_4(2p_y)$	0.0	0.0	0.00004	0.01364	0.08014	0.11560	0.28388	0.03796	1.16986	1.70112
$N(2p_z)$	0.0	0.0	0.0	0.05148	0.03748	0.48572	0.24706	0.41460	0.00008	1.23642
$C_1(2p_z)+C_2(2p_z)$	−0.00012	0.0	−0.00002	0.00240	0.36376	−0.00002	−0.00034	0.29072	0.27066	0.92704
$C_3(2p_z)+C_4(2p_z)$	0.0	0.0	0.0	0.07786	0.01662	0.35292	0.00888	0.72772	0.00552	1.18952
$H_1(1s)+H_2(1s)$	0.0	−0.00002	0.0	0.02566	0.00180	0.08692	0.40844	0.01750	0.23456	0.77486
$H_3(1s)+H_4(1s)$	0.0	0.0	−0.00002	0.00888	0.08144	0.14746	0.04660	0.30424	0.17042	0.75902
$H_5(1s)$	−0.00022	0.0	0.0	0.05740	0.09678	0.24510	0.06200	0.19748	0.00244	0.66098
Total	2.00000	2.00002	2.00002	2.00000	1.99990	2.00004	2.00000	1.99986	2.00004	17.99988

[6] This point will require further analysis, whereby the lowering in energy due to delocalization in the π MO's is compared to the lowering in energy in the valency σ MO's. We have initiated such analysis by using the bond-energy analysis concepts previously reported in this series of papers [E. Clementi, J. Chem. Phys. 46, 3842 (1967)].

TABLE X. Pyrrole—gross population analysis (B_2).

	$1b_2$	$2b_2$	$3b_2$	$4b_2$	$5b_2$	$6b_2$	Total
$N(2p_x)$	0.0	0.0	0.21860	0.47370	0.26210	0.18958	1.14398
$C_1(1s) - C_2(1s)$	1.99676	0.00044	0.00178	0.00016	0.0	0.00008	1.99922
$C_1(2s) - C_2(2s)$	0.00276	−0.00004	0.96008	0.08428	0.00078	0.02378	1.07164
$C_3(1s) - C_4(1s)$	0.00044	1.99682	0.00072	0.00070	0.00006	0.00004	1.99878
$C_3(2s) - C_4(2s)$	0.0	0.00282	0.34680	0.33742	0.01026	0.05590	0.75320
$C_1(2p_y) + C_2(2p_y)$	−0.00002	0.0	−0.01558	−0.00004	1.05212	0.00396	1.04044
$C_3(2p_y) + C_4(2p_y)$	0.0	−0.00004	0.12384	0.17368	0.00246	0.06090	0.36084
$C_1(2p_x) - C_2(2p_x)$	0.0	0.00002	0.04066	0.61552	0.02064	0.32090	0.99774
$C_3(2p_x) - C_4(2p_x)$	0.00006	0.0	0.13082	0.00570	0.06042	0.76172	0.95872
$H_1(1s) - H_2(1s)$	0.0	0.0	0.14564	0.08342	0.58696	0.00150	0.81752
$H_3(1s) - H_4(1s)$	0.0	0.0	0.04656	0.22544	0.00420	0.58166	0.85786
Total	2.00000	2.00002	1.99992	1.99998	2.00000	2.00002	11.99994

TABLE XI. Pyrrole—gross population analysis (π) A_2 and B_1.

	$1b_1$	$2b_1$	$1a_2$
$N(2p_z)$	0.89596	0.76294	
$C_1(2p_z) \pm C_2(2p_z)$	0.71820	0.02808	1.40416
$C_3(2p_z) \pm C_4(2p_z)$	0.38580	1.20896	0.59580
Total	1.99996	1.99998	1.99996

TABLE XII. Pyrrole—gross charges—summary.

		N	
$1s = 1.9975$	$1s = 1.9975$	$1s = 1.9975$	$\delta(\sigma) = -0.7488$
$2s = 1.3709$	$2s = 1.3709$	$2s = 1.3709$	$\delta(\pi) = 0.3411$
$2p_x = 1.2364$	$2p_\sigma = 2.3804$	$2p = 4.0393$	$\delta = -0.4077$
$2p_y = 1.1440$	$2p_\pi = 1.6589$		
$2p_z = 1.6589$			
		C_1 (or C_2)	
$1s = 1.9992$	$1s = 1.9992$	$1s = 1.9992$	$\delta(\sigma) = -0.0302$
$2s = 1.0469$	$2s = 1.0469$	$2s = 1.0469$	$\delta(\pi) = -0.0752$
$2p_x = 0.9624$	$2p_\sigma = 1.9841$	$2p = 3.0593$	$\delta = -0.1054$
$2p_y = 1.0217$	$2p_\pi = 1.0752$		
$2p_z = 1.0752$			
		C_3 (or C_4)	
$1s = 1.9991$	$1s = 1.9991$	$1s = 1.9991$	$\delta(\sigma) = -0.1602$
$2s = 1.0560$	$2s = 1.0560$	$2s = 1.0560$	$\delta(\pi) = -0.0953$
$2p_x = 1.0741$	$2p_\sigma = 2.1051$	$2p = 3.2004$	$\delta = -0.2555$
$2p_y = 1.0310$	$2p_\pi = 1.0953$		
$2p_z = 1.0953$			
		H_5	
$1s = 0.6610$	$1s = 0.6610$	$1s = 0.6610$	$\delta(\sigma) = 0.3390$
		H_1 (or H_2)	
$1s = 0.7962$	$1s = 0.7962$	$1s = 0.7962$	$\delta(\sigma) = 0.2038$
		H_3 (or H_4)	
$1s = 0.8084$	$1s = 0.8084$	$1s = 0.8084$	$\delta(\sigma) = 0.1916$

atom) distribution

$$1s^2 2s^2 2p_\sigma^{\,1} 2p_\pi^{\,2}.$$

Therefore the nitrogen has gained 0.41 of an electron. This gain is the sum of two effects: a gain of 0.75 of an electron from the σ orbital, and a loss of 0.34 of an electron from the π orbitals. The charge transfer acts two ways: *The nitrogen is π donor and σ acceptor*, with the net result of a gain of 0.41 of an electron. This two-way charge transfer brings about the problem on how reasonable are the charge distributions with the π electron approximation, where one assumes in general an undistorted core (σ electrons).

For the carbon atoms and the hydrogen atoms we have only one-way charge transfer. The carbon atoms are both σ and π acceptors, whereas the hydrogen are σ donors.

Finally the hybridization of the nitrogen atom is $s^{1.37}p^{2.38}$; the hybridization of the C_1 and C_2 is $s^{1.05}p^{1.98}$ and the hybridization of C_3 and C_4 is $s^{1.06}p^{2.10}$. These values are not too different from the s^1p^2 of a triagonal hybrid, exception made for the nitrogen atom.

ACKNOWLEDGMENTS

It is my pleasure to thank Professor R. S. Mulliken and Professor N. H. Nachtrieb for their hospitality during my stay in Chicago. It is my pleasure to thank the IBM Computer Center, Yorktown Heights, New York, for their cooperation in this work. Finally, it is my pleasure to thank Mr. K. Busse of the Department of Physics, University of Chicago, for the typing and proofreading of this manuscript.

VII-5

VII-5a

Subminimal ab initio Calculations

ARTHUR A. FROST

In most simple LCAO-MO calculations a minimal basis set of functions is employed. Such a set includes inner-shell and valence-shell orbitals of each atom. For example, for NH_3 there would be 1s, 2s, $2p_x$, $2p_y$, and $2p_z$ orbitals on N and a 1s orbital on each of the three H atoms, making eight orbitals in all. On the other hand, it is possible to place the 10 electrons in five doubly occupied localized orbitals. Such a set, which is smaller in size than the minimal set, is referred to here as subminimal.

The possibility of representing the electronic structure of certain molecules by electron pairs in localized orbitals may be considered to be a logical quantum mechanical development of the Lewis[1] electron-pair valence theory. Lewis stated that a double dot in a typical "Lewis structure" diagram represents the average position of the electrons in a distribution and that the average position can shift toward one atom or another of a bonded pair depending upon the relative electronegativities of the two atoms. The simplest probability distribution of a single localized electron in a spherical Gaussian distribution, which in turn is the square of a spherical Gaussian orbital. By placing electron pairs in a subminimal set of such orbitals, forming an antisymmetrized product equivalent to a single Slater determinantal wave function, and minimizing the energy with respect to parameters identifying the location and spread of the basis orbitals, a floating spherical Gaussian orbital (FSGO) model[2,3-7] is obtained. A more general floating localized orbital model can be created by varying the shape of each localized orbital.

The FSGO model is, of course, crude and cannot be expected to yield energies. The energies of a number of molecules turn out to be approximately 84 per cent of the Hartree-Fock energies, which are the best that can be obtained with a single closed-shell Slater determinant. The energy deficiency is primarily due to the lack of cusps in the wave function at the nuclei. The principal value of the model is in predicting the ground-state geometry of simple molecules. By minimizing the total energy, electronic plus nuclear repulsion, with respect to the variation of nuclear positions, the geometric form is obtained semiquantitatively. Surprisingly, the bond lengths and bond angles in a variety of molecules are predicted to within a few per cent of the experimental values without resorting to the use of semiempirical parameters.

As examples, consider the hydrocarbons[3,7] C_2H_6, C_2H_4, and C_2H_2. The carbon-carbon single, double, and triple bonds decrease in that order, in agreement with experiment within about 1 per cent. Similarly, in the same molecules with corresponding accuracy, the CH bonds show a slight trend downward. This latter effect is usually explained in terms of changing hybridization of the carbon orbitals, a concept that is unnecessary in the FSGO model.

The papers listed in the References give more detail and representative results of the calculations.

References

1. G. N. Lewis, *J. Am. Chem. Soc.* **38,** 762 (1916); *Valence and the Structure of Atoms and Molecules*, Chemical Catalog Co., New York, 1923, and Dover, New York, 1966.
2. G. E. Kimball and G. F. Neumark, *J. Chem. Phys.* **26,** 1285 (1957).
3. A. A. Frost, B. H. Prentice, III, and Robert A. Rouse, *J. Am. Chem. Soc.* **89,** 3064 (1967) (preliminary publication).
4. A. A. Frost, *J. Chem. Phys.* **47,** 3707 (1967) (Paper I).
5. A. A. Frost, *J. Chem. Phys.* **47,** 3714 (1967) (Paper II).
6. A. A. Frost, *J. Phys. Chem.* **72,** 1289 (1968) (Paper III).
7. A. A. Frost and R. A. Rouse, *J. Am. Chem. Soc.*, **90,** 1965 (1968) (Paper IV).

VII-5b

Reprinted from *The Journal of the American Chemical Society* 89, 3064 (1967)

A Simple Floating Localized Orbital Model of Molecular Structure

Sir:

Most molecules have an even number of electrons which are generally paired off to create a ground state which is a spectroscopic singlet. For such molecules and states the following quantum mechanical model is proposed.

Let there be a minimal set of n floating localized orbitals φ_i which are, in general, nonorthogonal and real, and let each one be occupied by a pair of electrons with opposing spin. The $2n$-electron wave function can then be written as a single normalized Slater determinant

$$\psi = |\varphi_1(1)\bar{\varphi}_1(2)\varphi_2(3)\bar{\varphi}_2(4)\ldots\varphi_n(2n-1)\bar{\varphi}_n(2n)| \quad [1/(\sqrt{(2n)!}\ \det S)]$$

where the bars over certain orbitals indicate β spin as opposed to α for the others, and det S is the determinant of the orbital overlap matrix S with elements

$$S_{ij} = \int \varphi_i{}^*\varphi_j\, dv$$

Given the set of orbitals and the appropriate nonrelativistic Hamiltonian operator H, the mean energy E is calculated according to a formula adapted from one derived by Löwdin[1]

$$E = \int \psi^* H\psi\, d\tau = 2\sum_{i,j}(i/j)T_{ij} + \sum_{i,j,k,l}(ij/kl)(2T_{ij}T_{kl} - T_{il}T_{jk})$$

where

$$(i/j) = \int \varphi_i{}^* h\varphi_j\, dv$$

are the kinetic and potential energy integrals with the one-electron operator h, and

$$(ij/kl) = \int \varphi_i{}^*(1)\varphi_j(1)\varphi_k{}^*(2)\varphi_l(2)(1/r_{12})\, dv(1)dv(2)$$

are the electron repulsion energy integrals. T_{ij}'s are elements of the reciprocal orbital overlap matrix

$$T = S^{-1}$$

For a given set of nuclear coordinates E is minimized, according to the variation method, by a variation in parameters defining the orbitals. This will generate a potential energy surface. If a "full minimization" of E with respect to nuclear coordinates as well as orbital parameters is carried out, then the equilibrium configurations of the molecule will be predicted. The calculation is strictly *ab initio* with no semiempirical parameters.

In this simple model the orbitals are taken to be floating spherical Gaussian functions[2]

$$\varphi_i = (2/\pi\rho_i{}^2)^{3/4} \exp[-(r_i/\rho_i)^2]$$

where r_i is the radial distance from the center of the orbital and ρ_i is an "orbital radius" parameter which defines a sphere which includes about 74% of the orbital charge density. For each orbital the coordinates of the center as well as the orbital radius are parameters to be varied.

Minimization of E with respect to all parameters will automatically lead to a result which will satisfy both the virial theorem and the Hellmann–Feynman theorem.[3]

Table I presents typical results for a series of diatomic and polyatomic molecules by the full minimization procedure.

The calculated energies are, of course, well above experimental values since no electron correlation is included other than that between electrons of like spin due to the antisymmetrization inherent in the determinantal wave function. Also the energies must be higher than those of Hartree–Fock calculations since the latter are by definition the values obtained by all possible variations of the orbitals in a single determinantal wave function. Because the total energies are crude, it would be expected that dissociation energies would be unsatisfactory and no attempt has been made to calculate them.

(1) P.-O. Löwdin, *J. Chem. Phys.*, **18**, 365 (1950).
(2) S. F. Boys, *Proc. Roy. Soc.* (London), **A200**, 542 (1950), introduction of gaussian orbitals; H. Preuss, *Z. Naturforsch.*, **11a**, 823 (1956); **19a**, 1335 (1964); **20a**, 18, 21, 1290 (1965); J. L. Whitten, *J. Chem. Phys.*, **39**, 349 (1963); **44**, 359 (1966); J. L. Whitten and L. C. Allen, *ibid.*, **43**, S170 (1965), use of off-center spherical Gaussian "pure" or "lobe" functions to simulate nonspherical atomic orbitals.
(3) A. C. Hurley, *Proc. Roy. Soc.* (London), **A226**, 170, 176, 193 (1954).

Table I. Calculated Energies and Bond Lengths According to the Floating Localized Orbital Model with Spherical Gaussian Functions

Molecules	Negative total energy (hartrees) This work	Hartree–Fock	Bond length, A Calcd	Obsd[h]
H_2	0.956	1.1336[a]	0.780	0.741
LiH	6.572	7.9851[b]	1.712	1.595
Li_2	12.282	14.8718[c]	2.807	2.672
HF	84.635	100.0580[d] (min at R = 0.920 A)	0.779	0.917
BeH_2				
linear	13.214		1.412	
BH_3				
planar D_{3h}	22.297	26.2358[e] (min at R = 1.16 A)	1.245	1.19 (av)
CH_4				
tetrahedral	33.992	39.8660[f] (min at R = 1.10 A)	1.115	1.093
NH_3				
pyramidal C_{3v}	47.568	55.9748[f] (min. at R = 1.04 A)	1.008	1.012
planar D_{3h}	47.141		1.489	. . .
H_2O				
angular C_{2v}	64.290	75.9224[f] (min at R = 0.963 A)	0.880	0.957
linear	64.203		1.621	. . .
C_2H_2		C≡C	1.210	1.205
linear, sym	64.684	76.7916[g]		
		C—H	1.073	1.059
C_2H_4		C═C	1.350	1.337
planar, D_{2h}	65.836	78.0012[g]		
		C—H	1.104	1.085

[a] W. Kolos and C. C. J. Roothaan, *Rev. Mod. Phys.*, **32,** 205 (1960). [b] D. D. Ebbing, *J. Chem. Phys.*, **36,** 1361 (1962). [c] P. E. Cade and A. C. Wahl, quoted by G. Das, *ibid.*, **46,** 1568 (1967). [d] E. Clementi, *ibid.*, **36,** 33 (1962). [e] B. D. Joshi, *ibid.*, **46,** 875 (1967). [f] R. Moccia, *ibid.*, **37,** 910 (1962); **40,** 2164, 2176, 2186 (1964). [g] R. J. Buenker, S. D. Peyerimhoff, and J. L. Whitten, *ibid.*, **46,** 2029 (1967). [h] Observed values are taken from L. E. Sutton, Ed., "Interatomic Distances," Special Publication No. 18, The Chemical Society, London, 1965.

The calculated bond lengths are surprisingly close to the experimental values, on the average within 4.4%. Considering the crudeness of the model no such direct quantitative similarity would be expected. However, the model should give general trends which certainly are present.

Bond angles are not as successfully calculated although NH_3 and H_2O are properly predicted to be pyramidal and bent, respectively. The angles are: H–N–H, 88.0° (obsd 106.6°); and H–O–H, 89.5° (obsd 104.5°).

Dipole moments (Debyes) are calculated to be: LiH, 6.56 (obsd 5.882); HF, 1.66 (obsd 1.98); H_2O, 1.92 (obsd 1.84); and NH_3, 1.71 (obsd 1.46).

LiH is a simple example which shows how the orbital parameters behave. One orbital turns out to have a small "radius," ρ, equal to 0.707 bohr and is located 0.0076 bohr from the Li nucleus on the side opposite from the proton. This can be considered an inner-shell Li orbital. The other orbital has a radius of 2.44 bohrs and is located about 89% of the way from Li to H. The bond could therefore be interpreted to be predominantly ionic.

This model has a simpler relation to the original electron pairing and shared pair concepts of Lewis[4] than does the quantum mechanical valence bond method since the present model uses only one orbital per electron pair bond instead of two. It is also related to molecular orbital theory through the use of a single determinantal wave function. Localized molecular orbitals have been discussed particularly by Lennard-Jones and co-workers[5] and by Edmiston and Ruedenberg.[6]

This model constitutes an extension of the Kimball–Neumark[7] spherical Gaussian orbital model which was applied by Neumark to the simple systems He and H_2. The "charge cloud" model of Kimball[7] which conceives of uniformly charged spheres for electron pairs resembles the present model but does not allow for overlap of the spheres and is only pseudo-quantum mechanical. Likewise the tangent-sphere model of Bent[8] and related ideas of King,[9] although giving considerable qualitative insight into molecular structure, are not sufficient for quantitative calculations. Details of the calculations and additional results will be published elsewhere.

Acknowledgment. This research has been supported by a grant from the National Science Foundation to Northwestern University.

(4) G. N. Lewis, *J. Am. Chem. Soc.*, **38,** 762 (1916); "Valence and the Structure of Atoms and Molecules," Chemical Catalog Co., New York, N. Y., 1923.

(5) J. E. Lennard-Jones, *Proc. Roy. Soc.* (London), **A198,** 1, 14 (1949); G. G. Hall and J. E. Lennard-Jones, *ibid.*, **A202,** 155 (1950); J. E. Lennard-Jones and J. A. Pople, *ibid.*, **A202,** 166 (1950); A. C. Hurley, J. E. Lennard-Jones, and J. A. Pople, *ibid.*, **A220,** 446 (1953).

(6) C. Edmiston and K. Ruedenberg, *Rev. Mod. Phys.*, **35,** 457 (1963).

(7) Ph.D. Dissertations by G. F. Neumark, L. M. Kleiss, H. R. Westerman, and J. D. Herniter, Columbia University; reviewed by J. R. Platt in "Encyclopedia of Physics," Vol. 37, Part 2, S. Flügge, Ed., Springer-Verlag, Berlin, 1961, p 258.

(8) H. A. Bent, *J. Chem. Educ.*, **40,** 446, 523 (1963); **42,** 302, 348 (1965).

(9) L. C. King, *Chemistry*, **37,** 12 (1964).

Arthur A. Frost, Bryant H. Prentice, III, Robert A. Rouse
Department of Chemistry, Northwestern University
Evanston, Illinois
Received March 29, 1967

VII-5c Reprinted from *The Journal of the American Chemical Society* **90**, 1965 (1968)

A Floating Spherical Gaussian Orbital Model of Molecular Structure. IV. Hydrocarbons[1]

Arthur A. Frost and Robert A. Rouse

Contribution from the Department of Chemistry, Northwestern University, Evanston, Illinois. *Received October 19, 1967*

Abstract: The FSGO model is applied to a series of hydrocarbons: methane, ethane, ethylene, acetylene, and cyclopropane. Bond lengths and bond angles are obtained with an average absolute deviation of 1.7 and 1.0%, respectively, from observed values. The barrier to internal rotation of ethane is calculated.

The floating spherical Gaussian orbital (FSGO) model is discussed in detail in paper I[2a] of this series. As currently applied, the model predicts the electronic and geometric structure of singlet ground states of molecules with localized orbitals without the use of any arbitrary or semiempirical parameters. The localized orbitals are constructed by using single normalized spherical Gaussian functions

$$\phi_i(\vec{r} - \vec{R}_i) = \left(\frac{2}{\pi\rho_i^2}\right)^{3/4} \exp[-(\vec{r} - \vec{R}_i)^2/\rho_i^2]$$

with orbital radius, ρ_i, and position, $\boldsymbol{R}_i$. A single Slater determinant represents the total electronic wave function. If S is the overlap matrix of the set of nonorthogonal localized orbitals $\{\phi_i\}$ and $T = S^{-1}$, then the energy expression for a molecule is

$$E = 2\sum_{j,k}(j|k)T_{jk} + \sum_{k,l,p,q}(kl|pq)[2T_{kl}T_{pq} - T_{kq}T_{lp}]$$

where $(j|k) = \int\phi_j h\varphi_k \, dv$ (h = one-electron operator) and $(kl|pq) = \int\phi_k(1)\phi_l(1)(1/r_{12})\phi_p(2)\phi_q(2) \, dv_1 dv_2$. The energy is minimized by a direct search procedure with respect to all parameters: orbital radii, ρ_i, orbital positions, $\vec{R}_i$, and nuclear positions.

Previous work with the FSGO model[2b,3] indicated that the model works best for elements in the middle of the second row of the periodic table. So in choosing larger and more complicated molecules to which to apply the model, the hydrocarbons were a natural selection. Simple hydrocarbons present a variety of molecular structure, double and triple bonds, ring compounds, and several interesting energetic quantities.

Results

Methane provides a simple example for detailed consideration of the application of the FSGO model to the hydrocarbons. In order to make the calculation most efficient, tetrahedral symmetry is imposed, thus allowing identification of symmetrically related integrals which are calculated only once. This in effect places a symmetry constraint on the minimization; *i.e.*, while the orbital positions and radii are varied, they are varied in such a way that the symmetry is maintained. Parameters were defined so that the four C–H orbital radii are varied together; the twelve orbital positions (x, y, and z for four C–H bonding orbitals) formed another parameter. The carbon 1s orbital radius was another parameter, but the orbital was held at the origin to maintain symmetry. The 12 hydrogen positions were defined by the fourth and final parameter with the carbon being held at the origin.

This symmetry constraint is not as serious as one might suspect. Several calculations were made with relaxed symmetry with LiH and BH_3, and the results were essentially the same as corresponding symmetry-constrained calculations. The remainder of the results reported here have the indicated symmetry imposed and presumably no error is introduced by such tactics.

The results for methane were presented in paper III[3] along with other first-row hydrides but are reproduced here for comparison with the other hydrocarbons. As-

(1) Portions of this paper were presented at the Computers in Chemistry Symposium, San Diego, Calif., June 1967, and at the 154th National Meeting of the American Chemical Society, Chicago, Ill., Sept 1967.

(2) (a) A. A. Frost, *J. Chem. Phys.*, **47**, 3707 (1967) (paper I); (b) **47**, 3714 (1967) (paper II).

(3) Paper III: A. A. Frost, *J. Phys. Chem.*, **72**, 1289 (1968); also see the preliminary communication: A. A. Frost, B. H. Prentice, III, and R. A. Rouse, *J. Am. Chem. Soc.*, **89**, 3064 (1967).

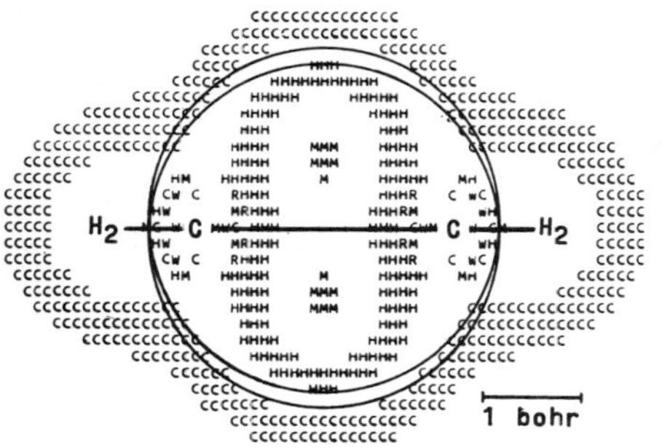

Figure 1. Electron density diagram for ethylene in a plane perpendicular to plane of the molecule. Contours, if drawn, would increase successively at outside and inside edges of the lettered regions of the computer output. Outside edge of C's, 0.05 au; inside of C's, 0.10 au; outside of H's, 0.15 au, etc. The circles represent the double-bond orbitals drawn with radii equal to the "orbital radii," ρ. The centers of these orbitals are only 0.2 bohr apart, but the maxima of the electron density lie in regions M which are about 1.5 bohrs apart because of the orbital overlap effect in the electron density.

suming a tetrahedral structure, the bond length was calculated to be 2.107 bohrs. The bonding orbitals had an orbital radius of 1.694 bohrs and were located 1.256 bohrs from the carbon nucleus. The orbital radius of the carbon inner shell was 0.328 bohr. The total energy for methane was −33.992 hartrees.

Both staggered and eclipsed conformations of ethane were investigated and Tables I and II present the results.

Table I. Ethane (Staggered, D_{3d}, Atomic Units)

Nuclear positions		x	y	z
C_a		0.0	0.0	1.418
C_b		0.0	0.0	−1.418
H_{a1}		0.0	1.979	2.166
H_{a2}		1.714	−0.990	2.166
H_{a3}		−1.714	−0.990	2.166
		(three other H's at z = −2.166 with x and y interchanged)		
Orbitals	**Radii**	x	y	z
C_a 1s	0.328	0.0	0.0	1.418
C_b 1s	0.328	0.0	0.0	−1.418
C_a–C_b	1.646	0.0	0.0	0.0
C_a–H_{a1}	1.695	0.0	1.201	1.869
C_a–H_{a2}	1.695	1.040	−0.601	1.869
C_a–H_{a3}	1.695	−1.040	−0.601	1.869
		(three other orbitals at z = −1.869 with x and y interchanged)		

Bond lengths, bohrs	Bond angles, deg	Energy, hartrees
C–C, 2.837 C–H, 2.116	C–C–H, 110.7 H–C–H, 108.2	−67.005

In ethylene the double bond is encountered for the first time. It is constructed by placing two identical spherical Gaussian functions at equal distances above and below the plane of the molecule at the midpoint of the C–C axis as shown in Figure 1. When the minimization procedure is applied to ethylene, the bonding orbitals tend to coalesce similarly to the cases mentioned previously in connection with lone pairs in paper III[3]

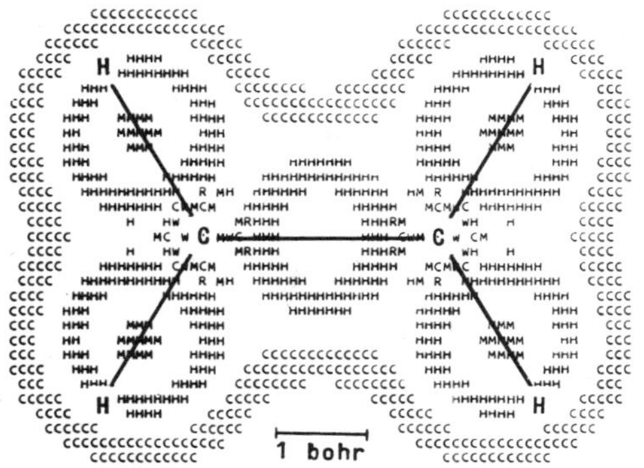

Figure 2. Electron density diagram of ethylene in the plane of the nuclei: contour intervals, 0.05 au.

in this series. So the positions of these orbitals were fixed and the other parameters varied to attain a minimum. The results for ethylene are presented in Table III. Figure 2 shows the electron density in the plane of the nuclei.

Table II. Ethane (Eclipsed, D_{3h}, Atomic Units)

Nuclear positions		x	y	z
C_a		0.0	0.0	1.430
C_b		0.0	0.0	−1.430
H_{a1}		0.0	1.9071	2.195
H_{a2}		1.707	−0.985	2.195
H_{a3}		−1.707	−0.985	2.195
		(three other H's at z = −2.195)		
Orbitals	Radii	x	y	z
C_a 1s	0.328	0.0	0.0	1.430
C_b 1s	0.328	0.0	0.0	1.430
C_a–C_b	1.651	0.0	0.0	0.0
C_a–H_{a1}	1.695	0.0	1.193	1.892
C_a–H_{a2}	1.695	1.033	−0.597	1.892
C_a–H_{a3}	1.695	−1.033	−0.597	1.892
		(three other orbitals at z = −1.892)		

Bond lengths, bohrs	Bond angles, deg	Energy, hartrees
C–C, 2.859 C–H, 2.114	C–C–H, 111.2 H–C–H, 107.7	−66.996

Table III. Ethylene (D_{2h}, Atomic Units)

Nuclear positions		x	y	z
C_a		0.0	0.0	1.277
C_b		0.0	0.0	−1.277
H_{a1}		1.791	0.0	2.338
H_{a2}		−1.791	0.0	2.338
		(two other H's at z = 2.338)		
Orbitals	Radii	x	y	z
C_a 1s	0.328	0.0	0.0	1.277
C_b 1s	0.328	0.0	0.0	−1.277
C_a=C_b	1.794	0.0	(0.100)	0.0
	1.794	0.0	(−0.100)	0.0
C_a–H_{a1}	1.642	1.084	0.0	1.937
C_a–H_{a2}	1.642	−1.084	0.0	1.937
		(two other orbitals at z = −1.937)		

Bond lengths, bohrs	Bond angles, deg	Energy, hartrees
C=C, 2.554 C–H, 2.081	C–C–H, 120.7 H–C–H, 118.7	−65.835

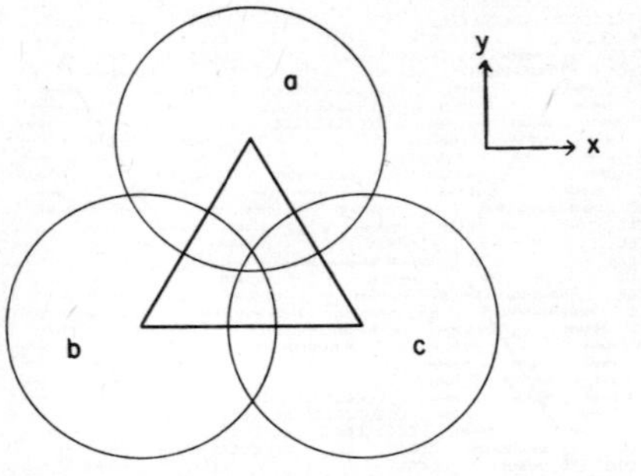

Figure 3. Schematic diagram of construction of the triple bond in acetylene.

The triple bond in acetylene is represented by three identical orbitals placed at the corners of an equilateral

Table IV. Acetylene (Nuclei Linear; Orbitals D_{3h}, Atomic Units)

Nuclear positions		x	y	z
C_a		0.0	0.0	1.148
C_b		0.0	0.0	−1.148
H_a		0.0	0.0	3.187
H_b		0.0	0.0	−3.187
Orbitals	Radii	x	y	z
C_a 1s	0.328	0.0	0.0	1.147
C_b 1s	0.328	0.0	0.0	−1.147
C≡C	1.781	0.0	(0.200)	0.0
	1.781	(0.173)	(−0.100)	0.0
	1.781	(−0.173)	(−0.100)	0.0
C_a–H_a	1.581	0.0	0.0	2.360
C_b–H_b	1.581	0.0	0.0	−2.360
Bond lengths, bohrs				Energy, hartrees
C≡C, 2.295				−64.678
C–H, 2.039				

Table V. Cyclopropane (D_{3h})

Nuclear positions		x	y	z	
C_a		0.0	1.672	0.0	
C_b		1.448	−0.836	0.0	
C_c		−1.448	−0.836	0.0	
H_{a1}		0.0	2.814	1.771	
H_{a2}		0.0	2.814	−1.771	
Orbitals	Radii, bohrs	x	y	z	
C_a 1s	0.3278	0.0	1.672	0.0	
C_b 1s	0.3278	1.448	−0.836	0.0	
C_c 1s	0.3278	−1.448	−0.836	0.0	
C_a–C_b	1.770	0.781	0.451	0.0	
C_b–C_c	1.770	0.0	−0.901	0.0	
C_c–C_a	1.770	−0.781	0.451	0.0	
C_a–H_{a1}	1.683	0.0	2.367	1.076	
C_a–H_{a2}	1.683	0.0	2.367	−1.076	
Bond length	Calcd, bohrs	Calcd, Å	Exptl nmr[a]	Exptl electron density	
C–C	2.897	1.533	(1.510)	1.510[b]	
C–H	2.108	1.115	1.123	1.089[b]	
Bond angle, H–C–H		114.4°	114.4°	115.1°	
Energy	−98.895 hartrees				

[a] L. C. Snyder and S. Meiboom, *J. Chem. Phys.*, **47**, 1480 (1967).
[b] J. Bastiansen, F. N. Fritsch, and K. Hedberg, *Acta Cryst.*, **17**, 538 (1964).

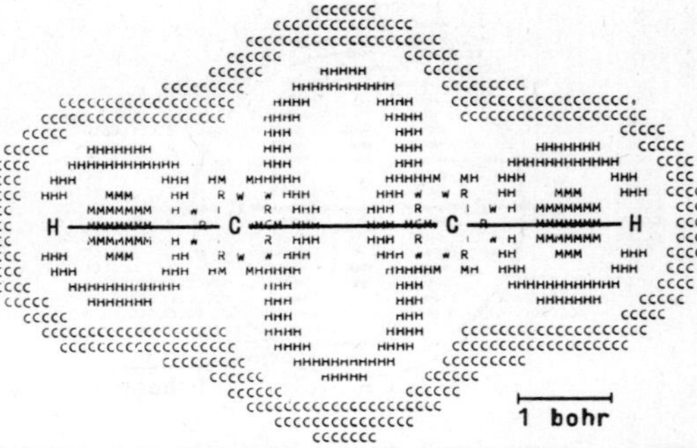

Figure 4. Electron density diagram of acetylene in the *xz* plane. The density in the triple bond is not quite exactly symmetrical with respect to reflection across the internuclear axis due to the finite separation of the orbital centers. Results for the *yz* plane are essentially the same: contour interval, 0.05 au.

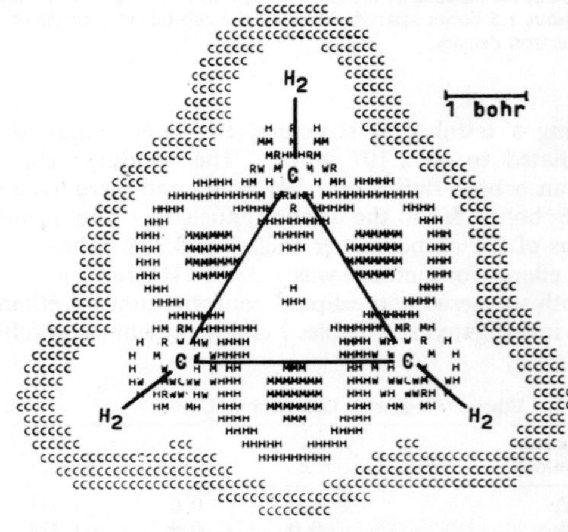

Figure 5. Electron density diagram of cyclopropane in the plane of the carbons: contour interval, 0.05 au.

triangle located at the midpoint of the C–C axis; this is shown schematically in Figure 3. Coalescence of the bonding orbitals again occurs and their positions are fixed and the minimization carried forth. Table IV and Figure 4 show the acetylene results.

The simplest ring hydrocarbon is cyclopropane and its predicted geometry, compared to observed geometry, is shown in Table V. Figure 5 shows the electron density map found for cyclopropane.

Discussion

Table VI compares the calculated bond lengths with experimental values. Examination of the results shows excellent agreement between experimental and calculated bond lengths and bond angles. Bond-length errors range from 0.7 to 2.5% with an average absolute error of 1.7%. Bond angle errors are found between 0.58 and 1.2% with an average absolute error of 1.0%. While little can be added to the discussion of methane, several points should be emphasized in the ethane investigation. The FSGO model yields an H–C–H angle of 108.2° within each methyl group in the staggered form that is smaller than the tetrahedral angle, in qualitative agreement with experiment. Also in the

Table VI. Comparison of FSGO Structural Results with Other Calculations and Experiment (Å)

	This calcn	SCF	Extended Hückel[b]	Observed[c]
Methane				
C–H	1.115	1.09[a]	1.02	1.093
Ethane				
C—C	1.501	...	1.92	1.534
C—H	1.120	...	1.0	1.093
∠H–C–H	108.2°	...	(109.45°)	109.1°
Ethylene				
C=C	1.351	1.333[d]	1.47	1.337
C—H	1.101	1.056[d]	0.95	1.086
∠H–C–H	118.7°	...	125°	117.3°
Acetylene				
C≡C	1.214	1.215[e]	0.85	1.205
C—H	1.079	1.085[e]	1.0	1.059
Cyclopropane				
C—C	1.533	1.54[f]		1.510[g]

[a] M. Krauss, *J. Res. Natl. Bur. Std., A*, **68**, 635 (1964). [b] R. Hoffmann, *J. Chem. Phys.*, **39**, 1397 (1963). [c] L. E. Sutton, Ed., "Interatomic Distances," Special Publication No. 18, The Chemical Society, London, 1965. [d] J. W. Moscowitz and M. C. Harrison, *J. Chem. Phys.*, **42**, 1726 (1965). [e] H. Preuss, *et al.*, "Arbeitsbericht der Gruppe Quanten Chemie," No. 5, Max-Planck-Institut, Munich, 1967, p 68. [f] H. Preuss and H. Diercksen, *Intern. J. Quantum Chem.*, **1**, 361 (1967). [g] Reference *b*, Table V.

eclipsed form the angle is still smaller, and in addition the C–C bond length is greater. All of these effects as well as the energy barrier can be interpreted by assuming repulsions between the two methyl groups. It would be tempting to interpret these effects in terms of repulsions between filled orbitals, but this does not seem to be possible in this model where the orbitals are definitely nonorthogonal. Although in paper III the bond angles for H_2O and NH_3 as calculated were too small by 17%, more faith can be put in calculations of hydrocarbons where lone-pair electrons are absent.

The energy of the eclipsed conformation is higher by 5.7 kcal/mole, yielding a barrier to internal rotation about twice the observed value, 3.03 ± 0.30 kcal/mole.[4] This energy is very poor compared to most other calculations.[5]

Ethylene serves as an example for the double bond. As was previously discussed, a "banana bond" approach was taken in constructing the double bond, and, upon minimizing, the bonding orbitals coalesced and the calculation became impossible as the off-diagonal elements of the overlap matrix converged toward unity, making the inverse matrix results void. It should be pointed out that the failure lies ultimately in an inherent violation of the Pauli exclusion principle manifested through the Slater determinant: with coalescence two electrons in the same molecule have the same set of quantum numbers, hence two rows in the Slater determinant are identical and the wave function is nonexistent. Following the appendix of paper III,[3] the orbitals in the limit of coalescence for ethylene are

$$\lim_{R \to 0}(a + b) = \text{s-type Gaussian}$$

$$\lim_{R \to 0}(a - b) = \text{p}_y\text{-type Gaussian}$$

where R = the distance between the two Gaussians.

(4) D. R. Lide, *J. Chem. Phys.*, **29**, 1426 (1958).

(5) R. M. Pitzer, *ibid.*, **47**, 965 (1967). This author summarized other calculations in addition to his own.

So the model gives the simple molecular orbital picture of a double bond (a σ bond and a π bond), which is equivalent to the banana bond through a transformation of the Slater determinantal wave function. It should also be noted that the H–C–H angle is less than 120°, possibly due to interaction of the C–H bonding orbitals with the double-bond orbitals as predicted by Gillespie.[6]

The triple bond of acetylene is interpreted in a similar manner. In the limit of $R \to 0$ (the distance from the center of the equilateral triangle to a vertex), two p- and one s-type orbital are obtained (see Figure 3).

$$\lim_{R \to 0}(a + b + c) = \text{s-type Gaussian}$$

$$\lim_{R \to 0}(b - c) = \text{p}_x\text{-type Gaussian}$$

$$\lim_{R \to 0}(2a - b - c) = \text{p}_y\text{-type Gaussian}$$

Again the simple molecular orbital picture of a triple bond is found, a σ bond and two π bonds. To check how close to the limit of $R = 0$ are the calculations for ethylene and acetylene, the POLYATOM program[7] was employed to calculate the energy using actual p orbitals. These orbitals were centered on the C–C axis, perpendicular to it, and had the identical exponential parameter as the spherical Gaussians. The results are (in hartrees)

	This calcn	POLYATOM
Ethylene	−65.835	−65.836
Acetylene	−64.678	−64.682

The energies are within 0.2% of each other and indicate that, in holding the bonding orbitals fixed at a small separation, no large error was introduced in the minimization process.

Cyclopropane exhibits several interesting features. Careful examination of the positions of the C–C bonding orbitals discloses that they lie outside of the C–C axis, essentially forming a bent bond. When cyclopropane was first synthesized, the existence of a three-membered ring was rationalized by saying the C–C bond must be bent in order to make the compound stable. The FSGO model reinforces this initial view. The H–C–H angle is greater than tetrahedral and is in excellent agreement with the experimental results shown in Table V.

Table VI also sets forth a comparison of the FSGO structural predictions with other calculated results. The FSGO findings are much better than the extended Hückel results as one expects and are comparable with accurate SCF results. In comparison to the extended Hückel method, the FSGO model is strictly quantum mechanical and involves no approximations or semiempirical parameters.

Figure 6 shows a comparison of calculated and experimental bond lengths, the FSGO predictions falling very close to the theoretical 45° line. In addition the C–C bond lengths decrease with increasing unsaturation,

(6) R. T. Gillespie and R. S. Nyholm, *Quart. Rev.* (London), **11**, 339 (1957); R. T. Gillespie, *Can. J. Chem.*, **38**, 818 (1960); R. A. Rouse and A. A. Frost, paper presented at 154th National Meeting of the American Chemical Society, Chicago, Ill., Sept 1967.

(7) I. G. Csizmadia, M. C. Harrison, J. W. Moskowitz, and B. T. Sutcliffe, *Theoret. Chim. Acta*, **6**, 191 (1966).

ethane the longest and acetylene the shortest. The C–H bond lengths also follow the correct trend with the C–H bond in ethane being the longest, then ethylene, and finally acetylene. This result comes automatically from the calculation without any necessary reference to changing hybridization of orbitals. The hybridization concept has no particular significance in the present model. Inspection of the carbon 1s orbital positions and radii discloses that these orbitals are always found very close to the carbon nuclei and with the same orbital radii. This indicates the inert behavior of this pair of electrons in the hydrocarbons.

Comparisons of FSGO and SCF energies are shown in Table VII. As pointed out previously[3] the FSGO energies are typically about 85% of the SCF values. This deviation is principally due to the lack of cusps in the inner-shell orbitals at the nuclei.

Table VII. Comparison of FSGO Energy Results (in Hartrees) to Hartree–Fock SCF Calculations

	This calcn	SCF
Methane	−33.992	−40.198[a]
Ethane		
Staggered	−67.005	−79.09797[b]
Eclipsed	−66.996	−79.09233[b]
Ethylene	−65.835	−78.0012[c]
Acetylene	−64.678	−76.7916[c]
Cyclopropane	−98.895	−116.02[d]

[a] C. D. Ritchie and H. F. King, *J. Chem. Phys.*, **47**, 564 (1967). [b] R. M. Pitzer, *ibid.*, **47**, 965 (1967). [c] R. J. Buenker, S. D. Peyerinnhoff, and J. L. Whitten, *ibid.*, **46**, 2029 (1967). [d] Reference *f*, Table VI.

The time required on the CDC3400 for computing each of these molecules varied from about 1 min for methane to 4 min for two-carbon species to 8 min for cyclopropane. These times include the search for the energy minimum for various nuclear configurations.

In a typical run the internuclear distances and bond angles not determined by symmetry are set at initial values which are within only about 10–20% of values that are known experimentally or just guessed from previous experience. Starting positions and radii of orbitals are estimated from experience with simpler molecules. The pattern search minimization procedure is then begun with 10% changes in the various parameters and allowed to proceed with decreasing step sizes until final parameter changes are of the order of only 0.001% of the original values. The cyclopropane calculation was accomplished with a variation of nine parameters in which the energy was computed 268 times requiring about 1.7 sec for each fixed set of parameters.

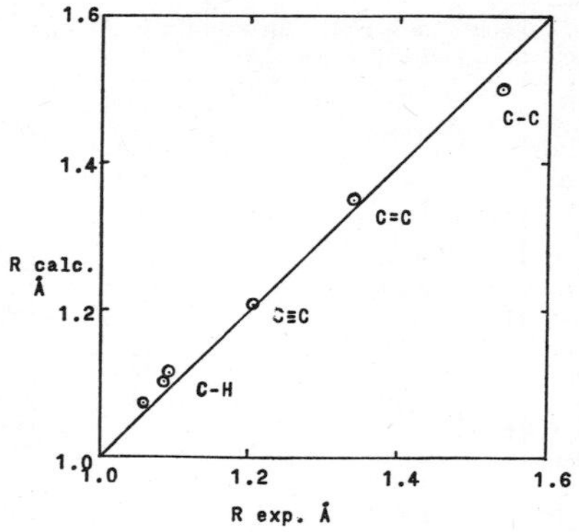

Figure 6. Plot comparing calculated and experimental CC and CH bond lengths.

Conclusion

The hydrocarbon results presented here have enlarged the range of applicability of the floating spherical Gaussian orbital model. Structural predictions are very good and encourage the expansion of the investigation to systems with a larger number of electrons.

The basis set in the FSGO model contains the fewest number of orbitals and yet in a sense contains an infinite variety. With all orbital parameters being varied, the very best orbitals are found while still keeping the number of orbitals low. We have chosen to call this optimized basis set a Lewis basis set in deference to G. N. Lewis, who suggested the electron dot model to have pairs of electrons representing bonds. It is important to notice the difference between a Lewis basis set and a minimal basis set in SCF jargon. For example, in a methane calculation, the SCF minimal basis set consists of five carbon orbitals: 1s, 2s, $2p_x$, $2p_y$, and $2p_z$, and four 1s H orbitals for a total of nine. The Lewis basis set would have just five orbitals: a 1s on carbon and four C–H bonding orbitals. Only in such cases as the He and Ne atoms would the basis sets be the same. The Lewis basis set through application of the FSGO model allows manageable programming and yields good results in reasonable times.

Finally we wish to emphasize that calculations with the FSGO model are strictly *ab initio* involving no arbitrary or semiempirical parameters.[8]

Acknowledgments. The National Science Foundation supported this research with a grant to Northwestern University.

(8) The present model is equivalent to the variation of a single "constellation" as defined by H. Preuss, *Mol. Phys.*, **8**, 157 (1964). For earlier history see paper I and ref 24 therein.

VII-6

Electron Correlation and How It Supplements Molecular Orbital Theory

VII-6a Reprinted from *The Journal of Chemical Physics* **36**, 706 (1962)

Many-Electron Theory of Atoms and Molecules.* I. Shells, Electron Pairs vs Many-Electron Correlations

OKTAY SINANOĞLU
Sterling Chemistry Laboratory, Yale University, New Haven, Connecticut

(Received July 24, 1961)

A theory is developed (a) to see what the physically important features of correlation in atoms and molecules are; (b) based on this to obtain a quantitative scheme for N-electron systems as in He and H_2; (c) to see what happens to the "chemical" picture, to semiempirical theories, and to shell structure, when correlation is brought in. It is shown why, unlike in an electron gas, in many atoms and in molecules mainly pair correlations are significant. In configuration-interaction, multiple excitations arise not as three or more electron "collisions," but as several binary "collisions" taking place separately but at the same time. The validity of theory is shown on Be, LiH, and boron. The theory does *not* depend on any perturbation or series expansion and the r_{12}-coordinate method can now be used for an N-electron system as in He and H_2.

INTRODUCTION

SHELL structure in atoms and molecules is due to nuclear wells and the exclusion principle. Long-range Coulomb repulsions between electrons have a disrupting influence on this structure.

The best one-electron orbits are obtained by letting each electron move in a potential averaged over the motions of all others (Hartree-Fock method). Based on orbitals such as these, much chemical and spectroscopic behavior of atoms and molecules is explained qualitatively or semiempirically.

Yet, full calculations of energy end up with so much error that often even molecule formation is not accounted for. This is usually attributed to electron correlation, the importance of which has been stressed in recent reviews.[1,2]

As the large errors show, electrons affect one another through their instantaneous potentials, not just by their average potentials as in the Hartree-Fock method.[1] Thus, one is faced with a many-electron problem. But what sort of a many-body problem? Does the long range of Coulomb repulsions cause all electrons to correlate with one another in a complicated way? Were this the case to the extent of overcoming the effects of the exclusion principle, shell structure would be wiped out; an atom or molecule would be more like a drop of electron liquid. The actual shell structure and radial electron densities, e.g., from x-ray results[3] are close to those of Hartree-Fock method.

This and *semiempirical success of orbitals must be reconciled with large correlation errors.*

The only accurate calculations on atoms and molecules so far[4] are still those on two-electron systems based on the classical work[1] of Hylleraas on He, and James and Coolidge on H_2, using trial functions containing the interelectron coordinate r_{12} directly. This method has not been extended to N-electron system. For such systems configuration-interaction[1] (C.I.) is used. This is slowly convergent[5] except when used to remove degeneracy or "resonance." As N and/or the finite orbital basis set used is increased, the number of multiply-excited configurations increases very rapidly into thousands, soon making an atom or molecule too large even for a computer. Best C.I. on small molecules so far has given only about 1/3 to 1/2 of the binding energies.

A method is needed whose difficulty does not increase rapidly with the number of electrons.

OBJECTIVES OF THE THEORY

The objectives of the theory developed here are:

(a) to examine the physically important features of correlation in atoms and molecules;

(b) to try to obtain a quantitative scheme based on this picture for the N-electron system as in the He, H_2 case;

(c) to see what happens to the "chemical picture," to semiempirical theories, and to shell structure when correlation is brought in.

Previously[6,7] we used perturbation theory, solving the first-order equations rigorously and starting from Hartree-Fock solutions as unperturbed. The first-order wave function contains correlations in an electron pair at a time. Here we shall go beyond perturbation theory and complete the picture in several directions along objectives (a), (b), and (c). The new results and the over-all scheme that come out were sum-

* For an over-all view of the scope and main results of this series of articles see reference 8.

[1] P. O. Löwdin, in *Advances in Chemical Physics* (Interscience Publishers, Inc., New York, 1959), Vol. II.

[2] K. S. Pitzer, in *Advances in Chemical Physics* (Interscience Publishers, Inc., New York, 1959), Vol. II.

[3] See, e.g., R. Daudel, R. LeFebvre, and C. Moser *Quantum Chemistry* (Interscience Publishers, Inc., New York, 1959).

[4] The next best calculation is the recent one on Be (reference 19), where $(2s)^2-(2p)^2$ "resonance" is of utmost importance and already corrects for much of the correlation.

[5] For a discussion of the r_{12} method vs C. I. and status of the correlation problem see J. C Slater, *Quantum Theory of Atomic Structure* (McGraw-Hill Book Company, Inc , New York, 1961), Vol. II.

[6] O. Sinanoğlu, Proc. Roy. Soc. (London) **A260**, 379 (1961).

[7] O. Sinanoğlu, Phys. Rev. **122**, 491, 493 (1961).

marized in a communication.[8] This series of articles presents the detailed theory.

Here we look at the physical characteristics of correlation in atoms and molecules as compared to electron gas and nuclear "matter." We show why significant correlations occur only in electron pairs despite long-range Coulomb forces. There are separate correlations in many pairs at the same time, but correlations due to the "collisions" of three or more electrons are unlikely, as shown on Be, LiH, boron, etc. We obtain a variational wave function exact in pairs, and find the correlation energy. The pairs are uncoupled if the molecule does not contain many nearly "metallic" electrons delocalized in the same region. For objective (b), the theory is expressed in terms of correlation functions which are in closed form and not limited to C.I., although C.I. may be used in short steps to estimate three or more electron correlation effects.

In article II we give several methods for obtaining pair correlations exactly. A variation method is applied to electron pairs in the Hartree-Fock "sea" just as in the He or H_2 problem. *The r_{12}-coordinate can now be used directly in the N-electron system.* Only about N two-electron problems need be solved. There is no "nightmare of innershells." One looks at each pair physically and then selects the appropriate method, r_{12} coordinate, or C.I. "open shell," etc. for it. Perturbation theory comes out as a special case. We derive the Schrödinger equations of electron pairs in the Hartree-Fock "sea" which (for example in ethylene) have precisely the form of the "Π-electron Hamiltonian." By simple transformations, correlation energies are converted into those in bonds and into accurately defined *intra*molecular Van der Waals' interactions. Both quantitative and semiempirical uses of the theory will be discussed.

COMPARISON WITH OTHER MANY-BODY PROBLEMS

To bring out the main physical features of the many-electron problem in atoms and molecules, we first discuss some related many-body problems. Another reason for doing this is to show that it is neither appropriate nor necessary to apply the theories devised for these to most atoms or molecules.

A. Nuclear Matter

The forces between nucleons are short range. This and the exclusion principle cause only local nucleon pair correlation to dominate inside nuclei. The strong short-range repulsions are described by a hard-core potential. With this type of potential one cannot start with the Hartree-Fock method because orbital energies become infinite.

In the Brueckner theory,[9] nucleons, while correlating pairwise, move in a medium that they constantly polarize. This is analogous to a charged particle moving in a dielectric medium with its polarization cloud around it, whereas in the Hartree-Fock method a particle moves in the potential of the undisturbed charge distribution.

Brueckner theory is applied mainly to infinite nuclear "matter" in which single nucleon orbitals are plane waves. In finite systems[10] difficulties arise. The polarized medium potential is strongly dependent on the orbital upon which it acts, so that orthogonal ground-state orbitals cannot be obtained easily. Also, the basis set is now discrete, and the simple momenta integrations with plane waves must be replaced by slowly convergent infinite sums.[11] As in ordinary perturbation theory,[1] these are difficult to evaluate. Moreover, the self-consistency procedure in getting orbitals, medium potential, pair correlations, back to orbitals, ··· becomes much too complicated and impractical.

B. Infinite Electron Gas

This idealized model of a metal replaces the periodic lattice (hardly seen by electrons, e.g., in Na) by a smeared out, uniform positive charge distribution. Then Hartree-Fock orbitals are plane waves.

Electron correlation is determined by the deviation of the instantaneous Coulomb potential $g_{ij}=1/r_{ij}$ between two electrons, say i and j, from the average [i.e., Hartree-Fock (H.F.)] potential they exert on one another. This deviation will be referred to as "fluctuation potential" (see below).

Let two electrons with opposite spins be in plane waves, ϕ_i and ϕ_j, normalized in a large spherical box of radius L. The fluctuation potential is

$$g_{ij}-S_i(j)-S_j(i), \tag{1}$$

with the H.F. part of the potential of, for example, i acting on j

$$S_i(j)=\langle\phi_i(\mathbf{r}_i),\, g_{ij}\phi_i(\mathbf{r}_i)\rangle_{r_i}$$
$$=\frac{3}{4\pi L^3}\int(\exp i\mathbf{k}_i\cdot\mathbf{r}_i)\exp(-i\mathbf{k}_i\cdot\mathbf{r}_i)\frac{1}{|\mathbf{r}_{ij}|}d\tau_i. \tag{2}$$

Changing the origin to j, so that $\mathbf{r}_i=\mathbf{r}_{ij}$, the simple integral above gives

$$S_i(j)=S_j(i)=3/2L, \tag{3}$$

which is uniform and tends to zero for $L\to\infty$.

Thus, in the electron gas, the fluctuation potential is the full g_{ij} itself with its long range. Moreover, each electron is totally delocalized and equally likely to be found anywhere. Therefore, *many* electrons "see" each other with their fluctuation potentials at the same time and thus correlate all at once, giving rise to collective screening and oscillation effects.

[8] O. Sinanoglu, Proc. Natl. Acad. Sci. **47**, 1217 (1961).

[9] K. A. Brueckner, in *The Many Body Problem, Grenoble, École d'été de physiqué théoriqué, Les Houches* (John Wiley & Sons, Inc., New York, 1959).

[10] H. A. Bethe, Phys. Rev. **103**, 1353 (1956).

[11] In article II we shall obtain some results that eliminate this difficulty also for the finite nucleus problem.

PHYSICAL ASPECTS OF ELECTRON CORRELATION IN ATOMS AND MOLECULES

Fortunately, in atoms and molecules (except in organic dyes, etc., where metallic behavior is approached) the problem is very different from that of an electron gas. The symmetry, nuclear wells, and exclusion principle tend to localize electrons (often in pairs) in different regions already in the orbital approximation.

The Hartree-Fock wave function is an ideal starting point for atoms and molecules. It takes care of the long-range effects of Coulomb repulsions, as we shall see; that is why it gives good radial densities. It has other advantages too, which will become apparent. Fortunately, even for molecules, H.F. solutions are rapidly becoming available.

Before examining how electrons correlate, we introduce some definitions. Let

$$\Psi=\phi_0+\chi; \qquad \langle\phi_0, \chi\rangle=0 \tag{4}$$

be the exact wave function of an N-electron system with a single-determinant H.F. solution ϕ_0, and with χ correcting for correlation. (This article discusses, mainly, closed shells or systems with a single determinant ϕ_0. For nonclosed shells, degeneracies must be removed first; the qualitative conclusions of this article then apply to what is left over. See II.) Then energy is separated into a H.F. and a correlation part.

$$E=\frac{\langle\psi, H\psi\rangle}{\langle\psi,\psi\rangle}$$

$$=E_{\mathrm{H.F.}}+\frac{2\langle\phi_0, (H-E_{\mathrm{H.F.}})\chi\rangle+\langle\chi, (H-E_{\mathrm{H.F.}})\chi\rangle}{1+2\langle\phi_0,\chi\rangle+\langle\chi,\chi\rangle},$$

$$E_{\mathrm{H.F.}}=\langle\phi_0, H\phi_0\rangle, \tag{5}$$

and

$$H=\sum_{i=1}^{N}h_i^0+\sum_{i>j}^{N}g_{ij}=H_0+H_1. \tag{6}$$

g_{ij} is the bare-nuclei Hamiltonian of electron i and $h_i^0=1/r_{ij}$. Atomic units (1 a.u.$=27.2$ ev) are used. We separate H into a part

$$H_0=\sum_{i=1}^{N}(h_i^0+V_i) \tag{7}$$

for independent electrons in the total H.F. potential V_i of the "sea" ϕ_0 and the residual *fluctuation potential* part[6]

$$H_1=\sum_{i>j}^{N}g_{ij}-\sum_{k=1}^{N}V_k(\mathbf{x}_k)=\sum_{i>j}^{N}[g_{ij}-\bar{S}_i(j)-\bar{S}_j(i)]. \tag{8}$$

In Eq. (7), V_i acts on electron i; $V_i=V_i(\mathbf{x}_i)$.

We denote spin-orbitals by numerals 1, 2, $\cdots$, N starting with the lowest orbital with spin α; e.g., $1\equiv(1s\alpha)$, $2\equiv(1s\beta)$, $3\equiv(2s\alpha)$, etc. Then

$$\phi_0=\mathcal{A}(123\cdots N); \qquad \langle\phi_0,\phi_0\rangle=1, \tag{9}$$

where $\mathcal{A}$ is the antisymmetrizer

$$\mathcal{A}=\frac{1}{(N!)^{\frac{1}{2}}}\sum_P(-1)^P P. \tag{10}$$

In Eq. (9), before $\mathcal{A}$ is applied, *electron i* with space, spin coordinates $\mathbf{x}_i$ occupies *orbital i*; e.g., $1(\mathbf{x}_1)$, $2(\mathbf{x}_2)$, etc. *Thus in Eq.* (8), *i, j can refer equally to orbitals or to electrons*, where we separated the total H.F. potentials into those of orbitals.[6]

$$\sum_{k=1}^{N}V_k(\mathbf{x}_k)=\sum_{i>j=1}^{N}[\bar{S}_i(j)+\bar{S}_j(i)]. \tag{11}$$

$\bar{S}_i(j)$ is the Coulomb (S_i) plus the exchange (R_i'') potential of orbital i acting on electron j.

$$\bar{S}_i(j)=S_i(j)-R_i''(j), \tag{12}$$

$$S_i(j)\equiv S_i(\mathbf{x}_j)=\langle i(\mathbf{x}_i), g_{ij}i(\mathbf{x}_i)\rangle_{\mathbf{x}_i}, \tag{13a}$$

$$R_i''(\mathbf{x}_j)j(\mathbf{x}_j)=\langle i(\mathbf{x}_i), g_{ij}j(\mathbf{x}_i)\rangle_{\mathbf{x}_i}i(\mathbf{x}_j). \tag{13b}$$

$\langle\ \rangle_{\mathbf{x}_i}$ means integration over coordinates $\mathbf{x}_i$ only. It is important to include "self-potentials" ($i=j$) in V_i to make it the same acting on any electron in ϕ_0.

From

$$H_0\phi_0=E_0\phi_0, \tag{14a}$$

H.F. spin-orbitals satisfy

$$e_i i=0,$$

where

$$e_i=h_i^0+V_i-\epsilon_i \tag{14b}$$

and ϵ_i are orbital energies. Then

$$E_{\mathrm{H.F.}}=E_0+E_1,$$

$$E_0\equiv\langle\phi_0, H_0\phi_0\rangle=\sum_{i=1}^{N}\epsilon_i. \tag{15a}$$

and

$$E_1\equiv\langle\phi_0, H_1\phi_0\rangle=-\sum_{i>j}^{N}(J_{ij}-K_{ij}'') \tag{15b}$$

J_{ij} and K_{ij} are the Coulomb and exchange integrals.

$$J_{ij}=\langle ij, g_{ij}ij\rangle; \qquad K_{ij}''=\langle ij, g_{ij}ji\rangle.$$

Writing

$$H-E_{\mathrm{H.F.}}=\sum_{i=1}^{N}e_i+\sum_{i>j}^{N}m_{ij}. \tag{16}$$

and defining[12]

$$m_{ij}=g_{ij}-\bar{S}_i(j)-\bar{S}_j(i)+J_{ij}-K_{ij}'', \tag{17}$$

Eq. (5) becomes

$$E-E_{\mathrm{H.F.}}=\frac{2\langle\phi_0, \sum_{i>j}m_{ij}\chi\rangle+\langle\chi, (\sum_i e_i+\sum_{i>j}m_{ij})\chi\rangle}{1+2\langle\phi_0,\chi\rangle+\langle\chi,\chi\rangle}. \tag{18}$$

[12] This notation is different from that used in reference 6. There m_{ij} denoted only part of Eq. (17).

m_{ij} is the complete fluctuation potential since Eq. (15b) is the expectation value of Eq. (8). Our potential must not be confused with the "*correlation potential*" t_{ij} of the Brueckner theory which need not be introduced for atoms and molecules (see below). It would unnecessarily complicate this problem.

Two things in Eq. (18) determine the nature of electron correlation in atoms and molecules: (i) charge distribution in the "sea" ϕ_0, (ii) pairwise fluctuation potentials m_{ij}.

i. Charge Distribution

Although Hartree-Fock orbitals (in molecules SCF MO's) are suitable in describing electronic excitation and ionization, they do not necessarily reflect the probabilities of finding electrons at various places relative to one another. The exclusion principle, $\mathcal{A}$ in Eq. (9), tends to arrange electrons with like spin in ϕ_0 to make them as far apart as possible. Linnett and Põe[13] looked at $|\phi_0|^2$ as a function of $3N$ coordinates and determined the positions of electrons that make it a maximum. For example, in Ne eight electrons are found arranged on two tetrahedra with arbitrary orientation (since Ne is ^1S). *Localization of electron pairs due to antisymmetry goes beyond the rule: two electrons per orbital.*

Directional bonding and shapes of molecules are mainly due to these electron distributions and are better represented by transforming the H.F. orbitals into "*equivalent orbitals*," leaving ϕ_0 unchanged.[14] The new orbitals are often localized and look like bonds or inner shells except in some Π systems containing electrons that are too free. Our theory will show that qualitative conclusions drawn on these orbitals should be valid in spite of correlation.

ii. Fluctuation Potential

Coulomb wells of nuclei and resulting geometric features of H.F. orbitals make also the fluctuation potentials, m_{ij} in Eq. (17), very different from those in an electron gas. In atoms and molecules with no strong delocalization, m_{ij} is usually of short range in directions going from orbital to orbital.

This is shown in Fig. 1 for the Be atom. Take a $(1s)$ electron and place a second arbitrary electron with opposite spin ar r_2. The $(1s)$ electron [taking a $(1s)$ Slater orbital with exponent 3.70 as sufficiently close to H.F.] is most likely to be found at $r_1=0.27$ a.u., the Bohr orbit of Be^{2+}. Were it there instantaneously, electron-two would see the full repulsion $g_{12}=r_{12}^{-1}$, not just the average Coulomb potential $S_1(r_2)$. Both potentials are shown in Fig. 1. The g_{12} is for the two electrons and the nucleus placed on the same line.

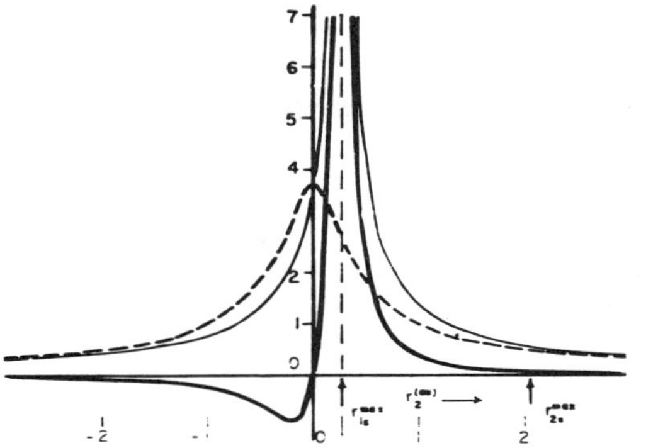

FIG. 1. Beryllium atom. Fluctuation potential as seen by an electron at r_2 due to a $(1s)$ electron ($\delta_1=3.70$) with opposite spin instantaneously at its Bohr radius (r_1^{max}). The nucleus, r_1 and r_2 are on the same line. ——— $=g_{12}=1/r_{12}$ (a.u.); - - - $=S_1(r_2)=$ H.F. potential of electron 1 acting on 2; ——— $=g_{12}-S_1(2)=$ fluctuation potential (a.u.).

The deviation, $g_{12}-S_1(2)$ differing from m_{12} by the constant value, 0.39 a.u. is the fluctuation potential (also in Fig. 1), which shows where electron-two (r_2) wants to be if electron-one is at $r_1=0.27$ a.u. Both g_{12} and $S_1(2)$ are of long range, but $m_1(2)$ has a short range. It has died off before reaching a distance of 2.1 a.u., about the Bohr radius of $(2s)$ in Be. The singularity of g_{12} at $r_{12}=0$ is the main feature left in m_{12}. Note also the minimum in $m_1(2)$ on the other side of the nucleus. Similar curves and potential surfaces can be drawn by varying r_1 and θ_{12} (the angle between r_1 and r_2) in addition to r_2. These would aid the selection of trial functions to give maximum charge density at the dips and zero at the singularities.

The short range of m_{12} in the radial direction explains, of course, why correlation energy between a $(1s)$ and a $(2s)$ electron is much smaller than that in the $(1s)^2$ shell even though Coulomb force is long range. More important, this also makes three- and four-electron correlations negligible. While correlating with one another, two $(1s)$ electrons do not see the third and fourth electrons that the exclusion principle has put in a different shell, *a safe distance away from the fluctuation potential* of $(1s)$.

The exclusion principle and fluctuation potentials operate this way generally. Even in many-electron shells (single determinant) pairs stay apart and cannot quite see other's m_{ij} potential. Out of any three electrons in a system, at least two will have the same spin and will not come together.

Thus in atoms and molecules with a single determinant ϕ_0, (except large and highly conjugated ones with nearly free electrons) we expect local pair correlations to dominate strongly over three or more electron "collisions." This is substantiated with more evidence below.

[13] J. W. Linnett and A. J. Põe, Trans. Faraday Soc. 47, 1033 (1951).

[14] See, e.g., J. A. Pople, Quart. Revs. 11, 291 (1957). These orbitals are due to Lennard-Jones.

PREVIOUS PERTURBATION THEORY RESULTS

The first-order wave function χ_1 of the perturbation theory contains just these pair correlations if we start with the H.F. "sea" ϕ_0 as unperturbed. Previously[6] we solved the first-order equation with operator techniques and got

$$\chi_1=\sum_{i>j}^{N}\frac{\mathcal{A}}{\sqrt{2}}\left\{123\cdots N\frac{\hat{u}_{ij}^{(1)}}{(ij)}\right\}, \tag{19}$$

which gives for second-order energy (the first term in correlation energy)

$$E_2=\sum_{i>j}^{N}\langle\mathcal{B}(ij),\, g_{ij}\hat{u}_{ij}^{(1)}\rangle. \tag{20}$$

The pair correlation in (ij) is given by $\hat{u}_{ij}^{(1)}$, which satisfies the first-order part[15] of the Schrödinger equation[6] of two electrons in the H.F. "sea." $\mathcal{B}$ is the two electron antisymmetrizer

$$\mathcal{B}=(1/\sqrt{2})(1-P_{ij}). \tag{21}$$

One crucial effect of the H.F. "sea" on a pair is that two electrons, say i and j with opposite spins, correlating with each other, keep away from any other electron k which will have the same spin as either i or j. Thus i and j, not only in H.F. orbital motion, but also while correlating, are prevented from going into spin-orbitals, $k\neq i, j$, already occupied in ϕ_0.

Hartree-Fock orbitals i and j are solutions of Eq. (14b), and so are already adjusted to the polarizing influence of the average "sea" potential V_i of ϕ_0. Therefore, in the language of C.I., single excitations which would have otherwise resulted from this "orbital average polarization" of individual electrons[16] are also missing from $\hat{u}_{ij}^{(1)}$.

The rigorous solution[6] reflects these effects in the *orthogonality* of $\hat{u}_{ij}^{(1)}$ to the *orbitals* k.

$$\langle\hat{u}_{ij}^{(1)}, k\rangle=\langle\hat{u}_{ij}^{(1)}(\mathbf{x}_i,\mathbf{x}_j),\, k(\mathbf{x}_i)\rangle_{\mathbf{x}_i}=0, \tag{22}$$

where $k=1, 2, 3, \cdots N$ includes i and j. Integration is again over coordinates of only one electron. How pair functions are made orthogonal to orbitals was shown in detail in reference 6 (see also article II). The procedure is quite similar to ordinary Schmidt orthogonalization, except that we have scalar products over only one electron space, as in Eq. (22).

χ_1 and E_2 are often sufficient approximations (see II), as expected physically from the importance of pair correlations. But sometimes the strength of m_{ij} may require the pair functions to be determined to greater accuracy than first order.

Here we shall go beyond perturbation theory in two respects and (i) examine many electron correlations; (ii) use in part a variational approach to take pair correlations into account exactly, and not just to first order.

CLUSTERS OF ELECTRONS IN THE HARTREE-FOCK "SEA"

The exact wave function χ will contain not just binary "collisions" in the Hartree-Fock "sea" as does χ_1, Eq. (19), but also the "collisions" of successive 3, 4, $\cdots$, and N electrons at a time. By "collisions" we mean, of course, electrons coming within reach of the fluctuation potentials m_{ij} of each other. χ may be written *exactly* as

$$\chi=\mathcal{A}\left\{(1234\cdots N)\left[\sum_{i=1}^{N}\frac{\hat{f}_i}{(i)}+\frac{1}{(2!)^{\frac{1}{2}}}\sum_{i>j}^{N}\frac{\hat{u}_{ij}}{(ij)}+\frac{1}{(3!)^{\frac{1}{2}}}\sum_{i,j,k}^{N}\frac{\hat{U}'_{ijk}}{(ijk)}+\cdots+\frac{1}{(N!)^{\frac{1}{2}}}\frac{\hat{U}'_{123\cdots N}}{(123\cdots N)}\right]\right\}. \tag{23}$$

Each term replaces some of the initial H.F. orbitals by a function involving only the corresponding electrons. These functions are antisymmetric with respect to odd permutations; i.e.,

$$\hat{u}_{ij}(\mathbf{x}_i,\mathbf{x}_j)=-\hat{u}_{ij}(\mathbf{x}_j,\mathbf{x}_i), \tag{24}$$

$$\hat{U}'_{ijk}(\mathbf{x}_i,\mathbf{x}_j,\mathbf{x}_k)=-\hat{U}'_{ijk}(\mathbf{x}_j,\mathbf{x}_i,\mathbf{x}_k),\ \text{etc.} \tag{25}$$

The caret indicates orthogonality to all the orbitals of the H.F. "sea" as in Eq. (22).

Equation (23) is *not* an infinite series for a system with a finite number N of electrons. But it can be obtained from the configuration-interaction (C.I.) series by summations. The infinite C.I. expansion[1] contains all unique Slater determinants formed from a complete one-electron basis set. Equation (23) replaces all single, double, etc., excitations from different H.F. orbitals by closed, as yet undetermined, functions. One could substitute this *exact* χ into the variational expression, Equation (18), and obtain a finite set of coupled integro-differential equations. But this would only amount to rewriting the original Schrödinger equation. Instead we look at each term and pick out the important effects.

The $\hat{f}_i$ in Eq. (23) adjusts the orbital i to the field of all other electrons to an extent beyond the H.F. potential. $\hat{u}_{ij}$ are pair correlations as in Eq. (19), but exact to all orders. U'_{ijk} contains all triple excitations from the orbitals $i, j,$ and k in ϕ_0. In $\hat{u}_{ij}$, electrons i and j "collide" (via m_{ij}) while each one is moving in the H.F. potential $V_i=V_j$. Part of $\hat{U}'_{ijk}$, on the other hand, changes this V_i so that while i and j "collide," i, say, will be moving in a "sea" in which k is being polarized.[9,17] This effect and $\hat{f}_i$ would come in if we went from the H.F. "sea" to the Brueckner "sea." However, in terms of perturbation theory, which is rapidly convergent if one starts with H.F., both effects alter the energy only in the fourth and higher orders, and so are both negligible. That electron densities come out well

---

[15] In reference 6 we denoted the first-order pairs, $\hat{u}_{ij}^{(1)}$ by $\hat{u}_{ij}$. From here on $\hat{u}_{ij}$ will be used to denote a general pair function not restricted to any order or to perturbation theory.

[16] See, e.g., O. Sinanoğlu, J. Chem. Phys. **33,** 1212 (1960).

[17] R. Brout, Phys. Rev. **111,** 1324 (1958).

with H.F. orbitals is also evidence that $\hat{f}_i$ may be neglected.

Parts of $\hat{U}'_{123}$, $\hat{U}'_{1234}\cdots$ correct for three, four, $\cdots$, n, $\cdots$ electron "collisions." Such n electron correlations will be large only if the probability of finding all the n electrons within reach of each other's fluctuation potential is appreciable. Above, the nature of these potentials and exclusion effects led us to believe that this probability is small. Then we expect χ to be well approximated (see, however, below) just by

$$\chi_s = \frac{\mathcal{A}}{\sqrt{2}} \sum_{i>j}^{N} (123\cdots N)\frac{\hat{u}_{ij}}{(ij)}, \qquad (26a)$$

$$\langle \hat{u}_{ij}(\mathbf{x}_i, \mathbf{x}_j), k(\mathbf{x}_i)\rangle_{\mathbf{x}_i} = 0, \qquad (k=1, 2, 3, \cdots N)$$

simple generalization of χ_1, Eq. (19), to *exact pairs* still in the H.F. "sea."

Terms in Eq. (23) involving progressively larger numbers of electrons will be referred to as "clusters," just as in imperfect gas theory.[18] We shall show below that there is still an important physical effect missing from Eq. (26). In this problem, too, we will find two types of clusters not apparent from C.I.

RECENT CONFIGURATION-INTERACTION RESULTS ON Be AND LiH

Recently Watson[4,19] studied the Be atom by C.I. by starting with the Hartree-Fock solution, and examining the effect of adding single, double, triple, and quadruple excitations. Single (f_i) and triple (U_{ijk}') excitations were found totally negligible, as we anticipated. ***But appreciable quadruple excitations appeared.*** At first sight this is a surprising result; for, if three-electron correlations are negligible, there ought to be even less chance for four electrons to "collide."

Ebbing[20] too, did such a study on the LiH molecule, transforming his orbitals to H.F. SCF MO's with similar results. (There are also some single excitations. These still ought to vanish in energy to second order so long as the approximate H.F. molecular orbitals are obtained from Roothaan's self-consistent field procedure.[3] Nevertheless, if there is a nonzero effect on energy—no separate energy estimate was available—it must be due to the relative crudeness of molecular Hartree-Fock orbitals compared to atomic H.F.)

We show below that important four-excitations in Be and LiH found in these studies are *not* due to four-electron "collisions."

"UNLINKED CLUSTERS"—SIMULTANEOUS CORRELATIONS IN SEPARATE PAIRS

According to χ_s, Eq. (26), only one pair of electrons correlate at any one time. Actually, it is very likely that when two electrons are "colliding" in some part of the atom or molecule, other binary "collisions" should be taking place in other regions separately but at the same time.[17] We can talk about "time" here since the theory could have been based equally on a time-dependent formalism.[21] To account for these simultaneous pairs, all possible products of $\hat{u}_{ij}$'s which have no electrons in common must be added to χ_s

$$\chi \cong \chi_s + \chi_{ss} + \chi_{sss} + \cdots$$

$$\equiv \mathcal{A}\left\{\frac{1}{\sqrt{2}}\sum_{i>j}(123\cdots N)\frac{\hat{u}_{ij}}{(ij)} + \sum_{\substack{i>j \\ (i,\, j\neq k,\, l)}}^{N}\sum_{k>l}^{N}\frac{\hat{u}_{ij}\hat{u}_{kl}}{(ijkl)} + \frac{1}{2\sqrt{2}}\sum_{\substack{i>j \\ i,\, j\neq k,\, l\neq m,\, n}}^{N}\sum_{k>l}^{N}\sum_{m>n}^{N}(123\cdots N)\frac{\hat{u}_{ij}\hat{u}_{kl}\hat{u}_{mn}}{(ijklmn)}\cdots\right\}. \qquad (27)$$

For instance, in a 4-electron system (e.g., Be or LiH)

$$\chi_{ss} = (\mathcal{A}/\sqrt{2})(\hat{u}_{12}\hat{u}_{34} + \hat{u}_{13}\hat{u}_{24} + \hat{u}_{14}\hat{u}_{23}). \qquad (28)$$

Such products[22] are "unlinked clusters" as in imperfect gas theory[18] and also play an important role in other many-body problems.[9,17,21] They are automatically taken care of in Rayleigh-Schrödinger perturbation theory,[23] but in, for example, variational methods,[17] must be specially introduced.

Then in Eq. (23), "clusters" should be separated into "linked" and "unlinked" parts; e.g.,

$$\frac{2}{(3!)^{\frac{1}{2}}}\hat{U}'_{1234} = \hat{u}_{12}\hat{u}_{34} + \hat{u}_{13}\hat{u}_{24} + \hat{u}_{14}\hat{u}_{23} + \hat{U}_{1234}, \qquad (29)$$

where now $\hat{U}_{1234}$ is the true four-electron correlation ("collision"), the one expected to be small. Drawing straight lines between each pair joined by a "U,"

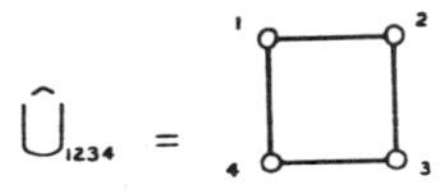

(Linked cluster)

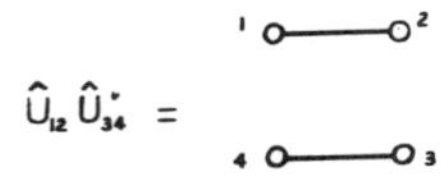

(Unlinked cluster)

FOUR-ELECTRON EXCITATIONS IN Be AND LiH ARE MAINLY UNLINKED CLUSTERS

Watson's[19] final wave function for Be contains 37 selected configurations with coefficients obtained by solving the 37×37 secular equation. We renormalize his function so as to have $\langle\phi_0, \Psi\rangle = \langle\phi_0, \phi_0\rangle = 1$ as in

---

[18] See, e.g., T. L. Hill, *Statistical Mechanics* (McGraw-Hill Book Company, Inc., New York, 1956).
[19] R. E. Watson, Phys. Rev. **119**, 170 (1960).
[20] D. D. Ebbing, Ph.D. thesis, Department of Chemistry, Indiana University, June, 1960.

[21] Such as used by J. Goldstone, Proc. Roy. Soc. (London) **A239**, 267 (1957).
[22] Just the first term in Eq. (28) is similar to a product of group functions in the "Atoms and molecules" method of M. Moffitt [Proc. Roy. Soc. London) **A210**, 245 (1951)]; but Eq. (28) is much more accurate, containing all the other pairs, and has many advantages such as a direct relation to C. I. Other group function methods will be discussed in II.
[23] K. A. Brueckner, Phys. Rev. **100**, 36 (1955).

TABLE I. Unlinked clusters vs four-electron excitations in Be atom.

Quadruply excited configuration[a]	Coefficient from 37-configuration wave function[a]	Coefficient calculated from double excitations as unlinked clusters [Eq. (28)]
$p_1^2(^1S)p_{11}^2(^1S)$	0.007063	0.0073
$p_1^2(^1S)s_1^2(^1S)$	0.005651	0.00647
$p_1^2(^1S)d_{11}^2(^1S)$	0.001585	0.00168
$p_{11}^2(^1S)d_1^2(^1S)$	0.0004639	0.000478
Energy contribution (Table III)	−0.075 ev	−0.074 ev

[a] R. E. Watson, Phys. Rev. 119, 170 (1960).

Eq. (4) instead of $\langle\Psi, \Psi\rangle = 1$, and use his notation for his orbitals.[19] Only configuration $1s^2\ p_1^2$ has an appreciable coefficient, −0.29706, compared to unity. This results from the near-degeneracy[19] of $2s^2$ with $2p^2$ and accounts for most of the correlation error in the outer shell. Such an important "resonance" is to be expected since Be is not an inert gas, but is capable of forming a metal. Most of the other configurations have coefficients less than | −0.027 | and contribute to the inner shell.

The C.I. wave function has the form

$$\Psi = \mathcal{A}(1234) + \sum_{k>l>4}^{\infty} a_{kl}\mathcal{A}(kl34) + \sum_{m>n>4}^{\infty} a_{mn}\mathcal{A}(12mn) + \cdots + \sum_{k>l>m>n>4}^{\infty} \gamma_{klmn}\mathcal{A}(klmn). \quad (30)$$

If the linked cluster $\hat{U}_{1234}$ in Eq. (29) is negligible, then coefficients of quadruple excitations must be simply obtainable from Eq. (28), i.e., by multiplying together already determined coefficients, a_{kl}, a_{mn}, etc. of double excitations ($\hat{u}_{ij}$'s in C.I. form).

Table I compares coefficients calculated from Eqs. (28) and (30) taking all contributing pair products, with the complete[19] γ_{klmn} from the full C.I. secular equation. The agreement is indeed satisfactory. Thirty-seven configuration results lie somewhat lower, indicating, as expected, small *negative* coefficients for $\hat{U}_{1234}$. One can now estimate coefficients for many more four-excited configurations from the available double excitations and Eq. (28) without augmenting the C.I. secular equation any further.

We found similar results (Table II) upon examining the Hartree-Fock C.I. coefficients by Ebbing[20] on LiH. Unlinked cluster coefficients are again calculated by multiplying coefficients of appropriate pair excitations and taking the right spin components. In Ebbing's notation ([11] is the inner, [22] the outer shell and 3, 4, 5, ··· are some excited orbitals), the main contribution to the coefficient of [3377] comes from the product of those of [1133] and [2277]; similarly [3357] comes from [1133] and [2257] (see calculation "D" in his[20] Table 6). There is no appreciable contribution from unlinked clusters to [3456] and [3457]. Their C.I. values and the difference between full C.I. and Eq. (28) values for [3357] and [3355] which are of the same magnitude, must be due to the linked cluster $\hat{U}_{1234}$. They are all small.

THREE OR MORE ELECTRON CORRELATIONS ARISE IN ENERGY EVEN WITH JUST PAIR FUNCTIONS

From physical considerations we arrived at the hypothesis that in atoms and in molecules with no near-metallic electrons, the wave function needs to correct only for pair correlations. Supported also by perturbation theory, this led us to Eq. (27), which we confirmed on Be and LiH.

Even with just pair functions, however, the energy will contain three or more electron correlations. The same thing happens in perturbation theory, since the first-order wave function χ_1 [Eq. (19)] determines the energy to third order.[6]

In the following sections, we obtain the complete energy given by χ_s, Eq. (26), and then by Eq. (27) using the variation method. We show that many electron correlation terms arising this way should be small too. Thus we end up with uncoupled pair energies reducing the problem to several He or H_2-like problems [objective (b)].

CORRELATION IN ONE PAIR AT A TIME

Take first χ_s, Eq. (26), which represents only one binary "collision" ($\hat{u}_{ij}$) in the entire system at a time. Use it as a trial function for χ in Eq. (5) or (18). In article II we show that, depending on what portion of this energy is minimized, one obtains different approximations for $\hat{u}_{ij}$. In particular, if just the

$$2\langle\phi_0, (H_1 - E_1)\chi\rangle + \langle\chi, (H_0 - E_0)\chi\rangle \quad (31)$$

part of Eq. (5) is minimized one gets χ_1. Then $\langle\chi_1, (H_1 - E_1)\chi_1\rangle$ becomes E_3, the third-order energy. Thus the analysis in this section includes perturbation

TABLE II. Unlinked clusters vs four-electron correlations in LiH.

Configuration[a]	Coefficient[a] from full C. I.	Coefficient calculated from unlinked clusters; [Eq. (28)]
[3377][b]	0.00113	0.00113
[3357]	−0.00086	−0.00078
[3355]	0.00062	0.00052
[3456][c]	{−0.00023, 0.00006	~0
[3457][c]	{0.00014, 0.00007	~0

[a] From calculation "D", Table 6 of D. D. Ebbing, Ph.D. thesis, Department of Chemistry, Indiana University, June, 1960.

[b] These are excited orbitals without spin in Ebbing's nomenclature, not our spin orbitals.

[c] These give two independent "codetors."

theory as a special case, but due to the generality of χ_s is by no means limited to it.

To analyze $E - E_{\mathrm{H.F.}}$ substitute χ_s, Eq. (26), and $\phi_0 = \mathfrak{A}(123\cdots N)$ into Eq. (18) and note that $\mathfrak{A}$ leaves

$$\sum_{i=1}^{N} e_i \quad \text{and} \quad \sum_{i>j}^{N} m_{ij}$$

unchanged. Also $\mathfrak{A}^2 = (N!)^{\frac{1}{2}}\mathfrak{A}$. Thus each term is reduced to a form having only one $\mathfrak{A}$ in it on the left side; e.g.,

$$\langle \phi_0, \sum_{i>j} m_{ij}\chi_s \rangle$$

$$= \left(\frac{N!}{2}\right)^{\frac{1}{2}} \left\langle \mathfrak{A}(123\cdots N), \left(\sum_{i>j}^{N} m_{ij}\right) \sum_{k>l}^{N} (12\cdots N) \frac{\hat{u}_{kl}}{(kl)} \right\rangle. \tag{32}$$

DIAGRAM TECHNIQUE

To proceed further, a diagram technique is very convenient. In a matrix element, such as in Eq. (32), we look at the direct term (without the $\mathfrak{A}$), draw a solid line for each $\hat{u}_{ij}$ and a dotted line for an m_{ij}. Thus, e.g.,

$\langle \mathfrak{B}(12), m_{12}\hat{u}_{12} \rangle$ =  (33)

Instead of doing complicated algebra, one now determines all possible diagrams arising from a matrix element. The exchange parts are then obtained by applying $\mathfrak{A}$ in the submatrix element corresponding to each diagram. This is done easily since a diagram is nonzero only for a subgroup of permutations that does not put electrons into orbitals so as to come out with $\langle \hat{u}_{ij}, k \rangle = 0$. One first finds this small subgroup, then applies only those permutations.

ENERGY WITH χ_s

With χ_s alone, the energy has a particularly simple form when ϕ_0 is the Hartree-Fock wave function. The first term in Eq. (18) gives

$$\langle \phi_0, (H_1 - E_1)\chi_s \rangle = \sum_{i>j}^{N} \langle \mathfrak{B}(ij), m_{ij}\hat{u}_{ij} \rangle. \tag{34}$$

This has the same *form* as E_2, Eq. (20), but now it is general. One point should be emphasized. When $\mathfrak{A}$ is applied in Eq. (32) to get Eq. (34), all the exchange terms (except $\mathfrak{B}$) drop out because of $\langle \hat{u}_{ij}, k \rangle = 0$, Eq. (26b). This does not mean however, that these exchange terms have been dropped by some artificial orthogonality condition. If we had had a u_{ij} in Eq. (32), nonorthogonal to $k \neq i, j$, after applying $\mathfrak{A}$ the new exchange terms would have been just the ones to Schmidt-orthogonalize u_{ij} to k to make it $\hat{u}_{ij}$. Thus just as in the perturbation result,[6,7] Eq. (34), is a shorthand form containing all the exchange effects.

In the other part of Eq. (18),

$$\chi_s, \left(\sum_i e_i + \sum_{i>j} m_{ij}\right)\chi_s,$$

the only nonvanishing diagrams, including their exchange parts, are

 $\equiv \langle \hat{u}_{ij}, (e_i - e_j + m_{ij})\hat{u}_{ij} \rangle$ (35a)

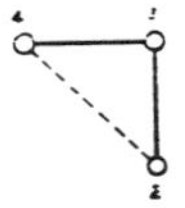 $= 2\{\langle 2_2\, \hat{u}_{34}(3,4), m_{24}\, \hat{u}_{23}(2,3)\, 4_4 \rangle - \langle 2_4\, \hat{u}_{34}(3,2), m_{24}\, \hat{u}_{23}(2,3)\, 4_4 \rangle\}$ (35b)

Also:

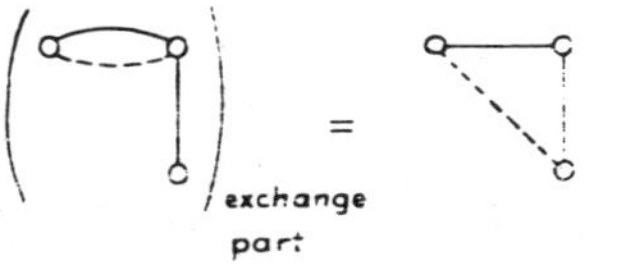

(35c)

In Eq. (35b), e.g., 2_4 means $2(\mathbf{x}_4)$ and $\hat{u}_{34}(3,4) \equiv \hat{u}_{34}(\mathbf{x}_3, \mathbf{x}_4)$. Only the triplet (234) is shown. There is one such term for every (ijk) and for each orientation of the triangle with numbering fixed on paper.

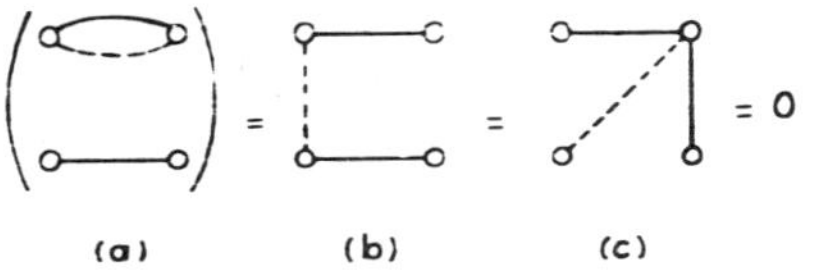
(36)

including their exchange parts vanish due to $\langle \hat{u}_{ij}, k \rangle = 0$.

The only other diagrams left have a dotted line with an open end.

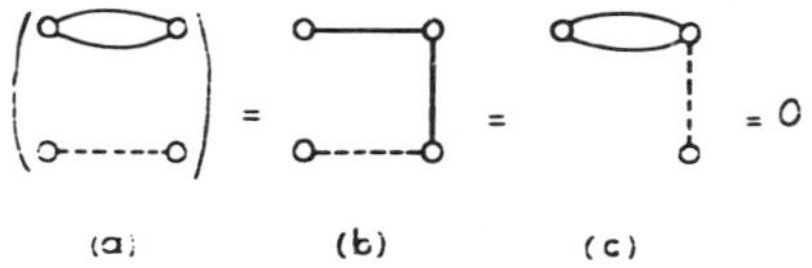
(37)

All these vanish, because (- - -o j) corresponds to the average of the fluctuation potential m_{ij} over orbital j. For example, Eq. (37c) has in it

$$\langle (1 - P_{13})1\hat{u}_{134}(3,4), m_{13}1_1\hat{u}_{34}(3,4) \rangle = 0, \tag{38}$$

which vanishes because $\langle (1 - P_{13})1_1, m_{13}1_1 \rangle_{\mathbf{x}_1} = 0$ from Eqs. (12), (13), and (17). Such open-end dotted lines indicate polarization by the average potential of orbitals. The H.F. choice for ϕ_0 and V_s eliminates the effects of this average polarization not only on single excitations, but also on diagrams such as Eq. (37).

Finally, using Eqs. (34) to (37), Eq. (18) becomes

$E \leq E_{\mathrm{H.F.}}$ [illegible] $+ 4\sum$ ) (39a)

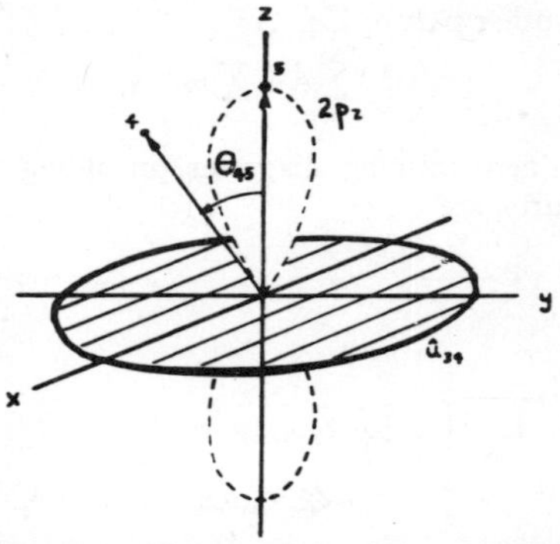

FIG. 2. The exclusion effect in boron atom $(1s^2 2s^2 2p_z)$. Due to $(2p_z)$, the $(2s)^2$ correlation $\hat{u}_{34}$ is confined to the vicinity of the xy plane. Electron 5 is shown at a most probable position.

where

$$\bar{\epsilon}_{ij} \equiv 2\langle B(ij), m_{ij}\hat{u}_{ij}\rangle + \langle \hat{u}_{ij}, (e_i + e_j + m_{ij})\hat{u}_{ij}\rangle \quad (39b)$$

and

$$D \equiv 1 + \langle \chi_s, \chi_s \rangle = 1 + \sum_{i>j} \langle \hat{u}_{ij}, \hat{u}_{ij} \rangle. \quad (39c)$$

The energy contains the pair correlations, but also the normalization change, discussed in the next section, and three-body correlations. The (ijk) triangle in Eq. (39a) means that the fluctuation potential of j causes simultaneous fluctuations in the orbital charge distributions i and k. Instantaneously induced distortions in the two charge clouds than interact electrostatically.

Now we see more specifically why such three-electron correlations should be small. Because of the exclusion principle j cannot be close both to i and k at the same time. When its fluctuation disturbs i, in general it will not reach k [see Fig. 1]. Another way of putting it: in

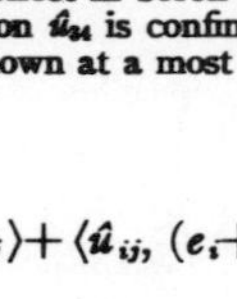

[see also Eq. (35b)],

$$\langle i\hat{u}_{kj}, g_{ki}\hat{u}_{ij}k\rangle, \quad (40)$$

the extensions of $\hat{u}_{kj}$ and $\hat{u}_{ij}$ in space are determined by the ranges of m_{kj} and m_{ij} and by where orbitals (ijk) are. Because of the exclusion principle, at least one of the electrons is in a different orbital. This, supplemented by Eq. (26b), means $\hat{u}_{kj}$ and $\hat{u}_{ij}$ will not both be large in the same region of space; therefore their product will be small everywhere.

This is easily seen in Be or LiH, where two pairs are localized in radially different regions. But what about cases with three or more electrons radially in close proximity?

FLUCTUATION POTENTIAL AND EXCLUSION EFFECT IN BORON

The boron atom $(1s^2 2s^2 2p_z)$ is the simplest example of such a case. Consider the three-electron correlation in $(2s_\alpha 2s_\beta 2p_z\alpha) \equiv (345)$. Almost all of $(2s)^2$ correlation, $\hat{u}_{34}$ in the Be atom[19] and presumably also in Boron is due to strong mixing ("resonance") with $(2p)^2$. In Be and B^+, $\hat{u}_{34}$ is ^1S, a combination of three $(2p)^2$ determinants containing p_x, p_y, and p_z. When $(2p_z\alpha)$, i.e., spin-orbital 5 is placed on top of the $(2s)^2$ shell to construct the boron atom; however, $\hat{u}_{34}$ gets changed to become orthogonal to p_z; $\langle \hat{u}_{34}, 5\rangle = 0$. The new $\hat{u}_{34}$ therefore is now a combination of only p_x and p_y and instead of over the whole sphere is large only on the xy plane as shown in Fig. 2.

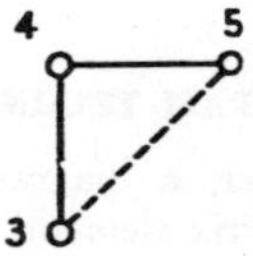

will be appreciable if the product $\hat{u}_{34}\hat{u}_{45}$ is large. But $\hat{u}_{45}$ is large only where electrons 4 and 5 [orbitals $\mathcal{B}(45)$] come together, i.e., around p_z and where m_{45} is large. The fluctuation potential $m_{45} \sim [g_{45} - S_5(4)]$ that electron 4 sees along the fixed radius as a function of angle θ_{45} from z when 5 is placed at its most probable place is shown in Fig. 3. Note that m_{45}, hence $\hat{u}_{45}$ is very small just where ($\sim 90°$) $\hat{u}_{34}$ is large. Thus, their product and the three-electron correlation will be small everywhere. The same arguments apply to other triplets; e.g., $\hat{u}_{35}\hat{u}_{45}$ is small because, for the $\alpha\alpha$ spins pair, $\hat{u}_{35}$ is small.

In *neon*, $\hat{u}_{34}$ for $(2s)^2$ should practically vanish, since no $2p$ is left available. There is only *one* triplet in Ne, $(2s_\alpha 2s_\beta 2p_z\alpha)$, about which little may be said *a priori*. We expect to make direct calculations on Ne comparing many electron correlations with just pairs.

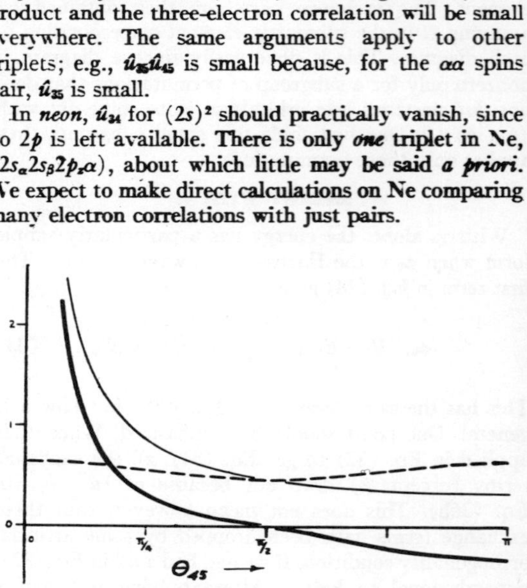

FIG. 3. Boron atom. Fluctuation potential ——— $= g_{45} - S_5(4)$ seen by the $(2s)$ electron, 4, due to electron 5 at the maximum of $(2p_z)$ (see Fig. 2); ——— $= g_{45} = 1/r_{45}$ (a.u.); - - - $= S_5(4) =$ H.F. potential (a.u.) of 5 acting on 4. Both electrons are at the same distance from the nucleus.

ESTIMATE OF TRIANGLES BY C.I.

In doubtful cases triangles may be estimated by C.I. (see II for other methods). Calculations in three stages would be illuminating:

(a) Take a trial function like χ_s, Eq. (26a), but with just one pair, say $\hat{u}_{34}$ in it. Then the entire energy is

$$E \leq E_0 + E_1 + \frac{\tilde{\epsilon}_{34}}{1+\langle \hat{u}_{34}, \hat{u}_{34}\rangle}. \quad (41)$$

Hitherto C.I. has been done mainly this way on molecules (e.g., Li_2).[24] Take another pair, say $\hat{u}_{23}$ and calculate it separately this way.

(b) Use the pairs from (a) and calculate the triangles they enter into.

(c) Take a χ_s with both $\hat{u}_{34}$ and $\hat{u}_{23}$ in. Determine them together by solving the big secular equation. Compare the C.I. coefficients from (a) and (c); also energies from (a), (b), and (c).

From Ebbing's C.I. (his calculations[20] "A", "B" and "C", and Table 5) on LiH, we estimate total triangle energies less than 0.01 ev. Coefficients obtained minimizing first Eq. (41), then Eq. (39a), (cf. Ebbing,[20] Table 6, "B" and "C") differ only by about 0.2%.

In general, we expect that Eq. (39a) can be approximated by

$$\tilde{E} = E_0 + E_1 + \frac{1}{D}\sum_{i>j}^{N} \tilde{\epsilon}_{ij}. \quad (42)$$

DIAGRAMS DUE TO UNLINKED CLUSTERS

To see the effect of unlinked clusters on energy, we substitute all of χ from Eq. (27) into Eq. (18). Terms coming from χ_{ss}, χ_{sss}, etc., will then be added to Eq. (39a).

All of $\langle \phi_0, \sum m_{ij}(\chi_s + \chi_{ss} + \cdots \rangle$ vanishes with diagrams like Eqs. (36a) and (36b). For simplicity consider a $4e^-$ case so that $\chi = \chi_s + \chi_{ss}$ only, where χ_{ss} is given by Eq. (28). This is a much more realistic wave function than χ_s alone since it allows for several separate binary "collisions" at the same time.

χ_{ss} has two effects on E; (i) it tends[25] to cancel the normalization change, $(1+\langle\chi, \chi\rangle)$ although with a finite number of electrons this cancellation is incomplete; (ii) it introduces some new linked clusters, many electron correlations analogous to the effect of $\hat{U}_{1234}$ etc.

For the $4e^-$ case, Eq. (28), we get in Eq. (18)

$$\langle \chi_s, (\sum e_i + \sum m_{ij})\chi_{ss}\rangle$$
$$= \sum_{\substack{i>j\ k>l \\ i,j\neq k,l}}^{N}\sum^{N} \langle \mathcal{B}(ij), n_{ij}\hat{u}_{ij}\rangle\langle \hat{u}_{kl}, \hat{u}_{kl}\rangle + R'. \quad (43)$$

The first terms in product form are unlinked clusters, Eq. (44a). R' contains the nonzero four-particle effects, Eqs. (44b), (44c), and (44d).

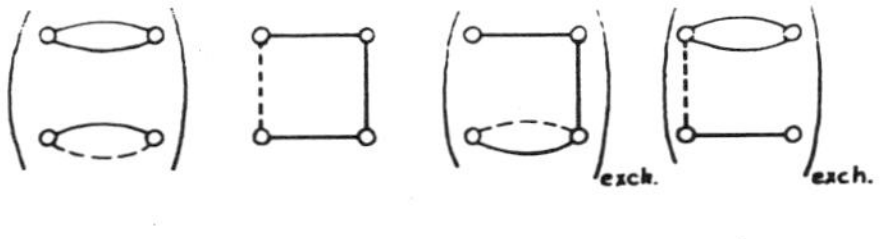

(a) (b) (c) (d) (44)

also

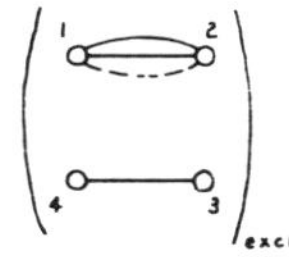

$$= \langle 34, m_{12}\hat{u}_{12}\rangle\langle \hat{u}_{12}, \hat{u}_{34}\rangle \quad (45)$$

and

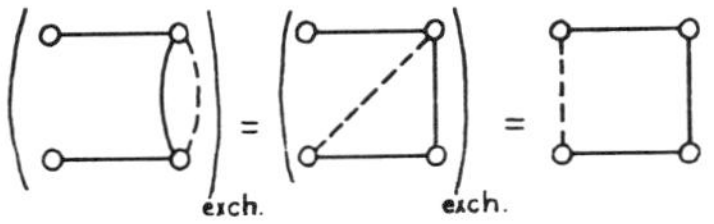

(46)

For example, Eq. (44b) is $\langle \mathcal{B}(12)\hat{u}_{34}, m_{12}\hat{u}_{14}\hat{u}_{23}\rangle$.

Equation (45) represents an exclusion effect not considered by $\langle \hat{u}_{ij}, k\rangle = 0$. Both pairs of electrons in $\hat{u}_{12}$ and $\hat{u}_{34}$ make excitations to the same states while correlating. However, electrons of one pair will be excluded from an excited state if that state happens to be occupied instantaneously by an electron of the other pair. This effect should be very small, since the fraction of time spent by each pair in any one excited state is already very slight. It is also dropped in the Brueckner method and also with generalized orthogonality conditions devised previously for many-electron group functions.[26]

The other term in E, Eq. (18), is

$$\langle \chi_{ss}, (\sum e_i + \sum m_{ij})\chi_{ss}\rangle$$
$$= \sum_{i>j}^{N}\sum_{k>l}^{N} \langle \hat{u}_{ij}, (e_i + e_j + m_{ij})\hat{u}_{ij}\rangle\langle \hat{u}_{kl}, \hat{u}_{kl}\rangle + 2R'', \quad (47)$$

where

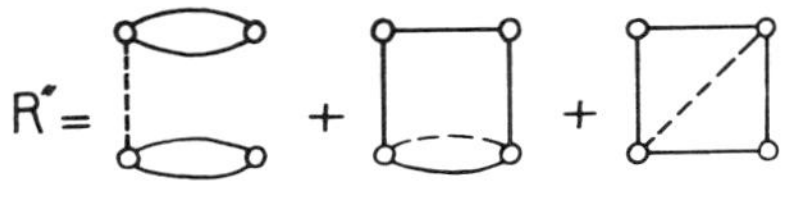

(48)

For a system with more than 4 electrons R' and R'' in Eqs. (43) and (47) will contain larger linked clusters in addition to Eqs. (44), (45), (46), and (48). All these represent many electron effects much smaller than the triangles discussed after Eq. (39a) and are expected to be negligible.

[24] E. Ishiguro, K. Kayama, M. Kotani, and Y. Mizuno, J. Phys. Soc. (Japan) 12, 1355 (1957).

[25] With the exact χ, Eq. (4), $(1+\langle\chi, \chi\rangle)$ in Eq. (5) would be canceled completely giving just $E = E_{H.F.} + \langle\phi_0, H_1\chi\rangle$.

[26] A. C. Hurley, J. E. Lennard-Jones, and J. A. Pople, Proc. Roy. Soc. (London) A220, 446 (1953).

TABLE III. Unlinked cluster effects in Be and LiH.

Be	
$D=1+\sum_{i>j}^{4}\langle\hat{u}_{ij},\hat{u}_{ij}\rangle$	1.0905
D' [Eq. (50)]	1.0906
$D_{12}=1+\langle\hat{u}_{34},\hat{u}_{34}\rangle$	1.08861
$D_{34}=1+\langle\hat{u}_{12},\hat{u}_{12}\rangle$	1.00172
$D_{13}=1+\langle\hat{u}_{24},\hat{u}_{24}\rangle$	1.0000
$D_{14}=1+\langle\hat{u}_{23},\hat{u}_{23}\rangle$	1.0001
$(\bar{\epsilon}_{12}/D)'$ (Configurations[a] "3" to "5" only)	−0.831 ev
Energy of four-excited configurations (6, 9, 19, 32)[a] from 37×37 C.I.	−0.075 ev
$(\bar{\epsilon}_{12}/D)'\times(D_{12}-1)$ = unlinked cluster energy.	−0.0074 ev
Unlinked cluster energy of $(p_1{}^2 d_{11}{}^2)$	−0.00019 ev
Full C.I. energy[a] of $(p_1{}^2 d_{11}{}^2)$	−0.00019 ev
LiH[b]	
$D'=D$	1.020
D_{12}	1.018
D_{34}	1.001
D_{13}, D_{24}, D_{23}, D_{14}	1.000

[a] R. E. Watson, Phys. Rev. **119**, 170 (1960); (configuration nos. refer to his Table III).
[b] Calculation "D", Table 6 in reference 13.

ENERGY WITH UNLINKED CLUSTERS

The total energy resulting from a χ with all pairs and their unlinked clusters, Eq. (27), is now obtained from Eqs. (18), (39a), (43), and (47) as

$$E\leq E_0+E_1+\sum_{i>j}^{N}\bar{\epsilon}_{ij}(D_{ij}/D')+\frac{R}{D'}. \qquad (49)$$

R is the sum of all true many electron correlation effects, i.e., linked clusters, e.g., triangles, squares, $\cdots$, Eqs. (44) to (46), (48) and larger ones for $N\geq 5$. The pair correlation energy $\bar{\epsilon}_{ij}$ was defined in Eq. (39b)

$$D'=1+\sum_{i>j}^{N}\langle\hat{u}_{ij},\hat{u}_{ij}\rangle+\langle\chi_{ss},\chi_{ss}\rangle+\cdots. \qquad (50)$$

D_{ij}'s come from factoring out the $\langle\hat{u}_{kl},\hat{u}_{kl}\rangle$ in Eqs. (43) and (47).

$$D_{ij}=1+\sum_{\substack{k>l\\ k,l\neq i,j}}\langle\hat{u}_{kl},\hat{u}_{kl}\rangle+\cdots. \qquad (51)$$

For $N>4$ terms like

$$\sum_{k>l\neq m>n}\sum\langle\hat{u}_{kl},\hat{u}_{kl}\rangle\langle\hat{u}_{mn},\hat{u}_{mn}\rangle$$

are added to D_{ij} corresponding to χ_{ss}, χ_{sss}, etc. D_{ij} is essentially D' from which the normalization effect of all pairs, which involve electrons i or j or both are missing. Since $\langle\hat{u}_{ij},\hat{u}_{ij}\rangle$'s are already small (see Table III) their products, $\langle\chi_{ss},\chi_{ss}\rangle$, etc. are entirely negligible and so using Eq. (39c)

$$D'\cong D \quad\text{and}\quad D_{ij}\cong 1+\sum_{k>l\neq i,j}^{N}\langle\hat{u}_{kl},\hat{u}_{kl}\rangle. \qquad (52)$$

The last equation is exact for $N=4$.

For reasons given in previous sections, R/D in Eq. (49) is expected to be negligible. Then in going from simple pairs, Eq. (42), to Eq. (49) with unlinked clusters the main difference will be just the replacement of $(1/D)$ by (D_{ij}/D). That unlinked clusters affect the energy only in this way is confirmed below on Be and LiH.

With N electrons, the number of pairs missing from D to give D_{ij} is $(2N+1)$, whereas D itself contains $N(N-1)/2$ pairs. Thus D_{ij}/D is of the order

$$D_{ij}/D\sim 1-2(2N+1)/N(N-1) \qquad (53)$$

so that as $N\rightarrow\infty$, $(D_{ij}/D)\rightarrow 1$, giving[9,17,21]

$$E\sim E_0+E_1+\sum_{i>j}^{N}\bar{\epsilon}_{ij}.$$

In atoms and molecules N is not large enough for such a complete cancellation of D's in Eq. (49).

UNLINKED CLUSTERS AND C.I.

As in the estimate of triangle energies, various effects of unlinked clusters can easily be tested by doing C.I. in several steps:

(i) Obtain the coefficient of double excitations for one electron pair at a time [see Eq. (41)].

(ii) Do all the pairs together with full C.I. This is the step in the estimation of triangle energies [Eq. (39a)]. Presumably, the coefficients will not change appreciably as compared to (i).

(iii) As Tables I and II showed, e.g., quadruple excitation coefficients can be obtained just from those of double excitations assuming only unlinked clusters are important. This is checked by adding these multiple excitations to C.I. and doing the full C.I. as in Be. The new energy is Eq. (49).

(iv) Calculate the D_{ij}/D's from the pair coefficients. The main unlinked cluster contribution should be [compare Eqs. (49) and (42) or (39a)]

$$\sum_{i>j}^{N}\frac{\bar{\epsilon}_{ij}}{D}(D_{ij}-1) \qquad (54)$$

which can be checked by comparing it to the full C.I. energy in (iii). The difference is due to linked clusters R.

MAGNITUDES IN Be AND LiH

Watson's[19] 37-configuration wave function for Be gives a total correlation energy −2.30 ev vs experimental −2.57 ev. In Table III we show the D_{ij}'s calculated from his coefficients renormalized to $\langle\phi_0,\phi_0\rangle=1$. The main contribution to D comes from the coefficient of $(1s^2p_1^2)$, the $(2s^2-2p^2)$ mixing term in $\hat{u}_{34}$. The largest D_{ij} too is $D_{12} \doteq 1+\langle \hat{u}_{34}, \hat{u}_{34} \rangle$, which therefore is close to D. Thus $D_{12}/D \sim 1$, making Eq. (49)

$$E_{Be} \cong E_0 + E_1 + \bar{\epsilon}_{12} + (\bar{\epsilon}_{34} + \bar{\epsilon}_{13} + \bar{\epsilon}_{24} + \bar{\epsilon}_{14} + \bar{\epsilon}_{23} + R)/D. \quad (55)$$

When four excitations (Table I) are added to a wave function with just double excitations and full C.I. repeated,[19] the energy is lowered by 0.075 ev (sum of D_{ij}/D and R effects).

On the other hand, just from renormalization, D_{ij}/D [Eq. (54)], and by looking at double excitations (Watson's[19] configurations "3" to "5") which when multiplied together give the four-excitations, we obtain 0.074 ev. Thus in their energy effect too, four excitations represent not four electron "collisions," but two binary ones. Especially[27] the agreement for $p_1^2 d_{11}^2$ indicates the negligibility of R.

Similar results are found upon examining LiH (Table III), although the C.I. here was less complete and gave only a third of the correlation energy. In Ebbing's calculations we observe that

(i) Double excitation coefficients are hardly changed after four-excitations are added to C.I. (cf. "C" and "D" in his Table 6) indicating the smallness of D_{ij} and R.

(ii) His four excitations lower the energy by 0.001 ev. This is only part of $\bar{\epsilon}_{12}(D_{12}/D-1)$; very few terms were taken in $\hat{u}_{12}$ (inner shell) so that $\bar{\epsilon}_{12}/D$ came out only about 0.3 ev. With $(\bar{\epsilon}_{12}/D) \cong 1$ ev and $D_{12} \cong 1.02$ (Table III), the unlinked cluster energy will be increased to $\sim$0.02 ev.

CONCLUSION

The results and physical arguments given indicate that in single determinant atoms and molecules with no nearly "metallic" electrons, mainly pair correlations are important. The fluctuation potential acts locally; in three or more electron "collisions" its effect is overshadowed by the Fermi holes of pairs. A realistic wave function, Eq. (27), includes many separate binary "collisions" at the same time. Multiple excitations of significance arise in C.I. this way. The energy given by Eqs. (27), (39b) and (51) neglecting the three or more electron collisions in Eq. (49) is

$$E \cong E_0 + E_1 + \sum_{i>j}^{N} \bar{\epsilon}_{ij}(D_{ij}/D'). \quad (56)$$

Neglected "linked clusters" R can be estimated too, and an upper limit to E obtained once the pair functions, $\hat{u}_{ij}$'s, are known (see II). This is especially easy with C.I. done stepwise, as indicated above.

Equation (56), where the last sum is close to just

$$\sum_{i>j}^{N} \bar{\epsilon}_{ij},$$

has significance for the following reasons:

(1) The energy is the sum of uncoupled pair (except trivially by D_{ij}/D) energies. Thus by methods developed in II, each pair correlation $\hat{u}_{ij}$ can be obtained separately. Now the r_{12} coordinate can be introduced into trial functions directly as in He and H_2.

(2) For an N-electron problem, there are just $\sim N$ pair functions to be determined, not $N(N-1)/2$ (see II). Thus even a large atom or molecule can be calculated quite easily.

(3) Had multiple excitations other than "unlinked clusters," or many-electron collisions been important, triplets, quadruplets, etc. of electrons would have had to be calculated. The number of these increase as $\sim N^3$, N^4 etc. with N so that the problem would have become huge very soon.

(4) Equation (56) has already the "chemical picture" and the reconcilation of the qualitative vs. quantitative situation of orbital theories in it. This picture to be given later will in particular (a) justify the "Π-electron" theory; (b) give a precise way of defining and calculating intramolecular van der Waals forces and bond energies.

(5) Adding the correlation energy $\bar{\epsilon}_{ij}$ to $(J_{ij}-K_{ij}'')$, Eq. (15b), we find the same *form* for the energy as given by the Hartree-Fock method. Correlation does not disturb the H.F. radial density much either, since it acts locally and has no long-range effects. Thus it is expected that, e.g., the shapes of molecules are determined[13,14] mainly by the orbitals and the exclusion principle (equivalent orbitals).

(6) The results of this article are easily generalized to, say, an outer shell which has either degeneracy (see II) or significant n-electron correlations in it just by introducing some $\hat{U}_{123\cdots n}$ in Eq. (27) in addition to $\hat{u}_{ij}$'s. With, say, organic dye molecules one then gets the cores in terms of u_{ij}'s and also a separate, "metallic" shell which can be dealt with in a number of special ways such as C.I. or free-electron or plasma models.

I wish to thank Professor Raymond Fuoss and Professor Lars Onsager for the kind interest they showed in this work. This research was supported partly by a grant from the National Science Foundation.

[27] Configurations "6" and "9" do not give any R's. Their effect on R can be checked by mixing them in towards the end of 37 C.I.

VII-6b

Reprinted from *Chemical Physics Letters* **1,** 337 (1967)

REDUCIBLE AND IRREDUCIBLE PAIR CORRELATIONS IN BENZENE

O. SINANOĞLU *
Sterling Chemistry Laboratory, Yale University, New Haven, Conn., USA

and

J. ČÍŽEK
Institute of Physical Chemistry of Czechoslovak Academy of Sciences, Prague, Czechoslovakia

Received 13 September 1967

Molecular orbital pair correlation energies in benzene are evaluated by two methods within a framework limited to a finite 2pπ-basis. Both the MO-pairs ("reducible") and irreducible symmetry pair correlations are obtained.

The ground state wave-function of benzene is expected to be well approximated by

$$\Psi \simeq \phi_0 + \chi_S' = \left[(0\bar{0}\ 1\bar{1}\ 5\bar{5}) \left(1 + \frac{1}{\sqrt{2}} \sum_{i<j}^{\bar{5}} \frac{\hat{u}_{ij}}{(ij)} + \frac{1}{2} \sum_{\substack{i<j\\(i<k;\,i,j \neq k,l)}}^{\bar{5}} \sum_{k<l}^{\bar{5}} \frac{\hat{u}_{ij}\hat{u}_{kl}}{(ijkl)} + \frac{1}{2\sqrt{2}} \sum_{\substack{i<j\\(i,j \neq k,\, l \neq m,n)}}^{\bar{5}} \sum_{k<l}^{\bar{5}} \sum_{m<n}^{\bar{5}} \frac{\hat{u}_{ij}\hat{u}_{kl}\hat{u}_{mn}}{(ijklmn)} \right) \right], \tag{1}$$

according to many electron theory (MET) [1], if the general considerations of the theory also apply to this molecule. The ϕ_0 is the Hartree-Fock (H. F.) determinant with the occupied molecular spin-orbitals 0, 1 and 5 with α-spin, and $\bar{0}$, $\bar{1}$ and $\bar{5}$ with β-spin**. A given $\hat{u}_{ij}$ is the correlation function between spin-MO's i and j. The last two terms are "unlinked pair clusters".

Complete configuration-interaction (CCI) involving the π-MO's (0, 1, 5, 2, 4, 3) provides an interesting model for examining correlation effects in benzene [2]. This wave-function indicates the three-electron excitations to contribute negligibly to the ground state. Also four-excitations are reproduced rather well by the products of two-electron excitations in accordance with eq. (1). One should note however that a ground state contains both non-dynamical correlations ("near-degeneracy" effects) and dynamical ones (slowly convergent, infinite CI) [3]. The π-system CCI may be viewed as only the non-dynamical correlations with the six π-MO's arising from the initially degenerate six $2p_z$ OAO's (Orthogonalized Atomic Orbitals). Nevertheless, whether viewed as a model, or as a starting $\phi_0 + \chi_S'$ (non-dynamical) to which dynamical correlations may be added later, the CCI should exhibit some of the basic features of π-electron correlations.

Two of the types of pair correlations introduced, related to each other by unitary transformations, are:
a) "$B(ij)$-type", i.e. spin-MO pairs (referred to here as "reducible", and
b) "irreducible pairs" [4].
The latter belong to irreducible representations of the molecular point group G as well as spin SU(2).

In this paper, we evaluate the individual pair correlations of both types for benzene at the CCI level. One calculation (CCI) uses the full CCI wave-function which is of the form

$$\Psi(\mathrm{CCI}) = \sum_{n,m} a_m^n G_m^n \tag{2}$$

to start with. The G_m^n is the mth symmetry adapted configuration involving only n-excited determinants. The second calculation (CMET) is

* Work supported in part by the Alfred P. Sloan Foundation and the U. S. National Science Foundation.
** The notation of benzene MO's used is as follows: a_{1u}: 0; e_{1g}: 1,5; e_{2u}: 2,4; b_{2g}: 3.

Table 1
The coefficients in $\hat{u}_{ij}$'s of eqs. (3).

c_i	MATNI[a]		STOAO[b]	
	CCI[c]	CMET[d]	CCI[c]	CMET[d]
1	-0.0501	-0.0552	-0.0744	-0.0921
2	-0.0452	-0.0422	-0.0536	-0.0441
3	-0.1407	-0.1391	-0.2166	-0.2168
4	-0.0177	-0.0200	-0.0339	-0.0422
5	-0.0699	-0.0663	-0.1144	-0.1064
6	-0.0763	-0.0774	-0.1118	-0.1173
7	-0.0586	-0.0574	-0.0779	-0.0751
8	-0.1305	-0.1312	-0.1840	-0.1892
9	-0.0606	-0.0649	-0.0696	-0.0828
10	-0.0452	-0.0422	-0.0536	-0.0441

[a] Using Mataga-Nishimoto parametrization [6]
[b] Using parametrization based on Slater-type orthogonalized atomic orbitals [7]
[c] Obtained from six π-MO complete configuration interaction wave function [2]
[d] Obtained from "coupled pair many-electron theory" [5].

based on eq. (1). The best $\hat{u}_{ij}$'s would be obtained by the use of the full eq. (1) in the variation method. This procedure would be cumbersome. One of us has developed a non-variational method however whereby the function eq. (1) is obtained conveniently with results very close to the CCI ones [5]. This method will be referred to as "coupled pair many-electron theory" (CMET).

The calculations are within the ZDO framework and with two kinds of parametrization: i) Mataga and Nishimoto [6] semi-empirical parameters (MATNI) and ii) McWeeny [7] parameters calculated from Slater-type OAO's (STOAO).

As obtained from determinants which mix with ϕ_0, the $\hat{u}_{ij}$'s are given by:

$$\begin{aligned}
\hat{u}_{0\bar{0}} &= c_1\, B(3\bar{3}) + c_2\, B(2\bar{4}) + c_2\, B(4\bar{2}) \\
\hat{u}_{1\bar{1}} &= c_3\, B(4\bar{4}) \\
\hat{u}_{01} &= c_4\, B(43) \\
\hat{u}_{15} &= c_5\, B(24) \\
\hat{u}_{0\bar{1}} &= c_6\, B(4\bar{3}) + c_7\, B(3\bar{4}) \\
\hat{u}_{1\bar{5}} &= c_8\, B(2\bar{4}) + c_9\, B(4\bar{2}) + c_{10}\, B(3\bar{3}).
\end{aligned} \tag{3}$$

B is the two-electron anti-symmetrizer. $(\sqrt{2})^{-1}(1-p_{12})$. There are $\frac{1}{2}N(N-1) = 15$ $\hat{u}_{ij}$-functions. The ones not shown in eq. (3) differ from the ones shown, only by spin or spatial orientation and involve the same coefficients.

Table 2
MO-pair ("reducible pair") correlation energies (in eV).

$-\epsilon_{ij}$	MATNI[a]		STOAO[b]	
	CCI[c]	CMET[d]	CCI[c]	CMET[d]
$0\bar{0}$	0.1154	0.1140	0.2179	0.2103
$1\bar{1}$	0.1043	0.1031	0.2181	0.2183
01	0.0022	0.0025	0.0111	0.0137
15	0.0504	0.0478	0.1915	0.1782
$0\bar{1}$	0.1094	0.1095	0.2274	0.2319
$1\bar{5}$	0.2749	0.2766	0.6351	0.6496
E_{corr}[e]	1.421	1.417	3.261	3.285

a.b.c.d have the same meaning as in table 1.

[e] $E_{corr} = \sum_{ij} \epsilon_{ij}$ (eq. (5)).

Table 1 lists the coefficients obtained from the CCI wave-function, from the CMET method and each for the two sets of parameters mentioned. The corresponding pair correlations,

$$\epsilon_{ij} = \langle B(ij) \left| \frac{1}{r_{ij}} \right| \hat{u}_{ij} \rangle, \tag{4}$$

are given in table 2. The total correlation energy in both methods* is:

$$E_{corr} = \epsilon_{0\bar{0}} + 2\epsilon_{1\bar{1}} + 4\epsilon_{01} + 2\epsilon_{15} + 4\epsilon_{0\bar{1}} + 2\epsilon_{1\bar{5}}. \tag{5}$$

The values in table 2 depend strongly on the parametrization used. Though the STOAO set $E_{corr} = -3.26$ eV may be thought close to the actual E_{corr} of benzene π-system, this may be fortuitous. The STOAO γ_{pp} is 17 eV whereas the γ_{pp} calculated by a valence state Hartree-Fock (VHF) theory is about 11 eV [8]. The atomic $2p_z^2$ correlation is close to -1 eV [3]. Thus the STOAO value greatly overestimates the correlation. Since the semi-empirical γ_{pp} (e.g. the MATNI value) is closer to the VHF value, the MATNI values of E_{corr} may be quite realistic (cf. also discussion of non-dynamical E_{corr} above).

Now we construct the "irreducible pairs" [4] from the $\hat{u}_{ij}$-functions. We have

* The complete correlation energies resulting from CCI and from CMET are exactly equal to the sum of pair correlation energies given in eq. (5) and in table 2.

Table 3
Irreducible pair correlation energies corresponding to pair symmetries $^1A_{1g}$ and $^1E_{1u}$ (in eV) (cf. eq. (7)).

$-\epsilon$	MATNI[a]		STOAO[b]	
	CCI[c]	CMET[d]	CCI[c]	CMET[d]
$^1A'_{1g}$	0.4994	0.5053	1.0786	1.1210
$^1E'_{1u}$	0.2167	0.2166	0.4437	0.4502

a.b.c.d have the same meaning as in table 1.

$$\hat{u}(^1A_{1g}) = \hat{u}_{0\bar{0}}$$

$$\begin{cases} \hat{u}(^1E_{2g};\ M = 2) = \hat{u}_{1\bar{1}} \\ \hat{u}(^1E_{2g};\ M = -2) = \hat{u}_{5\bar{5}} \end{cases}$$

$$\begin{cases} \hat{u}(^3E_{1u};\ M = 1, M_S = \mp 1) = \hat{u}_{01}, \hat{u}_{\bar{0}\bar{1}} \\ \hat{u}(^3E_{1u};\ M = -1, M_S = \mp 1) = \hat{u}_{05}, \hat{u}_{\bar{0}\bar{5}} \\ \hat{u}(^3E_{1u};\ M = 1, M_S = 0) = \frac{1}{\sqrt{2}}(\hat{u}_{0\bar{1}} + \hat{u}_{\bar{0}1}) \\ \hat{u}(^3E_{1u};\ M = -1, M_S = 0) = \frac{1}{\sqrt{2}}(\hat{u}_{0\bar{5}} + \hat{u}_{\bar{0}5}) \end{cases} \tag{6}$$

$$\begin{cases} \hat{u}(^3A_{2g};\ M_S = \mp 1) = \hat{u}_{15}, \hat{u}_{\bar{1}\bar{5}} \\ \hat{u}(^3A_{2g};\ M_S = 0) = \frac{1}{\sqrt{2}}(\hat{u}_{1\bar{5}} + \hat{u}_{\bar{1}5}) \end{cases}$$

$$\hat{u}(^1A'_{1g}) = \frac{1}{\sqrt{2}}(\hat{u}_{1\bar{5}} - \hat{u}_{\bar{1}5})$$

$$\begin{cases} \hat{u}(^1E_{1u};\ M = 1) = \frac{1}{\sqrt{2}}(\hat{u}_{0\bar{1}} - \hat{u}_{\bar{0}1}) \\ \hat{u}(^1E_{1u};\ M = -1) = \frac{1}{\sqrt{2}}(\hat{u}_{0\bar{5}} - \hat{u}_{\bar{0}5}) \end{cases}$$

M is the quasi-angular momentum about the six-fold symmetry axis. M_S is total spin z-component. The $\hat{u}$'s transform in the same way as the occupied spin-MO pairs, $B(ij)$.

From eqs. (6) and eq. (4), one obtains the irreducible pair correlation energies. We have

$$\begin{aligned} \epsilon(^1A_{1g}) &= \epsilon_{0\bar{0}} \\ \epsilon(^1E_{2g}) &= \epsilon_{1\bar{1}} \\ \epsilon(^3E_{1u}) &= \epsilon_{01} = \epsilon_{0\bar{1}} + \epsilon_{0\bar{1};\bar{0}1} \\ \epsilon(^3A_{2g}) &= \epsilon_{15} = \epsilon_{1\bar{5}} + \epsilon_{1\bar{5};\bar{1}5} \\ \epsilon(^1A'_{1g}) &= \epsilon_{1\bar{5}} - \epsilon_{1\bar{5};\bar{1}5} \\ \epsilon(^1E_{1u}) &= \epsilon_{0\bar{1}} - \epsilon_{0\bar{1};\bar{0}1} \end{aligned} \tag{7}$$

where for example

$$\epsilon_{0\bar{1};\bar{1}0} \equiv B(0\bar{1})\ \frac{1}{r_{12}}\ \hat{u}_{\bar{0}1}\ .$$

The total E_{corr} is:

$$E_{corr} = \epsilon(^1A_{1g}) + 2\epsilon(^1E_{2g}) + 6\epsilon(^3E_{1u}) + + 3\epsilon(^3A_{2g}) + \epsilon(^1A'_{1g}) + 2\epsilon(^1E_{1u}). \tag{8}$$

As in the ϵ_{ij}-pairs, there are again fifteen pairs, of which only six are distinct (cf. eq. (5)).

The values of $\epsilon(^1A_{1g})$ to $\epsilon(^3A_{2g})$ in eq. (7) are already in table 2. The remaining two, $\epsilon(^1A'_{1g})$ and $\epsilon(^1E_{1u})$, are given in table 3.

The reader will note that:

a) By far the largest irreducible pair correlation is $\epsilon(^1A'_{1g})$. This involves one electron in each of the two degenerate e_{1g}-MO's.

b) In table 2, pair correlations between two different MO's are as important as within doubly occupied MO's.

REFERENCES

[1] O. Sinanoğlu, J. Chem. Phys. 36 (1962) 706.
[2] J. Čížek, J. Paldus, L. Šroubková, to be published. C.I. limited up to triply excited states has been published in: J. Koutecký, J. Čížek, J. Dubský and K. Hlavatý, Theoret. Chim. Acta 2 (1964) 462.
[3] For atomic examples see: V. McKoy and O. Sinanoğlu. J. Chem. Phys. 41 (1964) 2689.
[4] O. Sinanoğlu, Proc. Roy. Soc. (London) A260 (1961) 379.
[5] J. Čížek, J. Chem. Phys. 45 (1966) 4256.
[6] N. Mataga and K. Nishimoto, Z. physik. Chem. (Frankfurt) 13 (1957) 140.
[7] R. McWeeny, Proc. Roy. Soc. A277 (1955) 288.
[8] M. Orloff and O. Sinanoğlu, J. Chem. Phys. 43 (1965) 49.

Nonempirical Calculations of Correlation Effects in the Diatomic Hydride Molecules

ERNEST R. DAVIDSON

1. Introduction

During the past decade much has been learned about the representation of good wave functions by a sum of configurations not explicitly involving the interelectronic distances. For two-electron systems, natural orbitals have proved to be a convenient conceptual device for discussing the results.[1-10] If the charge density matrix is defined by

$$\rho(\mathbf{r}_1, \mathbf{r}_1') = N \int \psi(\mathbf{r}_1, s_1, \mathbf{x}_2 \ldots \mathbf{x}_N)\, \psi^*(\mathbf{r}_1', s_1, \mathbf{x}_2 \ldots \mathbf{x}_N)\, ds_1\, d\mathbf{x}_2 \ldots d\mathbf{x}_N$$

where $d\mathbf{x} = d\tau\, ds$ and s stands for the spin coordinates, then the natural orbitals f_i satisfy

$$\int \rho(\mathbf{r}_1, \mathbf{r}_1')\, f_i(\mathbf{r}_1')\, d\tau' = n_i f_i(\mathbf{r}_i)$$

and the n_i are called the occupation numbers. The f_i form the orthonormal set which gives the most rapidly convergent expansion of the density,

$$\rho(\mathbf{r}_1, \mathbf{r}_1') = \Sigma\, n_i f_i(\mathbf{r}_1)\, f_i^*(\mathbf{r}_1')$$

The wave function may be expressed using the f_i and the α and β spin functions as a linear combination of Slater determinants. The wave function may also be written in a Fourier expansion of the form

$$g_i(s_1, \mathbf{x}_2 \cdots \mathbf{x}_N) = \int f_i^*(\mathbf{r}_1)\, \psi(\mathbf{x}_1 \cdots \mathbf{x}_N)\, d\tau_1$$

$$\psi = \Sigma\, f_i(\mathbf{r}_1)\, g_i(s_1, \mathbf{x}_2 \cdots \mathbf{x}_N)$$

$$n_i = \int |g_i|^2\, ds_1\, d\mathbf{x}_2 \cdots d\mathbf{x}_N$$

This Fourier series for ψ is again an extreme series for which each additional term minimizes the residual error in the expansion.[11]

The space part of the wave function for a two-electron system becomes particularly simple in terms of natural orbitals:

$$\psi(\mathbf{r}_1 \mathbf{r}_2) = \Sigma\, \mu_i f_i(\mathbf{r}_1)\, h_i(\mathbf{r}_2)$$

where $|\mu_1|^2 = n_i$, and $h_i = \mu_i^{-1} \int f_i^*(\mathbf{r}_1)\, \psi(\mathbf{r}_1, \mathbf{r}_2)\, d\tau_1$. It is easily shown that the h_i also form an orthonormal set and satisfy the definition of the natural orbitals with the same eigenvalues n_i. If n_i is not a degenerate eigenvalue of ρ, the product $\mu_i f_i(1)\, h_i(2)$ is uniquely defined (although the individual factors are arbitrary within a phase factor). If n_i is degenerate, then only the *natural configuration*,[9]

$$\phi = \Sigma\, \mu_j f_j(1)\, h_j(2)$$
$$\{j\ |\mu_j|^2 = n_i\}$$

is unique, as the f_j are arbitrary within a unitary transformation. Thus, for a closed-shell state, the SCF function

$$\psi = u(1)\, u(2)$$

is already in normal form. Similarly, the open-shell triplet function

$$\psi = 2^{-1/2}\, [u(1)\, v(2) - v(1)\, u(2)]$$

where u and v are orthonormal, is also a natural configuration. For open-shell singlet states, the SCF function becomes

$$\psi = 2^{-1/2}\, (1 + s^2)^{-1/2}\, [u(1)\, v(2) + v(1)\, u(2)]$$

where $s = \int u^* v\, d\tau$. For this function the natural configuration form is

$$\psi = \mu_1 f_1(1)\, f_1(2) + \mu_2 f_2(1)\, f_2(2)$$

$$f_1 = [2\,(1 + s)]^{-1/2}\, [u + v]$$

$$f^2 = [2\,(1 - s)]^{-1/2}\, [u - v]$$

$$\mu_1 = (1 + s)\, [2\,(1 + s^2)]^{-1/2}$$

$$\mu_2 = -\,(1 - s)\, [2\,(1 + s^2)]^{-1/2}$$

Since s is about 0.1 for low-lying excited singlet states, two natural configurations are required in the simplest reasonable approximation. (For He and H_2, s is nonzero only if both u and v are of the same symmetry—otherwise, s is zero, $\mu_1 = \mu_2$, and both terms are included as one natural configuration.)

The correlation energy, normally defined as the error in the SCF energy, is very nearly the error in the one (or two) natural configuration expansion. This number is known approximately for many states of He and H_2. As Table 1 shows, the error decreases rapidly as one goes to excited states. It is possible to correlate this error fairly well with the degree of penetration of the two SCF orbitals,[9] where the penetrating θ is defined by

$$\theta = \int |u(1)v(1)|\, d\tau_1$$

Hence the SCF approximation leads to small errors for Rydberg states. One might expect similar magnitudes for intershell correlation between core and valence electrons in larger molecules. The correlation energy for H_2 is presumably larger than for He because the single configuration approximation does not dissociate correctly.

Table 2 gives an indication of the convergence rate of the correlation energy calculated with a series of natural configurations. Because there is great difficulty in estimating the total correlation energy and because better wave functions might give faster convergence, the individual numbers in this table are subject to a large uncertainty. Nevertheless, the trend seems to be clear. *Regardless of the total correlation energy, the convergence rate is essentially the same.* About 40 to 60 per cent of the correlation energy is obtained from the configuration required to make the wave function behave correctly at large R.

For the closed-shell case (with real orbitals f_i),

$$\psi = \Sigma\, \mu_i f_i(1) f_i(2)$$

and $\mu_1 \approx 1$. The effect of the other terms can best be understood by considering the probability density $\Gamma(1, 2)$ of finding one electron at $\mathbf{r}_1$ and the other at $\mathbf{r}_2$:

$$\Gamma(1, 2) = \sum_{i,j} \mu_i \mu_j f_i(1) f_j(1) f_i(2) f_j(2)$$

Table 1. Approximate Correlation Energies of H_2 and He (in Hartrees)

H_2 (near R_e)[a,b]		He[c]	
State	$-\epsilon$	State	$-\epsilon$
$1s^2\ ^1\Sigma g$	0.0408	$1s^2\ ^1S$	0.0420
$1s2s\ ^1\Sigma g$	0.0058	$1s2s\ ^1S$	0.0026
$1s2s\ ^3\Sigma g$	0.0035	$1s2s\ ^3S$	0.0010
$1s2p\ ^1\Sigma u$	0.0131	$1s2p\ ^1P$	0.0014
$1s2p\ ^1\Pi u$	0.0046		
$1s2p\ ^3\Sigma u$	0.0043	$1s2p\ ^3P$	0.0018
$1s3p\ ^3\Pi u$	0.0047		
$1s3s\ ^1\Sigma g$	0.0020	$1s3s\ ^1S$	0.0009
$1s3s\ ^3\Sigma g$	0.0012	$1s3s\ ^3S$	0.0002
$1s3p\ ^1\Sigma u$	0.0041	$1s3p\ ^1P$	0.0004
$1s3p\ ^3\Pi u$	0.0013		
$1s3p\ ^3\Sigma u$	0.0013	$1s3p\ ^3P$	0.0005
$1s3p\ ^3\Pi u$	0.0011		
$1s3d\ ^1\Sigma g$	0.0013	$1s3d\ ^1D$	0.0001
$1s3d\ ^1\Pi g$	0.0013		
$1s3d\ ^1\Delta g$	0.0005		
$1s3d\ ^3\Sigma g$	0.0005	$1s3d\ ^3D$	0.0001
$1s3d\ ^3\Pi g$	0.0008		
$1s3d\ ^3\Delta g$	0.0003		

[a]S. Rothenberg and E. R. Davidson, *J. Chem. Phys.* **45**, 2560 (1966).

[b]G. H. Dieke, *J. Mol. Spectry.* **2**, 494 (1958).

[c]E. R. Davidson, *J. Chem. Phys.* **42**, 4199 (1965).

Table 2. Convergence Rate of CI Energies[a,b]

	1	2	3	4	5	6
$1s^2\ ^1\Sigma g$	0	44	70	87	89	90
$1s2s\ ^1\Sigma g$		0	49	60	66	73
$1s2s\ ^3\Sigma g$	0	41	61			
$1s2p\ ^1\Sigma u$	0	38	73	77	78	
$1s2p\ ^1\Pi u$	0	48	79	88	90	
$1s2p\ ^3\Sigma u$	0	62	85	86		
$1s2p\ ^3\Pi u$	0	28	68	88	89	
$1s3s\ ^1\Sigma g$		0	40	48	55	62
$1s3s\ ^3\Sigma g$	0	43	45	47		
$1s3p\ ^1\Sigma u$	0	48	76	80		
$1s3p\ ^1\Pi u$	0	49	88	90		
$1s3p\ ^3\Sigma u$	0	60	70			
$1s3p\ ^3\Pi u$	0	27	52	66		
$1s3d\ ^1\Sigma g$		0	47	61	66	
$1s3d\ ^1\Pi g$	0	54	62			
$1s3d\ ^1\Delta g$	0	34	56	72		
$1s3d\ ^3\Sigma g$	0	42	71			
$1s3d\ ^3\Pi g$	0	51	63			
$1s3d\ ^3\Delta g$	0	33	66			

[a]S. Rothenberg and E. R. Davidson, *J. Chem. Phys.* **45**, 2560 (1966).

[b]Entries in the table give the percentage of correlation energy obtained with the number of natural configurations indicated at the top of the column.

The probability of finding one electron at $\mathbf{r}_1$ is, similarly,

$$\rho(1) = \sum_i \mu_i^2 f_i^2(1)$$

Now if $\mu_i \mu_j$ is negligible for terms with neither i nor j equal to 1, then ρ can be written approximately as

$$\rho(1) \approx f_1^2(1)$$
$$\Gamma(1,2) \approx \rho(1)\,\rho(2)\,\gamma(1,2)$$

where

$$\gamma(1,2) \approx 1 - 2 \sum_{i \neq 1} \mu_i f_i(1) f_i(2)/f_1(1) f_1(2)$$

or

$$\gamma(1,2) \approx \prod_{i \neq 1} (1 - 2\,\mu_i f_i(1) f_i(2)/f_1(1) f_1(2))$$

The correlation factor γ is smaller in regions where $f_i(1)/f_1(1)$ and $f_i(2)/f_1(2)$ have the same sign and larger in regions where they have opposite signs. Thus if f_1 is an 1 s orbital and f_2 is a p orbital for an atom, the effect of f_2 is to increase the probability of finding the electrons on opposite sides of the nucleus.

For a many-electron system the above formulas become more complicated.[12] The probability density Γ_0 for the SCF wave function (with real orbitals f_i) is

$$\psi_0 = (N!)^{-1/2}\, |f_1(1)\,\alpha(1) f_1(2) \beta(2) \cdots f_{N/2}(N)\,\beta(N)|$$

takes the form

$$\Gamma_0 = \sum_{i,j} \rho_i(1)\rho_j(2) + \sum_{i \neq j} \rho_i(1)\rho_j(2) \left[1 - \frac{f_j(1)/f_i(1)}{f_j(2)/f_i(2)}\right]$$

where $\rho_i = f_i^2$.

The first of these sums in Γ_0 arises from electrons with opposite spins. This sum shows that their positions are uncorrelated in the SCF approximation. The second sum arises from the electron pairs with equal spins and includes the correlation factor from the Pauli exclusion principle. For H_2 and He this natural correlation between electrons of the same spin reduces the error in the energy to about 40 to 80 per cent of the error found for electrons with opposite spins. The addition to ψ_0 of a Slater determinant with the excitation $f_i\alpha \longrightarrow g_i\alpha$ and $f_j\beta \longrightarrow g_j\beta$ introduces a correlation factor γ_{ij} for the (i, j) pair into the equation for Γ_0. When all such pair excitations are considered, Γ takes the approximate form (to first order in the coefficients),

$$\Gamma(1,2) \approx \sum_{i,j} \rho_i(1)\rho_j(2)\gamma_{ij}^s(1,2) + \sum_{i \neq j} \rho_i(1)\rho_j(2)\gamma_{ij}^t(1,2) \left[1 - \frac{f_j(1)/f_i(1)}{f_j(2)/f_i(2)}\right]$$

The physical effect of any doubly excited configuration may thus be discussed from its effect on the distribution function of the corresponding pair.

Sinanoğlu (see, e.g., Chapter VII-6a and other references) has shown that an N-electron system may be treated in terms of N (N – 1)/2 pair correlations. He has developed a number of methods for the calculation of such pair correlations and of some residual effects.

The 1s2s correlation effects in the Be atom, which is a rather special case, are relatively small. As several people have shown,[13, 14] the Be wave function is nearly factorable into the product of two strongly orthogonal geminals. This means that in terms of natural orbitals, the wave function is approximated by

$$\psi \approx \mathcal{A} \left\{[\Sigma\, \mu_i f_i(1) f_i(2)]\ [\Sigma\, \mu_i' f_i'(3) f_i'(4)]\ \alpha(1)\beta(2)\alpha(3)\beta(4)\right\}$$

where the functions $\{f_1, f_2, \cdots, f_1', f_2', \cdots\}$ form an orthonormal set. This will work for Be because the penetration of the 1s and 2s orbitals is small. Probably 95 per cent of the correlation energy of Be is obtainable by a function of this form.

The form of this function for Be gives rise to another problem. The expansion of this form into Slater determinants gives rise to quadruple excitations. In most calculations it would be preferable to be able to ignore quadruple excitations. There are some results which indicate that this may be done if pair energies are calculated separately for each pair.[15,16] Thus, if the correlation energies obtainable with

$$\psi_1 = \mathcal{A} \left\{g_1(1, 2)\, f_1'(3)\, f_1'(4)\alpha(1)\beta(2)\alpha(3)\beta(4)\right\}$$

and

$$\psi_2 = \mathcal{A} \left\{f_1(1) f_1(2) g_2(3, 4)\alpha(1)\beta(2)\alpha(3)\beta(4)\right\}$$

are added together, the result seems to be a good approximation to the correlation energy obtained with

$$\psi = \mathcal{A} \left\{g_1(1, 2) g_2(3, 4)\alpha(1)\beta(2)\alpha(3)\beta(4)\right\}$$

Strangely, it is not a good approximation to using

$$\psi = \psi\ \ + \psi_2 + \psi_0$$

which would involve only double excitations. The quadruple excitations which arise naturally in multiplying the two geminals are called unlinked clusters, and their energy contribution is implicitly counted in adding individual pair correlation energies.[15] This argument can be justified by perturbation theory as well as numerical results. It cannot be proved by the variation method, however, and the sum of the pair energies could give an overestimate of the correlation energy. Further, the answer is not independent of a unitary transformation among the occupied orbitals, which leaves ψ_0 unchanged, as discussed in the Sinanoğlu theory.

2. Method

Two types of studies have been carried out on the diatomic hydride molecules. In one type, a variational calculation using excitations from one pair (i, j) was done to find the pair correlation energy. The occupied and virtual orbitals (f_i) were all expanded in a fixed basis set. As this set was large, many more configurations were formed than it seemed possible to handle at one time. Consequently, a procedure was developed in which (a) approximate natural orbitals were constructed, (b) configuration interaction with the most important configurations was used to get ψ_0, (c) perturbation theory was used for the rest of the configurations, and (d) new natural orbitals were calculated by diagonalizing the first-order density matrix of the resulting wave function. This procedure was then iterated until convergence was reached. The actual calculation was carried out using excitations from space orbitals into other space orbitals. All configurations were eigenfunctions of S^2 and S_z. Hence only average pair correlation energies for $f_i\alpha f_j\beta$, $f_i\alpha f_j\alpha$, etc., could be found.

In the second type of calculation an attempt has been made to get good correlated wave functions. All single and double excitations from occupied space orbitals were considered simultaneously in a calculation similar to that described above. This is a more complicated calculation, because many more configurations are involved and convergence is slower. Also, it seems that the perturbation-theory corrections to ψ_0 are less reliable. For some molecular properties of interest, good wave functions are necessary, and improvements in this type of calculation are still underway.

3. Results

Results using the above method for LiH have been published in detail elsewhere.[17] Since this is a relatively simple system with two sets of nonpenetrating pairs, the wave function was nearly of the form of the product of two strongly orthogonal geminals. The correlation energy was almost the sum of the correlation energies of the Li^+ and H^- ions. The efficacy of this method is demonstrated by the fact that, although LiH was run mostly as a problem for checking the program, the energy obtained was better than any result then in the literature. The error, 0.3 eV, in the total energy could be improved with the current version of the program.

The hydrogen fluoride molecule has also been discussed in detail elsewhere.[18] This molecule has several strongly interpenetrating valence orbitals. Hence a large fraction of the total correlation energy is due to intershell effects. The pair correlation energies resemble those of the neon atom. For this molecule, the individual correlated geminals, obtained by studying one pair at a time, are not strongly orthogonal, nor can the wave function be written as an antisymmetrized product of geminals alone. All pair correlations are needed (see Chapter V-2).

The hydrogen fluoride calculation did demonstrate that an expansion in Slater determinants based on natural orbitals was rapidly convergent. With 39 configurations, 60 per cent of the correlation energy was obtained. Since a study of individual pairs accounted for 80 per cent of the correlation energy, only 20 per cent (at most) lies outside the function space spanned by the basis set. The individual pair results also show that more than 100 configurations would be needed to get 90 per cent of the correlation energy, even with an improved basis set.

Detailed calculations on several other diatomic molecules are now available. From these calculations the pair correlation energies of individual pairs can be estimated. Tables 3 and 4 summarize the results of these calculations for the hydride series. In these tables some results are obtained with elliptical basis sets and other results are for Slater sets. Most results are based on 50 configuration variational calculations, but some are based on the partitioned-perturbation theory with a few hundred configurations. The least accurate numbers are based on a partitioned-perturbation theory calculation with all pairs considered simultaneously (but energy contributions from each kind of excitation correctly accounted for). In this worst calculation, approximate natural orbitals for the whole wave function were used. The trends in these numbers can best be understood by an examination of the contributing configurations. The σ orbital is essentially a 1s orbital on the heavy atom. The b orbital is the bonding orbital and varies from being essentially an H atomic orbital in LiH to being a $2p\sigma$ orbital on F in HF. The ℓ orbital is the lone-pair orbital and varies from being 75 per cent 2s in BeH to being almost pure 2s in HF. The π orbital is the molecular orbital which is mainly $2p\pi$ on the heavy atom but contains an appreciable amount of $2p\pi$ on H in HF.

The correlation orbitals for the σ^2 pair are of the same size as a 1s orbital. They are affected only slightly by the requirement of orthogonality to the valence-shell orbitals. Hence the configurations contributing to the $1\sigma^2$ pair are the same in all these molecules. The constancy of the correlation energy can be understood qualitatively by perturbation theory. The energy contribution from the configuration $\sigma^2 \longrightarrow g^2$ is roughly $|K_{\sigma g}|^2/\Delta E$. As the nuclear charge, Z, increases, both σ and g shrink. Hence $|K_{\sigma g}|^2$ and ΔE both increase proportional to Z^2 and the ratio is nearly constant.

The cross-shell correlation σ-b is negligible for chemical purposes. As pointed out previously, there is some correlation between the amount of correlation energy and the penetration of the orbitals. This penetration is quite small between the σ orbital and any valence orbital.

The b^2 correlation appears to decrease as Z increases. This is reasonable, because the correlation energy for H^- is about 0.040 and for $2p\sigma^2$ in Ne is about 0.026. The main contribution to the b^2 correlation energy comes from the antibonding orbital. The angular correlation energy is smaller and comes mostly from a $3d\pi$ atomic orbital on the heavy atom and $2p\pi$ on hydrogen. Consequently, the configurations contributing to the b^2 pair energy are not excluded as the other valence-shell orbitals are filled.

The ℓ^2 pair behaves differently, however. Since for atoms it is known that $2s^2 \longrightarrow 2p^2$ provides most of the correlation energy of $2s^2$, the correlation energy is subject to an "exclusion effect" as the 2p orbitals become filled, an effect first discussed in the Sinanoğlu theory[15] (see also Chapter III-2d). This is most apparent in the drop in ℓ^2 correlation from CH to NH as the excitation $\ell^2 \longrightarrow \pi\bar{\pi}$ becomes forbidden. Since this excitation is allowed in the $^1\Sigma^+$ state of NH, the correlation in that state would be comparable to CH. The other expected effect, that the correlation energy of $2s^2$ should increase with Z if the number of electrons is held constant, is not apparent from Table 4.[19] This latter effect arises because the previous perturbation argument does not hold for $2s^2 \longrightarrow 2p^2$. A better approximation is to say that ΔE is small compared with K, so the contribution to the energy from solving the 2×2 secular equation is $-|K|$, which increases linearly with Z.

The correlation of the pair $\ell\pi$ arises mostly from the excitation $\ell\pi \longrightarrow \bar{\pi}\delta$. This type of excitation is subject to the "exclusion effect" noted for ℓ^2 and hence its energy contribution should increase with Z somewhat and then disappear.

The correlation energies of $b\pi$, $\pi\bar{\pi}$, and π^2 are all related by symmetry in the Ne atom.[20,21] Hence the large value for $b\pi$ in HF is reasonable. The dominant configuration in π^2 correlation is $\pi^2 \longrightarrow (3p\pi)^2$, where $3p\pi$ is the atomic orbital on the heavy atom. Similarly, $\pi\bar{\pi} \longrightarrow (3p\pi)(3p\bar{\pi})$ is the dominant effect for that pair. The dominant effect for $b\pi$ is $b\pi \longrightarrow a(3p\pi)$, where a is the antibonding valence-shell orbital. This latter term introduces a correlation between the in-out position of the π electron and the left-right position of the b electron. The π^2 and $\pi\bar{\pi}$ correlation effects are not like the Dewar vertical correlation.[2] To represent vertical correlation of the Dewar type an excitation of the form $(2p\pi)^2 \longrightarrow 2s^2$ would be required, and this is forbidden in systems with occupied 2s orbitals. Instead, the correlation introduced is an in-out correlation within each lobe.

The change in correlation energy as the molecule dissociates introduces an error in the dissociation energy, D, resulting from an SCF calculation. This error ΔD may be expressed as $D = \Sigma\, n_{ij}\bar{\epsilon}_{ij} - \Sigma\, n'_{ij}\bar{\epsilon}'_{ij}$, where the first sum is over all district pairs in the molecule, the second is over all pairs in the atoms, and n_{ij} is the number of equivalent pairs. If there is an easy correspondence between the separated atoms and the molecule (as there is for the hydrides), this may be rewritten

$$\Delta D = \sum (n_{ij} - n'_{ij})\,\bar{\epsilon}_{ij} + \sum n'_{ij}(\epsilon_{ij} - \epsilon'_{ij})$$

For the hydrides, the last term tends to be small, so that ΔD may be estimated from the changes in pair populations.[23] It is clear that for complicated molecules the bonding pair will probably account for less than half of the correlation error D, and the remainder will be due to the change in the number of cross-shell correlations as the molecule dissociates.

Table 3. Calculated Average Pair Correlation Energies ($-\epsilon_{ij}$ in Hartrees)[a,b]

	HeH $^2\Sigma^+$	LiH $^1\Sigma^+$	BeH $^2\Sigma^+$	BH $^1\Sigma^+$	CH $^2\Pi$	NH $^3\Sigma^-$	OH $^2\Pi$	FH $^1\Sigma^+$
σ^2	0.0363	0.0372	0.0371	0.0373	0.0378	0.0378	0.0371	0.0316
σb	0.0003	0.0004	0.0006	0.0008	0.0008	0.0005	0.0005	0.0007
b^2		0.0347	0.0353	0.0295	0.0293	0.0309	0.0297	0.0265
σl			0.0008	0.0009	0.0013	0.0008	0.0006	0.0009
bl			0.0030	0.0073	0.0081	0.0070	0.0058	0.0048
l^2				0.0383	0.0250	0.0135	0.0130	0.0061
$\sigma\pi$					0.0014	0.0015	0.0015	0.0012
bπ					0.0094	0.0102	0.0101	0.0096
lπ					0.0016	0.0102	0.0056	0.0054
$\pi\bar{\pi}$						0.0101	0.0159	0.0145
π^2							0.0173	0.0159

[a] E_{ij} is obtained by considering all equivalent pairs simultaneously and dividing the energy improvement by the number of pairs. For example, for HF, $\epsilon_{b\pi}$ is $\frac{1}{8}$ of the energy gain from all configurations arising from b$\pi \longrightarrow$ fg or b$\bar{\pi} \longrightarrow$ fg. The orbitals π and $\bar{\pi}$ are treated as equivalent even if they are not.

[b] C. F. Bender and E. R. Davidson, *Phys. Rev.* **183**, 23 (1969).

Table 4. Estimated Pair Correlation Energies ($-\epsilon_{ij}$ in Hartrees)[a]

	HeH $^2\Sigma^+$	LiH $^1\Sigma^+$	BeH $^2\Sigma^+$	BH $^1\Sigma^+$	CH $^2\pi$	NH $^3\Sigma^-$	OH $^2\pi$	FH $^1\Sigma^+$	Ne ^{1}S
σ^2	0.038	0.040	0.040	0.040	0.040	0.040	0.040	0.040	0.040
σb	0.0004	0.0005	0.0008	0.0009	0.0009	0.0008	0.0008	0.0008	0.0016
b^2		0.041	0.045	0.032	0.033	0.036	0.036	0.032	0.026
σl			0.0017	0.001	0.001	0.001	0.001	0.001	0.0013
bl			0.004	0.008	0.009	0.008	0.007	0.006	0.007
l^2				0.042	0.028	0.016	0.016	0.007	0.011
$\sigma\pi$					0.0016	0.0017	0.0018	0.0015	0.002
bπ					0.0105	0.0120	0.0122	0.0116	0.012
lπ					0.013	0.012	0.0068	0.0065	0.007
$\pi\bar{\pi}$						0.012	0.019	0.0175	0.017
π^2							0.021	0.019	0.017

[a] See footnote a, Table 3.

References

1. P.-O. Löwdin and H. Shull, *Phys. Rev.* **101**, 1730 (1956).
2. P.-O. Löwdin, *Phys. Rev.* **97**, 1474 (1955).
3. E. R. Davidson, *J. Chem. Phys.* **37**, 577 (1962).
4. E. R. Davidson and Leon L. Jones, *J. Chem. Phys.* **37**, 2966 (1962).
5. H. Shull, *J. Am. Chem. Soc.* **82**, 1287 (1960).
6. H. Shull, *J. Chem. Phys.* **30**, 1405 (1959).
7. H. Shull and P.-O. Löwdin, *J. Chem. Phys.* **30**, 617 (1959).
8. E. R. Davidson, *J. Chem. Phys.* **39**, 875 (1963).
9. S. Rothenberg and E. R. Davidson, *J. Chem. Phys.* **45**, 2560 (1966).

10. S. Hagstrom and H. Shull, *Rev. Mod. Phys.* **35**, 624 (1963).
11. A. J. Coleman, *Rev. Mod. Phys.* **35**, 668 (1963).
12. R. McWeeny, *Rev. Mod. Phys.* **32**, 335 (1960).
13. K. J. Miller and K. Ruedenberg, *J. Chem. Phys.* **43**, 388 (1965).
14. T. L. Allen and H. Shull, *J. Chem. Phys.* **35**, 1644 (1961).
15. O. Sinanoğlu, *J. Chem. Phys.* **36**, 706 (1962).
16. R. K. Nesbet, *Phys. Rev.* **155**, 56 (1967).
17. C. F. Bender and E. R. Davidson, *J. Phys. Chem.* **70**, 2675 (1966).
18. C. F. Bender and E. R. Davidson, *J. Chem. Phys.* **47**, 360 (1967).
19. J. Linderberg and H. Shull, *J. Mol. Spectry.* **5**, 1 (1960).
20. K. F. G. Paulus, *J. Chem. Phys.* **46**, 3078 (1967).
21. H. J. Silverstone and O. Sinanŏglu, *J. Chem. Phys.* **46**, 854 (1967).
22. M. J. S. Dewar and C. Wulfman, *J. Chem. Phys.* **29**, 158 (1958).
23. See also C. Hollister and O. Sinanoğlu, *J. Am. Chem. Soc.* **88**, 13 (1966).

VII-7 Reprinted from *Modern Quantum Chemistry*, Vol. I, Academic Press, New York, 1965, p.221

6. Critical Examination of Pi-Electron Theories

f. The Pi-Electron Approximation and Coulomb Repulsion Parameters

O. SINANOĞLU AND M. K. ORLOFF

1. The Pi-Electron Approximation

One of the crucial approximations in semi-empirical π-electron theory is the "π-electron approximation" itself (see, e.g., reference [1]). To examine the validity of this approximation, let us consider the form of the exact energy of a conjugated molecule.

For a closed shell system, such as the ground state of ethylene or benzene, the form of the exact wave function (w.f.) and energy (E) can be written down [2]. After the first term, a Hartree-Fock (SCF) molecular orbital (MO) determinant the w.f. contains the one-electron effects of correlation ($\hat{f}_i$) on both the σ- and π- orbitals and the (linked cluster) correlation functions $\hat{U}_{ijk\ldots n}$ involving progressively larger numbers of only σ-, or π-, or some σ- and π-electrons (spin-orbitals). In addition, there are also the unlinked clusters of all these—the products of independent $\hat{f}_i$'s and $\hat{U}$'s. This w.f. substituted into $\langle\psi, H\psi\rangle/\langle\psi, \psi\rangle$ yields the detailed form of the exact energy. Where there are no strong near-degeneracies, the significant portion of this energy is [3],

$$E = E_\Sigma + E_\Pi + \varepsilon_{\sigma\pi} + R_{\sigma\pi} \tag{1}$$

E_Σ is the energy of the Σ-framework including correlation; $\varepsilon_{\sigma\pi}$ is the sum of the core-polarization energies [4] of π-electrons; and $R_{\sigma\pi}$ is the sum of all true many (more than two) electron correlation effects between σ- and π-electrons including linked clusters such as triangles, squares, etc. [5].[1]

[1] The right-hand side of Eq. (1) is an upper limit to E if normalization is included completely; see reference [5] for details.

In terms of Hartree-Fock (H.F.) MO's pertinent to the entire molecule, the Σ-core energy is

$$E_\Sigma = \sum_{i \geqslant j}^{n_\sigma} I_i + \sum_{i>j\geqslant 1}^{n_\sigma} \mathcal{J}_{ij} + E_\Sigma^{\text{corr}} \tag{2}$$

where I_i is the customary bare nuclei energy, $\langle i, h_i^0 i \rangle$, of the σ-spin orbital i; $\mathcal{J}_{ij}$ includes the usual Coulomb (J_{ij}) and exchange (K_{ij}) integrals. The core correlation energy E_Σ^{corr} consists mainly of individual σ-bond correlations and inter-bond attractions [2].

The π-electron energy E_Π for a closed shell state of the molecule, is given by

$$E_\Pi = \sum_{k>n_\sigma}^{N} \langle k, \mathscr{H}_k^{\text{core}} k \rangle + \sum_{l>k>n_\sigma}^{N} \mathcal{J}_{kl} + E_\Pi^{\text{corr}} \tag{3a}$$

where[2]

$$\langle k, \mathscr{H}_k^{\text{core}} k \rangle = \langle k, h_k^0 k \rangle + \sum_{i\geqslant 1}^{n_\sigma} \mathcal{J}_{ik} \tag{3b}$$

$$\mathscr{H}_k^{\text{core}} = h_k^0 + \sum_{i\geqslant 1}^{n_\sigma} \mathcal{J}_i(\mathbf{x}_k) \equiv h_k^0 + V_k^{\text{core}} \tag{4}$$

V_k^{core} is the sum of Coulomb and exchange potentials of σ-spin-orbitals acting on the π-spin-orbital k.

The E_Σ, E_Π, and $\varepsilon_{\sigma\pi}$ parts of the total energy Eq. (1) may be obtained individually using variation-perturbation techniques [2]. If the coupling term $R_{\sigma\pi}$ is small compared to E_Π, then E can be minimized by itself with a w.f. [3],

$$\hat{\psi}_\Pi = \mathscr{A}_n(\pi_1\pi_2 \ldots \pi_n) + \hat{U}_\Pi \tag{5}$$

subject to the rigorous condition, orbital-orthogonality to all the σ- and π-H.F. MO spin-orbitals m,

$$\langle \hat{U}_\Pi(\mathbf{x}_1, \mathbf{x}_2, \ldots, \mathbf{x}_n), m(\mathbf{x}_1) \rangle_{\mathbf{x}_1} = 0$$
$$(m = 1, 2, \ldots, n_\sigma, \pi_1\pi_2 \ldots \pi_n) \tag{6a}$$

The condition (6a) ensures that the π-energy approaches E_Π and does not go below into the energy of sigma.

[2] The terms in the first sum of Eq. (131) in reference [2] should read $\langle k, \mathscr{H}_k^{\text{core}} k \rangle$. Compare with Eqs. (135) and (136), same reference and Eq. (9b), reference [3].

As far as the ground state is concerned, the "π-electron approximation" [1] is justified in the sense that E_Π can be minimized separately if $R_{\sigma\pi}$ is small, and $R_{\sigma\pi}$ is expected to be small. What happens however to E_Σ and $\varepsilon_{\sigma\pi}$ in the calculation of electronic spectra? The π-electron approximation further assumes that E_Σ and $\varepsilon_{\sigma\pi}$ remain the same as additive constants (or that $\varepsilon_{\sigma\pi} \sim 0$) and cancel out in comparing π-energy levels. A rigorous discussion of whether this is so, requires a nonclosed shell H.F. MO theory along with a nonclosed shell many-electron theory. The latter (see the Silverstone-Sinanoğlu lecture in Part II of this book; [6]; also [14]) has not been applied as yet to a π-molecule in any detail. Indications of what happens to E_Σ and $\varepsilon_{\sigma\pi}$ are obtained, however, from a very simple model, atomic valence state reactions, as will be discussed in the next section of this lecture. Before we get onto that topic however, let us continue with further aspects of the ground state problem.

The E_Π^{corr} above, in Eq. (3) contains the correlations of various numbers of π-electrons. For a closed π-shell, the major terms in the exact π-correlation function $\hat{U}_\Pi$ should be the π-pair functions $\hat{u}_{kl}$ with again

$$\langle \hat{u}_{kl}, m \rangle = 0 \qquad (m = 1, 2, \ldots, n_\sigma, \pi_1, \pi_2, \ldots, \pi_n) \tag{6b}$$

Koutecký [7] has found the effect of three electron terms in CI (triple excitations) to be small on the ground state energy of benzene. With only $\hat{u}_{kl}$ and neglecting also the $\hat{f}_k$, one then obtains

$$E_\Pi^{\text{corr}} \leqslant \sum_{l>k>n_\sigma}^{N}{}' \tilde{\varepsilon}_{kl}' + \frac{R_\Pi'}{D_\Pi'} \tag{7}$$

ε_{kl}' is a pair correlation between two π-spin-MO's k and l; R_Π' is the remainder involving several different $\hat{u}_{kl}$, i.e., three or more π-electron energy effects [5]; D_Π' is the π part of normalization. The primes indicate that unlinked clusters arising from within the π-shell are included. (Note that this is how E_Π^{corr} has been defined from Eq. (1). The total normalization and unlinked cluster effects involving both σ- and π-electrons are implied in Eq. (1). Compare Eq. (82) of reference [2]; also [3]).

The remainder terms, R_Π'/D_Π', may be small for π-electron molecules like benzene, but their magnitudes have not been checked. It would be interesting to estimate them, for example by comparing partial CI's, which give the individual $\tilde{\varepsilon}_{kl}$ (subject to symmetry relations), with full CI which includes all the $\hat{u}_{kl}$'s and gives Eq. (7). If R_Π'/D_Π' is quite small, only the $\tilde{\varepsilon}_{kl}'$ terms may be minimized alone. Each $\tilde{\varepsilon}_{kl}'$ then reduces to (see [2])

$$\varepsilon_{kl} = \langle B(kl), g_{kl}\hat{u}_{kl} \rangle \tag{8}$$

with B the two-electron antisymmetrizer, and $g_{kl} = r_{kl}^{-1}$. This reduces Eq. (3) to

$$E \cong \sum_{k>n_\sigma}^{N} \langle k, \mathscr{H}_k^{\text{core}} k\rangle + \sum_{l>k>n_\sigma}^{N} (\bar{J}_{kl} + \varepsilon_{kl}) \tag{9}$$

Let us now consider the ground state of the ethylene molecule. The two π-MO's are $\pi_{\text{I}} = 2^{-1/2}(a + b)$ and $\pi_{\text{II}} = 2^{-1/2}(a - b)$, where a and b are the $2p_z$ atomic orbitals (AO's, or more properly, OAO's as discussed in previous lectures) centered on the two carbon atoms in the molecule.

It has been shown previously [3] that for the two π-electrons $k(\pi_{\text{I}}\alpha)$ and $l(\pi_{\text{I}}\beta)$ in ethylene, the effective Hamiltonian comes out of the many-electron theory as $H_\Pi = \mathscr{H}_k^{\text{core}} + \mathscr{H}_l^{\text{core}} + 1/r_{kl}$. This is just the well-know π-electron Hamiltonian (see for example [1]), formerly postulated on intuitive grounds.

From Eq. (9), the π-electron energy of the ground state of ethylene is

$$E_\Pi = \langle k, \mathscr{H}_k^{\text{core}} k\rangle + \langle l, \mathscr{H}_l^{\text{core}} l\rangle + \bar{J}_{kl} + \varepsilon_{kl} \tag{10}$$

With the standard notation, $\gamma_{ab} = \langle ab, (1/r_{12})ab\rangle$, and the approximation of zero differential overlap [1] (ZDO), Eq. (10) becomes

$$\begin{aligned} E_\Pi = I_k^c + I_l^c + \tfrac{1}{4}(\gamma_{aa} + \gamma_{bb}) \\ + \tfrac{1}{2}\gamma_{ab} + \langle B(kl), g_{kl}\hat{U}_{kl}\rangle \end{aligned} \tag{11}$$

We have denoted, e.g., $\langle k, \mathscr{H}_k^{\text{core}} k\rangle$ by I_k^c. Approximations analogous to ZDO may be made in ε_{kl} after writing $B(kl)$ in terms of a and b. Equation (11) then takes the form

$$E_\Pi = I_k^c + I_l^c + \tfrac{1}{2}\gamma_{aa}' + \tfrac{1}{2}\gamma_{ab}' \tag{12}$$

where

$$\gamma_{aa}' = \gamma_{aa} + \varepsilon_{aa} \tag{13a}$$

$$\gamma_{ab}' = \gamma_{ab} + \varepsilon_{ab} \tag{13b}$$

The Coulomb repulsion parameters of semi-empirical π-electron theory, thus incorporate correlation conveniently. The semi-empirical γ may incorporate other effects too, however, such as σ-changes [8] and $\varepsilon_{\sigma\pi}$ not explicit in the theory.

In comparing the semi-empirical and theoretical values of γ's, the question arises now as to what $2p_z$ AO's, $a, b, \ldots$, are appropriate for the π-electron molecule. Semi-empirical theory circumvents this question through parametrization. Integrals involving AO's are not calculated directly from specific AO w.f.'s, but are determined semi-empirically with the aid of some experimental data.

In the rest of this lecture, we shall examine the γ parameters in detail. Our purpose is (a) to see what $2p_z$ AO's are implied for π-electron theory by the semi-empirical values of γ, (b) to see if the discrepancy between the values of the conventional (STO) theoretical and experimental [1], [8] γ's is resolved, (c) to find a method by which the "semi-empirical γ_{aa}" of heteroatoms may be predicted, and (d) by examining the energy changes of atomic valence state reactions—models for π-electron systems—to find out more about σ and π orbital and correlation changes with π-electron number and state.

2. Coulomb Repulsion Parameters

Semi-empirical π-theory [9] assumes that a given MO is the same whatever state it appears in. For example in ethylene, π_{I} is assumed unchanged in the states N, V, T, and Z [1]. There is another assumption: the a in $\pi_{\text{I}} = 2^{-1/2}(a + b)$ and in $\pi_{\text{II}} = 2^{-1/2}(a - b)$ should be the same $2p_z$-orbital (OAO). These assumptions, added to the others above, seem to work quite well, though there are recent attempts to relax them and to use a nonclosed shell SCF MO theory where MO's change [10].

The semi-empirical theory has been fitted to the states of benzene, 6.76, 5.96, 4.71, and 3.59 eV, above the ground, assigned as ${}^1E_{1u}$, ${}^1B_{1u}$, ${}^1B_{2u}$ and ${}^3B_{1u}$ respectively [1]. For benzene the fit seems unambiguous. For the two coulomb parameters of main interest to us here, γ_{pp} and the nearest neighbor γ_{pq}, (i.e., γ_{01}), it yields

$$\begin{aligned} \gamma_{pp}(\text{spect.}) &= 11.35\ \text{eV} \\ \gamma_{pq}(\text{spect.}) &= 7.19\ \text{eV} \end{aligned} \tag{14}$$

Conventional theoretical values of these, calculated from $2p_z$-STO's (Slater orbital) with $Z^* = 3.18$, are [1]

$$\begin{aligned} \gamma_{pp}(\text{STO}) &= 16.93\ \text{eV} \\ \gamma_{pq}(\text{STO}) &= 9.03\ \text{eV} \end{aligned} \tag{15}$$

The discrepancy between the two sets of values has been attributed to (a) the failure of an STO calculation to incorporate the effects of σ- and

π-orbital changes [8, 11a, b], and (b) neglect of electron correlation [11a, 12a–e, 13a, b].

Pariser [8] noted that an MO w.f. includes ionic structures in valence bond language and that the σ-orbitals may change from those of neutral valence state carbon in such ionic structures. He then introduced the valence state carbon charge transfer reaction

$$\dot{C} + \dot{C} \rightarrow \ddot{C}^- + C^+ \tag{16}$$

as a model for such changes. The carbons in this reaction are in the valence states: $C(1s^2\eta^3 2p_z)$, $C^-(1s^2\eta^3 2p_z^2)$, and $C^+(1s^2\eta^3)$, where η^3 represents three electrons with random spins in the three sp^2-hybrid orbitals (the σ-core). Pariser showed that in the π-electron approximation, i.e., neglecting σ- and π-orbital changes with state (ionization) and neglecting correlation, the energy ΔE of reaction (16) would simply equal γ_{aa}. If instead one set

$$\text{``}\gamma\text{''}_{aa}\,(\text{``experimental''}) = \Delta E \tag{17}$$

and used this "γ"$_{aa}$ value for the molecule, σ-changes between covalent and ionic structures in the MO w.f. and similar to those in Eq. (16) may be automatically included [8].

The ΔE is given by

$$\Delta E = I_c - A_c \tag{18}$$

where I_c and A_c are the carbon valence state ionization potential and electron affinity. Though ΔE is referred to as "experimental," the evaluation of I and A for valence states requires special theory as well as data. Valence states as in Eq. (16) are combinations of several spectroscopic states. If the energies of all these spectroscopic states are known experimentally, an experimental valence state energy can be obtained. Usually only some of the states are available, so that theory must be used. We shall not go into it here, but theory of valence states including both Hartree-Fock and correlation is available; the reader may consult reference [14] (see also the chapter in Part II of this book [6] on nonclosed shell atomic states). In the literature, estimates for valence states have usually been obtained from semi-empirical I, F, and G parameters (see, e.g., reference [15]). For the carbon case, Eq. (16), all the methods yield about the same ΔE, 10.53 eV [8] to 10.69 eV [14].

It is remarkable that the value of ΔE, Eq. (18), is so close to the spectroscopic (semi-empirical) γ_{pp} of benzene, 11.35 eV. Is this a coincidence?[3] It would be difficult of course to calculate benzene completely and see why

[3] We thank Dr. Harris J. Silverstone for a stimulating discussion on this point.

the energy levels require the γ_{pp} value that they do. But we can calculate ΔE, Eq. (16), quite completely and see why it comes out 10.7 eV.

The exact form of ΔE for the carbon valence state reaction, Eq. (16) is

$$\Delta E = \Delta E_{\text{H.F.}} + \Delta E_{\text{corr}} \tag{19}$$

The valence state theory just mentioned [14] shows explicitly what orbital and correlation effects contribute to the energy of a valence state atom. Using this

$$\Delta E_{\text{H.F.}} = \gamma_{aa}^{\text{H.F.}} + \Delta E_{\text{H.F.}}^{\text{orbitals}} \tag{20}$$

$$\Delta E_{\text{H.F.}}^{\text{orbitals}} = \Delta E_{\text{H.F.}}^{\sigma\text{-}\pi} + \Delta E_{\text{H.F.}}^{\sigma} \tag{21}$$

and for the correlation part

$$\Delta E_{\text{corr}} = \Delta \varepsilon_{\sigma} + \Delta \varepsilon_{\sigma\text{-}\pi} + \Delta \varepsilon_{\pi} \tag{22}$$

For the carbon valence states, the σ refers to $(1s^2\eta^3)$ and π to $2p_z$ parts. The value for $\gamma_{aa}^{\text{H.F.}}$ (the coulomb integral $\langle 2p_z 2p_z, (1/r_{12}) 2p_z 2p_z \rangle$) is that calculated non-empirically with H.F. orbitals appropriate to the valence state $1s^2\eta^3 2p_z^2$ of negative carbon ion. $\Delta E_{\text{H.F.}}^{\text{orbitals}}$ is that part of the H.F. energy changes of the reaction due to changes in both the σ- and π-orbitals

$$\Delta E_{\text{H.F.}}^{\text{orbitals}} = \Delta E_{\text{H.F.}}^{\sigma\text{-}\pi} + \Delta E_{\text{H.F.}}^{\sigma} \tag{23}$$

$\Delta E_{\text{H.F.}}^{\sigma}$ involves the I_{σ}, $J_{\sigma\sigma'}$, and $K_{\sigma\sigma'}$ integrals of only the σ-electrons; $\Delta E_{\text{H.F.}}^{\sigma\pi}$ involves the energy of each π-electron in the field of the σ-core alone, $\langle 2p_z, \mathscr{H}_k^{\text{core}}\, 2p_z \rangle$, i.e., the $I(2p_z)$ and the $\sigma - \pi$ interactions $J(2p_z, \sigma)$ and $K(2p_z, \sigma)$; (cf. Eqs. (1)–(4)).

The results of H.F. calculations [14] for the valence states of carbon reaction are shown in Table I. These yield for Eqs. (16), (20), and (21),

$$\begin{aligned} \gamma_{aa}^{\text{H.F.}} &= 12.72 \text{ eV} & \Delta E_{\text{H.F.}}^{\sigma\text{-}\pi} &= -4.19 \text{ eV} \\ \Delta E_{\text{H.F.}}^{\sigma} &= +3.16 \text{ eV} & \Delta E_{\text{H.F.}}^{\text{orbitals}} &= -1.03 \text{ eV} \end{aligned} \tag{24}$$

The change in the correlation energy, Eq. (22), is given by the nonclosed shell many-electron theory [6] applied to valence states [14]. In atoms, the correlation energy of $1s^2 2s^2$ cores may change by several electron volts with changes in the number and state of $2p$-electrons [2, 16, 17]. Such core changes

TABLE I

HARTREE-FOCK RESULTS FOR VALENCE STATES IN THE CARBON REACTION (in eV).

	Valence states		
	$C(1s^2\eta^3 2p_z)$	$C^-(1s^2\eta^3 2p_z^2)$	$C^+(1s^2\eta^3)$
$E_{\text{H.F.}}$	-1019.047	-1017.110	-1009.291
$E^{\sigma}_{\text{H.F.}}$	-1007.900	-1003.349	-1009.291
$E^{\sigma-\pi}_{\text{H.F.}}$	-11.147	-13.241×2	0
$\gamma^{\text{H.F.}}_{pp}$	—	12.721	—

may be large in general, but cancel out in the carbon reaction. Calculations give [14]

$$\Delta\varepsilon_{\sigma} = -\ 0.04\ \text{eV} \tag{25}$$

In the rest of Eq. (22), $\Delta\varepsilon_{\sigma-\pi}$ is the change in the correlation energy ("core-polarization") of $2p_z$-electrons with the σ-core. This interaction in a given atom may be large; the correlation of a $2p$-electron with $1s^2 2s^2$, $\varepsilon(1s^2 2s^2 \to 2p)$ is about -1 eV [16, 17]. But again in the carbon reaction it tends to cancel out since the number of $2p_z$-electrons on both sides of Eq. (16) is the same. The $\Delta\varepsilon_{\pi}$ is the change in the correlation energy of $2p_z$-electrons among themselves. This comes simply from C^-, thus it is $\varepsilon(2p_z^2)$, a dynamical, transferable pair correlation with a value -1.0 eV ([16], [17]; see also [14]). Thus

$$\Delta E_{\text{corr}} = \Delta\varepsilon_{\pi} = \varepsilon(2p_z^2) = -\ 1.0\ \text{eV} \tag{26}$$

Before interpreting these results, let us note that the above equations give an accurate way for the prediction of other ΔE's, for heteroatoms. If the correspondence between such ΔE's and the nominal "γ_{pp}" of π-spectra is general, one can use these for heteromolecules. The net equation is

$$\text{"}\gamma_{pp}\text{"} \leftarrow \Delta E = \Delta E_{\text{H.F.}} + \varepsilon(2p_z^2) \cong \Delta E_{\text{H.F.}} - 1.0\ \text{eV} \tag{27}$$

ΔE values calculated [14] from this equation for several types of heteroatom are shown in Table II. The valence state reactions are taken analogous to the carbon reaction, Eq. (16). The atoms in the first column are the ones that disproportionate. For example, for the pyrolle-type nitrogen, ΔE refers to

$$\dot{N}^+ + \dot{N}^+ \to \ddot{N} + N^{+2}$$

The predicted ΔE values are quite accurate, but do they represent the "γ_{pp}"'s in heteromolecules? This requires further investigation. Let us examine the carbon results further.

The source of the discrepancy between the original STO value of γ_{pp}, Eq. (15) and the "experimental" one from Eq. (18) is clear from Eqs. (19)–(26). Correlation contributes only -1 eV; $\Delta E_{\text{H.F.}}^{\text{orbitals}}$ only -1.03 eV. There are indeed large σ-changes [8, 9], $+3.2$ eV in Eq. (24), but these are mostly canceled by equally large π-orbital changes, $\Delta E_{\text{H.F.}}^{\sigma-\pi}$. Thus the discrepancy is in the value of $\gamma_{pp}^{\text{H.F.}}$ itself. In the literature, the γ_{pp} integral has been usually calculated either from Slater orbitals, or from H.F. orbitals of neutral

TABLE II

Calculated Values of ΔE for Valence State Disproportionation Reactions of Carbon, Nitrogen, and Oxygen[a]

Atom	Number of electrons contributed to π-framework (e.g., in molecule)	$\Delta E(\rightarrow \text{"}\gamma_{pp}\text{"})$ [b]
$C(1s^2\eta^3 2p_z)$	1 (benzene)	10.69
$N(1s^2\eta^4 2p_z)$	1 (pyridine)	12.91
$N^+(1s^2\eta^3 2p_z)$	2 (pyrolle)	16.47
$O(1s^2 2s^2 2p_x^2 2p_y 2p_z)$	1 (formaldehyde)	15.38
$O^+(1s^2\eta^4 2p_z)$	2 (furan)	18.85

[a] in eV.
[b] Calculated from Eq. (27) in text.

carbon [13a]. In the reaction (16), $\gamma_{pp}^{\text{H.F.}}$ refers to valence state C^- H.F. orbitals. The value of $\gamma_{pp}^{\text{H.F.}}$ with these orbitals is already very close to the total ΔE, and to the benzene γ_{pp} (spect.) in Eq. (14). The formula

$$\gamma'_{pp} = \gamma_{pp}^{\text{H.F.}} + \varepsilon_{pp} = \gamma_{pp}^{\text{H.F.}} + \varepsilon(2p_z^2) \tag{28}$$

as in Eq. (13a), comes out even closer to the benzene value. From Eqs. (24) and (26), for C^- valence state,

$$\gamma'_{pp} = 11.7 \text{ eV} \tag{29}$$

compared to the benzene value γ_{pp} (spect.) $= 11.4$ eV. The $\gamma_{pp}^{\text{H.F.}}$ integral is very sensitive to charge on the carbon atom, less so to the state of a given ion. Neutral C values vary about 0.5 eV around 15 eV (15.7 eV in $1s^2 2s^2 2p^2\ {}^3P$,

16.1 eV in $1s^2 2s 2p^3\ ^5S$, and 15.3 eV in $1s^2 2p^4\ ^1D$). They are around 12.2 ∓ 0.5 eV for different states of C^-, and around 18 ∓ 0.5 eV for those of C^+. Thus the crucial point in the result above is the negative charge on carbon. But why the correspondence of these C^- values with the value of "γ_{pp}" in benzene?

As Pariser [8] noted, even the ground state of the benzene MO w.f. includes ionic valence bond structures with C^-C^+, but the carbon $2p\pi$ occupation is far from 2. The spectroscopic γ's however are fitted to several excited states, in fact to energy increments from the ground. They are weighted averages over the types of carbon occurring in these states. But excited states accentuate ionic structures. Thus the average, effective $2p\pi$-AO in the several states of benzene may be weighted more in the direction of C^-. This may be why—if it is not sheer coincidence—benzene yields the C^- γ'_{pp} value.

To examine this point further, we calculated the two-center $\gamma_{pq}^{\text{H.F.}}$ integral $\langle 2p_z^a 2p_z^b, g_{12} 2p_z^a 2p_z^b \rangle$ for benzene ($R = 1.39$ Å), with the C^- H.F. AO orbital.[4] The result was

$$\gamma_{pq}^{\text{H.F.}} = 8.02 \text{ eV} \tag{30}$$

Neutral carbon H.F. AO ($1s^2 2s^2 2p^2\ ^3P$) yields, on the other hand, 8.75 eV[5] which is quite close to the conventional STO value, 9.03 eV, in Eq. (15).

From Eq. (13b)

$$\gamma'_{pq} = \gamma_{pq}^{\text{H.F.}} + \varepsilon_{pq} \tag{31}$$

with the diffuse and expanded C^- $2p\pi$-orbitals, the two center correlation ε_{pq} is probably also close to -1 eV ($|\varepsilon_{pq}|$ versus R has a plateau up to about $R \cong 2R_e$ before it drops off in H_2, for example, as Fischer-Hjalmars has shown [11a]), in any case no less than 0.5 eV, in magnitude. Thus γ'_{pq} from Eqs. (30) and (31) is about 7.0 eV and most likely no more than 7.5 eV. This value is again remarkably close to the benzene value γ_{pq} (spect.) = 7.19 eV in Eq. (14).

The calculations above on both γ_{pp} and γ_{pq} indicate that the average, effective AO making up the MO's of the several electronic states of benzene is close in charge distribution to that of the C^- valence state. Aside from

[4] Six-term Roothaan open shell H.F. AO of ($1s^2 2s 2p^4\ ^4P$) of C^-. This orbital should be very close to the average one occurring in C^- valence state. It yields a $\gamma_{pp}^{\text{H.F.}} = 12.9$ eV versus the valence state value 12.7 eV. The diatomic molecule integral program of F. J. Corbato and A. C. Switendick was used.

[5] Our thanks are due to Miss Charlotte Hollister who calculated this value.

the rationalization above, concerning ionic states, it is not clear why the effective AO should be so. It could be that the similarity of the ΔE of the carbon disproportionation reaction, to the fitted "γ_{pp}" of benzene is just a coincidence. An examination of atomic populations and the kind of averaging over states that goes into the values of γ_{pp} and γ_{pq} may help in clarifying this situation. The ΔE's of analogous reactions for heteroatoms are calculable accurately (Eq. (27)), but again do these ΔE's correspond to the "γ_{pp}" of heteromolecules in a fundamental way?

For the carbon reaction, we found that both σ and π H.F. orbital changes are large [14],[6] but they nearly cancel in the ΔE. The σ and $\sigma - \pi$ correlation energies and changes are also large usually, but again tend to cancel out in the carbon reaction. Will all these sizable effects cancel out however in general for molecules? If the cancellations can be shown to be more general than in the carbon reaction, Eqs. (9), (13a) and (13b) can be used to define the γ'_{pp} and γ'_{pq} 's of π-electron theory (with average MO's) and to calculate them nonempirically from Eqs. (28) and (31).

Certain aspects of π-electron theory are justified formally and numerically, some of the questions above remain. A more detailed treatment would require nonclosed shell theory.

Acknowledgment

We thank H. J. Silverstone, C. Hollister, and A. H. Lowrey for helpful discussions. This work was supported by a grant from the National Science Foundation.

References

1. R. G. Parr, "Quantum Theory of Molecular Electronic Structure." Benjamin, N. Y., **1963**.
2. O. Sinanoğlu, "Advances Chemical Physics," Vol. VI, p. **315**. Wiley (Interscience), London, **1964**.
3. O. Sinanoğlu, *J. Chem. Phys.* **36**, 3198 (1962).
4. O. Sinanoğlu, *J. Chem. Phys.* **33**, 1212 (1960).
5. O. Sinanoğlu, *J. Chem. Phys.* **36**, 706 (1962).
6. H. J. Silverstone and O. Sinanoğlu, this book, Section A.7, Part II.
7. J. Koutecký, this book, Section B.6.e, Part I.
8. R. Pariser, *J. Chem. Phys.* **21**, 568 (1953).
9. R. Pariser and R. G. Parr, *J. Chem. Phys.* **21**, 466, 767 (1953).
10. L. Goodman and J. R. Hoyland, *J. Chem. Phys.* **39**, 1068 (1963).
11a. I. Fischer-Hjalmars, this book, Section B.6.d, Part I.

[6] See also Section II C–1 (b) and (c) of reference [18] for a review of recent findings on the importance of σ- and π-changes.

11b. K. Ohno, *Theoretica Chimica Acta* **2**, 219 (1964).
12a. W. Kolos, *Acta Phys. Polon.* **16**, 257, 299 (1957).
12b. M. J. S. Dewar and N. L. Hojvat, *J. Chem. Phys.* **34**, 1232 (1961).
12c. M. J. S. Dewar and N. L. Hojvat, *Proc. Roy. Soc.* **A264**, 431 (1961).
12d. M. J. S. Dewar and N. L. Sabelli, *J. Phys. Chem.* **66**, 2310 (1962).
12e. A. Julg, *J. Chim. Phys.* **57**, 19 (1960).
13a. T. Arai and P. G. Lykos, *J. Chem. Phys.* **38**, 1447 (1963).
13b. F. O. Ellison and N. T. Huff, *J. Chem. Phys.* **38**, 2444 (1963).
14. M. K. Orloff and O. Sinanoğlu, "Sigma and Pi Changes in Valence States of Pi-electron Theory and One-center Coulomb Repulsion Parameters", *J. Chem. Phys.* (in press).
15. G. Pilcher and H. A. Skinner, *J. Inorg. Nucl. Chem.* **24**, 937 (1962).
16. V. McKoy and O. Sinanoğlu, *J. Chem. Phys.* **41**, 2689 (1964).
17. V. McKoy and O. Sinanoğlu, this book, Section A, Part II.
18. O. Sinanoğlu and D. F. Tuan, *Ann. Rev. Phys. Chem.* **15**, 251 (1964).

Addendum to Chapter VII-7

The nonempirical Hartree-Fock values of the one-center ($\gamma_{aa}^{H.F.} = \gamma_0^{H.F.}$) and nearest-neighbor two-center ($\gamma_{ab}^{H.F.} = \gamma_1^{H.F.}$) Coulomb repulsions were given in this paper. With the same Hartree-Fock orbitals the other two next-nearest-neighbor ($\gamma_{ac}^{H.F.} = \gamma_2^{H.F.}$ and $\gamma_{ad}^{H.F.} = \gamma_3^{H.F.}$) values for benzene (as calculated by Ö. Pamuk at Yale) are also given below:

$$\gamma_0 = 12.72 \text{ eV}$$
$$\gamma_1 = 8.01 \text{ eV}$$
$$\gamma_2 = 5.37 \text{ eV}$$
$$\gamma_3 = 4.76 \text{ eV}$$

These values do not include any correlation corrections and therefore could be used to obtain ZDO-simplified Hartree-Fock MO results.

VII-8 Reprinted from *The Journal of Chemical Physics* **47**, 4494 (1967)

Sigma and Pi Electronic Reorganization in Acetylene*

MARTIN G. GRIFFITH† AND LIONEL GOODMAN‡

School of Chemistry, Rutgers University, New Brunswick, New Jersey and *Department of Chemistry, The Pennsylvania State University, University Park, Pennsylvania*

(Received 16 February 1967)

Accurate SCF–LCAO–MO wavefunctions have been computed for acetylene neutral ground state, singly excited $\pi\rightarrow\pi^*$ states, and for the acetylene anion and cation. A limited basis set of Slater orbitals was used, the carbon $2s$, $2p\pi$, and $2p\sigma$ orbital exponents were independently optimized for each state, and hydrogen $1s$ for the ground state. All integrals were computed accurately. For the ground state the $C2p\pm1$ optimized orbital ($\xi=1.59$) is very different from the $C2p\sigma$ optimized orbital ($\xi=2.03$) with the result that the Π cloud does not penetrate the Σ cloud as greatly as would be predicted by Slater or best-atom AO. However, the penetration remains sizeable (Fig. 2). The ground-state energy -76.678 a.u. (1 a.u. $=27.2098$ eV) is improved by 0.134 a.u. over McLean's best-atom calculations, but is 0.176 a.u. poorer than McLean's Hartree–Fock calculations. Reorganizations of the sigma and pi electronic shells with change in state of the pi system are computed to be large. A gain or loss of a pi electron is accompanied by a sigma energy change of 0.2 a.u. partially due to deformation of the C2s AO and partially due to sigma polarization away from the H atoms in the cation and vice versa in the anion. The large reorganization in the pi distribution is illustrated by the $2p\pi$ exponents: cation 1.68, anion 1.45, ground state 1.59. The total reorganization energy in acetylene (-0.033 a.u. G state$\rightarrow$cation; -0.069 a.u. G state$\rightarrow$anion) is only 6% of the largest reorganization term: the change in sigma–pi interaction (0.586 and 1.262 a.u., respectively). The success of conventional pi-electron theory then basically depends on cancellation of the sigma and pi reorganization energy terms with the change in the sigma–pi interaction. The H_{ij}^{core} parameters are found to be nearly independent of pi density providing justification for pi-electron approaches which adjust the pi atomic orbitals according to pi density.

I. INTRODUCTION

This paper reports a study of the molecular-orbital (MO) reorganizations which occur in a model unsaturated hydrocarbon when a change is made in configuration of the pi electronic shell. The objective is to provide a test calculation in a sufficiently rigorous framework to examine a basic tenet of pi-electron theory, that the sigma wavefunction remains unaltered by a change in configuration of the pi wavefunction.

For a closed-shell ground state the wavefunction is

$$\Psi_g=[(\Sigma_g)(\Pi_g)],\tag{1}$$

where (Σ_g) is a single antisymmetrized product of paired one-electron spin orbitals having σ symmetry. The brackets indicate that the product is made antisymmetric for exchange of electrons between the two groups. The tenet requires that for $(\Pi_g)\rightarrow(\Pi')$,

$$\Psi'=[(\Sigma_g)(\Pi')],\tag{2}$$

and that concommitantly that H^{core} matrix elements are not altered in going from Ψ_g to Ψ' (in order that the variational procedure for the pi-molecular orbitals remains valid). For perspective in the pi-electron approximation, see Fueno.[1] The representations given by Eqs. (1) and (2) are not the most general statements of pi-electron theory, cf. Lykos and Parr[2] and McWeeny.[3]

For a closed-shell system of N_σ doubly occupied sigma MO's and N_π pi MO's, the SCF effective one-electron Hamiltonian operator is

$$H_{\text{eff}}=-\frac{\hbar^2}{2m}\nabla^2+V+\sum_{i=1}^{N_\sigma}(2J_{\sigma_i}-K_{\sigma_i})+\sum_{i=1}^{N_\pi}(2J_{\pi_i}-K_{\pi_i}),$$

$$H_{\text{eff}}=H^{\text{core}}+\sum_{i=1}^{N_\pi}(2J_{\pi_i}-K_{\pi_i}),\tag{3}$$

where V is the nuclear field potential and J_{σ_i} and K_{σ_i} are the Coulomb and exchange operators[1] in terms of sigma molecular orbitals. The total energy is

$$Eg=\Sigma_{1e}+\Sigma_{2e}+\Pi_{1e}+\Pi_{2e}+\Sigma\Pi_{2e},\tag{4}$$

where for example Σ_{1e} indicates one-electron terms derived from σ functions alone and $\Sigma\Pi_{2e}$ comprises the Coulomb and exchange interactions between sigma functions and pi functions.

In practice only the pi MO's are computed in pi-electron theory. The sigma electrons enter the calculation as a potential in H^{core}, but the matrix elements of H^{core} over basis functions are treated as empirical parameters and carried from one molecule to another.

* Supported by a grant from the National Science Foundation, and The American Petroleum Institute.

† Taken from the Ph.D. thesis of M.G.G., PSU, April, 1967.

‡ John Simon Guggenheim Fellow.

[1] T. Fueno, Ann. Rev. Phys. Chem. **12**, 303 (1961).

[2] P. G. Lykos and R. G. Parr, J. Chem. Phys. **24**, 1166 (1956).

[3] R. McWeeny, Proc. Roy. Soc. (London) **A253**, 242 (1959); **A223**, 306 (1954).

Such handling of H_{ij}^{core} matrix elements is not strictly a part of pi-electron theory, but of its application.

If reference is made to the success of pi-electron theory in accounting for the electronic spectra of conjugated hydrocarbons, Eq. (2) is apparently a good approximation when the density $(\Pi')^2$ approximates $(\Pi_g)^2$. However, the sigma wavefunction certainly must change if electrons are added to or substracted from the pi system. Lykos and Parr[2] noted that computed pi-electron ionization energies are too high and suggested that for such computations pi-electron theory should be ammended to allow an "adjustable core." Since all MO's are expected to reorganize, every energy term in Eq. (4) will change. Changes in $\Pi_{1e}+\Sigma\Pi_{2e}$ will be altered by changes in the H_{ij}^{core} parameters. Changes in Π_{2e} are explicitly computed, but changes in $\Sigma_{1e}+\Sigma_{2e}$ do not even enter a pi-electron calculation.

In 1960 Parks and Parr[4] made a serious attempt to explicitly consider the adjustment of the sigma electrons in formaldehyde to the new electronic environments caused by pi-electron ionization and by $n\rightarrow\pi^*$ promotion. They found polarization of the sigma core upon ionization of a pi electron, but detailed analysis of their results is impeded by the many-integral approximations used in their calculations.

Shortly after, Hoyland and Goodman[5] attempted in a semiempirical framework to allow for deformation in both the sigma and pi shells. Their method is based on Lykos and Parr's observation that the calculated pi ionization potentials by Koopmans' theorem are usually greater by 2–3 eV than the experimental ones for conjugated organic molecules, so that making the ionized state wavefunction approach more closely a Hartree–Fock one should lower the computed ionization energy. They suggested that the poor ionized state wavefunction arises from neglect of (a) changes in the $2p\pi$ basis functions on ionization (pi deformation), (b) the effect of pi-electron ionization on the sigma framework (sigma deformation), (c) changes in the MO's through construction of a new self-consistent field for the ionized state (polarization). They attempted to correct for (a) by altering H_{ii}^{core} and Π_{2e}. The correction for (b) was brought about by first-order perturbation theory utilizing the change in Z_p at Atom p as a perturbation and utilizing hydrogenlike orbitals as the zeroth-order wavefunction. Linear dependence of Zp on the change in pi density at Atom p was assumed. This correction predicted for any conjugated hydrocarbon a constant change in $\Sigma_{1e}+\Sigma_{2e}$ of approximately 1 eV upon pi ionization. The correction for (c) was carried out (only for the pi MO's) theoretically by the Roothaan open-shell equations.[6] Ionization potentials about 2 eV smaller than those derived from direct application of Koopmans' theorem were obtained for a sizeable number of conjugated hydrocarbons mainly from the (a)- and (b)-type reorganizations. A major criticism of Hoyland and Goodman's approach is that the partitioning of reorganization effects may be erroneous even if the total reorganization is accurate, because nonlocal interactions were neglected in the Σ reorganization, and because of the empiricism.

This paper examines accurately computed SCF wavefunctions for states of acetylene. Acetylene was chosen because it is the simplest hydrocarbon having sufficient complexity to allow study of reorganization effects with reference to large molecules. Two types of deformation are accounted for: (1) molecular deformation, i.e., changes in the MO coefficients: (2) atomic deformation, i.e., contraction or expansion of the Slater atomic orbitals (AO's) which comprise basis functions for the MO's. There are, in a sense, two SCF problems: the molecular one, and the atomic one. The MO's are permitted to rearrange by polarization (change in coefficients) or scale deformation (change in Slater exponents). Because of the limited basis set (see Sec. II), the pi MO's can deform only at the atomic level, but in the sigma system, charge can also circulate between the C atoms and the H atoms by changes in coefficients on AO's. In this study $C2s$, $C2p\sigma$, and $C2p\pm1$ exponents are individually optimized for each state and $H1s$ for the ground state so as to allow evaluation of the relative importance of deformation versus polarization in sigma shell reorganization.

These calculations include all integrals, accurately computed, and should serve as models for the evaluation of approximate methods.

II. OPTIMIZED BASIS SET

Molecular wavefunctions for the neutral ground-state acetylene molecule and the cationic, anionic, and the three lowest triplet excited states $^3E_{2u}$, $^3A_{1u}$, $^3A_{2u}$ were computed according to methods given by Roothaan for closed-[7] and open-shell systems.[6,8] The geometry for all species was taken as the experimental conformation of the ground state, i.e., linear with H–C and C–C bond distances equal to 1.059 and 1.203 Å, respectively.[9] For studying electronic reorganization only vertical processes need be considered.

A minimal basis set of Slater orbitals was used with one $1s$ orbital on each hydrogen and $1s$, $2s$, $2p\sigma$, and $2p\pm1$ orbitals on each carbon. The carbon $1s$ orbital exponent: ξ $(C1s)$ was taken as 5.70. The three carbon and one hydrogen exponents were obtained by simultaneous minimization of the total energy for the ground

[4] L. M. Parks and R. G. Parr, J. Chem. Phys. **32,** 1657 (1960).
[5] J. R. Hoyland and L. Goodman, J. Chem. Phys **33,** 946 (1960).
[6] C. C. J. Roothaan, Rev. Mod. Phys. **32,** 179 (1960).
[7] C. C. J. Roothaan, Rev. Mod. Phys. **23,** 69 (1951).
[8] C. C. J. Roothaan and P. Bagus, Methods Computational Phys. **2,** 47 (1963).
[9] M. T. Christensen, D. R. Eaton, B. A. Green, and H. W. Thompson, Proc. Roy. Soc. (London) **A238,** 15 (1956).

TABLE I. Optimized orbital exponents for acetylenic species.[a]

Orbital	Neutral ground state[d]	Cation	Anion	Singly[b] excited
C2s	1.80	1.84	1.76	1.80
C2pσ	2.03	2.03	2.03	2.03
$C2p_{\pm1}$	1.59	1.68	1.46	1.59
H1s	1.20	c	c	c

[a] C1s = 5.70.
[b] Giving rise to $^3E_{2u}$, $^3A_{2u}$, $^3A_{1u}$ states.
[c] Not optimized. H1s has been set equal to 1.20.
[d] Slightly improved exponents obtained by use of a finer integration mesh are C2s = 1.75, C2pσ = 2.01, $C2p_{\pm1}$ = 1.59, H1s = 1.19. See Footnote f, Table II.

state with respect to each exponent by a "brute force approach." Preliminary minimization was carried out over a four-dimensional grid. In the first dimension the 2s exponent was varied over intervals of 0.04 from 1.60 to 1.88. In the second dimension the 2pσ exponent ranged from 1.48 to 2.08 at intervals of 0.05. In the third dimension 2p±1 varied from 1.43 to 1.73 also at 0.05 intervals, and in the fourth H1s ranged from 1.10 to 1.30 at 0.02 intervals. The 736 integral sets were sufficient to establish approximate minima and final optimization was obtained in a four-dimensional grid at 0.02 intervals. Optimization was considered accomplished when a change in any exponent of ±0.02 produced an energy change of less than 0.001 a.u. For the ionized species the three carbon exponents were simultaneously optimized for each species as described above. Lack of available computer time prevented optimization of H1s, and computations were carried out for only two values of $\xi(\mathrm{H}1s)=1.00$ and 1.20 with qualitatively the same reorganization results (see Table IV). [$\xi(\mathrm{H}1s)=1.00$ is utilized in Tables III–VI and Figs. 1 and 2.] The minor improvement in Eg in going to 1.20 from 1.00 (−76.649 a.u. for the latter), the minor change in σ charge migration for any combination of $\xi(\mathrm{H}1s)$ in going to the charged species, and the good agreement with the virial theorem (Table VI) suggests little error. As is discussed in Sec. IV, the principal sigma reorganization occurs in the MO coefficients.

Table I lists the optimized exponents for the various species. As would be expected, the sigma exponents are less affected than the pi exponents by the number of electrons in the pi shell. The exponents for the ground state and the singly excited states are identical to within ±0.02. The C2pσ exponent varies by less than 0.02 over all the states considered.

The Σ electronic configuration for all the acetylene states studied is

$$\Sigma=\sigma_{1g}^2\sigma_{1u}^2\sigma_{2g}^2\sigma_{2u}^2\sigma_{3g}^2.$$

The Π configurations are as follows:

Neutral ground state:

$$\Pi_g=\pi_u^4,$$

Cation:

$$\Pi_+=\pi_u^3,$$

Anion:

$$\Pi_-=\pi_u^4\pi_g^1,$$

Excited states:

$$\Pi_*(^3E_{2u},\,^3A_{2u},\,^3A_{1u})=\pi_u^3\pi_g^1.$$

Treatment of the excited states requires additional comment. The four possible 14-electron single-determinant triplet functions formed by single excitations are

$$^3\psi_1=|\Sigma(1\text{–}10)\bar{\pi}_{1u}(11)\pi_{1u}(12)\pi_{-1u}(13)\pi_{-1g}(14)|,$$

$$^3\psi_2=|\Sigma(1\text{–}10)\bar{\pi}_{-1u}(11)\pi_{-1u}(12)\pi_{1u}(13)\pi_{1g}(14)|,$$

$$^3\psi_3=|\Sigma(1\text{–}10)\bar{\pi}_{1u}(11)\pi_{1u}(12)\pi_{-1u}(13)\pi_{1g}(14)|,$$

$$^3\psi_4=|\Sigma(1\text{–}10)\bar{\pi}_{-1u}(11)\pi_{-1u}(12)\pi_{1u}(13)\pi_{-1g}(14)|.$$

A bar indicates β spin. π_{1u} and π_{-1u} are the π_u functions with M_z eigenvalues $+\hbar$ and $-\hbar$, respectively.

The linear combinations having the correct symmetry are[10]

$$^3\Sigma_u^+(A_{2u})=\sqrt{2}^{-1}(\psi_1+\psi_2),$$

$$^3\Sigma_u^-(A_{1u})=\sqrt{2}^{-1}(\psi_1-\psi_2),$$

$$^3\Delta_u(E_{2u})=\psi_3 \text{ or } \psi_4.$$

The SCF single-determinant functions ψ_1, ψ_2, ψ_3, and ψ_4 singlets and triplets, were obtained by Roothaan's open-shell formulation. The energy of the $^3A_{2u}$ and $^3A_{1u}$ states was computed from the Hamiltonian matrix elements of $^3\psi_1$ and $^3\psi_2$. Since in the restricted Hartree–Fock method, the Σ SCF wavefunctions in the four single-determinants are identical it follows that $\langle{}^3\psi_1H\,{}^3\psi_2\rangle=-\langle\pi_{1u}\pi_{-1u}|1/r_{12}|\pi_{-1g}\pi_{1g}\rangle$.

Integrals were evaluated to at least fifth-decimal-place accuracy. One- and two-center integrals were computed principally by Switendick and Corbato's[11] diatomic molecular integral program. Two-center exchange integrals were computed by a program of Pitzer, Wright, and Barnett,[12a] and a small number of one- and two-center integrals were computed by a program based on Roothaan's integral expressions.[12b] Programs of Pitzer, Wright, and Barnett and by Shavitt, Stevens, and Dleinman based on the method of Shavitt and Karplus[13] were used to evaluate the three- and four-center integrals. The major part of the calculations were carried out with an IBM 7074 computer at the Pennsylvania State University Computation Center. Some integral computations were carried out at the Indiana University computing center with a 709 com-

[10] I. G. Ross, Trans. Faraday Soc. **48**, 973 (1952).
[11] A. C. Switendick and F. J. Corbato, Quantum Chemistry Program Exchange, Indiana University, QCPE 29, 1966.
[12] (a) R. M. Pitzer, J. P. Wright, and M. P. Barnett, Quantum Chemistry Program Exchange, Indiana University, QPCE 22, 23, 24, 25, 1966. (b) C. C. J. Roothaan, J. Chem. Phys. **19**, 1445 (1951).
[13] I. Shavitt and M. Karplus, J. Chem. Phys. **43**, 398 (1965).

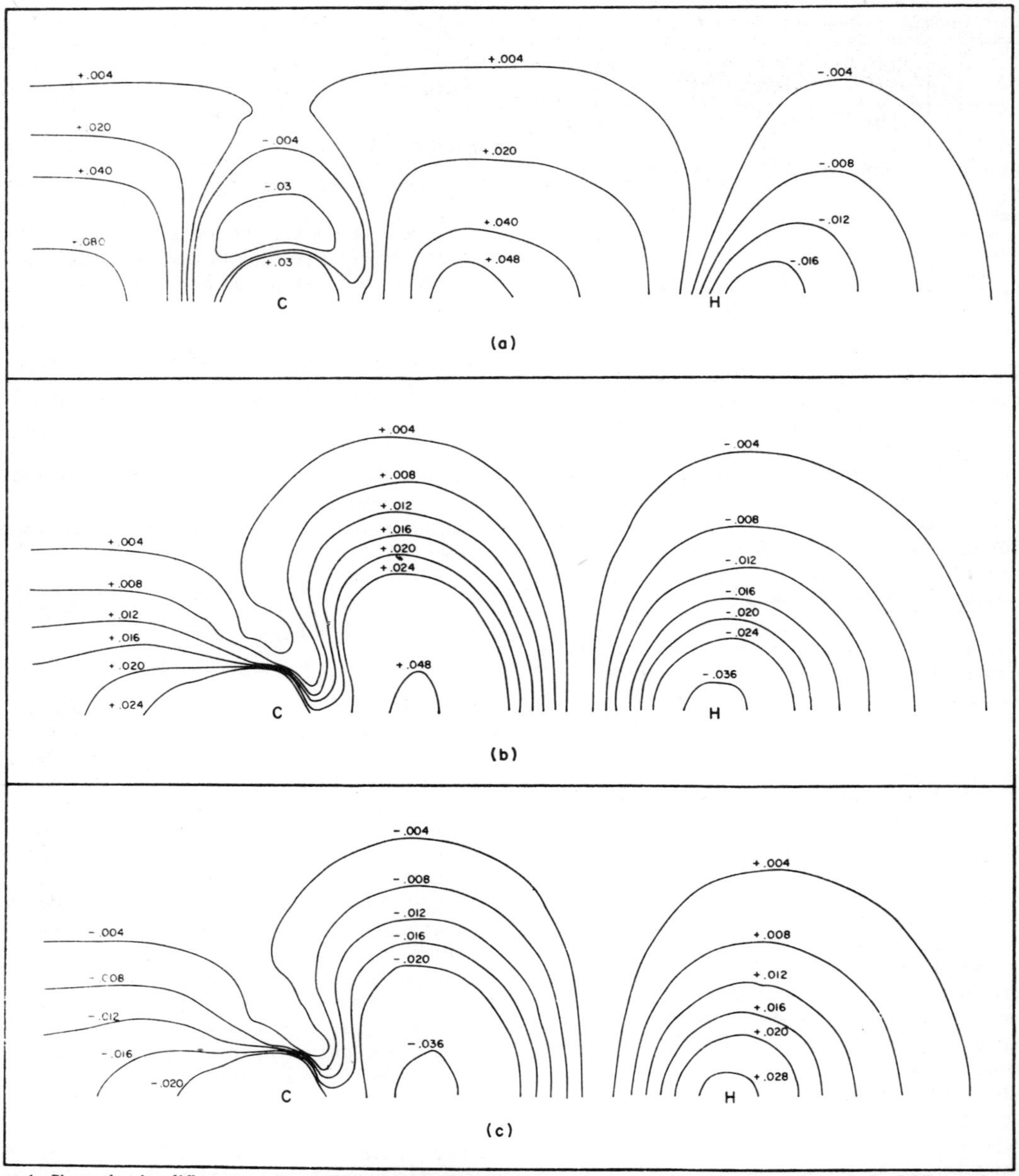

FIG. 1. Sigma density difference contours. The function plotted is $\delta(\sigma_B, \sigma_A) = \Sigma_B{}^*\Sigma_B - \Sigma_A{}^*\Sigma_A$. For (a), Σ_B is the SCF wavefunction for the sigma electrons of acetylene neutral ground state, Σ_g, and $\Sigma_A{}^*\Sigma_A$ is the sigma density arising from the isolated atoms in the state $Ks^2(s-p\sigma)\ p\pi$. Figure 1(a) demonstrates the effect of bond formation on the Σ density. For (b) Σ_B refers to the cation. For (c) Σ_B refers to the anion. Σ_A is the neutral ground-state function, Σ_g, in both cases. Figures 1(b) and 1(c) represent sigma density reorganization contours. The contours are carried to the molecular center in each case.

puter, and at Columbia University with a 7094 computer.

III. NEUTRAL GROUND STATE

Table II summarizes the current status of accurate SCF MO calculations for the acetylene neutral ground state. All these calculations are for single-determinant wavefunctions, and all except that of Moskowitz[14] and McLean's Hartree–Fock calculation[15] are in terms of

[14] J. W. Moskowitz, J. Chem. Phys. **43,** 60 (1965); **45,** 2338 (1966).

[15] A. D McLean, J. Chem. Phys. **32,** 1595 (1960); and private communication.

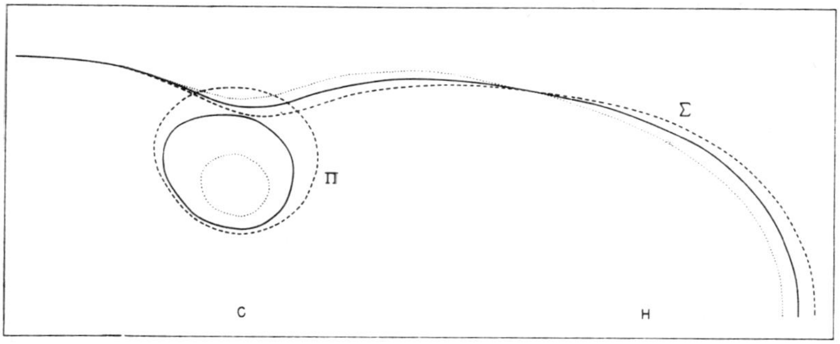

FIG. 2. The 0.06 contours for acetylene neutral ground state, ——; cation, • • •; and anion, – – –. The contours are given individually for total sigma density and total pi density.

a minimal basis set of Slater orbitals. Moskowitz used an extended Gaussian basis set. McLean used an extended Slater basis set chosen to closely converge to the Hartree–Fock energy. McLean's calculation comes closest to the experimental total energy for acetylene of −77.3605 a.u.

For the Slater orbitals computations, the optimization of the $2s$, $2p\sigma$, $2p\pm1$, and hydrogen $1s$ exponents carried out in this calculation produces an energy improvement in excess of 0.13 a.u. over the "best-atom" exponents used by McLean.[15] There is less improvement relative to the Palke and Lipscomb calculation[16] where normal Slater rule exponents were used except for H$1s$ ($\xi=1.20$). Palke and Lipscomb reported that an orbital exponent of 1.2 on H yields consistently lower energies for molecular hydrides than the value 1.0, and the optimized exponents of Table I confirm this. The disparity in computed values for the gross atomic H population is of interest especially in view of the well-known acidity of acetylene. The optimized minimal basis set gives almost zero net charge on H, and this result is not highly dependent upon the scale deformation of H$1s$. Since acetylene is acidic one assumes that an extended basis set calculation would go back to giving a net positive charge on the hydrogen (as is indicated by Moskowitz's calculation reported in Table II). It is probable that the lack of any functions on H which can adequately introduce polarization force the carbon functions to include polarization effects that should be associated with the hydrogen. In a population analysis it all counts as belonging to the carbon, however. The effect of optimization is to cluster electrons about the nuclei and also to transfer electrons to the CH bond region. As far as population analysis results are concerned the result is to transfer electrons to the hydrogen. The important point is that conclusions drawn from a minimal basis set, even with optimization, must be viewed cautiously. In a sense, this is an example of an unbalanced basis set, as discussed by Mulliken,[17] although in a slightly different way. In more accurate calculations with larger basis sets the simple analysis of polarization changes vs scale change which we apply in Secs. IV and V may not be applicable because of the large interdependence of the two effects. These remarks are intended to emphasize that our work is a first approximation to the reorganization problem though more rigorous than heretofore.

The ground-state SCF coefficients and eigenvalues are given in Table III. Figure 1(a) is a plot of the difference function[18]

$$\delta(\sigma_n,\sigma_v)=\Sigma_B{}^*\Sigma_B-\Sigma_A{}^*\Sigma_A,$$

TABLE II. Accurate SCF calculations for the acetylene ground state.

Total energy	McLean[a,b]	Moskowitz[c,b]	Palke and Lipscomb[d,b]	Hartree–Fock (McLean)[e]	This calculation[f,g]
(a.u. 1 a.u. = 27.2098 eV)	−76.544	−76.760	−76.617	−76.854	−76.678
N(H)[h]	9.79	0.79	0.81		0.998
KE[i]	...	...	76.316		77.440

[a] A. D. McLean, J. Chem. Phys. **32**, 1595 (1960). H$1s$ (orbital exponent, $\xi=1.00$), C$1s$ ($\xi=5.6726$), C$2s$ ($\xi=1.6082$), C$2p\sigma$, C$2p\pi$ ($\xi=1.56875$).

[b] $R_{CH}=2.002$ a.u., $R_{CC}=2.281$ a.u. Nuclear repulsion energy is 24.737 a.u.

[c] J. W. Moskowitz, J. Chem. Phys. **43**, 60 (1965); **45**, 2338 (1966) used extended basis set of Gaussian functions.

[d] W. E. Palke and W. N. Lipscomb, J. Am. Chem. Soc. **88**, 2384 (1966). H$1s$ ($\xi=1.20$), C$1s$ ($\xi=5.7$), C$2s$, C$2p\sigma$, C$2p\pi$ ($\xi=1.625$).

[e] A. D. McLean (private communication) used 46 optimized Slater orbitals to attain the Hartree–Fock function.

[f] Optimized exponents given in Table 1. A slightly improved energy (−76.682) is obtained with the exponents given in Footnote d of Table I.

[g] $R_{CH}=2.001$ a.u., $R_{CC}=2.273$ a.u. Nuclear repulsion energy is 24.798 a.u.

[h] H$1s$ gross atomic population.

[i] Total kinetic energy.

[16] W. E Palke and W. N. Lipscomb, J. Am. Chem. Soc. **88**, **2384** (1966).
[17] R. S. Mulliken, J. Chem. Phys. **36**, 3428 (1962).
[18] M. Roux, M Cornille, and L. Burnelle, J. Chem. Phys. **37**, 933 (1962).

TABLE III. Acetylene optimized basis set SCF molecular orbitals.

Molecular orbitals	State	Coefficients of symmetry orbitals				Orbital energies[a] (a.u.)
		H1s+H′1s	C1s+C′1s	C2s+C′2s	C2pσ+C′2pσ	
$1\sigma_g$	Ground	0.0040	−0.7031	−0.0135	0.0012	−11.2282
	Cation	0.0038	−0.7029	−0.0135	0.0007	−11.6241
	Anion	0.0040	−0.7033	−0.0131	0.0012	−10.8762
	Excited[b]	0.0040	−0.7031	−0.0135	0.0009	−11.1980
$2\sigma_g$	Ground	−0.1058	0.1927	−0.4973	−0.1737	−1.0091
	Cation	−0.0692	0.1954	−0.4926	−0.2205	−1.3956
	Anion	−0.1338	0.1870	−0.4886	−0.1535	−0.6610
	Excited[b]	−0.0986	0.1921	−0.4928	−0.1898	−0.9997
$3\sigma_g$	Ground	0.4228	−0.0025	−0.0084	−0.4047	−0.6692
	Cation	0.3738	−0.0307	0.0924	−0.4223	−1.0170
	Anion	0.4593	0.0137	−0.0738	−0.3841	−0.3430
	Excited[b]	0.4268	−0.0081	0.0059	−0.3986	−0.6567
$4\sigma_g$	Ground	0.8408	0.0967	−0.6684	0.5805	0.5432
	Cation	0.8593	0.1011	−0.6657	0.5500	0.2249
	Anion	0.8263	0.0923	−0.6707	0.6011	0.8534
	Excited[b]	0.8397	0.0979	−0.6717	0.5797	0.5603
		H1s−H′1s	C1s−C′1s	C2s−C′2s	C2pσ−C′2pσ	
$1\sigma_u$	Ground	0.0045	−0.7033	−0.0210	−0.0019	−11.2252
	Cation	0.0041	−0.7031	−0.0212	−0.0029	−11.6211
	Anion	0.0048	−0.7035	−0.0212	−0.0018	−10.8737
	Excited[b]	0.0045	−0.7032	−0.0219	−0.0026	−11.1955
$2\sigma_u$	Ground	0.4041	−0.1217	0.3371	−0.2232	−0.7489
	Cation	0.3574	−0.1332	0.3797	−0.2463	−1.0787
	Anion	0.4369	−0.1137	0.3135	−0.1989	−0.4356
	Excited[b]	0.4036	−0.1230	0.3461	−0.2153	−0.7349
$3\sigma_u$	Ground	0.8181	0.1735	−1.0386	0.1621	0.4344
	Cation	0.8196	0.1753	−1.0094	0.1274	0.1217
	Anion	0.8154	0.1713	−1.0821	0.1633	0.7310
	Excited[b]	0.8145	0.1749	−1.0507	0.1423	0.4418
$4\sigma_u$	Ground	−0.2824	0.0936	−0.7898	−1.1331	1.3449
	Cation	−0.3150	0.0977	−0.7551	−1.1314	0.9464
	Anion	−0.2681	0.0853	−0.7994	−1.1385	1.6873
	Excited[b]	−0.2932	0.0903	−0.7697	−1.1373	1.3453
		C2pπ+C′2pπ				
$1\pi_u$	Ground	0.6086				−0.3941
	Cation	0.6165				−0.7879
	Anion	0.5968				−0.0607
	Excited[b]	0.6086				−0.4633
		C2pπ−C′2pπ				
$1\pi_g$	Ground	0.8769				0.3080
	Cation	0.8545				−0.0311
	Anion	0.9159				0.1953
	Excited[b]	0.8769				−0.1655

[a] 1 a.u. = 27.2098 eV. [b] $^3E_{2u}$ $(\pi_u^3\pi_g^1)$ state.

where Σ_B is the neutral-ground-state sigma wavefunction and Σ_A is the total density due to the free atoms. In computing Σ_A each carbon is in the state $Ks^2p_\sigma p_\pi$; with Roothaan's best-atom exponents for a neutral carbon atom.[19] The results of Σ-bond formation are apparent in the depletion of density in the atomic regions and the buildup in the bond regions.

[19] C. C. J. Roothaan, Lab. Mol. Struct. Spectra, University of Chicago TR, 1955.

IV. Σ REORGANIZATION

Sigma electronic shell reorganization occurs in both the AO exponents and the MO coefficients. As seen in Table I, the atomic reorganization as measured by the exponents is small. The $2p\sigma$ exponents are identical for all species and the $2s$ exponents for the ionic species differ by only 0.04 from the neutral ground-state exponent. The $2s$ atomic orbitals are contracted in the cation and expaned in the anion. However, the small

TABLE IV. Gross atomic populations.

Orbital		Ground state	Cation	Anion	Excited[a]
C1*s*	b	1.995	1.995	1.995	1.995
	c	1.996	1.995	1.996	
C2*s*	b	1.084	1.139	1.047	1.085
	c	1.124	1.160	1.101	
C2*p*σ	b	0.869	0.962	0.794	0.863
	c	0.883	0.964	0.819	
H1*s*	b	1.052	0.904	1.164	1.057
	c	0.998	0.875	1.085	
$C2p_{\pm 1}$		1.000	0.750	1.250	1.000

[a] For the configuration $\sigma_{1g}^2\sigma_{1u}^2\sigma_{2g}^2\sigma_{2u}^2\sigma_{3g}^2\pi_u^3\pi_g^1$ giving rise to $^3E_{2u}$, $^3A_{2u}$, $^3A_{1u}$ states.
[b] $\xi(\mathrm{H}1s)=1.00$.
[c] $\xi(\mathrm{H}1s)=1.20$.

2*s* contraction sizeably increases the $2s$–$2p_\pi$ repulsion and even more sizeably increases the nuclear attraction for a 2*s* electron. For example, the one-center Coulomb integral $(2s, 2s \mid 2p_{+1}, 2p_{+1})$ increases from 0.612 to 0.637 a.u. when cation exponents are substituted for ground-state exponents. The accompanying $2p_{+1}$ contraction accounts for about half of this increase. At the same time, the 2*s* atomic kinetic energy increases from 0.540 to 0.564 a.u., but the over-all 2*s* one-electron energy (KE+NA) decreases from −8.185 to −8.287 a.u. Hence the largest effect of the contraction of the 2*s* orbital is to increase the nuclear attraction integral. The net effect is that decreased pi density in the cation allows the 2*s* functions to contract and pick up nuclear attraction energy. Similarly, increased pi density in the anion causes 2*s* expansion by diminishing repulsive interactions at the expense of nuclear attraction.

The 2*p*σ atomic orbital is expected to be less affected than the 2*s* orbital by changes in pi density, since the 2*p*σ distribution is concentrated more along the bond axis. The exponent value of 2.03 is not far from 1.97 where the two-center overlap integral [C(1), 2*p*σ | C(2), 2*p*σ] is maximum.

Reorganization at the molecular level is displayed in the Mulliken population analysis of Tables IV and V.

A decrease of 1*e* in total pi-electron density produces changes in gross atomic populations of approximately −0.05 in carbon 2*s*, −0.09 in carbon 2*p*σ, and +0.14 in hydrogen 1*s* as shown in the following diagram:

$$\underset{\substack{0.14\sigma e \\ \longrightarrow}}{[\mathrm{H}\text{—}\mathrm{C}}\text{—}\underset{\substack{0.14\sigma e \\ \longleftarrow}}{\mathrm{C}\text{—}\mathrm{H}]}^{+1\pi e}$$

σ-electron flow resulting from π-electron ionization.

Total carbon–carbon and carbon–hydrogen sigma overlap populations (Table V) change by about 0.03 and the change is in the direction opposite to flow of charge. The difference between an ionic acetylene sigma distribution and that of the neutral molecule is attributed to partial densities having net antibonding character. The net antibonding effect arises from the C(1) 2*s*–C(2) 2*s* overlap population which decreases with increasing sigma carbon density, while the C(1) 2*p*σ–C(2) 2*p*σ population behaves oppositely. In the acetylene cation the 2*p*σ orbitals act to increase sigma density in the C–C bond region, while any buildup of density in the 2*s* atomic region serves to place density in the vicinity of the diminished pi density. The over-all effect, on the Σ density, as shown in Fig. 1(b), is that sigma density shifts towards the center of the molecule, i.e., it increases in the C–C region at the expense of density near and at the hydrogen atoms.

The increased repulsion interactions in the anion cause an analogous sigma alteration but in the opposite direction, i.e., towards the hydrogens, Fig. 1(c).

The contour lines are somewhat denser for the cation, indicating a larger charge transfer. This is probably rationalized by there being $\frac{1}{3}$ more electrons in the neutral pi system than in the cation, but only $\frac{1}{4}$ more in the anion than in the neutral system.

Redistribution of sigma density in acetylene, and presumably in other small molecules, upon ionization or electron capture can be seen to be as important as the changes in electron distribution engendered by bond formation. [See the reorganization contours of Figs. 1(b) and 1(c) which are of the same magnitude as those of the differential density for formation of the neutral ground-state σ bonds given in Fig. 1(a).] The principal structural difference between the two types of differential densities is that the bond formation function has its density contours in the bond region while the reorganization functions are concentrated more in the region of the nuclei.

As can be seen from Table III, the occupied sigma eigenvalues are raised or lowered by 0.36±0.04 a.u. upon addition or deletion of a pi electron. This effect is mainly due to increased or diminished σΠ repulsion interactions corresponding to the number of pi electrons. The value for $(\pi_u\pi_u \mid \sigma_g\sigma_g)$ is 0.54±0.06 a.u. for all species where σ_g is a C1*s*, C2*s*, or C2*p*σ normalized symmetry orbital. The $(\pi_u, \pi_u \mid \mathrm{H}1s, \mathrm{H}1s)$ integral is 0.31±0.01 a.u. The repulsion contribution of a π_u electron to the sum of occupied eigenvalues can be estimated as the Coulombic repulsion between π_u and the 4 σ_g symmetry orbitals: 3(0.54)+0.31=1.9 a.u. which can be compared with 5(0.36)=1.8 a.u., the average total eigenvalue change.

Of the three reasons for σ eigenvalue alteration, (1) change in the basis functions, (2) change in MO coefficients, (3) change in σΠ interaction, the third is judged most important. Some σ exponents do undergo significant changes but the nuclear attraction terms approximately cancel the repulsion and kinetic-energy changes. Computation shows that the MO coefficients cannot account for the energy differences.

Even though $\pi\to\pi^*$ promotion to $\pi_u^3\pi_g^1$ leaves the

TABLE V. Overlap populations.

Orbitals[a]		Ground state	Cation	Anion	Excited[d]
C(1)2s	C(1)1s	−0.117	−0.134	−0.104	−0.117
C(1)1s	C(2)1s	0.000	0.000	0.000	0.000
C(1)1s	C(2)2s	−0.014	−0.012	−0.015	−0.013
C(1)1s	C(2)2pσ	−0.020	−0.021	−0.019	−0.020
C(1)2s	C(2)2s	0.235	0.182	0.266	0.216
C(1)2s	C(2)2pσ	0.249	0.247	0.249	0.250
C(1)2p+1	C(2)2p+1	0.518	0.360	0.237	0.120
C(1)2pσ	C(2)2pσ	0.191	0.221	0.174	0.197
H(1)1s	C(1)1s	−0.024	−0.025	−0.024	−0.025
H(1)1s	C(1)2s	0.408	0.442	0.376	0.419
H(1)1s	C(1)2pσ	0.364	0.346	0.364	0.358
H(1)1s	C(2)1s	0.001	0.001	0.001	0.001
H(1)1s	C(2)2s	−0.040	−0.029	−0.050	−0.040
H(1)1s	C(2)2pσ	−0.022	−0.019	−0.024	−0.023
H(1)1s	H(2)1s	0.004	0.003	0.006	0.004
C(1)σ	C(2)σ[b]	0.857	0.830	0.871	0.847
H(1)σ	C(1)σ[b]	0.748	0.763	0.761	0.752
H(1)σ	C(2)σ[b]	−0.061	−0.048	−0.073	−0.062
H_σ	C_σ[c]	0.687	0.716	0.644	0.690

[a] X(i) mr means Slater orbital mr on Atom X, center i.
[b] Indicates total sigma density between the two atoms.
[c] Sum of H(1)σC(1)σ+H(1)σC(2)σ.
[d] See Footnote a, Table IV.

gross Π-atomic population unchanged, the Σ populations are affected slightly. In the population analysis of Tables IV and V the largest change is a shift of 0.01 from the C(1)2s–C(2)2s overlap population to the C(1)2s–H(1)1s population. The occupied sigma eigenvalues are raised by an average 0.015 a.u., and this is due to the change in the pi distribution, which because of occupation of an antibonding orbital, is slightly more concentrated in the region of the carbons. The increase in ΣΠ repulsion amounts to 0.2 a.u., and as discussed in Sec. VII, this interaction is well accounted for in pi-electron theory.

V. Π REORGANIZATION

Because of symmetry and the limited basis set, pi reorganization engendered by a change in the number of pi electrons is observed only in the 2p Slater exponents. However, as seen in Table I, these changes are substantial, amounting to 7% for the cation and (curiously) 13% for the anion. They represent large electron density redistributions. For each of the exponents 1.68, 1.59, and 1.46, the distances of maximum electron density in a $2p\pm1$ atomic orbital from the nucleus are 1.19, 1.26, and 1.37 a.u., respectively. The total one-electron energies (kinetic-energy nuclear attraction) are −6.67, −6.56, and −6.25 a.u. for each case, and the Coulomb repulsion integrals $(2p_{+1}, 2p_{+1} \mid 2p_{+1}, 2p_{+1})$ are 0.66, 0.62, and 0.57, respectively. As with the sigma electrons, removal of a pi electron allows the pi MO's to contract so as to increase nuclear attraction energy because of the reduced electron repulsion interactions. Similarly, addition of a pi electron causes AO expansion so as to diminish the repulsion interactions, at the expense of nuclear attraction. The integral $(\pi_m, \pi_m \mid \pi_n, \pi_n)$, where π_m or π_n is π_u or π_g, is 0.47 ± 0.03 a.u. for all species. The orbital energy changes of π_u for the neutral and charged species reflect essentially this value multiplied by the change in the number of pi electrons, and modified by changes in other integrals due to different exponents. The energy changes due to the change in exponent are discussed in Sec. VI.

As with the sigma eigenvalues, the changes in pi eigenvalues reflect principally the number of pi electrons. The exponent changes are larger for the pi orbitals since, with the limited basis set, this is the only means of MO adjustment for the change in balance of nuclear field versus electronic field.

In Fig. 2 the 0.06 electron density contours for the pi system and the sigma system are plotted for each of the species: neutral ground state, cation, and anion. An increase in density in the sigma system is indicated by an increase in distance of the contour line from the bond axis. Similarly, an increase in area encompassed by a pi contour indicates greater density. Changes in the pi contours are most noticeable since one full electron is being added to or subtracted from the pi system. The approximate centers of the contours, representing highest pi density, can also be seen to shift closer to the bond axis according to the exponent on the atomic orbitals. The shift of sigma electron density from the carbons to the hydrogens, in the anion, is seen by the expansion of the anion sigma contour near the hydrogen, contraction near the carbons, and subsequent crossing in an intermediate area. In a similar manner, the cation contour expands near the carbons, and contracts near the hydrogens as sigma density moves to the center of the molecule.

Figure 2 also reveals strong penetration of pi density into sigma density. For the most part, pi contour lines cross sigma contours of greater density. The reason for this interpenetration is partly due to the larger number of sigma electrons (hence greater total sigma density), but also to the $2s$ maximum atomic density being in the vicinity of the maximum pi atomic density. For the ground state the maximum density is 1.18 a.u. from the carbon nucleus for $2s$, vs 1.26 a.u. for $2p\pi$.

VI. ENERGIES OF EXCITED AND CHARGED STATES

For the following discussion we define the *total reorganization energy* for a particular molecular state as the difference between the total state energies computed from (1) the self-consistent-field eigenvectors found for the state, and (2) the ground-state SCF eigenvectors (populated to form the state). For example, this quantity would correspond to the discrepancy between the ionization potential found by (1) Koopman's theorem, and (2) the difference in SCF energies of the cation and ground states.

For acetylene cation, the total electronic energy from ground-state orbitals is −101.053 a.u., and from reminimized orbitals, −101.086 a.u. The corresponding ionization potentials are 0.394 and 0.361 a.u. and the total reorganization energy is −0.033 a.u. (0.9 eV). (The zero-point correction does not exceed 0.02 a.u.[20]) The experimental vertical ionization potential is 0.4193 a.u.,[21] 0.058 a.u. higher than the value predicted by minimizing both states and only 0.025 a.u. higher than the value predicted by Koopmans' theorem. The error should be due to higher correlation energy in the ground state relative to the cation, rationalizing the low predicted I.P. The better Koopmans' value would then be explained following Mulliken[22] by the error in the ionized state energy (as calculated by ground-state orbitals) offsetting the correlation energy in the ground state. The difference in correlation energy should be approximately that due to one electron pair. The error of 0.058 a.u. may be compared to the approximate value of 0.1 a.u. which Moskowitz deduced as the average correlation energy per electron pair in the acetylene triple bond.

For the acetylene anion, the ground state and reminimized orbitals give −101.139 and −101.208 a.u., respectively, for the total electronic energy. The corresponding electron affinities are −0.308 and −0.239 a.u. and the total reorganization energy is −0.069 (1.9 eV). The negative electron affinity is in accord with the fact that the acetylene anion is experimentally unknown.

[20] G. Glockler and C. E. Morrell, J. Chem. Phys. **4**, 15 (1936).
[21] V. H. Dibeler and R. M. Reese, J. Chem. Phys. **40**, 2034 (1964).
[22] R. S. Mulliken, J. Chim. Phys. **46**, 497 (1949).

The first observed excited state of acetylene has been assigned $^1A_{2u}(\pi_u{}^3\pi_g)$.[20] The triplet states as yet have not been assigned. We therefore apply SCF theory which is formally correct for the triplets to the lowest excited singlet state. Although the excited-state optimization was carried out with respect to $^3E_{2u}$, all excited states turn out to have identical sigma exponents, and the $2p\pi$ exponent is 1.59 ± 0.02 for all states to an energy accuracy of 0.002 a.u. Both ground-state and excited-state molecular orbitals give the same total electronic energy, −101.127, and therefore the total reorganization energy upon excitation is less than 0.001 a.u.

The computed and experimental $^1A_{2u}\leftarrow{}^1A_{1g}$ vertical transition energies are 0.32 a.u. (the zero-point correction is minor[20]), and 0.21 a.u.,[23] respectively, and hence an *ab initio* approach to the excitation energy with a minimal basis set is a poor one. Huzinaga[24] has shown for ethylene that if the exponents in the bonding and antibonding orbitals are optimized separately excitation energies are improved greatly. However, this is equivalent to going beyond the minimal basis set.

Table VI lists the contribution to the total reorganization energy, e.g., $\Delta(\Sigma_{1e}+\Sigma_{2e})$, for each species. Changes due to reorganization are greater for the terms involving pi orbitals. Intragroup reorganization energy changes (sigma orbitals only and pi orbitals only) are largely offset by intergroup reorganization energy change (sigma–pi repulsive interactions). The balance between nuclear attraction and electron repulsion is apparent from Table VI. Removal of a pi electron from the neutral molecule allows both the sigma and pi shells to contract and gain nuclear attraction energy. Intershell electron repulsion is *greater* by 0.586 a.u. using cation SCF orbitals rather than ground-state SCF orbitals, but the net result is an energy *lowering* of 0.033 a.u.! A similar analysis holds for the anion. In this case the additional electron causes an over-all expansion and loss of nuclear attraction energy which however is more than offset by the decrease in sigma–pi repulsions.

In the excited $^3E_{2u}$ state, one pi electron occupies an antibonding π_g orbital instead of a π_u orbital. The resultant pi distribution is slightly more concentrated near the nuclei which increases the sigma–pi interaction by ~0.2 a.u. The result is similar to a much reduced ground state→anion reorganization: sigma density expands slightly, reducing nuclear attraction energy, and decreasing sigma–pi repulsion. The changes in sigma shell energy and sigma–pi interaction energy actually cancel out to an accuracy of 0.001 a.u.!

VII. PI-ELECTRON THEORY

Comparison of the total reorganization energy in the last line of Table VI with the preceding four lines

[23] C. K. Ingold and G. W. King, J. Chem. Soc. **1953**, 2702.
[24] S. Huzinaga, J. Chem. Phys. **36**, 453 (1962).

TABLE VI. Energies.[a]

	Ground state	Cation	Anion	$^3E_{2u}$
Total energy	−101.447	−101.086	−101.208	−101.127
Total SCF energy	−76.649	−76.288	−76.410	−76.419
Total kinetic energy	77.011	77.000	77.317	77.616
$-\frac{1}{2}V/\mathrm{KE}$[b]	0.9976	0.9954	0.9941	0.9923
Σ_{1e}[c]	−125.3226	−126.1647	−124.6503	−125.3124
Σ_{2e}[d]	28.2239	28.8849	27.7207	28.2152
$\Sigma_{1e}+\Sigma_{2e}$	−97.0988	−97.2798	−96.9295	−97.0972
Π_{1e}	−26.2326	−20.1689	−31.3388	−26.0763
Π_{2e}	2.7714	1.4421	4.2309	2.6442
$(\Sigma\Pi)_{2e}$[e]	19.1132	14.9207	22.8297	19.3119
$\mathrm{KE}(\pi_u)$[f]	1.061	1.191	0.892	1.061
$\mathrm{NA}(\pi_u)$[g]	−7.619	−7.913	−7.188	−7.619
$\Sigma-\Pi(\pi_u)$[h]	4.778	4.974	4.527	4.778
$H^{\mathrm{core}}(\pi_u)$[i]	−1.780	−1.749	−1.769	−1.780
$\mathrm{KE}(\pi_g)$	1.685	1.835	1.476	1.685
$\mathrm{NA}(\pi_g)$	−8.087	−8.392	−7.631	−8.087
$\Sigma-\Pi(\pi_g)$	4.979	5.176	4.722	4.978
$H^{\mathrm{core}}(\pi_g)$	−1.423	−1.381	−1.433	−1.424
$\Delta(\Sigma_{1e}+\Sigma_{2e})$[j]	...	−0.181	0.169	0.002
$\Delta(\Pi_{1e}+\Pi_{2e})$	...	−0.438	1.024	0.000
$\Delta(\Pi_{2e})$	...	0.056	−0.272	0.000
$\Delta(\Sigma\Pi_{2e})$	...	0.586	−1.262	−0.002
$\Delta(E_{\mathrm{total}})$[k]	...	−0.033	−0.069	0.000

[a] Energies in atomic units, 1 a.u. = 27.2098 eV.
[b] Check of virial theorem: ratio of $-\frac{1}{2}$ total potential energy to total kinetic energy.
[c] Kinetic energy + nuclear attraction energy of sigma electrons.
[d] Total two-electron interactions of sigma electrons.
[e] Total two-electron interactions between sigma and pi electrons.
[f] Kinetic energy of an electron in π_u.
[g] Nuclear attraction energy of an electron in π_u.
[h] Total two-electron interactions of an electron in π_u with sigma electrons.
[i] $H^{\mathrm{core}}(\pi_u) = \mathrm{KE}(\pi_u)\mathrm{NA}(\pi_u)+\Sigma-\Pi(\pi_u)$.
[j] The subsequent energies are reorganization energies—the change in energy contributions upon substituting SCF state molecular orbitals for ground-state SCF molecular orbitals, see Sec. VI.
[k] The difference in total state energy computed from (1) SCF molecular orbitals for that state and (2) ground-state SCF molecular orbitals.

(representing its partition) clearly illustrate that the "success" of conventional pi-electron theory basically depends on the pi and sigma reorganization nearly balancing the change in sigma–pi repulsion. In acetylene the total reorganization energy is only 6% of the largest reorganization term, the change in sigma-pi repulsion.

The energy terms of Eq. (4) are also tabulated in Table VI along with the matrix elements of H^{core}, and the terms which enter into their evaluation (in terms of π_u and π_g symmetry orbitals, for convenience in discussion). As noted in the Introductión, these matrix elements represent the only interaction of pi electrons with sigma electrons, and it would be very convenient in the application of pi-electron theory if they remain relatively invariant to a change in state of the pi system. In fact, for acetylene, $H^{\mathrm{core}}(\pi_u)$ undergoes changes of less than 2%, and $H^{\mathrm{core}}(\pi_g)$ less than 3%, for 25% changes in total pi density. The range is small because of counterbalancing changes in the individual energy terms. In the cation, the flow of sigma density to the center of the molecule, and the contraction of the pi system, causes an increase in sigma–pi repulsions and pi kinetic energy, which is largely offset by an increase in nuclear attraction energy. The use of the ground-state $H^{\mathrm{core}}(\pi_u)$ value, −1.780 a.u., in the energy calculation for the cation would introduce an error of $3(-1.780+1.749)=-0.092$ a.u. in the cation energy. This error is about one-half the error of +0.181 which would be made by assuming that the sigma energy $(\Sigma_{1e}+\Sigma_{2e})$ is unchanged. The only other contribution to the total reorganization energy is that from $\Pi_{2e}=-0.056$ a.u. which arises from reoptimization of the pi molecular orbitals for the cation. $\Delta(\Pi_{2e})$ could be accounted for in "ordinary" pi-electron theory (with $\Sigma'=\Sigma_g$).

The total reorganization energy for the cation is

$$3\Delta H^{\mathrm{core}}(\pi_u)+\Delta(\Sigma_{1e}+\Sigma_{2e})+\Delta(\Pi_{2e})=0.092-0.181+0.056=-0.033 \text{ a.u.},$$

and the total energy error between the cation SCF calculation and a pi-electron calculation which takes into account $\Delta(\Pi_{2e})$ but does not take into account $\Delta(\Sigma_{1e}+\Sigma_{2e})$ and $\Delta H^{\mathrm{core}}(\pi_u)$ is +0.089 a.u. The analogous error for the anion calculation is

$$-4\Delta H^{\mathrm{core}}(\pi_u)-\Delta H^{\mathrm{core}}(\pi_g)-\Delta(\Sigma_{1e}+\Sigma_{2e})=4(0.011)+(0.010)-0.169=-0.203 \text{ a.u.}$$

In summary, a pi-electron calculation on the acetylene neutral ground state, anion and cation, which uses ground-state core integrals and neglects energy changes of the sigma electrons, but recomputes SCF pi molecular orbitals for each state, will predict from a reorganization view the cation to be approximately 0.1 a.u. too high in energy, and the anion to be 0.2 a.u. too low. The discrepancies are due to (1) using approximate core integrals and (2) neglecting sigma energy changes. These two errors tend to balance each other, with the larger error being that due to sigma reorganization.

The energy term $\Sigma_{1e}+\Sigma_{2e}$, arising from sigma electrons alone, increases by about 0.2 a.u. from cation to neutral and from neutral to anion. Clustering (polarization) of the sigma electrons near the center of the molecule, as in the cation, produces an increase in nuclear attraction energy which overrides increases in kinetic energy and electron repulsion. The large sigma energy change gives fundamental support to Hoyland and Goodman's[5] conclusion that sigma reorganization upon pi-electron ionization or capture is important. Our *ab initio* calculated value is actually a great deal larger than the value of $\sim$0.05 a.u. computed by them. Cancellation of the various reorganization terms empirically partitioned in their method causes the (first-order) sigma energy reorganization to be only partially accounted for, and it does not allow for polarization effects. But clearly, as Parks and Parr[4] suggested from their early study of formaldehyde, the polarization effect in sigma electronic shell reorganization is important. One would expect that in ionized benzene, for example, 0.05e in sigma electron density would migrate to the carbon nuclei, and a net decrease of sigma energy by as much as 0.2 a.u. would result!

The idea of adjusting the pi-orbital exponent according to pi density,[25] and thus Π_{2e} in a pi-electron calculation, is also verified. The exponent changes for larger molecules should be smaller since (1) the change in total pi density per center is less, and (2) pi-molecular-orbital reorganization through change in coefficients may be of greater importance.

In larger molecules, changes in H^{core} matrix elements might be expected to be less than 2%, because of the smaller fractional charge at each center. It would seem best to determine these elements by a customary empirical procedure, and apply them as constants for a specific application.

For ionization potentials or electron affinities, an approximate sigma energy correction could be adopted for a given class of compounds. A simple procedure for approximately correcting for Π_{2e} would be to adjust the one-center repulsion integral (pp/pp)[26,27] for ionic species.

Nonalternate hydrocarbons, having nonuniform pi density distributions in the ground state, offer an interesting application. For these compounds, $\Delta(\Sigma_{1e}+\Sigma_{2e})$ should be about zero since there is no over-all change in the number of pi electrons. Changes in H^{core} matrix elements are probably best neglected, but the pi MO's should be optimized (deformed and polarized) at both the atomic (exponent) and the molecular (coefficients) levels.

ACKNOWLEDGMENTS

We are very grateful to Dr. A. D. McLean for checking our calculations, and for critically reading and commenting on the manuscript. We also thank Dr. Franklin Prosser and Professor Martin Karplus for computing some of the integrals and Professor J. A. Dixon for his continued interest and support.

[25] R. D. Brown and M. L. Hefferman, Trans. Faraday Soc. **54**, 757 (1958).

[26] R. Pariser and R. G. Parr, J. Chem. Phys. **21**, 466, 767 (1953).

[27] J. A Pople, Trans. Faraday Soc. **49**, 1375 (1953).

Chapter VIII

Sigma Molecular Orbital Theory in Inorganic Chemistry

VIII-1a

Sigma Molecular Orbital Theory: An Inorganic Chemist's Perspective

J. W. FALLER

1. Introduction

Modern MO theories of inorganic compounds, for the most part, are extensions of the methods described in other chapters devoted to organic compounds and began to come into widespread acceptance and use around 1963. Hence only a brief survey of the methods has been included to provide a source of leading references to the inorganic literature. Principally, I have addressed myself to a discussion of the MO theory applied to larger molecules, particularly transition metal complexes, with the understanding that the methods are equally applicable to a much broader range of inorganic compounds. Specifically, recent developments in inorganic chemistry are included to suggest possible areas of application and possible limitations of MO methods.

2. Semiempirical Molecular Orbital Theories

The MO theory commonly used for transition metal complexes has evolved from LCAO-MO theory and incorporates Moffitt's atoms-in-molecules approach.[1] The complexity of the systems suggests making approximations from the outset, rather than attempting *ab initio* calculations, and distinctive styles of computation have developed among the artisans of approximations for semiempirical MO calculations of these metal compounds. Although the error introduced by the approximations used in each method (as well as their validity) may be argued as more or less serious by those espousing another technique, there are features, nevertheless, which are general to all and which in some respects are peculiar to metal complex calculations. These differences, which may be delineated by consideration of some representative papers,[2-9] will not be pursued but only the nature of the calculations outlined.

Frequently, the metal complexes are highly symmetric (D_{4h}, T_d, O_h), and the energy matrices, which follow the same general form as expected for Hückel or extended Hückel theory, may often be partially diagonalized to blocks by group theory. Diagonal elements are generally approximated via a self-consistent procedure from valence-state ionization potentials (VSIP)[10,11] or valence orbital ionization potentials (VOIP).[2] The VSIP's or VOIP's are obtained from atomic spectroscopic data and presumably are representative of the energy required to remove an electron from a given basis orbital on the atom or ion. Since unit charges may not be present on the atoms in the molecule under consideration, intermediate values are interpolated for fractional charges. After an initial determination of the form of the MO's a Mulliken population analysis[12] gives the charge on each atom and these fractional charges are used to calculate the appropriate VSIP's or VOIP's for that charge distribution. The computation is repeated to obtain a new charge distribution and new VSIP's, and so on to self-consistency (SCCC method, self-consistent in charge and configuration).

Off-diagonal elements are usually evaluated by a Wolfsberg-Helmholz approximation[13] or variation thereof. Often the computations are limited by the availability of sufficiently accurate, yet computationally simple set of basis functions. Double-ζ STO approximations[14-16] to the Hartree-Fock solutions[17-23] for the atoms and ions have been used frequently for bases in the computations, although some use has been made of many-term Hartree-Fock functions (e.g., by Fenske and Radtke[6]).

The method suggested by Basche et al.[2] has been criticized as yielding molecular orbitals with excessive covalent character[7]; nevertheless, it is representative of the methods used by most groups. Furthermore, it has the advantage of having been applied to a large number of systems; hence its value in practical applications may be assessed.

In the SCCC-MO method of Basche et al., inner shells are neglected (core approximation); diagonal elements, H_{ii}, approximated by VOIP's; and off-diagonal elements, H_{ij}, approximated by a modified Wolfsberg-Helmholz method. Whatever accuracy may be obtained in the results will follow from the adequacy of these approximations (or the propitious offsetting of errors). It has generally been found advisable to supply an adjustable parameter or two to get reasonable agreement with experiment. In this method the parameters are introduced in the form of a proportionality constant, F, in the Wolfsberg-Helmholz approximation. A different F is used for σ- and π-type orbitals,

$$H_{ij} = FG_{ij}(VOIP_i + VOIP_j)/2$$

(F_σ and F_π). The overlap integral, G_{ij}, is computed from the overlap of the functions ϕ_i and ϕ_j, which may be basis atomic orbital functions or linear combinations of them determined by symmetry. The computations generally allow the F's to vary and those F's giving the lowest energy are selected.

It is obvious that for this method and, for the most part, the other methods, a technique for "normalization" or "calibration" is necessary to have a practically useful system of calculations. This is usually done by selecting a value of F appropriate to reproduce an experimental observable in a model compound. This value of F is then retained throughout the computations on a series of compounds. If one accepts the limitations which are imposed by this "calibration" procedure, then the computations can be valuable in the interpretation of the spectra of the remainder of the series. In particular, these calculations have been able to reproduce trends in the electronic spectra and the nuclear quadrupole resonance spectra of octahedral and tetrahedral complexes of the first-row transition metals. (The reader is referred to Chapter VIII-1b to 1d for details.)

Although these methods show promise and improvements are constantly forthcoming, the ability to accurately predict physical properties, such as electronic spectra, nuclear quadrupole resonance frequencies, or Mössbauer splittings, still exists only as a hope for the future. I believe at this stage that it is fair to comment that many inorganic chemists view semiempirical MO theory with suspicion and generally as *ex post facto* agreement with and rationalization of experimental fact.

The present semiempirical molecular orbital theories at best can indicate rough trends in the physical properties of transition metal complexes. Nevertheless, it is pertinent to consider what an inorganic chemist might expect or want from a theory. A variety of answers come to mind, but probably leading the list would be the ability to predict molecular geometry, spectra, and/or magnetic properties. In this respect it is interesting to note that it is the simple arguments, some naive, some well-founded on symmetry considerations, all drastic oversimplifications, that are most successful and practically useful.

3. Valence-Bond Theories

Even if a single theory or method would do both, it does not follow that the most convenient and practical method of predicting structure would be the most useful for spectroscopic properties of vice versa. In fact, valence-bond approaches are still the quickest and most reliable methods of predicting structure of a molecule with a central atom (particularly nontransition elements) surrounded by ligands. Methods of predicting molecular shape and stereochemistry by consideration of lone pair-bonding pair interactions, such as the valence-shell-electron-pair-repulsion model (VSEPR)[24–26], are quite successful. New simple models frequently appear; for instance, a novel but not widely accepted (for good reason) method of discussing structure in terms of tangent spheres representing electron pairs has been proposed,[27] but it is not competitive with VSEPR methods. For coordination numbers 1 through 5, the VSEPR method will predict, practically without exception, the correct stable structure for compounds of nontransition elements. Only the solid-state structure of $Sb(C_6H_5)_5$ does not appear to obey the rules for coordination numbers 1 to 5, whereas several exceptions are known for numbers greater than 6 (e.g., $SeC\ell_6^{2-}$ is octahedral).

Furthermore, it is quite productive to use the simple effective atomic number rule (EAN) to predict the number of ligands which are likely to be attached to a transition metal in an organometallic compound or inorganic complex. The stereochemistry can then be conveniently predicted as the most stable geometric configuration of the ligands about the metal on the basis of steric considerations. The noble gas or EAN rule suggests that the ideal number of electrons donated by ligand atoms should be the quantity necessary to bring the total number of electrons around the metal to the atomic number of the noble gas at the end of that row of the periodic table. This is merely a way of keeping track of the number of orbitals of appropriate energy to interact strongly and form strong-bonding MO's with ligand orbitals. Or in valence-bond terms, there is a natural propensity toward utilization of all available energetically favorable orbitals in hybridization and bonding (full hybridization theory). Hence considering carbon monoxide as a two-electron donor, six CO ligands would be expected to coordinate to a chromium atom, so that it might reach the noble gas quota of 36 electrons ($24 + 2 \times 6 = 36$). Therefore, one would expect a stable octahedral carbonyl of chromium of the formula $Cr(CO)_6$. Likewise one would predict stability of tetrahedral $Ni(CO)_4$ and trigonal bipyramidal $Fe(CO)_5$. Metals with odd numbers of electrons would be expected to take up as many ligands as possible without exceeding the "magic number" of noble gas electrons and then dimerize, if possible, forming a metal-metal bond; i.e., $[Mn(CO)_5]_2$ and $[Co(CO_4]_2$ are diamagnetic dimers, whereas $V(CO)_6$ is a paramagnetic monomer. It is important to note that the structures of transition metal complexes do not require the same rules as the nontransition elements and, specifically, lone pairs do not seem to have the same influence upon stereochemistry. That is, in the parlance of VSEPR theory, the full hybridization potential is not used and the lone pairs occupy symmetrical s orbitals. Nevertheless, within the transition metals, only steric interactions of the ligands about the metal are considered; hence one expects the same structures for all isoelectronic species. For example, $Co(CO)_4^-$ is tetrahedral like $Ni(CO)_4$; $Mn(CO)_5^-$ has the same structure as $Fe(CO)_5$; $Mn(CO)_6^+$ the same structure as $Cr(CO)_6$; and $Cr_2(CO)_{10}^{2-}$ the same structure as $Mn_2(CO)_{10}$. This method of electron counting to predict structures is extremely efficient in analyzing possible structures and compositions arising in most organometallic compounds, such as derivatives containing nitrosyls, π-cyclopentadienyls, π-benzenes, π-cycloheptatrienyls, π-allyls, phosphines, halides, hydrides, etc. Some exceptions are noted with Pd, Pt, Ir, and Rh complexes, which often end up two electrons short of an inert gas configuration. Nevertheless, most transition metal complexes, even if stable without a complete noble gas complement of electrons, will either reduce readily [$V(CO)_6 \longrightarrow V(CO)_6^-$] or will add additional ligands to reach the magic number.

4. Ligand Field Theory

On the other hand, attempting to understand the electronic spectra or magnetic properties of transition metal complexes immediately brings one to the necessity of applying different theories. Crystal field theory, which attempts to describe the origin of the splittings of the d orbitals or the Russell-Saunders states of a metal ion and the resulting consequences of an environment of point charges, has met with limited success. The adjusted crystal field theory (ACFT) or ligand field theory in allowing for a limited amount of covalent bonding by modification of the free-ion term values of the metal ion has more widespread applications. If the amount of covalent bonding is not excessive (i.e., the amount of metal orbital–ligand orbital overlap is small), one may quite adequately predict the spectra, some thermodynamic properties, and magnetic properties of metal complexes, particularly with the aid of Tanabe-Sugano or Orgel diagrams. Discussions of these methods and abundant examples are available in standard texts.[28-33]

If the metal is in an unusually high oxidation state (e.g., MnO_4^-) or there is appreciable covalent character in the bonding, usually with extensive π bonding [e.g., $Mo(CO)_6$] one must rely on MO theory. The extensive calculations of a semiempirical nature are not generally practicable (vide supra). Often, however, by merely considering the symmetry properties, one can resolve seeming anomalies by observing qualitatively the effects of covalence on the ligand and metal orbitals involved. Hence, by using partial MO diagrams, approximate relative energies, and estimating extents of overlap, one can often satisfactorily explain or predict effects. Nevertheless when one comes to attempting the explanation of the details of the electronic spectra of these compounds, he is forced to resort to the semiempirical MO approach and accept the consequences of the inaccuracies.

5. Multicenter Two-Electron Bonding

As with Tanabe-Sugano diagrams in ligand field theory, the most servicable results of MO theory derive from correlation diagrams based on symmetry arguments. Walsh diagrams[34] or modifications thereof still provide one of the most useful ways of predicting structures of ground and excited states of small molecules. When dealing with molecules containing larger numbers of atoms, the complete MO methods often become too cumbersome and consideration of the orbital overlap in only a portion of the molecule can often provide useful information.

Valence-bond theory enjoyed such an overwhelming acceptance in predicting structure because of the widespread occurrence of compounds, the structures of which could be interpreted within a framework of two-center two-electron bonds. Perhaps, one of the most useful qualitative concepts arising from MO theory is that of the stability of multicenter two-electron bonds. Outside the realm of organic chemistry, structures that can be readily interpreted within a framework of three-center and four-center bonds are quite common. Certainly the use of localized multicentered two-electron bonds does not preempt the value of a complete MO picture of the entire molecule, but it does allow a convenient flexibility in rapid consideration of structures in much the same way as drawing lines between carbon atoms in organic compounds. The simple MO diagram for the localized three-center bond is probably the most useful. The overlap of suitable valence orbitals will produce a bonding orbital, and, depending on the symmetry, either two antibonding or nonbonding (slightly antibonding) and an antibonding orbital (Fig.1).

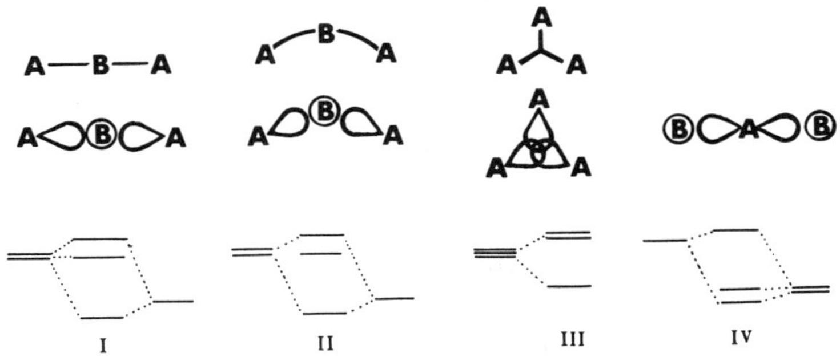

Fig. 1. Three-center bonds. These labels are not generally used, but are only for convenience in this discussion. Types I and II are "open" three-center bonds and type III is a "closed" three-center bond.

Diborane (B_2H_6) is the most obvious example of a compound in which there can be considered a bent three-center bond (type II) between two boron atoms and the hydrogen atom. This type of bonding is widespread in higher boron hydrides and has been discussed in detail by Lipscomb.[35] The closed three-center BBB bond is particularly common on boron cage compounds such as the icosahedral ion $B_{12}H_{12}^{2-}$ or the tetrahedral molecule $B_4C\ell_4$. These "electron-deficient" structures lend themselves well to interpretation via three-center bonds; however, cages do not always require three-center bonds. For instance, the tetrahedral P_4 molecule can be treated with "old-fashioned" two-center two-electron bonds along each edge of the tetrahedron. Multicenter bonding is not limited to boron and is common throughout the periodic table, in particular in transition metal complexes. For example, bridging metal hydrides of type I having the structure $(CO)_5M{-}H{-}M(CO)_5^-$ (where M = Cr, Mo, or W) have been prepared.[36] A bent Fe–C–Fe three-center bond (type II) is found[37] in $C_8H_8Fe_2(CO)_5$. Type IV bonding with either two or four electrons has been suggested as more appropriate for axial bonds in certain trigonal bipyramidal complexes.[33]

Some quite novel structures with metal atom clusters can be explained using three- and four-center two-electron bonds.[33, 38] The most common four-center two-electron bonds are derived from type III three-center bonds by considering another orbital perpendicular to the center of the triangle (Fig. 2). Methyllithium is a tetramer and consists of a tetrahedron of lithium

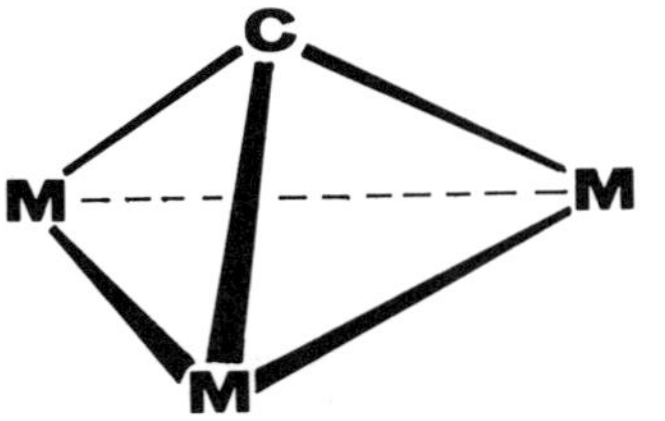

Fig. 2. Four-center bonds often are formed by a triangle of metal atoms and a triply bridging two-electron donor.

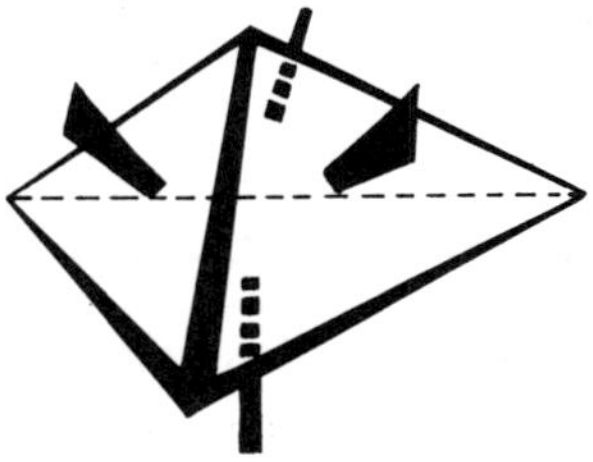

Fig. 3. Structure of the methyllithium tetramer.

atoms with a methyl group attached to the center of each face of the tetrahedron via a Li–Li–Li–C four-center two-electron bond (Fig. 3). A similar type of bonding may be invoked in $(C_4H_9)_3PCuI$ tetramer, which exists as a tetrahedron of copper atoms with iodine atoms at the center of each face and a phosphine ligand at each vertex. Transition metals generally cluster in groups of three and often contain multicenter bonds, rhenium halide derivatives being of particular note.[39] Transition metal organometallics can produce rather exotic combinations, such as $(C_5H_5)_3Ni_3(CO)_2$, which contains a triangular arrangement of nickel atoms with cyclopentadienyl groups at each corner and carbonyl groups attached to both faces of the triangle (Fig. 4). Hexarhodiumhexadecacarbonyl, $Rh_6(CO)_{16}$, contains a melange of multicenter

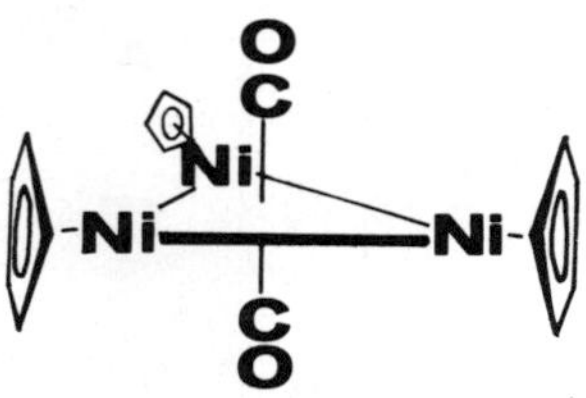

Fig. 4. Structure of π-$(C_5H_5)_3Ni_3(CO)_2$.

bonds, has T_d symmetry, and contains an octahedron of rhodium atoms with a triply bridging carbonyl on alternating faces and two carbonyls at each vertex[40] (Fig. 5). The dicarbollide of tetraphenylcyclobutadiene-palladium has a little bit of everything (Fig. 6).[41]

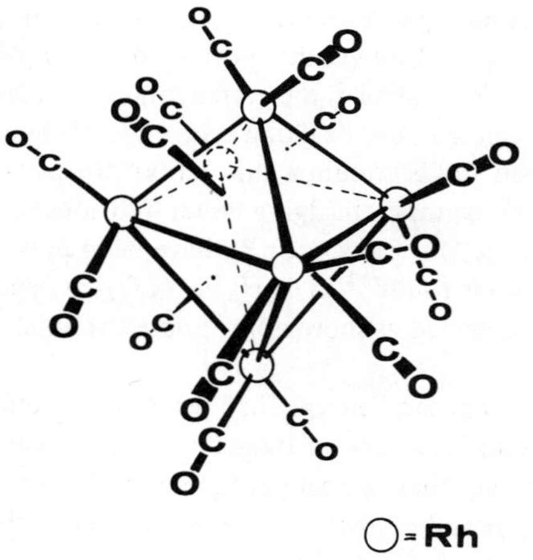

Fig. 5. Structure of hexarhodiumhexadecacarbonyl. (Modified from Ref. 40.)

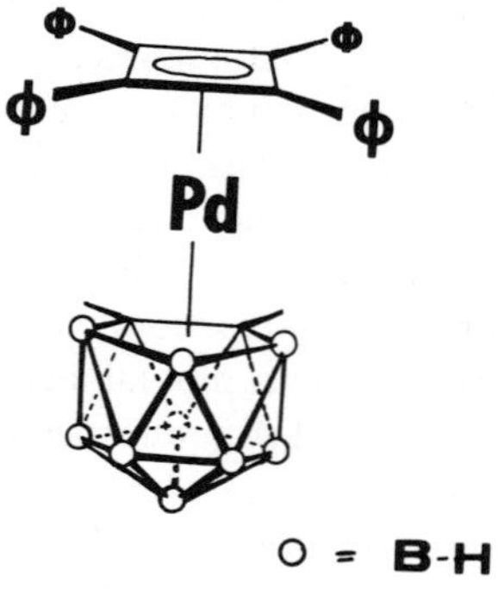

Fig. 6. Structure of $[\pi\text{-}C_4(C_6H_5)_4]Pd[\pi\text{-}(3)\text{-}1,2\text{-}B_9C_2H_9(CH_3)_2]$. (Modified from Ref. 41.)

6. Symmetry Aspects of Bonding

A further extremely useful result of symmetry aspects of MO pictures of bonding are the implications of the bonding of small molecules, such as N_2, O_2, CO, CS_2, and C_2H_4,[42-44] as ligands to transition metal atoms. The bonding is "synergic"; i.e., electrons are donated "forward" via a σ bond from the ligand to the metal and simultaneously electrons are donated "back" from the metal to the ligand via π bonding into antibonding orbitals (Figs. 7 and 8).

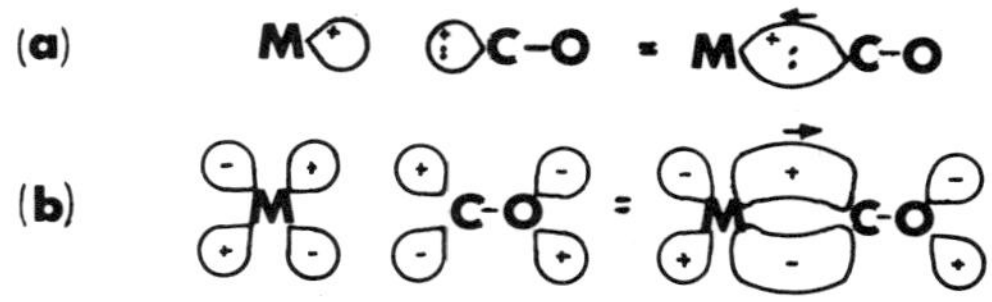

Fig. 7. (a) The formation of a carbon⟶metal σ bond by donation of an unshared pair on the C atom. (b) The formation of a metal⟶carbon π bond by back-donation from the metal. (Modified from Ref. 33, p. 728.)

Fig. 8. (a) Donation from filled π orbital to vacant metal orbital. (b) Back-bonding from filled metal orbital into vacant π^* orbital. (Modified from Ref. 33, p. 776.)

This conceptualization has allowed the correlation of a vast amount of data on metal carbonyl, nitrosyl, and olefin derivatives. The applications of this approach are treated in detail in Cotton and Wilkinson,[33] but of particular note is the correlation of infrared stretching frequencies of the carbonyl ligands. The amount of forward and back donation generally tends to be comparable, so there is no buildup of charge on the metal and it maintains a relatively small positive value. Hence, in considering the isoelectronic series $Mn(CO)_6^+$, $Cr(CO)_6$, and $V(CO)_6^-$, the greater negative charge on the metal spreads out into the antibonding orbitals of the carbonyl, lowering the C–O bond order, weakening the bond, and lowering the respective ν (T_{1u}) stretching carbonyl frequencies to 2090, 2000, and 1860 cm^{-1}.

These ideas have been used extensively and although x-ray crystal structures are now available for a large number of carbonyls, perhaps the most interesting observations have come from recent structures of other coordinated small molecules. The back-donation concept suggests that in an extreme viewpoint an electron is removed from a bonding orbital on the ligand and another placed into an antibonding orbital of the ligand. In other words, an excited electronic state of the ligand has been produced upon being attached to the metal. Although giving any credence to this view might appear to be completely absurd at first glance, the following observations may suggest that one had best not be too quick to judge the idea.

The mean C–S bond length is 1.63 Å and S-C-S angle is 136° in the coordinated carbon disulfide molecule in $[C_6H_5)_3P]_2Pt(CS_2)$. In the first excited ($^3A_2$) state of carbon disulfide the mean C–S bond length is 1.64 Å and the bond angle is 135°. In $[C_6H_5)_3P]_2Pt$ (diphenylacetylene) the ligand is cis-bent with angles of 140° and C–C bond length of 1.39 Å. Kroto and Santry suggest that a cis-bent excited state of acetylene should have a bond angle of 142° and a C–C bond length of 1.38 Å. These and other similar correlations with excited-state molecules, such as oxygen and butadiene, have been made by Mason.[44] It will be interesting to examine the success of this interpretation as more x-ray structures are completed.

Further applications of this interpretation would arise from the recent studies of molecular nitrogen [42,43] and oxygen[43,44] complexes. At this point it appears that most of the nitrogen complexes are bound in an end-on fashion as CO in metal carbonyls, whereas oxygen is bound in a sideways mode, as in ethylene–metal complexes. Furthermore, there is the interesting point that many organic moieties which are unstable as isolated species form quite stable metal adducts. For example, large numbers of cyclobutadiene–metal complexes are known, e.g., $C_4H_4Fe(CO)_3$, or further, the trimethylenemethane radical is relatively stable in the complex $C_4H_6Fe(CO)_3$.

Another application of symmetry properties which promises to be rewarding is the extension of the Woodward-Hoffmann rules to inorganic and organometallic systems. Bicyclo [2.2.1]-heptadiene undergoes dimerization to cyclobutane derivatives in the presence of zerovalent iron, nickel, and cobalt catalysts. This dimerization, as well as the facile valence tautomerism of quadricyclene to norbornadiene, has been rationalized by an extension of the Woodward-Hoffmann rules.[45] The explanation requires consideration of the correlation diagram for the metal and olefin orbitals within the point-group symmetry which is retained throughout a concerted reation. Hence the olefin dimerization, which is thermally forbidden by the Woodward-Hoffmann rules in the absence of a catalyst, becomes allowed in the presence of the metal. The correlations are illustrated in Fig. 9, in which two olefins attached to the same metal atom are considered to approach and form σ bonds while preserving C_{2v} symmetry throughout the concerted reaction.

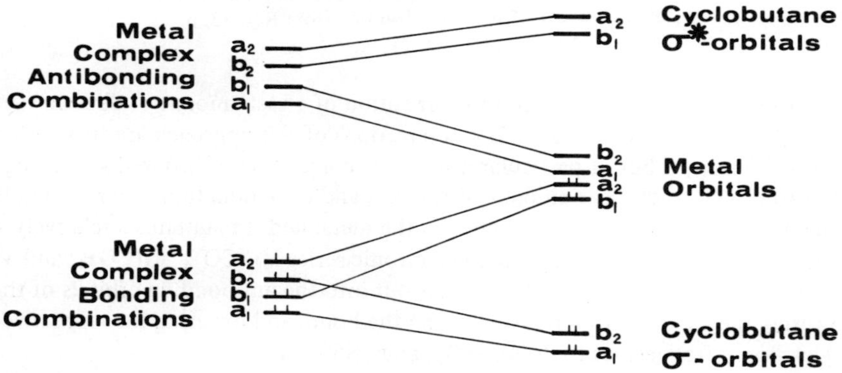

Fig. 9. Correlation diagram illustrating the orbital symmetry conservation in the thermally allowed process of dimerization of two olefins in the presence of a metal atom. (Modified from Ref. 45.)

In a modified approach, consideration of conversation of d orbital symmetry has allowed Eaton[46] to explain some of the great variations in rates of isomerization and bimolecular substitution reactions in four- and five-coordinate transition metal complexes. These arguments are considerably more involved, and a reprint of Eaton's paper included as Chapter VIII-1d.

7. Stereochemically Nonrigid Compounds

Finally I wish to point out the increasing awareness of time scales in inorganic research, particularly in view of their implications to the fundamental ideas of directed valence and static molecular configurations. Most of the ideas of directed valence derive from the empirical fact that most compounds with coordination numbers from 2 to 6 occur in a relatively small number of geometric configurations. Hence a six-coordinate compound should be octahedral, a four-coordinate compound tetrahedral or square planar, and a five-coordinate compound probably trigonal bipyramidal. Those that do not fall into these categories usually can be explained readily by VSEPR theory in considering the placement of lone pairs of electrons. Even the discovery of noble gas compounds did not phase the theories of bonding.[47,48] The following comments are noted to suggest that although the current bonding theories are fundamentally adequate to handle these situations, modifications may be appropriate to provide a viewpoint which is not so entangled with that of directed valence.

Werner's proposal that six-coordinate transition metal complexes were octahedral based on the number of isomers and optical activity studies holds about the same stature in coordination chemistry that the notions of planar hexagonal benzene rings do in organic chemistry. Within the past several years, however, some complexes of ethylenedithiolate ligands [S–C(R)=C(R)–S] have been shown to be trigonal prismatic and others intermediate between octahedral and trigonal prismatic.[49]

Within seven-coordinate compounds, several structures are found and differ little in energy. The solid-state molecular configurations which are adopted by the eight ligands about certain dodecahedral complexes appear to be determined by the size of the counterions.[50] The point to be emphasized is that the different configurations only differ by movements of the ligands by several tenths of an angstrom and the energy differences between the various conformations are quite low (0.05 to 1 eV). Furthermore, there are not high barriers (often less than 0.5 eV) to interconversion of these isomers. Consequently, even Hartree-Fock calculations are not likely to lead to adequate predictions of structure, because the errors due to neglect of correlation may be of the same order of magnitude. (See, however, Allen's conformational computations.[51]) More important, however, is that the low barriers to interconversion of conformations often allow very rapid interconversions at room temperature (first-order rate constants from about 1 to 10^6 sec^{-1} or faster. That is, at any instant one may expect a large percentage of the molecules to be distorted from the idealized configurations, and significantly more than would normally be anticipated may be in a configuration which might be designated as a transition state. Hence in some applications it may be appropriate to consider a potential surface surrounding the central atom on which the ligands move, rather than considering idealized bonds following the ligands around. The latter approach forces one to consider certain changes of ligand positions being correlated with changes in other ligand positions in a way which may not be the most general, realistic, or, for that matter, correct.

Although there are many different types of these *stereochemically nonrigid* or *fluxional* systems,[52,53] several representative examples will be discussed here. These intramolecular processes which interchange ligands can be conveniently considered under two divisions: those which apparently involve only changes in bond angles and lengths, and those which require bond breaking. A further subdivision might also be made if the starting and final configurations are the same (degenerate), or if the interchange involves interconversion of isomers.

The rates of these intermolecular processes are such that on the NMR time scale ($k \simeq 10^{-1}$ to 10^5 sec^{-1}) one will observe an averaged spectrum, whereas on the time scale of infrared or electronic spectra (10^{13} to 10^{15} sec^{-1}) one observes the spectrum of the individual components. Frequently, by variation of the temperature, the interchange rate may be varied over a range which will allow both an averaged spectrum and a "static" spectrum to be observed in the NMR. Subsequently NMR line-shape analysis can be used to obtain rates of the interchange as a function of temperature–hence allowing computation of barriers to interconversion.

The "classic" example of ligand interchange is that of the PF_5 molecule, which has trigonal bipyramidal geometry in the lowest energy conformation. The axial fluorine and equatorial fluorine atoms should show different chemical shifts in the NMR; yet, owing to rapid interchange of axial and equatorial ligands, only a single averaged chemical shift is observed. Consequently, a pseudorotation mechanism has been proposed for the interchange of ligands.[54]

$PF_3C\ell_2$, SF_4, and ReH_9^{2-} are but a few of the large numbers of compounds which have been found to undergo this type of fluxional rearrangement. Pseudorotation in the hydrolysis of phosphate esters has been suggested in organic and biochemical reactions.[55] With coordination numbers less than 6, one does not find that the different isomers are of comparable stability, but with higher coordination numbers isomer equilibria may be fairly common. The isomers of bis-substituted molybdenum complexes have been shown to be stereochemically nonrigid and presumably cis and trans isomers interconvert via an intramolecular pathway similar to that shown in Fig. 10.[56]

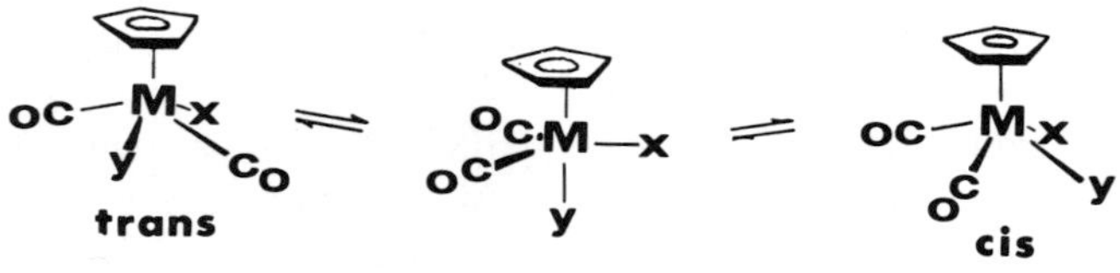

Fig. 10. Stereochemical nonrigidity in bis-substituted π-cyclopentadienylcarbonylmolybdenum alkyls.

Intramolecular interconversion via a bond-breaking mechanism is typified by the most well-known example—bullvalene ($C_{10}H_{10}$), which undergoes rapid degenerate Cope rearrangements internally at room temperature.[57] Exotic examples exist in the organometallic literature where the $M(CO)_3$ (M = Fe or Ru) group has been shown to "run around the edge" of a cyclooctatetraene ring in a series of 1,2 shifts (Fig. 11).[58] Several other examples of "ring whizzing" about

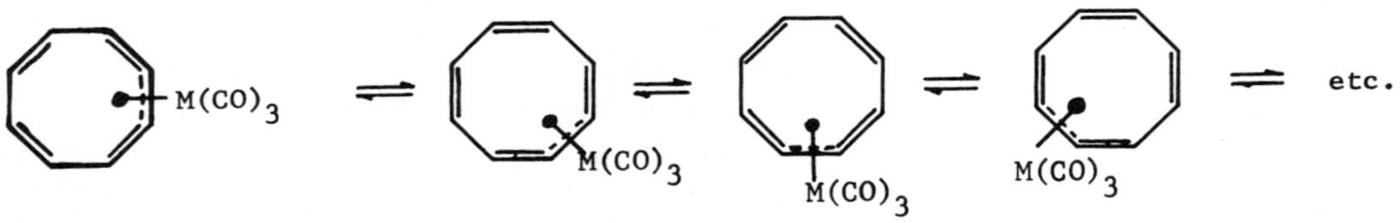

Fig. 11. Nature of the intramolecular rearrangement of the fluxional cyclooctatetrenemetal tricarbonyls.

five-, seven-, and eight-membered rings have been reviewed by Cotton.[52] Substituents on the rings can provide more than one pathway of rearrangement and, consequently, there may be more than one barrier to interchange[59] and/or several different isomers in rapid equilibrium.[60]

As I have attempted to indicate in the above examples, throughout the development of inorganic bonding theory, it has been the qualitative and semiquantitative aspects—the Walsh, Orgel, and Tanabe-Sugano diagrams, which developed from application of symmetry principles—which have had the greatest impact and have proved to have the most general utility. In this view, along with improvements expected in quantitative calculations, it appears that some of the most important developments which may be anticipated are extensions of the Woodward-Hoffman rules to inorganic systems, such as those suggested by Eaton[46] and Mango and Schachtschneider.[45]

References

1. W. Moffitt, *Proc. Roy. Soc.* (*London*) **A210,** 245 (1951).
2. H. Basch, A. Viste, and H. B. Gray, *J. Chem. Phys.* **44,** 10 (1966).
3. C. K. Jorgensen, S. M. Horner, W. E. Hatfield, and S. Y. Tyree, Jr., *Intern. J. Quant. Chem.* **1,** 191 (1967).
4. L. Oleari, G. DeMichelis, and L. DiSipio, *Mol. Phys.* **10,** 97, 111 (1966).
5. P. Ros and G. C. A. Schuit, *Theoret. Chim. Acta* **4,** 1 (1966).

6. R. F. Fenske and D. D. Radtke, *Inorg. Chem.* **7**, 479 (1968).
7. R. F. Fenske, K. G. Caulton, D. D. Radtke, and C. C. Sweeney, *Inorg. Chem.* **5**, 951 (1966).
8. F. A. Cotton and C. B. Harris, *Inorg. Chem.* **6**, 369, 376, 924 (1967).
9. W. C. Nieuwpoort, *Philips Res. Rept.* (Suppl. No. 6), 1 (1965).
10. G. Pilcher and H. A. Skinner, *J. Inorg. Nucl. Chem.* **24**, 937 (1962).
11. J. Hinze and H. H. Jaffe, *J. Phys. Chem.* **67**, 1501 (1963).
12. R. S. Mulliken, *J. Chem. Phys.* **23**, 1833 (1955).
13. M. Wolfsberg and L. Helmholz, *J. Chem. Phys.* **20**, 837 (1952).
14. E. Clementi, *J. Chem. Phys.* **40**, 1944 (1964).
15. J. W. Richardson, W. C. Nieuwpoort, R. R. Powell, and W. F. Edgell, *J. Chem. Phys.* **36**, 1057 (1962).
16. J. W. Richardson, R. R. Powell, and W. F. Edgell, *J. Chem. Phys.* **36**, 796 (1963).
17. F. Herman and S. Skillman, *Atomic Structure Calculations*, Prentice-Hall, Englewood Cliffs, N.J., 1963.
18. E. Clementi, C. C. J. Roothaan, and M. Yoshimine, *Phys. Rev.* **127**, 1618 (1962).
19. E. Clementi, A. D. McLean, D. L. Raimondi, and M. Yoshimine, *Phys. Rev.* **133**, A1274 (1964).
20. E. Clementi, *J. Chem. Phys.* **41**, 295, 303 (1964).
21. E. Clementi, Tables of Atomic Functions, *IBM J. Res. Develop.* (Suppl.) 9, 2 (1965).
22. R. E. Watson and A. J. Freeman, *Phys. Rev.* **124**, 1117 (1961).
23. R. E. Watson, *Phys. Rev.* **118**, 1036; **119**, 1934 (1960).
24. R. J. Gillespie and R. S. Nyholm, *Quart. Rev.* (*London*), **11**, 339 (1957).
25. L. S. Bartell, *J. Chem. Ed.* **45**, 754 (1968).
26. R. J. Gillespie, *J. Chem. Ed.* **40**, 295 (1963).
27. H. A. Bent, *J. Chem. Ed.* **45**, 768 (1968).
28. J. S. Griffith, *The Theory of Transition Metal Ions*, Cambridge University Press, New York, 1964.
29. C. J. Ballhausen, *Introduction to Ligand Field Theory*, McGraw-Hill, New York, 1962.
30. L. E. Orgel, *An Introduction to Transition Metal Chemistry*, *Ligand-Field Theory*, Wiley, New York, 1960.
31. C. S. G. Phillips and R. J. P. Williams, *Inorganic Chemistry*, Oxford University Press, New York, 1966.
32. M. C. Day, Jr., and J. Selbin, *Theoretical Inorganic Chemistry*, 2nd ed., Reinhold, New York, 1969.
33. F. A. Cotton and G. Wilkinson, *Advanced Inorganic Chemistry*, Wiley-Interscience, New York, 1966.
34. A. Walsh, *J. Chem. Soc.*, **1953**, 2260.
35. W. N. Lipscomb, *Boron Hydrides*, Benjamin, New York, 1963.
36. R. G. Hayter, *J. Am. Chem. Soc.* **88**, 4376 (1966).
37. E. B. Fleischer et al., *J. Am. Chem. Soc.* **88**, 3158 (1966).
38. F. A. Cotton, *Rev. Pure Appl. Chem.* **17**, 25 (1967).
39. E. Weiss and E. A. C. Lucken, *J. Organometal. Chem.* **2**, 197 (1964).
40. E. R. Corey, L. F. Dahl, and W. Beck, *J. Am. Chem. Soc.* **85**, 1202 (1963).
41. M. F. Hawthorne, *Accounts Chem. Res.* **1**, 281 (1968).
42. L. Vaska, *Accounts Chem. Res.* **1**, 335 (1968).
43. A. D. Allen and F. Bottomley, *Accounts Chem. Res.* **1**, 360 (1968).
44. R. Mason, *Nature* **543 (1968).**
45. F. D. Mango and J. H. Schachtschneider, *J. Am. Chem. Soc.* **89**, 2484 (1967).
46. D. R. Eaton, *J. Am. Chem. Soc.* **90**, 4272 (1968).
47. C. A. Coulson, *J. Chem. Soc.* **1964**, 1442.
48. G. J. Malm, H. Selig, J. Jortner, and S. A. Rice, *Chem. Rev.* **65**, 199 (1965).
49. S. I. Shupack, E. Billig, R. J. H. Clark, R. Williams, and H. B. Gray, *J. Am. Chem. Soc.* **86**, 4594 (1964).
50. E. L. Muetterties and C. M. Wright, *Quart. Rev.* (*London*) **21**, 109 (1967).
51. L. C. Allen, Chap. VI-2, this volume.
52. F. A. Cotton, *Accounts Chem. Res.* **1**, 257 (1968).
53. E. L. Muetterties, *Inorg. Chem.* **4**, 769 (1965).
54. R. S. Berry, *J. Chem. Phys.* **32**, 933 (1960).
55. F. H. Westheimer, *Accounts Chem. Res.* **1**, 70 (1968).
56. J. W. Faller and A. S. Anderson, *J. Am. Chem. Soc.* **91**, 000 (1969).
57. W. von E. Doering and W. R. Roth, *Angew. Chem., Intern. Ed. Engl.* **2**, 115 (1963).
58. W. K. Bratton, F. A. Cotton, A. Davison, A. Musco, and J. W. Faller, *Proc. Natl. Acad. Sci. U.S.* **58**, 1324 (1967).
59. F. A. Cotton, J. W. Faller, and A. Musco, *J. Am. Chem. Soc.* **90**, 4272 (1968).
60. F. A. L. Anet, H. D. Kaesz, A Maasbol, and S. Winstein, *J. Am. Chem. Soc.* **89**, 2489 (1967).

VIII-1b Reprinted from *The Journal of Chemical Physics* **44**, 10 (1966)

Molecular Orbital Theory for Octahedral and Tetrahedral Metal Complexes

HAROLD BASCH, ARLEN VISTE,* AND HARRY B. GRAY
The Department of Chemistry, Columbia University, New York, New York
(Received 4 May 1965)

Self-consistent charge and configuration (SCCC) molecular orbital calculations are reported for 32 selected octahedral and tetrahedral first-row transition-metal complexes containing halide and chalcogenide ligands. It is found that for the range of metal oxidation states II through IV, F_σ, chosen to fit the experimental Δ, is a function of only the metal atomic number for constant F_π. In the range of formal metal oxidation numbers V through VII, F_σ is also a function of oxidation number.

Calculated and observed trends in covalency, Δ values, and first L→M charge-transfer energies are compared. The conclusion is drawn that the molecular orbital method, in its present formulation, gives a reasonable account of the ground states and low excited states in simple metal complexes.

INTRODUCTION

THE semiempirical molecular orbital theory in the Wolfsberg–Helmholz framework[1] has been revived in recent calculations on a wide variety of molecules.[2–18] Such diverse molecular properties as geometric configurations,[4,5,9] electric-dipole moments,[10] transition moments,[1,3,11] electronic spectra,[1–8,10–13,15–18] charge distributions, and ionization potentials have been calculated with varying degrees of success. In particular, this reasonably simple MO approach has been of great value in interpreting various electronic structural properties of transition-metal complexes. Although the initial results are encouraging, there have been too few attempts at a critical evaluation of the method. A good start in evaluation of the generality of the simple MO method has been made by Cotton and Haas,[13] who studied the variation of the so-called F factor in exactly fitting the Δ values in a series of octahedral ammine complexes. Also, Fenske and Sweeney[11,12] have studied MnO_4^- and TiF_6^{3-} in detail. Recently, Boer, Newton, and Lipscomb[14] have calculated theoretical F factors for the B_2 and BH molecules.

It is clear that the steady development of a satisfactory MO theory for transition-metal complexes depends upon making detailed studies and improvements in the existing approximate schemes. In particular, there is the obvious need for a standard and consistent method of choosing the parameters of the calculation that can be transferred from one molecule to another. For this purpose we have chosen to calculate 32 octahedral and tetrahedral complexes containing first-row transition metals and one-atom ligands. The complexes were selected to explore the possibility that, using essentially the same set of initial assumptions, the variation of Δ and ligand-to-metal (L→M) charge transfer with (1) geometry, (2) ligand, (3) metal oxidation number, and (4) metal atomic number could be made the function of a single parameter, the F factor.

METHOD

The method as used in this investigation is similar to that applied in the calculation of MnO_4^-.[7] As usual, the secular equation to be dealt with is of the form,

$$|H_{ij}-G_{ij}E|=0. \quad (1)$$

The matrix elements are between properly normalized (with ligand–ligand overlap) symmetry basis functions, $\chi_i^{\mu\alpha}$. The $\chi_i^{\mu\alpha}$'s include atomic orbitals of the metal and linear combinations of ligand atomic (or molecular) orbitals; G_{ij} is the group overlap matrix,

$$G_{ij}=(\chi_i^{\mu\alpha}\mid\chi_j^{\nu\beta})=G_{ij}^{\mu}\delta_{\mu,\nu}\delta_{\alpha,\beta}. \quad (2)$$

* Present address: Augustana College, Sioux Falls, South Dakota.

1 M. Wolfsberg and L. Helmholz, J. Chem. Phys. **20**, 837 (1952).
2 (a) R. Hoffman and W. N. Lipscomb, J. Chem. Phys. **36**, 2179, 3489; **37**, 520 (1962); (b) R. Hoffman, *ibid.* **39**, 1397 (1963); **41**, 2474, 2480, 2745 (1964).
3 C. J. Ballhausen and H. B. Gray, Inorg. Chem. **1**, 111 (1962).
4 L. L. Lohr, Jr., and W. N. Lipscomb, J. Chem. Phys. **38**, 1607 (1963); J. Am. Chem. Soc. **85**, 240 (1963); Inorg. Chem. **2**, 911 (1963); *ibid.* **3**, 22 (1964).
5 T. Jordan, H. W. Smith, L. L. Lohr, Jr., and W. N. Lipscomb, J. Am. Chem. Soc. **85**, 846 (1963).
6 H. Bedon, S. M. Horner, and S. Y. Tyree, Jr., Inorg. Chem. **3**, 647 (1964).
7 A. Viste and H. B. Gray, Inorg. Chem. **3**, 1113 (1964).
8 S. I. Shupack, E. Billig, R. J. H. Clark, R. Williams, and H. B. Gray, J. Am. Chem. Soc. **86**, 4594 (1964).
9 E. A. Boudreaux, Inorg. Chem. **3**, 506 (1964); J. Chem. Phys. **40**, 246 (1964).
10 H. A. Pohl, R. Rein, and K. Appel, J. Chem. Phys. **41**, 3385 (1964).
11 R. F. Fenske and C. C. Sweeney, Inorg. Chem. **3**, 1105 (1964).
12 R. F. Fenske, Inorg. Chem. **4**, 33 (1965).
13 F. A. Cotton and T. E. Haas, Inorg. Chem. **3**, 1004 (1964).
14 F. P. Boer, M. D. Newton, and W. N. Lipscomb, Proc. Nat. Acad. Sci. (U.S.) **52**, 889 (1964).
15 A. Latham, V. C. Hascall, and H. B. Gray, Inorg. Chem. **4**, 788 (1965).
16 P. T. Manoharan and H. B. Gray, J. Am. Chem. Soc. **87**, 3340 (1965).
17 W. E. Hatfield, H. D. Bedon, and S. M. Horner, Inorg. Chem. **4**, 1181 (1965).
18 J. Halper, W. D. Closson, and H. B. Gray, Theoret. Chim. Acta (to be published).

H is the Hamiltonian matrix,

$$H_{ij}=(\chi_i^{\mu\alpha}\mid \mathcal{H}\mid \chi_j^{\nu\beta})=H_{ij}^{\mu}\delta_{\mu,\nu}\delta_{\alpha,\beta}, \qquad (3)$$

with $\mathcal{H}$ as the one-electron Hamiltonian operator.[19] The superscripts α and μ label the row and irreducible representation, respectively, to which the $\chi_i^{\mu\alpha}$'s belong. The δ function shows that there are no nonzero matrix elements between basis functions belonging to different representations, or different rows of the same representation.

(a) H_{ii}. The diagonal-Hamiltonian matrix elements H_{ii} are approximated as the negative values of valence orbital ionization potentials (VOIP's). The VOIP for the transition $M_a{}^q \rightarrow M_b{}^{q+1}$, corresponding to the ionization of an electron from configuration M_a to form configuration M_b, is computed from the formula,

$$\text{VOIP}(q)=I(q)+E(M_b)-E(M_a), \qquad (4)$$

where q is the charge on the atom, $I(q)$ is the ground-state ionization potential of $M_a{}^q$ (not necessarily resulting in $M_b{}^{q+1}$); $E(M_b)$ and $E(M_a)$ are the E_{Av}'s of the two configurations, M_b and M_a, where a particular E_{Av} is the weighted mean of the energies of all the multiplet terms arising from a given configuration relative to the ground state of the atom or ion in question. The weighting factor is equal to the total degeneracy (spin×orbital) of the term. After averaging over the J components, we have, for example,

$$E_{\text{Av}}(p^2)=\tfrac{1}{15}[9E(^3P)+5E(^1D)+E(^1S)]. \qquad (5)$$

The evaluation of E_{Av}'s for the pertinent configurations, where missing terms prevent a direct averaging can be done straightforwardly by using the equations of Slater[20] relating the energies of the individual multiplets to E_{Av} in terms of the Slater–Condon parameters. For example,

$$E_{\text{Av}}(p^2)=\begin{cases} E(^3P)+\ 3F_2(pp) \\ E(^1D)-\ 3F_2(pp). \\ E(^1S)-12F_2(pp) \end{cases} \qquad (6)$$

In these cases one can derive an E_{Av} for each Russell–Saunders term which theoretically should be identical for all the terms within one given configuration. However, the E_{Av}'s actually differ in some configurations by a few thousand cm^{-1}. In all such instances the E_{Av}'s derived from the different terms were averaged. Term energies and ionization potentials were taken from Moore.[21] Slater–Condon parameters were taken from the literature[22,23] or obtained by a least-squares fit of the relevant spectral data.

For the metal, three different orbital H_{ii}'s are required: $3d$, $4s$, and $4p$. The VOIP for a given metal orbital is considered as a linear combination of three different configurations in order to adequately represent fractional populations. For instance, for $4s$ the three configurations are taken to be $d^{n-1}s$, $d^{n-2}s^2$, and $d^{n-2}sp$. All the configurations are calculated as functions of charge and from the resulting nine VOIP curves, the metal $3d$, $4s$, and $4p$ H_{ii}'s can easily be evaluated.[7]

For the ligands, with valence orbitals s and p, the same procedure can be followed, requiring four VOIP curves. However, we have considered only the configuration s^2p^n; i.e., fractional population was considered only in the p level. The additional constraint was imposed that n be equal to the occupancy of the p level in the ground-state configuration of the neutral ligand, as in one calculation of MnO_4^-.[7] Richardson and Rundle[24] also found that the ligand H_{ii}'s were relatively insensitive functions of charge.

Valence state ionization energies of potentials (VSIE's or VSIP's) have been computed by Skinner, Pritchard, and Pilcher,[25–27] and Hinze and Jaffé.[28] These methods assume a hybridized state of the atom, implying a particular spatial orientation of the orbital favorable for bond formation. Such a concept is not necessary in MO theory. It is for this reason that we suggest the name valence-orbital ionization potentials for the energies calculated from E_{Av} values.[29] In addition, the VOIP's can be derived as naturally continuous functions of charge on the atom and, using Mulliken's prescription for summing populations,[30] the H_{ii}'s can be adjusted for variations of charge on the metal atom. Thus, the major advantage of the VOIP's is perhaps their simplicity. Finally, a clear relationship between H_{ii}'s, VOIP's, and the concept of electronegativity has recently been established.[31,32]

---

[19] C. C. J. Roothaan, Rev. Mod. Phys. **23**, 69 (1951).
[20] J. C. Slater, *Quantum Theory of Atomic Structure* (McGraw-Hill Book Company, Inc., New York, 1960), Vol. 1, Chap. 14 and Appendix 12a; Vol. 2, Appendix 22.
[21] C. E. Moore, "Atomic Energy Levels," Natl. Bur. Std. Circ. No. 467 (1958), Vols. 1, 2, and 3.
[22] J. Hinze and H. H. Jaffé, J. Chem. Phys. **38**, 1834 (1963). The F_k and G_k taken from Hinze and Jaffé differ by a multiplicative constant from the F^k and G^k used by Slater (Ref. 20).
[23] J. S. Griffith, *The Theory of Transition Metal Ions* (Cambridge University Press, Cambridge, 1961).
[24] J. W. Richardson and R. E. Rundle, U.S. Atomic Energy Commission, ISC-830, 1956.
[25] H. A. Skinner and H. O. Pritchard, Trans. Faraday Soc. **49**, 1254 (1953).
[26] H. O. Pritchard and H. A. Skinner, Chem. Rev. **55**, 745 (1955).
[27] G. Pilcher and H. A. Skinner, J. Inorg. Nucl. Chem. **24**, 937 (1962).
[28] J. Hinze and H. H. Jaffé, J. Am. Chem. Soc. **84**, 540 (1962); Can. J. Chem. **41**, 1315 (1963); J. Phys. Chem. **67**, 1501 (1963).
[29] H. Basch, A. Viste and H. B. Gray, Theoret. Chim. Acta (to be published).
[30] R. S. Mulliken, J. Chem. Phys. **23**, 1833 (1955).
[31] C. K. Jørgensen, *Orbital in Atoms and Molecules* (Academic Press Ltd., London, 1962), Chap. 7.
[32] G. Klopman, J. Am. Chem. Soc. **86**, 1463, 4550 (1964).

TABLE I. Average energies of configurations[a,b] (in 1000 cm^{-1}).

Metal	n	d^n	$d^{n-1}s$	$d^{n-1}p$	$d^{n-2}s^2$	$d^{n-2}sp$	$d^{n-2}p^2$
Sc	3	38.3	16.3	36.1	0.0	20.2	46.1
Ti	4	40.7	18.1	39.4	5.8	27.4	55.3
V	5	40.3	20.9	46.0	12.2	40.5	76.4
Cr	6	55.3	36.5	59.2	30.8	56.1	
Mn	7	58.2	42.9	71.3	39.3	69.3	104.3
Fe	8	40.4	25.1	50.8	27.1	50.8	
Co	9	27.9	13.1	42.2	19.3	46.8	
Ni	10	14.7	1.4	30.2	10.4	41.5	
Cu	11	···	0.0	30.7	12.0	46.5	
Zn	12	···	···	···	0.0	36.1	80.6

[a] Relative to ground-state multiplet.
[b] For neutral atom first-row transition metals.

(b) H_{ij}. The off-diagonal Hamiltonian matrix elements are approximated using Eq. (7),

$$H_{ij}=-\tfrac{1}{2}F_mG_{ij}[\text{VOIP}_i+\text{VOIP}_j], \qquad (7)$$

which contains an adjustable parameter F_m, called the F factor. The subscript m is used to differentiate among σ, π, ···, etc. —type interactions. In most applications of the method to date, it has been necessary to use F_m greater than one and also to use separate values for F_σ and F_π in order to achieve reasonable results.

(c) Implicit in the previous discussion is the neglect of all nonvalence atomic orbitals. The designation, "valence" in this context is necessarily arbitrary. For example, the metal $3p$ orbital has a finite, albeit small, overlap with the ligand functions and, if included in the calculation, would probably have some effect, particularly on the antibonding MO levels. Omitting the nonvalence atomic orbitals may be unduly optimistic,[14] but nevertheless seems necessary if simplicity is to be preserved. The valence orbitals used in these calculations are, therefore, $3d$, $4s$, and $4p$ for the metal, and ns and np on the ligand.

(d) The treatment of group overlap integrals, normalizations, and correction factors relating the H_{ii}'s to the H'_{ii}'s has been given in detail in a previous paper.[7]

RESULTS

VOIP's for the $3d$, $4s$, and $4p$ valence orbitals of the metals Sc through Zn have been determined from the average energies of the pertinent configurations.[29] The data have been smoothed by a least-squares quadratic fit across the transition series. For more efficient computer programming the resulting VOIP's are represented quadratically,

$$\text{VOIP}(q)=Aq^2+Bq+C, \qquad (8)$$

where q is the charge. The A, B, and C parameters, together with some neutral atom E_{Av}'s, are presented in Tables I and II. For the ligands, only the configuration s^2p^n of the neutral atom was used and these VOIP's are listed in Table III. In all calculations the ligand p_σ was assumed more stable (by 10 000 cm^{-1}) than p_π, and the latter was set equal to the value found in Table III. This adjustment is derived from inspection of theoretically calculated H'_{ii},[14,33,34–36] and from charge-transfer spectra.[37,38]

The ligands investigated were F, Cl, Br, S, and O. Interatomic bond distances were either taken from the

TABLE II. VOIP curves for first-row transition metals[a] (in 1000 cm^{-1}).

VOIP curve[b]		Ti	V	Cr	Mn	Fe	Co	Ni
A	1	17.15	15.8	14.75	14.1	13.8	13.85	14.2
	2	18.45	14.0	9.75	5.5	13.8	13.85	14.2
	3	18.45	14.0	9.75	5.5	13.8	13.85	14.2
	4	9.3	8.55	8.05	7.6	7.35	7.25	7.35
	5	9.3	8.55	8.05	7.6	7.35	7.25	7.35
	6	9.3	8.55	8.05	7.6	7.35	7.25	7.35
	7	7.8	7.45	7.25	7.2	7.3	7.55	7.95
	8	7.8	7.45	7.25	7.2	7.3	7.55	7.95
	9	7.8	7.45	7.25	7.2	7.3	7.55	7.95
B	1	60.85	68.0	74.75	80.8	86.2	91.15	95.5
	2	77.85	87.0	95.95	105.0	101.5	106.25	110.7
	3	76.75	87.3	96.95	106.0	101.9	105.55	108.2
	4	50.4	54.15	57.55	60.9	63.85	66.65	69.05
	5	58.5	62.95	66.85	70.3	73.05	75.25	77.05
	6	55.0	57.55	60.45	63.8	67.35	71.35	75.65
	7	35.6	45.45	47.55	49.3	50.8	51.95	52.85
	8	48.9	50.85	52.85	55.2	57.8	60.65	63.75
	9	48.9	50.85	52.85	55.2	57.8	60.65	63.75
C	1	27.4	31.4	35.1	38.6	41.9	44.8	47.6
	2	44.6	51.4	57.9	64.1	70.0	75.6	80.9
	3	55.4	61.4	67.7	74.3	81.2	88.4	95.9
	4	48.6	51.0	53.2	55.3	57.3	59.1	60.8
	5	57.2	60.4	63.3	65.9	68.3	70.5	72.3
	6	66.0	70.6	74.7	78.3	81.4	84.0	86.0
	7	26.9	27.7	28.4	29.2	29.9	30.7	31.4
	8	35.9	36.8	37.8	38.8	39.7	40.7	41.6
	9	34.4	36.4	38.1	39.4	40.3	40.8	40.9

[a] From Ref. 29.
[b] The type of electron being ionized, and the configurations, are as follows for the nine VOIP curves. 1: d, d^n. 2: d, $d^{n-1}s$. 3: d, $d^{n-1}p$. 4: s, $d^{n-1}s$. 5: s, $d^{n-2}s^2$. 6: s, $d^{n-2}sp$. 7: p, $d^{n-1}p$. 8: p, $d^{n-2}p^2$. 9: p, $d^{n-2}sp$. Successive points on a curve differ only in the number of d electrons.

[33] S. Sugano and R. G. Shulman, Phys. Rev. **130**, 506, 512, 517 (1963).
[34] R. E. Watson and A. J. Freeman, Phys. Rev. **134**, A1526 (1964).
[35] H. Basch and H. B. Gray, unpublished calculations on MnO_4^-.
[36] K. D. Carlson and R. K. Nesbet, J. Chem. Phys. **41**, 1051 (1964).
[37] C. K. Jørgensen, *Absorption Spectra and Chemical Bonding in Complexes* (Addison-Wesley Publishing Company, Inc., Reading, Massachusetts, 1962).
[38] C. J. Ballhausen and H. B. Gray, *Molecular Orbital Theory* (W. A. Benjamin, Inc., New York, 1964) Appendix 8.

literature[39-49] or estimated.[50,51] Day and Jørgensen[52] independently estimated the bond distances in $CoBr_4^{2-}$ (2.43 Å), $MnCl_4^{2-}$ (2.40 Å), and $FeCl_4^{2-}$ (2.30 Å) which differ from ours (Tables IV and V) by +0.05, +0.07, and +0.03 Å, respectively. This gives a good indication of the accuracy of estimated bond distances. The parameter F_π was consistently taken to be 2.10 for all H_{ij} and F_σ was varied in order to fit the observed[53-68] Δ exactly. All ligand–ligand correction factors[38] were calculated using $F_m=2.00$ for both σ- and π-type interactions. Wavefunctions for the ligands (neutral atom) were taken from Clementi[69] or (for Br) Watson *et al.*,[70] and for the metals from Richardson *et al.*[71,72] Neutral-atom functions were chosen for 4*s* and 4*p*, while the 3*d* function was taken from the d^n configuration for M^+. The effect of choice of wavefunctions on the calculation was not investigated, but it is expected that the *trends* under consideration here are not altered for any consistent choice. Also, it has recently been claimed that the one-electron LCAO–MO model is relatively insensitive to such changes.[73]

TABLE III. Ligand VOIPs[a,b] (in 1000 cm^{-1}).

Ligand	*s*	*p*
O	260 8	127 4
F	323.6	150.4
Cl	203.8	110.4
Br	193.8	99.6
S	166.7	93.4

[a] From Ref. 29.
[b] For configuration s^2p^n.

Once the elements of the secular equations had been calculated as input, the secular equations were solved for eigenvalues and eigenvectors, and a Mulliken population analysis[31] was carried out. The output metal charge (*q*), *s*, and *p* character were compared with the input *q*, *s*, and *p* (the latter may be regarded as the independent variables of charge and configuration in the input). The input *q*, *s*, and *p* populations were then adjusted by an iterative procedure until input and output *q*, *s*, and *p* agreed to 0.005 at which point the calculation was considered self-consistent. The whole process was programmed in FORTRAN to iterate automatically, and run on an IBM 7094 computer. Input consists essentially of the total number of electrons, metal oxidation state, metal and ligand VOIP's, group overlap matrix, *F*-factor matrix, and initial guess values for *q*, *s*, and *p*; the program then iterates to a self-consistent charge and configuration (SCCC). Overlap integrals are evaluated by a separate machine program.

The results of the calculations on 16 octahedral and 16 tetrahedral one-atom-ligand transition-metal complexes are presented in Tables IV and V. Table VI summarizes the composition of the relevant secular determinants for both octahedral and tetrahedral geometries.

DISCUSSION

The molecular orbital energy-level diagrams which represent these calculations of tetrahedral and octahedral complexes are shown schematically in Figs. 1 and 2. The MO's are connected to the atomic orbitals from which they are mainly derived and the MO schemes are most easily understood in terms of the ordering of the atomic energy levels. The brackets enclose levels of comparable energies derived from essentially the same atomic orbitals; in some cases there is a consistent ordering within a bracketed set (shown explicitly in the figures), in others not.

[39] A. F. Wells, *Structural Inorganic Chemistry* (Oxford University Press, London, 1962), Chap. 8.
[40] V. W. H. Baur, Acta Cryst. **11**, 488 (1958).
[41] C. K. Jørgensen, Acta Chem. Scand. **12**, 1539 (1958).
[42] H. W. Smith and M. Y. Colby, Z. Krist. **103**, 90 (1940).
[43] B. N. Figgis, M. Gerlock, and R. Mason, Acta Cryst. **17**, 506 (1964).
[44] M. Lister and L. E. Sutton, Trans. Faraday Soc. **37**, 393 (1941).
[45] W. N. Lipscomb and A. G. Whittaker, J. Am. Chem. Soc. **67**, 2019 (1945).
[46] B. Zaslow and R. E. Rundle, J. Phys. Chem. **61**, 490 (1957).
[47] P. Pauling, Ph.D. thesis, University College, London, 1960; quoted by A. B. Blake and F. A. Cotton, Inorg. Chem. **3**, 5 (1964).
[48] *Tables of Interatomic Distances and Configurations in Molecules and Ions*, Special Publication No. 11 (Chemical Society, London, 1958).
[49] R. W. G. Wyckoff, *Crystal Structures*, Interscience Publishers, Inc., New York, 1951), Vol. 1.
[50] M. J. Sienko and R. A. Plane, *Physical Inorganic Chemistry* (W. A. Benjamin, Inc., New York, 1963), p. 68.
[51] N. S. Hush and M. H. L. Pryce, J. Chem. Phys. **26**, 143 (1957).
[52] P. Day and C. K. Jørgensen, J. Chem. Soc. Suppl. 2, **1964**, 6226.
[53] C. J. Ballhausen and F. Winther, Acta Chem. Scand. **13**, 1729 (1959).
[54] V. O. Schmitz-DuMont, H. Brokopf, and K. Burkhardt, Z. Anorg. Chem. **295**, 7 (1958).
[55] C. K. Jørgensen, Advan. Chem. Phys. **5**, 62, 85 (1962).
[56] Reference 37, pp. 110–111 and pp. 284–289.
[57] Reference 37, p. 113.
[58] R. Englman, Mol. Phys. **3**, 48 (1960).
[59] W. Low, Phys. Rev. **109**, 247, 256 (1959).
[60] S. Siegel, Acta Cryst. **9**, 684 (1956).
[61] R. J. H. Clark, J. Chem. Soc. **1964**, 417.
[62] N. S. Gill and R. S. Nyholm, J. Chem. Soc. **1959**, 3997.
[63] C. J. Ballhausen, *Introduction to Ligand Field Theory* (McGraw-Hill Book Company, Inc., New York, 1962), Chaps. 4, 5, and 7.
[64] Reference 63, p. 228.
[65] F. A. Cotton, D. M. L. Goodgame, and M. Goodgame, J. Am. Chem. Soc. **84**, 167 (1962).
[66] C. Furlani, E. Cervone, and V. Valenti, J. Inorg. Nucl. Chem. **25**, 159 (1963).
[67] G. P. Smith, C. H. Liu, and T. R. Griffiths, J. Am. Chem. Soc. **86**, 4796 (1964).
[68] N. K. Hamer, Mol. Phys. **6**, 257 (1963).
[69] E. Clementi, J. Chem. Phys. **40**, 1944 (1964); IBM Research Paper RJ-256.
[70] R. E. Watson and A. J. Freeman, Phys. Rev. **124**, 1117 (1961).
[71] J. W. Richardson, W. C. Nieupoort, R. R. Powell, and W. F. Edgell, J. Chem. Phys. **36**, 1057 (1962).
[72] J. W. Richardson, R. R. Powell, and W. C. Nieupoort, J. Chem. Phys. **38**, 796 (1963).
[73] R. G. Shulman, Bell Telephone Laboratories, Memorandum for File, Case 38140-13.

TABLE IV. Results of SCCC calculations of various octahedral complexes (all energies in 1000 cm^{-1}).

	Complexes							
	TiF_6^{3-}	$TiCl_6^{3-}$	$TiBr_6^{3-}$	VF_6^{3-}	VCl_6^{3-}	VF_6^{4-}	CrF_6^{3-}	$CrCl_6^{3-}$
Bond distances, Å	1.97[a]	2.45[b]	2.56[c]	1.94[d]	2.39[e]	2.15[e]	1.93[e]	2.38[e]
3*d* VOIP	−120.3	−88.6	−84.0	−127.3	−94.5	−116.7	−134.1	−99.2
4*s* VOIP	−100.6	−78.9	−76.9	−101.6	−80.4	−95.3	−103.7	−81.4
4*p* VOIP	−55.7	−44.0	−41.2	−70.7	−50.6	−65.3	−70.4	−49.4
Metal charge	+1.12	+0.66	+0.68	+1.02	+0.59	+0.93	+0.97	+0.53
3*d* population	2.81	2.95	3.14	3.77	3.95	3.88	4.76	4.95
4*s* population	0.00	0.15	0.00	0.01	0.10	0.05	0.06	0.16
4*p* population	0.07	0.24	0.18	0.20	0.36	0.14	0.21	0.36
t_{2u} eigenvalue	−149.4	−108.4	−96.0	−149.2	−107.9	−149.8	−149.2	−108.9
$2t_{2g}$ eigenvalue	−90.4	−71.4	−70.2	−100.8	−80.2	−104.8	−110.3	−87.1
$3e_g$ eigenvalue	−73.1	−58.0	−57.1	−84.8	−66.5	−92.7	−94.7	−73.5
$2t_{2g}$ occupation	1	1	1	2	2	3	3	3
$3e_g$ occupation	0	0	0	0	0	0	0	0
Δ	17.5[e]	13.8[b]	13.0[c,a]	15.9[f]	13.9[b]	12.0[c,g]	15.2[h]	13.8[h]
First allowed L→M charge transfer: calculated orbital energy[i]	59.0	37.0	25.8	48.4	27.7	45.0	38.9	21.8
Observed band	>50[e]	≳38[e,j]	<38[e,j]	>40[e,j]	≳40[e,j]	≫40[e,j]	>37[e,j]	>37[e,j]
F_σ (±0.005)	1.53	1.60	1.51	1.55	1.57	1.59	1.60	1.61

	$CrBr_6^{3-}$	CrO_6^{9-}	MnF_6^{2-}	MnF_6^{4-}	FeF_6^{3-}	CoF_6^{3-}	CoO_6^{10-}	NiF_6^{4-}
Bond distances, Å	2.53[c]	2.00[e]	1.74[k]	2.12[d]	1.92[d]	1.89[d]	2.10[l]	2.00[m]
3*d* VOIP	−92.6	−114.7	−150.9	−129.6	−145.2	−153.4	−122.8	−153.4
4*s* VOIP	−79.0	−91.6	−112.4	−99.9	−108.2	−112.3	−93.5	−112.1
4*p* VOIP	−50.1	−59.3	−75.1	−65.0	−69.0	−69.7	−54.6	−67.4
Metal charge	+0.64	+0.75	+1.00	+0.85	+0.87	+0.84	+0.56	+0.78
3*d* population	5.26	4.91	5.65	5.89	6.76	7.72	7.96	8.79
4*s* population	0.10	0.08	0.06	0.10	0.14	0.17	0.22	0.20
4*p* population	0[n]	0.26	0.29	0.16	0.24	0.26	0.26	0.23
t_{2u} eigenvalue	−95.8	−125.5	−147.9	−149.7	−149.1	−148.9	−126.1	−149.4
$2t_{2g}$ eigenvalue	−83.0	−93.6	−111.0	−120.1	−125.6	−132.0	−112.5	−139.1
$3e_g$ eigenvalue	−69.7	−77.5	−89.1	−111.7	−111.9	−118.9	−103.0	−131.8
$2t_{2g}$ occupation	3	3	3	3	3	4	5	6
$3e_g$ occupation	0	0	0	2	2	2	2	2
Δ	13.2[o]	16.2[g,p]	21.8[g]	8.4[g]	14.0[g]	13.1[g,r]	9.6[p,h]	7.3[g]
First allowed L→M charge transfer: calculated orbital energy[i]	12.8	31.9	36.9	29.6	23.5	16.9	13.6	17.6
Observed energy	<37[c,j]	≳37[g,i]	>28[r]	>43[e,j]	>30[r]		~60[e,j]	>60[e,j]
F_σ (±0.005)	1.58	1.61	1.61	1.62	1.67	1.69	1.71	1.68

a From Ref. 60.
b From Ref. 61.
c Estimated value.
d From Ref. 39.
e From Ref. 6.
f From Ref. 53.
g From Ref. 56.
h From Ref. 59.
i Uncorrected for electronic repulsions.
j The values quoted are for hydrated salts from Ref. 58.
k From Ref. 48.
l From Ref. 62.
m From Ref. 40.
n $CrBr_6^{3-}$ was calculated without the metal 4*p* since it was found that a t_{1u} level interceded between the $2t_{2g}$ and $3e_g$ levels.
o From Ref. 57.
p In MgO.
q From Ref. 41.
r From Ref. 55.

Only a few of the levels appear to be important in discussing electronic transitions. For tetrahedral complexes of one-atom ligands, the most important levels are $3t_2$, t_1, $2e$, and $4t_2$. For these four levels, the schematic energy level diagram of Fig. 1 agrees with the conventional Ballhausen–Liehr ordering.[74] The $2e$ and $4t_2$ levels roughly correspond to the *d* orbitals of crystal-field theory, though with an admixture of ligand orbitals. The t_1 is a strictly ligand π level while $3t_2$ tends to be predominantly composed of ligand π functions.

For octahedral complexes, the most important levels are $3e_g$ and $2t_{2g}$ for the $d-d$ spectra, and, in addition, the two highest levels of odd parity, the $3t_{1u}$ and t_{2u}, for the ligand-to-metal charge transfer. The t_{1g}, although

[74] C. J. Ballhausen and A. D. Liehr, J. Mol. Spectry. 2, 342 (1958).

TABLE V. Results of SCCC calculations of various tetrahedral complexes (all energies in 1000 cm^{-1}).

	Complexes							
	$TiCl_4$	$TiBr_4$	VCl_4	VCl_4^-	CrO_4^{2-}	MnO_4^-	MnO_4^{2-}	MnO_4^{3-}
Bond distance, Å	2.18[a]	2.31[a]	2.03[b]	2.14[c]	1.60[d]	1.59[d]	1.63[e]	1.67[e]
3*d* VOIP	−90.8	−82.2	−95.1	−88.7	−128.3	−143.4	−137.2	−130.6
4*s* VOIP	−79.5	−74.3	−80.6	−76.7	−97.5	−104.5	−101.5	−98.2
4*p* VOIP	−45.3	−41.2	−50.7	−47.1	−61.8	−65.2	−63.3	−61.0
Metal charge	+0.64	+0.52	+0.58	+0.51	+0.67	+0.68	+0.68	+0.67
3*d* population	2.84	2.93	3.92	3.98	4.45	5.35	5.50	5.63
4*s* population	0.19	0.15	0.13	0.13	0.29	0.34	0.28	0.24
4*p* population	0.33	0.41	0.37	0.39	0.58	0.63	0.54	0.47
t_1 eigenvalue	−106.1	−93.5	−103.8	−105.6	−121.0	−120.8	−121.6	−122.3
2*e* eigenvalue	−65.9	−61.5	−66.1	−68.6	−72.8	−84.5	−87.0	−88.6
$4t_2$ eigenvalue	−57.2	−54.0	−57.1	−63.0	−46.8	−58.8	−67.8	−73.8
2*e* occupation	0	0	1	2	0	0	1	2
$4t_2$ occupation	0	0	0	0	0	0	0	0
Δ	8.7[c]	7.6[c]	9.0[e]	5.6[c]	26.0[f]	26.0[f]	19.0[f]	14.8[f]
First allowed L→M charge transfer: calculated orbital energy[g]	40.2	32.0	37.7	37.0	48.2	36.3	34.6	33.7
Observed energy	34.8[h]	29.0[h]	24.2[h]		26.8[f]	18.3[f]	22.9[f]	30.8[f]
F_σ (±0.005)	1.64	1.62	1.60	1.59	1.95	1.96	1.78	1.81
	$MnCl_4^{2-}$	$FeCl_4^-$	$FeCl_4^{2-}$	$CoCl_4^{2-}$	$CoBr_4^{2-}$	CoO_4^{6-}	CoS_4^{6-}	$NiCl_4^{2-}$
Bond distance, Å	2.33[c]	2.19[i]	2.27[e]	2.25[j]	2.38[c]	1.95[k,l]	2.52[m,n]	2.27[l]
3*d* VOIP	−94.3	−105.6	−98.2	−104.3	−95.4	−121.4	−89.3	−112.4
4*s* VOIP	−76.9	−83.2	−78.8	−81.3	−76.8	−92.5	−71.4	−85.0
4*p* VOIP	−43.4	−46.4	−43.1	−43.9	−40.6	−53.5	−35.5	−46.3
Metal charge	+0.37	+0.38	+0.33	+0.32	+0.30	+0.54	+0.13	+0.34
3*d* population	6.03	6.92	7.01	8.01	8.19	7.95	8.06	8.99
4*s* population	0.24	0.27	0.27	0.28	0.18	0.22	0.36	0.34
4*p* population	0.36	0.42	0.38	0.38	0.33	0.29	0.45	0.33
t_1 eigenvalue	−107.7	−106.3	−107.1	−106.9	−94.5	−125.4	−89.8	−107.1
2*e* eigenvalue	−86.2	−92.6	−89.9	−95.7	−88.2	−108.7	−84.5	−102.3
$4t_2$ eigenvalue	−82.6	−87.6	−85.8	−92.0	−85.1	−104.7	−81.3	−98.7
2*e* occupation	2	2	3	4	4	4	4	4
$4t_2$ occupation	3	3	3	3	3	3	3	4
Δ	3.6[o]	5.0[p]	4.0[q]	3.7[p]	3.1[r]	3.8[p]	3.2[p]	3.5[s]
First allowed L→M charge transfer: calculated orbital energy[g]	21.5	13.7	17.2	14.9	9.4	20.7	8.5	8.4
Observed energy	30[o]	27.2[m]		42.5[m]	23.4[t]			35.8[s]
F_σ (±0.005)	1.65	1.66	1.68	1.66	1.56	1.71	1.72	1.68

[a] From Ref. 44.
[b] From Ref. 45.
[c] Estimated value.
[d] From Ref. 42.
[e] From Ref. 64.
[f] From Ref. 7.
[g] Uncorrected for electronic repulsions.
[h] From C. Dyjkgraaf, Spectrochim. Acta 21, 769 (1965).
[i] From Ref. 46.
[j] From Ref. 43.
[k] ZnO lattice.
[l] From Ref. 48.
[m] ZnS lattice.
[n] From Ref. 49.
[o] From Ref. 65.
[p] From Ref. 56.
[q] From Ref. 66.
[r] From Ref. 68.
[s] From Ref. 67.
[t] From Ref. 62.

a strictly ligand π level as is the t_{2u}, is relatively unimportant since both $t_{1g}\rightarrow 2t_{2g}$ and $t_{1g}\rightarrow 3e_g$ are parity-forbidden transitions. The $3e_g$ and $2t_{2g}$ roughly correspond to the orbitals e_g and t_{2g} of crystal-field theory. The $3t_{1u}$ level is mainly composed of ligand π functions.

In calculations on the 16 octahedral complexes, permutations of levels occurred within some of the bracketed sets. Only one of these permutations is of importance, that interchanging $3t_{1u}$ and t_{2u}. Since they are both triply-degenerate orbitals of odd parity and of similar composition it is not likely that the correct order can be established in these cases. For consistency, the first L→M charge transfer was calculated from the t_{2u}.

However, with the possible exception of the t_{2u}–$3t_{1u}$ comparison, these calculations, for reasonable choices

TABLE VI. Metal and Ligand Orbital Contributions to the Secular Equations.[a]

Irreducible representation	Contribution: Metal	Contribution: Ligand	Dimension	Degeneracy
Octahedral				
a_{1g}	s	s, p_σ	3	1
e_g	d	s, p_σ	3	2
t_{1g}	...	p_π	1	3
t_{1u}	p	s, p_σ, p_π	4	3
t_{2g}	d	p_π	1	3
t_{2u}	...	p_π	1	3
Tetrahedral				
a_1	s	s, p_σ	3	1
e	d	p_π	2	2
t_1	...	p_π	1	3
t_2	p, d	s, p_σ, p_π	5	3

[a] Coordinate systems, ligand basis sets, analytic expressions for normalizations, group overlaps, and H_{ii}' correction factors are given in Ref. 38 for octahedral and Ref. 7 for tetrahedral.

of the parameters, yield the expected relative orderings for the levels, which should be most important in the $d-d$ and L→M charge-transfer spectra of both tetrahedral and octahedral complexes. In particular, $4t_2 > 2e$ in tetrahedral, and $3e_g > 2t_{2g}$ in octahedral.

The significance that can be attached to the wavefunctions and energies calculated by the simple MO method has been the subject of considerable debate.[11,12]

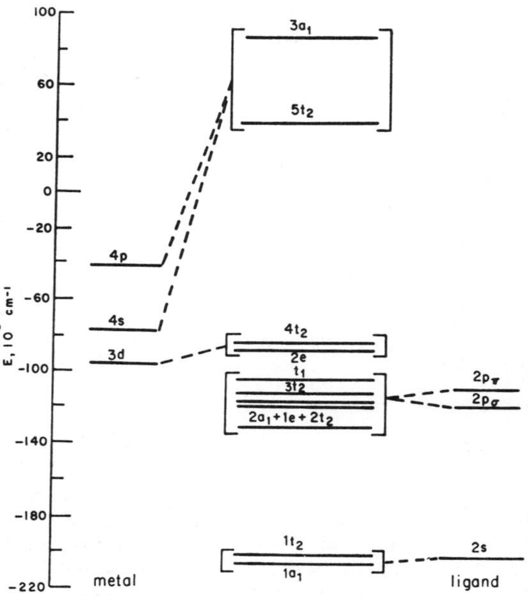

FIG. 1. Molecular orbital energy-level diagram for tetrahedral complexes. Drawn to scale for $FeCl_4^{2-}$.

Early investigators[1,3] placed great emphasis on the transition moments computed using the derived ground-state wavefunctions. In one case,[1] however, although the results compared satisfactorily with experimentally determined intensities, the band assignments subsequently have been shown to require revision.[7]

Direct experimental evidence of covalency in transition-metal complexes has come from magnetic resonance experiments which, under favorable conditions, can be interpreted to yield values for covalent mixing parameters.[63] In particular, one such work[38] reported the coefficients for the ligand $2s$ and $2p$ symmetry orbitals

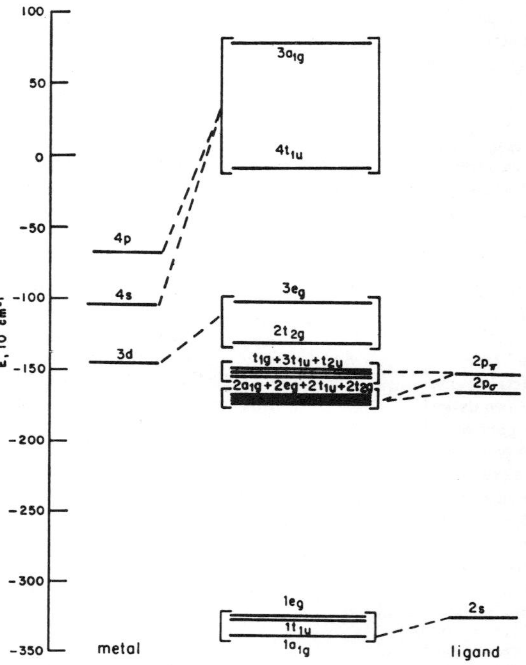

FIG. 2. Molecular orbital energy-level diagram for octahedral complexes. Drawn to scale for FeF_6^{3-}.

in $3e_g$ as 0.116 and 0.337, respectively, in $KNiF_3$. This can be compared with 0.007 and 0.386 calculated here for NiF_6^{4-}. Unfortunately, the comparison is not quantitatively valid since the "experimentally" determined coefficients are actually derived from ^{19}F hyperfine interactions using F^- $2s$ and $2p_\sigma$ atomic wavefunctions which differ from the neutral atom F wavefunctions used in this investigation.

Of greater use and reliability are the calculated charge distributions partitioning the electrons in the MO's into localized contributions of the participating atoms. Although arbitrarily defined, these gross atomic populations and resultant bond orders and bond polarities have long been used successfully to predict and confirm the existence of chemically reactive sites in complex organic molecules.[75] In addition, an examination of the dominating orbital contributions to a given calculated charge distribution has helped elucidate the factors governing the formation and properties of molecules. Of particular interest in the realm of transition-metal complexes are the relative magnitudes and trends in the derived charge distributions indicating degree and kind of covalency.

In the calculations reported here (Tables IV and V) the computed charge on the metal decreases and the occupation of the metal $4s$ and $4p$ orbitals increases with increasing atomic number (Z). This increased participation of $4s$ and $4p$ in the bonding reflects the increased stability of these orbitals in going across the transition series. The lowering of charge is indicative of increased covalency (for a given ligand) with increasing Z and is in qualitative agreement with the orbital electronegativities calculated for these metals by Hinze and Jaffé.[28]

Additional evidence is supplied by an examination of the transition-metal radial wavefunctions of Richardson *et al.*[71] or Clementi and Raimondi.[76] They show that the radii of maximum radial charge density of the free atom and ion wavefunctions shrink by approximately 0.04 Å on increasing Z by one unit. However, the computed ionic radii and observed bond distances in complexes (for a given oxidation state, ligand, and geometry) vary by a value somewhere between zero and 0.03 Å. This can be interpreted by assuming that the reduced charge on the metal due to increased covalency compensates for most of the natural contraction of the $3d$ radial function with increasing Z, yielding a slow net decrease in the observed bond distance. The calculations also reveal that, for a given ligand, the self-consistent charge changes slowly with increasing formal oxidation state. This is in qualitative agreement with a similar experimental observation made by Shulman and Sugano[77] for cyanide complexes, although their explanation may not be directly applicable here.

The variation of charge with ligand for a given metal is seen to follow F>O>Cl>Br>S, as is expected from the known trends in metal binding by these atoms. As outlined previously, transition-metal complexes are known to have substantial covalent character and this suggests that the ligand np ($n=2, 3, 4$) H'_{ii} are in the stability range of the metal $3d$ H'_{ii}. The actual choice of H'_{ii}, wavefunctions, and F factors determines the degree of covalency for an isolated complex, and all three choices are somewhat arbitrary; although one set may be superior to another. The simple MO method, however, offers no unambiguous way of finding this set in an absolute sense. In addition, since the numbers of interest to us here (electronic spectra) are of the same order of magnitude as the inherent error in the theory upon which the computations are based, significance can be attached only to consistent behavior within a group of molecules.

In both tetrahedral and octahedral geometries, Δ measures the splitting of the $d_\pi - d_\sigma$ (i.e., π and σ, respectively, with regard to interactions with ligand orbitals) metal orbitals. For a given set of VOIP's and wavefunctions, Δ is completely determined by the choice of F_σ and F_π [Eq. (7)]. Since F_π principally determines the position of the first L→M charge-transfer band, F_π was fixed at 2.10 to approximately fit the L→M band in the d^0 and d^1 cases, where the electronic repulsion corrections to the simple difference of the one-electron energy levels can be estimated [see later, Eq. (9)]. The results of fitting F_σ to reproduce the experimentally determined Δs are tabulated in Tables IV and V.

A detailed examination of F_σ values reveals the following:

(a) The F_σ's belonging to the 16 octahedral complexes (Table IV) follow the equation, $F_\sigma(n) = [0.027n+1.546]\pm 0.02$, where for Ti, $n=0$, for V, $n=1$, ··· etc., independent of the metal oxidation state or combining ligand.

(b) Nine of the 16 tetrahedral complexes (VCl_4, VCl_4^-, $MnCl_4^{2-}$, $FeCl_4^-$, $FeCl_4^{2-}$, $CoCl_4^{2-}$, CoO_4^{6-}, CoS_4^{6-}, $NiCl_4^{2-}$) in Table V follow the same equation with the same average deviation.

(c) For the high-oxidation-state (V, VI, VII) tetrahedral metal oxyanions (CrO_4^{2-}, MnO_4^-, MnO_4^{2-}, MnO_4^{3-}) the F_σ's show a definite oxidation-state dependence and lie outside the range found for the other complexes.

We now proceed to a comparison of the calculated and observed variations in Δ and L→M charge transfer.

(1) Δ with geometry. With the reservations specified above, it appears possible to transfer the F factor from octahedral to tetrahedral geometries quantitatively. It is important to note that this statement is limited to first-row transition metals containing one-atom ligands with a single π valence orbital. The inclusion of a second π valence orbital (as for example π^* in diatomic or complex unsaturated[8] ligands) could conceivably alter the whole character of this analysis. Extension to other geometries must be tested separately.

The variation of F_σ in the high-oxidation-state oxyanions seems to be a direct result of the unusually large Δs for these complexes.[7,78]

(2) Δ with metal-ion oxidation number. For metal complexes, which do not have appreciable π-acceptor ability, Δ is known to increase as the formal positive charge on the ion increases. For example, comparing

[75] C. A. Coulson, *Valence* (Oxford University Press, London, 1961), Chap. 9.

[76] E. Clementi and D. L. Raimondi, J. Chem. Phys. **38**, 2686 (1963).

[77] R. G. Shulman and S. Sugano, J. Chem. Phys. **42**, 39 (1965).

[78] A. Carrington and C. K. Jørgensen, Mol. Phys. **4**, 395 (1961).

experimental results for $V(H_2O)_6^{2+} \rightarrow V(H_2O)_6^{3+}$, Δ goes from 12 400 to 18 400 cm^{-1}.[56] Similarly, comparing $MnF_6^{4-} \rightarrow MnF_6^{2-}$, Δ goes from 8400 to 21 800 cm^{-1}; such large changes in octahedral Δs are accounted for by a nearly constant F_σ, as shown in Table IV.

(3) Δ with ligand (the spectrochemical series). The spectrochemical series of the halide ligands is F > Cl > Br. A glance at the results for octahedral Ti(III) and Cr(III), and tetrahedral Co(II) reveals a small scatter of F_σ values for the halide ligands in the calculations. This may be due to the inaccuracy of estimated bond distances for the complex or an ambiguity as to which value of Δ (which varies slightly with state and solvent) is appropriate to these calculations. In any event, this result gives an indication of the accuracy that can be expected from these calculations.

(4) Charge transfer with metal oxidation number. In discussing charge-transfer spectra, we utilize an equation given by Jørgensen[31]

$$E = I(\pi) - I(3d) - (q-1)J(a, a) + (q-k-1)J(a, b) + kJ(b, b) \quad (9)$$

approximately governing the energy (E) of the transition, $a^q b^k \rightarrow a^{q-1} b^{k+1}$, from a nonbonding π MO level, a, with eigenvalue $I(\pi)$ and occupation number q, to one of the "3d-orbital" MO levels, b, with eigenvalue $I(3d)$ and initial occupation number k. $J(a, b)$ is the two-electron integral $[a, a/(r_{12})^{-1}/b, b]$ between MO's a and b. Exchange integrals are neglected. In these calculations $q=6$ and k varies from 0 to 5. In an isoelectronic series of complexes with the same ligand, the first L→M charge-transfer band is expected to move to lower energy as the formal oxidation state of the metal increases.[31] Presumably, as the oxidation state of the metal increases, the metal 3d orbitals become more stable, giving a smaller energy separation between the (fully occupied) nonbonding ligand π levels and the lowest empty "d orbital" (2e or 4t_2 in tetrahedral, 2t_{2g} or 3e_g in octahedral).

In every case where an isoelectronic series can be compared (the interelectronic repulsion corrections to the simple differences of one-electron energy levels are approximately constant) these calculations give the expected trend. Thus we calculate $CrO_4^{2-} > MnO_4^-$, $MnCl_4^{2-} > FeCl_4^-$, $VF_6^{4-} > CrF_6^{3-} > MnF_6^{2-}$, and $MnF_6^{4-} > FeF_6^{3-}$.

As a rule of thumb, a similar movement of the first L→M charge-transfer band to lower energy with increasing formal positive charge is also expected for complexes of a given ligand and different oxidation states of a single metal. Here the comparison is not an isoelectronic one and so the expectation is qualitatively less clear cut. The calculations predict that the orbital energy of the first L→M transition is practically insensitive to oxidation number for such comparisons. However, the electronic repulsion correction factors are expected to increase directly proportional to the number of electrons in the upper ("d-orbital") MO. This corresponds to increasing k in Eq. (9) and shifts the calculated trend in the expected direction.

(5) Charge transfer with metal Z. It has been established experimentally that the position of the first L→M band in an analogous series of complexes shifts to lower energy in going from Ti to Co. For the series of octahedral fluorides and chlorides and tetrahedral chlorides, the calculated trend is as expected. The dominating factor seems to be the steady decrease in the metal 3d H'_{ii} with increasing Z, although the self-consistent charge on the metal decreases.

While the observed trend is generally followed by the calculated values, the magnitudes of the changes appear at first sight to be too large. It is probable that the VOIP curves exaggerate the increasing stability of the metal 3d H'_{ii} both with charge and with increasing Z.[4] This is not certain, however, since the interelectronic-repulsion corrections increase in a series from Ti to Co.

(6) Charge transfer with ligand. The energy of the first L→M band for a metal in a given oxidation state is expected to parallel the instabilities of the various ligand p levels. In these calculations the VOIP order is S < Br < Cl < O < F. For octahedral Ti(III) and Cr(III), and tetrahedral Co(II), the first L→M charge-transfer band follows the expected ordering, F > O > Cl > Br > S. The inclusion of electronic-repulsion terms in complexes with few d electrons is expected to reduce the calculated one-electron differences between adjacent ligands since, in terms of the quantities defined in Eq. (9), $J(a, a)$ is expected to increase in the order S < Br < Cl < O < F.

CONCLUSIONS

Two of the present results are particularly encouraging. First, for constant F_π and F_σ varied to obtain agreement with experimental values of Δ, it has been found that the optimum value of F_σ varies only slowly in going across the first transition series. This general conclusion agrees with the findings of Cotton and Haas.[13] Further, a single value of F_σ appears adequate to account for variations of Δ in the II through IV range of oxidation states for a given metal.

Secondly, a single value of F_σ is sufficient for both octahedral and tetrahedral complexes, with the same set of atomic orbitals. In previous calculations the transition from octahedral to tetrahedral has been particularly troublesome.[12] However, successful comparison seems to be limited to the II→IV range in oxidation states thus far.

There are two interesting complications. F_σ does seem to depend significantly on oxidation state, if the V→VII range in oxidation states is considered. Secondly, optimization of F_σ to bring agreement with Δ may not necessarily give close agreement with the observed first L→M charge-transfer band. However, in most cases the interelectronic-repulsion correction to the calculated one-electron charge-transfer energy brings theory and experiment into closer agreement. We also get a qualitative correlation of L→M band energy with formal oxidation number, ligand, and metal Z for a single F_{π}.

Thus the over-all results are sufficiently promising that the semiempirical molecular orbital method may be recommended for additional study, in hope of further refinement.

ACKNOWLEDGMENTS

We thank the National Science Foundation and the Public Health Service (Research Grant Number CA-07016-02, from the National Cancer Institute) for support of this research. The work was carried out during the tenure of a National Science Foundation Postdoctoral Fellowship to A. V., a Summer National Science Foundation Fellowship to H. B., and an Alfred P. Sloan Research Fellowship to H. B. G.

VIII-1c Reprinted from *Inorganic Chemistry* 7, 479 (1968)

Contribution from the Department of Chemistry,
University of Wisconsin, Madison, Wisconsin 53706

Parameter-Free Molecular Orbital Calculations

By Richard F. Fenske and Douglas D. Radtke[1]

Received July 11, 1967

A method is outlined for the calculation of electronic energy levels of transition metal complexes. The procedure is completely specified for a given choice of atomic wave functions and does not involve the use of arbitrary parameters for the evaluation of the matrix elements. The method is used to calculate the energies and molecular orbitals of four octahedral and seven tetrahedral chloro complexes. The results are evaluated in terms of Δ values, nephelauxetic B and β values, bond orders, overlap populations, stretching frequencies, and chlorine nqr results. The calculations substantiate increased covalency as a function of formal oxidation state by comparison of overlap populations with stretching frequencies and the correlation between calculated and experimental nephelauxetic β values. For different transition metals in the same oxidation state, an increase in covalent character with increasing atomic number is observed.

Introduction

In our recent publications,[2,3] we outlined a calculational procedure for estimation of the relative positions of the electronic energy levels of transition metal complexes. At that time, computational limitations necessitated the introduction of two parameters, R_σ and R_π, which were employed in the estimation of sums of two-center electrostatic interaction integrals. These two parameters were held constant for the series of fluoride complexes then under investigation. In this report, we shall indicate that not only was the assumption of constant R values a reasonable one, but also that it is possible to carry out the molecular orbital calculations without the introduction of these parameters. That is, for a proper basis set of atomic wave functions, we shall propose a calculational method which is completely specified. The procedure has been tested by application to four octahedral and seven tetrahedral chloride complexes of first-row transition metals. The accord achieved between experimental observables and the values calculated from the theoretical results are very satisfying and indicate that the calculational model holds promise for the elucidation of the electronic structures of metal complexes.

Matrix Elements of the Secular Determinant

The Diagonal Element.—As in our previous work,[2,3] the diagonal terms involving the metal wave functions, χ_i, are given by

$$(\chi_i|\mathcal{H}|\chi_i) = \epsilon_\chi(q_M) - \sum_{j=1}^{n} q_j(1/r_j|\chi_i\chi_i) \qquad (1)$$

where $\epsilon_\chi(q_M)$ is the orbital energy of the metal electron in the free ion of charge q_M, and the second term is the crystal field potential due to the ligand point charges, q_j. Both q_M and q_j are evaluated by means of the Mulliken electron population analysis[4] and it is required that self-consistency be established between the initial choices and final calculated values of q_M and q_j.

The ligand diagonal terms, exclusive of ligand–ligand interaction, have the form

$$(\phi_i|\mathcal{H}|\phi_i) = \epsilon_{\rho i1}(q_1) - q_M(1/r_M|\rho_{i1}\rho_{i1}) + \sum_{j=2}^{n} (1/r_j|\rho_{i1}\rho_{i1}) \qquad (2)$$

where $\epsilon_{\rho i1}$ is the orbital energy of the electron in the ith orbital of ligand 1 and of charge q_1, and the second and third terms constitute the crystal field potential due to the metal ion of charge q_M and the other ligands with charges q_j.

The concepts which led to the formulations given in eq 1 and 2 have been previously outlined.[2,3,5] It is worthwhile to note that while eq 1 and 2 are developed from consideration of the rigorous treatment of one-electron energies of closed-shell systems, the proposed method is semiempirical and deviates from rigor for the sake of computational simplicity. Furthermore, in our previous work we employed the commonly used simplification that the orbital energies, $\epsilon_{\chi i}$, could be obtained from the experimental values of the valence-state ionization energies (VSIE). This approximation has the undesirable feature that variations in the choice of the basis functions have no effect on the orbital

(1) Abstracted in part from the thesis submitted by D. D. Radtke in partial fulfillment of the requirements for the degree of Doctor of Philosophy at the University of Wisconsin.

(2) R. F. Fenske, K. G. Caulton, D. D. Radtke, and C. C. Sweeney, *Inorg. Chem.*, **5**, 951 (1966).

(3) R. F. Fenske, K. G. Caulton, D. D. Radtke, and C. C. Sweeney, *ibid.*, **5**, 960 (1966).

(4) R. S. Mulliken, *J. Chem. Phys.*, **23**, 1841 (1955).

(5) J. W. Richardson and R. E. Rundle, "A Theoretical Study of the Electronic Structure of Transition Metal Complexes," Ames Laboratory, Iowa State College, ISC-830, U. S. Atomic Energy Commission, Technical Information Service Extension, Oak Ridge, Tenn., 1956.

energies, as should be the case. Since the core functions, *e.g.*, 1s through 3p inclusive on the metal, do not vary appreciably as changes occur in the outer orbitals and since the wave functions for these orbitals are reasonably well known, it was decided to determine the orbital energies, $\epsilon_{\chi i}$, by the actual evaluation of the appropriate kinetic energy, nuclear attraction, and one-center electrostatic interaction integrals. Thus, as should be the case, when the wave function associated with an orbital is varied, not only the overlap and other multicenter integrals change, but the value of the orbital energy is also affected. Such considerations have not always been taken into account in previously reported calculational methods.[6-8]

The Off-Diagonal Elements.—The off-diagonal element involving the symmetry-adapted ligand wave function, ϕ_i, can be written in terms of the single-ligand wave functions, ρ_{il}. Thus, $(\phi_i|\mathcal{H}|\chi_i) = C(\rho_{il}|\mathcal{H}|\chi_i)$, where C is the same group factor which relates the diatomic overlap $S(\rho_{il}, \chi_i)$ to the group overlap, $G(\phi_i, \chi_i)$. In terms of the single-ligand wave function, the matrix element can be approximated by

$$(\rho_{il}|\mathcal{H}|\chi_i) = \epsilon_{\chi i}(q_M)S(\rho_{il}, \chi_i) - \sum_{j=2} q_j(1/r_j|\rho_{il}\chi_i) + (\rho_{il}|V_1|\chi_i), \quad (3)$$

where the middle term on the right-hand side of eq 3 consists of three-center nuclear attraction integrals multiplied by the self-consistent ligand charge, q_j. The final term in eq 3 is given by

$$(\rho_{il}|V_1|\chi_i) = \sum_k \bar{b}_k\{2(\rho_{k1}\rho_{k1}|\rho_{il}\chi_i) - (\rho_{k1}\rho_{il}|\rho_{k1}\chi_i)\} - Z_1(1/r_1|\rho_{i1}\chi_i) \quad (4)$$

In our previous work,[2,3] it was necessary to approximate eq 4 by the relationship

$$(\rho_{il}|V_1|\chi_i) = [R_a(6 - q_1) - Z_1](1/r_1|\rho_{i1}\chi_i) \quad (5)$$

where R_a was a "reduction factor" dependent upon ρ_{il} and χ_i but fixed for the complete series of fluoride complexes under study. In the present computations, we have been able to eliminate the use of arbitrary R values and thus free the calculations of all parameters. For a given choice of basis set wave functions, the computations and results are completely specified.

The elimination of the R parameters is made possible by the use of eq 4 rather than eq 5 for the evaluation of the off-diagonal elements. The former equation requires the determination of the two-center electrostatic interaction integrals.[9] Computationally, such integrals are time consuming and expensive to obtain. Fortunately, for a series of related complexes, we have obtained evidence[10] that the electrostatic interactions given in eq 4 can be quite accurately approximated by the relationship given by eq 5. That is, once the sum of two-center electrostatic interactions given by eq 4 has been determined for one complex, say $CrCl_6^{3-}$, it is possible to evaluate the R values which fit the relation

$$R_a(6 - q_1)(1/r_1|\rho_{il}\chi_i) = \sum_k \bar{b}_k\{2(\rho_{k1}\rho_{k1}|\rho_{il}\chi_i) - (\rho_{k1}\rho_{il}|\rho_{k1}\chi_i)\} \quad (6)$$

and these R values are essentially constant when one varies the central metal, internuclear distance, and even symmetry. While this is detailed more completely elsewhere,[10] it is worthwhile to reproduce some of the calculated R values to verify this important result. Table I displays the R values calculated by eq 6 for a group of chloride and fluoride complexes. Note the variations in symmetry and ligand species. The results indicate first of all that the assumption of constant R values from complex to complex which was made in our earlier calculations[2,3] was a reasonable one. Secondly, eq 6 provides a method for the approximation of the sums of two-center electrostatic interaction integrals in terms of the corresponding nuclear attraction integrals which are easier to compute. It should be emphasized that while eq 6 again incorporates R values into the computations, these are *not* parameters to be varied but are totally determined by the exact calculation of the interaction integrals for at least one complex in a series. Because the approximations of eq 6 hold, the reintroduction of R values is for computational simplicity only and is not an inherent part of the proposed method of calculation.

It might also be noted that while the values of $R(3d\sigma, 2p\sigma)$ and $R(3d\pi, 2p\pi)$ for the fluoride complexes in Table I are close to the value of $R_\sigma = R_\pi = 0.87$ used in our previous works,[2,3] they are not identical. Use of $R_\sigma = 0.84$ and $R_\pi = 0.89$ as suggested by Table I in a calculation identical with our previous work would yield similar wave functions but somewhat larger Δ values than previously obtained. However, the previous calculations did not involve the same size basis set as is used in the present work. This important factor is considered in detail in the next section.

TABLE I

R VALUES AS A FUNCTION OF THE COMPLEX[a]

	$TiCl_4$ (2.18 Å)	$CrCl_6^{3-}$ (2.34 Å)	$FeCl_6^{3-}$ (2.38 Å)	CrF_6^{3-} (1.93 Å)	NiF_3^- (2.10 Å)
$R(3d\sigma, n s)$	0.78	0.76	0.77	0.76	0.79
$R(3d\sigma, n p\sigma)$	0.84	0.83	0.84	0.83	0.86
$R(3d\pi, n p\pi)$	0.88	0.88	0.90	0.88	0.90

[a] All values calculated from eq 6 for the totally ionic case, *i.e.*, $q_1 = -1$ and $\bar{b}_k = 1$ for $k = n$s, np.

Calculational Considerations

As indicated by the results in Table IV of ref 2, our previous method required that the basis set used in the calculations consist of the 3d, 4s, and 4p orbitals for the first-row transition metals and the 2p fluorine ligand orbitals. At that time we showed that inclusion of the fluorine 2s orbitals increased $\Delta = 10Dq$,

(6) H. Basch, A. Viste, and H. B. Gray, *J. Chem. Phys.*, 44, 10 (1966).
(7) H. Basch and H. B. Gray, *Inorg. Chem.*, 6, 365 (1967).
(8) F. A. Cotton and C. B. Harris, *ibid.*, 6, 369 (1967).
(9) The two-center electrostatic interaction program for the CDC 3600 computer was made available to us through the kindness of Professor F. A. Matson, Molecular Physics Group, University of Texas, Austin, Texas.
(10) D. D. Radtke and R. F. Fenske, *J. Am. Chem. Soc.*, 89, 2292 (1967).

an effect which was not adequately corrected by inclusion of the free-ion 4d functions on the metal. We noted that the minimal reduction of Δ by the 4d orbitals was probably not because higher orbitals are ineffective but rather because of the form of the particular 4d functions obtained by minimization of the free-ion energy. That is, the *free-ion* 4d functions may not be the most appropriate functions for a *molecular* calculation.[11] Specifically, one notes in Table I of ref 2 that the free-ion 4d functions are so diffuse that the $4d\sigma$–$2p\sigma$ overlap is small and *negative*, while the $4d\pi$–$2p\pi$ overlap is very large.

The inability of our previous method to incorporate the ligand s orbitals imposes some limitations on its usefulness. For example, in dealing with NH_3 or CO ligands the inclusion of the σ-bonding ligand 2s orbitals is obviously desirable. Consequently, in the present work we decided not to restrict ourselves completely to the wave functions of the free ion at self-consistent charge. Of course, this raises the question as to how the wave functions should be chosen. Since this is a semiempirical calculation, there is no guarantee that the set which yields the lowest total energy, even if it could be easily calculated, is the proper choice. Thus, it was decided to formulate the wave functions in a systematic way on the basis of two considerations, both designed to maximize the contributions of a wave function to an occupied orbital.

Two factors influence the degree of participation of a metal wave function in a molecular orbital—the energy of the orbital relative to the ligand orbital energy and the overlap of the metal wave function with the ligand function. For illustrative purposes, assume an electron population analysis resulted in a metal configuration of $3d^5 4s^1 4p^1$. (Such a configuration might be apropos to the iron in $FeCl_6^{3-}$ with a self-consistent charge of $+1.0$.) Because of the high occupancy of the 3d orbitals, alteration of the 3d radial wave functions markedly changes the 3d orbital energy. Examination of calculated orbital energies shows that the lowest orbital energy is achieved when the wave function corresponds to that of the free ion, ignoring the outer-orbital 4s and 4p contributions. Attempts to vary the wave function in order to increase overlap with ligand orbitals result in rapidly increasing orbital energies and decreasing participation of the metal orbital in the bonding orbitals of the complex. Thus, as in previous work,[2] the chosen 3d functions were made to correspond to the effective charge on the metal. In general, as will be shown, the effective charge for the metal 3d orbitals was approximately $+2.0$.

Similarly, for the chloride complexes studied in this work the effective ligand charges are reasonably close to -1.0 so that chloride functions were used. For computational convenience, minimum basis set chloride functions were constructed such that they would have maximum overlaps with the many-term functions of Clementi, *et al.*[12,13] The final forms of these and the metal basis functions have been filed with the American Documentation Institute.[14]

Because of their lower occupancy, alteration of the radial functions for the 4s and 4p metal orbitals does not result in drastic changes in the calculated orbital energies but does cause substantial changes in the overlap integrals. Consequently, it was decided that the metal 4s, 4p, and 4d wave functions would be constructed such that they would have a maximum overlap interaction with the ligand p orbitals. Since the 4p and 4d functions are capable of both σ and π bonding, a compromise between the two was effected as indicated below.

Construction of the 4s, 4p, and 4d Functions.—The form of the outer 4s and 4p metal orbitals used in this work is similar to that given by Richardson, *et al.*,[15,16] except that Richardson's criterion was minimum free-ion energy whereas ours is maximum overlap with the ligand p functions. The metal functions are constructed from the smallest possible number of Slater-type orbitals which can be used for orthonormal functions. Thus

$$\psi(4s) = a_1\phi_{1s}(\alpha_1) + a_2\phi_{2s}(\alpha_2) + a_3\phi_{3s}(\alpha_3) + a_4\phi_{4s}(\alpha_4) \quad (7)$$

where $\phi_{ns}(\alpha_n) = N_n r^{n-1} e^{-\alpha_n r}$. The orbital exponents α_1, α_2, and α_3 are the same as those used for Richardson's[14] $\psi(1s)$, $\psi(2s)$, and $\psi(3s)$ core functions of the metal. For a given value of α_4, the coefficients a_1 through a_4 in $\psi(4s)$ are completely specified by the orthogonality of the function to the core functions and the normalization requirement. By variation of α_4, that $\psi(4s)$ function can be found such that $S(\psi_M(4s), \psi_{Cl}(3p\sigma))$ is a maximum. Where known, the internuclear distances used in the overlap calculations corresponded to the experimental distances for the complexes.[17] In those cases where the distances in the complexes have not been determined, the values for corresponding species were used.[18]

In a similar way, the 4p functions were constructed from

$$\psi(4p) = b_1\phi_{2p}(\beta_1) + b_2\phi_{3p}(\beta_2) + b_3\phi_{4p}(\beta_3) \quad (8)$$

Again, β_1 and β_2 are the same orbital exponents as the 2p and 3p core functions[15] and the coefficients b_1 to b_3

(11) See, for example, the Hartree–Fock calculations on HF by E. Clementi, *J. Chem. Phys.*, **36**, 33 (1962).

(12) E. Clementi, A. D. McLean, D. L. Raimondi, and M. Yoshimine, *Phys. Rev.*, **133**, A1274 (1964).

(13) E. Clementi, IBM Research Paper, RJ-256, 1963.

(14) Material supplementary to this article has been deposited as Document No. 9839 with the ADI Auxiliary Publications Project, Photoduplication Service, Library of Congress, Washington 25, D. C. A copy may be secured by citing the document number and by remitting \$2.50 for photoprints, or \$1.75 for 35-mm microfilm. Advance payment is required. Make checks or money orders payable to: Chief, Photoduplication Service, Library of Congress.

(15) J. W. Richardson, W. C. Nieuwpoort, R. R. Powell, and W. F. Edgell, *J. Chem. Phys.*, **36**, 1057 (1962).

(16) J. W. Richardson, R. R. Powell, and W. C. Nieuwpoort, *ibid.*, **38**, 796 (1963).

(17) (a) $TiCl_4$: M. Kimura, K. Kimura, M. Aoki, and S. Shibata, *Bull. Chem. Soc. Japan*, **29**, 95 (1956); (b) VCl_4: W. N. Lipscomb and A. G. Whittaker, *J. Am. Chem. Soc.*, **67**, 2019 (1945); (c) $FeCl_4^-$: B. Zaslow and R. E. Rundle, *J. Phys. Chem.*, **61**, 490 (1957); (d) $CoCl_4^{2-}$: B. N. Figgis, M. Gerlock, and R. Mason, *Acta Cryst.*, **17**, 506 (1964); (e) $NiCl_4^{2-}$: P. Pauling, Ph.D. Thesis, University College, London, 1960.

(18) (a) $MnCl_4^{2-}$ and $FeCl_4^{2-}$: see ref 6; (b) $TiCl_6^{3-}$ and VCl_6^{3-} from $TiCl_3$ and VCl_3: W. Klemm and E. Krose, *Z. Anorg. Chem.*, **253**, 218 (1947); (c) $CrCl_6^{3-}$ from $CrCl_3$: B. Morosin and A. Narath, *J. Chem. Phys.*, **40**, 1958 (1964); (d) $FeCl_6^{3-}$ from $FeCl_3$: N. Wooster, *Z. Krist.*, **83**, 35 (1932).

are fixed for a given value of β_3 by the orthonormality requirements. However, here β_3 will not be the same for maximum σ and π overlap. Since both of these interactions are important, it was decided to use an average β value, *i.e.*

$$\beta_3 = [\beta_3(\sigma) + \beta_3(\pi)]/2$$

where $\beta_3(\sigma)$ and $\beta_3(\pi)$ are the corresponding values for maximum σ and π overlaps.

The 4d wave functions were made orthogonal to the previously mentioned 3d functions. Since these latter functions are the sum of two Slater functions, *i.e.*

$$\psi(3d) = C_1\phi_{3d}(\gamma_1) + C_2\phi_{3d}(\gamma_2) \quad (9)$$

the 4d functions are of the form

$$\psi(4d) = C_3\psi(3d) + C_4\phi_{4d}(\gamma_3) \quad (10)$$

As in the case of the 4p functions, the value of γ_3 is the average of the ones required for maximum σ and π overlap

$$\gamma_3 = [\gamma_3(\sigma) + \gamma_3(\pi)]/2 \quad (11)$$

The general effect of using the average for β_3 and γ_3 is to contract the 4p and 4d wave functions relative to Richardson's values for the free ions. The final forms of all of the metal wave functions used for the complexes in this work have been deposited with the ADI.[14]

It should be noted that since the 4d wave function is orthogonal to the 3d, there is no cross-term between them in the G_{ij} matrix, *i.e.*, $S(\psi_{3d}, \psi_{4d}) = 0$. However, a term does appear in the H_{ij} matrix. Even though, in the approximations used, $(\psi(3d)|-{}^1/_2\Delta + V_M|\psi(4d)) = 0$, the interactions with the ligand centers must be evaluated. That is, $(\psi(3d)|\sum_{j=1} V_j|\psi(4d)) \neq 0$.

Effect of Basis Set Size.—Table II contains a comparison between two sets of energy levels calculated for $CrCl_6^{3-}$. Those in column 2 were computed using the metal 3d, 4s, and 4p and the chloride 3s and 3p orbitals. Those in column 3 were obtained using the same orbitals plus the specially constructed 4d orbital. The results are striking and their significance is apparent.

Note that the self-consistent charge and orbital coefficients in the two sets of calculations are quite similar. Indeed the only substantial differences between the two calculations occur in the relative positions of the $2t_{2g}$, $3e_g$, $3t_{2g}$, and $4e_g$ orbitals. However, the separation between the first two of these is $\Delta = 10Dq$, which is generally used as a major criterion for judging the worth of a calculational procedure! Note also the drastic alteration in the values of the $3t_{2g}$ and $4e_g$ levels. This reaffirms our earlier contention[2] that the value of the highest level for a given representation is subject to substantial variation. In the absence of the 4d functions, the highest levels for the t_{2g} and e_g representations are those associated with Δ. As previously mentioned, the sensitivity of these levels required that ligand s orbitals also be omitted from the calculational basis set. With the introduction of the appropriate 4d functions, such an omission is no longer necessary.

TABLE II

EFFECT OF 4d ORBITALS ON $CrCl_6^{3-}$ ENERGY LEVELS[a]

	Without 4d orbital interaction	With 4d orbital interaction
$4e_g(4d)$	22.21[b]	91.32
$3t_{2g}(4d)$	19.48[b]	49.58
$3e_g$	7.90	5.30
$2t_{2g}$	4.62	3.91
t_{2u}	−0.97	−1.00
t_{1g}	−0.97	−1.00
$3t_{1u}$	−1.19	−1.24
$1t_{2g}$	−1.59	−1.76
$2t_{1u}$	−2.28	−2.35
$2a_{1g}$	−3.63	−3.74
$2e_g$	−3.72	−4.20
$1e_g$	−17.45	−17.78
$1t_{1u}$	−18.06	−18.18
$1a_{1g}$	−18.46	−18.60
Charge	+1.75	+1.80
Δ[c]	3.28	1.39
3d coeff $2t_{2g}$	0.96	0.93
4d coeff $2t_{2g}$	0.00	−0.09

[a] All energies are in electron volts. [b] The 4d orbital energy without interaction. [c] The experimental Δ value is 1.57 ev.

This ability to include the ligand s orbitals should be substantially more important when one deals with ligand systems where the bonding to the metal might incorporate appreciable s character.

It should be emphasized that the introduction of the 4d orbitals into the calculation does not necessarily mean that there is substantial contribution by these atomic orbitals to the $2t_{2g}$ and $3e_g$ molecular orbitals. As indicated by the orbital coefficients given in Table II, the 4d contribution to either molecular orbital does not exceed 5%.

Electron Distribution.—As in our previous work, the charge on the metal ion was required to be self-consistent with that obtained *via* the Mulliken electron population analysis[4] of the occupied molecular orbitals. Two sets of calculations were performed. In the first, method A, only the participation of the metal 3d orbitals in the molecular orbitals was used to accumulate electron density on the metal while in the second, method B, all of the orbitals in the metal basis set were included. As with our fluoride results, the calculations indicate that the two methods of electron distribution give very similar results and that for most purposes either technique can be employed as long as the same method is used for all of the complexes in a series. Some typical values are given in Table III. The results reemphasize the fact that the charge on the metal is primarily a technique to control the calculation and is not an appropriate measure of covalent character. The 3d orbital coefficients obtained by the two distributions are quite similar, despite the drastic differences in self-consistent charges. However, because the metal 3d and 4d functions belong to the same irreducible representations, it is perhaps less permissible to say that only the 3d coefficient determines the metal participation within a molecular orbital. Furthermore, when only the 3d population density is used, a dilemma arises as to the distribution of the metal charge density

TABLE III

VALUES AS A FUNCTION OF ELECTRON DISTRIBUTION: METHOD A, 3d ORBITALS ONLY; METHOD B, ALL METAL ORBITALS

	$TiCl_4$	$MnCl_4^{2-}$	$CoCl_4^{2-}$	$TiCl_6^{3-}$	$CrCl_6^{3-}$
Metal charge					
A	1.70	1.67	1.55	1.86	1.80
B	0.33	0.20	0.09	0.60	0.44
Δ value[a]					
A	1.10	0.52	0.50	1.68	1.39
B	1.02	0.45	0.48	1.65	1.47
e coefficient[b]					
A	0.91	0.98	0.93	0.94	0.92
B	0.94	0.99	0.98	0.94	0.93
t_2 coefficient[b]					
A	0.90	0.97	0.93	0.96	0.93
B	0.92	0.98	0.95	0.98	0.96

[a] All Δ values are given in electron volts. [b] The e and t_2 coefficients refer to the coefficients of the 3d orbitals in those antibonding molecular orbitals generally associated with the metal orbitals.

associated with those higher metal orbitals not included in the population analysis. The division of this charge density between the ligand s and p orbitals is not uniquely specified and the manner in which one distributes the charge affects both the ligand electron configuration and possible charge asymmetry on the ligands. This latter consideration is important when one attempts to correlate ligand orbital charge densities with nqr data. The usual population analysis involving all orbitals avoids this problem. Consequently, it was decided to control the calculations by means of the self-consistent charge as determined by the Mulliken distribution involving all of the metal orbitals.

Ligand–Ligand Interaction.—In octahedral symmetry, the ligand π-bonding orbitals are basis functions for the irreducible representations t_{1u}, t_{2g}, t_{1g}, and t_{2u}. The latter two representations are nonbonding insofar as interactions with the metal orbitals are concerned, while the t_{1u} and t_{2g} functions can interact with the metal p and d orbitals, respectively. Under such considerations, a qualitative order of the four sets of ligand π-energy levels could be $t_{1g} = t_{2u} > t_{1u} > t_{2g}$. However, recent experimental results[19] on the Faraday effect in $IrCl_6^{2-}$ suggest that in that complex the order might be $t_{1g} > t_{1u} > t_{2u} > t_{2g}$. Such an order can most easily be explained as a consequence of the effects of ligand–ligand interactions. Hence, it would be desirable to be able to incorporate such interactions properly into the calculations.

Inclusion of ligand–ligand interactions have a twofold effect on the matrix elements of the secular determinant. First, it removes the accidental degeneracy of certain diagonal matrix elements. For example, without ligand–ligand interaction the ligand π orbitals t_{1u}, t_{2u}, t_{1g}, and t_{2g} have the same energy levels prior to interaction with the metal orbitals. Under the influence of ligand–ligand interaction, the values of the diagonal elements are altered as indicated by the partial energy level diagram for VCl_6^{3-} in Figure 1. Second, certain off-diagonal matrix elements between ligand orbitals of the same representation which were zero prior to ligand–ligand interaction can now possess nonzero values. For example, the matrix element in octahedral symmetry between the σ- and π-bonding p orbitals belonging to the t_{1u} representation becomes $\mathcal{H}(\sigma, \pi)_{t_{1u}} = \sqrt{2}[\mathcal{H}(\sigma_1, \sigma_2) - \mathcal{H}(\pi_1, \pi_2)]N_\sigma N_\pi$, where the scripts 1 and 2 refer to adjacent ligand atoms and the N's are the normalization factors for the ligand symmetry adapted wave functions.

The evaluations of the resultant terms for both the diagonal and off-diagonal elements were carried out in a manner analogous to that for metal–ligand interactions. Our early results were greeted with enthusiasm. The ordering of the calculated energy levels for VCl_6^{3-} were in accord with that suggested by the experimental results on $IrCl_6^{2-}$. As indicated by Figure 1, the only levels appreciably altered were those associated with the occupied ligand π orbitals. Such features as the self-consistent charge and the coefficients in the molecular orbitals were substantially the same for the calculations with and without ligand–ligand interaction.

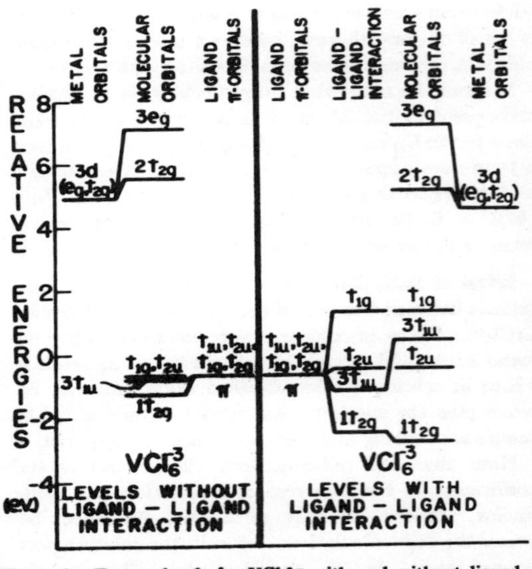

Figure 1.—Energy levels for VCl_6^{3-} with and without ligand–ligand interaction.

As additional computations were completed, our initial enthusiasm faded. One of the characteristic features of calculations carried out over a series of complexes MX_6^{3-}, where M has an increasing number of d electrons, is a slowly decreasing separation between the metal d orbitals and the ligand π orbitals prior to bonding. Inclusion of ligand–ligand interaction raised the filled nonbonding t_{1g} level closer and closer to the partially occupied antibonding $2t_{2g}$ level until, finally, in our calculation on $FeCl_6^{3-}$ the $2t_{2g}$ was *below* the t_{1g} level. This impossible situation has an obvious ex-

(19) P. N. Schatz, University of Virginia, Charlottesville, Va., private communication.

planation which becomes apparent on careful examination of the matrix elements. While the method qualitatively suggests the proper direction that the effect of ligand–ligand interaction should have, in its present technique of approximations it *overestimates* the importance of the interactions. For example, in VCl_6^{3-} the separation of the ligand diagonal terms for the t_{1g} and t_{2g} π orbitals *via* ligand–ligand interaction is calculated to be 3.78 ev, while the alteration of the metal 3d level as a result of σ-bonding interaction with the ligands is only 2.74 ev. It is extremely unlikely that ligand–ligand interactions should be more important than metal–ligand interactions in this respect.

The reason for the failure of this technique reasonably to assess ligand–ligand interaction lies in the approximational techniques required by the computational procedure. We have already asserted in connection with metal–ligand electrostatic interactions, eq 3, that point-charge approximations for charge densities on certain centers were reasonable only if the charge density was outside the region of overlap of the two wave functions.[20] In the evaluation of ligand–ligand interactions, terms of the sort $(\rho_{tl}|V_M|\rho_{tj})$ arise. As illustrated crudely in Figure 2, even though M and j are not on ligand center l and M $\neq$ j, it is still possible that the charge density V_M is in the region of overlap between ρ_{tl} and ρ_{tj}. In such cases, it is not reasonable to approximate the term by the integral $q_M(1/r_M|\rho_{tl}\rho_{tj})$ but it should be evaluated by the more complex relationship

$$(\rho_{tl}|V_M|\rho_{tj}) = \sum_k b_k\{2(\chi_{kM}\chi_{kM}|\rho_{tl}\rho_{tj}) - (\chi_{kM}\rho_{tl}|\chi_{kM}\rho_{tj})\} - Z_M(1/r_M|\rho_{tl}\rho_{tj}) \quad (12)$$

Equation 12 requires the evaluation of three-center two-electron electrostatic interaction integrals involving metal 3d and higher orbitals. At the present time, such computations are complex and time consuming. The necessity to sum such integrals over the metal orbitals of principal quantum numbers 3 and 4 makes their evaluation a prohibitive task. Consequently, we had no recourse but to eliminate the inclusion of ligand–ligand interaction from consideration in order to maintain the parameter-free character of the calculations. While this is an unfortunate limitation it does not seriously restrict the applicability of the method. As has already been mentioned, the inclusion of ligand–ligand interaction does not substantially alter the values of the orbital coefficients. The only energy levels affected by ligand–ligand interaction are those associated with the ligand π orbitals. This will affect the estimates of the energies associated with electronic transitions from these orbitals, *i.e.*, the ligand to metal charge-transfer transitions. Since such transitions involve a substantial alteration of charge distribution between the ground and excited states, it is uncertain that energy values based on one-electron separations in the ground state would achieve a high degree of correspondence

(20) For example, $(\rho_{il}|V_j|\chi_i)$ can be approximated by $q_j(1/r_j|\rho_{il}\chi_i)$ if $j \neq 1$ but requires the use of either eq 4 or 5 if $j = 1$.

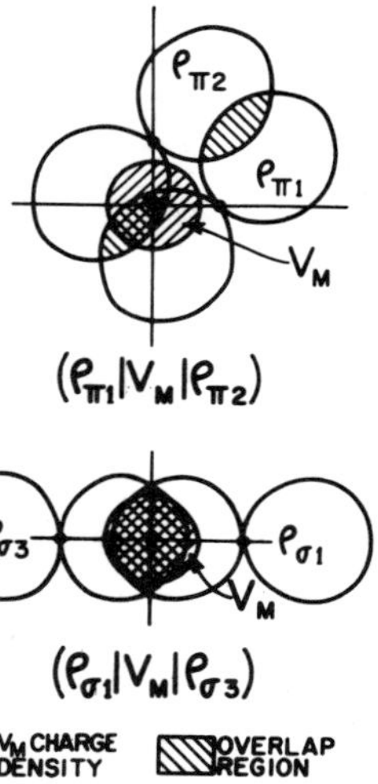

Figure 2.—Overlapping charge densities in ligand–ligand interactions.

with experimental values in any event. Even when ligand–ligand interaction is excluded, qualitative correlations between observed charge-transfer bands and the ground-state energy level separations can be made, as long as exact agreement is not expected.

Results and Discussion

Eigenvalues.—The method outlined in the previous sections was applied to four octahedral and seven tetrahedral chloride complexes of first transition row metals. The coefficients associated with the eigenfunctions have been deposited with the ADI.[14] The resultant eigenvalues along with the calculated and experimental values of $10Dq$ are given in Tables IV and V. The agreement between the $10Dq$ values is extremely gratifying particularly when one recalls that the method is devoid of arbitrary parameters which can be adjusted. Since the estimates of $10Dq$ are based upon approximate calculational methods of the ground-state energies, whereas a rigorous calculation would involve the difference between ground and excited states, it would be misleading to overemphasize the significance of the correlation between the t_2 and e energy levels and the $10Dq$ values. However, it does seem that, if appropriate basis sets for the ligand and metal atoms yield a good correlation for one complex, calculations with the same size basis sets properly reflect the trends in $10Dq$ values for similar complexes even when the symmetry is altered. Furthermore, in the halide complexes studied here, the t_2 and e orbitals involved in the $10Dq$ estimate are quite similar in metal character. Hence, one would not expect a substantial change in charge distribution between the ground and excited states nor a substantial alteration of electron repulsion integrals. Thus, estimates of $10Dq$ from the ground-state energy level separations in these cases are probably more realistic than they would be in

TABLE IV

EIGENVALUES FOR OCTAHEDRAL COMPLEXES[a]

	$TiCl_6^{3-}$	VCl_6^{3-}	$CrCl_6^{3-}$	$FeCl_6^{3-}$
	Distance,[b] Å			
	2.47	2.45	2.34	2.38
	Charge[c]			
	0.60	0.49	0.44	0.14
$1a_{1g}$	−17.47	−17.40	−17.23	−17.17
$1t_{1u}$	−17.40	−17.18	−16.87	−16.81
$1e_g$	−17.16	−17.02	−16.61	−16.60
$2e_g$	−2.75	−2.67	−2.44	−2.38
$2a_{1g}$	−2.33	−2.29	−1.87	−1.92
$2t_{1u}$	−1.40	−1.34	−0.87	−0.82
$1t_{2g}$	−1.20	−1.13	−0.82	−0.90
$3t_{1u}$	−0.91	−0.77	−0.34	−0.31
$1t_{2u}$	−0.86	−0.75	−0.32	−0.34
$1t_{1g}$	−0.86	−0.75	−0.32	−0.34
$2t_{2g}$	+7.00	+5.65	+5.62	+2.69
$3e_g$	+8.66	+7.14	+7.09	+3.94
$3t_{2g}$	+34.99	+35.62	+37.64	+38.99
$4t_{1u}$	+60.70	+54.78	+73.46	+65.72
$4e_g$	+87.66	+88.62	+90.46	+93.27
$3a_{1g}$	+109.60	+127.16	+370.52	+248.05
$10Dq$(calcd)	1.66	1.49	1.47	1.25
$10Dq$(exptl)	1.71	1.54	1.57	1.14

[a] All energies are given in electron volts. [b] Distance refers to the experimental metal–ligand distance used in the calculations. [c] Charge refers to the metal charge at self-consistency *via* the Mulliken population analysis.

Crudely, one might say that transitions from the ligand orbitals to the empty metal 3d orbitals are possible but transitions to the metal 4p orbitals are not.

The fact that in certain of the charged complexes the occupied orbitals are above zero does not mean that such complexes are unstable or the results are erroneous since it must be remembered that the basic calculation did not take into account the potential well caused by the cations. Thus, the observed correspondence between the charge on the species and the absolute values of the energy levels is to be expected. Note that if −3, −9, and −12 ev were added to the energy levels of the complexes with −1, −2, and −3 charges, respectively, not only would all occupied orbitals be below zero but the absolute values of the low-lying oribtals would be practically identical. The energies −3, −9, and −12 ev are those required to make the $1a_1$ and $1t_2$ energies of the −1, −2, and −3 tetrahedral species essentially equal to those of the neutral molecule.

Covalency.—In a previous paper,[3] we outlined a method for the calculation of the Racah electrostatic interaction parameter, B(complex), and the nephelauxetic β value from the 3d orbital coefficients in the

TABLE V

EIGENVALUES FOR TETRAHEDRAL COMPLEXES[a]

	$TiCl_4$	VCl_4	$FeCl_4^-$	$MnCl_4^{2-}$	$FeCl_4^{2-}$	$CoCl_4^{2-}$	$NiCl_4^{2-}$
	Distance,[b] Å						
	2.185	2.03	2.19	2.33	2.27	2.25	2.27
	Charge[c]						
	0.33	0.23	0.34	0.20	0.24	0.09	0.08
$1a_1$	−30.02	−30.27	−27.32	−21.37	−21.05	−21.14	−21.18
$1t_2$	−30.20	−30.35	−26.97	−21.23	−20.88	−20.89	−20.92
$2a_1$	−15.90	−16.02	−12.35	−6.26	−6.05	−6.16	−6.31
$2t_2$	−16.42	−16.65	−12.64	−5.88	−5.95	−5.89	−5.99
$1e$	−14.62	−14.96	−11.07	−4.78	−4.73	−4.69	−4.77
$3t_2$	−14.34	−14.51	−10.66	−4.86	−4.76	−4.69	−4.75
$1t_1$	−13.65	−13.69	−10.30	−4.67	−4.54	−4.47	−4.51
$2e$	−7.01	−6.85	−7.05	+3.07	+1.32	+0.21	−1.32
$4t_2$	−5.99	−5.89	−6.42	+3.52	+1.84	+0.68	−0.88
$5t_2$	+10.27	+13.59	+16.92	+20.40	+21.74	+21.55	+20.80
$3e$	+16.80	+20.57	+21.43	+25.22	+26.66	+26.98	+26.11
$6t_2$	+66.39	+77.04	+66.88	+65.95	+65.64	+66.73	+65.49
$3a_1$	+38.54	+64.60	+92.23	+62.03	+68.25	+76.40	+71.26
$10Dq$(calcd)	1.02	0.96	0.63	0.45	0.52	0.48	0.44
$10Dq$(exptl)	0.93	1.12	0.62	0.45	0.50	0.41	0.44

[a] All energies are in electron volts. [b] Distance refers to the experimental metal–ligand distance used in the calculations. [c] Charge refers to the metal charge at self-consistency *via* the Mulliken population analysis.

systems, such as carbonyl and cyanide complexes, where one would expect substantial differences in metal character between the two orbitals.

It will be noted that the absolute values of the energy levels fall into distinct regions depending upon the charge on the complex. The neutral species $TiCl_4$ and VCl_4 have all their occupied orbitals below zero. Of the unoccupied orbitals in these two complexes only the 2e and $4t_2$ levels are less than zero which suggests (but certainly does not prove) that transitions from filled orbitals such as $1t_1$, $3t_2$, etc., to these orbitals can occur but that ionization of an electron would result rather than transitions to the $5t_2$ and higher orbitals.

molecular orbitals and the appropriate 3d wave functions. It might be noted here that in the estimate of B and β a reduction results as a consequence of both the coefficients in the molecular orbitals and the wave functions used in the *atomic* basis set. This double correction is necessary for calculational consistency. As we have previously outlined,[3] the 3d wave function should reflect the effective charge seen by the 3d electron as determined by self-consistent charge calculation which is used to control the computational procedure. If a different wave function had been employed, *e.g.*, that of the formal charge on the free ion, it would affect the molecular orbital coefficients and they would not correspond to those obtained in this self-consistent calculation. It should be kept in mind that the method outlined for estimation of the B values is only an approximate one and that in principle the full molecular orbitals should be employed. The 3d metal wave functions corresponding to lower than the formal charge were used in the estimation of the B values because they were the atomic orbitals used in the molecular orbital calculation.

Application of the method to the 2e and $4t_2$ orbitals in tetrahedral symmetry and the $2t_{2g}$ and $3e_g$ orbitals in octahedral symmetry yields the calculated values given in Table VI. The results are in good accord with the experimental values. Since the β values can be used as a measure of the covalency of the specified orbitals, several features are noteworthy: (1) For a given formal oxidation state of the metal there is an increasing orbital covalency as the atomic numbers of the elements increase. This is illustrated by both the tetrahedral metal(II) complexes and the octahedral metal(III) complexes. (2) As shown by the isoelectronic pair $FeCl_4^-$ and $MnCl_4^{2-}$, the increased formal charge on the metal substantially increases the orbital covalency; that is, β is much smaller. (3) The calculated orbital covalencies of the two iron(III) complexes, $FeCl_4^-$ and $FeCl_6^{3-}$, are essentially identical.

TABLE VI

NEPHELAUXETIC EFFECT. B AND β VALUES

	$FeCl_4^-$	$MnCl_4^{2-}$	$CoCl_4^{2-}$	$NiCl_4^{2-}$	VCl_6^{3-}	$CrCl_6^{3-}$	$FeCl_6^{3-}$
B(calcd)	625[a]	830	880	875	619	649	640
B(exptl)	590	770	720	765	563	561	655
β(calcd)	0.58	0.95	0.88	0.82	0.70	0.66	0.59
β(exptl)	0.58	0.86	0.75	0.72	0.65	0.62	0.65

[a] All B values are in units of cm^{-1}.

Other commonly used measures of covalent character are bond orders and overlap populations. Coulson's[21] original definition of bond order concerned the π character between two orbitals on adjacent atoms with the simplifying assumption that the overlap integral could be neglected in computing the molecular orbital coefficients. With this assumption, two electrons in the bonding or antibonding orbital of a diatomic homopolar molecule have bond orders of +1.0 and −1.0, respectively, so the net bond order due to full occupation of both orbitals is zero. In adapting bond orders to include overlap considerations, Mulliken[22] defined the bond order in such a way that the upper limit is still +1.0. However, partial or complete occupation of the antibonding orbital can result in a negative value for the bond order, the limit being $-(1 + S)/(1 - S)$, where S is the overlap integral.

In dealing with the symmetry-adapted functions in the present work, it is convenient to specify a given molecular orbital belonging to the irreducible representation, a, by

$$\psi_{ia} = \sum_j C_{ij}\chi_j(a) + \sum_q C_{iq}\phi_q(a) \qquad (13)$$

(21) C. A. Coulson, *Proc. Roy. Soc.* (London), **A169**, 419 (1939).
(22) R. S. Mulliken, *J. Chem. Phys.*, **23**, 1841 (1955).

where C_{ij} and C_{iq} are the coefficients for the normalized symmetry-adapted wave functions on the metal, $\chi_j(a)$, and ligands, $\phi_q(a)$, respectively. Then the orbital bond order becomes

$$P_{ia}(jq) = N(ia)C_{ij}C_{iq}(1 + G_{jq}) \qquad (14)$$

where $N(ia)$ is the number of the electrons in the ith orbital and $G(jq)$ is the group overlap integral of $\chi_j(a)$ with $\phi_j(a)$. The total bond order is given by

$$P_a(jq) = \sum_i P_{ia}(jq)/D_a \qquad (15)$$

where D_a is the orbital degeneracy of the irreducible representation, a. Division by the orbital degeneracy is necessary to keep an upper limit of +1.0 for a fully covalent bond. Since this is an adaptation of Mulliken's definition, it is also possible that occupation of the antibonding levels can result in bond orders which are negative. A partial tabulation of the calculated bond orders for the eleven complexes is given in Table VII. The complete list of bond orders has been deposited with the ADI since the most informative trends concerning covalency can be obtained by examination of the bond orders involving the metal 3d orbitals. Several features are significant.

TABLE VII

BOND ORDERS

	3d–3s	3dσ–pσ	t_2 3dπ–pπ	e 3dπ–pπ
$TiCl_6^{3-}$	0.050	0.594	0.351	...
VCl_6^{3-}	0.036	0.638	0.283	...
$CrCl_6^{3-}$	0.046	0.664	0.230	...
$FeCl_6^{3-}$	−0.001	0.334	0.316	...
$TiCl_4$	0.052	0.544	0.434	0.716
VCl_4	0.060	0.546	0.484	0.540
$FeCl_4^-$	−0.018	0.268	0.181	0.346
$MnCl_4^{2-}$	0.006	0.130	0.034	0.062
$FeCl_4^{2-}$	0.012	0.167	0.069	−0.005
$CoCl_4^{2-}$	0.006	0.193	0.109	−0.114
$NiCl_4^{2-}$	0.008	0.132	0.056	−0.092

(1) The negligible values for the 3d–3s bond orders for all the complexes are indicative of the relative unimportance of the low-lying 3s chlorine orbital.

(2) A clear distinction must be made between covalency of *orbitals* as measured by the nephelauxetic β values and covalency of *bonds* as measured by the bond order. We have previously noted the increase of orbital covalency with increasing atomic number. However, whether the bond order increases or decreases depends upon the change, if any, in occupation of the antibonding orbitals in the sequence of compounds. Thus for the four octahedral chlorides, the 3dσ–3pσ bond order increases from Ti through Cr, reflecting the increasing orbital covalency, since the $3e_g$ orbital remains unoccupied. The 3dσ–3pσ bond order of $FeCl_6^{3-}$ shows a substantial decrease because of the two electrons in the $3e_g$ orbital. Similarly the 3dπ–3pπ bond order decreases from Ti through Cr because of the increased occupation of the $2t_{2g}$ (antibonding) orbital. From $CrCl_6^{3-}$ to $FeCl_6^{3-}$ there is no change in the $2t_{2g}$

orbital occupation so the increase in bond order reflects the increased orbital covalency.

(3) For the isoelectronic pair of complexes $FeCl_4^-$ and $MnCl_4^{2-}$ the bond orders reflect the increased covalency as a function of increased formal oxidation state. This is analogous to the conclusion reached on the basis of the β values.

(4) The negative values for the tetrahedral e $3d\pi$–$3p\pi$ bond orders are indicative of the nonbonding character of these interactions because of the partial or full occupation of the antibonding 2e orbitals.

Rather than using bond orders, Mulliken[22] expresses a strong preference for overlap populations as a measure of bond covalency. In terms of the wave function formulation given by eq 13, the overlap populations are given by

$$n(a, j, q) = 2\sum_i N(ia)C_{ij}C_{iq}G(jq) \qquad (16)$$

where $n(a, j, q)$ is the overlap population due to the interaction of $\chi_j(a)$ and $\phi_q(a)$. Once obtained, these overlap "densities" may be summed together, *e.g.*, σ density with π density, and/or divided between the number of metal to ligand bonds in thec omplex. The calculated overlap populations are given in Table VIII. Only the metal 3d orbital interactions are listed for two reasons: it is in these populations that the greatest variations occur, and, even more important, the higher metal orbitals are so diffuse that the resultant densities are not between the metal and ligand nuclei but are actually concentrated on the ligand centers.[23] In general, the covalency characteristics suggested by the overlap populations are analogous to those obtained from the bond order results.

One thought-provoking correlation exists between the overlap populations and the infrared stretching frequencies[24] for the metal–chlorine bonds. The relationship is illustrated by the results in Table IX. The stretching frequencies for a particular oxidation state are fairly constant, as are the overlap populations, but between oxidation states the populations and frequencies differ substantially. While a qualitative correlation exists, there does not appear to be a quantitative relation between overlap population and the stretching frequencies. This suggests that other factors such as different charge distributions on the atoms contribute to the stretching frequencies.

Correlation with Nqr Results.—Recently Cotton and Harris[25] proposed a method for the calculation of quadrupole coupling constants and frequencies from the LCAO–MO eigenvectors. Using their relationship we have calculated the frequencies for various complexes. The results are given in Table X. In general the agreement is quite good particularly when one considers the approximations in both calculational models. It should also be kept in mind that the nqr data for the solids were obtained on compounds in which the chlorines were bridging the metal atoms, and, since nqr results are a measure of charge asymmetry, the values for MCl_3 and MCl_6^{3-} need not be identical.

TABLE VIII

TOTAL OVERLAP POPULATIONS

			t_2	e				
	$3d\sigma$–3s	$3d\sigma$–$p\sigma$	$3d\pi$–$p\pi$	$3d\pi$–$p\pi$	$3d\sigma$–lig	$3d\pi$–lig	Total	Per bond
$TiCl_6^{3-}$	0.03	0.40	0.29	...	0.43	0.29	0.72	0.12
VCl_6^{3-}	0.02	0.38	0.22	...	0.40	0.22	0.62	0.10
$CrCl_6^{3-}$	0.03	0.38	0.16	...	0.41	0.16	0.57	0.09
$FeCl_6^{3-}$	0.00	0.15	0.15	...	0.15	0.15	0.30	0.05
$TiCl_4$	0.04	0.49	0.24	0.43	0.53	0.67	1.20	0.30
VCl_4	0.04	0.47	0.29	0.34	0.51	0.63	1.14	0.28
$FeCl_4^-$	−0.01	0.15	0.06	0.12	0.14	0.18	0.32	0.08
$MnCl_4^{2-}$	0.01	0.07	0.01	0.02	0.08	0.03	0.11	0.03
$FeCl_4^{2-}$	0.01	0.08	0.02	0.00	0.09	0.02	0.11	0.03
$CoCl_4^{2-}$	0.01	0.09	0.03	−0.03	0.10	0.00	0.10	0.02
$NiCl_4^{2-}$	0.01	0.06	0.01	−0.02	0.07	−0.01	0.06	0.01

TABLE IX

CORRELATION OF OVERLAP POPULATIONS WITH METAL–CHLORINE STRETCHING FREQUENCIES

Complex	Total overlap population	Exptl str freq, cm^{-1}
$TiCl_4$	1.20	490
VCl_4	1.14	482
$FeCl_4^-$	0.32	378
$MnCl_4^{2-}$	0.11	285
$FeCl_4^{2-}$	0.11	282
$CoCl_4^{2-}$	0.10	310
$NiCl_4^{2-}$	0.06	285

TABLE X

NQR FREQUENCIES FOR CHLORINE LIGANDS

	Mc/sec		
Complex	Freq, calcd	Freq, exptl	Exptl compd
$TiCl_4$	8.23	5.98	$TiCl_4$
$TiCl_6^{3-}$	7.40	7.39	$TiCl_3$(s)
VCl_6^{3-}	9.05	9.40	VCl_3(s)
$CrCl_6^{3-}$	9.32	12.81	$CrCl_3$(s)
$FeCl_6^{3-}$	7.95	10.02	$FeCl_3$(s)
Cl atom	...	54.8	Cl

Acknowledgment.—The authors wish to thank the National Science Foundation (Grant GP-3413) and the Wisconsin Alumni Research Foundation for support of this work.

(23) See Figure 3 of ref 2 for radial density curves of the diffuse metal orbitals.

(24) D. M. Adams, J. Chatt, J. M. Davidson, and J. Gerratt, *J. Chem. Soc.*, 2189 (1963).

(25) F. A. Cotton and C. B. Harris, *Proc. Natl. Acad. Sci. U. S.*, **56**, 12 (1966).

 Reprinted from *The Journal of the American Chemical Society* **90,** 4272 (1968)

Selection Rules for the Isomerization and Substitution Reactions of Transition Metal Complexes

D. R. Eaton[1]

Contribution No. 1408 from the Central Research Department, Experimental Station, E. I. du Pont de Nemours and Company, Wilmington, Delaware.
Received January 15, 1968

Abstract: The isomerization and bimolecular substitution reactions of four-coordinated transition metal complexes are considered from a point of view very similar to that previously adopted by Woodward and Hoffmann to treat a number of organic reactions. A combination of symmetry considerations and crystal field theory allows the classification of the reactions as either thermally allowed or forbidden. A method is suggested for classifying reactions for which there is insufficient symmetry to allow the above sharp distinction to be drawn.

Woodward and Hoffmann have developed the concept of symmetry-based selection rules for chemical reactions. They have applied their approach to electrocyclic reactions,[2a] sigmatropic reactions,[2b,3] and concerted cycloaddition reactions.[4,5] It is the purpose of the present work to extend the application of such selection rules to some reactions of transition metal complexes and to attempt some generalization of the concept.

Mango and Schachtschneider[6] have considered the participation of transition metals in concerted cycloaddition reactions using the Woodward–Hoffmann scheme. In the present case a slightly different approach will be adopted.

For clarity we will consider first an example from the work of Woodward and Hoffmann,[4] namely the addition of two molecules of ethylene to give cyclobutane. In essence their method is to draw a molecular-orbital energy-level diagram for the reactants (two molecules of ethylene) and then a molecular-orbital level diagram for the product (cyclobutane), and finally to correlate the individual initial molecular orbitals with the individual final molecular orbitals using symmetry restrictions imposed by an assumed geometry for the transition state. In the present case there are four molecular orbitals, each belonging to a different symmetry class in the relevant C_{2v} point group, and the correlation connects a filled bonding orbital in the initial state with an empty antibonding orbital in the final state. The reaction is therefore described as thermally disallowed but photochemically allowed. As a first step we wish to rephrase this description.

Consider a basis set of atomic orbitals, in this case four carbon p orbitals; apply first a Hamiltonian H_A to obtain the molecular orbitals of the initial state and then a Hamiltonian H_B to obtain the molecular orbitals of the final state, and finally correlate the orbitals according to the symmetry common to H_A and H_B. In this case H_A and H_B are simple Hückel Hamiltonians differing in that H_A has off-diagonal elements corresponding to two ethylene π bonds and H_B has off-diagonal elements corresponding to two cyclobutane σ bonds. For this problem the symmetry conserved corresponds to the geometry of the transition state, but for the more general case this is not necessarily so.

Consider now an analogous problem in which the basis set comprises the five d orbitals of a transition metal atom and H_A and H_B are crystal-field Hamiltonians corresponding to different arrangements of ligands around the central metal atom. If certain elements of symmetry are conserved as the ligands are rearranged from the situation corresponding to H_A to that corresponding to H_B, the eigenfunctions of these operators can be classified accordingly and allowed and disallowed reactions defined by the Woodward–Hoffmann criterion. Specifically let us consider the four-coordinate rearrangements illustrated in Figure 1. The reaction a → b or c → d corresponds to a tetrahedral → square-planar isomerization; a → c racemizes a tetrahedral complex and b → d accomplishes the *cis–trans* isomerization of a square-planar complex. The coordinate system is that shown in the diagram, and the crystal-field operators corresponding to these four arrangements of ligands must be defined in terms of this coordinate system. The conventional d orbitals d_{xy}, d_{xz}, d_{yz}, $d_{x^2-y^2}$, and d_{z^2} are eigenfunctions of the problem for a, c, and d but not for b. Note that in this coordinate system in structure d, d_{xy} not $d_{x^2-y^2}$ is the orbital pointing toward the ligands. For d^8 complexes the energy-level diagrams of Figure 1 show immediately that the square-planar to tetrahedral and tetrahedral racemization reactions are thermally allowed. In agreement with this in several cases where equilibria between square-planar and tetrahedral Ni(II) complexes have been observed, the lifetimes of the individual isomers have been shown to be less than 10^{-5} sec.[7,8] Ernst, O'Connor, and Holm have recently demonstrated that the racemization of optically active tetrahedral Ni(II) complexes is also very fast.[9] The *cis* → *trans* square-planar problem is a little less trivial and is illustrated in Figure 2. It is necessary to derive the eigenfunctions of an octahedral crystal-field potential corresponding to the axis of quantization being

(1) Address inquiries to the Department of Chemistry, McMaster University, Hamilton, Ontario, Canada.
(2) (a) R. B. Woodward and R. Hoffmann, *J. Am. Chem. Soc.*, **87,** 395 (1965); (b) R. B. Woodward and R. Hoffmann, *ibid.*, **87,** 2511 (1965).
(3) R. Hoffmann and R. B. Woodward, *ibid.*, **87,** 4389 (1965).
(4) R. Hoffmann and R. B. Woodward, *ibid.*, **87,** 2046 (1965).
(5) R. Hoffmann and R. B. Woodward, *ibid.*, **87,** 4388 (1965).
(6) F. D. Mango and J. H. Schachtschneider, *ibid.*, **89,** 2484 (1967).
(7) D. R. Eaton, W. D. Phillips, and D. J. Caldwell, *ibid.*, **85,** 397 (1963).
(8) R. H. Holm, A. Chakravorty, and G. O. Dudek, *ibid.*, **86,** 379 (1964).
(9) R. E. Ernst, M. J. O'Connor, and R. H. Holm, *ibid.*, **89,** 6104 (1967).

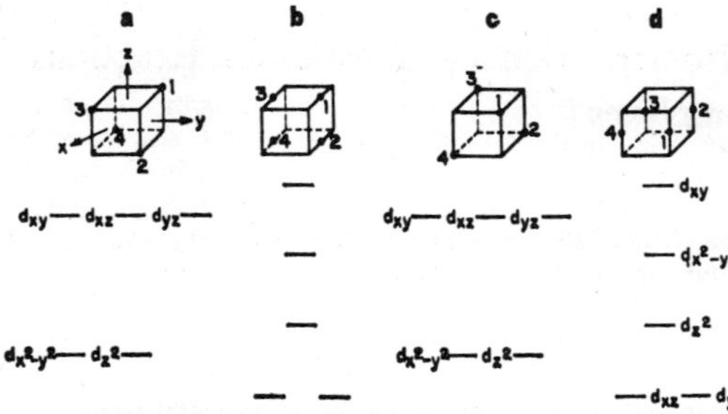

Figure 1. Four-coordinate isomerizations.

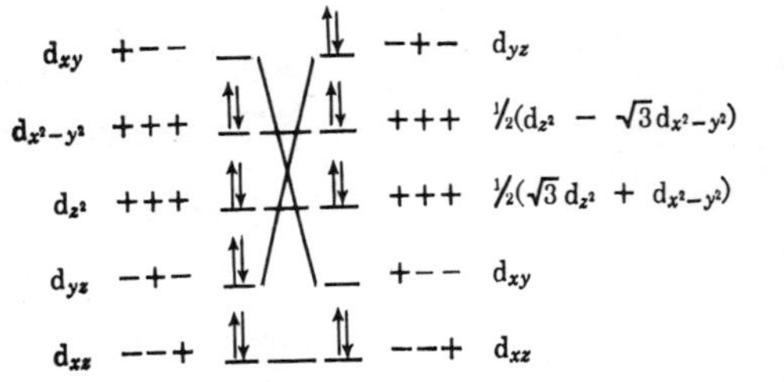

Figure 2. *cis–trans* isomerization for d^8 complex. C_{2z}, C_{2x}, and C_{2y} preserved ($\equiv D_2$).

the *z* axis of Figure 1b and then to apply the appropriate tetragonal distortion. The resulting energy levels are correlated with the conventional square-planar energy levels in Figure 2. Three twofold axes of symmetry (corresponding to point group D_2) are conserved and suffice to correlate the orbitals. It is predicted that the reaction is thermally disallowed but photochemically allowed for d^8 complexes. This agrees with the experimental observations of Haake and Hylton[10] and the more recent work of Perumareddi and Adamson[11] on the *cis–trans* isomerization of $Pt(PEt_3)_2Cl_2$. It should be particularly noted that these authors found the d–d transitions to be photochemically active. It might also be mentioned that this result could have been intuitively predicted from the observation that the transformation 1b → 1d involves rotation of the square by $\pi/2$ so that an orbital such as d_{xy}, which in 1d points toward the ligands and is antibonding, in 1b points out of the plane of the ligands and is nonbonding. Predictions based on arguments of this kind have certain limitations. Thus both the Woodward–Hoffmann approach and the present results give a qualitative description of the profile of the *orbital* energy of the system during a *concerted* transition from the initial state to the final state. Thus in the present case no account has been taken of interelectron repulsion which can in fact suffice to make the intermediate tetrahedral configuration a stable (triplet) state of the molecule for Ni(II) complexes. The presence of such an intermediate allows the possibility of carrying out the reaction in two distinct steps rather than in a concerted manner, and the argument is no longer valid. An analogous objection could be raised in the ethylene dimerization problem if the triplet

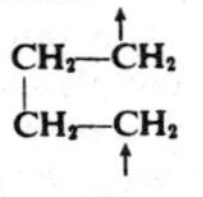

were a stable molecule. However, Pt(II) and Pd(II) compounds are "strong field" complexes in the sense that the interelectron repulsion energies are small compared with the bonding energies (as they are in organic compounds) so that the tetrahedral configuration is not stable and a description of the energy profile neglecting such effects is reasonably valid. For this reason the qualitative agreement between theory and experiment noted above is to be anticipated.

(10) P. Haake and T. A. Hylton, *J. Am. Chem. Soc.*, **84**, 3774 (1962).
(11) J. R. Perumareddi and A. W. Adamson, *J. Phys. Chem.*, **72**, 414 (1968).

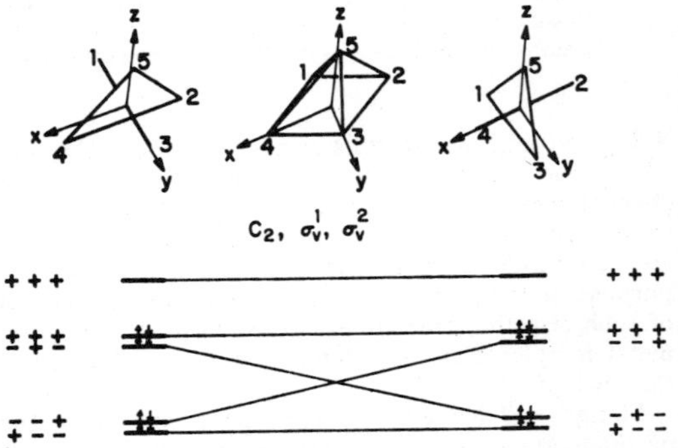

Figure 3. Axial–equatorial isomerization of trigonal bipyramid.

As a further example of this approach, Figure 3 illustrates the axial–equatorial exchange of a trigonal bipyramid through a square-pyramidal intermediate. C_{2v} symmetry is conserved. The reaction is thermally allowed for d^8 complexes but would be forbidden for spin-free d^2 and spin-paired d^3 and d^4 complexes. Bramley, Figgis, and Nyholm[12] inferred very fast intramolecular exchange for the d^8 trigonal bipyramid $Fe(CO)_5$ from nmr data. In this laboratory we have been unable to observe Sn nmr in $Pt(SnCl_3)_5{}^{3-}$ and have attributed this to the effects of a rapid exchange reaction.

We will now consider the substitution reactions of square-planar and tetrahedral d^8 complexes. Plausible mechanisms for these reactions are shown in Figure 4. The bimolecular tetrahedral substitution involves approach of the incoming ligand on one of the faces of the tetrahedron and leads to inversion of the tetrahedron. This process is a familiar one in organic chemistry. C_{3v} symmetry is conserved and the above considerations show immediately that it is thermally allowed. It is assumed that the bimolecular square-planar substitution also goes through a trigonal-bipyramidal intermediate. This is not a restrictive assumption since even if the intermediate is a square pyramid, apical and basal ligands must be interchanged through a trigonal bipyramid before any ligand exchange is accomplished. To go from dsp^2 (square planar) to dsp^3 (trigonal bipyramid) hybridization the additional metal orbital involved is p_z so that the obvious approach of the fifth ligand is toward this orbital, *i.e.*, in a direction perpendicular to the square plane. A trigonal bipyramid re-

(12) R. Bramley, B. N. Figgis, and R. A. Nyholm, *Trans. Faraday Soc.*, **58**, 1893 (1962).

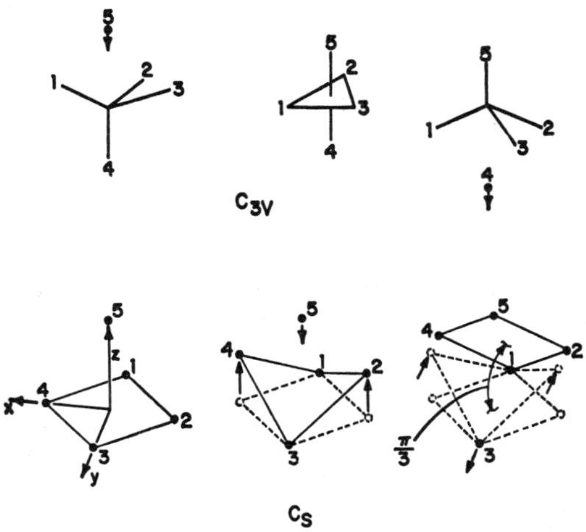

Figure 4. Bimolecular tetrahedral and square-planar substitution.

sults from dropping two diagonal corners of the square below the plane so that these two ligands, together with the incoming ligand, form the equatorial ligands of the trigonal bipyramid. Loss of a different equatorial ligand in the same manner leads to ligand exchange without *cis–trans* isomerization as demanded by experiment.[13] If the coordinate system is kept fixed, the over-all effect is to rotate the plane of the square by $\pi/3$. Only a single plane of symmetry is conserved, and it is predicted that the reaction is thermally allowed. Experimentally, square-planar d^8 substitutions are fast but appreciably slower than tetrahedral substitutions. Thus, phosphine ligand exchange of $Ni(PEt)_3Cl_2$ (square planar) has been observed by nmr methods. At high concentrations (~0.2 *M*) sharp multiplets are observed for the CH_2 and CH_3 resonances at 220 Mcps in positions corresponding to the average of free and complexed ligand. At lower concentrations (~0.05 *M*) there is sufficient broadening to lose the spin–spin splittings. This observation demonstrates that the ligand-exchange rate is slower than that of $Ni[P(C_6H_5)_3]_2Cl_2$[14] (tetrahedral) by a factor of around 10^2. Even this rate may be due to about 1% of paramagnetic tetrahedral isomer in $Ni(PEt)_2Cl_2$. Ligand exchange of $Pt[P(C_6H_5)_3]_2Cl_2$ and $Pd[P(C_6H_5)_3]_2Cl_2$ has been similarly studied by 220-Mcps pmr and found to be slower than that of $Ni[P(C_6H_5)_3]_2Cl_2$ by a factor of 10^4–10^5. Although slow by the nmr criterion, these are still fast reactions and occur in times of the order of seconds at nmr concentrations, *i.e.*, ~0.1 *M* in complex and ligand. Thus it has been shown that statistical equilibrium between triphenylphosphine and tri-*p*-tolylphosphine ligands is reached in less time than is required to mix the solutions and position the sample in an nmr spectrometer. This may be compared with the many hours needed for the thermal isomerization of $Pt(PEt_3)_2Cl_2$.[10] Ligand exchange of tetrahedral chelated nickel(II) aminotroponeimineates takes around 10^1 sec,[15] but the square-planar complexes are very much slower. Similar observations have been made on nickel(II) salicylaldimines.[16] For $Pd[(CH_3)_2PC_6H_5]_2Cl_2$ the rate of ligand exchange is faster than that of *cis–trans* isomerization in the absence of ligand by at least 10^2. In brief, the implication is that square-planar substitution occupies a position intermediate between the allowed tetrahedral substitution and the disallowed *cis–trans* isomerization.

The role of symmetry in formulating selection rules is one of convenience rather than of necessity. This point has been made explicitly in a recent paper by Woodward and Hoffmann.[17] What we really seek is some method of qualitatively predicting the energy profile of the system during the course of a smooth transition from the initial to the final state. Thus Longuet-Higgins and Abrahamson[18] have discussed selection rules in terms of state symmetries rather than orbital symmetries. This is undoubtedly the more elegant and appealing approach although the conclusions are identical with those obtained by the simpler Woodward–Hoffmann method. The resulting state correlation diagrams give a qualitative idea of how the energy of the system varies during the course of the reaction, and, even when the symmetry is insufficient to prevent levels from crossing, such diagrams demonstrate that a considerable activation energy will be required for reactions described as forbidden. There must obviously be a continuous spectrum of conditions ranging from "fully allowed" to "fully forbidden," and it would be desirable to have some criterion for determining the relative positions of two comparable reactions within such a spectrum. We wish to suggest the following approach to this objective.

In the ethylene dimerization problems, the four molecular orbitals φ_1, φ_2, φ_3, and φ_4 are eigenfunctions of both H_A and H_B. This arises because each belongs to a different symmetry class, and no mixing can occur as the reaction proceeds by smoothly changing H_A to H_B. The ground state of the reactants can be described by a wave function Ψ_A which is a product wave function involving the filled molecular orbitals φ_1 and φ_2. Similarly, the total Hamiltonians $\mathcal{H}_A$ and $\mathcal{H}_B$ can be written as the sum of the one-electron Hamiltonians $H_A(1)$, etc. Since the φ_A are eigenfunctions of H_B, it follows that $\Psi_A\mathcal{H}_B\Psi_A$ must be an eigenvalue of the product cyclobutane. It must therefore represent the energy either of the gound state or of an excited configuration. Inspection shows that it is an excited-state energy. The selection rule might therefore be stated in the form that the reaction is thermally forbidden if the application of the Hamiltonian corresponding to the reaction products ($\mathcal{H}_B$) to the ground-state wave function of the reactants (Ψ_A) leads to an excited-state energy of the products and is thermally allowed if it gives the ground-state energy. $\Psi_A\mathcal{H}_B\Psi_A$ could, if desired, be evaluated without using symmetry to factorize the determinants involved. Consider now the effect of making a small change in the Coulomb integral of one of the p orbitals such as might arise from replacing hydrogen

(13) F. Basolo and R. G. Pearson, "Mechanisms of Inorganic Reactions," 2nd ed, John Wiley and Sons, Inc., New York, N. Y., 1967.
(14) W. D. Horrocks, Jr., and L. H. Pignolet, *J. Am. Chem. Soc.*, **88**, 5929 (1966).
(15) D. R. Eaton and W. D. Phillips, *J. Chem. Phys.*, **43**, 392 (1965).
(16) A. Chakravorty and R. H. Holm, *J. Am. Chem. Soc.*, **86**, 3999 (1964).
(17) R. Hoffmann and R. B. Woodward, *Accounts Chem. Res.*, **1**, 17 (1968).
(18) H. C. Longuet-Higgins and E. W. Abrahamson, *J. Am. Chem. Soc.*, **87**, 2045 (1965).

by an alkyl substituent. Both elements of symmetry are lost, and a correlation diagram can no longer be constructed by the usual procedure. However, common sense dictates that the reaction must still be forbidden to a first approximation. In terms of the above approach some of the elements in H_A which were previously equal will now be slightly different and the φ_A will now be slightly different combinations of the original basis set atomic orbitals than the φ_B. However, since φ_A and φ_B are formed from the same basis set, a given φ_A can always be written as a linear combination of the φ_B's. Similarly, Ψ_A can be written as a linear combination of eigenfunctions Ψ_B. Evaluation of $\Psi_A \mathcal{H}_B \Psi_A$ will give $aE_0 + bE_1 + cE_2 + \cdots$ where E_0, $E_1, \cdots$ are the ground-state and excited-state energies of the products, respectively. In the present case the coefficient of one of the excited-state energies will be very much larger than all the other coefficients, and this could be interpreted to mean that the reaction is almost forbidden.

This approach can be applied particularly easily to the square-planar substitution reaction considered previously. It has been shown that this reaction leads to rotation of the plane by $\pi/3$. It is more convenient to consider $\Psi_B \mathcal{H}_A \Psi_B$, *i.e.*, to use the final wave functions and the initial Hamiltonian. The wave functions for a $\pi/2$ rotation have been given in Figure 2. Those for a $\pi/3$ rotation can be expressed in terms of the original wave functions and the $\pi/2$ rotated wave functions, *i.e.*

$$\varphi_B = \cos \pi/3\ \varphi_0 + \sin \pi/3\ \varphi_{\pi/2} = \frac{1}{2}\varphi_0 + \frac{\sqrt{3}}{2}\varphi_{\pi/2}$$

Since the $\pi/2$ rotation leads to a forbidden reaction, it is readily shown that

$$\Psi_B \mathcal{H}_A \Psi_B = \frac{1}{4}E_0 + \frac{3}{4}E_1$$

Thus qualitatively the reaction is expected to be slower than the fully allowed tetrahedral substitution (for which $\Psi_B \mathcal{H}_A \Psi_B = E_0$) but faster than the forbidden *cis–trans* isomerization (for which $\Psi_B \mathcal{H}_A \Psi_B = E_1$).

It is unlikely that arguments of this kind will have more than qualitative significance but they may find a place in rationalizing what are at first sight rather puzzling observations such as the greatly enhanced rate of ligand exchange in tetrahedral d^8 complexes compared to square planar. Application of similar arguments to the isomerization and substitution reactions of complexes of higher coordination number is under consideration and will be reported in due course.

Acknowledgment. The author is indebted to Drs. J. P. Jesson and E. L. Muetterties for helpful discussions.

VIII-2

Nonempirical SCF-MO Calculations on Transition Metal Complexes

HAROLD BASCH, C. HOLLISTER, and JULES W. MOSKOWITZ

In this article we shall demonstrate the feasibility of employing Gaussian functions as basis orbitals in molecular orbital computations on transition metal complexes. This calculation represents the first treatment of a transition metal complex ion involving all electrons and no approximations within the Roothaan SCF-MO formalism. The qualitative features of the calculation can be expected to have wide significance both for the understanding of the electronic structure of transition metal complexes and for the application of MO theory to large systems of this type.

We will report the results of calculations on the (hypothetical) square planar NiF_4^{2-} ion (the reasons for this particular choice will be explained later) and discuss those interesting ground- and excited-state molecular electronic properties which can be extracted from a computed wave function.

It is well known that individual Gaussians are much poorer representations of atomic orbitals (AO's) than are single Slater-type orbitals (STO's), especially in two very important areas: near the nucleus, and in the tail region of the AO. These deficiencies are very serious if one wants to use Gaussians as basis functions for molecular calculations. Of course, a single STO is also known to reproduce rather poorly the tail region of an AO, and it takes a linear combination of STO's to span adequately the whole space of an AO. This suggests the method whereby one can correct the deficiencies of the individual Gaussian-type orbitals (GTO's); use many of them in linear combinations to form Gaussian-type functions (GTF's) and then use the GTF's as the basis functions in the MO calculation. The proper linear combinations needed to form the GTF's are easily obtained either by directly fitting the Hartree-Fock AO's with GTO's, or by doing an atomic analytic SCF calculation in a GTO basis. We find that on the average only two to three GTO's per STO are required for quantitatively comparable results.

This is illustrated in Fig. 1, which compares the 3d AO of Ni^{2+} ($3d^8$ configuration) in the single Slater, Hartree-Fock, and Gaussian approximations. Note that the position of the maximum for the Hartree-Fock and GTF coincide and that they tail off pretty much together, in contrast to the single Slater function. The Gaussian curve is a linear combination of five GTO's and is identical, on the scale of this figure, with the two-term Slater representation computed by Richardson.[2]

This ratio of two to three Gaussians per Slater means that we must use a basis set two to three times larger for comparable results and evaluate many more integrals. Fortunately, the speed with which the necessary two-electron integrals over GTO's can be computed more than compensates for the handicap of the larger basis. This very favorable speed is due entirely to the fact that all the multicenter integrals over GTO's can be evaluated using simple analytic formulas which are easily coded in FORTRAN language, without recourse to complex numerical integration techniques, numerous slowly convergent expansions, or sophisticated programming structure.[3] This is the fundamental reason for using Gaussians.

To summarize, the MO wave function is an antisymmetrized product of MO's which are expressed as linear combinations of GTF's. The coefficients of the GTF's in the MO's are determined variationally via the well-known Roothaan Hartree-Fock equations.[4] The GTF's are themselves linear combinations of GTO's but with the coefficients in these linear combinations

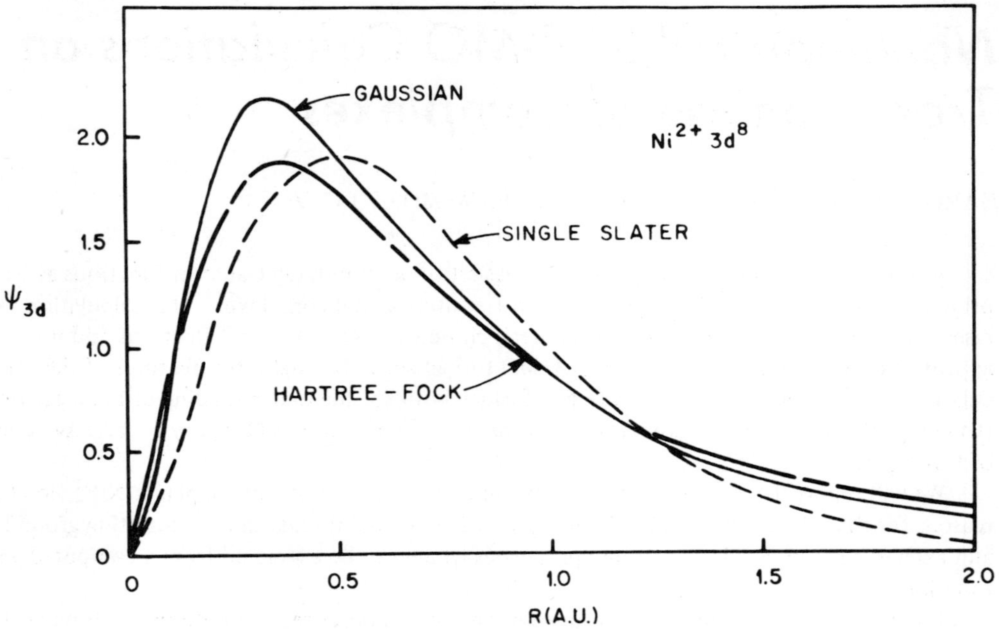

Fig. 1. 3d wave function of Ni^{2+} ($3d^8$ configuration) in the single Slater, Hartree-Fock, and Gaussian approximations.

fixed either directly or indirectly by previous atomic SCF calculation. Finally, and this is to be emphasized, no loss in accuracy in the computed MO's is necessary using Gaussians as basis functions provided they are used properly, in the sense that a sufficient number of them are used to span adequately the function space.

The transition metal ion complex whose ground-state wave function we have computed is the 66-electron NiF_4^{2-} ion in a square planar geometry. Actually, this ion has not been observed in a planar configuration. However, the planar $PdCl_4^{2-}$ complex ion is well known and has been very extensively studied.[5] Making reasonable extrapolations from the known shifts in behavior in going from Ni(II) to Pd(II) and F^- to Cl^- it was hoped to obtain reasonably good estimates of properties to compare with those computed here for NiF_4^{2-}. Unfortunately, this expectation was not realized in a quantitative manner.

Outside of a specific interest in the square planar situation, the reasons this particular ion was chosen for our initial study are threefold. Square planar Ni(II) complexes are predicted to be closed shell in the ground state. This avoids the conceptual and practical difficulties that are known to arise in rigorously applying the SCF method to open-shell electronic configurations. Further, since the D_{4h} symmetry has only two doubly degenerate representations, it is likely that the excited electronic states calculated in the virtual orbital approximation would not involve the separation of multiplet states arising from electrons distributed in two partially occupied, multiply degenerate MO's. Finally, square planar NiF_4^{2-} is planned to be part of a series of calculations on Ni(II)–F complexes which will include NiF_4^{2-} (tetrahedral), NiF_2 (linear), and NiF_6^{4-} (octahedral).

The coordinate system used in the calculation has the nickel atom at the origin and the four fluorine atoms in the xy plane. A Ni–F bond length of 2.00 Å was assumed.

The basis set for the square planar NiF_4^{2-} calculation is given in Table 1, which shows the distribution of both the GTF's and GTO's in the basis set with respect to the AO's on Ni and F. Notice that some of the AO's are represented by two basis functions. Especially important are

Table 1. Basis Set for NiF^{2-}

		GTF's	GTO's
Ni^{2+}	1s, 1s′, 2s, 2s′, 3s, 4s	6	12
	2p, 2p′, 3p, 4p	12	24
	3d, 3d′	12	30
F^-	1s, 2s	2	7
	2p, 2p′	6	12
		62	144

the 3d on Ni and 2p of F. This double-ξ representation allows sufficient freedom so that the free ion AO's may adjust to the molecular environment in the course of the SCF calculation. For example, the five-term Gaussian representation of the Ni^{2+} 3d AO was split up into a long-range and short-range component. This allows the 3d AO either to *expand* or *contract*; i.e., it allows the energetics of the calculation to determine the spatial extension of the 3d AO in the complex. An analogous argument applies to the F^- 2p, which is represented here by 2p and 2p′ basis functions.

The resultant ground-state ordering and orbital energies of the computed MO's are shown in Table 2. We would hasten to point out that all previous notions of the MO orderings in the

Table 2. Ground State of NiF_4^{2-} (D_{4h})

% Minority composition		% Majority composition		Orbital energy, cm^{-1}
4.2	2p	95.8	$4p_z$	141,040
7.3	2p	92.7	$3d_{x^2-y^2}$	139,980
8.5	2p	90.1	4s	114,790
			2p: 24 electrons	−17,790 −23,360
22.9	2p	77.1	$3d_{z^2}$	−46,530
5.7	2p	94.3	$3d_{xz, yz}$	−50,370
4.4	2p	95.6	$3d_{xy}$	−67,390
			Core: 24 electrons	

diamagnetic square planar complexes of necessity arose from absorption spectra measurements, whether optical or magnetooptical. Unfortunately, absorption spectra involve both a *ground* state and an *excited* state, and evidence is here presented that the energy ordering of the ground-state *MO's* does not necessarily determine the energy ordering of the excited *states*, and vice versa. We will return to this after a closer look at the ground state.

The calculation shows the electronic structure built up starting with the "core" electrons; 1s, 2s, 3s, 2p, 3p on Ni and 1s, 2s on F. The valence shell begins with the 3d orbital MO's, so called because Mulliken population analysis[6] shows them to be predominantly metal 3d. Then come all 24 F^- 2p electrons sandwiched within a narrow range of 6000 cm^{-1}. This completes the occupied levels. The first unoccupied MO is the Ni 4s, then the $3d_{x^2-y^2}$, and the 4p is last.

The population analysis shows that the MO's are rather completely segregated between the Ni and F orbitals, except for the $6a_{1g}$ ($3d_{z^2}$) MO, which has a substantial amount of F 2p character. Note also that z^2 is the highest occupied 3d orbital and that xy is lowest. From the orbital energy diagram presented here, we would expect the first electronic transition to be F 2p to Ni 4s, but in fact, as we will show, this expectation is not fulfilled: Although the orbital energy gap between occupied and unoccupied d-orbital MO's is over 190,000 cm^{-1}, the d-d bands are calculated to fall lowest, in the visible and near-UV regions of the spectrum.

The total orbital populations in this computation are Ni $(core)^{18.00} 3d^{8.13} 4s^{0.20} 4p^{0.49}$ and F $(core)^{3.96} 2p^{5.84}$. The point of interest here is that the total occupancy of the Ni 3d orbitals is almost identical with the free-ion configuration, $3d^8$. However, the 4s and 4p orbitals draw electrons from the ligands so that the computed net charge of +1.12 on the Ni is substantially reduced from the formal ionic value of +2, while each F has a computed net charge of −0.78.

The computed MO's can be interpreted to give two types of information. The first is covalency, which is the extent of mixing or interaction of the ligand and metal AO's and is reflected in the relative weights of metal and ligand coefficients in the computed MO's. The second type of information is the distortion of the basis set AO's themselves, i.e., the change in the spatial extension of an AO in the complex when compared to its spatial extension in the free atom or ion. It is unfortunate that heretofore all discrepancies in the properties computed assuming the 3d radial function of the free ion to be a valid representation of the 3d AO in the complex have been attributed entirely either to expansion (or contraction) of the free ion AO, which is the crystal-field view, or to covalency mixing, which is the traditional MO treatment of ESR, NMR, and optical data. In fact, both these mechanisms operate simultaneously, and with the use of the double-ζ representation we can examine the inherent expansion or contraction of AO's due to complex ion formation. This is illustrated in Table 3.

Table 3. Expansion and Contraction of Fluorine AO's

	Atomic (F^-)	Atomic (F^0)	$1b_{2u} (\pi^{n.b.})$	$7a_{1g} (\sigma^b)$
C_1	0.02652	0.02905		
C_2	0.17158	0.18408		
C_3	0.43180	0.49025	0.44212	0.35716
C_4	0.60428	0.53497	0.59240	0.51830
C_3/C_4	0.715	0.916	0.746	0.689

The double-ζ basis for the F^- 2p AO was obtained by breaking up a four-term Gaussian representation into a long-range and a short-range component. The expansion coefficients for atomic F° and the F^- ion using the same set of exponents are given in the first two columns and beneath them is the ratio in various MO's. C_4 is the coefficient of the Gaussian with the smallest exponent so that a smaller ratio means a more diffuse orbital. Note that the ratios for F° and F^- differ by about 30 per cent.

The corresponding coefficients in two ligand 2p MO's are shown: the π nonbonding $1b_{2u}$ and the σ-bonding $7a_{1g}$. The ratio of C_3/C_4 for both these MO's is extremely close to that found in the free F^- ion calculation.

An analagous result for the Ni^{2+} 3d AO is shown in Table 4. Here a five-term Gaussian expansion was broken up into a set of three and a set of two. Here we compare the ratio C_3/C_4,

Table 4. Expansion and Contraction of Nickel AO's

	Atomic (Ni^{2+})	xy	z^2	xz, yz
C_1	0.07302			
C_2	0.23864			
C_3	0.45326	0.44236	0.38260	0.43072
C_4	0.40293	0.39003	0.35841	0.39801
C_5	0.17167			
C_3/C_4	1.125	1.134	1.067	1.082

which does not depart significantly from that computed for the free metal ion in its formal +2 oxidation state, even for the $3d_{z2}$ MO, which contains a significant amount of F^- 2p character. Note, however, that the orbital deformation, as reflected in the C_3/C_4 ratios, is slightly different for each d-orbital MO. Thus, using a double-ζ representation is equivalent to optimizing the spatial extension of the basis function separately for each computed MO.

The excited electronic states of the NiF_4^{2-} ion are presented in Table 5. The qualitative features of the computed absorption energies are quite clear. The ordering of the states arising

Table 5. Electronic States of NiF_4^{2-}

One-electron transition	State	Energy, 1000 cm^{-1}	$PdC\ell_4^{2-}$, ϵ_{max}	$PtC\ell_4^{2-}$, ϵ_{max}
	$^1A_{1g}$	0		
xz, yz $\rightarrow x^2 - y^2$	^3E_g	4.0	?	18.0 (2)
xy $\rightarrow x^2 - y^2$	$^3A_{2g}$	11.0	?	20.7 (2)
xz, yz $\rightarrow x^2 - y^2$	^1E_g	19.0	22.8 (104)	29.5 (46)
xy $\rightarrow x^2 - y^2$	$^1A_{2g}$	19.5	20.0 (67)	26.3 (28)
$z^2 \rightarrow x^2 - y^2$	$^3B_{1g}$	26.0	17.5 (13)	24.0 (5)
$z^2 \rightarrow x^2 - y^2$	$^1B_{1g}$	40.0	29.5 (67)	36.5

from d-d transitions in the virtual orbital approximation gives those originating from xy, xz, and yz low in energy and that from z^2 high in energy, quite the opposite from that expected from the one-electron orbital energies in the ground state. The triplets are very low lying, so low as to fall in the near-IR region of the spectrum.

Since the xy, xz, yz, and $x^2 - y^2$ MO's are overwhelmingly of metal character we would expect that in going from NiF_4^{2-} to $PdC\ell_4^{2-}$ only the Ni to Pd shift should matter, giving a slight blue shift in the computed energies of these transitions. On the other hand, transitions originating from the z^2 MO, which contains 23 per cent F^- 2p, and considering the tremendous blue-shift effect of the F^- ion, should be red shifted in going to $PdC\ell_4^{2-}$. Thus the d-d bands in $PdC\ell_4^{2-}$ should be found in the 10 to 36,000 cm^{-1} range, with the singlet $z^2 \longrightarrow x^2 - y^2$ transition highest in energy. These conclusions are not inconsistent with the observed absorption spectrum of $PdC\ell_4^{2-}$, with the d-d bands reported in the range 17 to 30,000 cm^{-1} (Ref. 5). However, a more detailed assignment of the d-d states of $PdC\ell_4^{2-}$ in the light of a calculation on NiF_4^{2-} would probably tax the powers of the calculation beyond its present capabilities.

Two-high-intensity bands have also been observed in the absorption spectrum of the $PdC\ell_4^{2-}$ ion at 36,000 and 45,000 cm^{-1}, with oscillator strengths estimated at about 0.2 and 0.5, respectively. Anex[7] has measured the polarizations of these bands and finds them both to be *xy* or in-plane polarized, with a weak *z* polarized band hiding under the higher-energy *xy* polarized band.

The lowest-energy symmetry-allowed transitions calculated for NiF_4^{2-} are shown in Table 6

Table 6. Charge-Transfer States of NiF_4^{2-}

One-electron transition	State	Energy, cm^{-1}	$f(\nabla)$	f(r)
2p $\longrightarrow$ 4s	^1E_u(x,y)	69,400	0.055	0.045
2p $\longrightarrow$ 4s	^1E_u(x,y)	86,200	0.491	0.263
2p $\longrightarrow$ 4s	$^1A_{2u}$(z)	91,400	0.195	0.106
2p $\longrightarrow$ 3d	^1E_u(x,y)	100,100	0.081	0.212
2p $\longrightarrow$ 3d	$^1A_{2u}$(z)	101,600	0.000	0.000
2p $\longrightarrow$ 3d	^1E_u(x,y)	107,000	0.165	0.285
3d $\longrightarrow$ 4p	$^1A_{2u}$(z)	108,000	0.209	0.270

together with their oscillator strengths as computed from the dipole velocity and dipole length. These excited states are, of course, computed to fall at very high energy in NiF_4^{2-}, owing to the tremendous blue-shift effect of F^-. But, most surprising, the assignments of the lowest lying of these transitions are all 2p-4s rather than the usually assumed 2p-3d. The relative orderings and intensities of the three computed 2p-4s bands are consistent with the observations of Anex, although the same can be said about the three 2p-3d bands which are next highest in energy. The use of an unoptimized single-ζ representation for the Ni 4s orbital renders any conclusions about these assignments tentative.

In Table 7 we give a detailed breakdown of the computation of the excitation energies. The equation is given at the top and the values of the contributing terms for the three single

Table 7. Excitation Energies of NiF_4^{2-}

$$^1E(i \longrightarrow j) = \epsilon_j - \epsilon_i - J_{ij} + 2K_{ij}$$
$$1 \text{ a.u.} = 219{,}450 \text{ cm}^{-1}$$

Transition	$\epsilon_j - \epsilon_i$, a.u.	J_{ij}, a.u.	K_{ij}, a.u.	Energy, cm^{-1}
$xz, yz \longrightarrow x^2 - y^2$	0.86741	0.84913(0.95806)[a]	0.03455	19,200
$xy \longrightarrow x^2 - y^2$	0.94495	0.89496(0.99429)[a]	0.01940	19,500
$z^2 \longrightarrow x^2 - y^2$	0.84991	0.73132(0.94598)[a]	0.03264	40,300
$2p \longrightarrow 4s$	0.54432	0.22783	0.00738	69,400

[a]Computed from free-ion 3d functions.

d-d transitions and the lowest-lying charge-transfer transition are given in their respective columns. ϵ_i and ϵ_j are the orbital energies, J_{ij} and K_{ij} are the Coulomb and exchange integrals, respectively.

The importance of the electron repulsion correction of the Coulomb integral to the simple differences of ground-state orbital energies is immediately obvious. Note that for the d-d bands the Coulomb integral is very similar to that calculated from the free-ion metal 3d orbitals alone, given in parentheses next to the MO values. In contrast, note the small Coulomb integral value for the 2p-4s transition. Thus, although the orbital energy differences for the d-d transitions are considerably larger than those for the charge-transfer bands, the enormous release in energy obtained by splitting up a pair of d-orbital electrons more than compensates for this large difference; i.e. the enormous self-repulsion of the 3d electrons must be considered as a crucial factor in the gross ordering of the low-lying excited-state configurations.

The ratio of the Coulomb integral for the $z^2 \longrightarrow x^2 - y^2$ transition in the molecule to its free-ion value is much smaller than for the other d-d transitions. This is a direct outcome of the admixture of F^- 2p orbitals in the z^2 MO. Thus, although the orbital energy difference is smallest for the $z^2 \longrightarrow x^2 - y^2$ excitation, reflecting the high energy of the z^2 MO, the value for the Coulomb integral due to covalent mixing gives an excitation energy for $z^2 \longrightarrow x^2 - y^2$ highest of all the d-d transitions. Thus the amount of covalent mixing and the accompanying variable reduction in the electron repulsion integrals apparently determine the finer details of the ordering of the excited states in NiF_4^{2-}.

The purpose of this computation has been to demonstrate to those interested in the application of MO theory to problems in inorganic chemistry the feasibility of doing nonempirical SCF-MO calculations on transition metal complexes with Gaussian-type basis functions. The specific example discussed, square planar NiF_4^{2-}, although admittedly a hypothetical complex, was chosen as a prototype to illustrate certain general characteristics of the calculation. Since the presentation of these results, we have redone the computations using completely optimized atomic wave functions for both the Ni^{2+} and F^- ions. The results are in qualitative agreement with those reported above. In addition, SCF calculations have been completed on octahedral NiF_6^{4-} as part of a study of the interesting spin properties of this species, which give rise to transferred hyperfine interactions, spin-orbit coupling, and electronic g factors.

Acknowledgment

The authors would like to thank Dr. D. Neumann for many helpful discussions and programming assistance. This research was supported in part by the National Science Foundation under Grant GP-5677.

References

1. J. M. Schulman, J. W. Moskowitz, and C. Hollister, *J. Chem. Phys.* **46**, 2759 (1967).
2. J. W. Richardson, W. C. Nieupoort, R. R. Powell, and W. F. Edgell, *J. Chem. Phys.* **36**, 1057 (1962); J. W. Richardson, R. R. Powell, and W. C. Nieupoort, *J. Chem. Phys.* **38**, 796 (1963).
3. I. G. Csizmadia, M. C. Harrison, J. W. Moskowitz, and B. T. Sutcliffe, *Theoret. Chim. Acta* **6**, 191 (1966).
4. C. C. J. Roothaan, *Rev. Mod. Phys.* **23**, 69 (1951); **32**, 179 (1960).
5. H. Basch and H. B. Gray, *Inorg. Chem.* **6**, 365 (1967).
6. R. S. Mulliken, *J. Chem. Phys.* **23**, 1833 (1955).
7. B. Anex, unpublished results.